Dry Valley Drilling Project

American Geophysical Union

ANTARCTIC RESEARCH SERIES

American Geophysical Union

ANTARCTIC RESEARCH SERIES

FRONTISPIECE

Drill site (DVDP 12) west of the Canada Glacier in Taylor Valley. View looking northeast with the
Canada Glacier in the background. Photo by L. McGinnis.

Volume 33 | ANTARCTIC RESEARCH SERIES

Dry Valley Drilling Project

Lyle D. McGinnis, Editor

American Geophysical Union
Washington, D. C.
1981

Volume 33 | ANTARCTIC
RESEARCH
SERIES

DRY VALLEY DRILLING PROJECT

LYLE D. McGINNIS, EDITOR

Copyright © 1981 by the American Geophysical Union
2000 Florida Avenue, N. W.
Washington, D. C. 20009

Library of Congress Cataloging in Publication Data

Dry valley drilling project.

  (Antarctic research series; v. 33)
  Includes bibliographies.
  1. Geology—Antarctic regions—Victoria Land—
Addresses, essays, lectures.   2. Borings—
Antarctic regions—Victoria Land—Addresses, essays,
lectures.   I. McGinnis, Lyle David.   II. Series.
QE350.D79   559.8′9   81-3608

ISBN 0-87590-177-8   AACR2

Published by the
AMERICAN GEOPHYSICAL UNION
With the aid of grant DPP-7721859 from the
National Science Foundation

Printed in the United States of America

CONTENTS

Analyses of Crystalline Rocks

Analyses of Sedimentary Rocks

Glacial and Geologic History

DVDP Core Storage and Bibliography

THE ANTARCTIC RESEARCH SERIES:
STATEMENT OF OBJECTIVES

The Antarctic Research Series, an outgrowth of research done in the Antarctic during the International Geophysical Year, was begun early in 1963 with a grant from the National Science Foundation to AGU. It is a book series designed to serve scientists and graduate students actively engaged in Antarctic or closely related research and others versed in the biological or physical sciences. It provides a continuing, authoritative medium for the presentation of extensive and detailed scientific research results from Antarctica, particularly the results of the United States Antarctic Research Program.

Most Antarctic research results are, and will continue to be, published in the standard disciplinary journals. However, the difficulty and expense of conducting experiments in Antarctica make it prudent to publish as fully as possible the methods, data, and results of Antarctic research projects so that the scientific community has maximum opportunity to evaluate these projects and so that full information is permanently and readily available. Thus the coverage of the subjects is expected to be more extensive than is possible in the journal literature.

The series is designed to complement Antarctic field work, much of which is in cooperative, interdisciplinary projects. The Antarctic Research Series encourages the collection of papers on specific geographic areas (such as the East Antarctic Plateau or the Weddell Sea). On the other hand, many volumes focus on particular disciplines, including marine biology, oceanology, meteorology, upper atmosphere physics, terrestrial biology, snow and ice, human adaptability, and geology.

Priorities for publication are set by the Board of Associate Editors. Preference is given to research projects funded by U.S. agencies, long manuscripts, and manscripts that are not readily publishable elsewhere in journals that reach a suitable reading audience. The series serves to emphasize the U.S. Antarctic Research Program, thus performing much the same function as the more formal expedition reports of most of the other countries with national Antarctic research programs.

The standards of scientific excellence expected for the series are maintained by the review criteria established for the AGU publications program. The Board of Associate Editors works with the individual editors of each volume to assure that the objectives of the series are met, that the best possible papers are presented, and that publication is achieved in a timely manner. Each paper is critically reviewed by two or more expert referees.

The format of the series, which breaks with the traditional hard-cover book design, provides for rapid publication as the results become available while still maintaining identification with specific topical volumes. Approved manuscripts are assigned to a volume according to the subject matter covered; the individual manuscript (or group of short manuscripts) is produced as a soft cover 'minibook' as soon as it is ready. Each minibook is numbered as part of a specific volume. When the last paper in a volume is released, the appropriate title pages, table of contents, and other prefatory matter are printed and sent to those who have standing orders to the series. The minibook series is more useful to researchers, and more satisfying to authors, than a volume that could be delayed for years waiting for all the papers to be assembled. The Board of Associate Editors can publish an entire volume at one time in hard cover when availability of all manuscripts within a short time can be guaranteed.

PREFACE

Papers included in this volume represent final results of part of the research conducted under the auspices of the Dry Valley Drilling Project (DVDP), a coordinated effort by science groups from Japan, New Zealand, and the United States. Primary support for the project came from the U.S. National Science Foundation, Division of Polar Programs; the New Zealand Department of Scientific and Industrial Research, Antarctic Division; and the Japan National Institute of Polar Research. Responsibility for project operations included Japanese support of the Thiel Earth Science Laboratory at McMurdo, New Zealand's provision of drilling and other technical personnel at the drill sites, and the U.S. purchase of the drill rig and primary logistics such as helicopter airlift support and base shops at McMurdo, all staffed by U.S. Navy personnel and supported by the National Science Foundation. Project coordinators were D. Kear, L. McGinnis, T. Nagata, R. Thomson, T. Torii, and M. Turner. Project advisors were E. Barghoorn, P. Barrett, C. Bentley, R. Black, P. Damon, S. Goldich, H. Kurasawa, M. Murayama, N. Nakai, R. Roy, S. Treves, P. Webb, H. Wright, and Y. Yoshida.

Some preliminary papers, as well as papers covering aspects of DVDP not in this volume, have already been published in scientific periodicals and in other volumes, such as the SCAR Symposium Transactions, University of Wisconsin Press, Madison, and the proceedings volume of DVDP Seminar III held in Tokyo. Final reports of the efforts of U.S. scientists are included here, although final heat flow analyses and a synthesis of the geology of the dry valleys were not completed on time to appear on these pages.

Major emphasis in this volume is on core analysis; however, regional geophysical surveys and downhole geophysical logging amplify the three-dimensional view of the geologic setting that would not be available from core analysis alone. As the first rock drilling on the Antarctic continent, the 2 km of core retrieved from the 15 holes completed by DVDP are extremely important in developing an Antarctic geologic history of the past 10,000,000 years.

There was 93% core recovery during drilling; however, the deeper sections of glacial-marine strata were not penetrated because of technical difficulties associated with drilling. Geophysical measurements indicate over 2 km of sediment rest on the floor of McMurdo Sound, but only 65 m of that section was drilled at DVDP 15. New Zealand, with help from U.S. and Japanese scientists and the DVDP drill rig, made a second attempt to drill through glacial-marine sediments in McMurdo Sound in late 1979. This program, the McMurdo Sound Sediment and Tectonic Study (MSSTS), resulted in the coring of 225 m of sediment below the sea floor, stopping in glacial-marine sediments of Middle Miocene age. The wealth of specific data from DVDP, with the additional data from the MSSTS, has placed the geologic and glacial history of the McMurdo area of Antarctica for the Late Tertiary and Quaternary on solid ground. The very significant conclusions in this volume will be the basis for the next generation of studies and projects that will be carried out in Antarctica. It is apparent, for example, that a full and complete geologic history of preglacial to glacial successions in the McMurdo area will not be available until a drill rig capable of drilling holes several kilometers in depth is brought into use. Until then, estimates of age and environmental history of the transition from temperate to glacial conditions in Antarctica will remain conjectural.

MORT D. TURNER
Division of Polar Programs
National Science Foundation
Washington, D. C. 20550

THE ROLE OF THE DRY VALLEY DRILLING PROJECT IN ANTARCTIC AND INTERNATIONAL SCIENCE POLICY

PHILIP M. SMITH

Office of Science and Technology Policy, Executive Office of the President, Washington, D. C. 20024

INTRODUCTION

By the late 1960's the ice-free valleys west of McMurdo Sound were emerging as one of the greatest research centers in Antarctica. Geological and glaciological research in the dry valleys carried out by Wright, Priestley, Taylor, and others on the early exploring expeditions had set the stage for the exploration and research that were to follow. With the discovery of additional large, ice-free areas during Operation Highjump and in the early reconnaissance flights of Operation Deepfreeze prior to the International Geophysical Year (IGY) it was clear that these vast, virtually unexplored terrains would occupy the attention of scientists of several disciplines for many years. The first confirming evidence of scientific richness of the dry valleys stemmed from the geological reconnaissance by New Zealand parties and more limited U.S. field research during the IGY.

As we began to plan the post-IGY program at the National Science Foundation, the initiation of additional biological, geological, and glaciological research in the McMurdo Sound region and, in particular, the dry valleys, was an obvious necessity. Two steps were taken to facilitate the opportunity for dry valley research. Construction of well-equipped scientific laboratories at McMurdo Station became a high priority. Expansion of the helicopter support capability at McMurdo Station provided ready access to the dry valleys. With the new facilities and ease of access to the valleys, scientists quickly came forward with a variety of research proposals.

During the 1960's a pattern of international cooperation that influenced later development in the dry valleys also began to emerge. New Zealand and the United States, because of the proximity of their main Antarctic stations and the latter country's dependence on the former for the logistic staging of its expeditions, had begun to undertake an annual review of scientific programs and operations. Further, they initiated some joint research programs, whereas other research activities were complementary. For example, when New Zealand established its station at Lake Vanda, its year-long or seasonal meteorological and hydrological observations aided the intensive glaciological program carried out by the United States at the nearby Meserve Glacier. Cooperation between the United States and Japan owed part of its origin to collaboration under the terms of the Antarctic Treaty and through the International Council of Scientific Unions. Also important were the bilateral science relationships that had been established earlier between the two countries under the aegis of Harry Kelley, who had been on McArthur's staff and was subsequently Assistant Director of the National Science Foundation. Thus when Japan had temporarily closed its own Antarctic station, it was natural for Japanese officials to propose to the National Science Foundation that the United States provide some assistance for glaciological and limnological research in the dry valleys. This Japan–U.S. cooperation soon extended also to Japan–New Zealand contacts. By the late 1960's, Antarctic scientists of the three nations, as well as the officials of the three national programs, were in regular communication in the field and between their home laboratories and offices. The close international contact facilitated the planning of the Dry Valley Drilling Project (DVDP) that was to follow.

Of fundamental importance to the DVDP above all else was the growing awareness of the scientific complexity of the dry valleys. Each year's field investigations sought answers to questions raised by earlier observations. The complexity of the Cenozoic record, as it was known in the valleys at the end of the 1960's, went beyond the generalized first impressions reported by Berg, Black, Bull, Denton, McKelvey, Nichols, Pewe, Webb, and others. The record of the Cenozoic, when Antarctica first became glaciated, remained elusive. In order to eliminate the many speculative and diverse theories on the origins of the valleys, scientists in each of the three nations discussed with increasing intensity their desire to examine subsurface sediments through core drilling programs and downhole measurements, which could extend the geologic record on land and in the lakes of the valleys as well as in the McMurdo Sound. It was from this strong scientific interest that the separate discussions in each nation and the collective discussions of drilling in the dry valleys emerged.

ORGANIZATION OF THE DVDP

The organization, development, and management of the DVDP has been documented in the national polar publications of the three nations and in the *DVDP Newsletter* and *DVDP Bulletin* series. From my perspective as an official of the U.S. Antarctic Research Program at the time the DVDP was organized, there were several important considerations involved in the planning and organization of this effort. These were the following:

1. Most importantly, while the Antarctic Treaty provided the framework for international cooperation and such cooperation had been successful beyond most expectations, the DVDP offered an opportunity for a more complete multinational sharing of planning and program costs than was the usual circumstance. In general, the pattern of Antarctic cooperation during the 1960's had been one of planning national research activities within a generalized series of research objectives derived through the deliberations of the Scientific Committee on Antarctic Research. Where geophysical observations were of a synoptic nature or where the coordinated efforts led to an understanding of geologic processes over the entire continent, this pattern of cooperation was and remains an effective method of planning research. In the case of the dry valleys, however, it seemed that the scientific interests and the national program activities were so tightly interlocked that it would be worthwhile to propose a more ambitious form of international cooperation, one in which the three nations would together plan a common scientific

program and then share in the costs. If successful, such a cost sharing of research might serve as a model for future activities in Antarctica.

2. It was clear that the United States, because of its laboratories and helicopter support capability, had to assume a leadership role in organizing the activities of the three nations. The willingness of Japan and New Zealand to participate in a commonly planned program was understood from informal contacts. However, the central importance of the logistic air support and the laboratory facilities required the first informal and formal discussions to be initiated by U.S. officials.

3. The high costs of the program, including costs of the drilling equipment, suggested that drilling needed to be based on the best geological and geophysical data. While much had been done, especially in geology, additional geophysical data were desirable. Hence plans included early formulation of an aeromagnetic survey of the entire McMurdo Sound dry valley region.

4. Since this was the first project of its kind in Antarctica, it was desirable that it proceed in an environmentally acceptable way with on-site environmental monitoring of the project during the drilling. Between the New Zealand and U.S. programs, there had already been initiated a number of environmental planning and site use protocols limiting, for example, scientific campsites at certain penguin rookeries and planning for the cleanup and helicopter evacuation of debris from field activities in the dry valleys. The concern for the environment led to the consideration of air support as the primary means of moving the DVDP equipment, including the drilling rig, rather than the pioneering of overland transport routes from drilling site to drilling site in the dry valleys.

5. Beyond the scientific desirability of drilling in the volcanics of Ross Island, it was clear that this ambitious new endeavor would benefit from initial testing of the drilling equipment close to the logistic facilities at McMurdo Station and Scott Base. The planning group from the three nations was encouraged to think out a program strategy that would enable first field trials on Ross Island, with the drilling in the valleys taking place in the second field season.

The discussions of these and other issues took place among scientists and national program officials in Japan, New Zealand, and the United States from 1969 to early 1971. Parallel discussions were held by scientific advisory groups in the three countries. In the United States, for example, the scientific feasibility of a drilling program in the dry valleys was considered in detail by the Geology and Geophysics Panel, Committee on Polar Research, National Academy of Sciences. In January 1971, I wrote to R. B. Thomson, Superintendent of the Antarctic Division, New Zealand Department of Scientific and Industrial Research, and to M. Murayama, Director, Polar Research Center, Japan, National Science Museum, extending an official invitation to both national programs to join the United States in the DVDP. Both responded affirmatively. The DVDP was officially announced in the *Antarctic Journal of the United States* 2 months later. Scientific response in all three countries was swift and enthusiastic. The program planning then moved rapidly, with first field activities and then geophysical survey taking place in the 1971-1972 austral summer.

RESEARCH MANAGEMENT PRINCIPLES
IN THE DVDP

Each of the participants in the national programs has an impression of the overall scientific success of the DVDP and also observations concerning the less successful elements of the enterprise. From my own perspective of science policy and the management of research and development, however, I would offer the following observations derived from DVDP. They may be applicable to other similar international research projects in Antarctica.

1. *Organizers of the national Antarctic programs must provide vigorous leadership to bring about jointly planned and funded international scientific programs.* Even though the magnitude of the research problems in Antarctica often demands cooperation, such collaboration will not of its own accord come into being. National program managers must have the vision to see the opportunity for scientific research and the confluence of other forces such as availability of new technology or logistics. They must recognize particular patterns of interest and cooperation that can serve as a basis for proposing new cooperative endeavors. At the same time, national program organizers must not overreach the prevailing attitudes and research base capabilities, domestically or internationally, by proposing activities without major scientific substance. In the case of the DVDP, it was clear to a few of us in the late 1960's that the moment for such a program had arrived. Even though the timing seemed right, there remained in

each nation some scientific and fiscal skeptics who felt that DVDP was being promoted too rapidly. In 1970 and 1971, part of our planning challenge was to persuade these skeptics that the 'moment' should not be lost.

2. *A balance must be maintained between larger-scale scientific projects and individual research endeavors, and among the larger-scale projects themselves.* In retrospect, it is clear to me that some of the criticism of the proposal for the DVDP as it was voiced in the United States was centered in a fear that this team effort would consume virtually all the financial resources available to some disciplines. Other critics wondered whether the commitment to the DVDP would delay initiation of other large-scale international cooperative programs that were then under national discussion. One of these was the proposal to drill through the Ross Ice Shelf to conduct glaciological, oceanographic, and biological studies. Even though that project was less 'ready' technologically and international interest had not advanced to the point of commitment, there was a worry in some circles that the U.S. involvement in the DVDP would delay ice shelf drilling. In the intervening years since the start of the DVDP, the capability of maintaining a program balance seems to have been demonstrated by the three nations conducting the dry valley research.

3. *Scientific and logistic flexibility must be maintained so adjustment can be made for the vagaries of Antarctica.* In the case of DVDP, execution of the much desired drilling at the site on the McMurdo Sound sea ice proved the most vexsome problem. But throughout the program the plan had to be readjusted to accommodate discoveries at some drill sites and technical difficulties at others. In my view, the national coordinators and scientific teams exhibited ingenuity throughout the program in their ability to adjust to and often take advantage of the 'Antarctic factor.'

4. *International cost sharing is an effective mechanism for the funding of larger-scale Antarctic research projects.* Early in the organization of the DVDP a decision was made to identify methods of sharing the costs rather than to contribute funds from each nation to a common fund or 'pool.' The latter system might have worked, but it would have undoubtedly resulted in a delay of the project for a year or two. Each national program office would have had to develop a new method of budgeting some of their program resources. Further, an examination of the appropriation schedules of

the Japan Diet, the New Zealand Parliament, and the United States Congress revealed that development of a common budget would take some time. Hence each of the three nations identified as a share some of the overall DVDP program costs, agreeing to full responsibility for those costs throughout the life of the DVDP.

5. *Cooperation is facilitated by an effective communication network involving both official and unofficial contacts.* The DVDP well illustrates one of the important elements of Antarctic cooperation, that both official and informal communications take place through the formal arrangements under the Antarctic Treaty, through the Scientific Committee on Antarctic Research, and among both scientists and program officials of the various nations. In the period of formative discussion of the DVDP prior to the official start of the project, this network of official and informal contact was essential. It remained an asset throughout the life of the DVDP.

6. *Collaboration in program planning and field operations succeeds to a greater extent than does collaboration in research analysis and publication.* Not surprisingly, as the DVDP entered its research analysis and publication phase, the scientists in each nation tended to seek out methods of publishing research results in domestic journals. Much preliminary DVDP data has been made available by way of the three scientific symposia, one of which was hosted by an institution in each of the three participating nations. However, the further analysis and subsequent publication have tended to focus the scientists' attention on publications in their own nations, where recognition among their immediate peers is the highest. This pattern seems also to have existed in other Antarctic projects and in other cooperative international science projects such as the International Program of Ocean Drilling. It is an understandable pattern and should be planned for.

7. *The project organization must provide for clear lines of management.* In general, the management of the DVDP proceeded effectively, with clear lines of responsibility established. The decision to divide funding responsibility among the three nations undoubtedly aided in this regard. Logistics responsibility too was generally divided. Where there were some problems—for example, in the initial procurement and operation of the drilling equipment—part of the problem lay in the ambiguity of management responsibility. These

problems were recognized quickly by the national program leadership in each of the three nations and corrected.

FUTURE DIRECTIONS

Cooperation among nations working in Antarctica in the planning and execution of larger-scale scientific research programs has been demonstrated effectively by the DVDP. Several other projects have, in the last decade, also shown the benefits of such collaboration. These include the Ross Ice Shelf Project (RISP), the International Antarctic Glaciological Project, and the Antarctic phase of the Deep Sea Drilling Project. Nevertheless, few if any of these projects have so clearly established a method of international research planning and cost sharing as has the DVDP. In the 1980's we should expect this pattern of cooperation to continue and to intensify.

Moreover, as we look forward to a period in which the management of Antarctic's living and nonliving resources becomes an imperative, the models that have been developed for essentially scientific research projects such as the DVDP should be used as a basis for planning. In all respects, the resource assessment needs would benefit from joint planning and cost sharing. The research and assessment needs cover wide areas geographically, involve many disciplines of science, and are logistically complex. They are costly enterprises, beyond the resources of a single nation, and therefore prudent policy direction suggests that cooperative, cost-sharing methods of conducting Antarctic research should be pursued.

Costs of Antarctic operations including the spiraling costs of energy also suggest that nations should pursue more science along the model created by the DVDP. For industrial nations there seems to be no end in sight for general cost increases, for inflation, and for rapidly escalating energy costs. Each nation's Antarctic commitment has to be balanced against a broad range of budgetary and fiscal demands. One would hope that Antarctic program planners would be encouraged by the success of the DVDP and seek to identify other research problems where the sharing of costs would provide more benefit to all participants than any one nation could enjoy through its own conduct of similar research. In the future I would encourage experimentation with the pooling of funds contributed to a single project budget. This extension of the DVDP approach would be an appropriate step in the management of Antarctic research projects in the 1980's.

With the success of the DVDP, RISP, and other projects it would seem that national policy in all of the Antarctic Treaty nations should promote intensified international cooperation as the treaty enters its third and crucially important decade. The future management of Antarctica will surely benefit from an extension of the treaty's basic tenets beyond 1991 into the next century. A successful series of internationally funded and managed research and resource assessment programs will reveal new modes of collaboration that will form a basis for Antarctica's future wise and international management.

In the United States, most recent presidents have espoused international cooperation in science and technology. Impetus to the International Geophysical Year and the Antarctic Treaty itself was provided by President Eisenhower. Presidents Kennedy and Johnson promoted the Global Atmospheric Research Project and the International Decade of Ocean Exploration, respectively. Scientific cooperation is a U.S. policy that has existed with consistency over the last two decades.

President Carter, in a message to the U.S. Congress on science and technology, articulated the U.S. policy on science, technology, and international relations by noting that

> Science and technology is increasingly international in its scope and significance. This international dimension affects the planning and conduct of our research and development activities. Such activities, whether carried out by us or by others, serve to increase the fundamental stock of human knowledge. They can also foster commercial relationships, impact on the quality of life in all countries, and affect the global environment . . . Several themes have shaped my Administration's policy in this area. We are:
>
> —pursuing new international initiatives that advance our own research and development objectives;
>
> —developing and strengthening scientific exchanges that bridge political, ideological, and cultural divisions between countries;

> —formulating programs and institutions that help developing countries use science and technology; and
>
> —cooperating with other nations to manage technologies with global impact.

In the area of new international initiatives, President Carter said

> United States scientific and technological objects are advanced by cooperating with other nations. For example, we work together with many nations on large-scale scientific programs; joint funding of expensive research, development, and demonstration projects; and efforts to alleviate common problems. Two decades ago, the International Geophysical Year set a pattern for international cooperation on large-scale scientific problems. This model has been extended to most fields of science . . . . As the cost of large-scale research programs and research facilities rises, all countries find the financial support increasingly burdensome. We must join together to support the most expensive and significant projects. We are discussing with other nations a program to drill deeply into the offshore continental margins between the continental shelves and ocean basins. The program would provide new knowledge of the sea floor and help us assess the margins' potential for resources. Other large-scale scientific programs that could be pursued jointly include the next generation of high energy physics accelerators, telescopes and fusion energy research facilities.

The examples listed by President Carter could have equally well cited Antarctica. From the U.S. perspective, both the national research and development policy and the U.S. policy objectives for Antarctica suggest that in the next decades the United States would encourage more international projects like the DVDP.

As an experiment in policy as well as in science the DVDP has been eminently successful. Its success served as an illustration of innovation in science planning that seemed appropriate for the last decade. I would hope that all those currently involved in the planning of national Antarctic programs would be encouraged to draw on the DVDP experience as plans are formulated for the Antarctic research and resource assessment programs of the decade ahead.

AEROMAGNETIC SURVEY OF ROSS ISLAND, MCMURDO SOUND, AND THE DRY VALLEYS

D. R. PEDERSON

Chevron Oil Company, Harvey, Louisiana 70058

G. E. MONTGOMERY

Texaco, Bellaire, Texas 77401

L. D. McGINNIS AND C. P. ERVIN

Northern Illinois University, DeKalb, Illinois 60115

H. K. WONG

*Geologisch-Palaontologisches Institut, Universitat Hamburg, Bundesstrasse 55
D-2000 Hamburg 13, Federal Republic of Germany*

Magnetic anomalies of Ross Island, McMurdo Sound, and the dry valley area, Antarctica, are correlated with regional geology. Magnetic measurements were made from helicopters 300 m above ground surface along east-west flight lines spaced 2 km apart. Jurassic dolerite sills produce short-wavelength, high-gradient, primarily positive anomalies in the western dry valleys. Negative anomalies with amplitudes of -300 to -600 γ found north of Shapeless Mountain may be caused by reversely polarized dolerite. Several ground profiles over volcanic vents reveal short-wavelength anomalies having amplitudes up to several thousand gammas that were not detected from the air. The ground measurements indicate both normal and reversed magnetic fields associated with the vents. The eastern dry valleys and most of McMurdo Sound are characterized by a smooth magnetic field, indicating that the relatively nonmagnetic basement complex found in the eastern dry valleys continues beneath the sound. A positive magnetic anomaly with a wavelength of about 30 km, extending westward from Hut Point Peninsula, is probably caused by a layer of pyroclastics up to 2000 m thick; however, the broad, positive anomaly east of the Daily Islands which merges with the lower-amplitude anomaly west of Hut Point Peninsula is caused by a deep-seated, crystalline intrusive body. Extensive, low-amplitude, positive anomalies south of Ross Island may reflect a broad cover of pyroclastics or lava flows beneath the Ross Ice Shelf. Six north-northeast trending magnetic lineaments, each about 5 km wide, up to 50 km long, and with amplitudes up to 1000 γ or more, are observed over Ross Island. Several less distinctive lineaments trending north-south and east-west are also evident. Modeling, age dating, and magnetic measurements on core indicate that the magnetic lineaments are caused by alternating bands of reversed and normally polarized lavas several hundred meters thick. The number of the lineaments reflect the long time period involved in the growth of Ross Island.

INTRODUCTION

Analyses of magnetic fields supplement other investigations of the Dry Valley Drilling Project (DVDP). Models derived from the magnetic fields are inferred beneath ice, water, and nonmagnetic sediments. Magnetic fields are also used to determine lineaments in the crystalline rocks and to describe the configuration of volcanic and intrusive

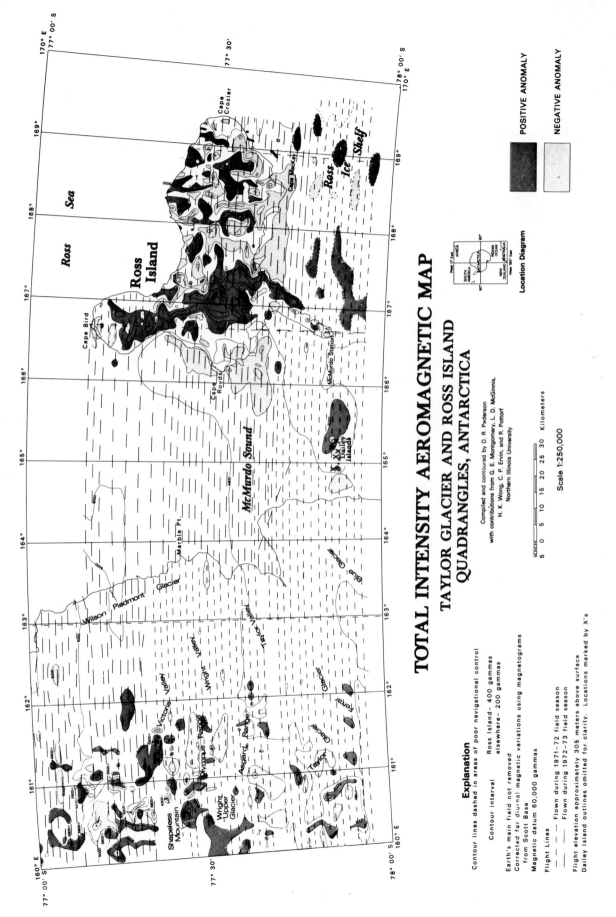

TOTAL INTENSITY AEROMAGNETIC MAP

TAYLOR GLACIER AND ROSS ISLAND
QUADRANGLES, ANTARCTICA

Compiled and contoured by D. R. Pederson
with contributions from G. E. Montgomery, L. D. McGinnis,
H. K. Wong, C. P. Ervin, and R. Pottorf
Northern Illinois University

Scale 1:250,000

Explanation

Contour lines dashed in areas of poor navigational control

Contour interval   Ross Island— 400 gammas
                    elsewhere— 200 gammas

Earth's main field not removed
Corrected for diurnal magnetic variations using magnetograms
    from Scott Base
Magnetic datum 60,000 gammas

Flight Lines
— — — Flown during 1971-72 field season
———— Flown during 1972-73 field season

Flight elevation approximately 305 meters above surface
Dailey Island outlines omitted for clarity. Locations marked by X's

POSITIVE ANOMALY

NEGATIVE ANOMALY

Location Diagram

Fig. 1.   Total field aeromagnetic map. Data are corrected for diurnal changes in the earth's magnetic field which was measured at
Scott Base.

structures and volcanogenic sediments. This study includes an aeromagnetic survey composed of east-west flight lines flown at 2-km intervals over most of the area covered by the U.S. Geological Survey topographic maps (1 : 250,000) of Ross Island and Taylor Glacier (Figure 1). In addition, 11 ground magnetic profiles were made across small volcanic vents in Taylor and Wright valleys for ground control.

An aeromagnetic survey of Ross Island with a spacing of 5 km between lines as well as single traverses flown through Wright and Taylor valleys was discussed by *Robinson* [1964a]. An unpublished, residual total intensity map was constructed by Behrendt [see *Robinson*, 1964a] using the data obtained in these surveys. The earlier surveys are consistent with the present study, considering the wider spacings.

Bull and Irving [1960] and *Bull et al.* [1962] studied the paleomagnetism of dolerite sheets and dikes in the dry valley region and found them to be normally polarized. In magnetic surveys of the Ross Sea, *Adams and Christoffel* [1962] and *Hayes and Davey* [1975] found only small-amplitude anomalies up to 600 γ. The most significant anomaly in the southwestern Ross Sea is an elongate feature trending northeast. *Bennett* [1964] noted that a similar magnetic character is displayed over the Ross Sea and the Ross Ice Shelf. He suggested that shorter-wavelength anomalies over the Ross Ice Shelf are due to a magnetic layer at a depth of about 2.5 km with a susceptibility between 0.0008 and 0.0011. *Ostenso and Thiel* [1964] proposed that this layer consists of Ferrar Dolerite, at or near the top of the Beacon Supergroup. The featureless magnetic field over the Ross Embayment has generally been attributed to magnetic source rocks underlying a thick sedimentary cover. *Cox* [1966] and *McMahon and Spall* [1974] studied the paleomagnetism of volcanic rocks from the southern tip of Ross Island and found some to be reversely polarized.

Rock units in the McMurdo Sound region have characteristic magnetic susceptibilities, and therefore the magnetic field is representative of the geologic framework in which the rocks are contained. The principal geologic units include (1) metasediments of the Ross Supergroup, (2) the Granite Harbor Intrusives, containing both syntectonic and posttectonic plutons, (3) the Beacon Supergroup

[*Barrett et al.*, 1972] composed of flat-lying, nonmarine sediments, (4) the Middle Jurassic Ferrar Dolerites, (5) the McMurdo Volcanics of late Cenozoic age, and (6) glacial, marine, and terrestrial sediments of Cenozoic age.

Prior to the magnetic study it was felt that the magnetic field could be used to interpret at least the major structures in the region. *Warren* [1969] reported that folding in the basement rocks in the dry valley area is predominantly about steeply plunging axes that trend north-northwest; however, the overlying Beacon strata are not folded, showing only a gentle southwest dip. *Schopf* [1969 and *Cullen* [1968] suggest that the Transantarctic Mountains, and hence the dry valley region, may be the margin of an ancient rift that has undergone some faulting. *Hamilton* [1965] proposed that the dry valley area is a large, north trending dome broken by a few normal faults. Both of these ideas are compatible with the widely accepted belief that the Transantarctic Mountains underwent extensive block faulting in the late Tertiary and Quaternary [*Nichols*, 1970]. *Calkin and Nichols* [1972] reported that the generally flat-lying Beacon Sandstone is found below the Ross Sea and beneath Ross Island and also at thousands of feet above sea level in the dry valleys, suggesting the presence of a major fault between the two areas.

Three large faults have also been found in the dry valleys proper. *Angino et al.* [1962] suggested the presence of a major right-lateral, strike slip fault, with a lateral movement of several kilometers, along the axis of Taylor Valley to the north of Nussbaum Riegel. *Hamilton* [1965] described a normal fault south of Taylor Glacier, near longitude 161°20'E, that strikes west-southwest, dips southward, and shows a downdrop of about 600 m on its southern side. He also reported a fault striking west-northwest in Pearse Valley with its north side downdropped about 500 m.

DATA COLLECTION AND ANALYSIS

The aeromagnetic survey was made during the 1971-1972 and 1972-1973 field seasons using a Varian M-50 nuclear precession magnetometer which gives discrete, absolute measurements of the total field intensity. Data were recorded with an Esterline Angus 'S' type portable strip chart recorder. The magnetic sensing head was towed behind

Navy and Coast Guard helicopters with a 60-m cable in 1971-1972 and a 46-m cable in 1972-1973. Both cables were long enough to render the magnetic effects of the helicopter on the sensor negligible.

During both seasons the flight lines were oriented east-west with a spacing of 2 km. Four north-south tie lines were flown across the flight lines to facilitate datum adjustment between lines. The positions of the flight and tie lines are shown on Figure 1. A total of 9390 line km were surveyed, 4958 in the 1971-1972 season and 4432 in the 1972-1973 season. An average air speed of 152 km/h and a surface clearance of 305 m were maintained whenever possible. However, where topography was irregular, it was impossible to meet these conditions. The surface clearance ranged from 600 m over deep, narrow valleys to 100 m over steep cliffs.

A magnetometer polarization rate of 3 s was used during the 1971-1972 season, so that a reading was taken approximately every 160 m of flight path. A 3-s interval was used during the 1972-1973 season until the Esterline Angus recorder broke down, after which the polarization rate was increased to 10 s to facilitate manual recording of the data. This interval produced a reading approximately every 530 m of flight path. Helicopters were equipped with a radar altimeter to determine the height of the flight path above the surface. In addition, a Tacan instrument, with a range of 30 km, was used in the 1971-1972 field season to measure the straight line distance between the helicopter and a radio source at Williams Field near McMurdo Station. Determination of the lateral position of the helicopter during both field seasons depended primarily on visual sighting of topographic features. This was done with the use of 1 : 250,000 scale topographic maps on which the flight and tie lines were marked; therefore the most accurate locations of the flight paths was over Ross Island and the dry valleys, where recognizable surface features frequently occur. In these areas the error in location did not exceed about 0.5 km. Over McMurdo Sound, flight path location was accurately known only at the ends of the lines, where they approached the coasts of Ross Island and Victoria Land. Navigation was poorest over the Ross Ice Shelf, where there are no landmarks; location errors there were probably of the order of 1 km.

The geomagnetic observatory at Scott Base (166°48′E, 77°51′S) within the surveyed area provides continuous variations of the three orthogonal components of the magnetic field. These were recorded at intervals corresponding to the time periods when aeromagnetic observations were being made. The vector sum of the components was subtracted from the average total intensity value for the period of the survey to give the magnetic variation for each observation time. This variation was then subtracted from the corresponding interval of the aeromagnetic record to complete the daily variation correction. With this method of correction the assumption was necessarily made that the magnetic variation measured at Scott Base is the same as at our aeromagnetic survey locations.

All readings of this survey were taken within 200 km of Scott Base and most of them were within 70 km. Therefore it appears that daily variation corrections based on the Scott Base magnetograms are probably valid to within at least 10 γ. Since the anomalies that are considered important in this investigation have amplitudes of several hundred gammas or more, a possible 10-γ error in data correction is not significant.

Locations were determined by computer and data were plotted on a 1 : 250,000 scale base map with the aid of a CalComp plotter. The corrected total magnetic intensities of both the flight and tie lines were marked on the map at their corresponding locations and hand contoured (Figure 1). In a few instances a particular flight line would not 'fit in' with adjacent flight and tie lines; the values along it were consistently too high or low. This most likely occurred because the diurnal variation correction determined from the Scott Base records differed from the field value. In such cases, all the values along the anomalous flight line were changed by a constant to make the values consistent with adjacent flight lines and crossing tie lines. The regional field intensity in Victoria Land in the area of the survey increases to the north-northwest, toward the south magnetic pole at an average rate of 5 γ/km [*Robinson*, 1964a]. This field was removed graphically to give the residual magnetic intensity map (Figure 2).

Magnetic susceptibility measurements (Table 1) were made on samples from DVDP holes 2 and 3, located near McMurdo Station, and hole 6, at Lake Vida. Additional susceptibilities were measured for rocks collected from volcanic vents during a

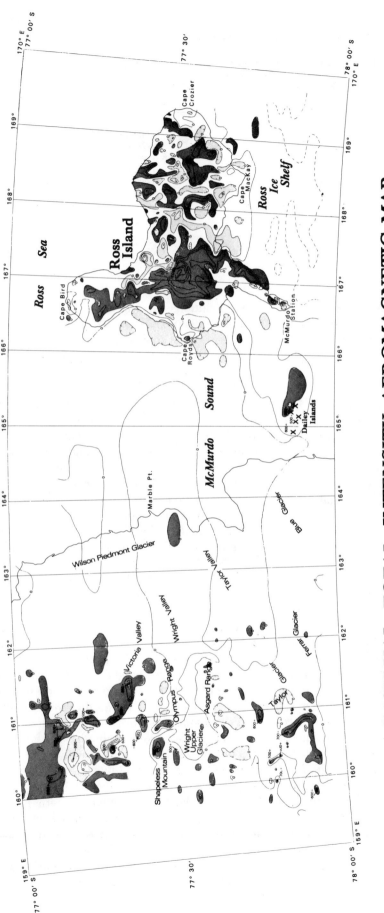

RESIDUAL TOTAL INTENSITY AEROMAGNETIC MAP

TAYLOR GLACIER AND ROSS ISLAND QUADRANGLES, ANTARCTICA

Compiled and contoured by D. R. Pederson
with contributions from G. E. Montgomery, L. D. McGinnis,
H. K. Wong, C. P. Ervin, and R. Pottorf
Northern Illinois University

Scale 1:250,000

Explanation

Contour lines dashed in areas of poor navigational control

Contour interval   Ross Island– 400 gammas
                  elsewhere– 200 gammas

Earth's main field removed graphically

Corrected for diurnal magnetic variations using
magnetograms from Scott Base

Aeromagnetic flight lines flown in the east–west direction
with a line spacing of 2 kilometers

Flight elevation approximately 305 meters above surface

Dailey Island outlines omitted for clarity. Locations marked by X's

POSITIVE ANOMALY

NEGATIVE ANOMALY

Fig. 2. Residual aeromagnetic map. The regional magnetic field was removed graphically to give the residual map.

TABLE 1.   Magnetic Susceptibilities in the Dry Valley Region

Location	Susceptibility $\overline{k}$	Number of Samples	Range	Sources
McMurdo Volcanic Group				
Ross Island lavas and flows*	0.003	12	0.0007-0.018	1, 3
Ross Island pyroclastics†	0.0001	4	0.00008-0.00038	1
Dry valley cones‡	0.001	44	0.00008-0.003	1
Ferrar Dolerites	0.001	5	0.0008-0.012	1, 2
Vida Granite	0.00006	2	0.00004-0.00006	1
Olympus Granite Gneiss	0.00005	2	0.00003-0.00006	1
Beacon Supergroup Sediment	0.00002	1	•••	1

*This paper.
†*Bull et al.* [1962].
‡*Robinson* [1964].

ground survey. Measurements were made in triplicate using a Soiltest MS-3 magnetic susceptibility bridge on samples crushed to 5-mm-sized fragments.

INTERPRETATION

To facilitate discussion of the magnetic field, the area of investigation is divided into four zones based on variations in magnetic anomaly characteristics. These zones are (1) the western dry valleys, from longitudes 160°E to approximately 162°30'E, (2) the eastern dry valleys and McMurdo Sound, (3) the Ross Ice Shelf, and (4) Ross Island.

Western Dry Valleys

This zone is characterized by a large number of relatively short wavelength (up to about 10 km) anomalies with amplitudes ranging from approximately 100 to 600 γ and gradients up to 400 γ/km (Figure 2).

Geologic maps of *Mirsky* [1964], *Warren* [1969], and *Lopatin* [1972] show that the Ferrar Dolerites occur extensively in the western dry valleys. The dolerites are the only rocks other than McMurdo Volcanics that have magnetic susceptibilities significantly higher than other rock types in the area (Table 1). Basalt cones, flows, and feeder dikes of McMurdo Volcanics are also present in the dry valleys [*Angino et al.*, 1962; *Calkin et al.*, 1970; *Fleck et al.*, 1972]. Anomalies caused by these features were modeled using a magnetic modeling program based on the polygonal lamina technique [*Talwani*, 1965]. The cones and flows are, in general, less than 10 m thick and 250 m in diameter. The susceptibility contrast used, 3.3×10^{-3}, is the difference between the minimum susceptibility of the dry val-

ley volcanic rocks and basement rocks. As measured at 300 m above a model having these characteristics, the theoretical anomaly would have an amplitude of less than 10 γ. For purposes of modeling, the largest volcanic neck was assumed to be about 100 m in diameter. Volcanic necks should theoretically cause an anomaly with an amplitude up to about 70 γ and wavelength of approximately 100 m, as measured at 300 m above the surface. An examination of the residual anomaly map (Figure 2) shows that magnetic anomalies are not found over locations where the cones, flows, and necks were described. It is probable that no flight lines crossed over an anomaly caused by one of these features, and if a line did cross, the magnetic sampling rate permitted observations at only 160-m intervals. Thus an anomaly could have been missed altogether, or it might have appeared as a one-point anomaly and thus would be ignored in the contouring.

After completion of the air survey a ground magnetic study was made over selected volcanic vents in both Taylor and Wright valleys. From the ground observations and from measurements made on samples it was found that the small cones, flows, and volcanic necks do indeed contribute magnetic anomalies of the wavelength and amplitude predicted from their physical appearance. Anomalies on the ground, directly over the cones, have very short wavelengths, but they do have amplitudes up to 1200 γ, including both normal and reversed polarity. A ground magnetic profile over a small cone near the Sollas Glacier in Taylor Valley is shown in Figure 3. It is concluded that magnetic anomalies measured during the air survey in the western dry valleys are due entirely to sills and dikes of the Ferrar Dolerite.

TSIA

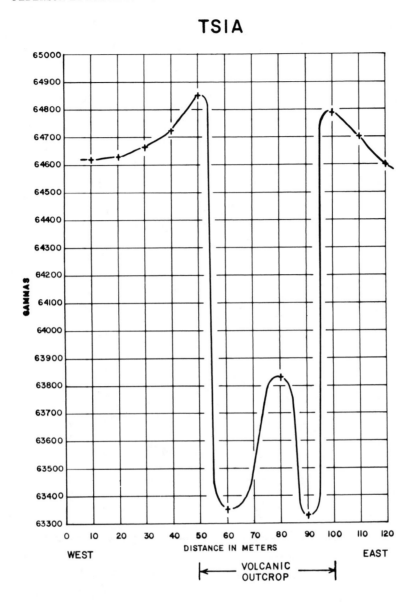

Fig. 3. Magnetic anomaly from a ground survey over Sollas Glacier volcanic center in Taylor Valley.

Ferrar Dolerite occurs mainly at three sills: (1) the basement sill, which intrudes granites of the basement complex, (2) the peneplain sill, which occurs along the unconformity between the basement complex and the overlying Beacon rocks, and (3) the Beacon sill, found within the Beacon strata. The two lower sills occur very extensively in the western dry valleys [*Hamilton*, 1964], whereas the uppermost sill shows irregular emplacement [*McKelvey and Webb*, 1961].

The average susceptibility value (Table 1) for the sills is 0.001, which is close to the average value of the McMurdo Volcanics. This value, however, does not take remanent magnetism into account. *Bull et al.* [1962] found that the remanent magnetization of the sills is polarized in the same direction as the present field of the earth, so that it can be added to the induced magnetization by using the equation of *Nagata* [1961]:

$$K_A = K(1 + Q) \qquad (1)$$

where K_A is apparent susceptibility, to be used in calculating the theoretical anomaly, K is true sus-

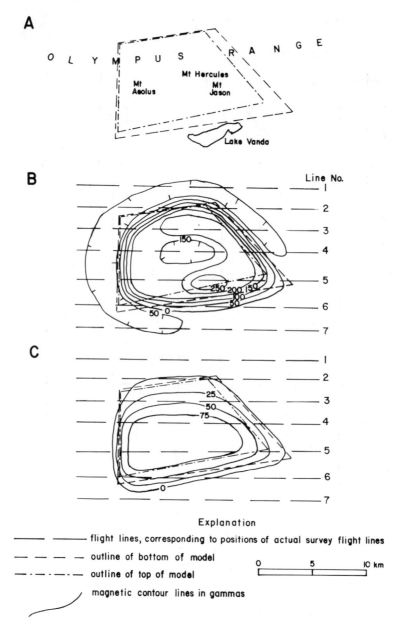

Fig. 4.   Theoretical anomalies for a model of the basement sill of Ferrar Dolerite north of Lake Vanda beneath the Olympus Range: (*a*) shows the location of the model, (*b*) gives the theoretical magnetic field at a height 300 m above the model, and (*c*) gives the theoretical magnetic field at a height 1700 m above the model.

ceptibility, and Q is Koningsberger factor equal to J/KF [*Garland*, 1971], where J is intensity of remanent magnetization and F is total intensity of the earth's field where the rock is located.

Bull et al. [1962] measured the ratio of remanent magnetization to susceptibility for the sills and obtained average values of 1.0 (basement sill), 1.7 (peneplain sill), and 2.0 (Beacon sill). Substituting these ratios in (1) produced apparent susceptibility

values of 0.0038 for the basement sill, 0.0054 for the peneplain sill, and 0.0061 for the Beacon sill, respectively. These values were used in calculating theoretical anomalies caused by the three sills.

In order to study the magnetic effects of the sills they were modeled individually. The basement and peneplain sills generally occur at great depths beneath the mountain ranges of the western dry valleys. Along the flight lines, over the peaks of the ranges, the magnetometer sensor was much farther from these sills than along the flight lines through the valleys. To approximate both cases, the theoretical field was calculated at different elevations above the sill models.

Basement sill. The basement sill occurs at approximately 1600 m beneath the peaks of Mount Aeolus, Mount Hercules, and Mount Jason in the Olympus Range, just to the north of Lake Vanda [*Warren*, 1969]. The lateral shape of this sill was approximated from several geologic maps [*Mirsky*, 1964; *Warren*, 1969,; *Lopatin*, 1972] and is shown in Figure 4. Its thickness was assumed to be 270 m as estimated by *Hamilton* [1964]. The theoretical magnetic field was calculated at elevations of 300 and 1700 m above the model. The magnetic contribution of the basement sill can be estimated along lines numbered 1, 2, 6, and 7 in Figure 4 above the valleys flanking the Olympus Range. The field intensity along lines 1 and 7 is negative and averages

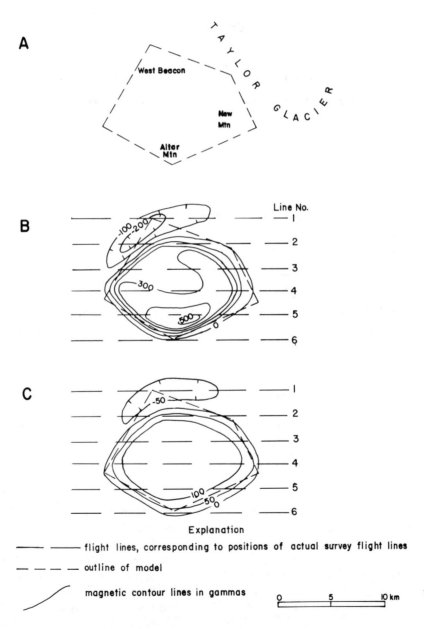

Fig. 5.   Theoretical anomalies for a model of the peneplain sill of Ferrar Dolerite at New Mountain southwest of Taylor Glacier: (*a*) shows the location of the model, (*b*) gives the theoretical magnetic field at a height 300 m above the model, and (*c*) gives the theoretical magnetic field at a height 1150 m above the model.

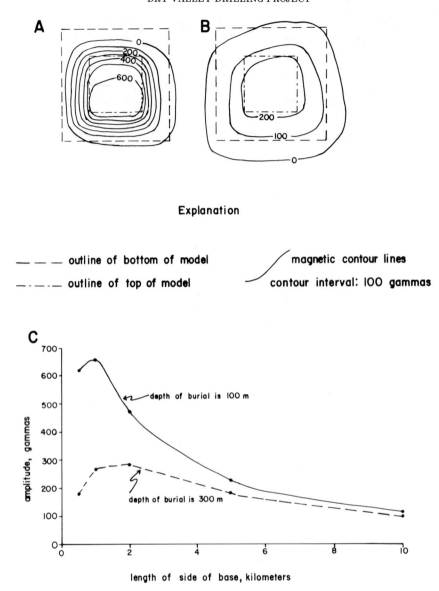

Fig. 6.   Theoretical anomalies for a model of the Beacon sill in the shape of a truncated pyramid with a square base 1 km on a side and a square top 0.5 km on a side: (*a*) gives the theoretical magnetic field at a height 100 m above the model, (*b*) gives the theoretical magnetic field at a height 300 m above the model, and (*c*) shows the relationship between lateral dimensions of the model and the amplitude of the theoretical anomaly.

about -20 to -80 γ. Along line 2 the intensity varies from about 100 γ near the center of the northern boundary to about -100 γ just outside the boundaries. The intensity along flight line 6 ranges from 50 to -80 γ. Along flight lines 3, 4, and 5, 100 m above the Olympus Range, the maximum contribution of the magnetic field is 80 γ. The magnetic intensity drops rapidly along all three lines toward the east and west edges of the model and becomes slightly negative beyond the model boundaries.

In summary, the basement sill should contribute less than 100 γ to the magnetic field as measured directly over the sill and about -20 to -80 γ as measured outside the boundaries of the sill.

Peneplain sill.   Theoretical models were likewise developed for the peneplain sill beneath New Mountain, East and West Beacon, and Altar Mountain (Figure 5). The thickness of the sill was estimated to be 300 m from a geologic cross section of New Mountain [*Hamilton*, 1965]. Its depth of burial beneath the peak of New Mountain was assumed to be 1050 m. The theoretical magnetic field was then calculated at elevations above the model of 1150 and 300 m, respectively. The former may be compared to measurements along flight lines 2, 3, 4, and 5, where the magnetometer sensor was towed over the mountains at a terrain clearance of about 100 m. The maximum contribution is 180 γ. The theoretical intensity decreases away from the center of the model and becomes slightly negative (of the order of -30γ) outside of the model boundaries.

Figure 5 shows the theoretical intensity produced by the sill model along lines 1 and 6. The values are all negative along line 1, reaching a minimum of about -200γ at the northern edge of the model. Along line 6 the values are also negative and are of the order of -40 to -100γ except at the southern edge of the model where the magnetic values approach zero.

In summary, the peneplain sill at this location should contribute up to 180 γ to the field as measured directly over the sill, and up to -200γ as measured immediately outside the boundaries of the sill.

Beacon sill. The Beacon sill forms the top layer of many of the peaks in the western dry valleys because it is more resistant to erosion than the surrounding sandstone. Along flight lines over these peaks the magnetometer sensor was, in some cases, no more than 100 m above the peaks. When modeling this sill therefore, the theoretical field was calculated at 100 m above the model (Figure 6) as well as at 300 m, the usual surface clearance. To approximate a horizontal slab of the sill that has been eroded along its upper edges, the shape of the model was chosen to be a truncated pyramid, with a square base 1 km on a side and a square top 0.5 km on a side. The model consisted of five layers, with a total thickness of 150 m. At 100 m above the model the theoretical anomaly has steep gradients and an amplitude of more than 600 γ. At 300 m above the model the theoretical amplitude is considerably less, of the order of 250 γ.

The Beacon sill in the western dry valleys undoubtedly occurs in a wide variety of shapes and sizes. To investigate the effects of size, the lateral dimensions of the truncated pyramid were altered variously to 0.5, 2, 5, and 10 times those of Figure 6, while keeping the thickness, magnetic susceptibility, and elevations above the model constant. The smallest model, with a 0.5-km-square base, was found to cause a theoretical anomaly with gradients and amplitude similar to those produced by

the model in Figure 6. The anomalies produced by the larger models, however, had lower amplitudes. As the ratio of the lateral dimensions to the thickness increases, the anomaly shape flattens out, and the amplitude decreases sharply. The amplitude of a particular anomaly can be used to estimate, within an order of magnitude, the lateral extent of the slab of Beacon sill causing the anomaly. Assuming a thickness of the order of 150 m and apparent susceptibility contrast of 0.0061, the slab must be smaller than 2 km across to produce an anomaly with an amplitude greater than about 500 γ. It must be kept in mind, of course, that the lower two sills are also contributing 200-300 γ to these positive anomalies.

The fact that the amplitude of the anomaly decreases as the lateral dimensions increase occurs because the anomaly is caused by the superposition of the magnetic effects from the sides of the slab and the top and bottom of the slab [*Jakosky*, 1950; *Vacquier et al.*, 1951]. As the sides become farther from the point of observation above the slab, the effects of the opposite sides overlap less, and the peak amplitude of the anomaly decreases.

Eastern Dry Valleys and McMurdo Sound

Only rocks of the basement complex having low susceptibilities (Table 1) outcrop in the eastern dry valleys. This is borne out by an examination of Figure 2, as the magnetic field over this area is very smooth. Although several large faults have been found in the eastern dry valleys [*Angino et al.*, 1962; *Hamilton*, 1965], they are not evident from the magnetics because they occur in rocks that have very little magnetic contrast.

McMurdo Sound, south of approximately 77°45'S, is characterized by (1) steep gradient, short-wavelength anomalies associated with the volcanic Dailey Islands and Tent Island and (2) a broad magnetic high extending westward from Hut Point Peninsula. The magnetic high between the Dailey Islands and Hut Point Peninsula might be caused by (1) sills and dikes of the Ferrar Dolerites, (2) lava flows on the bottom of the sound, or (3) a deep intrusion association with the McMurdo Volcanics. Modeling studies show that the body causing the anomaly is buried at depths greater than 4 km and has a thickness of at least several kilometers.

A number of models were tried, with different shapes, depths of burial, and susceptibilities. The theoretical field, observed field, model, and inter-

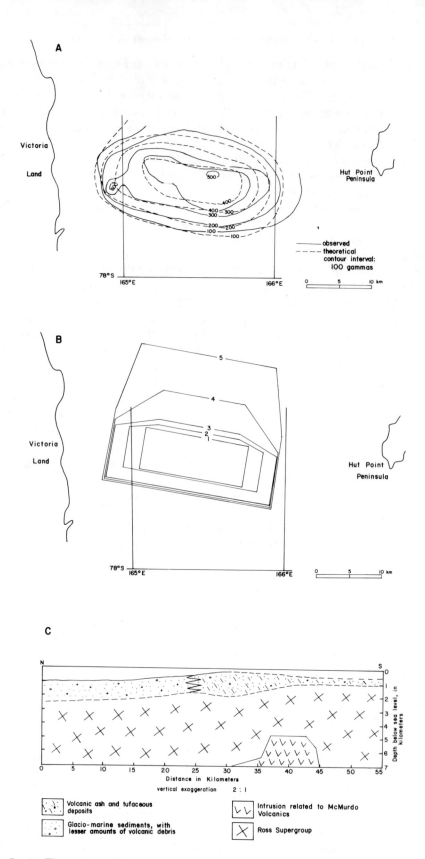

Fig. 7. (a) Theoretical and observed magnetic anomalies west of Hut Point Peninsula. (b) Proposed model from which the theoretical field shown in 7a was derived. (c) Hypothetical geologic cross section through the model in the north-south direction in McMurdo Sound.

preted profile are shown in Figure 7. A susceptibility contrast of 0.0054 was calculated from (1), using a true susceptibility of 0.003, the average for Ross Island rocks, and a Q factor of slightly less than 1. The largest deviation between the theoretical and observed anomalies is found on the west side of the anomaly and is caused by the volcanic rocks of the Dailey Islands.

The top of the model is 4.7 km below sea level. A bathymetric chart of McMurdo Sound [*McGinnis*, 1973] shows that the water depth is between 200 and 600 m. Therefore the body causing the anomaly is interpreted to lie more than 4 km beneath the floor of McMurdo Sound. The total thickness of the model was 3 km. However, the modeling studies showed that additional layers could be added to the bottom of the model with only a slight increase in the amplitude of the anomaly. The vertical extent of the body causing the anomaly could therefore be much greater than 3 km.

The magnetic high extending westward from Hut Point Peninsula has low gradients and an amplitude of about 200 γ (Figure 2). *Robinson* [1964*b*] discussed seismic refraction and gravity profiles made in this area and concluded that the subsurface of the sound consists locally of two layers. The uppermost layer was found to have a thickness of about 1500 to 2000 m and a seismic velocity of about 3.4 km/s. Robinson suggested that it is composed of volcanic ash and tuffaceous deposits similar to material on Ross Island. The second layer, with a velocity of about 4.79 km/s, may be basalt [*Robinson*, 1964*b*]. Alternatively, this layer could be an extension of the basement complex as found in the dry valleys area, which has velocities ranging from 4.8 to 5.8 km/s [*Clark*, 1972].

Using the susceptibility of pyroclastics (Table 1) and assuming a water depth of about 500 m, a model was developed to approximate the uppermost layer in this area. The model produced theoretical gradients and amplitude similar to those observed, supporting Robinson's suggestion that this layer consists of volcanic debris.

North of approximately 77°45'S, McMurdo Sound is characterized by a smooth magnetic field, except for several broad lows flanking Ross Island (Figure 2). These lows can be correlated with bathymetric depressions adjacent to the Island [*McGinnis*, 1973].

The smoothness of the magnetic field over most of McMurdo Sound and the seismic velocities determined by *Robinson* [1964*b*] in the sound and by

Clark [1972] in the dry valleys suggest that most of the sound is floored by a continuation of the basement complex found in the dry valleys. Extensive deposits of sediments mixed with randomly oriented volcanic debris overlie these rocks. Modeling studies show that a 600-m thick upper layer of pyroclastics would produce an anomaly of only about 50 γ. On the basis of bathymetric [*McGinnis*, 1973], magnetic, and gravity and seismic [*Robinson*, 1964*b*] data, a north-south geologic profile in the sound was constructed (Figure 7*c*). The profile extends along longitude 165°30'E between latitudes 77°30'S and 78°00'S.

Ross Ice Shelf

South of Ross Island a number of short-wavelength (up to 10 km) anomalies were found superposed on a magnetic field that is consistently about 200 or 300 γ higher than the regional field (Figure 2). This high field could be due to a layer or pyroclastics or submarine lava flows over the entire area. The short-wavelength anomalies have low gradients and amplitudes. Depth estimates were made for several of these anomalies by the method given by *Grant and West* [1965]. Because the anomalies are very poorly defined, the computed depths are probably accurate only to an order of magnitude. The depth estimates range from 2 to 4 km. These anomalies perhaps indicate the locations of vents for the proposed flows.

Bennett [1964] determined that the shorter-wavelength anomalies found over much of the Ross Ice Shelf are due to bodies at a depth of about 2.5 km with computed susceptibilities between 0.008 and 0.0011. *Ostenso and Thiel* [1964] proposed that these bodies consist of volcanic rocks at or near the top of the Beacon Supergroup. It is possible that the short-wavelength anomalies found south of Ross Island in the present investigation have a similar cause.

Ross Island

Magnetic fields over the volcanic islands of Hawaii, Kamchatka, and Japan contain topographic features that produce bipole anomalies [*Malahoff*, 1969]. In the magnetic latitude of Ross Island such bipole anomalies should consist of a magnetic low, located north-northwest of a high, which is shifted toward the south magnetic pole. Although Ross Island is also composed of volcanic rocks, the ob-

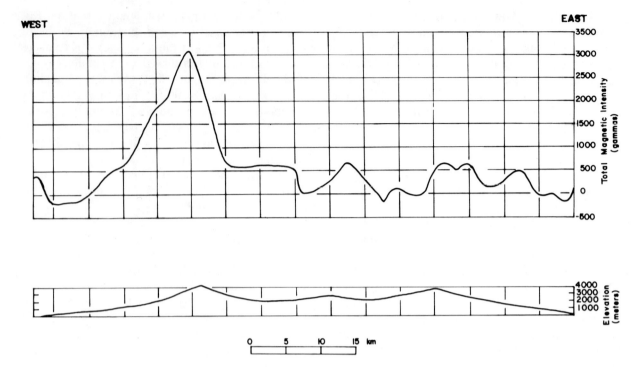

Fig. 8.   Topography and residual magnetic profile over Ross Island along latitude 77°30′S.

served field appears to be more complex than the fields discussed by Malahoff. Figure 8 shows profiles of the topography and magnetic field across Ross Island. The profiles are located along the latitude line 77°30′S. It is apparent that there is only a very general correlation between the topography and the magnetic intensity, suggesting that the anomalies are caused primarily by variations in the polarity of the remanent magnetization and the susceptibility of the rocks. The remanent magnetization is very important, since it almost always predominates over the induced magnetization in volcanic rocks [*Doell and Cox*, 1965].

In studies of the intensity of remanent magnetization of Hawaiian basalts, *Tarling* [1965] found wide variations ranging from 0.00096 to 0.13730 cgs units. *Malahoff and Woollard* [1966] measured the susceptibilities of Hawaiian basalts and obtained values ranging from 0.0005 to 0.0133. Similar wide variations were found for volcanic rocks of Kamchatka [*Steinberg and Rivosh*, 1965]. The comparatively low number of susceptibility measurements made on rocks from Ross Island (Table 1) also show considerable variations.

DISCUSSION

On the basis of the theoretical modeling of anomalies in the western dry valleys the following statements can be made:

1. The basement sill, where it is present at depths of about 1500 m or more beneath the mountain ranges, contributes less than 100 γ to the magnetic field as measured above the ranges. This sill makes a significant contribution to the negative anomalies found over some of the valleys.

2. The peneplain sill also contributes to both the positive anomalies over the mountain ranges and the negative anomalies over the valleys. However, since the peneplain sill occurs above the basement sill and has a higher apparent susceptibility, the magnitude of the contributions is larger. Many of the comparatively broad, low-amplitude (100 to 300 γ) anomalies in the western dry valleys area are caused by the peneplain and basement sills.

3. The uppermost sill is primarily responsible for most of the steep gradient, high-amplitude (600 γ or more) positive anomalies in the dry valleys.

Anomalies with amplitudes greater than about 500 or 600 γ can be largely attributed to slabs of Beacon sill less than 2 km across.

4. Volcanic cones are associated with magnetic anomalies that are of too short a wavelength to be seen using the techniques described in this study.

Specific anomalies in the dry valley region were not modeled. Many of the anomalies are essentially defined by only one flight line, for which three-dimensional modeling obviously could not be done. Even the anomalies defined by more than one line were not delineated well enough for modeling purposes. *Vacquier et al.* [1951] state that a flight line spacing of half the depth of burial delineates an anomaly adequately for modeling. By this criterion, modeling was not justified for any of the dry valley anomalies.

Depth estimates were made on 14 anomalies in the western dry valleys using the method discussed by *Grant and West* [1965, p. 344], as modified from *Vacquier et al.* [1951]. The method uses the maximum gradient and amplitude of an anomaly and the two widths of the anomaly pattern parallel to and transverse to its 'strike' direction at an amplitude which is halfway between the maximum and minimum values. These quantities are used with complementary curves [*Grant and West,* 1965, p. 348] to determine the depth of the prism. This depth estimation method is considered most reliable for sources having dimensions which are large compared to the depth of burial [*Dobrin,* 1960], a condition which is satisfied for the dry valley anomalies.

Inaccuracies in the depth estimates are probably caused by the following:

1. The method assumes that the source extends to infinite depths. It can, however, be used for horizontal plates as in the present study, but the depths of burial will be overestimated [*Grant and West,* 1965]. According to *Ervin* [1972] the field for a prism of finite thickness is only 76% of the field above an infinite prism when the thickness equals the depth of burial, as is true for many of the dry valley anomalies.

2. The method assumes that the sides of the source are vertical. This is undoubtedly not the case for many of the anomalies in the dry valleys, which are probably caused by small slabs of the uppermost sill and considerably larger slabs of the basement and peneplain sills.

3. The method becomes progressively less reliable as the anomalies become more poorly defined [*Vacquier et al.,* 1951].

Using the results from modeling hypothetical dolerite sills and the depth estimates, anomalies found in the western dry valleys were grouped according to their probable causes. The first group is made up of positive anomalies with steep gradients (about 400 γ/km) and generally high amplitudes (up to 600 γ or more). They occur over the Beacon Supergroup and the Ferrar Dolerites and are associated with topographic highs. It is likely therefore that their primary source is the Beacon sill.

The second group consists of positive anomalies with gradients ranging from 100 to 300 γ/km and low to medium amplitudes (from 100 to 300 γ). These are also associated with the Beacon sandstones and the Ferrar Dolerites and in many cases occur very near the trace of the Kukri surface, the unconformity between the basement complex and the Beacon Supergroup. They are commonly observed along the sides of valleys and glaciers and are thus caused by the basement and peneplain dolerite sills.

The third group consists of positive anomalies found over ice-covered areas. They indicate the locations of topographic highs, capped by the uppermost sill, that have been buried by ice. Depth calculations were made for five of them to estimate the thickness of the ice over the topographic highs. Depths range from 200 to 450 m.

The fourth group is composed of generally broad, low-gradient (less than 100 γ/km) negative anomalies with amplitudes of −200 to −300 γ. With one exception they are all associated with valleys or glaciers, suggesting that they may be attributed to the basement and peneplain dolerite sills, which occur at the sides of the valleys and glaciers. The exception is located over the southern part of the Asgard Range near the trace of the Kukri surface and probably also is caused by the two lowermost sills.

The fifth group consists of negative anomalies with amplitudes that appear to be too large to be caused by edge effects of the dolerite sills, as in group four. The anomalies are located about 30 km north of Shapeless Mountain (Figure 2). Reversely polarized dolerite is proposed to account for these negative anomalies. A number of positive anomalies were also found in the area, indicating that normally polarized dolerite may also be present. Only normally polarized dolerite has been found in Victoria Land to date, but the number of samples studied paleomagnetically is small [*Bull et al.,* 1962]. Both normal and reverse polarization of Jurassic dolerite has been found elsewhere in Antarctica, at longitude 30°W and latitude 80°S near the Theron Mountains [*Blundell and Stephenson*, 1959].

McDougall [1963] dated samples of dolerite from the dry valleys and found ages ranging from 147 to 163 m.y. *Jones et al.* [1973] reported ages from these dolerites ranging from 150 to 180 m.y. While the geomagnetic field was predominantly normal from 147 to 180 m.y. ago, reversed events did occur from approximately 147 to 150, 168 to 169, and 175 to 176 m.y. [*McElhinny and Burek*, 1971]. Some of the dolerites north of Shapeless Mountain therefore could have been formed during one of these time intervals.

The complex magnetic field over Ross Island is also explained as being composed of both reversely and normally polarized rocks; however, they are much younger than the Ferrar Dolerites, ranging in age from slightly greater than 4 m.y. to very recent [*Armstrong*, 1978]. Much of the volcanic material therefore could have formed during the Matuyama reversed epoch, from approximately 2.43 to 0.69 m.y., or the Gilbert reversed epoch, from about 4.5 to 3.32 m.y. According to *Giggenbach et al.* [1973], Mt. Erebus and its associated lava flows were erupted since 1 m.y. ago. Most of Mt. Erebus therefore consists of Brunhes normally polarized rock. This is supported by the relatively simple character and high amplitude (over 3400 γ)of the anomaly (Figure 2).

According to *Cole and Ewart* [1968], Cape Bird is made up of seaward dipping basalt flows erupted from the main Mt. Bird cone and later penetrated by Trachyte and basalt plugs and cones. K-Ar age determinations [*Armstrong*, 1978] indicate that the rocks were formed during the Gauss normal and the Matuyama reversed epochs. Since both normal and reversed rocks occur in the area and because of the structural complexity, a complex and irregular magnetic field is observed over Cape Bird (Figures 1 and 2). *Malahoff* [1969] found that the magnetization of intrusive basalts is usually of greater intensity than the magnetization of flow basalts; therefore it is suggested that the high-intensity, positive anomalies over Cape Bird mark the locations of basaltic plugs of normal polarity.

The basaltic shield volcanoes of Mt. Terror and Mt. Terra Nova [*Dort*, 1972] are also associated with a very complex magnetic field. Again, the presence of strong magnetic lows and highs indicates the occurrence of both normally and reversely polarized rocks.

Cole et al. [1971] state that Cape Crozier consists of a sequence of alternating basalt and trachyte. K-Ar age determinations [*Armstrong*, 1978] indicate that much of the volcanic material making up Cape Crozier was formed during the Matuyama reversed epoch, which agrees with prominent magnetic lows shown on Figures 2 and 3. The Knoll, near Cape Crozier, consists of basanite and phonolite [*Cole et al.*, 1971] and is also characterized by a magnetic low. Magnetic highs are located over outcrops of olivine-augite basalt north of Post Office Hill also near Cape Crozier.

From the character of the magnetic field over Cape Royds (Figures 1 and 2), it appears that almost the entire area, with the exception of the tip of Cape Royds, consists of reversely polarized rock. Age determinations [*Treves*, 1968; *Armstrong*, 1978] indicate that formation of these rocks occurred near the end of the Matuyama reversed epoch. The probable occurrence of normally polarized rock near the tip of Cape Royds suggests that these rocks are younger than surrounding rocks of the area.

Hut Point Peninsula consists of a number of coalesced cones which lie in two trends striking northnortheast, with the more prominent trend lying along the western coast. Basalt is the dominant rock type [*Cole et al.*, 1971]. Phonolite from Observation Hill at the tip of Hut Point Peninsula was found to be reversely polarized by *Cox* [1966]. K-Ar dates of about 1.18 m.y. [*Forbes et al.*,1974] indicate that the hill was formed during the Matuyama reversed epoch. The large negative anomaly over the hill, with an amplitude over -600 γ (Figures 1 and 2), agrees with these findings. The core recovered from DVDP holes 1 and 2 near McMurdo Station was also found to be reversely polarized [*McMahon and Spall*, 1974].

Rock formed during the Brunhes normal epoch is also present on the peninsula, as shown by K-Ar dates of 0.43 for samples obtained about 1 km north of McMurdo Station. Positive anomalies were found over Crater Hill and Castle Rock, suggesting that these volcanic features are probably younger than 0.69 m.y., although *Kyle* [this volume] argues for an age of 1.18 m.y. for Castle Rock.

Careful examination of Figures 1 and 2 reveals a number of major magnetic lineaments that trend north-northeast. There also appears to be minor north-south and east-went trends. Several of the north-northeast lineaments can be traced over as great a distance as 50 km or more, from the southern tip of Hut Point Peninsula to the northern coast of Ross Island. The lineaments appear to terminate south of Ross Island and may be explained by the general lack of magnetic correlation between subaerial and submarine volcanism [*Ward*, 1971; *Marshall and Cox*, 1971; *Barth*, 1962; *Moore and Fiske*, 1969; *Serson, et al.*, 1968].

The north-south and north-northeast trending lineaments resemble those produced by sea floor spreading. A strip of normally polarized material containing Mt. Erebus is flanked by bands of magnetic lows that are probably caused by reversely polarized rock. To the east as many as five alternating positive and negative stripes can be identified. Gradients between the north-northeast trending lineaments of magnetic highs and lows are about 300-600 γ/km. Models were developed to estimate the thickness of the volcanics causing these gradients. The models consisted of two adjacent, oppositely polarized horizontal slabs, each having lateral dimensions of 5×20 km. The true susceptibility was assumed to be 3×10^{-3}, which is the average susceptibility for Ross Island rocks. The Q factor was varied from 1 to 4. *Watkins and Richardson* [1968] found Q factors for volcanic rocks ranging from 0.61 to 36.2 with an average about 4. The apparent susceptibility of the rocks was then

calculated using (1). It was found (Figure 9) that theoretical gradients similar to those observed between the lineaments on Ross Island could be produced by slabs 150 m thick if the Q factor is 4. Assuming a Q factor of about 2 or 3, the thickness should be of the order of 200 to 400 m to produce gradients of 300 to 600 γ/km.

On the basis of magnetic models it is proposed that the lineaments are due to normal and reversed flows which emanate from fractures beneath Ross Island. *Dort* [1972] suggested that Mt. Erebus, Mt. Bird, Mt. Terra Nova, and Mt. Terror, along with the smaller cones of Hut Point Peninsula, lie along large orthogonal fractures. *Wellman* [1964] proposed several west-northwest and north-northeast trending faults in Hut Point Peninsula following lines of volcanic cones and necks, whereas *Kyle and Cole* [1974] believe 120 radial faults extend outward from Mt. Erebus through Hut Point Peninsula, Cape Bird, and Cape Crozier. From the magnetic map shown in Figure 2 a Hut Point Peninsula anomaly extends north-east, completely across the island. Northeast of the peninsula the anomaly is negative, whereas on the peninsula the magnetic lineament contains both positive and negative anomalies. *Kyle and Cole*'s [1974] speculation of radial symmetry might be true in part, but it does not explain the lack of a strong east-trending

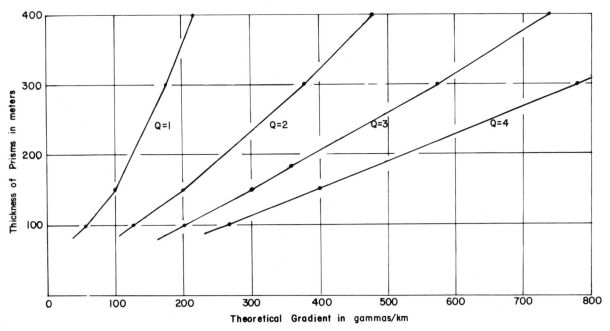

Fig. 9.   Theoretical gradients over Ross Island. The curves show the relationship of the Q factor, model thickness, and theoretical gradient between two adjacent, oppositely polarized, 5×20 km prisms, each with a true susceptibility of 0.003.

lineament through Mt. Erebus and Cape Crozier. It should be noted, however, that the map is constructed from east-west flight lines from which it would be extremely difficult to identify even a strong east-west lineament. As many normal and reversed anomalies are present, it is clear that the island has been constructed over a long period of time during several normal and reversed magnetic intervals. An easterly trending magnetic lineament could have been obliterated by later eruptions along secondary fractures. The fractures may be related to tectonism of the Transantarctic Mountains, to crustal tilting and fracture due to glacial bending, or to a combination of the two. Mt. Erebus and Mt. Bird are located on a northwest-southeast lineament, whereas a weak, east-west lineament can be correlated with Mt. Erebus, Mt. Terra Nova, and Mt. Terror. Well-defined lineaments are found above the north-northeast trend of volcanic cones and necks of Hut Point Peninsula and are in line with Mt. Discovery to the south, which is also volcanic and shows a north-northeast topographic trend.

SUMMARY

The Ferrar Dolerites are the major source of the magnetic anomalies observed in the dry valleys. Positive anomalies with gradients of about 400 γ/km and amplitudes up to 600 γ or more are caused by slabs of the uppermost sill that cap many of the peaks. Positive anomalies with lower amplitudes and gradients and negative anomalies over glaciers and valleys can be attributed to the basement and peneplain sills. Negative anomalies observed north of Shapeless Mountain probably indicate occurrences of reversely polarized Ferrar Dolerite.

McMurdo Sound is floored by a continuation of the basement complex found in the dry valleys. A layer of pyroclastics or submarine lava flows extends westward from Hut Point Peninsula along the bottom of the sound. A broad positive anomaly between the Dailey Islands and Hut Point Peninsula is caused by an intrusive body, probably related to the McMurdo Volcanics, approximately 5 km below sea level and probably extending to the base of the crust. Expression of a major fault between the Transantarctic Mountains and McMurdo Sound is not apparent on the magnetic map. The lack of a magnetic signature may be explained by low magnetic susceptibility in the basement rocks involved in the faulting.

The consistently high residual magnetic field found over the Ross Ice Shelf south of Ross Island is caused by a layer of pyroclastics or lava flows. The short-wavelength anomalies superposed on this field may mark the locations of volcanic vents.

Mt. Erebus is characterized by a relatively simple, high-amplitude positive anomaly, unlike the other volcanic peaks of Ross Island. North-northeast trending magnetic lineaments that are subparallel to Hut Point Peninsula, a northwest-southeast trend through Cape Bird, and a minor east-west trend through Mt. Terra Nova and Mt. Terror may reflect major fractures beneath Ross Island.

Acknowledgments. This study was supported by the Dry Valley Drilling Project C-642 under contract to the National Science Foundation. Helicopters were provided by Antarctic Development Squadron Six, U.S. Navy Antarctic Support Force, and by the U.S. Coast Guard Icebreaker, *Northwind*. The manuscript was critically reviewed by Phil Kyle and early phases of the work were reviewed by John Stuckless and I. Edgar Odom.

REFERENCES

Adams, R. D., and D. A. Christoffel, Total magnetic field surveys between New Zealand and the Ross Sea, *J. Geophys. Res., 67(2)*, 805-814, 1962.

Angino, E. E., M. D. Turner, and E. J. Zeller, Reconnaissance geology of lower Taylor Valley, Victoria Land, Antarctica, *Geol. Soc. Amer. Bull., 73*, 1553-1561, 1962.

Armstrong, R. L., K-Ar dating: McMurdo Volcanics and dry valley glacial history, Victoria Land, Antarctica, *N.Z. J. Geol. Geophys., 21*, 687-698, 1978.

Barrett, P. J., G. W. Grindley, and P. N. Webb, The Beacon Supergroup of east Antarctica, in *Antarctic Geology and Geophysics*, edited by R. J. Adie, pp. 319-358, Universitetsforlaget, Oslo, 1972.

Barth, T. F. W., *Theoretical Petrology*, 2nd ed., John Wiley, New York, 1962.

Bennett, H. G., A gravity and magnetic survey of the Ross Ice Shelf area, Antarctica, *Res. Rep. Ser. 64-3*, Univ. of Wis., Madison, 1964.

Blundell, D. J., and P. J. Stephenson, Paleomagnetism of some dolerite intrusions from the Theron Mountains and Whichaway Nunataks, Antarctica, *Nature, 184*, 1880, 1959.

Bull, C., and E. Irving, Paleomagnetism of some hypabyssal intrusive rocks from south Victoria Land, Antarctica, *Geophys. J. Roy. Astron. Soc., 3*, 211-224, 1960.

Bull, C., E. Irving, and I. Willis, Further paleomagnetic results from south Victoria Land, Antarctica, *Geophys. J. Roy. Astron. Soc., 6*, 320-336, 1962.

Calkin, P. E., and R. L. Nichols, Quaternary studies in Antarctica (review), in *Antarctic Geology and Geophysics*, edited by R. J. Adie, pp. 625-644, Universitetsforlaget, Oslo, 1972.

Calkin, P. E., R. E. Behling, and C. Bull, Glacial history of Wright Valley, southern Victoria Land, Antarctica, *Antarct. J. U.S., 5*, 22-27, 1970.

Clark, C. C., Geophysical studies of permafrost in the dry valleys, M.S. thesis, N. Ill. Univ., DeKalb, 1972.

Cole, J. W., and A. Ewart, Contributions to the volcanic geology of the Black Island, Brown Peninsula, and Cape Bird areas, McMurdo Sound, Antarctica, *N.Z. J. Geol. Geophys.*, *11*, 793-828, 1968.

Cole, J. W., P. R. Kyle, and V. E. Neall, Contributions to the Quaternary geology of Cape Crozier, White Island, and Hut Point Peninsula, McMurdo Sound region, Antarctica, *N. Z. J. Geol. Geophys.*, *14*, 528-546, 1971.

Cox, A. V., Paleomagnetic research on volcanic rocks of McMurdo Sound, *Antarct. J. U.S.*, *1*(4), 136, 1966.

Cullen, D. J., Movement of sialic blocks oblique to the direction of sea floor spreading *Earth Planet. Sci. Lett.*, *5*, 123-126, 1968.

Dobrin, M. B., *Introduction to Geophysical Prospecting*, McGraw-Hill, New York, 1960.

Doell, R. R., and A. Cox, Paleomagnetism of Hawaiian lava flows, *J. Geophys. Res.*, *70*, 3377-3405, 1965.

Dort, W., Late Cenozoic volcanism in Antarctica, in *Antarctic Geology and Geophysics*, edited by R. J. Adie, pp. 645-652, Universitetsforlaget, Oslo, 1972.

Ervin, C. P., Automated analysis of aeromagnetics, Ph.D. thesis, Univ. of Wis., Madison, 1972.

Fleck, R. J., L. M. Jones, and R. E. Behling, K-Ar dates of the McMurdo Volcanics and their relation to the glacial history of Wright Valley, *Antarct. J. U.S.*, *7*(6), 245-246, 1972.

Forbes, R. B., D. L. Turner, and J. R. Carden, Age of trachyte from Ross Island, Antarctica, *Geology*, *2*(6), 297-299, 1974.

Garland, G. D., *Introduction to Geophysics*, W. B. Saunders, Philadelphia, Pa., 1971.

Giggenbach, W. F., P. R. Kyle, and G. L. Lyon, Present volcanic activity on Mt. Erebus, Ross Island, Antarctica, *Geology*, *1*(3), 135-137, 1973.

Grant, F. S., and G. F. West, *Interpretation Theory in Applied Geophysics*, McGraw-Hill, New York, 1965.

Hamilton, W. B., Diabase sheets differentiated by liquid fractionation, Taylor Glacier region, south Victoria Land, Antarctica, in *Antarctic Geology*, edited by R. J. Adie, pp. 442-454, John Wiley, New York, 1964.

Hamilton, W. B., Diabase sheets of the Taylor Glacier region, Victoria Land, Antarctica, *U.S. Geol. Surv. Prof. Pap. 456-B*, 1965.

Hayes, D. E., and F. J. Davey, A geophysical study of the Ross Sea, Antarctica, in *Initial Reports of the Deep Sea Drilling Project*, vol. 28, edited by D. E. Hayes and L. A. Frakes, pp. 887-908, U. S. Government Printing Office, Washington, D. C., 1975.

Jakosky, J. J., *Exploration Geophysics*, Trija, Los Angeles, Calif., 1950.

Jones, L. M., R. L. Walker, B. A. Hall, and H. W. Borns, Jr., Origin of the Jurassic dolerites and basalts of southern Victoria Land, *Antarct. J. U.S.*, *8*(5), 268-270, 1973.

Kyle, P. R., Glacial history of the McMurdo Sound area as indicated by the distribution and nature of McMurdo Volcanic Group rocks, this volume.

Kyle, P. R., and J. W. Cole, Structural control of volcanism in the McMurdo Volcanic Group, Antarctica, *Bull. Volcanol.*, *38*, 16-25, 1974.

Lopatin, B. G., Basement complex of the McMurdo 'oasis,' south Victoria Land, in *Antarctic Geology and Geophysics*, edited by R. J. Adie, pp. 287-292, Universitetsforlaget, Oslo, 1972.

Malahoff, A., Magnetic studies over volcanoes, in *The Earth's Crust and Upper Mantle, Geophys. Monogr. Ser.*, vol. 13, edited by P. J. Hart, pp. 436-446, AGU, Washington, D. C., 1969.

Malahoff, A., and G. P. Woollard, Magnetic surveys over the Hawaiian Islands and their geologic implications, *Pac. Sci.*, *20*, 265-311, 1966.

Marshall, M., and A. Cox, Magnetism of pillow basalts and their petrology, *Geol. Soc. Amer. Bull.*, *82*, 537-552, 1971.

McDougall, I., Potassium-argon age measurements on dolerites from Antarctica and South Africa, *J. Geophys. Res.*, *68*, 1535-1546, 1963.

McElhinny, M. W., and P. J. Burek, Mesozoic paleomagnetic stratigraphy, *Nature*, *232*, 98-108, 1971.

McGinnis, L. D., McMurdo Sound—A key to the Cenozoic of Antarctica, *Antarct. J. U.S.*, *8*(4), 166-169, 1973.

McKelvey, B. C., and P. N. Webb. Geological reconnaissance in Victoria Land, Antarctica, *Nature*, *189*, 545-547, 1961.

McMahon, B., and H. Spall, Results of paleomagnetic investigation of selected cores recovered by Dry Valley Drilling Project, paper presented at Dry Valley Drilling Project Seminar-I, Seattle, Wash., May 29-31, 1974.

Mirsky, A., Reconsideration of the 'Beacon' as a stratigraphic name in Antarctica, in *Antarctic Geology*, edited by R. J. Adie, pp. 364-376, North-Holland, Amsterdam, 1964.

Moore, J. G., and R. S. Fiske, Volcanic substructure inferred from dredge samples and ocean-bottom photographs, Hawaii, *Geol. Soc. Amer. Bull.*, *80*, 1191-1202, 1969.

Nagata, T., *Rock Magnetism*, Maruzen, Tokyo, 1961.

Nichols, R. L., Geomorphic features of Antarctica, in *Geologic Maps of Antarctica, Antarct. Map Folio Ser., Folio 12*, edited by V. C. Bushnell and C. Craddock, plate XXII, pp. 2-6, American Geographical Society, New York, 1970.

Ostenso, N.A., and E. C. Thiel, Aeromagnetic reconnaissance of Antarctica between Byrd and Wilkes stations, *Res. Rep. Ser. 64-6*, Univ. of Wis., Madison, 1964.

Robinson, E. S., Correlation of magnetic anomalies with bedrock geology in McMurdo Sound area, Antarctica, *J. Geophys. Res.*, *69*, 4319-4325, 1964a.

Robinson, E. S., Geophysical investigations in McMurdo Sound, Antarctica, *J. Geophys. Res.*, *69*, 257-262, 1964b.

Schopf, J. M., Ellsworth mountains—Position in west Antarctica due to sea-floor spreading, *Science*, *164*, 63-66, 1969.

Serson, P. H., W. Hannaford, and G. V. Haines, Magnetic anomalies over Ireland, *Science*, *162*, 355-357, 1968.

Steinberg, G. S., and L. A. Rivosh, Geophysical study of the Kamchatka volcanoes, *J. Geophys. Res.*, *70*, 3341-3369, 1965.

Talwani, M., Computation with the help of a digital computer of magnetic anomalies caused by bodies of arbitrary shape, *Geophysics*, *30*(5), 797-817, 1965.

Tarling, D. H., Paleomagnetic studies of the Hawaiian lavas, *Geophys. J.*, *10*, 93-98, 1965.

Treves, S., Volcanic rocks of the Ross Island area, *Antarct. J. U.S.*, *3*(4), 108, 1968.

Vacquier, V. N., C. Steenland, R. G. Henderson, and I. Zietz, *Interpretation of Aeromagnetic Maps, Mem., 47*, Geological Society of America, New York, 1951.

Ward, P. L., New interpretations of geology of Iceland, *Geol. Soc. Amer. Bull.*, *82*(11), 2991-3012, 1971.

Warren, G., Terra Nova Bay-McMurdo Sound area geologic map in *Geologic Maps of Antarctica*, edited by C. Craddock, Folio 12, sheet 14, American Geographical Society, New York, 1969.

Watkins, N. D., and A. Richardson, Paleomagnetism of the Lisbon volcanics, *Geophys. J. Roy. Astron. Soc.*, *15*, 287-304, 1968.

Wellman, H. W., Later geological history of Hut Point Peninsula, Antarctica, *Trans. Roy. Soc. N.Z. Geol.*, *2*(10), 149-154, 1964.

SEISMIC REFRACTION STUDY IN WESTERN McMURDO SOUND

L. D. McGinnis

Department of Geology, Northern Illinois University, DeKalb, Illinois 60115

Three reversed seismic refraction profiles in western McMurdo Sound were shot from sea ice in water depths ranging from less than 100 m to over 200 m. Velocity and depth interpretations indicate abnormally high sea floor velocities of 2.7 to 2.9 km/s, which are explained as being caused by submarine permafrost sediment. A layer having an unfrozen velocity of 1.9 km/s and porosity of 27% would increase in velocity to 2.9 km/s upon freezing; therefore it is believed that the sea floor velocities observed here are the frozen equivalents of the lower-velocity sea floor sediments found farther out in McMurdo Sound. Basement depth varies from 0.48 km below sea level in New Harbor to ~1.75 km about 15 km offshore at line S-78-79-1. A fault, having a vertical displacement between 2 and 3 km, lies between lines S-78-79-1 and S-78-79-3. Basement velocities range from 4.97 km/s at line S-78-79-1 to 5.62 km/s at New Harbor. In areas of deep water (~ 200 m) and thick sediment, bottom fractions are attenuated due to thin, high-velocity, bonded submarine permafrost resting upon lower-velocity, unfrozen sediments. The combination of low, ocean water temperature ($-1.8°C$) and low pore water salinity, at times less than one fifth that of sea water, is sufficient to explain the presence of frozen beds near the sea floor. Submarine, freshwater sediments are probably due to a combination of sea floor lowering as observed on the Atlantic Continental shelf of the United States and ponding marginal to a retreating McMurdo Ice Shelf. Intermediate velocities ranging from 3 to 3.6 km/s observed below the frozen sea floor may represent sediment of Late Mesozoic to Early Cenozoic age.

INTRODUCTION

Reversed seismic refraction measurements in western McMurdo Sound were made from sea ice in late November-early December 1978 (Figure 1). The study was designed to determine the character and thickness of sediment underlying western McMurdo Sound and to obtain preliminary information on the lithology and structure of the crystalline basement surface.

Seismic reflection studies (unpublished) on sea ice in McMurdo Sound were first conducted on January 25-31, 1958, during the International Geophysical Year by J. Cook with this author's assistance. Refraction seismic studies were first attempted in 1959 [see *Crary et al.*, 1962]. Results of these studies and additional work by Robinson were summarized by *Robinson* [1963]. Principal results were obtained from four refraction profiles shot and recorded from sea ice and the McMurdo Ice Shelf in the vicinity of Hut Point Peninsula (see Figure 1). Robinson reports the presence of a single velocity horizon 2.0 km thick, having a velocity of 3.14 km/s lying above basement with a velocity of 4.79 km/s. Marine seismic profiling in McMurdo Sound, completed in 1975 [*Northey et al.*, 1975; *Wong and Christoffel*, this volume], consisted of single-channel air gun and sonobuoy refraction measurements from the U.S. Coast Guard ice breaker *Burton Island*. *Wong and Christoffel* [this volume] indicate the presence of four seismic layers having refraction velocities of 1.9, 2.4, 2.8-3.1, and 3.9-4.2 km/s and sedimentary thicknesses over 1.4 km. These studies were made northeast of Dry Valley Drilling Project (DVDP) 15. Borehole 15, drilled from sea ice 3 m thick [see *Barrett et al.*, 1976], penetrated only 65 m below the sea floor

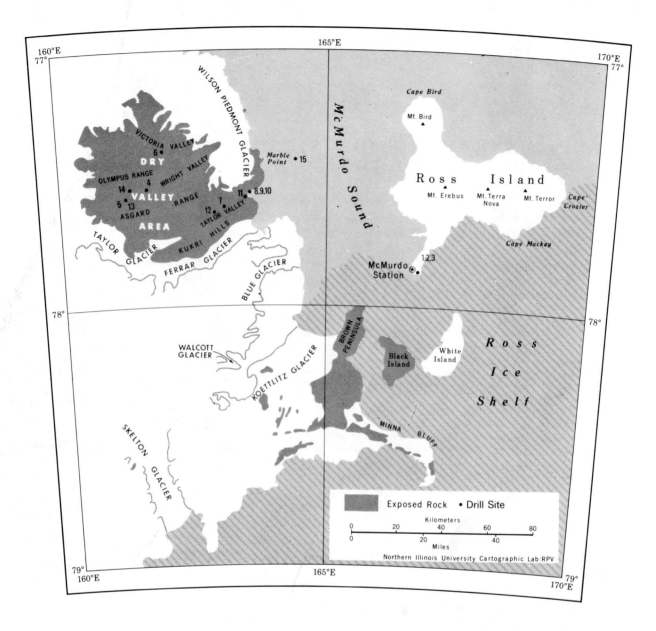

Fig. 1.   Location of seismic refraction lines 1, 2, and 3 in western McMurdo Sound. Lines labeled R are from *Robinson* [1963]. Those labeled W are from *Wong and Christoffel* [this volume]. Robinson observed bottom velocities averaging 3.14 km/s, and Wong and Christoffel report 1.9 km/s; whereas in the three stations reported here they are 2.90 km/s for line 1, 2.74 km/s for line 2, and 2.96 km/s for line 3. The high velocities observed here and by Robinson are believed to be caused by frozen submarine sediments.

which was at a depth of 122 m. The hole was drilled 16 km east of Marble Point.

Instrumentation consisted of a 12-channel Dresser SIE, Inc., RS-4. Frequency response is 5-125 Hz and input sensitivity is such that 100 μV (root mean square) produces a 1.27-cm peak to

trough deflection. Geophones had a natural frequency of 26 HZ, and paper speed is 28 cm/s. Timing lines are recorded at 0.010-s intervals on Kodak dry write paper which is developed in indirect sunlight, fluorescent or incandescent light. The seismograph is designed to operate within the temper-

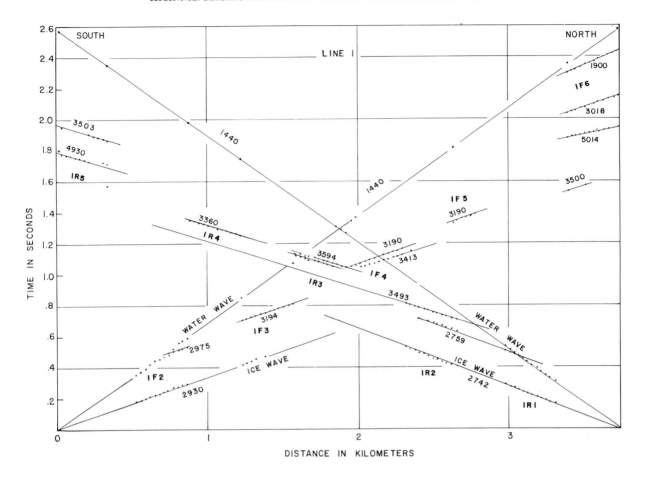

Fig. 2. Reversed, time-distance curve for line 1 in western McMurdo Sound. The direct, compressional waves through ice and water form an envelope of energy arrivals within which refracted energy is returned from the sea floor and below the sea floor. From the intercept time method the basement lies at a depth below sea level between 1.5 and 1.75 km.

ature range −20° to +50°C. Recordings were made in a Scott Polar Tent where ambient temperatures were generally above −15°C. About the only malfunctions occurred when a full roll of film was loaded in the seismic recorder and the paper drive motor had to be manually assisted.

Layout of the geophone cable was in line, and shot detector distances ranged from 13 m to 6 km. Shot size ranged from 1 kg of Nitramon S dynamite to 18 kg, depending on shot detector distances. Shot depths were at first placed directly under the sea ice which was 2.8 m thick; however, to increase shot efficiency, later shots were placed at 9.1-, 16.2-, and 32.4-m depths. The best depth was found to be that where the explosion bubble just breaks the water or ice surface [see *Dobrin*, 1976, p. 118]. This prevents a second arrival from being generated by the collapsing bubble and per-

mits greater resolution of refractions immediately following first arrivals. Radios were used to coordinate the shot and recording. Distances were chained and are known to the nearest meter.

PROCEDURE

Seismic refraction procedures used are the standard, in-line, forward and reversed shot to detectors. The detector spreads containing 12 geophones were laid out along a cable with geophone separations of 30.49 m. The spread was held stationary while shots were placed in line away from the spread at intervals of 500 to 1000 m. Shot distances were increased until the author was assured that a basement arrival had been recorded. The criteria used was simply the detection of a velocity segment on the order of 5 km/s. This stationary detector versus moving shot procedure is similar to

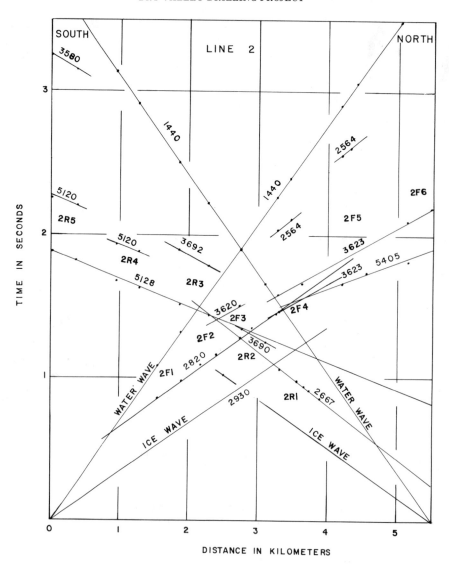

Fig. 3.   Reversed, time-distance curve for line 2 in western McMurdo Sound. Basement is at a depth of 1.26-1.33 km.

that used in marine refraction profiling using sono-buoys. Although a shipboard refraction profile can be completed in a relatively short time, it does have the disadvantage in that normally the sono-buoy is not fixed but will drift with ocean currents. In conducting refraction profiling from sea ice, detectors remain fixed, and interpretations are therefore more certain. Additionally, because of the use of 12 channels a velocity can be determined for each shot. An advantage of marine profiling, especially when using an air gun, is that shots are fired at very short time intervals, and therefore energy arrivals are more continuous over the time-distance curve. Phase changes, amplitude, and segment character (i.e., en echelon, continuous-discontinuous, etc.) can often be better identified as a result. To acquire the same kind of data from sea ice would require shot spacings equivalent to the geophone spread. Depending upon the information required, this may or may not be warranted.

As in marine seismic profiling using sonobuoys, the water arrival in shooting from sea ice plays a significant interpretational role. At long distances it is the arrival of greatest energy and can be followed out to great distances with no change in velocity. Therefore it becomes fundamentally important in calculating shot detector distances or times if one or the other is not known. In ship-borne

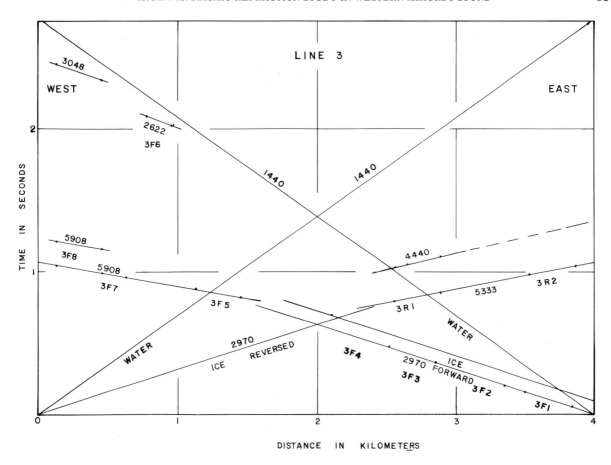

Fig. 4. Reversed, time-distance curve for line 3 at New Harbor in western McMurdo Sound. Basement is at a
depth of 0.48 km.

work the water arrival is used to calculate distance, since interval time is known from radio signals. In the study reported here a radio shot time was not recorded on the seismic record; however, distances were measured to the nearest meter, and therefore the water arrival time can be calculated. All other times are measured relative to the water arrival. The velocity of sound through water measured in this study was 1440 ± 2 m/s. *Robinson* [1963] and *Wong and Christoffel* [this volume] also used a value of 1440 m/s in McMurdo Sound.

In addition to the *P* wave velocity measured directly through water, refraction seismology conducted from sea ice can also utilize the *P* wave and plate wave through sea ice, although these velocities are not as invariant as the water *P* wave [*Oliver et al.*, 1954]. At short distances the ice waves must be used, since they obliterate the water wave arrivals. Ice wave velocities measured in this study range from 2700 to 3100 m/s. Ice thicknesses ranged from 2.8 to 4.1 m. In thin ice the ice

wave attenuates rapidly and will not obscure seismic events arriving through the water or from the sea floor. The water and ice arrivals form an envelope bounded by straight lines passing through the origin of a time-distance plot. Refracted arrivals from sea bottom and subbottom occur within the water-ice envelope at the distances shot in this study.

TIME-DISTANCE CURVES

Data collected during the present study are incorporated in Figures 2, 3, and 4. Line 1 (Figure 2), shot forward and reversed, is located 22 km southeast of Marble Point and is oriented northwest-southeast on 2.8 m of annual ice over 204 (mean depth) m of water. Ice velocities range from 2.93 km/s for the southernmost shot (forward) to 2.74 km/s for the northernmost shot (reversed). Sea bottom velocities range from (apparent) 2.98 km/s on the south to (apparent) 2.76 km/s on the north. The bottom slopes up about 5 m from north

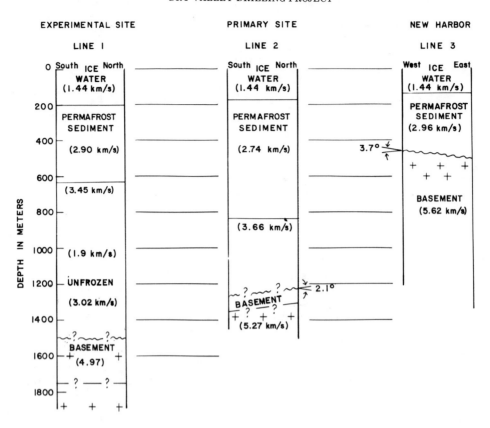

Fig. 5   Interpretation of seismic refraction data in western McMurdo Sound. The 3.45-km/s layer beneath line 1 represents unfrozen sediment of late Mesozoic to early Cenozoic age.

to south over the spread length of 335 m for the south shot. Correcting the velocity for bottom slope gives a true velocity of 2.90 km/s. Conversely, the spread for the reverse shot was down dip 9 m, which gives a corrected velocity of 2.89 km/s, a difference of only 10 m/s. Calculating water depth from refraction arrivals gives a mean depth of 196 m, which indicates that the ~2.89-km velocity is at the sea floor and therefore little or no low-velocity marine sediment is present above the 2.9 km/s layer. P. J. Barrett (personal communication, 1978) conducted a grab-sampling program as part of on-going geological studies in McMurdo Sound. The sampling program indicates an abundant faunal assemblage living on the sea floor which is composed of unfrozen, basaltic sands and silts. From the seismic data it would appear that this unfrozen bottom layer is a thin veneer, being at most only several meters thick.

At shot detector distances greater than 1 km a persistent arrival that averages 3.45 km/s is observed. From the intercept times this layer lies at a depth 435 m below the sea floor for a total depth

below sea level of 631 m, and it could not therefore represent frozen sediment, since *Decker and Bucher* [1980] indicate only 240 m of permafrost near the shoreline at New Harbor. *Wong and Christoffel* [this volume] state that McMurdo Sound sediments having velocities of 1.9 and 2.4 km/s have a total thickness of approximately 1 km in water depths less than 370 m. If these layers were present, they would not be seen because of the high-velocity bottom. The seismic boundary at 631 m below sea level is representative of late Oligocene to early Miocene sandstones [see *Barrett and Froggatt*, 1978].

Time-distance curves for line 2 (Figure 3) are interpreted in the same manner used for line 1. Mean velocities at line 2 are 1.44 km/s for water, 2.74 km/s for the first refractor, 3.66 km/s for the deep refractor, and 5.27 km/s for basement. Depth to the first refractor is 233 m, whereas the mean water depth below the profile determined from bottom sounding is only 168 m. The disparity in depths is probably due to the fact that an unfrozen, low-velocity layer, approximately 65 m thick, lies

TABLE 1. Velocity and Depth Data for Three Reversed Seismic Refraction Profiles in Western McMurdo Sound

Line	Mean Water Depth, m	Depth to Refractor 1, m	Depth to Refractor 2, m	Depth to Basement, km	Line Orientation	Apparent Dip	Mean Basement Velocity, km/s	Mean Sea Floor Velocity, km/s
	204	196	631	1.5 (1.75)	NW-SE	0°	4.97	2.90
2	168	233	835	1.26 (1.33)	NW-SE	2.1°S	5.27	2.74
3	118	∙ ∙ ∙	∙ ∙ ∙	(0.48)	E-W	3.7°E	5.62	2.96 (?)

*Numbers in parentheses are depths to basement, assuming mean sediment velocity of 2.96 km/s.

above frozen sediment. Depth to the second layer (3.66 km/s) is 835 m, and depth to basement is 1260 m. Lower-velocity strata are inferred beneath the 2.74 km/s layer on the basis of high-energy, secondary arrivals. Using a mean sediment velocity of 2.96 km/s and mean water depth beneath the profile of 168 m, the depth to basement is 1330 m, a difference of 70 m. If the 2.74-km/s layer were the frozen equivalent of the 1.9-km/s layer, it would have a porosity of 23%.

Average depth to the sea floor beneath the seismic profile (Figure 4) at New Harbor (line 3) is 118 m. The center of line 3 is located about 6 km east of DVDP 10. The time-distance profile is marked by the absence of a refraction off the sea floor or, for that matter, a refractor near the sea floor. Sea ice velocity at New Harbor is 2.97 km/s, and the ice wave arrives at high energy out to several kilometers. It is likely that the sea ice wave, with a velocity near that of submarine permafrost, obliterates arrivals off the sea floor. At the New Harbor line, shots were placed at depths greater than at the preceding lines in an attempt to increase the energy into the subsurface. As a result, a large secondary arrival due to the bubble effect, the relatively thick sea ice, and the similarity in wave velocity of the ice and sea floor all contributed to making the bottom refractor indistinguishable from the direct wave and later phases through the sea ice. A good refraction was obtained off basement with an average velocity of 5.62 km/s, and this arrival, along with known ocean depths and an assumed mean sediment velocity of 2.96 km/s, gives a basement depth below sea level near the center of the spread of 477 m. From the difference in apparent velocities shooting east and west the basement is sloping up to the west toward borehole 10 at an angle of 3.7°. Since hole 10 was drilled to 206 m through Late Cenozoic sediments, the slope on the basement surface must decrease to-

ward the borehole. Geological interpretations of lines 1, 2, and 3 are illustrated in Figure 5. Data are summarized in Table 1.

DISCUSSION

Pore water salinity becomes a predominant factor in velocity control on polar continental shelves, since it controls the degree of freezing. From data collected in the dry valley region and other data collected from the literature, McGinnis et al. [1973] found the porosity of sediment to be given by the equation

$$P = 3.63 + 0.4438 \times V_c \qquad (1)$$

where V_c is the change in velocity in percent after freezing or

$$V_c = \frac{V_F - V_u}{V_u} \times 100 \qquad (2)$$

Upon substitution of the relationship shown in (2) into (1) we get

$$P = \left(\frac{V_F - V_u}{V_u}\right) 44.38 + 3.63 \qquad (3)$$

where

P porosity in percent;
V_u unfrozen velocity;
V_F frozen velocity.

Thus it is apparent that at the higher porosities encountered in near-bottom sediments (say 50%, i.e., see McGinnis and Otis [1979]) the increase in velocity with freezing can be over 100%. The 1.9-km/s layer, proposed as the sea floor velocity for McMurdo Sound by Wong and Christoffel, would have a porosity near 32% [see McGinnis and Otis, 1979] which, when frozen, would increase in velocity by 63.9%, or from 1.9 to 3.11 km/s. Salinities in the pore ice of some sediments beneath McMurdo Sound were found to be quite variable and in some cases less than sea water [*McGinnis et al.*, this volume]; therefore velocities will probably range from normal frozen to normal unfrozen.

From (3), if we assume that the 2.9-km/s layer is the frozen equivalent of the 1.9-km/s layer, we can solve for the porosity and find it to be 27%, which is very low for normal sea floor sediments. The low porosities may be explained by overconsolidation due to glacial loading or to the exposure of over-consolidated sediment due to glacial scour. Given the probability for the presence of error limits to be expected in refraction seismology, the sea floor velocities, as reported by *Wong and Christoffel* [this volume], *Northey et al.* [1976], *Robinson* [1963], and here in this paper, are probably all representative of equivalent strata except that the higher velocities are caused by submarine permafrost.

Intermediate velocities of about 3.45 km/s observed here are similar to values observed from laboratory samples by *Barrett and Froggatt* [1978] for late Oligocene to early Miocene rocks obtained in Deep Sea Drilling Project holes 270 and 272. It is possible that the similarity in velocity does not reflect a similarity in age but only a similarity in consolidation (compaction) history.

Between 3 and 4 km, on both forward and reversed shots of line 1, the basement arrival having a mean velocity of 4.97 km/s is observed as the first strong arrival. Its arrival follows weak, low-amplitude waves refracted through frozen sediment. The rapid attenuation of the wave propagating through frozen sediments probably indicates high-speed layering. On the last forward shot, two strong arrivals having velocities of 1.9 and 3.02 km/s were recorded between basement and water arrivals. From an inspection of travel paths it is possible that these velocities represent the low-velocity strata reported in deeper water. No velocities in the range 3.9 to 4.2 km/s were observed as reported by Wong and Christoffel farther out in the sound.

It should be pointed out here that basement velocities for line 1 were determined for reversed spreads 4 km apart. In averaging velocities one assumes that the basement is a plane surface over the distance measured. This assumption is certainly not rigorously met, and therefore the mean velocity is only a 'best' estimate of the true velocity. A 'true' velocity could be determined if the lo-

cation of the geophone spread were held in place and the shots moved out from either end of the spread. This procedure was considered but rejected because it would have resulted in a complete lack of velocity control beneath each shot. Where the array is reversed, assumptions which must be made are more limited. It should also be pointed out here that the procedure used in plotting the time-distance curves is done for interpretational purposes only. If plotted correctly, the times would increase with increasing shot distance, but the position of the geophone array on the distance axis would be held stationary. Assuming a plane surface, the basement would have an apparent dip slightly down to the south of ($<<<1°$).

To estimate depth to basement, I assumed low-velocity strata were present beneath the frozen layers having velocities of 1.9 and 3.02 km/s. Not knowing the individual layer thicknesses, I assumed a mean velocity of 2.96 km/s from the sea floor to basement and using the basement intercept times found the thickness of the sediment layer to be 1.31 km. Thus the basement depth below sea level at line 1 is 1.5 km, and it has little or no dip along the profile. Interpreting the time-distance curves as though no low-velocity layering were present below the 2.90-km/s layer, the depth to basement would be 1.75 km. Since line 1 is only about 10 km east of the southward extension of the coastline of the dry valley region, which is also composed of crystalline basement, a major vertical fault must exist between line 1 and the coast. It is assumed that the deep basement in New Harbor is caused by glacial scour. Displacement on the fault must be of the order of 2-3 km.

CONCLUSIONS

Three reversed seismic refraction profiles in western McMurdo Sound provide data from which it is concluded that submarine permafrost is present to water depths over 200 m [*McGinnis*, 1979]. Compressional velocities in the submarine permafrost range from 2.7 to 3.1 km/s. Bottom sediments in deeper waters of McMurdo Sound and the Ross Sea have velocities near 1.9 km/s. It is assumed here that the ~3-km/s bottom layer is the frozen equivalent of the 1.9-km/s layer. The submarine layers remain frozen because they contain pore water having salinity less than sea water [*McGinnis et al.*, this volume].

The basement surface in McMurdo Sound ranges from 3 km below sea level near Ross Island to 1.7

km in the western sound. From all available data it appears that the crystalline basement floor is relatively flat, rising gently to the west. A major displacement on the basement floor lies between stations 1 and 2 of this study and the coastline. Total displacement in the form of vertical faulting is of the order of 2-3 km.

Acknowledgments. This study was supported by the U.S. National Science Foundation, Division of Polar Programs (grant DPP-7821112), and the New Zealand Department of Scientific and Industrial Research, Antarctic Division, as part of the McMurdo Sound Sediment and Tectonic Study. The author was assisted in the field by K. Power and D. Grund.

REFERENCES

Barrett, P. J., and P. C. Froggatt, Densities, porosites, and seismic velocities of some rocks from Victoria Land, Antarctica, *N. Z. J. Geol. Geophys.*, *21*, 175-187, 1978.

Barrett, P., S. Treves, C. Barnes, H. Brady, S. McCormick, N. Nakai, J. Oliver, and K. Sillars, Initial report of DVDP 15, western McMurdo Sound, Antarctica, *Dry Val. Drilling Proj. Bull.*, *7*, 1-100, 1976.

Crary, A. P., E. S. Robinson, H. F. Bennett, and W. W. Boyd, Glaciological studies of the Ross Ice Shelf, Antarctica, 1957-60, *IGY Glaciol. Rep. 6*, 34 pp., IGY World Data Center A: Glaciol, American Geographical Society, New York, 1962.

Decker, E. R., and G. J. Bucher, Preliminary geothermal studies in the Ross Island-Dry Valley region, in *Antarctic Geoscience*, edited by C. Craddock, University of Wisconsin Press, Madison, in press, 1980.

Dobrin, M. B., *Introduction to Geophysical Prospecting*, 3rd ed., 630 pp., McGraw-Hill, New York, 1976.

McGinnis, L. D., Seismic refraction detection of submarine permafrost in McMurdo Sound, Antarctica (abstract), *Eos Trans. AGU*, *60*, 287-288, 1979.

McGinnis, L. D., and R. M. Otis, Compressional velocities from multichannel refraction arrivals on Georges Bank—Northwest Atlantic Ocean, *Geophysics*, *44*, 1022-1033, 1979.

McGinnis, L. D., K. Nakao, and C. C. Clark, Geophysical identification of frozen and unfrozen ground, Antarctica, in *Permafrost, 2nd International Conference, North American Contribution*, pp. 136-146, National Academy of Sciences, Washington, D. C. 1973.

McGinnis, L. D., J. S. Stuckless, D. R. Osby, and P. R. Kyle, Gamma ray, salinity, and electric logs of DVDP boreholes, this volume.

Northey, D. J., C. Brown, D. A. Christoffel, H. K. Wong, and P. J. Barrett, A continuous seismic profiling survey in McMurdo Sound, Antarctica—1975, *Dry Val. Drilling Proj. Bull.*, *5*, 167-179, 1975.

Oliver, J., A. P. Crary, and R. Cotell, Elastic waves in arctic pack ice, *Eos Trans. AGU*, *35*(2), 282-292, 1954.

Robinson, E. S., Geophysical investigations in McMurdo Sound, Antarctica, *J. Geophys. Res.*, *68*, 257-262, 1963.

Wong, H. K., and D. A. Christoffel, A reconnaissance seismic survey of McMurdo Sound and Terra Nova Bay, Ross Sea, this volume.

A RECONNAISSANCE SEISMIC SURVEY OF McMURDO SOUND AND TERRA NOVA BAY, ROSS SEA

How Kin Wong

Geologisch-Paläontologisches Institut, Universität Hamburg, 2000 Hamburg 13
Federal Republic of Germany

David A. Christoffel

Physics Department, Victoria University of Wellington, Wellington, New Zealand

The sea floor of McMurdo Sound may be described as a north-south trending, eastward dipping slope incised by two submarine, fjordlike valleys thought to be extensions of Taylor and Ferrar valleys, respectively. Sediments subparallel to and underlying this slope continue beneath the flat-lying, stratified sequence in the deep (over 900 m) Erebus Basin and may persist uninterrupted to underlie Ross Island. Continuous seismic profiling in McMurdo Sound has demonstrated the presence and pervasiveness of the angular unconformity (a glacial erosional surface, here labeled *T*) first mapped elsewhere in the Ross Sea. By assuming that this unconformity is contemporaneous with that at sites 270-272 of the Deep Sea Drilling Project, an age of 4-5 m.y. may be assigned, and from this an uncorrected, average sedimentation rate in McMurdo Sound of 18 m/m.y. since mid-Pliocene follows. The total sedimentary sequence exceeds 1.4 km in thickness in the central part of the sound. Four north-south sonobuoy refraction profiles provide information on the sedimentary structure in the sound. Four layers with refraction velocities of 1.9, 2.4, 2.8-3.1, and 3.9-4.2 km/s have been resolved. They are interpreted as marine, pebbly, muddy sand, a coarse, nearshore facies of Miocene-Oligocene mudstone, older preglacial sandstone and mudstone, and metasediments (?), respectively. The depth to basement may be estimated by substituting the measured basement velocity v to the linear equation $v = 2.11 + 1.23\ h$, where h is the basement depth measured from the sea floor. Terra Nova Bay is characterized by a series of northeast-southwest trending troughs and depressions cut into a slope, the most prominent of which is Drygalski Basin with a depth exceeding 1100 m. The area in Terra Nova Bay over which the angular unconformity *T* may be clearly discerned is more restricted than in McMurdo Sound, but the angularity and acoustic characteristics of the stratifications are similar. In the western part of the bay is a large, stratified depositional feature interpreted as a delta moraine. This is a fluvioglacial deposit formed when the expanded Ross Ice Shelf invaded Terra Nova Bay and caused local or general ice grounding. The moraine is deposited where the stationary ice front remained for some time, ending in a lake or sea. The single continuous seismic profile between Terra Nova Bay and McMurdo Sound shows, in addition to the angular unconformity and morainal and meltwater deposits, at least two near-vertical faults. An offset cannot be observed at the sea floor because of erosion, but throw is at a maximum in the immediate subbottom. This throw appears to decrease with depth, suggesting that the faulting activity is youthful and is probably continuing.

INTRODUCTION

A single-channel, air gun seismic reflection and sonobuoy refraction survey was carried out aboard the U.S. Coast Guard icebreaker *Burton Island* in McMurdo Sound and Terra Nova Bay, southeastern Ross Sea, during a two-week period in January-February 1975. Approximately 1800 km of survey tracks were run. This includes a detailed grid in McMurdo Sound, a similar grid in Terra Nova Bay, a long line from Terra Nova Bay to Beaufort Island, and another long survey track extending from Beaufort Island northeastward for about 120 km (Figures 1-3).

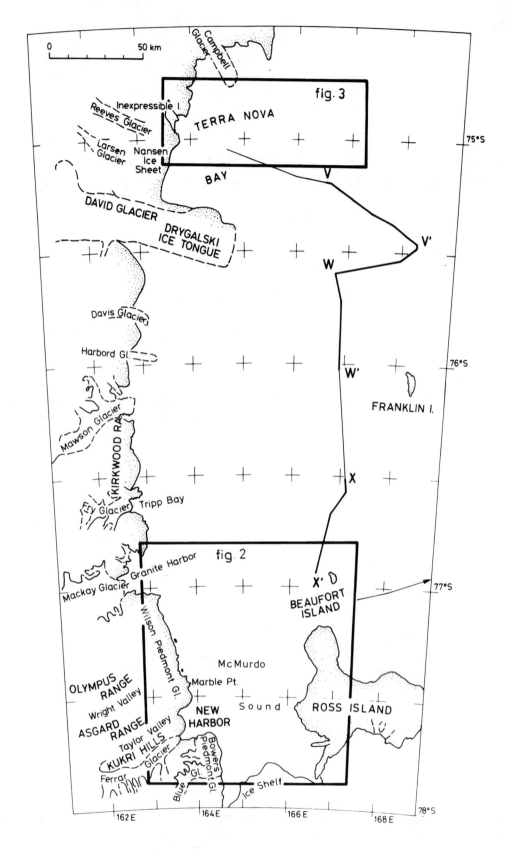

Fig. 1. The survey areas in southeastern Ross Sea. Rectangles outline the areas of detailed survey in McMurdo Sound and Terra Nova Bay, the survey grids of which are shown in Figures 2 and 3. Shown also is the long line from Terra Nova Bay to Beaufort Island and part of the track extending from Beaufort Island northeastward.

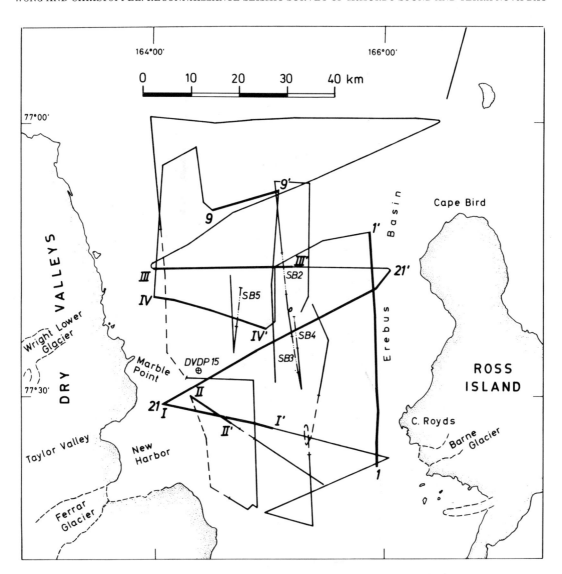

Fig. 2.   Track chart for the survey area in McMurdo Sound. Along the thin solid survey lines, both bathymetric and profiler data are collected, while only bathymetric data are available on the dashed tracks. Continuous seismic profiles along the thick solid track lines are presented in this paper. Sonobuoy runs are labeled SB2-SB5.

The survey was carried out with two sets of interchangeable equipment, so that loss of ship time due to equipment malfunction was minimal. The sound sources used include a 490-cm^3 (30 in^3) Bolt-PAR air gun and a 230-cm^3 (14 in^3) air gun similar to the Lamont-Doherty Geological Observatory design. These were operated at a pressure of 7-14 MPa (1000-2000 psi) and a repetition rate of 6-10 s.

Two separate multielement hydrophone arrays were towed from the port and starboard sides of the ship to function as receivers. The profiling signal was band-pass filtered over bandwidths of 40-100 Hz and 100-400 Hz and was then registered on two separate EPC model 4100 graphic recorders. The sonobuoy signal was received via a Yaggi antenna and a phase-locked receiver, after which it

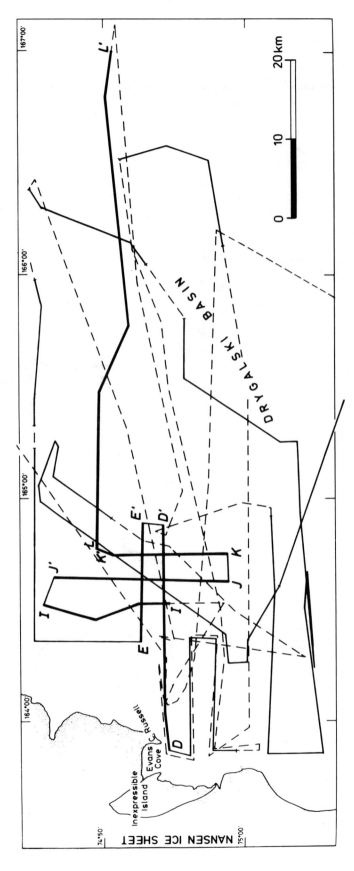

Fig. 3. Track chart for the survey area in Terra Nova Bay. Symbols are the same as in Figure 2.

was recorded wide band on a 4-channel FM tape recorder.

Navigational control was provided by satellite fixes at approximately hourly intervals. Radar fixes and occasional visual sightings were obtained every 10 min and at course changes. The ship's track was then plotted by hand. Subsequently, minor adjustments were made so as to match the bathymetric data at track crossings. The accuracy of the final track charts is estimated to be ±1 km.

Operational difficulties include occasional severe ice conditions which caused freezing up of lines between the various stages in the compressor and of the air supply line to the air gun. On occasions, deployment of any towed equipment had to be interrupted to facilitate maneuverability of the ship. When ice cover was heavy (up to 75%), the hydrophone arrays often had to ride over blocks of floating ice already broken up by the ship. The acoustic noise so produced would then completely mask the desired signal.

Earlier work in the Ross Sea includes the sediment thickness studies of *Crary* [1961], studies on glacial marine sedimentation by *Chriss and Frakes* [1972], and the profiler and sonobuoy measurements of *Houtz and Meijer* [1970] and *Houtz and Davey* [1973]. In the last two investigations the bathymetry and structure of the shelf and continental slope were established. The *Glomar Challenger* drilled three holes (sites 270, 271, and 272) in the south central part of the Ross Sea at water depths ranging from 550 to 650 m during Leg 28 of the Deep Sea Drilling Project (DSDP) [*Hayes et al.*, 1975]. For the first time, cores from as much as 410 m below the sea floor became available for unraveling the geological history of the area. In McMurdo Sound itself, in addition to shallow refraction and gravity measurements [*Robinson*, 1963], a preliminary profiler survey was carried out over about 100 km of track, but the usefulness of these latter results was limited because of equipment malfunction [*Northey and Sissons*, 1974]. Nonetheless, sediment stratification across the submarine extension of Taylor Valley was demonstrated for the first time [*Barrett et al.*, 1974; *Northey and Sissons*, 1974].

Preliminary results of the McMurdo Sound portion of our survey have already been reported with the purpose of recommending drill sites for the Dry Valley Drilling Project (DVDP) [*Northey et al.*, 1975]. In this paper, we present a detailed interpretation of our data and its implications on the structure of the southwestern Ross Sea.

McMURDO SOUND

Bathymetry

The sea floor beneath McMurdo Sound may be described as a north-south trending, eastward dipping slope with an average gradient of 0.7°, incised by two well-developed fjordlike submarine valleys (Figure 4, based on our bathymetric data only). The more northerly valley, possibly a scouring result of an expanded Taylor Glacier [*McGinnis*, 1973], is about 150-200 m deeper than the surrounding sea floor (Figure 10). It measures nearly 400 m wide at its deepest and exhibits a typical U shape. The southerly valley, thought to be eroded by an expanded Ferrar Glacier [*McGinnis*, 1973], is less prominent but possesses an equally well-developed U-shaped cross section. These two submarine valleys are also evident in the bathymetric chart of *McGinnis* [1973], but our chart shows a much more distinct north-south morphologic trend and a smoother sea floor. Perhaps many of the meanders in the depth contours of *McGinnis* [1973] are due to the nonuniformity of data base inherent in HO Chart 6666 (Franklin Island to McMurdo Sound), the soundings of which (plus three other profiles) constitute the basis for the contours of his chart.

The depression to the northwest of Ross Island, named Erebus Basin by *Vanney and Johnson* [1976], is part of a much larger system lying to the north, west, and southeast of the island [*Hayes and Davey*, 1975a; *Vanney and Johnson*, 1976]. It may be the result of isostatic adjustment in response to flexural loading by the Cenozoic volcanics erupted on Ross Island over the last 15 m.y., or it could be formed by the collapse along concentric fractures of a magma chamber at depth, such as is postulated for the Shasta depression east of the Klamath Mountains of northern California [*Heiken*, 1976]. To choose between these possibilities, it would be crucial to determine the accurate shape of the depression, to measure its associated gravity field, and to investigate whether concentric or radial fractures or faults exist within the depressions.

Sonobuoy Refraction Data

Five sonobuoy stations were occupied in McMurdo Sound. Because of heavy seas and occasionally a thick ice cover the quality of the records is not very good. Sonobuoy 1 did not yield any usable results, and for all five runs the wide angle reflection data are not reliable enough to provide inter-

TABLE 1. Sonobuoy Refraction Data From McMurdo Sound

Sonobuoy Number	v_1, km/s	v_2, km/s	v_3, km/s	h_1, km	h_2, km	Warer Depth, km	Location
2	2.09	2.84	4.18	0.57	0.87	0.554	77°16.6'S,165°0.72'E
3	2.10*	2.83	4.06	0.50	0.71	0.660	77°25.0'S,165°13.4'E
4	2.49*		3.86	1.20		0.625	77°23.1'S,165°14.0'E
5	1.90–2.37	3.14	4.01	0.44–0.61	1.03	0.375	77°20.6'S,164°43.5'E

Here v is the refraction velocity, and h is the layer thickness.
* Velocity assumed.

val velocities within the sedimentary column because of air gun ringing and propagation noise.

Contrary to the results reported by *Houtz and Davey* [1973] in the Ross Sea, our refraction arrivals from sonobuoys 2–5 can be closely approximated by segments of straight lines, suggesting that the subbottom structure in the sound is stratified. A mean velocity of 1.44 km/s is assumed for the water layer, corresponding to a mean temperature of $-1.5°C$ (measured by expendable bathythermographs) and a salinity of 34.7°/oo (measured by a conductivity bridge). Whenever a refraction interface can be directly correlated with a reflector in the profiler data, its dip is approximated by fitting a line through this reflector. Where the interface lies deeper, it is assumed to have the same dip as the deepest reflector observable on the corresponding continuous reflection profile. The velocities are then corrected for dip of the interface. Table 1 summarizes the sonobuoy refraction results based on a constant velocity layer model. Asterisks indicate assumed velocities.

Houtz and Davey [1973] deployed 23 sonobuoys on the shelf part of the Ross Sea, none of which, however, are located in or near McMurdo Sound. Of these, only four recorded well-defined refracted arrivals in the form of straight line segments. They all lie in the central Ross Shelf between 170°W and 180°W. The remainder, except for a few from areas where the sediment layer is very thin, display pronounced curvature in the refraction data. By approximating the curved arrivals with straight lines they obtained a velocity-depth distribution off Victoria Land that could be fitted by the straight line:

$$v = 2.10 + 1.78\ h$$

(constant velocity layer model) [from *Houtz and Davey*, 1973, Figure 4]. They pointed out that this relationship and the curvature of the refraction

data suggest strongly that the velocity is predominantly a function of overburden pressure, and hence the analysis should proceed by way of computing velocity gradients from the original travel time curves. By following this procedure they found that the depth to basement may be estimated from the equation

$$v = 2.10 + 1.27\ h$$

(velocity gradient analysis), where h is depth in kilometers from the sea floor to the instantaneous velocity v. The depth to basement is obtained by substituting the maximum observed velocity in the sediment to this equation.

In contrast, our four sonobuoy profiles in McMurdo Sound (locations in Figure 2) suggest that models with constant velocity layers are applicable (Table 1 and Figure 5). To compare our results to those of Houtz and Davey, a plot of depth versus instantaneous velocity at that depth is necessary. For a model with a stepwise increase in velocity with depth the problem arises as to which depth the velocity of a constant velocity layer with finite thickness should be assigned. The obvious choice would be to attribute the velocity to a depth corresponding to the middle of the layer. This choice would also obviate the difficulty which arises when two layers cannot be resolved. In such a case, an intermediate velocity would be measured, which would still be representative of the middle of the two combined layers. Assigning this intermediate velocity to the top of the combined layers, however, would bias the resulting curve to the high-velocity side. In Figure 6, measured layer velocities are plotted against depth from the sea floor to the middle of the layer. The least squares regression line is $v = 1.83 + 0.71\ h$, which differs significantly from the results of Houtz and Davey. If, however, we plot layer velocity versus depth to

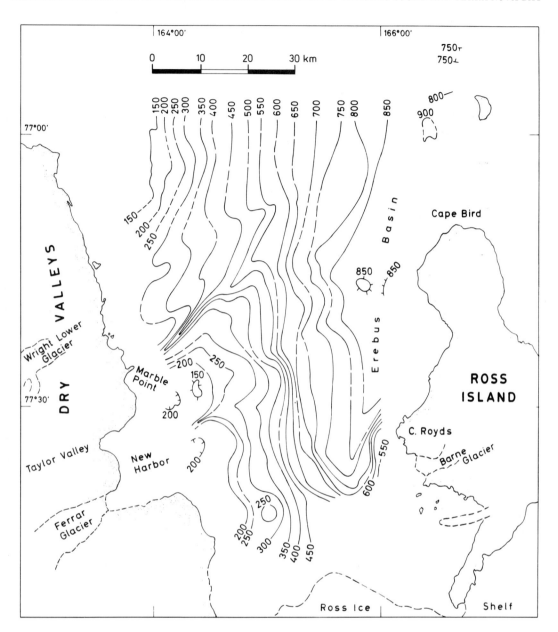

Fig. 4. Bathymetry of McMurdo Sound. Depth contours in meters at 50-m intervals. A sound velocity of 1440 m/s is assumed, corresponding to a mean temperature of −1.5°C and a salinity of 34.7‰. See Figure 2 for data coverage.

the top of the layer (Figure 6), then a least squares line of $v = 2.11 + 1.23\,h$ is obtained, which is nearly identical to the equation $v = 2.10 + 1.27\,h$ of *Houtz and Davey* [1973]. The latter plot provides perhaps a more reasonable comparison, since the equation of Houtz and Davey is considered valid for the estimate of depth to top of basement. We recognize that in evaluating a regression line from our data we are determining a velocity function that increases linearly with depth, contrary to

our structural model of a stepwise increasing velocity function. This is done solely for purposes of comparison, as has just been discussed.

Our data suggest that four layers with refraction velocities of 1.9, 2.4, 2.8-3.1, and 3.9-4.2 km/s can be recognized in McMurdo Sound, although they are not always separately resolvable. Where the 1.9- and 2.4-km/s layers cannot be resolved, a layer velocity of 2.1 km/s results (sonobuoys 2 and 3, Table 1); where resolution of the first three layers

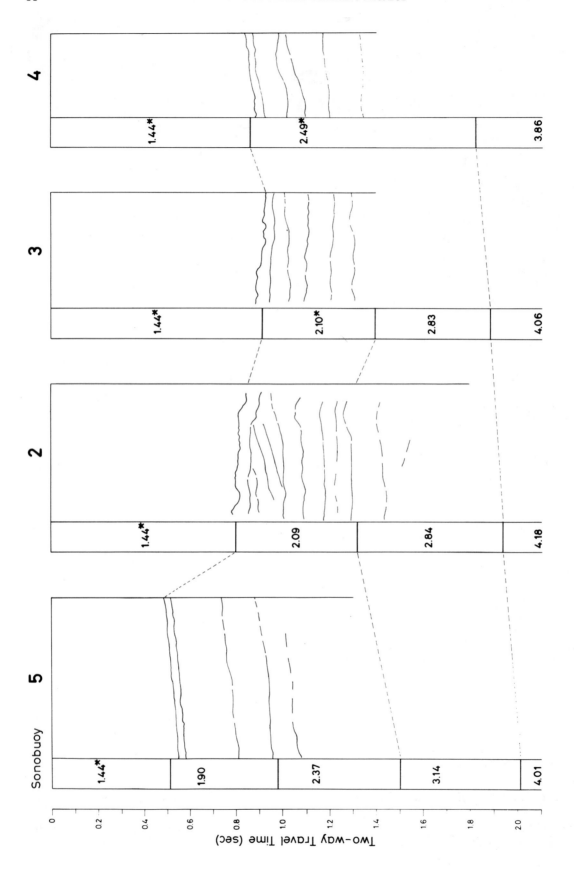

Fig. 5. Summary of sonobuoy refraction results in McMurdo Sound. Velocities are in kilometers per second and are accompanied by an asterisk when assumed. Normal incidence reflection profile obtained during each sonobuoy run is shown alongside the sonobuoy results. Tentative correlations are indicated by dotted straight lines.

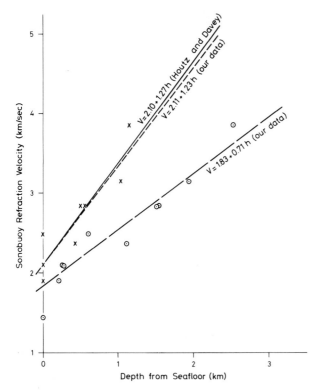

Fig. 6.   Plot of measured layer velocities from sonobuoy refraction results versus depth from the sea floor to the middle of the layer (points with circles; least squares regression line is denoted by long dashes). Analogous plot of layer velocity versus depth to top of layer (crosses; least squares line indicated by short dashes). Solid line is from the equation of *Houtz and Davey* [1973] for the estimate of depth to basement in the eastern Ross continental shelf.

(1.9, 2.4, and 2.8-3.1 km/s) is not attained, the intermediate velocity of 2.49 km/s is measured (sonobuoy 4). At a water depth of about 600 m the first two layers together constitute the top 500 m of the sedimentary column, while the third layer averages about 750 m in thickness. With decreasing water depth these thicknesses increase, both reaching approximately 1 km in value under 370 m of water.

Table 2 summarizes seismic velocity measurements reported in McMurdo Sound and the adjacent Ross Sea and Victoria Land. Of particular significance are the refraction profiles of *McGinnis* [this volume], the laboratory seismic velocity measurements of *Barrett and Froggatt* [1978], and the results from sites 270-272 of the Deep Sea Drilling Project [*Shipboard Scientific Party*, 1975]. *McGinnis* [this volume] reported three reversed refraction profiles from New Harbor to about 15 km southeast of Marble Point, shot from sea ice in water depths of 118-204 m. He measured a sea floor

velocity of 2.9 km/s, which was ascribed to late Cenozoic permafrost sediments. This ascription is supported by the fact that the salinity of pore ice in some sediments of McMurdo Sound is less than that of normal seawater [*McGinnis et al.*, this volume], so that the occurrence of frozen sediments at these localities cannot be excluded. Once a sediment becomes frozen, its seismic velocity increases in accordance with its porosity [*Timur*, 1968; *McGinnis et al.*, 1973]:

$$v_f = v_u \left[1 + \left(\frac{P - 3.63}{44.38} \right) \right]$$

where

v_f velocity of frozen sediment;
v_u velocity of same sediment when unfrozen;
P porosity, %.

For a 48% porosity, the sediment velocity becomes doubled upon freezing. With this in mind, the discrepancy between our sea floor velocity (1.9 km/s) and the one reported by *McGinnis* [this volume] (2.9 km/s) may be explained: we are dealing with essentially the same material but under different conditions, unfrozen in our case, partially or completely frozen in the other.

Barrett and Froggatt [1978] made laboratory measurements on compressional wave velocities of 75 samples from the major rock units of Victoria Land and the adjacent sea floor. Pertinent to our present study are the values of 1.6-2.6 km/s for late Cenozoic glacial sediments and 2.2-4.4 km/s for the Beacon Supergroup. The first result brackets our sea floor velocity of 1.9 km/s; the second brackets the range of highest velocities we were able to record (3.9-4.2 km/s). *Barrett and Froggatt* [1978] noted that field measurements on seismic velocities of the Beacon sandstones are about 0.9 km/s higher than laboratory values, this being a result of overburden pressure. Taking their average of 3.26 km/s for Beacon sandstones (calculated from *Barrett and Froggatt* [1978, p. 181, Table 2]), and adding 0.9 km/s, we expect a typical value of 4.16 km/s to be measured in the field. This suggests that we cannot dismiss Beacon sandstone as a possible candidate for our 3.9-4.2 km/s material.

The results of sites 270-272 of the Deep Sea Drilling Project in the south central Ross Sea are crucial to the interpretation of our data. These sites cluster around 77°S, 177°W, and lie about 500 km offshore from our survey area. At site 270, early Paleozoic marble and calc-silicate gneiss (4.3-

TABLE 2.   Summary of Seismic Velocity Measurements Reported in McMurdo Sound and the Adjacent Ross Sea and Victoria Land

Location	Velocity (Thickness)	Interpretation	Reference
DVDP 15: W. McMurdo Sound, 77°26'14.2"S, 164°22'49.4"E	2.0–2.2 2.0–3.2 5.1	unfrozen sand partially or completely frozen sand drilled cobble of schist	*Barrett et al.* [1976a, b]
DSDP sites 270, 271, and 272 south central Ross Sea	1.6–2.0 2.1–2.5 3.5–4.2 4.3–5.5	soft Plio-Pleistocene sandy mud marine mudstone, semilithified glacial deposits of Miocene to Oligocene age coarse sediment breccia, Oligocene Paleozoic basement of marble and calc-silicate gneiss	*Shipboard Scientific Party* [1975]
W. McMurdo Sound, New Harbor to 15 km SW of Marble Point	(thin) 2.9 3–3.6 4.97–5.62	unfrozen low-velocity layer permafrost sediment unfrozen late Mesozoic-early Cenozoic sediment basement	*McGinnis* [this volume]
Victoria Land and adjacent sea floor (laboratory measurements)	1.6–2.6 3.0–4.2 2.6 4.6 2.2–4.4 5.8 4.7 5.3	late Cenozoic glacial sediments early Cenozoic shallow marine, calcareous sediments pyroclastic sequence, McMurdo volcanics flow-dominated sequence, McMurdo volcanics Beacon Supergroup (Devonian to Triassic) Jurassic dolerite sills granitic basement rocks (Precambrian to early Paleozoic) metamorphic basement rocks (Precambrian to early Paleozoic)	*Barrett and Froggatt* [1978]
McMurdo Station	3.1 (1.2) 4.8 5.9		*Kaminuma* [1978]
E. McMurdo Sound (77°50'S,167°E)	3.1 (2.0) 4.8	volcanic ash and tuffaceous material	*Robinson* [1963]
E. McMurdo Sound near Hut Point (77°50'S, 167°E)	2.9 (1.3) 3.7	local volcanics	*Robinson* [1963]
E. McMurdo Sound (77°50'S,167°E)	3.1 (2.4) 5.0	local volcanics	*Robinson* [1963]
E. McMurdo Sound (77°50'S,167°E)	? (1.7) 4.1		*Crary and Van der Hoeven* [1961]
Ross Sea continental rise	1.6–2.0 2.3 3–4 4.4–4.7	basement	*Houtz and Davey* [1973]
Ross Sea shelf	5.2–6.3	basement, pre-Cretaceous rocks of Marie Byrd Land or Precambrian-lower Paleozoic metamorphics of Victoria Land	*Houtz and Davey* [1973]
Don Juan Pond	1.8 5.12	unfrozen sediments basement	*Clark* [1972]
Don Quixote Pond	4.59	basement	*Clark* [1972]
Little America Station	2.4 (1.33) 4.24 (0.65) 6.4	 Beacon sandstone basement	*Crary* [1961, 1963]
Skelton Glacier, Victoria Land	4.30 5.43 6.54 6.8	Beacon sandstone granite or gneiss gabbro Ferrar dolerite sills	*Crary* [1961, 1963]
Ohio Range	4.4	Permian coal measures, lower part of Victoria Group (Beacon Supergroup)	*Bentley and Clough* [1972]

TABLE 3. Interpretation of Sonobuoy Refraction Velocities McMurdo Sound

Velocity, km/s	Interpretation
1.9 2.1	marine pebbly muddy sand
2.4	Miocene-Oligocene pebbly mudstone, coarse nearshore facies
2.8–3.1	older preglacial sandstone and mudstone
3.9–4.2	Beacon sandstone? metasediments? low-velocity basement?

5.5 km/s) is overlain by 30 m of sedimentary breccia (3.5-4.2 km/s), which is in turn overlain by 360 m of glacial marine deposits (2.1-2.5 km/s) ranging in age from Oligocene to early Miocene. The topmost Plio-Pleistocene sandy mud is characterized by velocities of 1.6-2.0 km/s. In view of the coarser surficial sediments in McMurdo Sound [Glasby et al., 1975] we interpret the 1.9-km/s layer from our sonobuoy results as a sequence of marine, pebbly, muddy sand similar to that found on the floor of the sound today, the unfrozen equivalent of the 2.9-km/s material of McGinnis [this volume] (Table 3). Analogously, our 2.4-km/s layer could represent a coarse, nearshore facies of the Oligocene to Miocene pebbly mudstone of site 270. The 2.8- to 3.1-km/s material may then be correlated with older (early Cenozoic), preglacial sandstones and mudstones. Interpretation of the highest velocity layer encountered (3.9-4.2 km/s) is problematic. The lack of magnetic anomalies in McMurdo Sound [Wong, 1973; Pederson et al., this volume] makes basalt or McMurdo volcanics unlikely candidates. Typical basement velocities lie in the range 4.5-5.5 km/s, somewhat higher than what has been measured. Could this material be Beacon sandstone? Or some variety of metasediment? Or perhaps even low-velocity basement?

Our lithologic interpretation of layer velocities is consistent with results of DVDP hole 15, drilled 16 km east of Marble Point through 122 m of water [Barrett et al., 1976a, b]. Here the top 13 m consist of fine to coarse, unconsolidated, poorly sorted silty sand, with angular to subrounded pebbles scattered throughout. This unit is apparently deposited from melting ice and wind and is of Recent age. The underlying unit, extending to at least 65-m subbottom, consists of well-stratified, moderately sorted fine sand to poorly sorted medium sand. It is dominated by basaltic material similar to the top unit but lacks pebbles. Possible age is Plio-Pleistocene.

In contrast to our interpretation is that of Robinson [1963] for his refraction and gravity data, in which a sequence of volcanic ash and tuffaceous material, 2 km in thickness and extending two thirds of the way from Hut Point across McMurdo Sound, is inferred to overlie basalt. In view of the smooth magnetic field within the sound this interpretation appears unlikely.

Profiler Data

Shallow sediment structures are revealed by profiler data obtained in the sound. The single most important horizon that may be correlated over a large part of the survey area is the angular unconformity T first recognized by Houtz and Meijer [1970]. This unconformity (profile 9-9' of Figure 7, for example) truncates the underlying gently dipping (1.5°-3°) sequence on a regional scale. The overlying section is in part acoustically homogeneous. Whatever sparse reflections that occur cannot be traced over long distances and are not indicative of progradation. They are often unconformable to each other as well as to the sea floor. The lower section appears to be coarser and exhibits a very distinct stratification, which in at least one instance suggests slumping.

Profile 1-1' (Figure 8), lying almost exactly along the 166°E meridian, shows in the south (near Cape Royds) a very strong acoustic reflector at the sea floor which effectively reduced penetration to zero. The acoustic character of this reflector led us to attribute it to the acoustic basement, which probably is composed of volcanic material from Ross Island. However, our interpretation is still open to discussion. Since the refraction profiles of McGinnis [this volume] suggest the occurrence of thick permafrost sediments in New Harbor and immediately offshore, that similar sediments may exist off Cape Royds cannot be dismissed offhand. Should frozen sediments actually occur, a high acoustic impedance contrast would be encountered at the sea floor because of its high (2.9 km/s) compressional wave velocity. The net effect is an increase in the amplitude of the reflected wave and a reduction in penetration. While we do not favor this alternative interpretation of ascribing the strong reflector to permafrost sediments, we do not exclude it as a possibility.

The angular unconformity T can be clearly traced across a large part of this profile (1-1'). As Erebus Basin is approached in the north, the truncated dipping layers grade into the flank of a broad

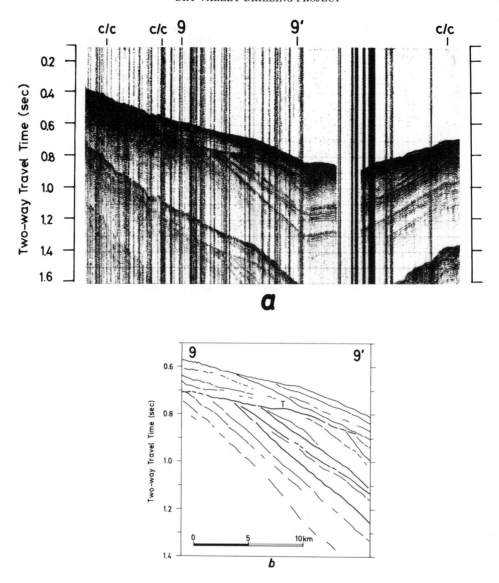

Fig. 7.   (a) Profiler record and (b) line drawing interpretation of profile 9-9'. Location in Figure. 2. Vertical exaggeration about 20 times. For all reflection profiles, water depth is approximately 720 m/s of two-way travel time, and sediment thickness is 1000 m/s. Black bands across the profile are due to noise produced when the hydrophone arrays were forced to ride over ice floes. Note occurrence of angular unconformity T under about 120 m of sediment. Ths unconformity, possibly of mid-Pliocene age, is believed to be of glacial erosional origin.

synclinal structure, whereby angularity appears to be lost. The overlying sequence is pierced by three diapirlike structures, which also cause an elevation of the sea floor. Reflections within these structures are lacking. Whether they are associated with evaporites, with shallow water diatomaceous clay [*Talwani and Eldholm*, 1972], or with intrusive plugs is not clear.

Profile 21-21' (Figure 9) extends from northern New Harbor northeastward to Erebus Basin. In the shallow waters of New Harbor to the west,

three ridgelike features, 50-100 m high and 1-5 km across, are recorded at the sea floor. That they are acoustically transparent suggests that they consist possibly of glacial till and could have been deposited as end moraines when the ice front remained stationary for some time. Horizontally stratified sediments are found within Erebus Basin. Underlying this is the continuation of the dipping beds that parallel the trend of the bottom slope. It appears that these layers may extend beyond Erebus Basin to continue under Ross Island, to underlie

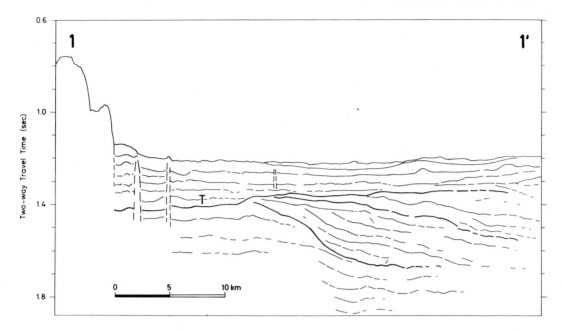

Fig. 8.   Line drawing interpretation of profile 1-1'. Location in Figure 2. Vertical exaggeration about 20 times. Note angular unconformity T, change in angularity of dipping layers below T, and the three diapirlike structures in the southern half of the profile.

the Cenozoic volcanics, the eruption of which established the island as a major topographic feature. An extension of similar layers has also been recorded on two other profiles off Cape Royds and one to the northeast of Cape Bird.

Profiles I-I' to IV-IV' (Figure 10) cut across, at various oblique angles, the fjordlike submarine valleys in the western part of the sound. A direct comparison of the shapes of these valleys is made difficult by the different vertical exaggerations and profile orientations relative to the valley axes. However, their locations suggest that they are extensions of the Taylor and Ferrar valleys, respectively.

The east Antarctic ice sheet is dammed to a considerable thickness by the Transantarctic Mountains in southern Victoria Land. Spill-overs from this ice sheet are observable today as glacier tongues. In the early glacial history of McMurdo Sound these glacier tongues have repeatedly expanded and invaded the lower reaches of the dry valleys and the nearby areas of the sound. At least three such invasions have occurred in Taylor Valley, for example [Denton and Armstrong, 1968]. We hypothesize that the observed U-shaped submarine valleys were eroded during one or more of these glacier invasions. The very rugged sea floor in the immediate vicinity of the valleys and the oc-currence on the east banks of distinct ice contact deposits whose steep sides face the valleys lend support to our hypothesis. In addition, in both instances, the valley floor corresponds approximately to the 4-5 m.y. angular unconformity T, implying that both submarine valleys have probably undergone a similar history of development.

An attempt has been made to correlate several of the strongly reflecting horizons across the different profiles by utilizing track crossings and similarities in acoustic character. However, the resulting correlations (except for the unconformity T) are so uncertain that they will not be presented here.

TERRA NOVA BAY

Bathymetry

The survey area in Terra Nova Bay consists of a series of northeast-southwest trending troughs and depressions cut into a southeastward dipping slope of 1° gradient (Figure 11). The trend of these troughs contrasts with the northwest-southeast structural trend (e.g., of fold structures) mapped on Victoria Land. The most prominent of these linear depressions is Drygalski Basin (over 1100 m, maximum 1157 m), which contains some of the greatest depths reported to date in the Ross Sea.

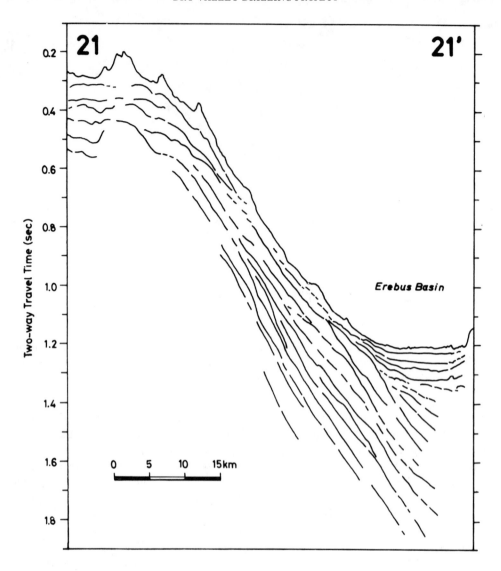

Fig. 9.   Line drawing interpretation of profile 21-21'. Location in Figure 2. Vertical exaggeration about 40 times. Of particular interest are the ridgelike features at the sea floor which may represent end moraines, the flat-lying sedimentary sequence in Erebus Basin, and the continuation of the dipping layers beneath this sequence.

Southeast of this basin, the sea floor slopes upward to less than 400 m at 170°E, where 'Crary Bank' is encountered [*Vanney and Johnson*, 1976].

The origin of these linear deeps remains speculative. They could be a product of glacial scour, they could be related to the tectonic movements that occurred along the Victoria Land coast in association with uplift of the Transantarctic Mountains [*Hayes and Davey*, 1975*b*], or they could be a result of crustal depression just beyond the edge of the Antarctic ice sheet.

Profiler Data

North-south profiles II', JJ', and KK' (Figure 12) and east-west profiles EE', DD' (Figure 13) in the western part of Terra Nova Bay traverse a large stratified depositional feature which measures about 300-500 m high and about 10 km in dimension. We interpret this feature to be a delta moraine, a fluvioglacial deposit formed when an expanded Ross Ice Shelf invaded the bay from the south and caused local or general ice grounding.

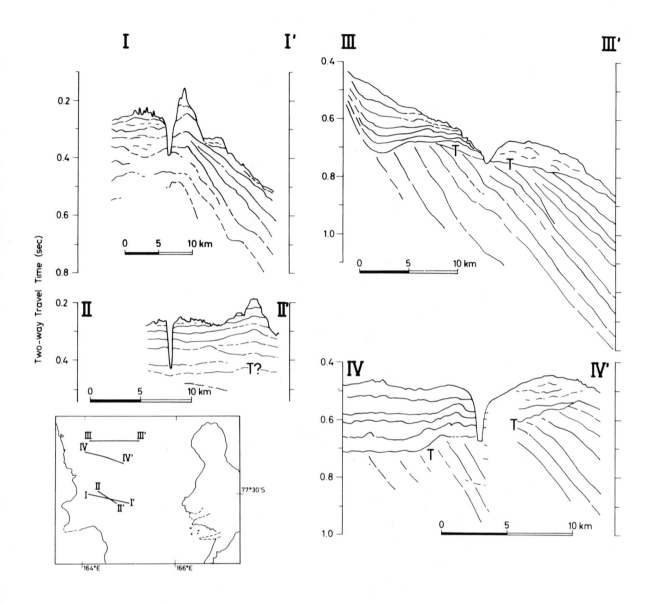

Fig. 10.   Profiles I to IV across the two fjordlike submarine valleys in McMurdo Sound. These valleys are believed to be extensions of the Taylor and Ferrar valleys, respectively. They were eroded during one or more invasions of glacier tongues into McMurdo Sound from the ice sheet dammed by the Transantarctic Mountains in southern Victoria Land. Note the unconformity T and the ice contact deposits on the east banks.

The moraine is deposited where the stationary ice front remained for some time, ending in a lake or sea. It shows deltaiclike bedding on the distal side, and on the proximal side, gentle folding and contortions in the structure provide some evidence of later glacial advance. The steep ice-contact slope faced the decayed and vanished ice margin which lay to the south.

Profile LL′ (Figure 14) runs approximately from west to east along latitude 74°50′S, crossing Drygalski Basin where water depth is greatest. In the west the two linear depressions seen as bathymetric lows in the profile appear to be a result of erosion. Many of the upper sedimentary layers lying conformably on older strata outcrop along the flanks of topographic highs (i.e., the sides of the

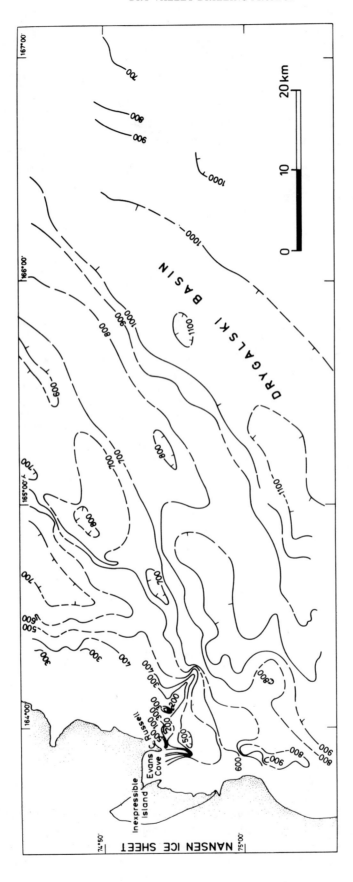

Fig. 11. Bathymetry of survey area in Terra Nova Bay. Depth contours in meters at 100-m intervals. A sound velocity of 1440 m/s is assumed. See Figure 3 for data coverage.

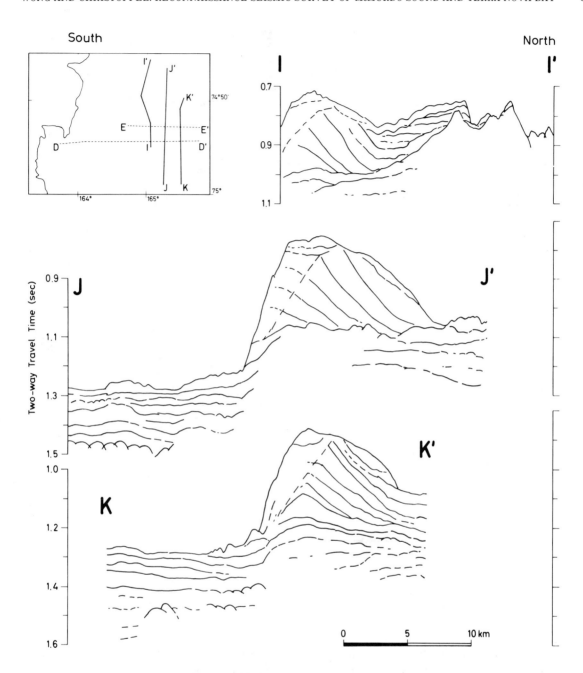

Fig. 12. Line drawing interpretation of profiles II′, JJ′, and KK′. Location in Figure 3. Vertical exaggeration about 22 times. The large, asymmetric, stratified depositional feature is interpreted as a delta moraine, deposited when the ice front of an expanded Ross Ice Shelf became stationary for some time, ending in a lake or in the sea.

depressions). Drygalski Basin itself is bounded to the east by a moraine, beneath which the angular unconformity T marked by truncated dipping layers can be observed. Sediments within the basin proper are well stratified and exceed 600 m in thickness.

TERRA NOVA BAY TO BEAUFORT ISLAND

Profile VV′WW′ (Figure 15) lies along the survey track from Terra Nova Bay to Beaufort Island, between 75°S and 76°S (Figure 1). At the northern

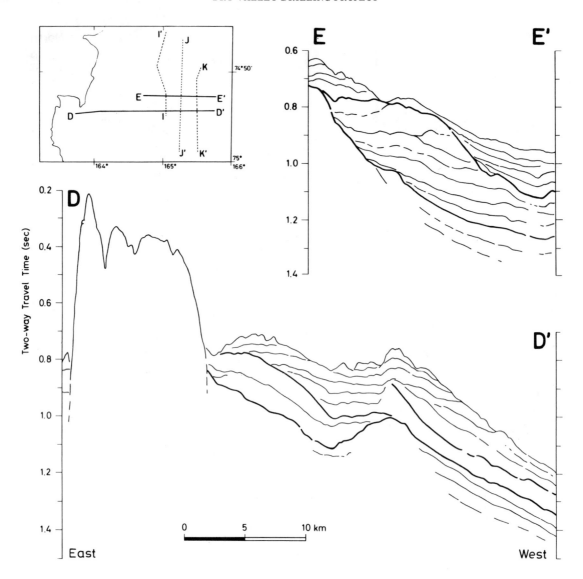

Fig. 13.   Line drawing interpretation of profiles EE' and DD'. Location in Figure 3. Vertical exaggeration about 22 times. The inferred delta moraine is again clearly visible.

end of the profile (near V) the truncated dipping layers can be clearly traced for about 25 km, after which their angularity begins to be lost, whereby they become subparallel to the sea floor. The overlying sediments (above unconformity T) average about 200 m in thickness and are characterized by poor stratification except for two pervasive, highly reflecting horizons. Small erosional channels, probably of current or iceberg scouring origin, occur at the sea floor.

About 25 km south of W, two near-vertical faults, each with a throw of approximately 100 m, have been mapped. A vertical displacement is not

observed at the sea floor, since the latter appears to represent an erosional surface at this location. These faults constitute the only peice of indisputable evidence of vertical tectonics in all the profiler data we have gathered throughout our survey. Could they be associated with the uplift of the Transantarctic Mountains along Victoria Land?

The profile XX' (Figure 16), located just to the north of Beaufort Island (Figure 1), shows a morainal deposit lying on what is otherwise a flat sea floor. Stratification within the glacial drift suggests that either the original till has been transported both by glacier ice and by meltwater, whose

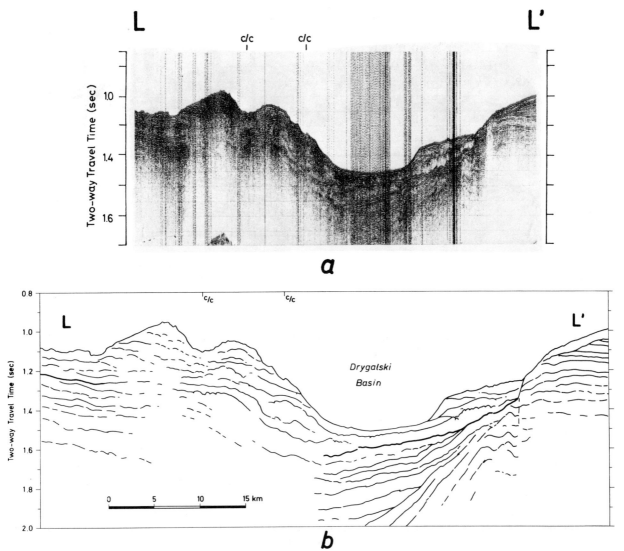

Fig. 14.   (a) Profiler record and (b) line drawing interpretation of profile LL'. Location in Figure 3. Vertical exaggeration about 20 times. Note that the sediments within Drygalski Basin are well stratified and exceed 600 m in thickness and that the eastern flank of the basin itself is bounded by a moraine.

release often accompanies the deposition of rock debris, or that redistribution of the till by bottom currents has taken place subsequent to deposition.

Near X' in the vicinity of Beaufort Island, the acoustic basement can be traced to or almost to the sea floor. This is characterized by the lack of penetration and the occurrence of numerous hyperbolic reflectors suggestive of ruggedness of the reflecting interface. To the north, the basement appears to be down faulted and deepens abruptly. Whether this sudden change in basement depth is related to the general sharp increase in depth to basement of about 2 km in the eastern part of the Ross continental rise reported by *Houtz and Davey* [1973] is not clear.

DISCUSSION

Angular Unconformity

The angular unconformity *T* widely observed both in McMurdo Sound and Terra Nova Bay has been interpreted to represent a major erosional surface caused by grounded shelf ice as it advanced north of its present position [*Hayes et al.*, 1975]. The underlying dipping sequence was truncated in a short interval of time, during which a major glacial pulse occurred, comprising a substantial northward displacement of the northern limit of grounded shelf ice followed by a rapid retreat of the ice shelf edge to a position similar to that of today.

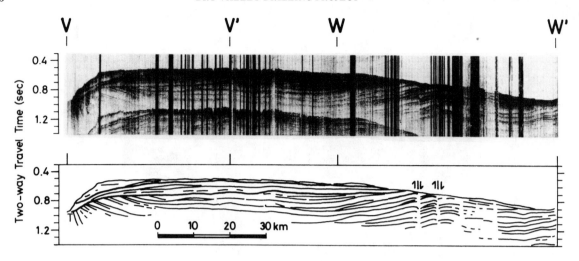

Fig. 15.   Profiler record and line drawing interpretation of profile VV'WW'. Location in Figure 1. Vertical exaggeration about 20 times. Of particular note are the unconformity T and the two near-vertical faults about 25 km south of W.

Melting and ice calving could cause meltwater to be ponded in front of the ice sheet if sea level were at a low stand, thus producing freshwater lakes in which lacustrine deposits and delta moraines would be laid down. This is particularly likely during the collapse phase of a particular ice advance. The distribution of areas from our profiler data where truncated dipping beds and the angular unconformity T can be clearly recognized (Figure 17) suggests that such freshwater conditions could have occurred locally in the southern part of McMurdo Sound, thus offering a possible explanation for the reported occurrence of freshwater diatoms in parts of the DVDP 15 cores [*Brady*, 1977]. In Terra Nova Bay the location of a mapped feature interpreted as a delta moraine is also consistent with the position of the ice front inferred from the angular unconformity.

At sites 270 and 272 of the Deep Sea Drilling Project a similar angular unconformity has been observed. It is accompanied by a marked increase in lithification and an early Pliocene to Miocene stratigraphic hiatus. Assuming that the unconformity we observed is contemporaneous with that drilled in DSDP, the sediment layer above it must have been deposited within the last 4-5 m.y. Figure 18 shows the thickness distribution of sediments above this unconformity T. The thickest section is encountered in Erebus Basin in the eastern part of the sound (over 300 m, assuming $v = 2$ km/s), in a small area northeast of Marble Point in the west (up to 220 m), and seaward of the ice shelf in

the south. In the center of the sound is a north-south trending belt where this layer averages only about 40 m. Thus the thickness of sediments deposited in McMurdo Sound since mid-Pliocene time (4-5 m.y. B.P.) reflects the proximity to a sediment source. It has been demonstrated that for DVDP site 15 the sediments of Recent and Plio-Pleistocene age have been derived primarily from the alkaline volcanic suite of the McMurdo area and partly from the basement complex of intrusives of the Transantarctic Mountains [*Barrett et al.*, 1976*b*]. The main transport agents are wind and melting ice. Our observation on the post mid-Pliocene sediment thickness distribution is thus consistent with this finding. It should be noted that if permafrost sediments exist at the sea bottom, particularly in nearshore areas, the estimate of 2 km/s for bottom velocity would be too low [*McGinnis*, this volume] and the thicknesses given would only represent lower limits.

Taking the mean sediment thickness above the angular unconformity to be 80 m, the sedimentation rate since mid-Pliocene would average about 18 m/m.y. in McMurdo Sound. However, this value must represent a gross underestimate, since a widespread disconformity separating sediments of Gauss age (>2.4 m.y.) from a thin discontinuous layer of Bruhnes sediments (<0.7 m.y.) has been well established from paleomagnetic, radiolarian, foraminiferal, and ice-rafted debris studies of *Eltanin* cores [*Fillon*, 1972, 1975, 1977]. The disconformity is thought to be a result of climatic cool-

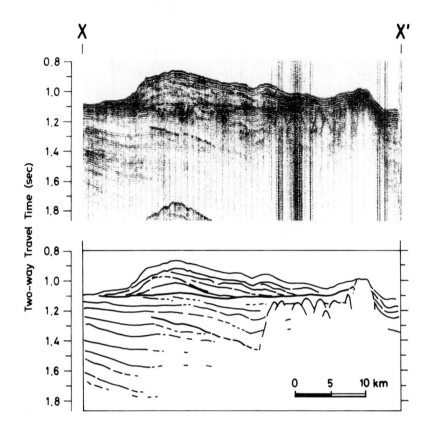

Fig. 16.   Profiler record and line drawing interpretation of Profile XX'. Location in Figure 1. Vertical exaggeration about 20 times. The stratification within the morainal deposit lying on what is otherwise a flat sea floor is due either to transport by glacier ice and meltwater or to redistribution of till by bottom currents subsequent to deposition.

ing, expansion of the Ross Ice Shelf, increased circulation of the Antarctic bottom water, and hence increased current erosion or nondeposition. In view of this disconformity our mean sedimentation rate may be too low by perhaps a factor of 2. The corrected value should thus lie somewhere around 36 m/m.y., which is low compared to the average of 53 m/m.y. for the DSDP Ross Sea drill holes, or to the rate of 75 m/m.y. for the complete sequence drilled at site 271. No doubt such a low value reflects ice scour effects as well as major ice advances in the early Pliocene and late Quaternary, when extensive ice shelves were built and dry base glaciation, slow sedimentation took place.

Palynological evidence suggests that vegetation persisted in the Ross Sea area until late Oligocene [*Kemp and Barrett*, 1975], while sedimentological data indicate that ice rafting of clastic debris commenced at about the same time [*Barrett*, 1975]. In addition, the species diversity of planktonic foraminifera was low in the Oligocene and Eocene

[*Margolis and Kennett*, 1970, 1971]. All of this evidence points to glaciation of Antarctica at least at times during the lower and middle Eocene and during the Oligocene [*Hayes and Frakes*, 1975]. Climate during lower and middle Miocene was considerably warmer. The cooling trend commenced again near the end of the Miocene, leading to expanded Pleistocene glaciations on the continent.

In the past 3 m.y. the Ross Ice Shelf has expanded on at least four occasions into an ice sheet that was grounded on the floor of the Ross Sea [*Denton, et al.*, 1970, 1975; *Denton and Borns*, 1974]. This ice sheet grounding within McMurdo Sound [*Brady*, 1977; *Stuiver and Denton*, 1977] has left numerous small erosional features such as scour channels that are either surficial or lie within the top several hundred meters of subbottom. Off the Victoria Land coast, small unstratified depositional features interpretable as morainal deposits are observed, marking halts in the advance of the ice sheet. However, we could not determine

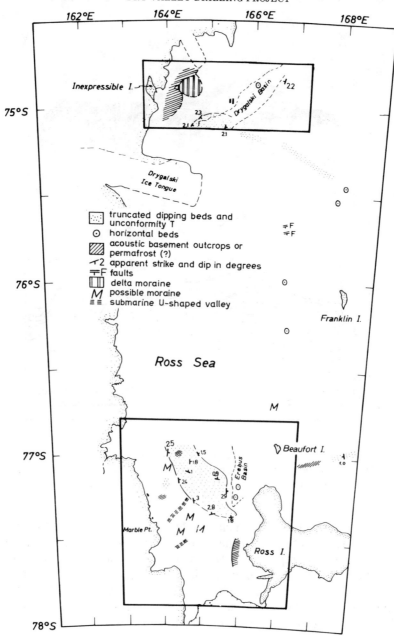

Fig. 17.   Near-surface structural features of McMurdo Sound and Terra Nova Bay from profiler data. Note distribution of moraines, truncated dipping beds, and possible basement outcrops.

whether they are a result of Ross Sea glaciations or of spill-overs from the east Antarctic ice sheet.

A current-scouring origin for these erosional channels cannot be dismissed with certainty. Bottom current velocities of 10-15 cm/s have been reported on shallow banks within the Ross Sea [*Jacobs et al.*, 1970], while velocities as high as 1 m/s have been measured just north of the Dailey Islands and around Hut Point Peninsula [*Barrett et*

al., 1976*b*]. Even stronger bottom currents could have operated during periods of an expanded Ross Ice Shelf, when circulation became more restricted and thermohaline interaction of seawater with the underside of the ice shelf more enhanced.

In the Terra Nova Bay area, *Denton et al.* [1975] concluded that at least two glaciations occurred. The Reeves and Priestley glaciers thickened considerably and grounded ice flowed eastward

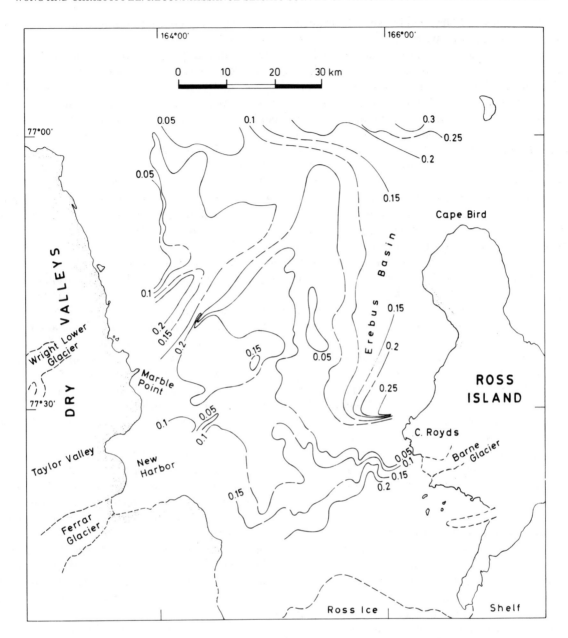

Fig. 18.   Thickness distribution of sediments overlying the conformity T, presumably of mid-Pliocene age. Isopachs in seconds of two-way travel time, or in kilometers, if an average sound velocity of 2 km/s in the sediments is assumed.

around and over Inexpressible Island. Thus, here again, glacial erosional features are common at the sea bottom and the immediate subbottom.

Near-Surface Structure

In Figure 17, near-surface structures of Mc-Murdo Sound and Terra Nova Bay from our profiler data are summarized. Stipled areas represent truncated bedding, where the angular unconformity T can be readily discerned. Strikes of the truncated layers are assumed to be normal to the profiles, except in four cases where intersection of profiler tracks allow the true strike to be computed. The apparent dip is determined by assuming a sound velocity in water of 1.44 km/s and a near-surface sediment sound velocity of 2 km/s. Horizontal bedding (double circles) is confined largely to Erebus and Drygalski basins. Here, conditions are presumably favorable for rapid deposition in a quiet milieu, while erosion is subdued.

The acoustic basement has been mapped at three locations in McMurdo sound and immediately offshore in Terra Nova Bay. While we cannot dismiss the possibility that it may correspond to permafrost sediments as discussed earlier, we prefer the interpretations which follow. North of Marble Point, the acoustic basement probably represents a glacially eroded surface cut into early Paleozoic-late Precambrian basement rocks similar to those exposed a short distance away in the dry valleys. Off Cape Royds in the eastern part of the sound, it probably corresponds to McMurdo volcanics such as those erupted on Ross Island. This is supported by the limited westward extension of magnetic anomalies observed over Cape Royds [*Pederson et al.*, this volume] and by the rough topography of the basement outcrops. In Terra Nova Bay we interpret the basement either to belong to the metamorphic complex of Precambrian age or to consist of granitic intrusives associated with the Ross Orogeny similar to the rocks exposed on Inexpressible Island and the adjacent coasts [*Ricker*, 1964]. The fact that the basement reflector has only been infrequently observed is in agreement with the profiler and sonobuoy results of *Houtz and Davey* [1973] for the Ross Sea, which show an abrupt thickening to the west of the sedimentary sequence to over 2.5 km, approximately along the 177°E meridian.

Total sediment thickness exceeds 1.4-2 km in the central part of McMurdo Sound, where our sonobuoy measurements were made. Profiler data suggest that the sediment sequence is thinner in the west and probably reaches its maximum thickness in Erebus Basin in the east. In Terra Nova Bay the lack of refraction data does not permit much to be said about the total sediment thickness distribution, but again, indications are that the thickest sequences are to be encountered in Drygalski Basin.

CONCLUSIONS

1.   The bathymetry of McMurdo Sound is characterized by a north-south trending, eastward dipping slope incised by two submarine, fjordlike valleys believed to be the subaqueous extensions of Taylor and Ferrar valleys, respectively. Off the west coast of Ross Island is a well-developed basin (Erebus Basin), over 900 m deep, whose stratified

sediments may continue uninterrupted to underlie the island.

2.   Sonobuoy refraction in the center of the sound reveals the existence of four layers with velocities of 1.9, 2.4, 2.8-3.1, and 3.9-4.2 km/s. These are interpreted lithologically to represent marine, pebbly, muddy sand, pebbly mudstone, preglacial sandstones and mudstones, and metasediments (?), respectively. The total sediment thickness exceeds 1.4 km.

3.   Terra Nova Bay consists of a series of northeast-southwest trending linear deeps cut into a southeastward dipping slope. Maximum depth (1157 m) is reached in Drygalski Basin in the east.

4.   East of an area of acoustic basement outcrops immediately offshore in Terra Nova Bay is a large, asymmetric, stratified depositional feature, interpreted to be a delta moraine. It is believed to have been deposited when the ice front of an expanded Ross Ice Shelf became stationary for some time, ending in a lake or in the sea.

5.   Two near-vertical faults of about 100-m throw each are mapped northwest of Franklin Island. They constitute the only evidence of vertical tectonics from our entire survey and may be associated with the uplift of the Transantarctic Mountains.

6.   Profiler data demonstrated the pervasive occurrence, both in McMurdo Sound and in Terra Nova Bay, of an angular unconformity *T* at or near the sea floor. This unconformity is interpreted as a major erosional surface caused by grounded shelf ice. Assuming that it is contemporaneous with the truncated dipping layers drilled at sites 270 and 272 of the Deep Sea Drilling Project about 500 km away in the south central Ross Sea, the ice erosion must have taken place in the mid-Pliocene (4-5 m.y. B.P.). From this, the post mid-Pliocene sediment distribution pattern may be determined, and an average uncorrected sedimentation rate of 18 m/m.y. in McMurdo Sound may be estimated.

7.   The northern limit of locally or regionally grounded ice certainly extends at least to Terra Nova Bay. It could have reached the shelf edge at some point in time, as has been reported by *Houtz and Davey* [1973].

Acknowledgments.   We thank the captain and crew of the U.S. Coast Guard icebreaker *Burton Island* for their cheerful support and cooperation during the field work in McMurdo Sound and Terra Nova Bay. Lyle D. McGinnis was instrumental in getting this project launched. Our special thanks go to Colin Brown, James J. Kohsmann, and Douglas J. Northey, who has provided us with invaluable help during data collection.

C. Brown and D. J. Northey have also helped during early stages of data reduction and interpretation. This paper has benefited from reviews by L. D. McGinnis, C. P. Ervin, and R. Sylwester. The work reported here has been supported to a large extent by the Division of Polar Programs, U.S. National Science Foundation, through the Dry Valley Drilling Project (C-642), and by the Antarctic Division of the Department of Scientific and Industrial Research, New Zealand.

REFERENCES

Barrett, P. J., Textural characteristics of Cenozoic preglacial and glacial sediments at site 270, Ross Sea, Antarctica, in *Initial Reports of the Deep Sea Drilling Project*, vol. 28, pp. 757-767, U. S. Government Printing Office, Washington, D. C., 1975.

Barrett, P. J., and P. C. Froggatt, Densities, porosities, and seismic velocities of some rocks from Victoria Land, Antarctica, *N. Z. J. Geol. Geophys.*, *21*(2), 175-187, 1978.

Barrett, P. J., D. A. Christoffel, D. J. Northey, and B. A. Sissons, Seismic profiles across the extension of Wright Valley into McMurdo Sound, *Antarctic J. U.S.*, *9*(4), 138-140, 1974.

Barrett, P. J., S. B. Treves, C. G. Barnes, H. T. Brady, S. A. McCormick, N. Nakai, J. S. Oliver, and K. J. Sillars, Dry Valley Drilling Project, 1975-1976: First core drilling in McMurdo Sound, *Antarctic J. U.S.*, *11*(2), 78-80, 1976a.

Barrett, P. J., S. B. Treves, C. G. Barnes, H. T. Brady, S. A. McCormick, N. Nakai, J. S. Oliver, and K. J. Sillars, Initial report on DVDP 15, western McMurdo Sound, Antarctica, *Dry Val. Drilling Proj. Bull.*, *7*, 1-100, 1976b.

Bentley, C. R., and J. W. Clough, Antarctic subglacial structure from seismic refraction measurements, in *Antarctic Geology and Geophysics*, edited by R. J. Adie, pp. 683-692, Universitetsforlaget, Oslo, 1972.

Brady, H. T., Freshwater lakes in Pleistocene McMurdo Sound, *Antarctic J. U.S.*, *12*(4), 117-118, 1977.

Chriss, T., and L. A. Frakes, Glacial marine sedimentation in the Ross Sea, in *Antarctic Geology and Geophysics*, edited by R. J. Adie, pp. 747-762, Universitetsforlaget, Oslo, 1972.

Clark, C. C., Seismic refraction and electrical resistivity investigations in the dry valleys, *Antarctic J. U.S.*, *7*, 91-92, 1972.

Crary, A. P., Marine-sediment thickness in the eastern Ross Sea area, Antarctica, *Geol. Soc. Amer. Bull.*, *72*, 787-790, 1961.

Crary, A. P., Results of United States traverses in east Antarctica, 1958-1961, *IGY Antarctic Geophys. Year Glaciol. Rep. Ser.*, *7*, 144 pp., World Data Center A, Amer. Geogr. Soc., New York, 1963.

Crary, A. P., and F. G. Van der Hoeven, Sub-ice topography of Antarctica, long 160°W to 130°E, Antarctic Glaciology, *Int. Ass. Sci. Hydrol. Publ.*, *55*, 125-131, 1961.

Denton, G. H., and R. L. Armstrong, Glacial geology and chronology of the McMurdo Sound region, *Antarctic J. U.S.*, *3*, 99-101, 1968.

Denton, G. H., and H. W. Borns, Jr., Former grounded ice sheets in the Ross Sea, *Antarctic J. U.S.*, *9*, 167, 1974.

Denton, G. H., R. L. Armstrong, and M. Stuiver, Late Cenozoic glaciation in Antarctica: The record in the McMurdo Sound region, *Antarctic J. U.S.*, *5*, 15-21, 1970.

Denton, G. H., H. W. Borns, Jr., M. G. Groswald, M. Stuiver, and R. L. Nichols, Glacial history of the Ross Sea, *Antarctic J. U.S.*, *10*(4), 160-164, 1975.

Fillon, R. H., Evidence from the Ross Sea for widespread submarine erosion, *Nature*, *238*(81), 40-42, 1972.

Fillon, R. H., Late Cenozoic paleo-oceanography of the Ross Sea, Antarctica, *Geol. Soc. Amer. Bull.*, *86*, 839-845, 1975.

Fillon, R. H., Ice-rafted detritus and paleotemperature: Late Cenozoic relationships in the Ross Sea region, *Marine Geol.*, *25*, 73-93, 1977.

Glasby, G. B., P. J. Barrett, J. C. McDougall, and D. G. McKnight, Localized variations in sedimentation characteristics in the Ross Sea and McMurdo Sound regions, Antarctica, *N.Z. J. Geol. Geophys.*, *18*, 605-621, 1975.

Hayes, D. E., and F. J. Davey, Bathymetry of the Ross Sea (1974), Plate 1, in *Initial Reports of the Deep Sea Drilling Project*, vol. 28, U.S. Government Printing Office, Washington, D. C., 1975a.

Hayes, D. E., and F. J. Davey, A geophysical study of the Ross Sea, Antarctica, in *Initial Reports of the Deep Sea Drilling Project*, vol. 28, pp. 887-907, U. S. Government Printing Office, Washington, D. C., 1975b.

Hayes, D. E., and L. A. Frakes, General synthesis, Deep Sea Drilling Project Leg 28, in *Initial Reports of the Deep Sea Drilling Project*, vol. 28, pp. 919-942, U. S. Government Printing Office, Washington, D. C., 1975.

Hayes, D. E., et al., *Initial Reports of the Deep Sea Drilling Project*, vol. 28, 1017 pp., U.S. Government Printing Office, Washington, D. C., 1975.

Heiken, G., Depressions surrounding volcanic fields: A reflection of underlying batholiths? *Geology*, *4*(9), 568-572, 1976.

Houtz, R., and F. J. Davey, Seismic profiler and sonobuoy measurements in Ross Sea, Antarctica, *J. Geophys. Res.*, *78*(17), 3448-3468, 1973.

Houtz, R., and R. Meijer, Structure of the Ross Sea shelf from profiler data, *J. Geophys. Res.*, *75*(32), 6592-6597, 1970.

Jacobs, S. S., A. F. Amos, and P. M. Bruchhausen, Ross Sea oceanography and Antarctic bottom water formation, *Deep Sea Res.*, *17*(6), 935-962, 1970.

Kaminuma, K., The upper crustal structure under McMurdo Station, Antarctica, obtained by blasts, *Dry Val. Drilling Proj. Bull.*, *8*, 29, 1978.

Kemp, E. M., and P. J. Barrett, Antarctic glaciation and early Tertiary vegetation, *Nature*, *258* (5535), 507-508, 1975.

Margolis, S. V., and J. P. Kennett, Antarctic glaciation during the Tertiary recorded in sub-Antarctic deep-sea cores, *Science*, *170*, 1085-1087, 1970.

Margolis, S. V., and J. P. Kennett, Cenozoic paleoglacial history of Antarctica recorded in subantarctic deep-sea cores, *Amer. J. Sci.*, *271*, 1-36, 1971.

McGinnis, L. D., McMurdo Sound—A key to the Cenozoic of Antarctica, *Antarctic J. U.S.*, *8*, 166-169, 1973.

McGinnis, L. D., Initial report on a seismic refraction study in western McMurdo Sound, this volume.

McGinnis, L. D., K. Nakao, and C. C. Clark, Geophysical identification of frozen and unfrozen ground, Antarctica, in *Permafrost, 2nd International Conference, North American Contribution* pp. 136-146, Academy of Sciences, Washington, D.C., 1973.

McGinnis, L. D., J. S. Stuckless, D. R. Osby, and P. R. Kyle, Gamma-ray, salinity, and electric logs of DVDP boreholes, this volume.

Northey, D. J., and B. A. Sissons, Preliminary seismic profiling survey in McMurdo Sound, *Dry Val. Drilling Proj. Bull.*, *3*, 234-239, 1974.

Northey, D. J., C. Brown, C. A. Christoffel, H. K. Wong, and P. J. Barrett, A continuous seismic profiling survey in Mc-Murdo Sound, Antarctica—1975, *Dry Val. Drilling Proj. Bull.*, *5*, 167-179, 1975.

Pederson, D. R., G. E. Montgomery, L. D. McGinnis, C. P. Ervin, and H. K. Wong, Magnetic study of Ross Island and Taylor Glacier quadrangels, Antarctica, this volume.

Ricker, J., Outline of the geology between Mawson and Priestley glaciers, Victoria Land, in *Antarctic Geology, Proceedings of the First International Symposium on Antarctic Geology*, edited by R. J. Adie, pp. 265-275, North-Holland, Amsterdam, 1964.

Robinson, E. S., Geophysical investigations in McMurdo Sound, Antarctica, *J. Geophys. Res.*, *68*, 257-262, 1963.

Shipboard Scientific Party, Shipboard site report: Sites 270, 271, 272, in *Initial Reports of the Deep Sea Drilling Project*, vol. 28, pp. 211-324, U.S. Government Printing Office, Washington, D. C., 1975.

Stuiver, M., and G. H. Denton, Glacial history of the McMurdo Sound region, *Antarctic J. U.S.*, *12*(4), 128-130, 1977.

Talwani, M., and O. Eldholm, Continental margin off Norway: A geophysical study, *Geol. Soc. Amer. Bull.*, *83*, 3575-3606, 1972.

Timur, A., Velocity of compressional waves in porous media at permafrost temperatures, *Geophysics*, *33*, 584-595, 1968.

Vanney, J. R., and G. L. Johnson, Floor of the Ross Sea and adjacent oceanic provinces, *Antarctic J. U.S.*, *11*(4), 231-233, 1976.

Wong, H. K., Aeromagnetic data from the McMurdo Sound region, *Antarctic J. U.S.*, *8*, 162-163, 1973.

THE LITHOLOGIC LOGS OF DVDP CORES 10 AND 11, EASTERN TAYLOR VALLEY

B. C. McKelvey

Department of Geology, University of New England, Armidale, New South Wales, Australia 2351

The 185 and 328 m cores recovered from DVDP sites 10 and 11, respectively, show the Cenozoic sequence in eastern Taylor Valley to consist of Late Miocene and Pliocene marine diamictites (tillites) overlain disconformably by a regressive sequence of Pleistocene tillites and shallower marine (strandline) conglomerates, pebbly sandstones, and fluvial sandstones.

INTRODUCTION

Taylor Valley extends from the Polar Plateau eastwards down through the southern Victoria Land coastal mountain ranges for a distance of 100 km to McMurdo Sound. Sites 10, 11, and 12 of the Dry Valley Drilling Project are all located within the deglaciated eastern end of Taylor Valley. Site 10 is at the valley mouth on the coast, at an altitude of 2.8 m and approximately 50 m from the edge of the sea ice at New Harbor. DVDP drilling sites 8 and 9 [*Chapman-Smith and Luckman,* 1974] occupied this same location. Site 11 is approximately 3 km inland, situated on undulating valley floor moraines at an altitude of 80.2 m, near the terminus of the Commonwealth Glacier. Site 12 [*Chapman-Smith,* 1975] is located 75 m above sea level, 15 km farther inland from DVDP 11.

The rocks of both the Kukri Hills and the Asgard Range, bordering the Taylor Valley, consist of late Precambrian or early Paleozoic Skelton Group metamorphics, and plutons and dike swarms of the lower Paleozoic Admiralty Intrusives. Farther inland, this crystalline basement complex [*Lopatin,* 1972] is overlain nonconformably by cratonic sediments of the Beacon Supergroup (Devonian or older to Jurassic), which is intruded by Jurassic Ferrar Dolerite sheets and sills [*Haskell et al.,* 1965]. The floor of the Taylor Valley is mantled by a hummocky surface of Cenozoic moraine diversified by several frozen glacial lakes and scattered basaltic cinder cones [*Péwé,* 1960; *McCraw,* 1967].

NOMENCLATURE AND STRATIGRAPHIC PROCEDURE

1. *Induration.* All core recovered was frozen, coming from within the permafrost interval. However, in the older horizons, appreciable diagenesis has resulted in a considerable degree of lithification. As all the core was examined and logged at below 0°C, it was not possible to distinguish between frozen sediment and partially lithified strata. For simplicity and convenience the nomenclature used is that for indurated rocks.

2. *Grain size.* Very few of the Cenozoic deposits are well sorted, and the great bulk of the sedimentary rocks display considerable textural immaturity. This has led to difficulties in applying size-grade nomenclature. In the case of sandstones these have been described as either coarse, medium, or fine (Wentworth scale) according to the estimated average or median grain size. However, it is emphasized that the considerable coarse and fine admixtures present make such hand specimen identification subject to considerable error. Similarly, most of the mudstones are texturally very immature and have to be qualified by either of the prefixes sandy or pebbly, as they almost invariably contain an abundant though variable dispersed sand-grade component and very occasional dropstones. Those mudstones containing more abundant dropstones (i.e., >1%) are logged as diamictites or tillites (see number 3 below).

3. *Textural variation.* Very poorly sorted sedimentary rocks, composed of dispersed granule or pebble frameworks set within finer and more abundant matrices, exhibit a wide spectrum of textural variation. Two contrasting end members are recognized, diamictites [*Flint et al.,* 1960] and pebbly sandstones. The former consist of widely differing amounts of granules and pebbles, irregularly dispersed throughout sandy and silty mud matrices. The latter are commonly laminated, though often only indistinctly. The fabric of the

dispersed pebble frameworks in the diamictites is largely random, and frequently, apparent long axes of clasts are highly inclined or vertical. It is assumed that these diamictites are tillites and represent a combination of suspension current sedimentation (matrices) combined with profuse glacial rafting (dispersed frameworks).

The pebbly sandstones contain scattered clasts set in variously sorted sand matrices. Many pebbles show a preferred orientation with their apparent long axes parallel to bedding or lamination. It is assumed that the pebbly sandstones, particularly those with medium or coarse sand matrices, are largely traction current deposits.

There exists every gradation between these two end members, and it was sometimes difficult during geologic logging to decide whether some sediments should be logged as pebbly sandstones or as diamictites. Fortunately, such intermediate textural types are mostly restricted to rather short stratigraphic intervals at both sites (for example, unit 3, DVDP 10).

4. *Graphic logging.* Particular features of the core such as bioturbation, diagenesis, stratal inclination, and deformation, etc. are highlighted alongside the lithologic log by the various symbols annotated in the legend preceding each log (Appendices A and B). Symbols qualified by a prime (e.g., f′, bt′, etc.) or underlined indicate that the particular feature occurs at that exact meterage. Symbols without a prime or not underlined indicate that the feature specified occurs widely throughout much of the stratigraphic unit portrayed alongside.

5. *Core recovery.* The percentage core recovery figures cited for both logs were obtained by measuring the total length of all the recovered core. However, only major intervals of complete nonrecovery are indicated graphically in both the logs. Frequently, small intervals of partially washed pebbles and brecciated or disaggregated core were retrieved. Where the original rock types could be determined, the length of the stratigraphic interval was estimated and shown as full recovery.

6. *Provenance.* Examination of granule-grade and coarser detritus in both cores shows derivation of the strata to have been from the southern Victoria Land basement complex of metamorphic rocks and associated lower Palaeozoic intrusive rocks. Smaller contributions were derived from the Jurassic Ferrar Dolerites and also from late Tertiary

basaltic rocks. The latter appear to be present throughout both cores in only very small quantities. No kenyte lava detritus such as occurs as lava flows on Ross Island was definitely identified. In addition, clasts of two rock types widely exposed throughout the Ferrar, Taylor, and Wright valley systems were almost entirely absent from the cores. These are the widespread and lithologically striking Vida or Irizar Granite and sediments of the Beacon Supergroup.

The majority of sandstones logged are distributed across a compositional range between lithic and litho-feldspathic end members. Few are dominantly feldspathic. For brevity, only the feldspathic varieties are specified petrographically in the geologic log descriptions.

7. *Lithostratigraphic subdivision.* The two cores recovered from DVDP 10 and 11 have been subdivided into 5 and eight major lithostratigraphic units, respectively (Figure 1). No formal stratigraphic status of any of these units is intended. For convenience of description and reference these units have been further subdivided, generally on the basis of a dominant sediment type into smaller intervals, whose upper and lower boundaries can be clearly determined (for example, units 6.2, 3.4, and so forth). Several of these smaller divisions, however, refer to relatively thin intervals characterized by rapid alternation of contrasting sediments, with no one sediment type being really dominant (for example, DVDP 11, units 3.1, 4.4, 5.2, and so forth). Unit 5 of DVDP 10 is the only example encountered in either hole of a thick interval characterized by such rapid alternation of sediment types. This has necessitated considerable subdivision of unit 5. The use of two hachure patterns in unit 5 of DVDP 10 (Figure 1) is intended to portray the progressive increase in abundance downwards of diamictite or marine tillite.

8. *Biostratigraphy.* In Figure 1 the positions of the Pleistocene-Pliocene disconformity and the Pliocene-Miocene boundary are approximate and are shown only to indicate the general stratigraphic span of the two cores. Detailed biostratigraphic data obtained from DVDP 10 and 11 are discussed elsewhere.

In DVDP 10 the Pleistocene-Pliocene disconformity is placed at 137 m for the following reasons.

1. *Brady* [1980] finds early Pliocene floras to extend from 185 m up to 137 m. None occur above 137 m. *Webb and Wrenn* [1980] report that in the

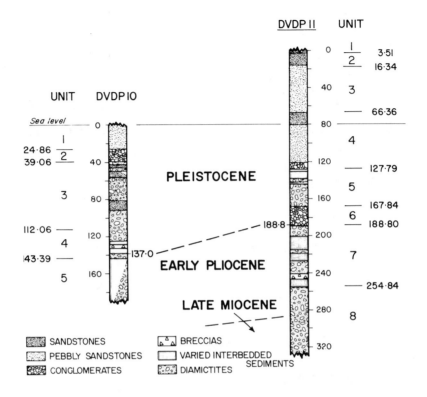

SUMMARY STRATIGRAPHIC LOGS — DVDP 10 AND 11
EASTERN TAYLOR VALLEY

Fig. 1.   Summary of the stratigraphic logs of DVDP 10 and 11 in eastern Taylor Valley. Time boundaries are based upon studies by *Webb and Wrenn* [1980] and *Brady* [1980].

overlying sequence their oldest Pleistocene faunas at 125 m overlie a barren interval, and their youngest Pliocene faunas extend down from 154 m.

2.   A stratal inclination of 30° was noted within unit 4.7 from 137 m down to 139 m. This suggests deformation by grounded ice perhaps associated with its erosion of the disconformity. In addition the somewhat different provenance (indicated by the incoming of abundant granitoid clasts) shown by unit 4.7, compared with strata above 137 m, also suggests the possibility of a disconformity.

In DVDP 11 the oldest Pleistocene faunas of Webb and Wrenn occur in unit 6.2 at 173.4 m, and the youngest Pliocene faunas occur at 205.96 m in unit 7.8. In Figure 1 the Pleistocene-Pliocene boundary has been somewhat arbitrarily placed at

188.8 m, at the base of unit 6.2, because the conglomerates of this unit appear to be tillites that have undergone winnowing contemporaneous with their deposition. As such, they reflect a major modification of the depositional environment when they are compared with the older tillites of units 7 and 8. However, subsequent reexamination of the core shows that unit 7.1 and 7.2 are somewhat similar lithologically to unit 6, and so the Pleistocene-Pliocene boundary, if it is to be placed at a horizon where à change in the depositional environment is indicated, could be lowered to approximately 195.2 m.

Brady [1980] has found Late Miocene floras beneath 291 m, and for this reason the Miocene strata are so indicated in DVDP 11.

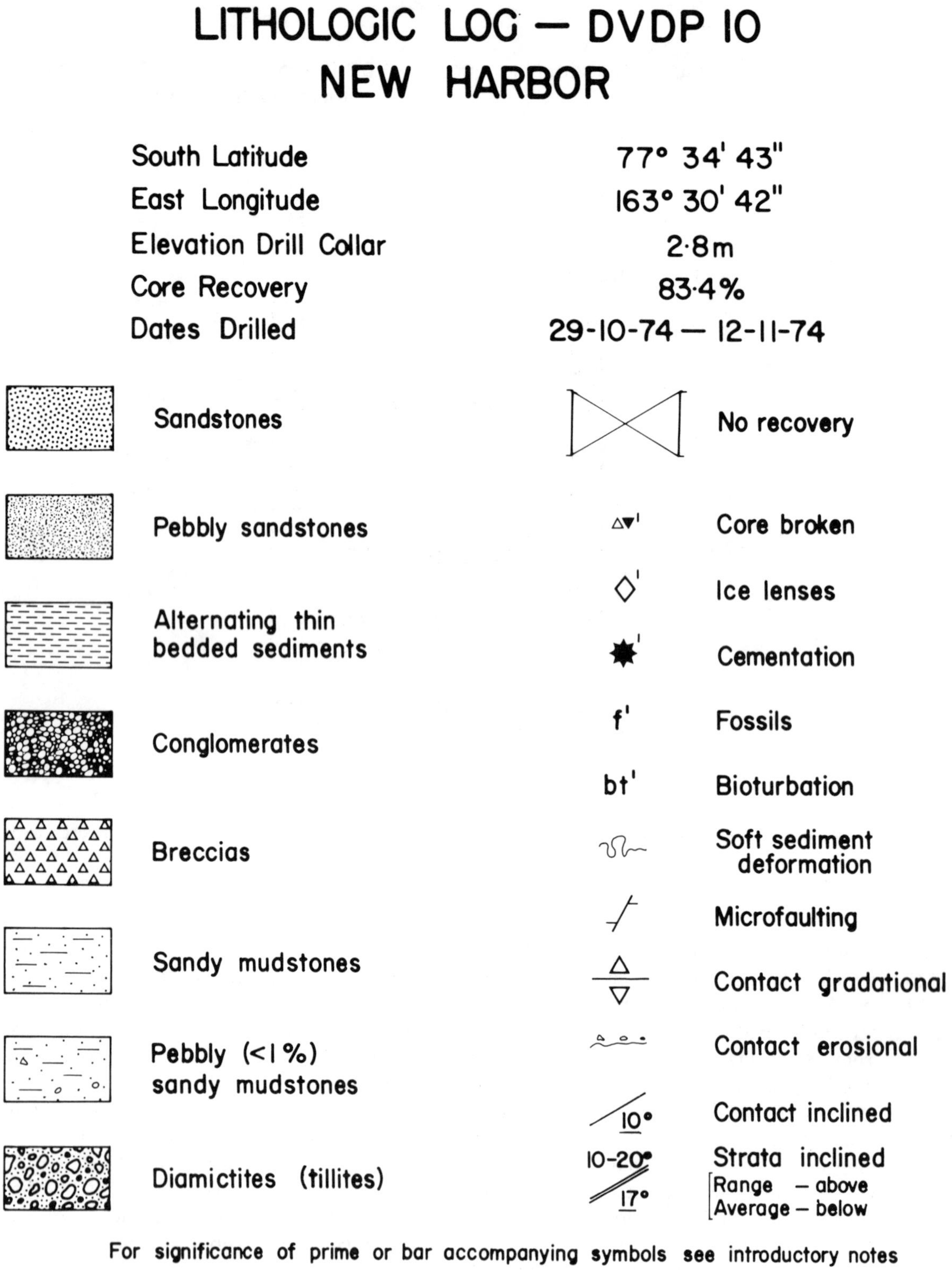

LITHOLOGIC LOG — DVDP 10
NEW HARBOR

South Latitude	77° 34' 43"
East Longitude	163° 30' 42"
Elevation Drill Collar	2·8 m
Core Recovery	83·4%
Dates Drilled	29-10-74 — 12-11-74

Sandstones

Pebbly sandstones

Alternating thin bedded sediments

Conglomerates

Breccias

Sandy mudstones

Pebbly (<1%) sandy mudstones

Diamictites (tillites)

No recovery

$\triangle \blacktriangledown'$ Core broken

$\diamondsuit'$ Ice lenses

Cementation

f' Fossils

bt' Bioturbation

Soft sediment deformation

Microfaulting

Contact gradational

Contact erosional

10° Contact inclined

10-20° Strata inclined
17° [Range — above / Average — below]

For significance of prime or bar accompanying symbols see introductory notes

Fig. A1. DVDP 10, legend.

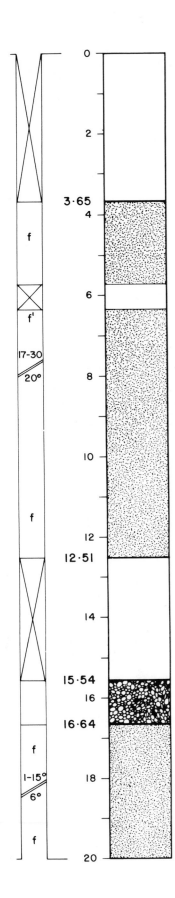

Unit 1.1. (3.65 - 12.51 m). Interbedded medium to very coarse sandstones
(10Y 4/2), with minor granule conglomerate beds (10 cm), e.g. 8.65 m.
Scattered subangular to subrounded pebbles (<2%) throughout many sandstones;
range up to 2 cm, average 1 cm. Medium sandstones (5Y 3/2) predominate
8.5 - 11.1 metres. Above 8.5 metres medium to fine sandstones (2-4 cm)
noted ca. 5.4, 7 and 7.26 m.
                          Stratification generally indistinct, inclined
17° to 30°. Averages 20°. Some beds show normal grading. Broken shell
debris noted, one valve relatively complete, convex upwards at 6.5 m.

Unit 1.2. (15.54 - 16.64 m). Pebble conglomerate. Average clast 4 mm,
maximum clast 2 cm. Moderately sorted, lacks interstitial fines. Subrounded to subangular. Intraformational semilithified fine sandstone
clasts (av. 3.5 mm). Stratification subhorizontal.

Unit 1.3. (16.64 - 24.86 m). Coarse to medium laminated sandstones
(5 6Y 4/2) interbedded with minor pebble conglomerates (latter subangular to subrounded). Sandstone beds and laminae 1 cm - 20 cm; show
both reverse and normal grading. Scattered pebbles (<1%), average clast
1.5 cm, maximum clast 2.5 cm. Medium sandstone predominates beneath ca.
22 m.
                          Pebble conglomerate 2-16 cm thick. Average
clast 1.5 cm, maximum clast 3 cm. Particularly abundant interbedded with
sandstones ca. 22-23 m.
                          Stratification distinct. Inclination
1-15°. Average 6°. Organic carbonate detritus throughout, one articulated
bivalve at 22-16 m. Single valve at 24.2 m.

Fig. A2. DVDP 10, 0-20 m.

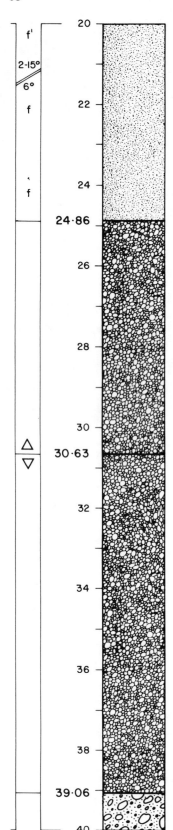

Unit 2.1. (24.86 - 30.63 m). Pebble and granule conglomerates associated
with minor pebbly sandstones and sandstones. Pebble conglomerate clasts
angular to rounded. Average clasts 1-3 cm, maximum clast 10 cm. Conglomerates contain up to 20% coarse sand matrix.
                          Minor lithologies common above 26 m and
below 29.54 m. Between 29.54 - 30.14 m, two 30 cm beds grade normally
from granule conglomerates to very coarse sandstones.
                          Stratification horizontal. Gradational base.

Unit 2.2. (30.63 - approximately 39.06 m). Pebble conglomerates. Clasts
subangular to subrounded, predominantly subangular. Moderate sorting,
lack sand grade matrix. Average clast size increases downwards from 4 mm
to 15 mm (ca. 36 m). Maximum clasts 12 cm to 17 cm; one of >46 cm noted
ca. 33 metres.
                                    Some conglomerate beds
exceed 60 cm. Normal grading common. Intraformational mudstone and fine
sandstone fragments form up to 5%, ca. 36 m. Reworked tillite clast ca.
31.5 m. Maximum size, 4 cm.

Fig. A3. DVDP 10, 20-40 m.

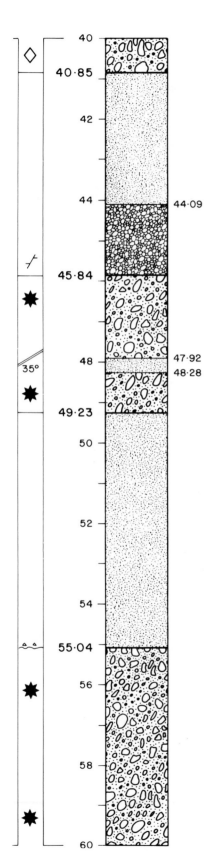

Unit 3.1. (39.06 - 40.85 m). Diamictite. Massive. Clasts rounded to
angular. Average clast 2.5 cm, maximum clast 15 cm. Clasts >3 cm comprise
less than 5%.
                        Matrix poorly sorted. Averages medium
sandstone (5Y 4/2). Numerous ice lenses (up to 5 mm) ca. 40.3 - 40.6
metres. Base of unit sharp.

Unit 3.2. (40.85 - 45.84 m). Medium to coarse massive sandstone (5 6Y 4/2)
grading down to granule and pebble conglomerates. Clasts subangular to
subrounded. Average clast ca. 5 mm. Varied intraformational mudstone -
fine sandstone fragments up to 15 mm (5Y 4/2 - 10YR 6/6).
                        Uppermost horizons (40.85 m - ca. 41.5)
consist of well sorted medium and coarse sandstone (5Y 4/2) with rare
pebbles (<1%), overlying approximately 25 cm of laminated (1 cm) granule
conglomerate.
                        At 45.47 m poorly sorted fine to coarse
sandstone (5 6Y 5/2) containing occasional pebbles (max. 4 cm) overlie
14 cm laminated (1 cm) mudstone (5Y 4/4) showing microfaulting.

Unit 3.3. (45.84 - 49.23 m). Two diamictites separated by 36 cm coarse
sandstones, passing with gradational contact down to granule conglomerate
(ca. 47.92 - 48.28 m). Diamictites massive. Clasts comprise 5%, angular
to rounded. Average clast 1 cm, maximum clast >16 cm. Matrix poorly
sorted, averages medium sandstone. Contains abundant dispersed feldspathic
coarse sand-granule debris. Some indistinct stratification (deformed)
in lower diamictite.
                        Diamictite - sandstone contacts irregular
(loading) and inclined 35° (deformation). Patches (up to 4 cm) of secondary
carbonate within diamictites and along diamictite - granule conglomerate
contact. Latter contact erosional.

Unit 3.4. (49.23 - 55.04 m). Interbedded coarse, medium and fine sandstones.
Contacts both sharp and gradational. Minor mudstones and granule-pebble
conglomerates.
                        Coarse sandstones massive, fine and medium
sandstones laminated (1 mm to 1 cm). Medium and coarse sandstones contain
scattered pebbles up to 2 cm. Fine sandstones, 1-3 cm thick, associated
with mudstones, e.g. 11 cm mudstone ca. 50.1 m; and calcareous mudstone
laminae, 1 cm, common ca. 52.6 - 53.6 m. Two mudstones 6 cm and 10 cm
thick, at ca. 53.9 m and 54.05 m.
                        Granule conglomerate 25 cm thick, average
clast 2 mm ca. 51.4 m. Graded pebble conglomerate, clasts subangular
to subrounded, average 3 cm 54.16 m to 55.04 m. Erosional contact with
unit 3.5.

Fig. A4.   DVDP 10, 40-60 m.

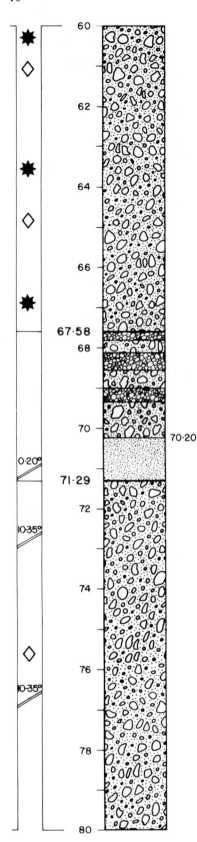

Unit 3.5. (55.04 - 67.58 m). Diamictite. Highly variable texturally.
Clasts average 5-10%, Angular to rounded, average subangular to subrounded.
Average clast ranges 2 to 4 cm. Maximum clast ranges 15 to 30 cm, one
clast exceeds 50 cm ca. 65.8 m. Two angular clasts (4 cm) of semi-
lithified older diamictites noted.

Matrix (10Y 6/2) variable, massive and
laminated. Latter noted particularly ca. 57-60 m. Between 57.1 - 57.45 m
matrix consists of alternating (1-2 mm) coarse and fine sand and mud.
Inclined (deformed) 25°.

Secondary carbonate patches (average 2 cm)
widespread and carbonate rinds (<2 mm) present on clasts. Carbonate postdates soft sediment deformation. Ice lenses, max. 5 mm, inclined, max.
45°, and subhorizontal widespread below 60.75 m.

Unit 3.6. (67.58 - 71.29 m). Highly variable interval. Three thin
diamictites interbedded with pebble conglomerates. Pass down at ca.
70.2 m to interbedded pebbly coarse, medium and fine sandstones.

Diamictites (matrix averages 5 GY 4/2)
18 cm - 1.10 m thick. Oldest is thickest. Clasts average 5%. Average
clast 2 cm. Matrices massive, average poorly sorted fine sand.

Conglomerates 20 to 30 cm thick. Some
normal grading. Average clast 1.5 cm. Very poorly sorted fine sand
matrices.

Pebbly sandstones and sandstones poorly
sorted. Laminated (<1 cm). Proportion of clasts in pebbly sandstones
varies. Maximum 2%. Average clast 1.5 cm. Maximum clast 3 cm. Alternating
fine sand silt laminae (1 cm) between 70.9 and 71.2 m. Stratification
gradational to sharp. Attitude 0-20°.

Unit 3.7. (71.29 - 81.91 m). Texturally variable unit. Diamictite -
pebbly sandstone. Clasts increase downwards from 5% to 20%. Average
clast 2 cm. Maximum clast 37 cm ca. 77.5 cm. Clasts predominantly
subangular to subrounded.

Matrix (averages 10Y 5/2) poorly sorted
coarse sand. Beneath 79 m grades down to granule conglomerate and very
coarse sand. Matrix generally massive beneath ca. 77 m. Above 77 m
frequently laminated, often inclined 10-35°. Some ice lenses ca. 75-76 m,
up to 2 mm.

Fig. A5. DVDP 10, 60-80 m.

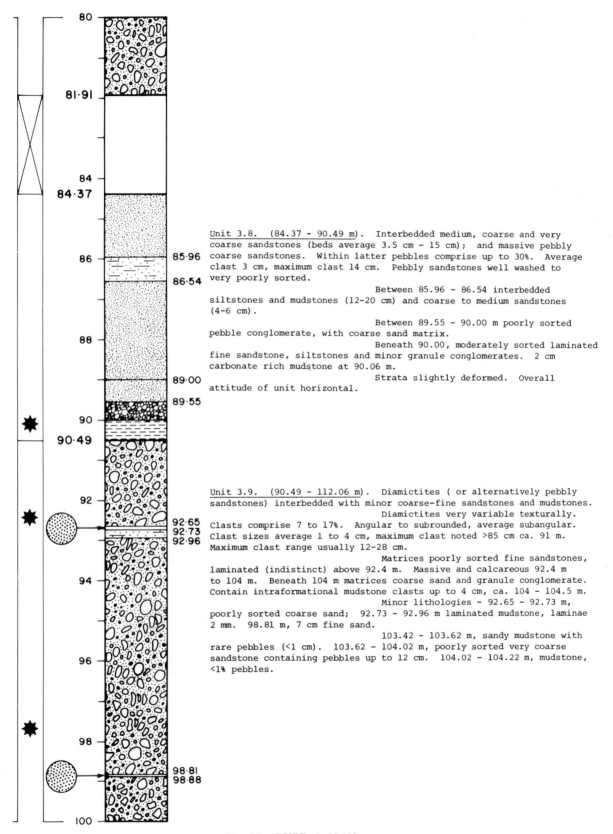

Unit 3.8. (84.37 - 90.49 m). Interbedded medium, coarse and very
coarse sandstones (beds average 3.5 cm - 15 cm); and massive pebbly
coarse sandstones. Within latter pebbles comprise up to 30%. Average
clast 3 cm, maximum clast 14 cm. Pebbly sandstones well washed to
very poorly sorted.
       Between 85.96 - 86.54 interbedded
siltstones and mudstones (12-20 cm) and coarse to medium sandstones
(4-6 cm).
       Between 89.55 - 90.00 m poorly sorted
pebble conglomerate, with coarse sand matrix.
       Beneath 90.00, moderately sorted laminated
fine sandstone, siltstones and minor granule conglomerates. 2 cm
carbonate rich mudstone at 90.06 m.
       Strata slightly deformed. Overall
attitude of unit horizontal.

Unit 3.9. (90.49 - 112.06 m). Diamictites (or alternatively pebbly
sandstones) interbedded with minor coarse-fine sandstones and mudstones.
       Diamictites very variable texturally.
Clasts comprise 7 to 17%. Angular to subrounded, average subangular.
Clast sizes average 1 to 4 cm, maximum clast noted >85 cm ca. 91 m.
Maximum clast range usually 12-28 cm.
       Matrices poorly sorted fine sandstones,
laminated (indistinct) above 92.4 m. Massive and calcareous 92.4 m
to 104 m. Beneath 104 m matrices coarse sand and granule conglomerate.
Contain intraformational mudstone clasts up to 4 cm, ca. 104 - 104.5 m.
       Minor lithologies - 92.65 - 92.73 m,
poorly sorted coarse sand; 92.73 - 92.96 m laminated mudstone, laminae
2 mm. 98.81 m, 7 cm fine sand.
       103.42 - 103.62 m, sandy mudstone with
rare pebbles (<1 cm). 103.62 - 104.02 m, poorly sorted very coarse
sandstone containing pebbles up to 12 cm. 104.02 - 104.22 m, mudstone,
<1% pebbles.

Fig. A6. DVDP 10, 80-100 m.

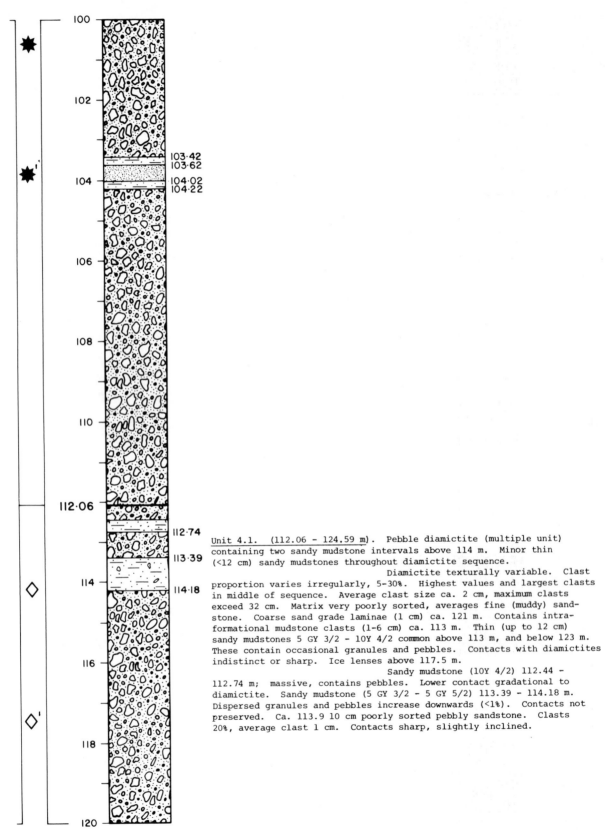

Unit 4.1. (112.06 - 124.59 m). Pebble diamictite (multiple unit)
containing two sandy mudstone intervals above 114 m. Minor thin
(<12 cm) sandy mudstones throughout diamictite sequence.

               Diamictite texturally variable. Clast
proportion varies irregularly, 5-30%. Highest values and largest clasts
in middle of sequence. Average clast size ca. 2 cm, maximum clasts
exceed 32 cm. Matrix very poorly sorted, averages fine (muddy) sandstone. Coarse sand grade laminae (1 cm) ca. 121 m. Contains intraformational mudstone clasts (1-6 cm) ca. 113 m. Thin (up to 12 cm)
sandy mudstones 5 GY 3/2 - 10Y 4/2 common above 113 m, and below 123 m.
These contain occasional granules and pebbles. Contacts with diamictites
indistinct or sharp. Ice lenses above 117.5 m.

               Sandy mudstone (10Y 4/2) 112.44 -
112.74 m; massive, contains pebbles. Lower contact gradational to
diamictite. Sandy mudstone (5 GY 3/2 - 5 GY 5/2) 113.39 - 114.18 m.
Dispersed granules and pebbles increase downwards (<1%). Contacts not
preserved. Ca. 113.9 10 cm poorly sorted pebbly sandstone. Clasts
20%, average clast 1 cm. Contacts sharp, slightly inclined.

Fig. A7.   DVDP 10, 100-120 m.

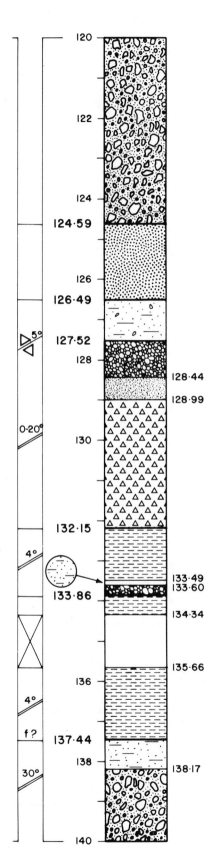

Unit 4.2. (124.59 - 126.49 m). Massive medium sandstone. Well washed.
Contains dispersed very coarse sand detritus. Some thin (1 cm) disrupted
pebble lenses. One 12 cm metadiorite clast. Circa 126.08 m, 15 cm
mudstone (intraformational mudstone breccia?) 5Y 4/1. Gradational
contacts. Drilling breccia? 126.08 - 126.16.

Unit 4.3. (126.49 - 127.52 m). Mudstone (5 GY 3/2) passing down
(ca. 127 m) to silty mudstone (5 GY 5/2). Contains >30 cm foliated
leucocratic migmatite clast. Base gradational, inclined (<5°).

Unit 4.4. (127.52 - 132.15 m). Breccias overlain at approximately
128.99 m by medium to coarse pebbly sandstone (maximum clast 10 cm)
and younger (sharp contact at 128.44 m) pebble conglomerate within
latter. Average clast .75 cm, maximum clast 5 cm. Coarse indistinct
lamination.

Breccia (128.99 - 132.15 m), poorly
sorted with coarse sand matrix. Average clast 1.5 cm, maximum 13 cm.
Clasts predominantly angular. Long axes indicate stratification
varies from horizontal to 20° inclination. Thin (max. 15 cm) laminated
sandstone interbeds. Contacts gradational.

Unit 4.5. (132.15 - 133.86 m). Interbedded laminated (.2 - 2.5 cm)
coarse, medium and fine sandstones. Strata inclined 4°. Overall
homogeneous interval. Gradational (interbedded) upper contact with
breccias of unit 4.4. Overlies interlaminated fine sandstones and
mudstones (133.47 - 133.60 m). Latter interval passes gradationally
to pebble conglomerate (av. clast .5 cm) containing scattered pebbles
up to 5 cm (133.60 - 133.86 m). Sharp base.

Unit 4.6. (133.86 - 137.44 m). Laminated sandy mudstone and fine
sandstones. Laminae 1-5 mm. Below 135.6 m four medium to coarse
sandstones (max. 10 cm). Mudstones and fine sandstones contain dispersed
coarse sand and granule grade detritus. Lamination indistinct in
oldest 60 cm. Carbonate detritus (sand grade) dispersed throughout
oldest 20 cm. Strata inclined 4°. N.B. No recovery 134.34 - 135.66 m.

Unit 4.7. (137.44 - 143.39 m). Diamictite clasts subangular to subrounded. Granitic lithologies prominent. Average clast 1 cm, maximum
clast exceeds 11 cm. Matrix, moderately sorted medium sand (no fines).
Abundant dispersed coarse sand and granule admixture. Two mudstone
laminae (2 cm) ca. 140.6 m. Pebble orientation and lamination suggests
strata inclination above 139 m of 30°.

137.44 - 138.17 m. Sandy mudstone
(5 GY 3/2) with few dispersed pebbles (<1%). Abundant dispersed granule
and coarse sand detritus.

N.B. 143.01 - 148.43. No recovery.
Washed pebbles only. Maximum size exceeds 13 cm.

Fig. A8. DVDP 10, 120-140 m.

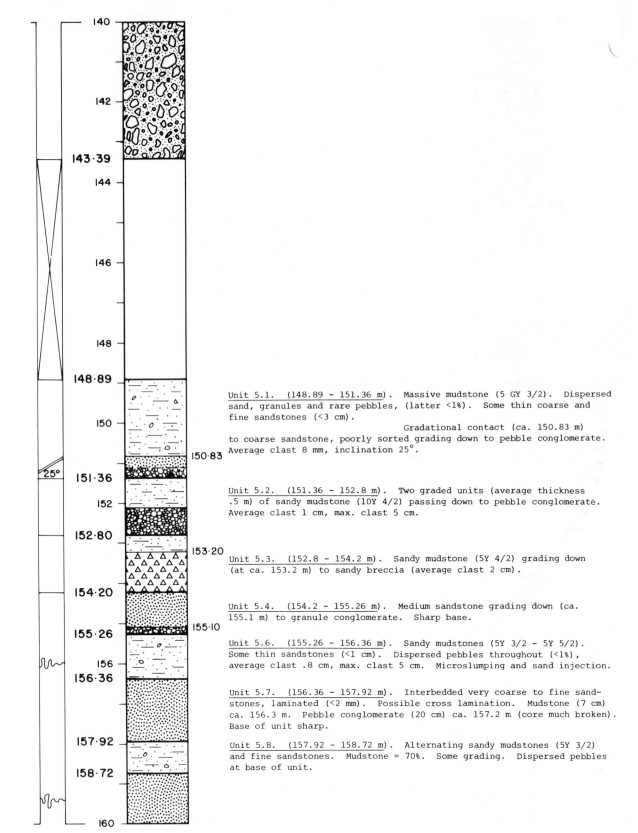

Unit 5.1. (148.89 - 151.36 m). Massive mudstone (5 GY 3/2). Dispersed sand, granules and rare pebbles, (latter <1%). Some thin coarse and fine sandstones (<3 cm).

Gradational contact (ca. 150.83 m) to coarse sandstone, poorly sorted grading down to pebble conglomerate. Average clast 8 mm, inclination 25°.

Unit 5.2. (151.36 - 152.8 m). Two graded units (average thickness .5 m) of sandy mudstone (10Y 4/2) passing down to pebble conglomerate. Average clast 1 cm, max. clast 5 cm.

Unit 5.3. (152.8 - 154.2 m). Sandy mudstone (5Y 4/2) grading down (at ca. 153.2 m) to sandy breccia (average clast 2 cm).

Unit 5.4. (154.2 - 155.26 m). Medium sandstone grading down (ca. 155.1 m) to granule conglomerate. Sharp base.

Unit 5.6. (155.26 - 156.36 m). Sandy mudstones (5Y 3/2 - 5Y 5/2). Some thin sandstones (<1 cm). Dispersed pebbles throughout (<1%), average clast .8 cm, max. clast 5 cm. Microslumping and sand injection.

Unit 5.7. (156.36 - 157.92 m). Interbedded very coarse to fine sandstones, laminated (<2 mm). Possible cross lamination. Mudstone (7 cm) ca. 156.3 m. Pebble conglomerate (20 cm) ca. 157.2 m (core much broken). Base of unit sharp.

Unit 5.8. (157.92 - 158.72 m). Alternating sandy mudstones (5Y 3/2) and fine sandstones. Mudstone = 70%. Some grading. Dispersed pebbles at base of unit.

Fig. A9. DVDP 10, 140-160 m.

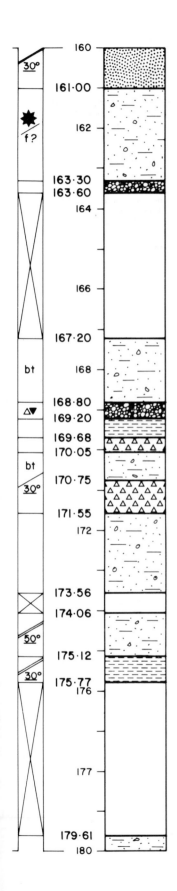

Unit 5.9. (158.72 - 161.0 m). Interbedded medium to fine sandstones and pebble conglomerates. Sandstones poorly sorted. Conglomerate beds up to 12 cm, average clast 1.5 cm. Soft sediment deformation. Strata horizontal except ca. 160.2 m, inclined 30°.

Unit 5.10. (161.0 - 163.3 m). Sandy mudstones (5Y 3/2). Calcareous, laminated. Dispersed granules and pebbles (<1%). Occasional medium sandstones (3 cm), sharp contacts.

Unit 5.11. (163.3 - 163.6 m). Pebble conglomerate, poorly sorted. Average clast 1.5 cm, max. clast 23 cm. Coarse sand matrix.

Unit 5.12. (167.2 - 168.8 m). Interbedded sandy mudstone and medium sandstone. Mudstone 65%, sandstone 30%, minor coarse sandstone. Mudstones average 6 cm, sandstones average 2 cm. Sharp horizontal contacts. Rare (<1%) pebbles (<1 cm). Bioturbation, - cylindrical burrows, diameter <4 mm.

Unit 5.13. (168.8 - 169.2 m). Pebble conglomerate grading up to cross-stratified medium sandstone.

Unit 5.14. (169.2 - 169.68 m). Laminated (1-5 mm) sandstones and mudstones (shales). Core much broken.

Unit 5.15. (169.68 - 170.05 m). Breccias. Average clast 1.5 cm, max. clast 3 cm.

Unit 5.16. (170.05 - 170.75 m). Sandy mudstone, includes (ca. 170.2 m) 2 cm coarse sandstone. Bioturbation ca. 170.5 m. Sharp base to unit.

Unit 5.17. (170.75 - 171.55 m). Pebble breccia. Clasts subangular (predominant) to subrounded. Average clast 1 cm, max. clast >6 cm. Upper contact inclined 30°, lower contact horizontal.

Unit 5.18. (171.55 - 173.56 m). Variable unit. Sandy mudstones (5-75 cm); interbedded with pebble conglomerates and coarse sandstones (2-25 cm). Average conglomerate clasts 1 cm. Occasional pebbles (<2 cm) in mudstones.

Unit 5.19. (174.06 - 175.12 m). Sandy mudstone (5Y 3/2); some fine sandstone laminae (1 mm - 2 cm). Strata inclined up to 50°. Circa 174.3 m dispersed pebbles abundant, max. clast 6 cm. Core much broken.

Unit 5.20. (175.12 - 175.77 m). Laminated (5 mm) fine sandstone, poorly sorted, interbedded with sandy mudstones (3-14 cm). Strata inclined up to 30°.

Fig. A10. DVDP 10, 160-180 m.

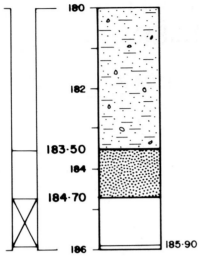

Unit 5.21. (179.61 - 183.5 m). Mudstone (5Y 3/2). Lamination 1-5 mm.
Scattered pebbles throughout <1%. Circa 180.3 pebble conglomerate
(4 cm), average clast 5 mm. Some medium to coarse sandstones, 2-12 cm.

Unit 5.22. (183.5 - 184.7 m). Poorly sorted fine sandstone (5Y 3/2).
Dispersed clasts <1%, average clast 5 mm, max. clast 2 cm. Circa 184.3 m,
sharp contact with coarse and very coarse sandstone. Contains pebbles
(<1%) up to 5 cm.
                              Below 184.7 no core recovery, washed
pebbles only, some exceed 8 cm.

Fig. A11.   DVDP 10, 180-185.9 m.

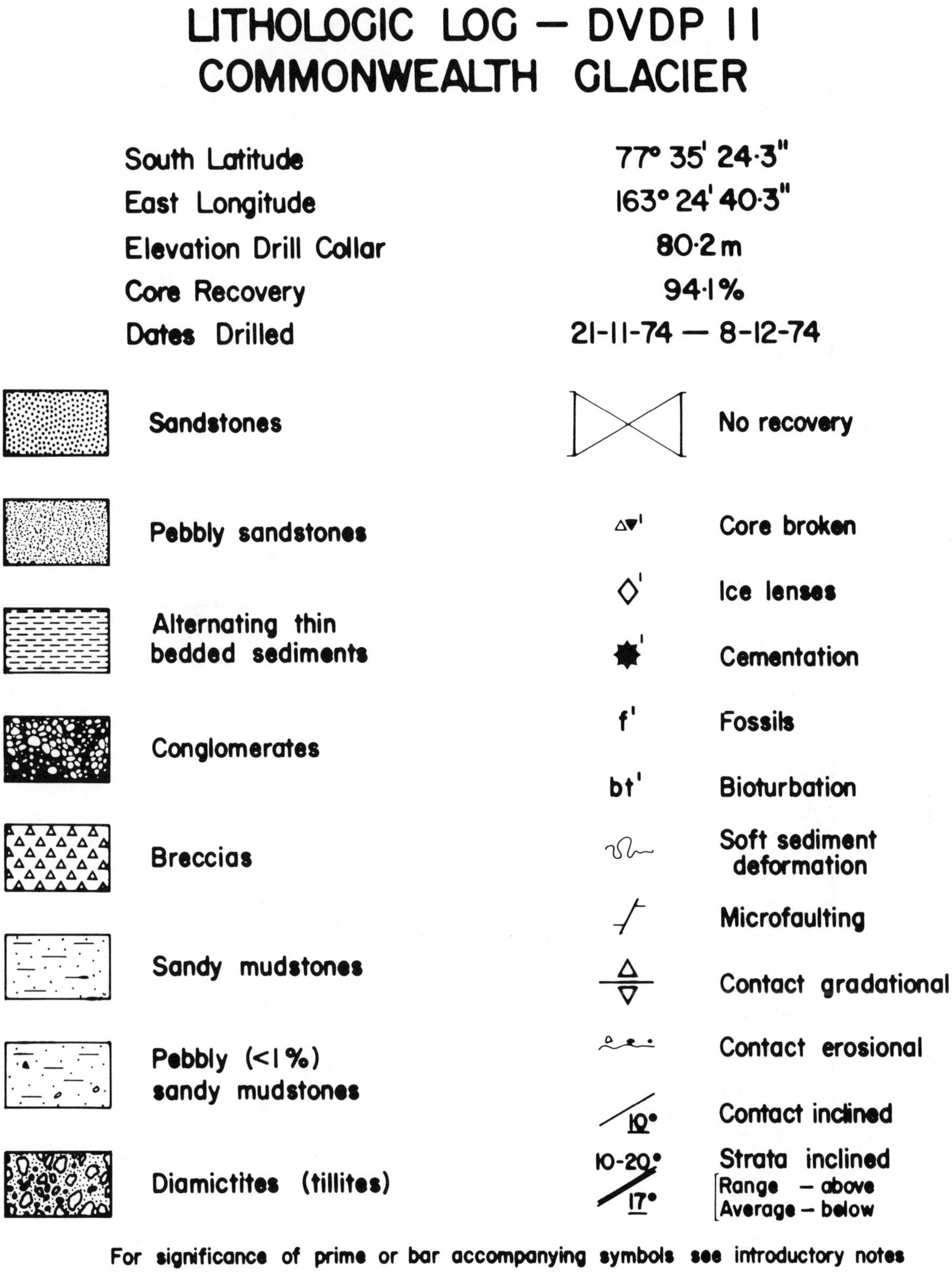

LITHOLOGIC LOG — DVDP 11
COMMONWEALTH GLACIER

South Latitude	77° 35' 24·3"
East Longitude	163° 24' 40·3"
Elevation Drill Collar	80·2 m
Core Recovery	94·1%
Dates Drilled	21-11-74 — 8-12-74

Sandstones

Pebbly sandstones

Alternating thin bedded sediments

Conglomerates

Breccias

Sandy mudstones

Pebbly (<1%) sandy mudstones

Diamictites (tillites)

No recovery

△▼' Core broken

◇' Ice lenses

✷' Cementation

f' Fossils

bt' Bioturbation

Soft sediment deformation

Microfaulting

Contact gradational

Contact erosional

10° Contact inclined

10-20° Strata inclined
17° [Range — above
[Average — below

For significance of prime or bar accompanying symbols see introductory notes

Fig. B1. DVDP 11, legend.

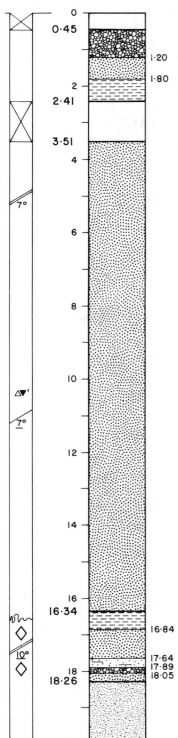

Unit 1. (0.45-2.41 m). 0.45-1.20 m; fragments (6-9 cm) of granule conglomerate (5Y 5/2), loose pebbles, and drilling breccia.
              1.20-1.80 m; fine sandstones to granule conglom-
erates (5Y 4/2). Laminated (1-5 mm), inclination up to 15⁰. Rare pebbles
(5 cm). Appreciable basaltic detritus. Ice lenses (1 mm - 2 cm).
              1.80-2.41 m; fine to medium sandstones (10Y
4/2), indistinct lamination, beds 2.5-10 cm. Calcareous. Interbedded
with structureless sandy mudstones (5Y 5/2). Intraformational mudstone
detritus in sandstones.

Unit 2. (3.51 - 16.34 m). Very uniform. Medium to medium coarse sandstones
(5Y 3/2 - 5Y 5/2). Angular to sub-angular. Moderate to good sorting.
Lamination indistinct. Above 11.4 metres lamination dips at up to 7°.
Beneath 11.4 metres, lamination horizontal.
              Circa 7.12 m, 2 cm very coarse sandstone
(lamellae 2 mm thick) overlies 2.5 cm of coarse sandstone and granule
conglomerate (lamellae 3 mm).
              Circa 10.82 m, 15 cm intraformational mudchip
granule conglomerate (average clast 3 mm). Upper contact sharp, slightly
load casted mudstone (0.5 - 4 cm).
              Circa 13.1 - 13.4 m, intraformational semi-
lithified mudchips common (max. size 12 mm, average 2 mm). Impersistent
laminae, up to 4 mm thick. Similar lithologies circa 14.4 - 14.48 m.
              Circa 10.38 - 10.6 m core much broken.

Unit 3.1. (16.34 - 18.26 m). 16.34 - 16.84 m; interbedded (2-6 cm)
fine to medium sandstones and silty mudstones (5Y 3/2 - 5Y 5/2). All
laminated (1-4 mm). Fine sandstones predominate. Includes 3 thin (<1.5
cm) granule conglomerates. Angular mudstone fragments (<5%) in sandstones
(and conglomerates). Slight soft sediment deformation. Abundant ice lenses,
16.62 - 16.76 m. 16.76 - 16.84 m; pebble conglomerate, max. clast 4 cm,
average clast 1 cm. Clasts subangular, ice cemented, well sorted.
              16.84 - 17.64 m; fine to coarse sandstones,
laminated (2.5 cm), inclined up to 10°. Thin mudstone laminae (up to 1 cm).
Some laminae graded.
              17.64 - 17.74 m; silty mudstone (5Y 3/2),
fragmented and ice cemented, disconformable basal contact. Ice lenses
(<4 mm).
              17.74 - 18.05 m; fine sandstone, cross
stratified overlying (at 17.89) pebble conglomerate (av. clast 6 mm,
subangular). Gradational base.
              18.05 - 18.26; medium sandstone (5Y 3/2).
Poorly sorted, scattered pebbles (max. 10 cm).

Fig. B2. DVDP 11, 0-20 m.

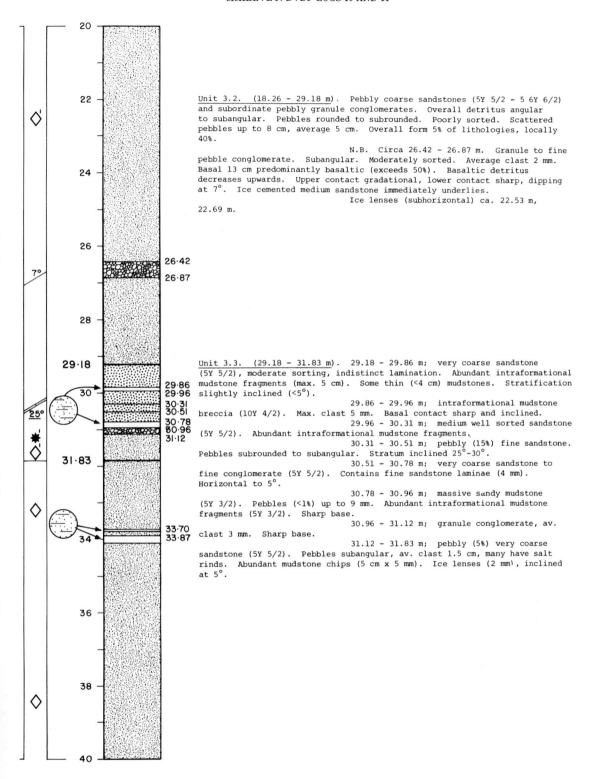

Unit 3.2. (18.26 - 29.18 m). Pebbly coarse sandstones (5Y 5/2 - 5 6Y 6/2)
and subordinate pebbly granule conglomerates. Overall detritus angular
to subangular. Pebbles rounded to subrounded. Poorly sorted. Scattered
pebbles up to 8 cm, average 5 cm. Overall form 5% of lithologies, locally
40%.

                         N.B. Circa 26.42 - 26.87 m. Granule to fine
pebble conglomerate. Subangular. Moderately sorted. Average clast 2 mm.
Basal 13 cm predominantly basaltic (exceeds 50%). Basaltic detritus
decreases upwards. Upper contact gradational, lower contact sharp, dipping
at 7°. Ice cemented medium sandstone immediately underlies.
                         Ice lenses (subhorizontal) ca. 22.53 m,
22.69 m.

Unit 3.3. (29.18 - 31.83 m). 29.18 - 29.86 m; very coarse sandstone
(5Y 5/2), moderate sorting, indistinct lamination. Abundant intraformational
mudstone fragments (max. 5 cm). Some thin (<4 cm) mudstones. Stratification
slightly inclined (<5°).
                         29.86 - 29.96 m; intraformational mudstone
breccia (10Y 4/2). Max. clast 5 mm. Basal contact sharp and inclined.
                         29.96 - 30.31 m; medium well sorted sandstone
(5Y 5/2). Abundant intraformational mudstone fragments.
                         30.31 - 30.51 m; pebbly (15%) fine sandstone.
Pebbles subrounded to subangular. Stratum inclined 25°-30°.
                         30.51 - 30.78 m; very coarse sandstone to
fine conglomerate (5Y 5/2). Contains fine sandstone laminae (4 mm).
Horizontal to 5°.
                         30.78 - 30.96 m; massive sandy mudstone
(5Y 3/2). Pebbles (<1%) up to 9 mm. Abundant intraformational mudstone
fragments (5Y 3/2). Sharp base.
                         30.96 - 31.12 m; granule conglomerate, av.
clast 3 mm. Sharp base.
                         31.12 - 31.83 m; pebbly (5%) very coarse
sandstone (5Y 5/2). Pebbles subangular, av. clast 1.5 cm, many have salt
rinds. Abundant mudstone chips (5 cm x 5 mm). Ice lenses (2 mm), inclined
at 5°.

Fig. B3.   DVDP 11, 20-40 m.

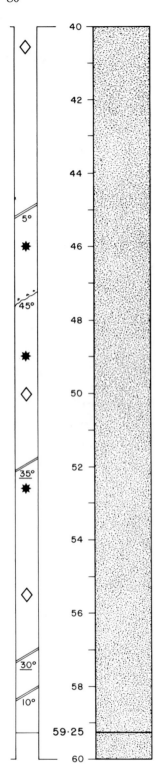

Unit 3.4. (31.83 - 59.25 m). Coarse (pred.) to fine pebbly sandstones
(5Y 5/2 - 5Y 3/2). Very poorly sorted with abundant silt-mud and very
coarse sand-granule grade admixtures. Pebbles angular to subangular.
(Fine pebbly sandstones predominate between ca. 39 - 46.3 m). Pebble
component averages 1-3%, maximum 7% ca. 40 m. Average clast size varies
between 1-3 cm; maximum clasts exceed 16 cm ca. 38-39 m. Pebbles predominantly medium-coarse granodioritic lithologies.
                                Stratification and/or lamination indistinct,
essentially horizontal. Inclined stratification noted ca. 45.62 m, 5°;
51.91 m, 35°; 57.17 m, 30°; between 56-59 m inclination averages 10°.
Erosional contact inclined at 45° noted at 47.65 m.
                                Sandy mudstones (5Y 5/2) 6 cm and 20 cm
respectively noted at ca. 33.7 m and 33.87 m. Contacts sharp. Older
mudstone contains semi-lithified mudstone fragments (intraformational)
up to 4 mm.
                                Ice lenses (max. 3 mm) every 15-50 cm.
Predominantly subhorizontal, some inclined, rarely vertical.
                                Between 46-53 m white salt rind (<.1 mm)
on pebbles. Drilling breccia 34.52 - 34.7 m.

Fig. B4.   DVDP 11, 40-60 m.

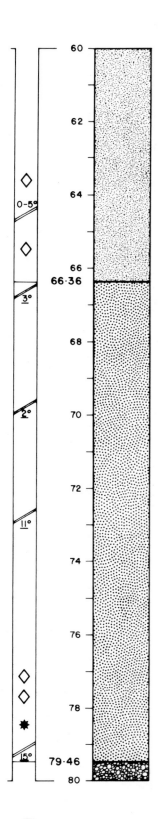

Unit 3.5. (59.25 - 66.36 m). Pebbly medium sandstones (5Y 5/2). Pebbles
angular to subangular. Well sorted - moderately sorted (cf. Unit 3.4).
Upper boundary gradational. Pebbles 1-4%; range up to 4 cm. Ca. 59.1 m,
pebbles comprise up to 10%, max. size 15 cm.
                    Lamination-stratification discernable (cf.
Unit 3.4). Beneath 64.2 m laminae .5 - 4 cm thick, attitude 0-5°.
                    Ice lenses (1-2 mm) particularly noted ca.
63.5 - 65.5 m.

Unit 4.1. (66.36 - 79.46 m). Variable unit. Very coarse to fine sandstones (5Y 3/2 - 5Y 5/2). Medium sandstones predominate. Minor pebble
conglomerates. Fine grained sandstones absent ca. 73.6 - 75.95 m.
Scattered pebbles (<.5%) throughout.
                    Pebble-granule conglomerates, angular,
poorly sorted and associated with coarse sands, noted ca. 71.82 m (7 cm)
75.57 m (20 cm) 76.62 cm (30 cm) and 79.31 m (8 cm). Clasts average 1.5 cm,
maximum noted 3.5 cm at 76.7 m. Basal contacts sharp. Some reverse grading.
                    Stratification generally indistinct above
70.3 m and sharp below. Attitude subhorizontal. Inclined strata noted
ca. 66.7 m (3°); 69.93 m (2° with enclosed cross-sets at 18°); 72.77 m
(11°) 79.3 (15°). Between 67-70 m strata range from 2-23 cm in thickness.
                    Circa 78.65 m patches (2.5 cm) of secondary
carbonate cement. Four inclined (20°-40°) ice lenses (up to 3 mm) noted
between 77 m and 77.92 m.

Fig. B5.   DVDP 11, 60-80 m.

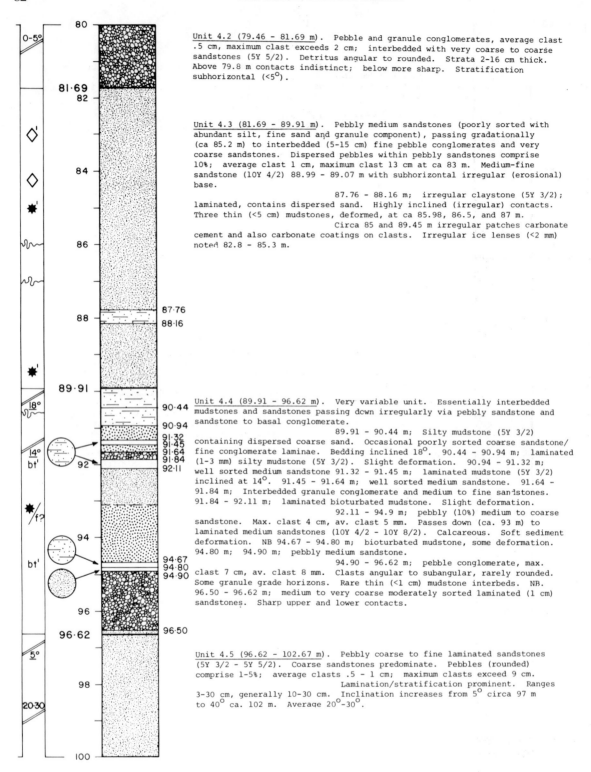

Unit 4.2 (79.46 - 81.69 m). Pebble and granule conglomerates, average clast
.5 cm, maximum clast exceeds 2 cm; interbedded with very coarse to coarse
sandstones (5Y 5/2). Detritus angular to rounded. Strata 2-16 cm thick.
Above 79.8 m contacts indistinct; below more sharp. Stratification
subhorizontal (<5°).

Unit 4.3 (81.69 - 89.91 m). Pebbly medium sandstones (poorly sorted with
abundant silt, fine sand and granule component), passing gradationally
(ca 85.2 m) to interbedded (5-15 cm) fine pebble conglomerates and very
coarse sandstones. Dispersed pebbles within pebbly sandstones comprise
10%; average clast 1 cm, maximum clast 13 cm at ca 83 m. Medium-fine
sandstone (10Y 4/2) 88.99 - 89.07 m with subhorizontal irregular (erosional)
base.
          87.76 - 88.16 m; irregular claystone (5Y 3/2);
laminated, contains dispersed sand. Highly inclined (irregular) contacts.
Three thin (<5 cm) mudstones, deformed, at ca 85.98, 86.5, and 87 m.
          Circa 85 and 89.45 m irregular patches carbonate
cement and also carbonate coatings on clasts. Irregular ice lenses (<2 mm)
noted 82.8 - 85.3 m.

Unit 4.4 (89.91 - 96.62 m). Very variable unit. Essentially interbedded
mudstones and sandstones passing down irregularly via pebbly sandstone and
sandstone to basal conglomerate.
          89.91 - 90.44 m; Silty mudstone (5Y 3/2)
containing dispersed coarse sand. Occasional poorly sorted coarse sandstone/
fine conglomerate laminae. Bedding inclined 18°. 90.44 - 90.94 m; laminated
(1-3 mm) silty mudstone (5Y 3/2). Slight deformation. 90.94 - 91.32 m;
well sorted medium sandstone 91.32 - 91.45 m; laminated mudstone (5Y 3/2)
inclined at 14°. 91.45 - 91.64 m; well sorted medium sandstone. 91.64 -
91.84 m; Interbedded granule conglomerate and medium to fine sandstones.
91.84 - 92.11 m; laminated bioturbated mudstone. Slight deformation.
          92.11 - 94.9 m; pebbly (10%) medium to coarse
sandstone. Max. clast 4 cm, av. clast 5 mm. Passes down (ca. 93 m) to
laminated medium sandstones (10Y 4/2 - 10Y 8/2). Calcareous. Soft sediment
deformation. NB 94.67 - 94.80 m; bioturbated mudstone, some deformation.
94.80 m; 94.90 m; pebbly medium sandstone.
          94.90 - 96.62 m; pebble conglomerate, max.
clast 7 cm, av. clast 8 mm. Clasts angular to subangular, rarely rounded.
Some granule grade horizons. Rare thin (<1 cm) mudstone interbeds. NB.
96.50 - 96.62 m; medium to very coarse moderately sorted laminated (1 cm)
sandstones. Sharp upper and lower contacts.

Unit 4.5 (96.62 - 102.67 m). Pebbly coarse to fine laminated sandstones
(5Y 3/2 - 5Y 5/2). Coarse sandstones predominate. Pebbles (rounded)
comprise 1-5%; average clasts .5 - 1 cm; maximum clasts exceed 9 cm.
          Lamination/stratification prominent. Ranges
3-30 cm, generally 10-30 cm. Inclination increases from 5° circa 97 m
to 40° ca. 102 m. Average 20°-30°.

Fig. B6. DVDP 11, 80-100 m.

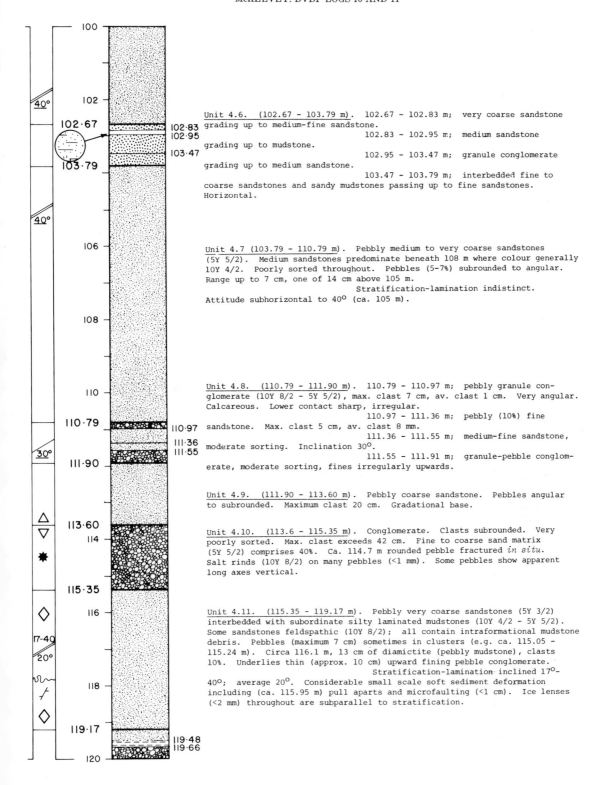

Unit 4.6. (102.67 - 103.79 m). 102.67 - 102.83 m; very coarse sandstone
grading up to medium-fine sandstone.
                              102.83 - 102.95 m; medium sandstone
grading up to mudstone.
                              102.95 - 103.47 m; granule conglomerate
grading up to medium sandstone.
                              103.47 - 103.79 m; interbedded fine to
coarse sandstones and sandy mudstones passing up to fine sandstones.
Horizontal.

Unit 4.7 (103.79 - 110.79 m). Pebbly medium to very coarse sandstones
(5Y 5/2). Medium sandstones predominate beneath 108 m where colour generally
10Y 4/2. Poorly sorted throughout. Pebbles (5-7%) subrounded to angular.
Range up to 7 cm, one of 14 cm above 105 m.
                              Stratification-lamination indistinct.
Attitude subhorizontal to 40° (ca. 105 m).

Unit 4.8. (110.79 - 111.90 m). 110.79 - 110.97 m; pebbly granule conglomerate (10Y 8/2 - 5Y 5/2), max. clast 7 cm, av. clast 1 cm. Very angular.
Calcareous. Lower contact sharp, irregular.
                              110.97 - 111.36 m; pebbly (10%) fine
sandstone. Max. clast 5 cm, av. clast 8 mm.
                              111.36 - 111.55 m; medium-fine sandstone,
moderate sorting. Inclination 30°.
                              111.55 - 111.91 m; granule-pebble conglom-
erate, moderate sorting, fines irregularly upwards.

Unit 4.9. (111.90 - 113.60 m). Pebbly coarse sandstone. Pebbles angular
to subrounded. Maximum clast 20 cm. Gradational base.

Unit 4.10. (113.6 - 115.35 m). Conglomerate. Clasts subrounded. Very
poorly sorted. Max. clast exceeds 42 cm. Fine to coarse sand matrix
(5Y 5/2) comprises 40%. Ca. 114.7 m rounded pebble fractured *in situ*.
Salt rinds (10Y 8/2) on many pebbles (<1 mm). Some pebbles show apparent
long axes vertical.

Unit 4.11. (115.35 - 119.17 m). Pebbly very coarse sandstones (5Y 3/2)
interbedded with subordinate silty laminated mudstones (10Y 4/2 - 5Y 5/2).
Some sandstones feldspathic (10Y 8/2); all contain intraformational mudstone
debris. Pebbles (maximum 7 cm) sometimes in clusters (e.g. ca. 115.05 -
115.24 m). Circa 116.1 m, 13 cm of diamictite (pebbly mudstone), clasts
10%. Underlies thin (approx. 10 cm) upward fining pebble conglomerate.
                              Stratification-lamination inclined 17°-
40°; average 20°. Considerable small scale soft sediment deformation
including (ca. 115.95 m) pull aparts and microfaulting (<1 cm). Ice lenses
(<2 mm) throughout are subparallel to stratification.

Fig. B7.   DVDP 11, 100-120 m.

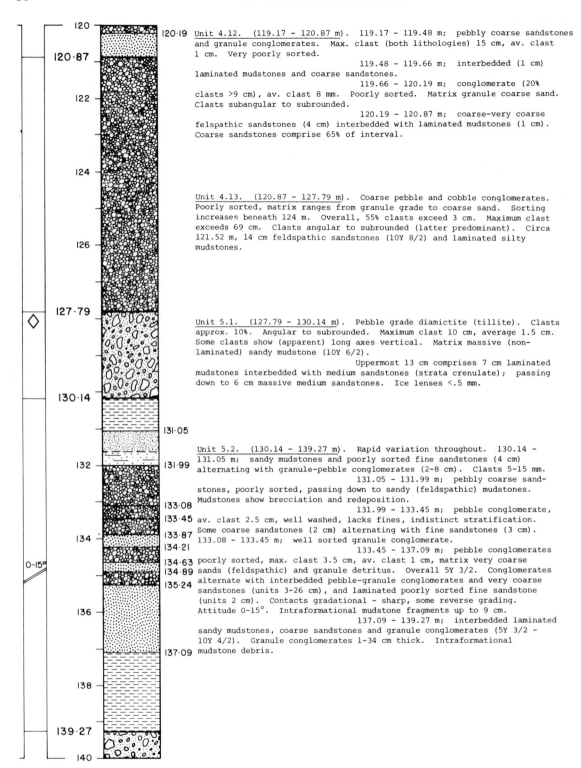

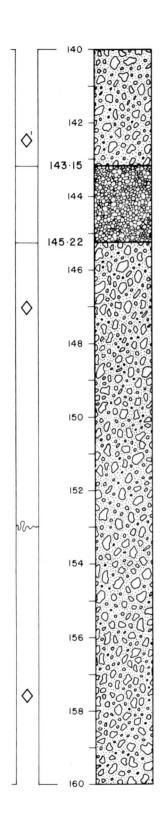

Unit 5.3. (139.27 - 143.15 m). Pebble grade diamictite (tillite). Clasts
approx. 7%. Subangular to subrounded (latter predominant). Maximum clast
5.5 cm, average clast 1 cm. Some clasts show apparent long axes vertical.
Matrix massive (non laminated). Matrix contains silty mud to granule grade
detritus, averages fine sand. Base of unit sharp. 1 cm ice lens noted
ca. 142.4 m.

Unit 5.4. (143.15 - 145.22 m). Pebble conglomerate. Angular to subrounded, latter predominant. Average clast 5 mm, maximum 5.5 cm. Moderate-
poor sorting, matrix averages medium sand grade, lacks extreme fines.
Stratification horizontal.

Unit 5.5. (145.22 - 166.62 m). Diamictite (tillite), highly variable.
Clasts (>4 mm) range from .5% to 25%. Lowest values in general above
150.7 m. Beneath 150.7 m, overall average 7%. Average clast size 1.5 to
3 cm. Maximum clasts generally between 10 to 20 cm. At top of unit, one
clast (hornblende biotite diorite) exceeds 1.04 m.
                    Matrix (10Y 4/2 - 5Y 3/2) poorly sorted.
Components range from mud to coarse sand. Averages generally medium to
fine sandstone. Beneath ca. 158 m matrix overall finer grained and
noticeably laminated (e.g. ca. 159 - 160.5 m). Considerable soft sediment
deformation with inclinations up to 45° (ca. 161.9 m). Ice lenses (<2 mm)
throughout. Base of unit gradational.

Fig. B9.   DVDP 11, 140-160 m.

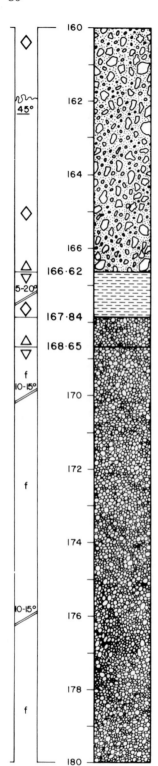

Unit 5.6. (166.62 - 167.84 m). Interbedded mudstones and medium sandstones (5Y 3/2). Minor coarse sandstones (<5%). Stratification
indistinct, inclined 5°-20°. Ice lens 4 mm.

Unit 6.1. (167.84 - 168.65 m). Pebble conglomerate, very poorly sorted.
Pebbles (average 7 cm) in stratified (average 6 cm) matrix of granule
conglomerate and coarse sandstone. Gradational base.

Unit 6.2. (168.65 - 188.8 m). Fine to medium pebble conglomerates.
Poorly sorted. Clasts sub-rounded to sub-angular, former predominate.
Average clast varies between 1 to 5 cm, maximum clast varies between 5
to 40 cm. Maximum clast noted exceeds 60 cm (ca. 180 m). Overall, coarsest
grades 176-188 m. At 176 m clasts greater than 20 cm comprise 40% of rock.
                         Stratification generally indistinct
(gradational contacts). Circa 170-176 m inclinations of 10°-15° noted.
Elsewhere pebble orientation suggests horizontal strata.
                         Silty fine sandstones, 5 cm and 3 cm,
noted at 172.05 and 173.5 m respectively.
                         Fossil problematica (2 cm x 1.5 mm) noted
throughout, while, slightly arcuate. ?Bivalve fragments. No reaction
with dil. HCl.

Fig. B10. DVDP 11, 160-180 m.

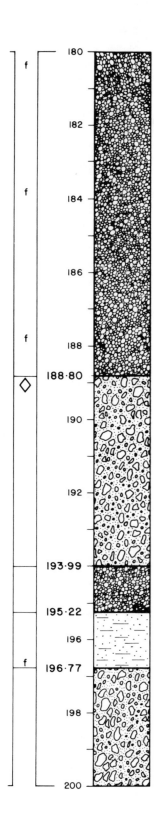

Unit 7.1. (188.8 - 193.99 m). Pebble grade diamictite. Clasts approx.
25%. Sub-angular to sub-rounded. Average clast 2.5 cm, maximum clast
25 cm (one of 50 cm noted). Matrix poorly sorted, averages medium-fine
sand (10Y 4/2). Contains appreciable dispersed coarse sand and granule
grade debris.

                Youngest 40 cm comprises sandy mudstones
containing a few pebbles, and coarse sandstone laminae (<3 mm) passing
down to 9 cm laminated diamictite (pebbles 25%; average clast 5 mm, maximum
clast 15 mm) overlying 2 cm mudstone. Slight soft sediment deformation.
Ice lens (5 mm) at 188.96 m.

Unit 7.2. (193.99 - 195.22 m). Pebble conglomerate. Subrounded, average
clast 2 cm, maximum clast 16 cm. Poorly sorted overall. Considerable
clast size variation.

Unit 7.3. (195.22 - 196.77 m). Sandy to silty mudstone (10Y 6/2).
Contains dispersed coarse sand detritus (<5%). Prominent lamination
(1-2 mm) lowermost 32 cm. Two very coarse sandstone laminae (<1 cm)
ca. 195.77 m. Bivalve fragment (convex upwards) at ca. 196.67 m, valve
thickness approx. 2 mm.

Unit 7.4. (196.77 - 200.81 m). Diamictite. Average clast 3 cm; maximum
clast above 199.28 m exceeds 20 cm, below 199.28 m exceeds 16 cm. Matrix,
laminated sandy mudstone (5Y 3/2). Contains appreciable dispersed granule
grade feldspathic debris.
                Some stratification at top of unit;
e.g. 10 cm of diamictite (max. clast 8 cm) overlies 15 cm of sandy mudstone
containing rare clasts (average clasts 3 mm, maximum clast 1 cm). 3 cm
of fine sand separates this sequence from underlying diamictite. Base
of unit gradational.

Fig. B11.   DVDP 11, 180-200 m.

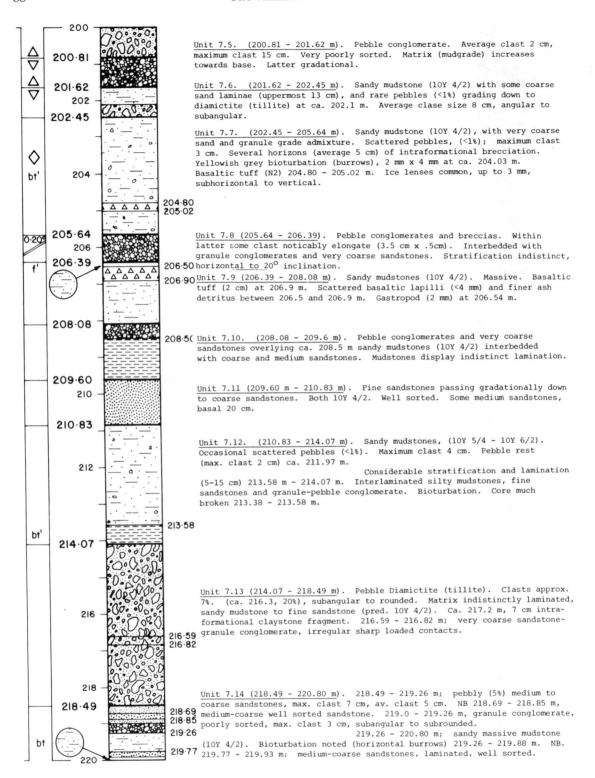

Unit 7.5. (200.81 - 201.62 m). Pebble conglomerate. Average clast 2 cm, maximum clast 15 cm. Very poorly sorted. Matrix (mudgrade) increases towards base. Latter gradational.

Unit 7.6. (201.62 - 202.45 m). Sandy mudstone (10Y 4/2) with some coarse sand laminae (uppermost 13 cm), and rare pebbles (<1%) grading down to diamictite (tillite) at ca. 202.1 m. Average clase size 8 cm, angular to subangular.

Unit 7.7. (202.45 - 205.64 m). Sandy mudstone (10Y 4/2), with very coarse sand and granule grade admixture. Scattered pebbles, (<1%); maximum clast 3 cm. Several horizons (average 5 cm) of intraformational brecciation. Yellowish grey bioturbation (burrows), 2 mm x 4 mm at ca. 204.03 m. Basaltic tuff (N2) 204.80 - 205.02 m. Ice lenses common, up to 3 mm, subhorizontal to vertical.

Unit 7.8 (205.64 - 206.39). Pebble conglomerates and breccias. Within latter some clast noticably elongate (3.5 cm x .5cm). Interbedded with granule conglomerates and very coarse sandstones. Stratification indistinct, horizontal to 20° inclination.

Unit 7.9 (206.39 - 208.08 m). Sandy mudstones (10Y 4/2). Massive. Basaltic tuff (2 cm) at 206.9 m. Scattered basaltic lapilli (<4 mm) and finer ash detritus between 206.5 and 206.9 m. Gastropod (2 mm) at 206.54 m.

Unit 7.10. (208.08 - 209.6 m). Pebble conglomerates and very coarse sandstones overlying ca. 208.5 m sandy mudstones (10Y 4/2) interbedded with coarse and medium sandstones. Mudstones display indistinct lamination.

Unit 7.11 (209.60 m - 210.83 m). Fine sandstones passing gradationally down to coarse sandstones. Both 10Y 4/2. Well sorted. Some medium sandstones, basal 20 cm.

Unit 7.12. (210.83 - 214.07 m). Sandy mudstones, (10Y 5/4 - 10Y 6/2). Occasional scattered pebbles (<1%). Maximum clast 4 cm. Pebble rest (max. clast 2 cm) ca. 211.97 m.
                              Considerable stratification and lamination (5-15 cm) 213.58 m - 214.07 m. Interlaminated silty mudstones, fine sandstones and granule-pebble conglomerate. Bioturbation. Core much broken 213.38 - 213.58 m.

Unit 7.13 (214.07 - 218.49 m). Pebble Diamictite (tillite). Clasts approx. 7%. (ca. 216.3, 20%), subangular to rounded. Matrix indistinctly laminated, sandy mudstone to fine sandstone (pred. 10Y 4/2). Ca. 217.2 m, 7 cm intraformational claystone fragment. 216.59 - 216.82 m; very coarse sandstone-granule conglomerate, irregular sharp loaded contacts.

Unit 7.14 (218.49 - 220.80 m). 218.49 - 219.26 m; pebbly (5%) medium to coarse sandstones, max. clast 7 cm, av. clast 5 cm. NB 218.69 - 218.85 m, medium-coarse well sorted sandstone. 219.0 - 219.26 m, granule conglomerate, poorly sorted, max. clast 3 cm, subangular to subrounded.
                              219.26 - 220.80 m; sandy massive mudstone (10Y 4/2). Bioturbation noted (horizontal burrows) 219.26 - 219.88 m. NB. 219.77 - 219.93 m; medium-coarse sandstones, laminated, well sorted.

Fig. B12. DVDP 11, 200-220 m.

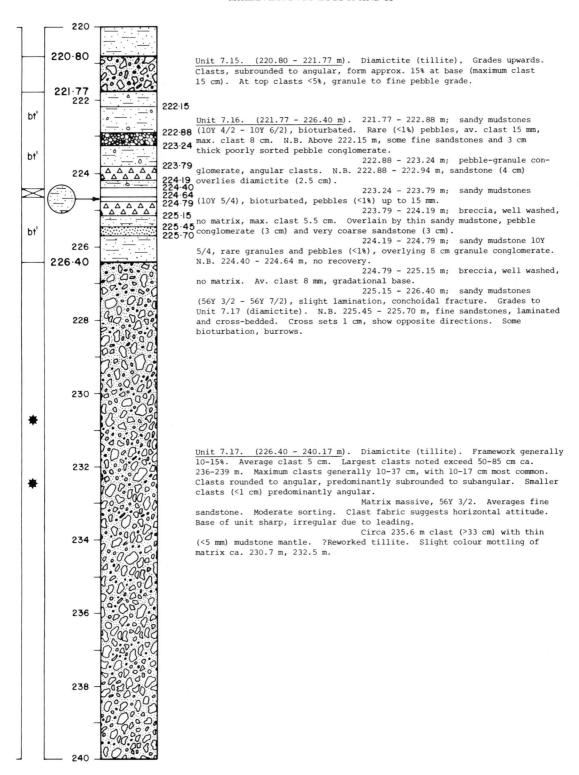

Unit 7.15. (220.80 - 221.77 m). Diamictite (tillite). Grades upwards. Clasts, subrounded to angular, form approx. 15% at base (maximum clast 15 cm). At top clasts <5%, granule to fine pebble grade.

Unit 7.16. (221.77 - 226.40 m). 221.77 - 222.88 m; sandy mudstones (10Y 4/2 - 10Y 6/2), bioturbated. Rare (<1%) pebbles, av. clast 15 mm, max. clast 8 cm. N.B. Above 222.15 m, some fine sandstones and 3 cm thick poorly sorted pebble conglomerate.
          222.88 - 223.24 m; pebble-granule con-glomerate, angular clasts. N.B. 222.88 - 222.94 m, sandstone (4 cm) overlies diamictite (2.5 cm).
          223.24 - 223.79 m; sandy mudstones (10Y 5/4), bioturbated, pebbles (<1%) up to 15 mm.
          223.79 - 224.19 m; breccia, well washed, no matrix, max. clast 5.5 cm. Overlain by thin sandy mudstone, pebble conglomerate (3 cm) and very coarse sandstone (3 cm).
          224.19 - 224.79 m; sandy mudstone 10Y 5/4, rare granules and pebbles (<1%), overlying 8 cm granule conglomerate. N.B. 224.40 - 224.64 m, no recovery.
          224.79 - 225.15 m; breccia, well washed, no matrix. Av. clast 8 mm, gradational base.
          225.15 - 226.40 m; sandy mudstones (56Y 3/2 - 56Y 7/2), slight lamination, conchoidal fracture. Grades to Unit 7.17 (diamictite). N.B. 225.45 - 225.70 m, fine sandstones, laminated and cross-bedded. Cross sets 1 cm, show opposite directions. Some bioturbation, burrows.

Unit 7.17. (226.40 - 240.17 m). Diamictite (tillite). Framework generally 10-15%. Average clast 5 cm. Largest clasts noted exceed 50-85 cm ca. 236-239 m. Maximum clasts generally 10-37 cm, with 10-17 cm most common. Clasts rounded to angular, predominantly subrounded to subangular. Smaller clasts (<1 cm) predominantly angular.
          Matrix massive, 56Y 3/2. Averages fine sandstone. Moderate sorting. Clast fabric suggests horizontal attitude. Base of unit sharp, irregular due to leading.
          Circa 235.6 m clast (>33 cm) with thin (<5 mm) mudstone mantle. ?Reworked tillite. Slight colour mottling of matrix ca. 230.7 m, 232.5 m.

Fig. B13.   DVDP 11, 220-240 m.

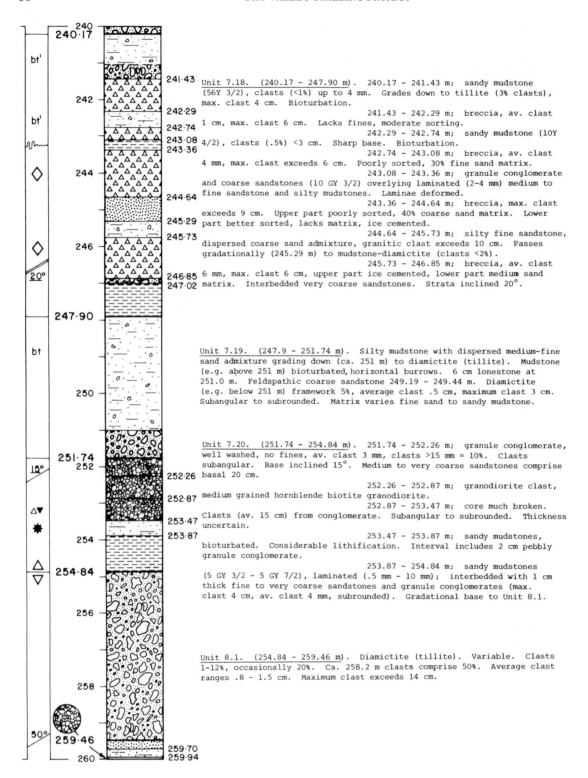

Unit 7.18. (240.17 - 247.90 m). 240.17 - 241.43 m; sandy mudstone (56Y 3/2), clasts (<1%) up to 4 mm. Grades down to tillite (3% clasts), max. clast 4 cm. Bioturbation.

241.43 - 242.29 m; breccia, av. clast 1 cm, max. clast 6 cm. Lacks fines, moderate sorting.

242.29 - 242.74 m; sandy mudstone (10Y 4/2), clasts (.5%) <3 cm. Sharp base. Bioturbation.

242.74 - 243.08 m; breccia, av. clast 4 mm, max. clast exceeds 6 cm. Poorly sorted, 30% fine sand matrix.

243.08 - 243.36 m; granule conglomerate and coarse sandstones (10 GY 3/2) overlying laminated (2-4 mm) medium to fine sandstone and silty mudstones. Laminae deformed.

243.36 - 244.64 m; breccia, max. clast exceeds 9 cm. Upper part poorly sorted, 40% coarse sand matrix. Lower part better sorted, lacks matrix, ice cemented.

244.64 - 245.73 m; silty fine sandstone, dispersed coarse sand admixture, granitic clast exceeds 10 cm. Passes gradationally (245.29 m) to mudstone-diamictite (clasts <2%).

245.73 - 246.85 m; breccia, av. clast 6 mm, max. clast 6 cm, upper part ice cemented, lower part medium sand matrix. Interbedded very coarse sandstones. Strata inclined 20°.

Unit 7.19. (247.9 - 251.74 m). Silty mudstone with dispersed medium-fine sand admixture grading down (ca. 251 m) to diamictite (tillite). Mudstone (e.g. above 251 m) bioturbated, horizontal burrows. 6 cm lonestone at 251.0 m. Feldspathic coarse sandstone 249.19 - 249.44 m. Diamictite (e.g. below 251 m) framework 5%, average clast .5 cm, maximum clast 3 cm. Subangular to subrounded. Matrix varies fine sand to sandy mudstone.

Unit 7.20. (251.74 - 254.84 m). 251.74 - 252.26 m; granule conglomerate, well washed, no fines, av. clast 3 mm, clasts >15 mm = 10%. Clasts subangular. Base inclined 15°. Medium to very coarse sandstones comprise basal 20 cm.

252.26 - 252.87 m; granodiorite clast, medium grained hornblende biotite granodiorite.

252.87 - 253.47 m; core much broken. Clasts (av. 15 cm) from conglomerate. Subangular to subrounded. Thickness uncertain.

253.47 - 253.87 m; sandy mudstones, bioturbated. Considerable lithification. Interval includes 2 cm pebbly granule conglomerate.

253.87 - 254.84 m; sandy mudstones (5 GY 3/2 - 5 GY 7/2), laminated (.5 mm - 10 mm); interbedded with 1 cm thick fine to very coarse sandstones and granule conglomerates (max. clast 4 cm, av. clast 4 mm, subrounded). Gradational base to Unit 8.1.

Unit 8.1. (254.84 - 259.46 m). Diamictite (tillite). Variable. Clasts 1-12%, occasionally 20%. Ca. 258.2 m clasts comprise 50%. Average clast ranges .8 - 1.5 cm. Maximum clast exceeds 14 cm.

Fig. B14. DVDP 11, 240-260 m.

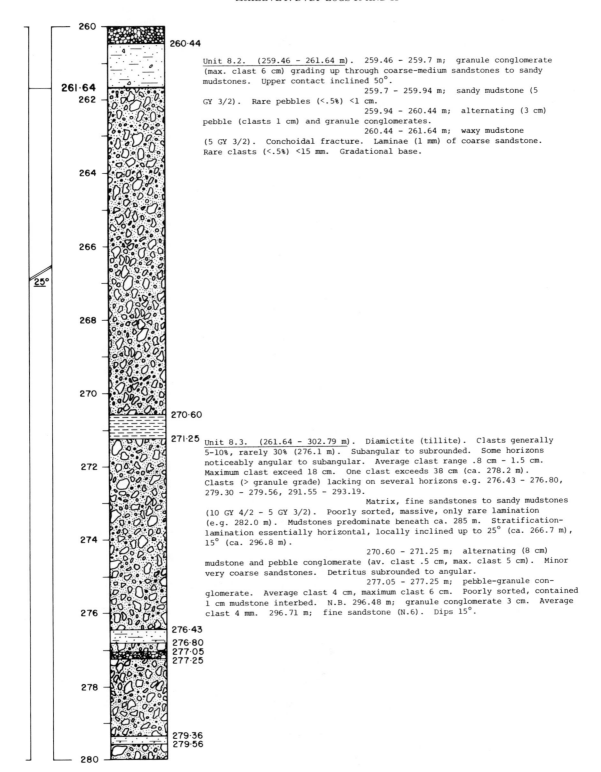

260.44

Unit 8.2. (259.46 - 261.64 m). 259.46 - 259.7 m; granule conglomerate (max. clast 6 cm) grading up through coarse-medium sandstones to sandy mudstones. Upper contact inclined 50°.
259.7 - 259.94 m; sandy mudstone (5 GY 3/2). Rare pebbles (<.5%) <1 cm.
259.94 - 260.44 m; alternating (3 cm) pebble (clasts 1 cm) and granule conglomerates.
260.44 - 261.64 m; waxy mudstone (5 GY 3/2). Conchoidal fracture. Laminae (1 mm) of coarse sandstone. Rare clasts (<.5%) <15 mm. Gradational base.

270.60

271.25 Unit 8.3. (261.64 - 302.79 m). Diamictite (tillite). Clasts generally 5-10%, rarely 30% (276.1 m). Subangular to subrounded. Some horizons noticeably angular to subangular. Average clast range .8 cm - 1.5 cm. Maximum clast exceed 18 cm. One clast exceeds 38 cm (ca. 278.2 m). Clasts (> granule grade) lacking on several horizons e.g. 276.43 - 276.80, 279.30 - 279.56, 291.55 - 293.19.
Matrix, fine sandstones to sandy mudstones (10 GY 4/2 - 5 GY 3/2). Poorly sorted, massive, only rare lamination (e.g. 282.0 m). Mudstones predominate beneath ca. 285 m. Stratification-lamination essentially horizontal, locally inclined up to 25° (ca. 266.7 m), 15° (ca. 296.8 m).
270.60 - 271.25 m; alternating (8 cm) mudstone and pebble conglomerate (av. clast .5 cm, max. clast 5 cm). Minor very coarse sandstones. Detritus subrounded to angular.
277.05 - 277.25 m; pebble-granule con-glomerate. Average clast 4 cm, maximum clast 6 cm. Poorly sorted, contained 1 cm mudstone interbed. N.B. 296.48 m; granule conglomerate 3 cm. Average clast 4 mm. 296.71 m; fine sandstone (N.6). Dips 15°.

276.43
276.80
277.05
277.25

279.36
279.56

Fig. B15.   DVDP 11, 260-280 m.

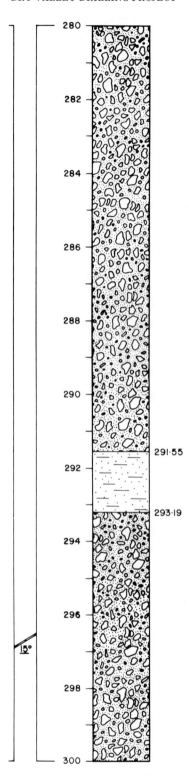

Fig. B16.   DVDP 11, 280-300 m.

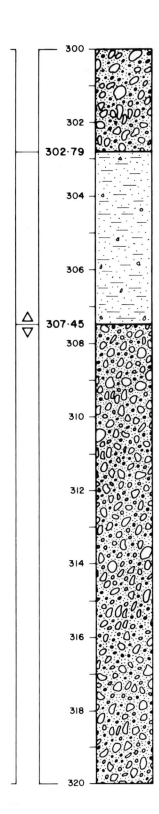

Unit 8.4. (302.79 - 307.45 m). Sandy mudstones containing fine sandstone
laminae (<.5 mm). Overall 5 GY 3/2. Dispersed coarse sand detritus
throughout. Clasts (>4 mm) less than .1%. Maximum clast >7 cm. Some
sandstone laminae feldspathic (N.7) e.g. ca. 302.0 m. Stratification
horizontal, local slight deformation. Base of unit gradational (with
unit 8.5), position approximate.

Unit 8.5. (307.45 - 328 m). Diamictite (tillite). Framework variable,
generally 5-7%, locally 1% (ca. 325.8 m). Average clast range .7 cm -
2.5 cm. Maximum clast range 7 - 16.5 cm (ca. 326.8 m, >28.5 cm). Clasts
subangular to rounded, predominantly subangular.
                    Matrix fine sandstones and sandy mudstones
(5 GY 3-5/2 - 5 Y 3/2). Minor fine to medium sandstones. Dispersed coarse
sand admixture. Stratification indistinct overall horizontal, locally
inclined (20° ca. 327.5 m). Some intraformational brecciation ca. 320 m.
                    Slickensiding and/or shearing at 328 m.

Fig. B17.   DVDP 11, 300-320 m.

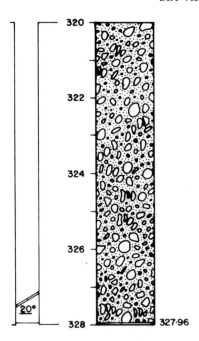

Fig. B18.   DVDP 11, 320-327.96 m.

Acknowledgments.   This research was supported by National Science Foundation contract C-642 and by a University of New England, Armidale, Australia, research grant. The author acknowledges the considerable assistance given both in the field and in the laboratory by other members of the Dry Valley Drilling Project.

REFERENCES

Brady, H. T., Late Cenozoic history of Taylor and Wright valleys and McMurdo Sound derived from diatom biostratigraphy and palaeoecology of DVDP cores, in *Antarctic Geoscience*, edited by C. Craddock, University of Wisconsin Press, Madison, in press, 1980.

Chapman-Smith, M., Geology of DVDP 12 and 14, *Antarctic J. U.S.*, *10*(4), 170-172, 1975.

Chapman-Smith, M., and P. G. Luckman, Late Cenozoic glacial sequences cored at New Harbor, Victoria Land, Antarctica (DVDP 8 and 9), *Dry Val. Drilling Proj. Bull.*, *3*, 120-137, 1974.

Flint, R. F. J. E. Sanders, and J. Rodgers, Diamictite, a substitute for the term symmictite, *Geol. Soc. Amer. Bull.*, *71*, 1809-1819, 1960.

Haskell, T. R., J. P. Kennett, W. M. Prebble, G. Smith, and I. A. G. Willis, The geology of the middle and lower Taylor Valley of south Victoria Land, Antarctica, *Trans. Roy. Soc. N.Z. Geol.*, *2*(12), 169-186, 1965.

McCraw, J. D., Some surface features of McMurdo Sound region, Victoria Land, Antarctica, *N. Z. J. Geol. Geophys.*, *10*(2), 394-417, 1967.

Lopatin, B. C., Basement complex of the McMurdo 'oasis,' south Victoria Land, in *Antarctica Geology and Geophysics*, edited by R. J. Adie, pp. 287-292, Universitetsforlaget, Oslo, 1971.

Péwé, T. L., Multiple glaciation in the McMurdo Sound region, Antarctica, A progress report, *J. Geol.*, *68* (5), 298-514, 1960.

Webb. P. N., and J. H. Wrenn, Late Cenozoic micropalaeontology and biostratigraphy of eastern Taylor Valley, Antarctica, in *Antarctic Geoscience*, edited by C. Craddock, University of Wisconsin Press, Madison, in press, 1980.

GAMMA RAY, SALINITY, AND ELECTRIC LOGS OF DVDP BOREHOLES

L. D. McGinnis

Department of Geology, Northern Illinois University, DeKalb, Illinois 60115

J. S. Stuckless

U.S. Geological Survey, Denver, Colorado 80225

D. R. Osby

Illinois Environmental Protection Agency, Springfield, Illinois 62706

P. R. Kyle

Polar Research Center, Ohio State University, Columbus, Ohio 43210

Natural gamma radiation measurements were made in eight boreholes on Ross Island and in the dry valleys. Total salinity of pore water was determined in five holes in Taylor and Wright valleys and in McMurdo Sound. Electrical resistivity measurements were made only in the Don Juan Pond hole which is uncased. Additional electrical measurements were made on core samples. Resistivities are used to estimate salinities at intervals where water could not be squeezed from a sample. Gamma logs, supplemented by laboratory measurements of radioelement contents of core samples, show that relative to rocks of similar silica contents, basement rocks of the dry valley region are anomalously low in uranium and thorium but that the volcanic rocks from Ross Island are anomalously high in these elements as well as potassium. We attribute anomalies in the two groups of rock to radioelement abundance present at the time of crystallization. The gamma ray log of DVDP 3 shows an increasing radioactivity with increasing silica. Below 90 m the hole is dominated by basanite and a low radioactivity. The upper portion of the hole is dominated by differentiated alkalic rocks of higher radioactivity and a few ice lenses with little or no radioelement content. Logs of holes 10-14 reflect the low radioelement content of reworked sediment. In general, diamictons are more radioactive than coarse-grained sands and gravels. In Wright Valley the gamma log through the Ferrar Dolerite reflects the low radioelement content usually associated with tholeiites. Total salinity is determined from water squeezed from sediment in DVDP holes 10, 11, 12, 14, and 15, and the data are plotted as logs along with gamma ray measurements. From salinity variations of pore ice in Taylor Valley it is inferred that connate water is present in marine and freshwater sediment in holes 10 and 12; whereas sediment in hole 11 has undergone postdepositional flushing. Resistivity logs in hole 13 at Don Juan Pond in Wright Valley indicate that groundwater in the Ferrar dolerite is less saline than that in the sands. The dolerite has been suggested as being the aquifer through which sub-ice sheet melt water is discharging to the periphery of the ice sheet, and the salinity is consistent with this supposition. Salinity of water discharged from the Don Juan Pond hole under natural artesian conditions is greater than 200 ppt. In North Fork sediment, pore ice salinity remains greater than sea water from the surface to basement. Only salinities were measured in the McMurdo Sound borehole, and they were found to vary from the salinity of sea water near the water-sediment interface to less than one fifth that of sea water from 138 to 165 m below sea level. From 165 m to the bottom of the hole, sea water salinities were encountered.

INTRODUCTION

The Dry Valley Drilling Project was formulated in 1971 in response to questions regarding the geological history of the dry valleys and adjacent regions. As part of the project it was proposed that access into the boreholes be maintained after drilling, wherever it was technically feasible to do so, in order to conduct various geological, geophysical, and hydrogeological downhole tests. The final borehole was drilled into McMurdo Sound in December 1975, and most of the downhole geophysical logs were completed in the following field season. The drilling and logging summary is shown in Table 1. This report is based primarily on downhole geophysical measurements made in January 1977. Additional logs were made in the Don Juan Pond borehole in December 1978. Salinity logs are derived from analyses of pore water squeezed from melted core from 9 of the 15 holes (Figure 1).

Natural gamma radiation logs were made in two boreholes on Ross Island, three in Taylor Valley, two in Wright Valley, and one in Victoria Valley. The gamma logs are calibrated from known amounts of radioactive elements contained in Ross Island, Don Juan Pond, and Lake Vida core. Salinity logs are determined for three holes in Taylor Valley, one in Wright Valley, and one in McMurdo Sound. Electrical resistivity logs were made only in the uncased Don Juan Pond borehole in Wright Valley. In addition to downhole electrical logging, several resistivity measurements were made on selected samples of core from which water could not be obtained.

Geophysical logs are constructed to provide data on the physical and chemical characteristics of subsurface permafrost sediments and crystalline rocks and to aid in a reconstruction of geological history. For a detailed description of the geology of the boreholes, see other papers in this volume.

INSTRUMENTATION AND PROCEDURE

Gamma Ray Logs

Natural gamma radiation measurements were made in January 1977 with a Johnson GR-73 gamma ray logging system. A Geiger-Mueller probe was lowered in the holes at 1.52-m intervals. At Don Juan Pond, both the Geiger-Mueller probe and a scintillation probe were used. Background noise at ground level both inside and outside the casing was measured before and after logging the holes to check calibration. To further verify calibration, borehole 3 on Ross Island was measured three times, in the beginning of the logging period, once near the middle, and finally at the end of the logging period. Deviations in the three logs were

TABLE 1. DVDP Drilling Summary

DVDP Borehole Number	Drill Penetration,* m	Collar Elevation, m	Core Recovery,† m	Drilling Dates	Log	Geographic Description
1	201	66.9	197	Jan. 1973	3	McMurdo
2	179	47.6	171	Feb. 1973	1, 4	McMurdo
3	381	47.6	341	Sept.–Oct. 1973	1, 3, 4	McMurdo
4‡	17	84.0	17	Nov. 1973	3, 4	Lake Vanda
5	4	116.7	3	Dec. 1973	· · ·	Don Juan Pond
6	306	349.2	303	Dec. 1973	1, 3, 4	Lake Vida
7	11	18.5	3	Dec. 1973	· · ·	Lake Fryxell
8	157	2.8	130	Jan. 1974	1, 3, 4	New Harbor
9	38	2.8	33	Jan. 1974	· · ·	New Harbor
10	185	2.8		Oct.–Nov. 1974	1, 3, 4	New Harbor
11	328	80.2	328	Nov.–Dec. 1974	1, 3, 4	Commonwealth Glacier
12	185	75.1	184	Dec.–Jan. 1974–1975	1, 3, 4	Lake Leon
13	75	118.4	68	Jan. 1975	1, 2, 3, 4	Don Juan Pond
14	78	68.4	77	Jan. 1975	1, 3, 4	North Fork
15¶	65	1.0	34	Nov. 1975	3, 4	McMurdo Sound

1 denotes gamma log, 2 denotes resistivity and spontaneous potential, 3 denotes salinity and resistivity measured on core, and 4 denotes temperature.

* Total drill penetration is 2231.

† Total core recovery is 2074 (recovery = 93%).

‡ DVDP number 4 was drilled from lake ice in 70 m of water and into bottom 17 m.

¶ DVDP number 15 was drilled from sea ice in 122 m of water and into bottom 65 m.

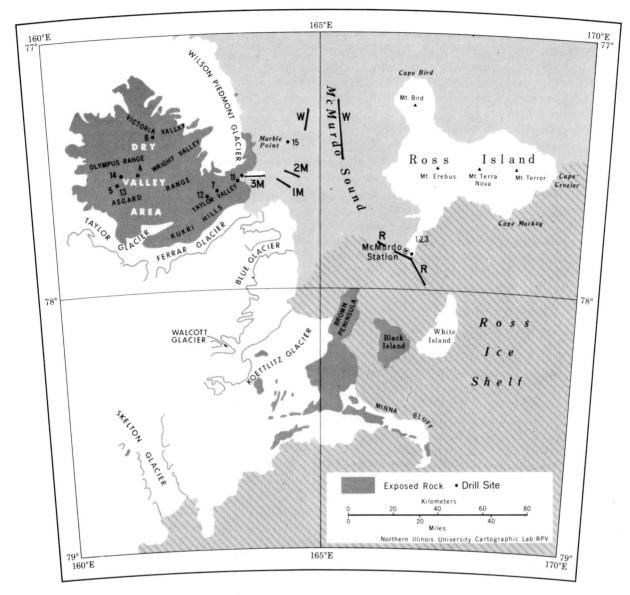

Fig. 1.   Geographic setting of the dry valley-McMurdo Sound-Ross Island region. Borehole 15 was drilled from sea ice which covers McMurdo Sound through most of the year.

within the limits of expected error; that is, gamma ray emission follows a Poisson distribution, which means the standard deviation is equal to the square root of the number of disintegrations observed.

The gamma log technique was selected because it detects radiation through casing and is temperature independent. The more sensitive scintillation probe was used only at Don Juan Pond because of its large diameter, which did not permit its insertion into the 3.175-cm-diameter black iron pipe placed in most other holes. The pipe was placed in

the holes immediately after drilling and was then filled with arctic diesel fuel to maintain an open hole for heat flow studies.

Keys and MacCary [1971] describe logging techniques especially designed for water resources investigations. These techniques are directly applicable to studies in boreholes in glacial sediment in permafrost terrain such as those drilled during DVDP, differing only in that the temperatures in DVDP holes are always below 0°C and that the holes are, with few exceptions, filled with diesel fluid. Gamma logging techniques may be used in either cased or uncased holes that are filled with any type of fluid.

Natural gamma radiation logs are related to the measurement of electromagnetic radiation emitted from the nucleus of atoms of potassium 40, uranium 235, uranium 238, and thorium 232 as part of the radioactive decay process. The energies of emitted gamma rays are characteristic of the nucleus and therefore of the isotope emitting them. With the instrumentation used in this study it is not possible to discriminate the variable radiation energy levels from the above isotopes. A radiation intensity was measured as the number of seconds per 100 counts and was then converted to counts per minute or in the case of Don Juan Pond, counts per second. Only a fraction of the total radioactive disintegrations per second from a source are ob-

served, with the fraction observed depending on the efficiency of the detecting and counting system, the distance from the source, and absorption by the well casing. One of the causes of error is what is referred to as 'dead time,' which occurs when two radiation pulses arrive during a time interval smaller than the resolving time of the equipment, so that only one pulse is recorded. The logs have been quantified by using measured abundance of radioactive elements in samples from boreholes 2, 6, and 13 (see Table 4).

Electric Logs

Apparent resistivities were measured below the casing in DVDP 13 which contains only natural waters encountered during drilling. Measurements

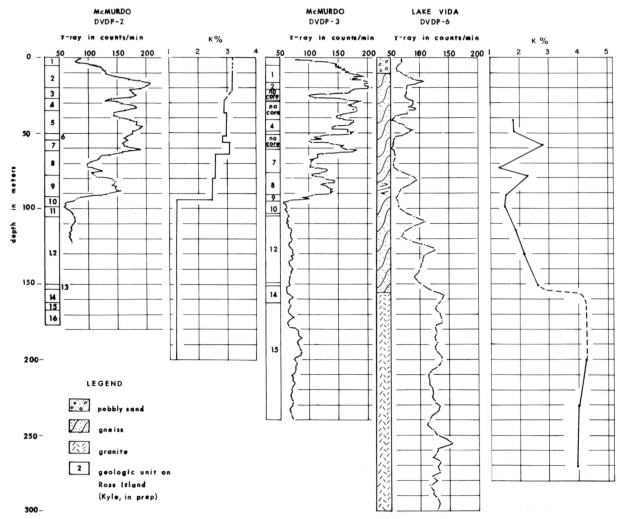

Fig. 2.   Gamma ray logs of holes 2, 3, and 6. K content is plotted for holes 2 and 3 between the logs of the holes. K content of hole 6 is plotted to the right of the log. Geologic units in holes 2 and 3 are listed in Tables 2 and 3. Lithology of hole 6 is from *Kurasawa et al.* [1974].

were made with a W. G. Keck electrical logging system DR-74. In January of 1977 a 'short normal' resistivity array was inserted in the hole and read at intervals of 1.52 m. Electrode spacing between the short normal potential and current electrodes is 0.0762 m, which provides an apparent resistivity of the fluids in the hole. Apparent resistivity of a down hole measurement is

$$R_a = (E/I)\, 4\pi L$$

where E is potential drop, I is current, and L is distance between the potential and the current electrode. In early December 1978, down hole electrical measurements were repeated using probes having potential-current electrode spacings of 0.0762, 0.7620, and 3.048 m in order to observe formation resistivity and to note possible differences from the 1977 measurements.

During phase 1 of DVDP and prior to drilling, surface resistivity measurements were made in the Don Juan Pond basis using the Wenner electrical depth sounding technique [McGinnis et al., 1973]. Electrical depth soundings indicate resistivities similar to those measured in down hole logging. To further document resistivities derived from logs and surface data, samples of core were prepared in the lab, and resistivities were measured on these [McGinnis et al., 1980].

Salinity Logs

Samples of sediment were taken at 5-m intervals from core drilled at Lake Vida, New Harbor, Commonwealth Glacier, Lake Leon, Don Juan Pond, North Fork, and McMurdo Sound. The samples, with an approximate volume of 150 cm^3 each, were shipped in dry ice to Northern Illinois University from the Antarctic Core Storage Facility at Florida State University, Tallahassee. Before analysis, the frozen samples were allowed to reach room temperature in an airtight container to minimize the loss of pore fluids.

Salinities were obtained by using a technique devised by the USGS [Manheim, 1966] which employs a cylinder and piston, core squeezing apparatus [McGinnis et al., 1980]. Fluids are multifiltered and drawn into a syringe from which drops are placed on a refractometer-salinometer. Two refractometers were American Optical, Goldberg T/C 10419 and Endeco Type 102 having

ranges of 0-160 and 0-48 ppt, respectively. Where an occasional salinity went slightly beyond the calibrated range of the refractometer, the salinity could be estimated by extrapolating the calibration scale out to the salinity reading.

DISCUSSION

Geophysical logs are shown in Figures 2-5. Because of their scatter the gamma ray logs are running averages of three adjacent measurements. Electrical and salinity logs are curves drawn through single measurements.

DVDP 2 and 3

Holes 2 and 3 are only 3 m apart, and therefore the logs are very similar (Figure 2). The first hundred meters in each hole has high gamma activity which fluctuates between 100 and 200 cpm (counts per minute). There is a general tendency for activity to decrease with depth. Below about 100 m the activity is low and uniform. The two low points at 25 and 55 m in DVDP 3 correspond to zones in which thick lenses of ice were encountered during drilling. The rapid decrease in activity at the very top of each hole is due to the shielding effect of double casing.

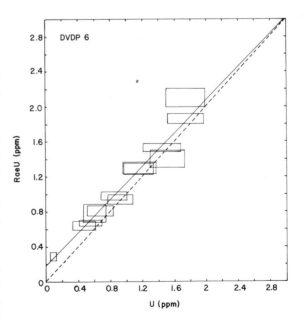

Fig. 3. Uranium versus radium equivalent uranium. Dashed curve shows a 1:1 correlation; solid curve shows the actual regression. The two curves agree within experimental error.

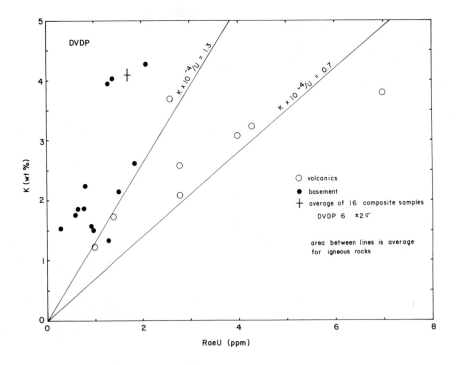

Fig. 4.   Radium equivalent uranium (or uranium) versus potassium. The lines show the field of normal K × 10⁻⁴/U ratios which vary from 0.7 to 1.3. Open circles are volcanic rocks. Solid circles are crystalline basement rocks from the dry valleys.

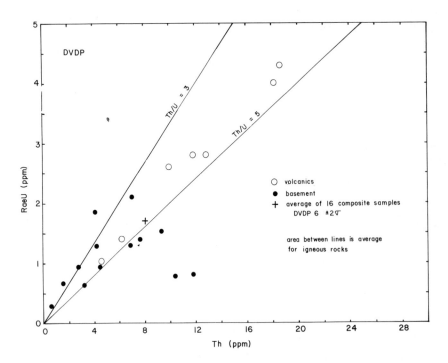

Fig. 5.   Thorium versus radium equivalent uranium (or uranium). The lines show the field of normal thorium/uranium ratios which vary from 3 to 5. Open circles are volcanic rocks. Solid circles are crystalline basement rocks from the dry valleys.

TABLE 2. Description of Geologic Units—DVDP 2

Unit	Description	Thickness, m	Depth Below Platform, m	Elevation Above/Below Sea Level, m
1	undifferentiated intermediate lava flow	3.51	6.81	40.82
2	Ne benmoreite flow	14.94	21.75	25.88
3	undifferentiated intermediate lava flow	7.11	28.86	18.77
4	Ne mugearite flow	8.13	36.99	10.64
5	Ne mugearite flow	15.05	52.04	−4.41
6	Ne hawaiite flow	4.35	56.39	−8.76
7	Ne benmoreite flow	7.33	63.72	−16.09
8	Ne hawaiite pyroclastic breccia	16.23	79.95	−32.32
9	Ne hawaiite flow	16.63	94.55	−46.92
10	basanite pyroclastic breccia	5.63	100.18	−52.55
11	basanite flow	7.53	107.71	−60.08
12	basanite flow	44.51	152.22	−104.59
13	basanite pyroclast	3.35	155.57	−107.94
14	basanite flow	9.26	164.83	−117.20
15	basanite pyroclastic breccia	4.41	169.25	−121.62
16	basanite hyaloclastite	10.16	179.41	−131.78

Depths of DVDP 2 core given in the preliminary log [*Treves and Kyle*, 1973] were based on measurement of the core only. The depths and unit thicknesses have been revised to conform with depths as indicated by the drill stem. The depths listed above are from the drill platform which was 47.63 m above sea level; coring commenced 3.30 m below this depth.

The gamma logs show excellent correlation with the geologic logs of DVDP 2 and 3. Cores from DVDP 2 and 3 represent a differentiation sequence from primitive, mantle-derived basanite (basanitoid) lavas and pyroclastic units at the bottom of the holes, which pass up through Ne hawaiite, Ne mugearite, and Ne benmoreite units (Tables 2 and 3). Note that although there is a general increase in the degree of fractionation with decreasing depth, the trend is irregular. Radioactive elements K, U, and Th are termed incompatible because during magmatic crystallization they do not enter mineral phases; rather they are enriched in the residual liquid. Therefore as the degree of differentiation increases in the DVDP 2 and 3 lavas, so do the K, U, and Th abundances. Analyses of surface

TABLE 3. Description of Geologic Units—DVDP 3

Unit	Description	Thickness, m	Depth Below Platform, m	Elevation Above/Below Sea Level, m
1	Ne benmoreite flow with pyroclastic breccia flow top (2.48 m)	11.03	20.48	27.15
2	undifferentiated intermediate lava flow, glassy	4.89	25.37	22.26
	no core	6.94	32.31	15.32
3	Ne mugearite flow	3.05	35.36	12.27
	no core	10.06	45.42	2.21
4	Ne mugearite flow, thin (0.68 m) pyroclastic flow base	6.99	52.41	−4.78
5	Ne hawaiite flow top	3.37	55.78	−8.15
6	Ne benmoreite flow base	1.00	64.40	−16.77
7	Ne hawaiite pyroclastic breccia	16.10	80.50	−32.87
8	Ne hawaiite flow; oxidized flow base	14.66	95.16	−47.53
9	basanite pyroclast, lapilli tuff (2.44 m) and pyroclastic breccia (1.20 m)	4.44	99.60	−51.97
10	basanite flow, scoriaceous brecciated flow top	8.18	107.78	−60.15
11	basanite pyroclastic breccia	1.36	109.14	−61.51
12	basanite flow, secondary minerals	44.03	153.17	−105.54
13	basanite pyroclastic breccia	2.25	155.42	−107.79
14	basanite flow, red oxidized base	11.41	166.83	−119.20
15	basanite hyaloclastite; mixed volcanic breccia with lapilli tuff and blocky lapilli tuff	214.17	381.00	−333.37

The depths listed above are from the drill platform which was 47.63 m above sea level; coring commenced 9.45 m below this depth.

samples from Hut Point Peninsula and DVDP 1 and 2 for K, U, and Th are given in Table 4. A generalized log of K content of DVDP 2 and 3 shows excellent correspondence with the gamma log (Figure 2).

The relative proportions of uranium, thorium, and potassium in the volcanic rocks are normal, but the concentrations of each element are high in relation to nonalkalic volcanic rocks of similar silica content [Rogers and Adams, 1969a, b]. Tilling et al. [1970] have noted the tendency for an increase in radioelement contents with increasing degree of alkalinity for any given silica content. The volcanic rocks of the Ross Island region are strongly alkalic and consist of a basanite to phonolite association [Goldich et al., 1975; Kyle and Rankin, 1976]. This feature and the normal proportionalities of radioelements suggests that the high radioelement contents are a primary feature.

DVDP 6

The gamma activity of DVDP 6 (Figure 2) falls into two natural divisions. From 0 to 150 m, activity is generally less than 100 cpm and shows large fluctuations. At 150 m activity increases to about 130 cpm and shows a more regular trend.

The geologic log of DVDP 6 [Kurasawa et al., 1974] recorded 10.52 m of glacial sediment overlying biotite granite gneiss, correlated with the Olympus granite gneiss. The gneiss is massive above a fault of 88 m, but below this it is cut by meter-thick tongues of massive biotite granite. Below 156 m the core consists of massive biotite-oligoclase-quartz-orthoclase granite believed to be a correlative of the Vida granite.

The gamma log correlates well with the geologic log. The increase in gamma activity at 156 m corresponds to the gneiss/granite boundary. Radioelement contents of DVDP 6 samples are listed in Table 4 and graphically presented in Figure 2.

In general, the gamma log, supplemented by laboratory measurements of radioelement contents of DVDP 6 core samples (Table 4), shows that relative to rocks of similar silica content [Rogers and Adams, 1969a, b], basement rocks from DVDP 6 are anomalously low in uranium and thorium. We attribute the anomaly to radioelement abundance present at the time of crystallization.

TABLE 4.   Radioelement Content of Rocks From DVDP Cores and One From a Surface Sample on Hut Point Peninsula

Hole Number and Sample Depth	SiO$_2$, wt %	K, wt %	U, ppm	Th, ppm	RaeU, ppm
Surface—25793	54.82	3.79	8.8	26.0	
1—90.72	42.15	1.74	1.7	6.5	
2—61.23	51.90	3.07	5.0	18.0	
2—76.38	42.27	2.58	3.6	13.0	
2—105.53	41.68	1.23	1.2	4.7	
6—41.6		1.80	0.47	3.30	0.64
6—49.2		1.84	0.55	1.57	0.67
6—59.2		2.80	0.92	0.94	1.56
6—72.6		1.34	1.15	4.28	1.29
6—78.6		2.24	0.67	11.83	0.81
6—91.4		1.53	0.09	0.61	0.29
6—99.8		1.51	0.84	4.55	0.98
6—113.0		. . .	1.01	. . .	. . .
6—117.1		1.87	0.60	10.4	0.78
6—130.1		2.15	1.44	9.4	1.53
6—150.4		2.62	1.74	4.15	1.86
6—200.2		4.27	1.74	7.1	2.1
6—230.3		4.03	1.51	7.7	1.4
6—271.1		3.95	1.16	6.96	1.30
6—298.7		. . .	2.16	. . .	. . .
6—298.7		4.09	. . .	8.07	1.71
		0.11		0.2	0.06

Surface sample is from Hut Point Peninsula. Uranium analyses for hole 6 are determined by delayed neutron; analysts are H. T. Millard, Jr., C. M. Ellis, and C. McFee of the U. S. Geological Survey, Denver. Th, K, and RaeU abundances are determined by gamma ray spectrometry. RaeU = radium equivalent uranium (the amount of uranium needed for secular equilibrium with the measured radium − 226). The last sample from hole 6 includes a composite of 8 samples of Vida granite supplied by E. Decker.

Several isotopic studies (for a summary, see *Stuckless and Nkomo* [1978]) have shown that low uranium contents in most granitic rocks are best attributed to uranium loss, usually in response to incipient weathering. This loss may extend to a depth of several hundred meters. Three lines of evidence suggest that uranium loss is not the only explanation for low contents in the antarctic basement rocks. These are as follows.

1. Movement of uranium in granite creates isotopic disequilibrium with the uranium decay chain [*Stuckless and Ferreira*, 1976] . . . most granitic samples from DVDP 6 show no evidence of isotopic disequilibrium (Figure 3).

2. Analyzed samples of the basement rocks are low in uranium relative to potassium (Figure 4), but thorium and uranium occur in approximately normal proportions (Figure 5) . . . hence if the low uranium is attributed to loss, thorium must have been lost in similar proportions . . . but thorium is much less mobile than uranium during weathering of granitic rocks in a cold climate [*Titayeva et al.*, 1973] . . .

3. If uranium and thorium were both mobilized, the character of the gamma ray log in DVDP 6 (Figure 2) would not be expected to closely reflect the geology, and the granite should exhibit large variations depending on the extent of uranium and thorium gain or loss.

DVDP 10, 11, and 12

Gamma ray logs of the Taylor Valley boreholes, drilled through primarily clastic sediments, reflect radioelement contents about equivalent to the average values in the Olympus granite gneiss in the Lake Vida hole or to the deeper, basanite units from the Ross Island holes (Figure 6). In the upper parts of holes 10 and 11 the gamma ray curves are unusually smooth and of low emission intensity, suggesting a radioelement leaching and homogenization that might be expected in multireworked sediments. Rocks in the upper 80 m in hole 11 have particularly low emission rates. Between 80 and 100 m the curve rises, where it again flattens out, corresponding with a similar curve and intensity in hole 10. At about 170 m the curve in hole 11 then rises in a cyclic manner to nearly 250 m, where the cyclicity continues to be constant but at a higher level than before. Emission intensity in hole 10 remains rather constant from about 25-m depth to the bottom of the hole, whereas in hole 11, gamma radiation gradually increases from the top to the

bottom of the hole. In general the diamictites are relatively more radiogenic than the coarse-grained sands and gravels.

The gamma log of borehole 12 bears no resemblance to the gamma logs of boreholes 10 and 11 except perhaps in the lower 140 m of hole 11. Amplitudes of the log fluctuate from well below 75 cpm to greater than 100 cpm, with a gradual increase in count rate with depth. Of the three holes in Taylor Valley, only hole 12 enters basement. This is reflected by the higher radioactivity in the bottom of the hole. Again, the diamictites exhibit the highest radioelement content of all the sedimentary units with the exception of the diamictite directly overlying basement. As the sedimentary column in hole 12 lacks the low-count homogeneous appearance contained in the upper sections in holes 10 and 11, it would be difficult to assume any rock-stratigraphic correlation from hole 12 to the other two holes. Either hole 12 sediments are immature and relatively unleached of their radioelement content or they have been enriched through some process. The first alternative is preferred. It should be noted that a large erratic of Ferrar dolerite in hole 12 displays a very low gamma count.

In addition to the gamma ray logs, salinities of the pore ice in the sediments of holes 10, 11, and 12 were determined at 5-m intervals. A discussion of the significance of the variations in salinity of these holes has been prepared by *McGinnis et al.* [1980] and will therefore only be reviewed briefly here. On the basis of the characteristic salinity curves in each hole, they suggest that the pore ice in hole 10 is connate, whereas the ice in hole 11 is not connate because the sediments, some of which are marine, contain freshwater ice to a depth of 110 m below sea level. They propose that the upper 200 m of sediment in hole 11 has been flushed by water draining through it from a freshwater lake. This interpretation would be modified if it could be shown that a large amount of freshwater sediment has been removed from the upper part of hole 10. In both hole 10 and hole 11 the salinities of the water squeezed from the deepest sediments is the same (72 ppt) as the salinity which flowed up into hole 10 and which caused termination of drilling.

Salinities in hole 12, as well as the radioelement content, cannot be correlated with those in holes 10 and 11. Salinities increase from freshwater ice at the surface to greater than 60 ppt just above basement. Since the sediments are nonmarine in origin, *McGinnis et al.* [1980] propose that the increasingly high salinities with depth are caused by sedi-

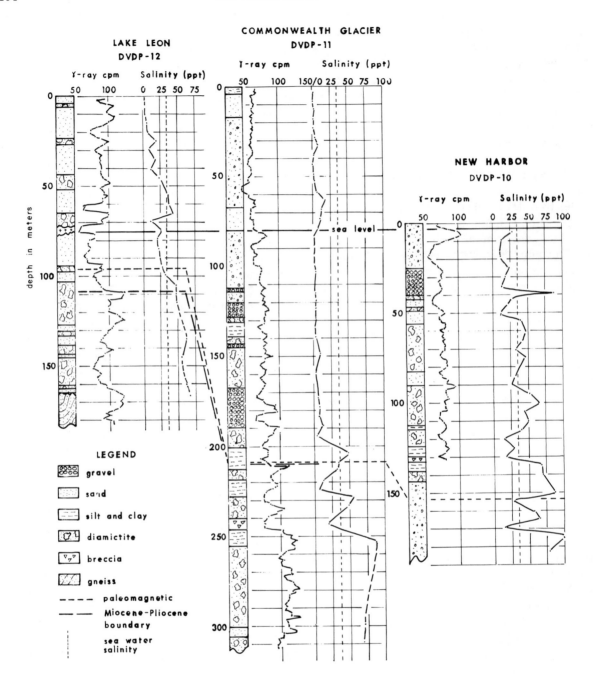

Fig. 6.   Gamma ray and salinity logs of holes 10, 11, and 12 in Taylor Valley. Stratigraphy of hole 10 is from New Harbor, hole 11 from Commonwealth Glacier, and hole 12 from Lake Leon.

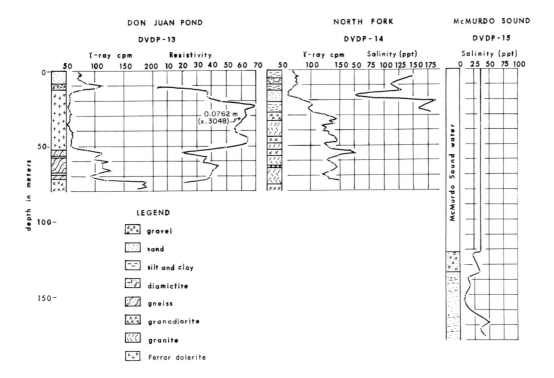

Fig. 7.   Gamma ray, salinity, and electric logs of holes 13, 14, and 15. Stratigraphy of hole 13 is from Don Juan Pond, hole 14 from North Fork, and hole 15 from McMurdo Sound.

ments deposited during polar desert conditions similar to those existing today, followed by downward freezing and enrichment in salts as the permafrost front descended.

DVDP 13, 14, and 15

Gamma logs were run only in holes at Don Juan Pond (DVDP 13) and North Fork (DVDP 14) in Wright Valley (Figure 7). *Mudrey et al.* [1975] have described the core DVDP 13 as follows.

> Continuous core recovery at Don Juan Pond provides a record of 12.67 m of sorted sands and silt, 39.53 m of the lowermost units of a Ferrar Sill, and 22.78 m of migmatite.
>
> Water-saturated sands contain no recognized secondary, saline minerals, and (the water) may be related to upward moving water that is less saline than found in Don Juan Pond. Subsurface, sub-zero water is migrating upward along fractures in the basement rocks, and through the sub-zero, yet nonfrozen, surficial sands.
>
> Diabase represents the lower third of a Ferrar Sill emplaced within the crystalline basement. Recognized sill stratigraphy from the bottom up includes 1.5 m of contact chill, 13 m of fine-grained microdiabase with abundant interstitial granophyre,

and 26 m of medium-grained diabase in which three kinds of pyroxene (hyperthene, augite, and pigeonite) contain abundant exsolution blebs of secondary pyroxenes.

> The crystalline basement includes bands of biotite-amphibolite gneiss, leuco-flaser gneiss, biotite-flaser gneiss, and hornblendegranodiorite. All of these rocks are cataclasized and recrystallized to some extent.

The gamma log in hole 13 corresponds with the geology. The surficial sands have intermediate radioelement content. A thin layer of radioelement-enriched sediment rests directly on the Ferrar dolerite. The zone of enrichment is probably caused by precipitation of radioactive minerals from waters moving upward through the Ferrar dolerite. The dolerites represent the geologic unit with the lowest concentrations of radioactivity of all the holes cored. This is to be expected for tholeiitic rocks, which are mantle derived, and have a low content of radioactive elements. The lower 13 m of fine-grained microdiabase is slightly more radiogenic than the upper medium-grained diabase. Beneath the dolerites the migmatites display erratically increasing radioelement concentrations, which suggests that they are unleached, normal basement rock, at least at depths 20 m below the dolerite-migmatite contact.

K measurements of 18 samples from the sill average 0.74 ± 0.20 wt % (H. Kurasawa, personal communication, 1979). The microdiabase has slightly higher K abundances (0.91 ± 0.15 wt %) compared with the medium-grained diabase (0.66 ± 0.17 wt %). The gamma activity of the underlying gneiss is higher than that found in gneiss from DVDP 6 but similar to the granite. Granodiorite at the bottom of the hole has high activity, suggesting radioelement contents similar to the upper parts of DVDP 2 and 3.

Borehole 14 in the North Fork was drilled through 28 m of sediment and approximately 52 m of migmatites [Chapman-Smith, 1975]. The coarser-grained sediments again display relatively lower radioelement concentrations, whereas the clays and clay-bearing diamictites have higher concentrations. The migmatites appear to be unleached with a rather steady gamma ray count of 125-150 cpm.

At the North Fork site, water was squeezed from the 28 m of sediment at 5-m intervals. Salinities ranged from 50 to 185 ppt, well above the salinity for sea water but less than in Don Juan Pond (Figure 7). Mechanisms for emplacement of brines in sediment of the North Fork are probably similar to those operating now in the Don Juan Pond basin.

Following standard procedures, salinities were also determined for pore water in the sediments cored from McMurdo Sound at site 15. Somewhat unexpectedly, at a depth of about 15 m below bottom, the salinity was found to drop from that for sea water to 7 ppt at a depth of about 30 m below the water-sediment interface. The low salinity zone is 30 m thick, extending from 15 to 45 m below bottom. At 45 m the salinity again rises to near that of sea water.

Salinities at Don Juan Pond were not measured directly in this study; however, downhole resistivity was measured at three electrode separations to give values for the fluid both in the borehole and in the formation. A resistivity log was made in January 1977 by using an electrode separation of 0.0762 m, 2 years after drilling (Figure 7). During this time interval and the subsequent 2 years before a repeat measurement, the well was capped, and flow from the well was not observed. In December 1978, after clearing soft ice from the upper 13.5-m

cased section of the well, it was observed to flow at a rate of 4 liters per minute [McGinnis, 1979]. The well was relogged at the same electrode spacing, and results were identical (see Figures 7 and 8) except for an increase of about 6 ohm-m between 23- and 33-m depth. It is suggested that the high resistivity zone in the depth interval 23-33 m is a zone of high permeability and that the higher resistivities, measured when the well flows, are caused by less diffusion of salts into the zone of high flow.

Maximum resistivities using the three electrode separations of 0.0762, 0.7620, and 3.0480 m are 27, 189, and 244 ohm-m, respectively. An increase in resistivity with an expanding array is caused by the increasing influence of adjacent rock. The resistivity values measured in the hole using the 3.0480-m array are about 5 times greater than those reported by McGinnis et al. [1973] using surface techniques. The reason for the difference is not known.

The resistivity curve is clearly dependent upon the geology, with the dolerite sill containing the highest values, up to 244 ohm-m. Immediately below the casing, resistivity has a minimum value of less than 30 ohm-m, probably owing to an accumulation of salts. In the migmatites below the sill the resistivities are intermediate in value, the lowest resistivity being measured in the basement rock at the bottom of the hole. The relatively high resistivity values for rocks saturated with brines over 200 ppt may be explained by the fact that they are near the freezing point. From the apparent resistivity curve, it may be inferred that the zone of lowest brine concentrations is in the dolerite sill. Low resistivities at the bottom of the hole may be explained by the sinking of heavier, more concentrated brines down into the open hole and through fractures in the migmatite.

SUMMARY AND CONCLUSIONS

The various geophysical logs reveal a geological history that is not immediately apparent from the geological description of the core. On Ross Island the younger volcanic rocks contain twice as much radioactive material as the older, more deeply buried rocks. The same pattern is evident in the basement rocks in the dry valleys, the older gneisses and granite gneisses being less radioactive than the younger granites. Although it might be inferred that declining radioactivity with age denotes radioelement loss through weathering or

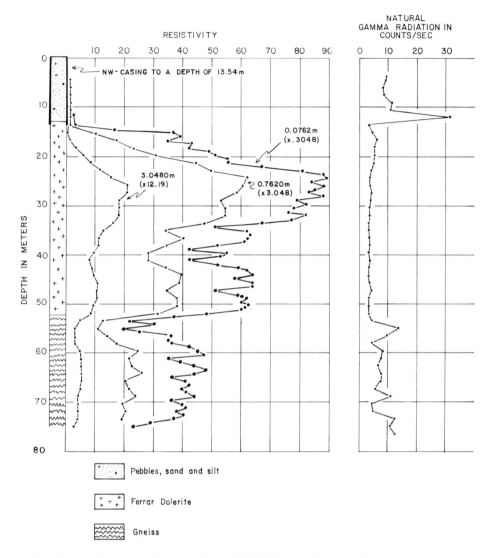

Fig. 8. Resistivity and natural radioactivity logs of DVDP 13 in the Don Juan Pond basin. The number in parentheses should be multiplied by the numbers on the horizontal scale to get resistivity in ohm-meters. The numbers 0.0762, 0.7620, and 3.0480 indicate the distance in meters between the potential and the current electrodes. The 0.0762-m readings are a measure of primarily the resistivity of the fluid in the borehole, whereas the 3.0480-m readings are primarily 'formation' resistivity.

leaching, isotopic ratios indicate that it is more probable that the relative abundances are present near the original amounts and proportions. From the large variations in the gamma ray intensity in the Olympus granite gneiss in hole 6 it is probable that the gneiss reflects a metamorphosed sedimentary sequence rather than a more homogeneous, metamorphosed granite. On Ross Island the more deeply buried mafic flows and pyroclastics have the lowest radioactivities.

In general, the gamma ray logs in sedimentary strata in the dry valleys display lower natural gamma radiation than the rocks from which they were derived. The coarse clastic sediments are especially low, being only slightly higher than the tholeiitic Ferrar dolerites. The more marine fraction of the sediments and the diamictites are about equal radiometrically. It is presumed that the clay fraction, through chemical or mechanical processes, is enriched in radioelements, while the coarse clastics and weathered zones in crystalline basement are depleted.

The salinity of water extracted from permafrost sediments reveals a varied hydrologic history in the dry valleys over short vertical and horizontal intervals. With pore ice salinities ranging from less than 1 ppt to over 200 ppt in the dry valleys it is clear that the single most important variable controlling the physical and biological properties of sediments in permafrost is the total salt content. Therefore before a permafrost terrane can be modeled and its response to climatic change and thermal regimen be predicted, some knowledge of the paleohydrology of the region must be ascertained. It is the paleohydrogeology that ultimately controls the distribution and redistribution of salts in permafrost. Depending on the salt concentrations and temperature, permafrost can be highly permeable as at Don Juan Pond or impermeable over millions of years as in the eastern regions of Taylor Valley. In the correct geological setting, saline permafrost of high permeability peripheral to ice sheets could permit the circulation of subglacial groundwater on a continentwide scale.

Acknowledgments. This study was supported by the National Science Foundation, Division of Polar Programs under contract C-642. The authors express their appreciation to the U.S. Navy VXE-6 Helicopter Squadron for repeated transport to drill sites.

REFERENCES

Chapman-Smith, M., Geologic log of DVDP 14, North Fork basin, *Dry Val. Drilling Proj. Bull.*, *5*, 94-99, 1975.

Goldich, S. S., S. B. Treves, N. H. Suhr, and J. S. Stuckless, Geochemistry of the Cenozoic volcanic rocks of Ross Island and vicinity, Antarctica, *J. Geol.*, *83*, 415-435, 1975.

Keys, U. S., and L. M. MacCary, Application of borehole geophysics to water-resources investigations, in *Techniques of Water-Resources Investigations of the United States Geological Survey*, U.S. Geol. Surv. Collect. of Environ. Data, Book 2, chap. E-1, 126 pp., U.S. Geological Survey, Reston, Va., 1971.

Kurasawa, H., Y. Yoshida, and M. G. Mudrey, Jr., Geologic log of the Lake Vida core—DVDP 6, *Dry Val. Drilling Proj. Bull.*, *3*, 92-108, 1974.

Kyle, P. R., and P. C. Rankin, Rare earth element geochemistry of Late Cenozoic alkaline lavas of the McMurdo volcanic group, Antarctica, *Geochim. Cosmochim. Acta*, *40*, 1497-1507, 1976.

Manheim, F. T., A hydraulic squeezer for obtaining interstitial water from consolidated and unconsolidated sediments, *U.S. Geol. Surv. Prof. Pap.*, *550-C*, c256-261, 1966.

McGinnis, L. D., Artesian well at Don Juan Pond, *Antarct. J. U.S.*, *14*, 26-27, 1979.

McGinnis, L. D., K. Nakao, and C. C. Clark, Geophysical identification of frozen and unfrozen ground, Antarctica, in *Permafrost, 2nd International Conference, North American Contribution*, pp. 137-146, National Academy of Sciences, Washington, D. C., 1973.

McGinnis, L. D., D. R. Osby, and F. A. Kohout, Paleohydrology inferred from salinity measurements on dry valley drilling project (DVDP) core in Taylor Valley, Antarctica, in *Antarctic Geophysics*, edited by C. Craddock, University of Wisconsin Press, Madison, in press, 1980.

Mudrey, M. G., Jr., T. Torii, and H. Harris, Geology of DVDP 13—Don Juan Pond, *Dry Val. Drilling Proj. Bull.*, *5*, 78-93, 1975.

Rogers, J. J. W., and J. A. S. Adams, Uranium, in *Handbook of Geochemistry*, vol. 2, edited by K. W. Wedepohl, pp. 92-B–92-O, Springer, New York, 1969a.

Rogers, J. J. W., and J. A. S. Adams, Thorium, in *Handbook of Geochemistry*, vol. 2, edited by K. W. Wedepohl, pp. 90-l-90-O, Springer, New York, 1969b.

Stuckless, J. S., and C. P. Ferreira, Labile uranium in granite rocks, in *International Symposium on Exploration of Uranium Ore Deposits, Proceedings Symposium*, pp. 717-730, International Atomic Energy Agency, Vienna, 1976.

Stuckless, J. S., and I. T. Nkomo, Uranium-lead isotope systematics in uraniferous alkali-rich granites from the Granite Mountains, Wyoming; Implications for uranium source rocks, *Econ. Geol.*, *73*, 427–441, 1978.

Tilling, R. I., D. Gottfried, and F. C. W. Dodge, Radiogenic heat production of contrasting magma series: Bearings on interpretation of heat flow, *Geol. Soc. Amer. Bull.*, *81*, 1447-1462, 1970.

Titayeva, N. A., V. A. Filonov, V. Y. Ovchenkov, T. I. Veksler, A. V. Orlova, and A. S. Tyrina, Behavior of uranium and thorium isotopes in crystalline rocks and surface waters in a cold wet climate, *Geochim. Int.*, *10*, 1146-1151, 1973.

Treves, S. B., and P. R. Kyle, Geology of DVDP 1 and 2 Hut Point Peninsula, Ross Island, Antarctica, *Dry Val. Drilling Proj. Bull.*, *2*, 11-82, 1973.

MAGNETIC STRATIGRAPHY OF LATE CENOZOIC GLACIOGENIC SEDIMENTS FROM DRILL CORES, TAYLOR VALLEY, TRANSANTARCTIC MOUNTAINS, ANTARCTICA

MICHAEL E. PURUCKER,[1] DONALD P. ELSTON, AND STEPHEN L. BRESSLER

U. S. Geological Survey, Flagstaff, Arizona 86001

Frozen glaciogenic sediments, deposited in a former fjord, were cored to depths as great as 325 m in ice-free Taylor Valley as part of the National Science Foundation Dry Valley Drilling Project. Holes 8 and 10 were drilled at the edge of the Ross Sea, hole 11 was drilled 3 km inland to the west, and hole 12 was drilled 16 km inland. About 750 samples, spaced at approximately 1-m intervals and oriented only with respect to the up direction, were obtained from material ranging from unstratified diamictite matrix to finely laminated sandstone and siltstone. The Koenigsberger ratio (remanent magnetization/induced magnetization) was used to discard material of low magnetic stability. Removal of material with Koenigsberger ratios less than 0.1 resulted in a significant refinement in the definition of polarity zones and allowed a refined temporal correlation between the sections in DVDP holes 8, 10, and 11. Intervals of increased susceptibility reflect increased influx of titanomagnetite-bearing sediment derived from the McMurdo Volcanic Group, subaerially erupted to the east in the area of McMurdo Sound. One of these susceptibility zones is time-transgressive in relation to the polarity zonation, perhaps because deposition of till from a grounded ice sheet at site 11 continued after deposition of till had ceased at sites 8 and 10. The finding of magnetite as the principal magnetic mineral in core 12 and in the lowermost susceptibility zones in cores 8, 10, and 11 indicates a different source area for this sediment. Magnetite in core 12 was derived from the Ferrar Dolerite, presumably from bedrock in western Taylor Valley. Titanohematite-bearing volcanic fragments in core 12 were erupted from a local source in Taylor Valley at a time when it was covered by ice.

INTRODUCTION

The dry valleys of the McMurdo Sound region, Transantarctic Mountains, though now free of throughflowing glaciers, display evidence of earlier glaciations. The limited and complex nature of the evidence at the surface led to the formation of the Dry Valley Drilling Project (DVDP), undertaken jointly by scientists of the United States, New Zealand, and Japan, to investigate the surficial and subsurface glacial deposits [*McGinnis et al.*, 1972, 1975]. The drill program succeeded in obtaining several cores, to depths as great as 325 m, of sediments underlying Taylor Valley (Figure 1), one of the dry valleys.

PREVIOUS WORK

Early paleomagnetic studies in southern Victoria Land [*Bull and Irving*, 1960; *Bull et al.*, 1962] con-centrated on the Paleozoic and Mesozoic basement complex. Paleomagnetic studies of basalt and associated extrusive rocks of the late Cenozoic Mc-Murdo Volcanic group of *Nathan and Schulte* [1968] by *Cox* [1966] and *McMahon and Spall* [1974] revealed the presence of both normally and reversely magnetized material. *Pederson et al.* [1979] measured the magnetic susceptibilities of many of the common rock types present in southern Victoria Land and determined that basalts of the McMurdo Volcanic Group had by far the highest magnetic susceptibilities. *Park and Barrett* [1975] first studied DVDP cores 11 and 10 for magnetic inclination and proposed that the sediments of hole 11 were deposited within the past 4 m.y.

The first samples for this study were taken from DVDP 11 by H. Spall and D. P. Elston in 1976. Measurement revealed the existence of several polarity reversals that led to a second sampling by D. P. Elston and S. L. Bressler in 1977. At that time, samples were collected from DVDP 8, 9, 10, and 12 as well as from DVDP 11. Preliminary results from

---

[1] Now at the Department of Geological and Geophysical Sciences, Princeton University, Princeton, New Jersey 08544.

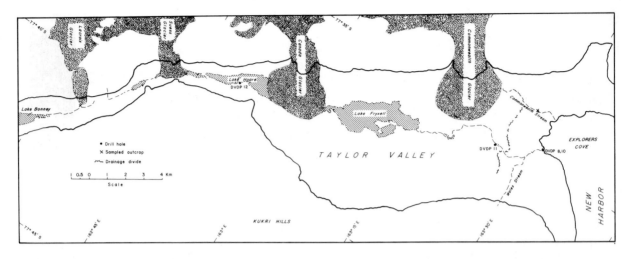

Fig. 1.   Index map showing location of Taylor Valley and drill sites for DVDP cores 8, 9, 10, 11, and 12.

these samples were reported by *Spall and Elston* [1977]. Polarity zonations for the DVDP cores from Taylor Valley, using results from all samples, were reported by *Elston et al.* [1979]. Purucker subsequently carried out a study of the magnetic properties and susceptibilities of the various sediments that resulted in the recognition of unstably magnetized samples and a means for their rejection. The polarity zonations reported here include only data points not rejected by the Koenigsberger, or Q, ratio technique; as a result, they differ in some detail from those reported by *Elston et al.* [1979]. The brief discussion given here to provide a base for evaluation of the magnetic properties of the sediment contains no new sedimentologic or geologic data.

GEOLOGIC SETTING

Taylor Valley (Figure 1) is one of the major ice-free valleys in the McMurdo Sound region of southern Victoria Land. The walls of the valley are mainly of Precambrian and early Paleozoic metasedimentary and granitic rocks [*Haskell et al.*, 1965]. Above this basement complex the Paleozoic Beacon Supergroup crops out in western Taylor Valley, 22 km or more west of New Harbor. Sheets of the Jurassic Ferrar Dolerite cut both the basement and the Beacon. Small outcrops of basalt assigned to the late Cenozoic McMurdo Volcanic Group [*Armstrong*, 1978] are present on the flanks of central and western Taylor Valley. This volcanic assemblage occurs extensively north, east, and

south of Taylor Valley. The floor of much of Taylor Valley is mantled with glacial and fluvioglacial sediments. Drill cores have revealed that interbedded diamictite, sandstone, conglomerate, and mudstone underlie the valley floor. Much of the sediment, to a depth of about 190 m in hole 11, is cemented by ice. The work of *Denton and Armstrong* [1968], *Denton et al.* [1970, 1971], *Webb and Wrenn* [1976, 1977], *McKelvey* [1977, 1979], and *Porter and Beget* [this volume] shows that Taylor Valley was the site of deposition of glaciomarine sediments in a fjord channel, overlain by fjord tillites, strandline sediments, and fluvioglacial deposits associated with advances of ice from the Ross Sea, the East Antarctic ice sheet, and local glaciers. From counts of lithic fragments and microfabric analysis, *Porter and Beget* [this volume] found evidence for both westerly land-based and easterly marine-based sources of detritus and therein for both land- and marine-based glaciation.

SAMPLING AND LABORATORY PROCEDURES

The cores studied are currently stored in a frozen state at the Antarctic Core Storage Facility at Florida State University, Tallahassee, where sampling for this study was carried out. The finest grained material was sampled at intervals of 1 m to less than ⅓ m. Core samples were obtained by using a drill press equipped with a diamond-tipped core barrel. Compressed air was used as a drilling medium. After drilling, the uphole directions of the individual samples were marked by using a non-

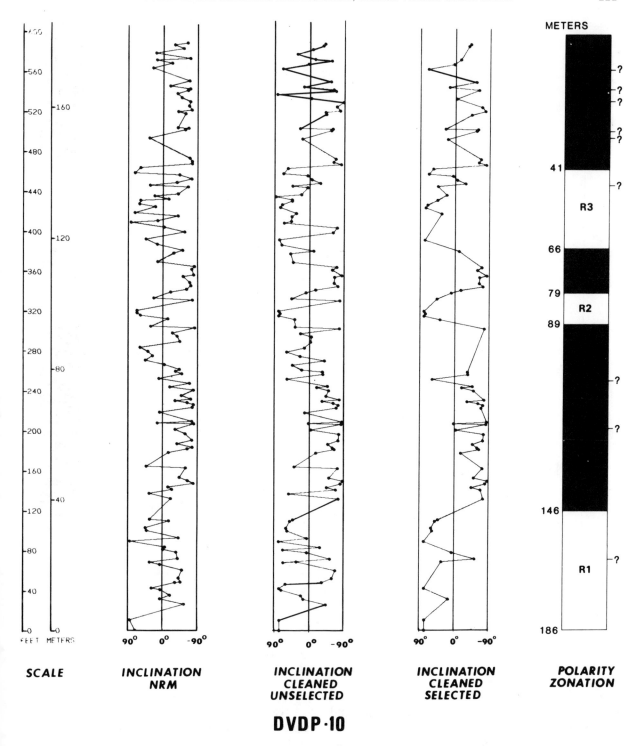

DVDP·10

Fig. 2a.   Magnetic directions (inclination only) for material from hole 10. Scale on right-hand column shows the distance from top (collar) of hole to polarity boundaries. Scale on the left shows distance from the lowest sample in feet and meters. Samples were subjected to peak fields of 100 to 150 Oe during partial alternating field demagnetization; results appear in the column labeled 'cleaned.' Criteria for data selection are discussed in text. Reverse polarity zones are identified numerically and in stratigraphic order for each core, beginning at the base of DVDP 10 and proceeding through DVDP 8, 11, and 12. Black denotes normal polarity intervals, and white reverse polarity intervals. Question marks indicate narrow intervals of uncertain magnetic polarity. NRM is natural remanent magnetization.

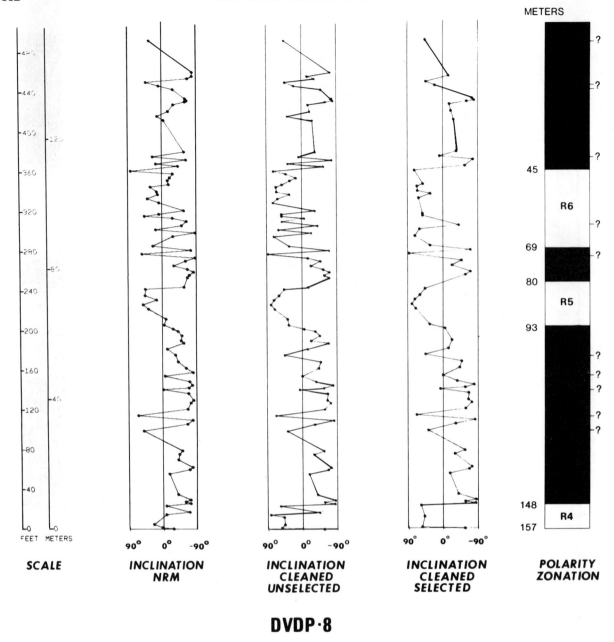

DVDP·8

Fig. 2b.   Same as Figure 2a for DVDP 8.

magnetic scribe. The samples were then trimmed with a diamond saw to nearly cubical shapes and placed in plastic boxes. Stratigraphic separations were measured to the nearest centimeter, and the location and depth of the samples marked on reference photographs of the parent cores. Because the parent drill cores are unoriented with respect to azimuth, only the remanent magnetic inclination could be determined. The high southern latitude of Antarctica results in inclinations that are either steeply negative (normal polarity) or steeply positive (reversed polarity). The present magnetic inclination and the inclination predicted from an axial dipole are essentially identical at $-83°$. Before and during the measuring process the samples were stored in a low magnetic field to allow viscous magnetizations to decay and to prevent the acquisition of viscous magnetizations during demagnetization analysis. Samples were kept frozen until the magnetic measurements were completed.

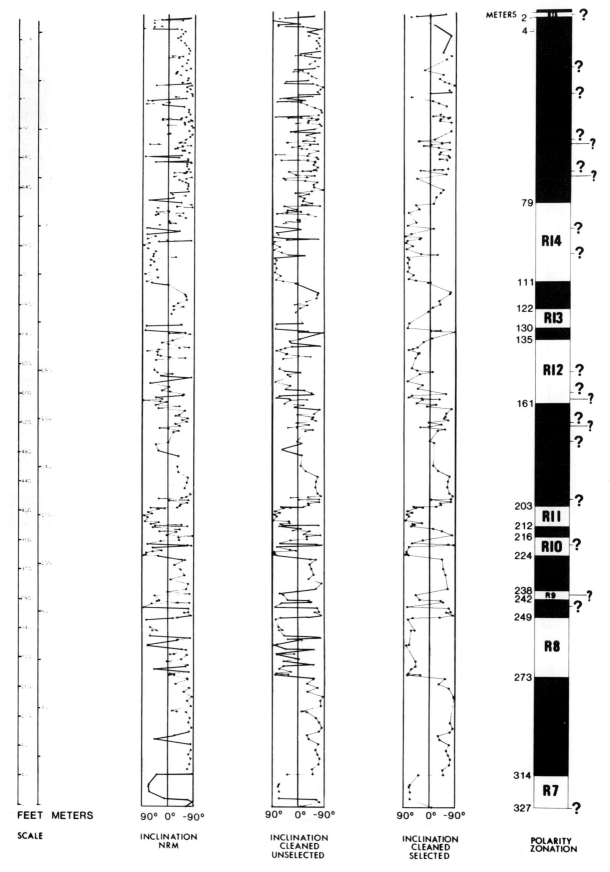

Fig. 2c.   Same as Figure 2a for DVDP 11.

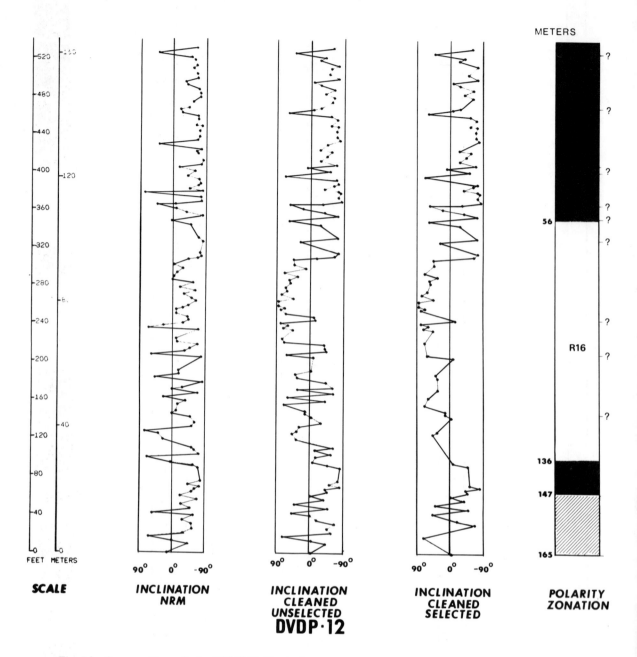

Fig. 2*d*.   Same as Figure 2*a* for DVDP 12. Hatched pattern denotes intervals of mixed or uncertain polarity.

A total of 749 samples from five drill cores (DVDP 8–12) were obtained for this study. The short length of DVDP 9 (36 m) and an undiagnostic magnetic signature made its paleomagnetic correlation with the other cores highly uncertain, and results from it are not considered further. A total of 99 samples were collected from the DVDP 8 core, 140 samples from DVDP 10, 352 samples from DVDP 11, and 141 samples from DVDP 12.

Nearly all samples were measured at least 3 times, and some samples were measured more than 10 times during progressive alternating field demagnetization analysis.

The natural remanent magnetization (NRM) of the samples was measured at the Paleomagnetics Laboratory of the U. S. Geological Survey at Flagstaff, Arizona, by using a superconducting cryogenic magnetometer as described by *Goree and*

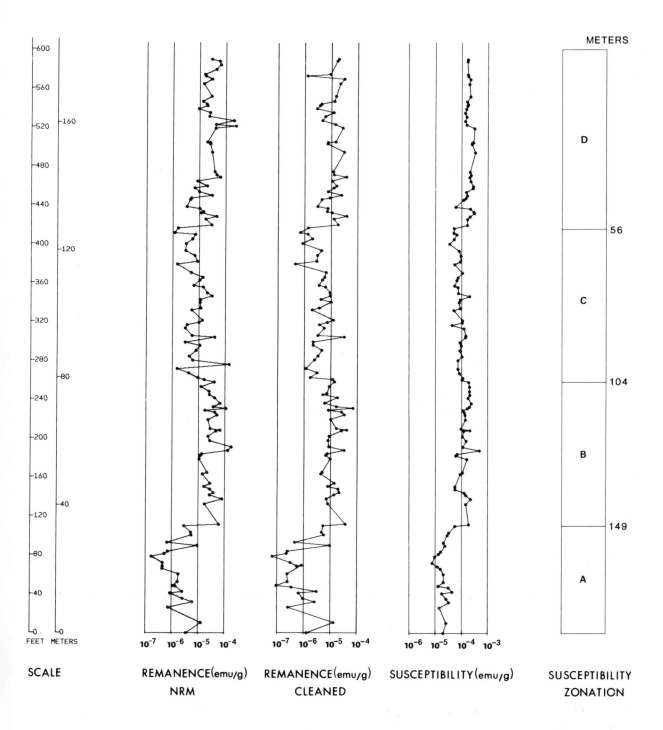

DVDP·10

Fig. 3a.   Magnetic intensities (remanence and susceptibility) for material from hole 10. The scale on the right-hand side shows the distance from the top (collar) of hole to susceptibility boundaries; emu/g is electromagnetic units per gram. See the caption for Figure 2a for an explanation of other abbreviations and conventions.

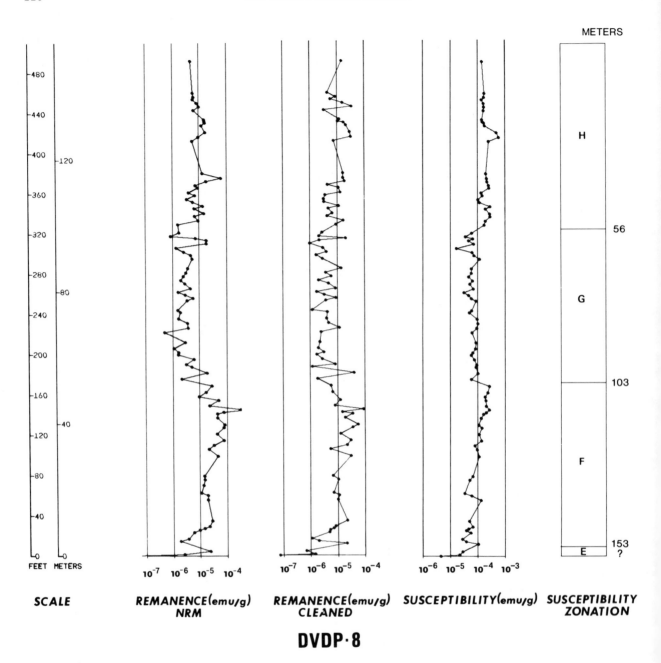

DVDP·8

Fig. 3b.   Same as Figure 3a for DVDP 8.

Fuller [1976]. All measurements were made in a near-zero field and analyzed simultaneously by computer. The NRM directions and intensities before partial demagnetization are shown in Figures 2 and 3, respectively.

Partial alternating field (af) demagnetization was carried out on all samples by using an af demagnetizer of commercial design, in part at the Paleomagnetics Laboratory of the University of Ari-

zona. Progressive af demagnetization was carried out on about 5% of the samples representing various lithologic types. Peak fields of 25 to 500 Oe were used. The peak field that most efficiently cleaned the secondary components without decreasing the intensity of magnetization below easily measurable levels or producing erratic directional changes was used to demagnetize the remaining samples. Peak fields of 100 to 150 Oe

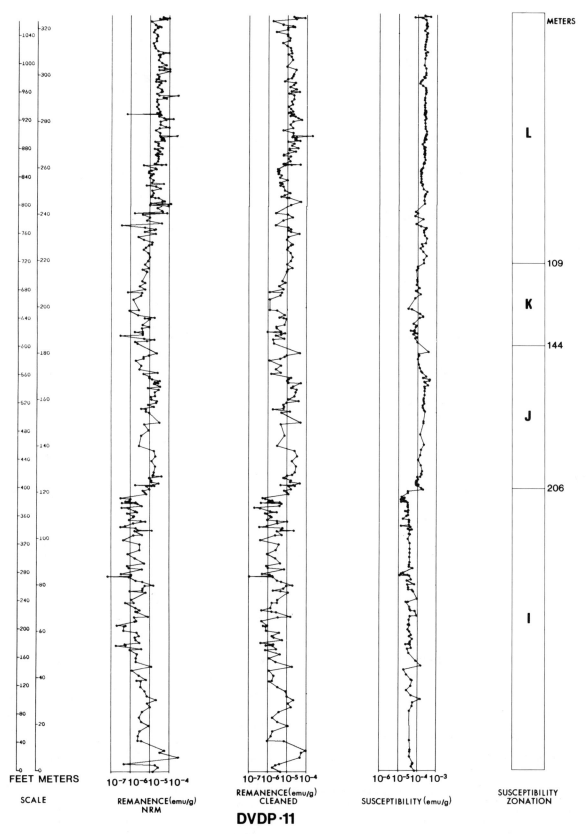

FEET METERS
SCALE

REMANENCE(emu/g)
NRM

REMANENCE(emu/g)
CLEANED

SUSCEPTIBILITY (emu/g)

SUSCEPTIBILITY
ZONATION

DVDP·11

Fig. 3c.   Same as Figure 3a for DVDP 11.

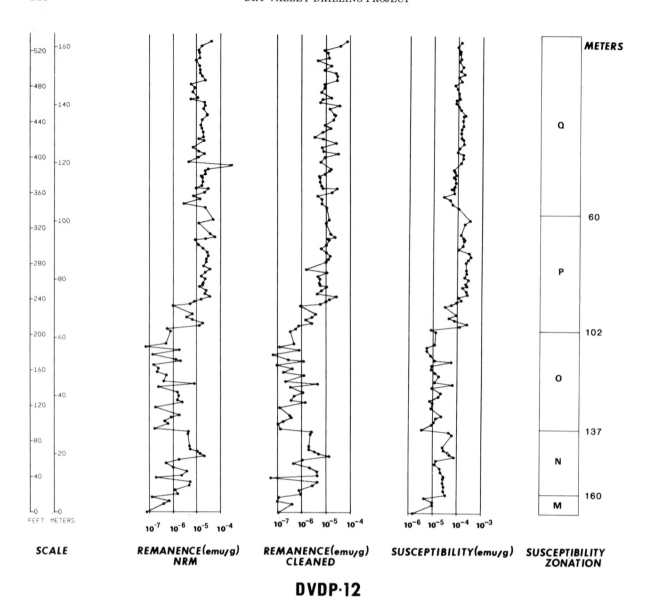

DVDP·12

Fig. 3d.   Same as Figure 3b for DVDP 12.

were found to produce optimum cleaning, typically resulting in a decrease in intensity of 10 to 50%. Directions and remanent intensities for the samples, following partial af cleaning, are shown in Figures 2 and 3, respectively.

The low-field bulk magnetic susceptibility of the sediments was measured by using the two-axis superconducting magnetometer in Flagstaff. A field of approximately 0.5 Oe was applied along the vertical axis, while the horizontal axis was maintained in an essentially field-free condition. Measurements were made along three mutually orthogonal

axes of each sample, and the remanent and induced components of magnetization then calculated for each of the three axes. Magnetic susceptibility, measured after partial af demagnetization, is shown in Figure 3. Magnetic remanence values obtained as a consequence of this procedure are not reported here but are in close agreement with the remanence measured previously in a near-zero field.

Magnetic mineralogy was investigated by determination of the Curie temperature and by the examination of selected magnetic separates in pol-

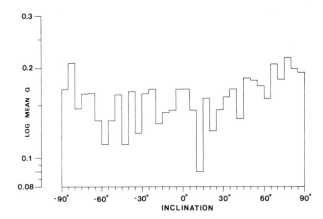

Fig. 4. Log mean Q (remanent magnetization/induced magnetization) as a function of magnetic inclination. Q has been calculated for 5° inclination increments. Samples which contained large magnetic clasts were omitted from the calculations.

ished thin sections. Curie temperature determinations were carried out in both oxidizing and partially reducing conditions. Mineralogic investigations were carried out at the Rock Magnetics Laboratory of the U.S. Geological Survey in Denver, Colorado.

SELECTION CRITERIA

The diamictons, abundant throughout the section, show a wide range of grain size. They are poor paleomagnetic recorders of the ancient magnetic field because both gravity and magnetic forces act on the magnetic particles during the process of deposition. When magnetic particles are large, gravitational forces dominate over magnetic forces, preventing the magnetic particles from orienting themselves closely with the earth's magnetic field. Magnetic particles larger than fine sand size are poor recorders of the magnetic field. A technique is needed to identify and objectively remove such samples from the data set. The method employed in this study relies on the ratio of the remanent to the induced magnetization, typically called the Koenigsberger ratio, or simply Q. If the effects of grain size are disregarded, a large Q (NRM values greater than 0.5) is indicative of high magnetic stability and small grain size of the magnetic particles [Stacey and Banerjee, 1974, pp. 115–116]. Conversely, a small Q is indicative of low magnetic stability and large grain size. An assemblage of very small magnetic particles will lie in directions such that their magnetic directions tend to parallel the earth's magnetic field during deposi-

tion. If larger magnetic particles are introduced, they will generally not cause an increase in the remanent intensity because they will be randomly oriented with respect to the ambient magnetic field. The quality of grouping of the magnetic particles within a sample can be assessed by normalizing the remanent intensity with respect to the induced intensity, considered to be a measure of the total amount of magnetic material present. The largest Q values are associated with steep positive or negative inclinations (Figure 4). Because partial af demagnetization commonly reduces the remanent magnetization by 10 to 50% without affecting

TABLE 1. Mean Magnetic Inclination by Polarity Zone

Zone	NRM	Cleaned—Unselected	Cleaned—Selected
		DVDP 10	
R1	1(43)	25(49)	43(38)
N1	−35(40)	−32(44)	−53(32)
R2	16(47)	36(49)	54(26)
N2	−57(28)	−36(46)	−56(25)
R3	8(49)	33(44)	38(33)
N3	−45(33)	−29(46)	−32(41)
		DVDP 8	
R4	−15(30)	26(50)	25(47)
N4	−48(35)	−39(38)	−37(40)
R5	25(24)	56(17)	60(16)
N5	−55(39)	−35(46)	−31(55)
R6	−6(41)	36(35)	48(30)
N6	−28(40)	−23(43)	−26(41)
		DVDP 11	
R7	2(62)	20(57)	49(29)
N7	−62(30)	−58(19)	−62(17)
R8	−8(51)	37(44)	63(14)
N8	−26(58)	−31(52)	−41(48)
R9	−1(54)	49(11)	42(7.1)
N9	−43(24)	−49(21)	−53(7.5)
R10	30(54)	33(57)	37(62)
N10	−29(34)	−28(27)	−44(28)
R11	36(42)	48(37)	59(18)
N11	−33(34)	−31(34)	−37(31)
R12	9(43)	22(41)	36(33)
N12	−40(31)	−34(34)	−34(34)
R13	23(55)	35(41)	52(32)
N13	−36(25)	−30(32)	−30(29)
R14	25(44)	38(43)	57(29)
N14	−43(39)	−32(41)	−36(35)
R15	38(39)	53(28)	56(9.6)
N15	−26(50)	−32(25)	−32(25)
		DVDP 12	
N16	−49(37)	−54(25)	−46(23)
R16	−27(41)	16(49)	32(45)
N17	−59(17)	−42(39)	−40(40)

Numbers in parentheses represent standard deviation of each mean inclination value.

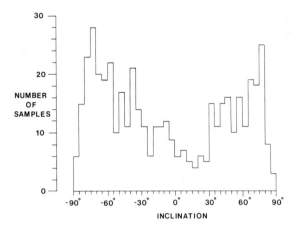

Fig. 5. Histogram showing distribution of remanent inclination for selected data after partial demagnetization of samples from cores 8, 10, 11, and 12.

the induced magnetization, Q values will be smaller when measured after af demagnetization, and therefore values of Q less than 0.5 can still indicate the presence of a stable magnetization. Samples with a Q less than 0.1 are associated with large angular dispersions and therefore were removed from the inclination data set (Figure 2). They were not removed from the susceptibility or remanence data set (Figure 3). The 0.1 cutoff point is arbitrary, but it works well in practice.

We have observed that very large values of Q (greater than 1) in sediments can indicate the presence of a single stably magnetized clast within a sample. Such magnetic clasts do not record the magnetic inclination at the time of deposition. A large value of Q could indicate the presence of needles of magnetite that would not orient themselves closely with the ambient magnetic field. Samples that were suspected, on the basis of magnetic properties or on physical examination, to contain a magnetic fraction of unusable size were disaggregated, and the magnetization of any small clast was determined separately. If an individual clast contained a significant part of the magnetization of the parent sample, the sample was discarded from the selected inclination, remanence, and susceptibility plots (Figures 2 and 3). Magnetic particles that have large values of Q because of their shape will still remain undetected. In all, 33% of the samples were rejected, 27% because of their low Q ratio and 6% because they contained large magnetic clasts. For the remaining samples the directions and intensities of remanence and the bulk magnetic susceptibilities are shown in Figures 2 and 3.

MAGNETIC INCLINATION AND POLARITY ZONATION

Establishment of a polarity zonation in each DVDP core ideally could make possible both correlations between cores and correlations with the polarity time scale. This paper attempts correlation between cores. Relation of the polarity zones to the geologic framework and possible correlations with the polarity time scale are the subject of a paper by *Elston and Bressler* [this volume].

The principal polarity intervals, or polarity zones, are identified numerically and in stratigraphic order for each core, beginning at the base of DVDP 10 and proceeding through DVDP 8, 11, and 12 (Figure 2). Individual data points that have inclinations of polarity opposite that of the enclosing polarity zone are indicated by question marks. Some of these may record real polarity reversals. Narrow intervals of opposite polarity defined by two or three adjacent samples that exhibit steep inclinations are shown as brief, but questionable, reversals. The polarity zonation in DVDP 12 is considered provisional. Because of the large

TABLE 2.   Log Mean Magnetic Susceptibility by Zone

Zone*	Susceptibility, emu/g	SD + †	SD − †
	DVDP 10		
A	2.08E-05‡	3.38E-05	1.28E-05
B	1.40E-04	2.25E-04	8.76E-05
C	8.10E-05	1.17E-04	5.62E-05
D	1.91E-04	2.65E-04	1.38E-04
	DVDP 8		
E	1.44E-05	3.84E-05	5.38E-06
F	9.92E-05	1.93E-04	5.10E-05
G	6.53E-05	9.54E-05	4.46E-05
H	1.99E-04	2.91E-04	1.36E-04
	DVDP 11		
I	3.64E-05	5.73E-05	2.32E-05
J	1.68E-04	2.57E-04	1.10E-04
K	8.69E-05	1.30E-04	5.83E-05
L	2.17E-04	2.92E-04	1.61E-04
	DVDP 12		
M	5.10E-06	1.22E-05	2.13E-06
N	3.06E-05	4.97E-05	1.88E-05
O	1.19E-05	2.28E-05	6.21E-06
P	1.64E-04	2.97E-04	9.11E-05
Q	1.06E-04	1.60E-04	7.09E-05

* There is standard deviation overlap between zones B and C, F and G, J and K, N and O, and P and Q.

† SD + and SD − are the high and low range of susceptibility obtained by respectively adding and subtracting the standard deviation from the mean susceptibility.

‡ Read 2.08E-05 as 2.08×10^{-5}.

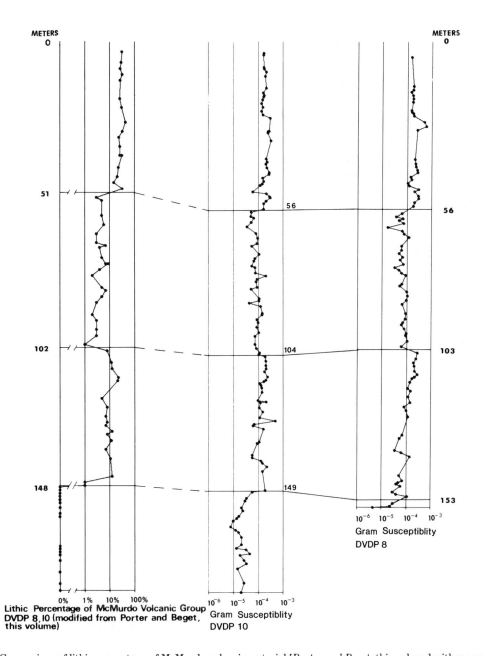

Fig. 6. Comparison of lithic percentage of McMurdo volcanic material [*Porter and Beget*, this volume] with magnetic susceptibility records. Numbers refer to depth in meters from top of core.

present-field overprints on these samples, the direction of movement during partial af demagnetization has been used in the definition of polarity intervals. Diffuse reversal boundaries such as the one that occurs at the base of R12 (161 m) in DVDP 11 may represent a time delay in the acquisition of a stable magnetization, and the delay may be related to the formation of ice in the sediment. Between the time of deposition of water-saturated sediments and the time the water freezes to form an ice cement, some magnetic particles may be able to rotate into the ambient field. If the polarity of the field reverses before the water turns to ice, the original magnetic record will be modified, resulting in a diffuse record of the field reversal. When this model is used, the position of a polarity switch is drawn at the top of the diffuse interval.

The improved definition of polarity zones can be observed by comparing plots of the inclination before and after selection on the basis of the Q ratio

(Figure 2). This selection procedure eliminates an interval of reversal polarity that had been proposed earlier for deposits in the upper 20 m of DVDP 10 and 8 [*Elston et al.*, 1979]. Many samples in this interval displayed an apparent reverse polarity. Because nearly all samples in this interval have now been rejected on the basis of their Q ratios, normal polarity has been assigned to the upper 20 m of DVDP 10 and 8, well documented to be of Holocene age. The great scatter in inclinations is possibly attributable to fluviatile deposition and reworking in the modern (Holocene) delta on which holes 10 and 8 are located, less than 5800 years old on the basis of surface geology, fauna, flora, and isotopic ages [*Chapman-Smith*, 1979; *Brady*, 1978; *Stuiver et al.*, 1976; J. H. Wrenn, oral communication, 1978]. The Q ratio selection procedure resolves a major discrepancy between an early interpretation of the paleomagnetic data and the ages assigned to the beds from other methods.

Mean magnetic inclinations (Table 1) have been calculated for the individual polarity zones shown in Figure 2. Reverse polarity inclinations can be seen to steepen after partial af demagnetization and Q ratio selection, reflecting removal of present field components of magnetization.

A histogram of inclinations (Figure 5) compiled from the selected inclination data shows strong bimodal peaks for the intervals $-70°$ to $-75°$ and $75°$ to $80°$. A bimodal distribution with peaks nearly equally spaced about $0°$, the one-dimensional analog of an antiparallel reversal test, indicates that the stable magnetization was acquired during or shortly after deposition. The difference ($3°-13°$) in inclination between the present value and the histogram peaks might be explained in at least three ways. One explanation rests on the observation that the area on a sphere occupied by near-polar inclination increments is considerably less than that occupied by near-equatorial inclination increments. This has the effect of decreasing the number of samples that will exhibit steep inclinations and consequently compresses the angular separation of the bimodal distribution (D. V. Kent, oral communication, 1979). A second explanation appeals to a depositional inclination error. Magnetic grains may roll as they encounter the sediment-water interface. Rolling of grains, if uncorrected by postdepositional rotations, results in inclinations that are statistically shallower than the steep applied field. A third possibility is that Taylor Valley was in lower latitudes at the time the sediments were deposited. However, since no systematic changes in inclination with depth have been

detected in the mean inclinations calculated in Table 1, continental drift was not important for the interval of time studied.

Finally, the close resemblance of the polarity zonations for DVDP 8 and 10 is an important indicator of reproducibility and hence of magnetic stability of the sediments. A lack of detailed lithologic similarity between DVDP 8 and 10 (see drill logs of *McKelvey* [1975]) has hindered previous correlations and made lithologic correlations to DVDP 11 uncertain if not impossible.

MAGNETIC SUSCEPTIBILITY

Magnetic susceptibility, normalized to account for differences in mass between samples, is an indicator of the kind and amount of magnetic material present. If the kind, amount, and grain size of magnetized material remain constant across a stratigraphic section, differences in remanent intensity could be interpreted as reflecting changes in the intensity of the earth's magnetic field at the time of deposition. Conversely, if it is assumed that the intensity of the earth's magnetic field has been more or less constant during deposition, abrupt changes in remanent intensity (for example, at 206 m in DVDP 11, 149 m in DVDP 10, and 153 m in DVDP 8, Figure 3) could be interpreted to reflect changes in the kinds and amounts of magnetic materials and hence changes in provenance. This second interpretation can be tested by comparing remanent intensity and susceptibility data. Changes in remanent intensity that correlate with changes in susceptibility are most simply interpreted as reflecting changes in the kind and amount of magnetic material present.

Magnetic susceptibilities were measured for all samples for the results shown in Figure 3. On the basis of the amplitude of the magnetic susceptibility, four susceptibility zones are recognized in sections penetrated by holes 8, 10, and 11 and five susceptibility zones in the section penetrated by hole 12. Susceptibility zones in each core have been given individual letter designations in an ascending stratigraphic order following the ordering of the polarity zonations. Mean magnetic susceptibilities and standard deviations for each zone (for selected data only) are summarized in Table 2.

The close correlation of susceptibility zones between neighboring holes 10 and 8 (Figure 6) indicates a stratigraphic reproducibility for the magnetic data despite a lack of detailed lithologic similarity (see drill logs by *McKelvey* [1975]). The only strongly magnetic materials present are basaltic and trachytic rocks of the McMurdo Volcanic Group. The percentage of these lithic constituents across the section at New Harbor (DVDP 8, 9, and 10) has been reported by Porter and Beget [this volume]. The close resemblance of their results plotted with respect to magnetic susceptibility (Figure 6) indicates that McMurdo volcanic material dominates the magnetization. Other magnetic minerals from the Ferrar Dolerite, Irizar Granite, and lamprophyre dike rocks contribute to the background susceptibility. *Porter and Beget* [this volume] report that these materials generally make up less than 10% of the material in cores 8, 9, and 10.

MAGNETIC MINERALOGY

Saturation magnetizations and Curie temperatures were determined for specimens taken from eight stratigraphic horizons in DVDP 10, 11, and 12. The specimens were measured in air and in a reducing mixture of hydrogen and nitrogen.

DVDP 11

Magnetic separates (silt size and finer) for a sample at a depth of 161.22 m in DVDP 11 show two distinct Curie temperatures, 150°C and 575°C (Figure 7a). The Curie temperatures and the shape of the heating and cooling curves indicate the presence of two phases of titanomagnetite. Continued thermal cycling raises the Curie temperature of the low-temperature component from 150° to 195°C, apparently reflecting the exsolution and (or) oxidation of Ti to TiO_2. This change has the effect of increasing the Fe/Ti ratio, thereby increasing the saturation magnetization. A magnetic fraction of larger grain size from the same horizon in DVDP 11 probably contains titanomaghemite (Figure 7b), possibly formed as a low-temperature oxidation product of titanomagnetite [*Prévot and Grommé*, 1975]. Another possible interpretation of the shape of this curve is that it represents titanomagnetite in which the titanium exsolved at high temperatures in the Curie balance. An abundance of pebble size material in the magnetic separates at 161.22 m probably accounts for the low magnetic inclination ($-22°$) found at this horizon. A Q of 0.18 at this horizon is consistent with a magnetic fraction dominated by comparatively large grains not oriented with the ambient field at the time of depo-

sition. *Porter and Beget* [this volume] found that samples from near this horizon in DVDP 11 contained 20 to 60% McMurdo volcanic material by lithic percentage. *Goldich et al.* [1975], in a geochemical study of the McMurdo Volcanic Group on Ross Island, found that the prominent rock type was an alkalic olivine basalt characterized by high TiO_2 content (3–4%).

A magnetic separate (silt size and finer) from a sample at 118.47 m in DVDP 11 shows a single Curie temperature of approximately 570°C (Figure 7c), which indicates the presence of a low-titanium magnetite. *Porter and Beget* [this volume] found that samples from this horizon contained, in addition to detritus from McMurdo volcanic material, approximately 2% Ferrar Dolerite. A whole rock sample of the Ferrar Dolerite from Beacon Heights (northern Victoria Land, 77.9°S, 160.6°E) yielded a Curie temperature of approximately 565°-570°C (Figure 7d), nearly identical to the Curie temperature from 118.47 m in DVDP 11. In addition, a vesicular volcanic fragment from 118.47 m in DVDP 11 shows two Curie temperatures, 240°C and 570°C. Close similarity to the shape of the heating and cooling curves of Figure 7a indicates the presence of titanomagnetite. Our results therefore support and complement the results of Porter and Beget.

A magnetic separate (silt size and finer) containing associated quartz from 262.05 m in DVDP 11 shows a single Curie temperature of 575°C (Figure 7e). The only possible ferrimagnetic material identified by *Porter and Beget* [this volume] from this horizon is the Irizar Granite. No material from the Ferrar Dolerite or the McMurdo Volcanic Group was identified within 60 m of this horizon.

DVDP 10

In DVDP 10 a magnetic separate with associated quartz at 174.70 m shows a single Curie temperature of 565°C (Figure 7f), indicative of a low-titanium magnetite. Lamprophyre dike rocks make up as much as 20% of the sediment at this horizon [*Porter and Beget*, this volume]. The only other material identified from this horizon is the Irizar Granite.

DVDP 12

Magnetic separates from 24.12 m, 65.10 m, 124.98 m, and 137.28 m in DVDP 12 show a single Curie temperature of approximately 575°C (Figure 7g), again indicative of a low-titanium magnetite. Examination of polished thin sections confirmed

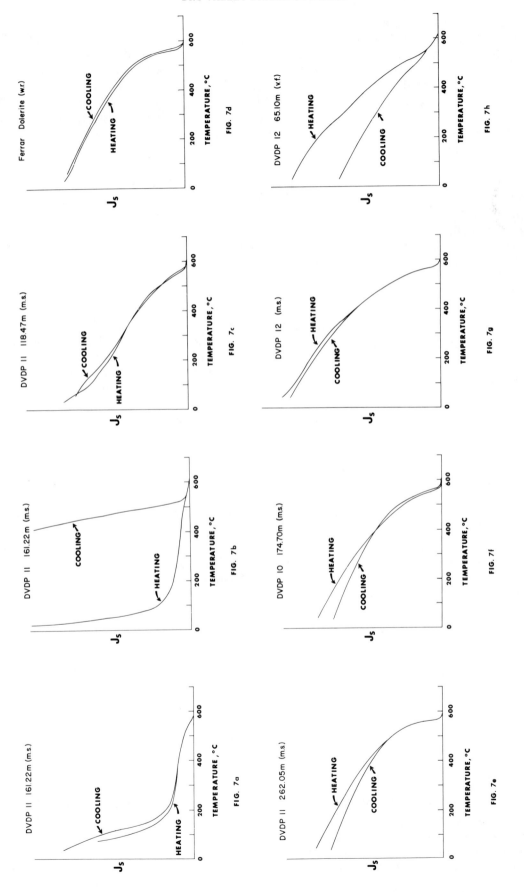

Fig. 7.   Saturation magnetization J_s as a function of temperature for DVDP holes 10, 11, and 12. Amplitude of saturation magnetization arbitrary. A field of 1800 Oe was applied to each sample, sufficient to saturate all common magnetic minerals except hematite. Small amounts of hematite may remain undetected; m.s. is magnetic separate; w.r. is whole rock; v.f. is volcanic fragment.

the existence of magnetite as a major magnetic mineral. Ferrar Dolerite identified in the lithologic logs of DVDP 12 [*Chapman-Smith*, 1975] appears to have been the source of this material. Magnetite is the only magnetic mineral we have identified in the Ferrar Dolerite (Figure 7d). Minor maghemite, a metastable oxidation product of magnetite, appears to be present in the magnetic separates from DVDP 12, a suggestion supported by the examination of polished thin sections of material from 24.12 m and 65.10 m.

Volcanic fragments have been identified in core from DVDP 12 [*Chapman-Smith*, 1975]. Analysis of a fine-grained vesicular volcanic fragment from 65.10 m revealed multiple Curie temperatures that range from approximately 520° to 615°C (Figure 7h). Curie temperatures below 578°C reflect the presence of titanomagnetite and magnetite, whereas the Curie temperature of 615°C probably is produced by a titanohematite. Titanohematite is represented by the chemical formula $Fe_{2-x}Ti_xO_3$, where x varies between 0 and 1. Presumably, hematite-rich components $(0 < x < 0.2)$ are responsible for the Curie temperature of 615°C. Volcanic fragments from this horizon typically are enclosed by oxidation rinds. When the oxidation rind has not been removed before analysis, an additional Curie temperature of approximately 300°C is seen. With removal of the oxidation rind, only a slight suggestion of a 300°C Curie temperature is seen (Figure 7h).

Volcanic fragments examined under the reflecting microscope commonly exhibit primary exsolution textures. A faint yellow-beige tint seen in the fragments possibly reflects oxidizing conditions. Examination of magnetite from the same horizon reveals only a faint oxidation rind, probably maghemite. The primary exsolution textures and the evidence for only minor oxidation indicate that the titanohematite was formed before its incorporation in the sediment. *Hamilton* [1972] proposed that oxidation by steam of lava erupted under ice may be responsible for the formation of reddish scoria commonly seen in volcanic rocks from the Hallett volcanic province, north of the dry valleys. Such a mechanism could account for the formation of primary titanohematite seen in volcanic fragments from the core of DVDP 12. The oxidation of Fe^{2+} to Fe^{3+} can occur either in steam or in air but is not associated with the eruption of submarine flows. The absence of titanohematite from McMurdo volcanic material found in the upper two thirds of the DVDP 11 section suggests that volcanic fragments in DVDP 12 were erupted from a local source in Taylor Valley at a time when Taylor Valley was covered by ice. Rocks from minor volcanic vents identified in Taylor Valley are considered to be part of the McMurdo Volcanic Group [*Armstrong*, 1978]. Our magnetic studies suggest that material from the vents in Taylor Valley can be distinguished from McMurdo volcanic material erupted from under the Ross Sea by the difference in the oxidation state of the iron.

CORRELATION AND DISCUSSION

The polarity zonations for the DVDP cores schematically summarized in Figure 8 provide time-stratigraphic information, whereas the susceptibility zonations provide rock-stratigraphic information. Correlations based on polarity patterns are widely used; magnetic susceptibilities have rarely been used for correlation [*Thompson*, 1973].

The close resemblance of the polarity zonations between DVDP 8 and 10 is an important indicator of reproducibility and thus of magnetic stability. The polarity zonations in DVDP 8 and 10 appear to correlate broadly with the DVDP 11 section at Commonwealth Glacier, as shown in Figure 8. Two polarity zones, between 2 and 4 m and between 122 and 130 m in DVDP 11, are not present in DVDP 8 and 10. These zones may indicate that DVDP 11 contains a more complete record of the magnetic field.

The close resemblance of the susceptibility records for neighboring holes 8 and 10 indicates magnetic reproducibility (Figure 6). Susceptibility zones for the New Harbor section not only correlate with one another but appear to correlate well with susceptibility zones of the Commonwealth Glacier section (Figure 8). Zones A, E, and I have low remanent intensities and susceptibilities and predate the appearance of highly magnetic material of the McMurdo Volcanic Group. Zones B, F, and J have high intensities and susceptibilities and reflect the deposition of material of the McMurdo Volcanic Group in abundance. Zones C, G, and K reflect a decrease in its abundance, and zones D, H, and L a substantial increase.

A correlation at the level that marks the appearance of material of the McMurdo Volcanic Group (206 m in DVDP 11) would, on cursory examination, appear to be extendable to the DVDP 12 section [*Elston et al.*, 1979]. Zones M, N, and O in

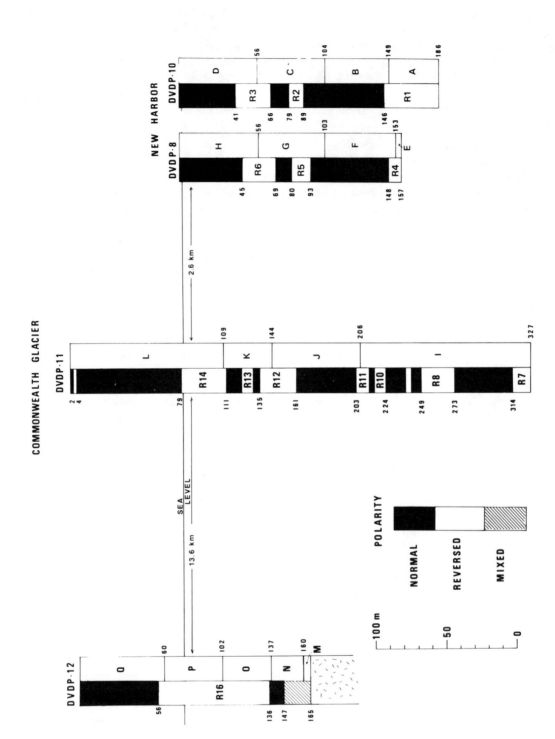

Fig. 8.   Magnetic polarity and susceptibility zonations for cores 8, 10, 11, and 12. Reverse polarity zones numbered correspond to those listed in Table 2. Numbers adjacent to columns indicate depth in meters from top of core.

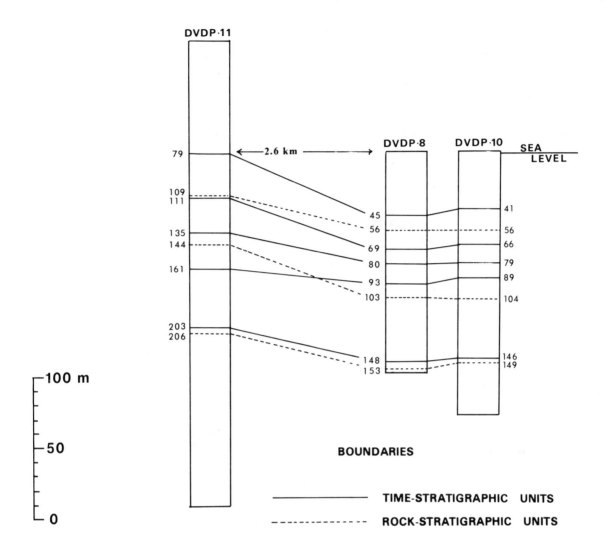

Fig. 9.   Correlations between cores 8, 10, and 11 based on polarity and susceptibility zonations. Numbers adjacent to columns indicate depth in meters from top of core. Boundaries between time-stratigraphic units shown by solid lines; boundaries between rock-stratigraphic units shown by dashed lines.

DVDP 12 have low susceptibilities and remanent intensities that characterize sediments deposited before the appearance of basalt. Zone P (60- to 102- m depth) appears to be magnetically characteristic in susceptibility and remanence of sediments east of this hole that contain abundant material of the McMurdo Volcanic Group. However, the considerable differences in the mineralogy of DVDP 12 and DVDP 8, 10, and 11 indicate differences in provenance that preclude a straightforward correlation on susceptibility and remanence.

Combining the magnetic mineralogy with the magnetic susceptibility zonation, we infer that the upper three susceptibility zones in DVDP 8, 10, and 11 contain titanomagnetite derived from the McMurdo Volcanic Group of the Ross Sea area. Subordinate magnetite derived from the Ferrar Dolerite and titanomaghemite derived by low-temperature oxidation of the McMurdo volcanic material are present in the upper three susceptibility zones. The lowermost susceptibility zones in DVDP 8, 10, and 11 contain magnetite as the primary magnetic mineral, the apparent source for which is lamprophyre dike rock and Irizar Granite, both of Taylor Valley provenance. In DVDP 12, magnetite and subordinate maghemite derived

from the Ferrar Dolerite of Taylor Valley provenance are the major magnetic minerals. Subordinate titanohematite and titanomagnetite(?) from
locally erupted volcanic material are present.

The temporal and stratigraphic correlations
shown in Figure 9 have implications for the geologic history of Taylor Valley and vicinity. The susceptibility change at 104 m in DVDP 10 and 103 m
in DVDP 8, in a zone of normal polarity, very
likely correlates with the susceptibility change
found at 144 m in DVDP 11. In DVDP 11, however, the susceptibility change in a zone of reverse
polarity implies that the susceptibility boundary
crosses time lines. Porter and Beget [1979] suggest, on the basis of microfabric evidence, that a
basal till occupies the interval between 104 and 125
m in DVDP 8 and 10. Oxygen isotope studies led
Stuiver et al. [1976] to report freshwater conditions, possibly indicative of a grounded ice sheet,
between 100 and 125 m in DVDP 8. In DVDP 11
the top of a massive till, present between 145 and
167 m, coincides with the change in susceptibility.
Our results suggest that the top of the till in
DVDP 8 and 10 (104 m) correlates with the top of
the till in DVDP 11. The time transgression shown
by the magnetic data indicates that deposition of
till at the hole 11 Commonwealth Glacier site continued for a time after such deposition had ceased
at the hole 8 and 10 New Harbor sites and thus
that grounded ice disappeared earlier at New Harbor.

Acknowledgments. This research was supported in part by
NSF grant DPP-77-13048. Careful management of the DVDP
cores by Dennis Cassidy assured maintenance of their stratigraphic integrity. Stephen Porter and James Beget allowed us
to use the graph of lithic percentage of McMurdo volcanic rocks
in DVDP 8 and 10. Our conclusions benefited from their careful
sedimentologic study of the DVDP cores. We thank Donald
Watson for use of the U.S. Geological Survey's Rock Magnetics
Laboratory and for help in interpretation of Curie balance
results. We thank Richard Reynolds for assistance with reflected light microscopy and for helpful discussions. Ed Kelley
and Stephen Gillett aided in the development of the susceptibility measurements. D. A. Christoffel, of the University of Wellington, provided samples of Ferrar Dolerite for mineralogic examination. We thank Robert Butler for the use of the
Paleomagnetics Laboratory at the University of Arizona. The
helpful suggestions of Rob Van Der Voo, Robert Hargraves,
and Warren Hamilton improved the manuscript.

REFERENCES

Armstrong, R. L., K-Ar dating: Late Cenozoic McMurdo Vol-
    canic Group and dry valley glacial history, Victoria Land,
    Antarctica, *N.Z. J. Geol. Geophys.*, *21*, 685-698, 1978.

Brady, H., The dating and interpretation of diatom zones in
    Dry Valley Project Holes 10 and 11, Taylor Valley, South
    Victoria Land, Antarctica, publication, 37 pp., Jap. Inst. of
    Polar Stud., Tokyo, 1978.
Bull, C., and E. Irving, The paleomagnetism of some hyp-
    abyssal intrusive rocks from South Victoria Land, Antarc-
    tica, *Geophys. J.*, *3*, 211-224, 1960.
Bull, C., E. Irving, and I. Willis, Further paleomagnetic results
    from South Victoria Land, Antarctica, *Geophys. J.*, *6*, 320-
    336, 1962.
Chapman-Smith, M., Geologic log of DVDP-12, Lake Leon,
    Taylor Valley, *Dry Val. Drilling Proj. Bull.*, *5*, 61-70, 1975.
Chapman-Smith, M., The Taylor Formation (Holocene) and its
    macrofaunas, Taylor Dry Valley, Antarctica, (abstract), in
    *Proceedings of the Seminar III on Dry Valley Drilling Pro-
    ject, 1978, Mem. Spec. Issue 13*, edited by T. Nagata, pp.
    9-10, National Institute of Polar Research, Tokyo, 1979.
Cox, A., Paleomagnetic research on volcanic rocks of McMurdo
    Sound, *Antarct. J. U.S.*, *1*, 136, 1966.
Denton, G. H., and R. L. Armstrong, Glacial geology and chro-
    nology of the McMurdo Sound Region, *Antarct. J. U.S.*, *3*,
    99-101, 1968.
Denton, G. H., R. L. Armstrong, and M. Stuiver, Late Ceno-
    zoic glaciation in Antarctica: The record in the McMurdo
    Sound Region, *Antarct. J. U.S.*, *5*, 15-21, 1970.
Denton, G. H., R. L. Armstrong, and M. Stuiver, The late Ce-
    nozoic glacial history of Antarctica, in *The Late Cenozoic Gla-
    cial Ages*, edited by K. K. Turekian, pp. 267-306, Yale Uni-
    versity Press, New Haven, Conn., 1971.
Elston, D. P., and S. L. Bressler, Magnetic stratigraphy of
    DVDP drill cores and late Cenozoic history of Taylor Valley,
    Transantarctic Mountains, Antarctica, this volume.
Elston, D. P., M. E. Purucker, S. L. Bressler, and H. Spall,
    Polarity zonations, magnetic intensities, and the correlation
    of Miocene and Pliocene DVDP cores, Taylor Valley, Antarc-
    tica (abstract), in *Proceedings of the Seminar III on Dry Val-
    ley Drilling Project, 1978, Mem. Spec. Issue 13*, edited by T.
    Nagata, pp. 12-15, National Institute of Polar Research, To-
    kyo, 1979.
Goldich, S. S., S. B. Treves, N. H. Suhr, and J. S. Stuckless,
    Geochemistry of the Cenozoic volcanic rocks of Ross Island
    and vicinity, Antarctica, *J. Geol.*, *83*, 415-435, 1975.
Goree, W. S., and M. Fuller, Magnetometers using R-F driven
    Squids and their application in rock magnetism and pa-
    leomagnetism, *Rev. Geophys. Space Phys.*, *14*, 591-608, 1976.
Hamilton, W., The Hallett Volcanic Province, Antarctica, *U. S.
    Geol. Surv. Prof. Pap.*, *456-C*, 62 pp., 1972.
Haskell, T. R., J. P. Kennett, W. M. Prebble, and I. Willis, The
    geology of the middle and lower Taylor Valley of South Victo-
    ria Land, Antarctica, *Trans. Roy. Soc. N. Z.*, *2*, 169-186,
    1965.
McGinnis, L. D., T. Torii, and P. N. Webb, Dry Valley Drilling
    Project, *Antarct. J. U.S.*, *7*, 53-56, 1972.
McGinnis, L. D., T. Torii, and R. Clark, Antarctic Dry Valley
    Project: Report on Seminar I, *Eos Trans. AGU*, *56*, 217-220,
    1975.
McKelvey, B. C., Preliminary site reports, DVDP sites 10 and
    11, Taylor Valley, *Dry Val. Drilling Proj. Bull.*, *5*, 16-60,
    1975.
McKelvey, B. C., Upper Cenozoic marine and terrestrial glacial
    sedimentation in Eastern Taylor Valley, Southern Victoria
    Land (abstract), in *Third Symposium on Antarctic Geology*

and Geophysics, *Volume of Abstracts*, University of Wisconsin Press, Madison, 1977.

McKelvey, B. C., The Miocene-Pliocene stratigraphy of eastern Taylor Valley (abstract), in *Proceedings of the Seminar III on Dry Valley Drilling Project, 1978, Mem. Spec. Issue 13*, edited by T. Nagata, pp. 60-61, National Institute of Polar Research, Tokyo, 1979.

McMahon, B. E., and H. Spall, Paleomagnetic data from unit 13, DVDP Hole 2, Ross Island, *Antarct. J. U.S.*, *9*, 229-232, 1974.

Nathan, S., and F. J. Schulte, Geology and petrology of the Campbell-Aviator Divide, northern Victoria Land, Antarctica, *N. Z. J. Geol. Geophys.*, *11*, 940-975, 1968.

Park, J. E., and P. J. Barrett, Paleomagnetic measurements on glacial sediment from DVDP 10 and 11, Taylor Valley, Antarctica, *Dry Val. Drilling Proj. Bull.*, *6*, 24-25, 1975.

Pederson, D. R., G. E. Montgomery, L. D. McGinnis, C. P. Ervin, and H. K. Wong, Magnetic study of Ross Island and Taylor Glacier quadrangles, Antarctica (abstract), in *Proceedings of the Seminar III on Dry Valley Drilling Project, 1978, Mem. Spec. Issue 13*, edited by T. Nagata, pp. 72-73, National Institute of Polar Research, Tokyo, 1979.

Porter, S., and J. Beget, Provenance and depositional environments of late Cenozoic sediments in permafrost cores from Lower Taylor Valley, Antarctica (abstract), in *Proceedings of the Seminar III on Dry Valley Drilling Project, 1978, Mem.*

Spec. Issue 13, edited by T. Nagata, pp. 74-76, National Institute of Polar Research, Tokyo, 1979.

Porter, S., and J. Beget, Provenance and depositional environments of late Cenozoic sediments in permafrost cores from Lower Taylor Valley, Antarctica, this volume.

Prévot, M., and S. Grommé, Intensity of magnetization of subaerial and submarine basalts and its possible change with time, *Geophys. J.*, *40*, 207-224, 1975.

Spall, H., and D. P. Elston, Magnetic stratigraphy in DVDP Hole 11, Taylor Valley, Antarctica (abstract), in *Third Symposium on Antarctic Geology and Geophysics, Volume of Abstracts*, University of Wisconsin Press, Madison, 1977.

Stacey, F., and S. Banerjee, *The Physical Properties of Rock Magnetism*, 195 pp., Elsevier, New York, 1974.

Stuiver, M., I. C. Yang, and G. H. Denton, Permafrost oxygen isotope ratios and chronology of three cores from Antarctica, *Nature*, *261*, 547-550, 1976.

Thompson, R., Paleolimnology and paleomagnetism, *Nature*, *242*, 182-184, 1973.

Webb, P. N., and J. H. Wrenn, Foraminifera from DVDP holes 8-12, Taylor Valley, *Antarct. J. U.S.*, *11*, 85-86, 1976.

Webb, P. N., and J. H. Wrenn, Late Cenozoic micropaleontology and biostratigraphy of Eastern Taylor Valley, Antarctica (abstract), in *Third Symposium on Antarctic Geology and Geophysics, Volume of Abstracts*, University of Wisconsin Press, Madison, 1977.

OXYGEN ISOTOPE RATIOS OF ANTARCTIC PERMAFROST AND GLACIER ICE

MINZE STUIVER AND IN CHE YANG

Quaternary Research Center, University of Washington, Seattle, Washington 98195

GEORGE H. DENTON AND THOMAS B. KELLOGG

Institute for Quaternary Studies, University of Maine, Orono, Maine 04469

Oxygen isotope records of permafrost waters of DVDP cores 10, 11, and 12 often appear to be defined by nonconnate waters infiltrated much later. For the Lake Hoare core (DVDP 12) the preserved [18]O record agrees with permafrost formation advancing downward from the surface after drainage of glacial Lake Washburn about 10,000 years ago. Replacement of fresh water by seawater took place fairly recently in the New Harbor core (DVDP 10). Perhaps only the main part of the Commonwealth Glacier core permafrost (DVDP 11) was formed from connate waters. The isotope ratios from the core ice are compared with oxygen isotope ratios of valley glaciers, polar plateau ice, ice-cored moraines, and McMurdo shelf ice. The McMurdo Ice Shelf isotope ratios show the main part of the shelf ice to be of seawater origin.

INTRODUCTION

A multitude of applications of oxygen isotope ratio studies in the fields of glaciology, climatology, and oceanography attests to the usefulness of this geochemical parameter for studying past changes in climate and the physical environment. Although this type of study is applied mostly to simple systems such as oceans (climate record in sediments) and ice sheets (changes in the isotope ratio of precipitation), we have attempted here to use [18]O isotope ratio studies for the more complicated environment in which permafrost waters are formed.

In addition to permafrost waters, the oxygen isotope composition of glacial ice was studied because it provides the present-day analog with which these permafrost ratios have to be compared.

The cores used for permafrost oxygen isotope ratio studies were obtained from three Dry Valley Drilling Project (DVDP) sites in Taylor Valley, Antarctica (Figure 1) [*Treves and McKelvey*, 1975; *McKelvey*, 1975]. Three cores (cores 8, 9, and 10 yielding an isotope record for a 185-m long composite core) were available from the New Harbor site. Borehole 11, 320 m deep, at an elevation of about 80 m, is 2.4 km to the west of the New Harbor site and 3 km from the terminus of Commonwealth Glacier. The Lake Hoare borehole (core 12) is 160 m deep and located 16.1 km from the New Harbor site.

For a description of the technical procedures, the reader is referred to the paper that also discusses the oxygen isotope distribution of the New Harbor core [*Stuiver et al.*, 1976].

The oxygen isotope ratios are expressed as

$$\delta^{18}O = \frac{(^{18}O/^{16}O)_S - (^{18}O/^{16}O)_{SMOW}}{(^{18}O/^{16}O)_{SMOW}} \times 1000\%o$$

where S denotes sample and SMOW, standard mean ocean water. Sample isotope ratios were measured relative to a laboratory standard. The laboratory standard has a $\delta^{18}O$ ratio of $-9.56\%o$ relative to SMOW, as determined through the use of (1) the NBS-1 Potomac reference sample and (2) *Craig*'s [1961] $-7.94\%o$ definition of NBS-1 relative to SMOW.

OXYGEN ISOTOPE COMPOSITION

Glacier Ice

The $\delta^{18}O$ values from -32 to $-34\%o$ [*Stewart*, 1975] and from -28.9 to $-31.9\%o$ [*Nakai et al.*,

131

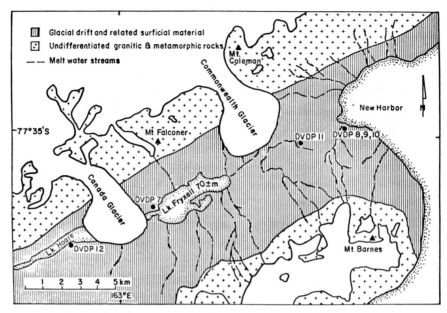

Fig. 1.   Location map of DVDP sites in lower Taylor Valley.

1975] have been reported for snow accumulating on the surface of the McMurdo Ice Shelf. Snow precipitated at higher elevations has a lower isotope ratio. The lowest $\delta^{18}O$ values are found in ice of polar plateau origin. The Taylor Glacier, at the head of Taylor Valley, originates at the polar plateau and has $\delta^{18}O$ values down to $-44.2‰$. Thirteen samples were collected from this glacier, of which ten had $\delta^{18}O$ values between -39.8 and $-44.2‰$. The remaining three samples were somewhat higher in ^{18}O content ($\delta^{18}O$ values of -34.0, -36.1, and $-37.4‰$), but contain on average less ^{18}O than ice derived from low elevation (alpine) glaciers on valley sides.

One of us (G.H.D.) collected samples from several alpine glaciers in the dry valleys during the 1975-1977 field seasons. The $\delta^{18}O$ determinations for these samples are given in Figures 2 and 3 and Table 1. The range in isotopic composition of alpine glacier ice is from -25 to $-35‰$.

'Fossil' glacier ice, collected from ice-cored moraines, has a wide spread in isotope ratios (Figure 2). Ice of marine origin was incorporated in the ice-cored moraines near Hobbs Glacier, Garwood Glacier, and the Blue Glacier. Therefore these moraines can be tied to the Ross Sea glaciation. This glaciation resulted from the grounding of an earlier ice sheet in McMurdo Sound [*Stuiver et al.*, 1981]. Thus the oxygen isotope ratios of ice-cored mo-

raines can be used as a diagnostic tool in determining the history of Ross Sea glaciations.

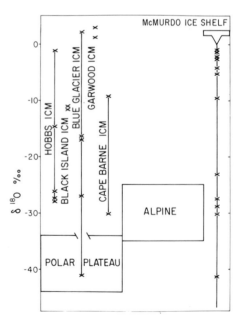

Fig. 2.   Schematic distribution of $\delta^{18}O$ values in polar plateau glacier ice, alpine glacier ice, McMurdo Ice Shelf ice, and selected ice-cored moraines (ICM). The blocks for polar plateau ice, alpine glacier ice, and McMurdo shelf ice give the range of $\delta^{18}O$ values for a large number of samples (see Tables 1 and 2). The $\delta^{18}O$ values for single samples are indicated by crosses. Horizontal scale has no significance, except for separating the four environments under discussion.

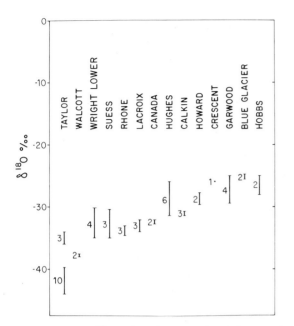

Fig. 3.   The $\delta^{18}O$ ratios relative to the SMOW standard of polar plateau and alpine glaciers. Numbers near the vertical bars give the number of samples analyzed; the vertical bar gives the range in $\delta^{18}O$ values.

McMurdo Ice Shelf

The McMurdo Ice Shelf, an extension of the Ross Ice Shelf, covers southern McMurdo Sound. In the east it is fed by ice inflow from the Ross Ice Shelf and from glaciers flowing south from Mount Erebus and Mount Terror; in the southwest it is fed by the Koettlitz Glacier.

Except for relatively clean ice derived from the Koettlitz Glacier, the upper surface of the shelf is covered by a thin debris mantle. Well-preserved marine macrofossils occur on the ice shelf surface and were discussed by *Debenham* [1920]. Marine microfossils also have been studied [*Kellog et al.*, 1977]. Debenham suggested that the ice shelf in the ablation area, which covers an extensive area of the McMurdo Ice Shelf [*Stuiver et al.*, 1981], was maintained by basal freezing and surface ablation, with marine sediments and macrofossils frozen onto the base of the shelf and subsequently moved upward through the shelf. The marine origin of part of the ice in the ablation area was confirmed by *Gow and Epstein* [1972], who found seawater $\delta^{18}O$ values for ice in two cores, in addition to freshwater ratios in a core near the Koettlitz Glacier.

TABLE 1.   The $\delta^{18}O$ Ratios, Relative to the SMOW Standard, of Glaciers and Ice-Cored Moraines in the Dry Valleys and Royal Society Range, Antarctica

	$\delta^{18}O$ Ratios
Taylor Glacier	−34.0, −36.1, −37.4, −39.8, −40.0, −40.4, −41.7, −42.0, −42.8, −42.9, −43.3, −43.5, −44.2
Commonwealth	−28.7
Hobbs ice-cored moraine	−1.1, −14.8, −26.2, 27.4, −28.0
Black Island ice-cored moraine	−10.9, −11.8
Blue Glacier ice-cored moraine	+2.3
Wright lower	−30.1, −34.2, −34.5, −34.9
Suess	−30.4, −30.6, −35.1
Sollas	−31.8
Garwood ice-cored moraine	+3.9, +1.4
Garwood	−27.6, −30.2, −30.8, −31.3, −32.1
Calkin	−30.9, −31.0, −31.1
Howard	−27.9, −29.8
Crescent	−25.9
Rhone	−33.0, −34.0, −34.7
Lacroix	−32.2, −32.8, −33.5
Canada	−32.2, −32.8
Hughes	−26.0, −28.5, −29.5, −30.4, −30.8, −31.2, −31.6, −32.4
Cape Barne ice-cored moraine	−9.2, −30.0
Adams	−26.7, −28.1
Miers Glacier ice-cored moraine	−27.9, −31.6
Cape Evans glacier tongue entering the sea	+0.5
Blue Glacier ice-cored moraine	−16.6, −17.0, −24.8, −27.0, −41.3
Blue Glacier	−24.8, −25.6
Hobbs, Ross Island	−41.7

Samples were collected by G. H. Denton during the 1975-1977 field seasons.

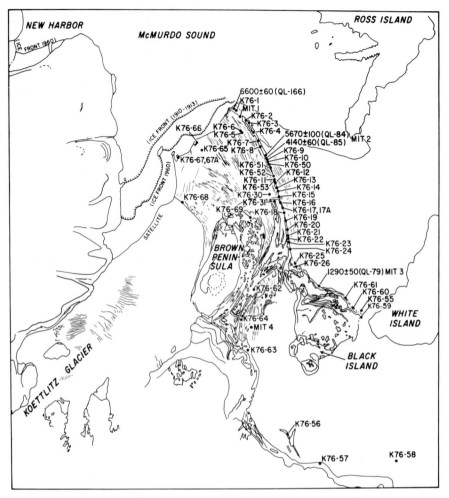

Fig. 4.   Collection sites on the McMurdo Ice Shelf of samples listed in Table 2. [14]C ages are given with the QL (Quaternary Isotope Laboratory) sample numbers.

The ^{14}C dating of shells found in debris bands on the shelf show that the shells are being incorporated into the bands near shore, and from there they are carried to the edge of the ice shelf [*Stuiver et al.*, 1981; *Kellogg et al.*, 1977].

Further evidence that McMurdo shelf ice is mainly of marine origin is provided by the $\delta^{18}O$ measurements reported here. In 1976, one of us (T.B.K.) collected samples from the debris band that runs northward from Black Island to the edge of the ice shelf as well as from a large area of the McMurdo Ice Shelf (Figure 4). Of the 45 samples analyzed, 33 had positive $\delta^{18}O$ ratios, indicating ice of pure marine origin (Table 2 and Figure 2). Several of the remaining samples are mixtures of seawater and fresh water. Only a few of these samples are entirely of pure freshwater origin. One of these samples, K76-59, with a $\delta^{18}O$ ratio of −41.5‰ is on

the edge of the shelf and should perhaps not be considered shelf ice.

The above data prove that almost all samples collected from the McMurdo Ice Shelf were formed from seawater, in full agreement with the mechanism described by Debenham and discussed above.

OXYGEN ISOTOPE RATIOS

Influence of Core Lab Storage

Water vapor is depleted in the heavy ^{18}O isotope when compared to the isotopic composition of the water from which it is derived. Thus tropical lakes, for instance, are enriched in ^{18}O because substantial portions of water are evaporated. Glacier ice also can be enriched in ^{18}O through evaporation losses (freeze drying).

TABLE 2.   Oxygen Isotope Ratios Relative to the SMOW Standard of Surface Samples of the McMurdo Ice Shelf

Site	$\delta^{18}O$
K76-1	+2.2
K76-2	+1.5
K76-3	+1.9
K76-4	+2.0
K76-5	+2.6
K76-6	+2.1
K76-7	+2.1
K76-8	−28.8
K76-9	+2.3
K76-10	−5.2
K76-11	+2.2
K76-12	+2.1
K76-13	+2.4
K76-14	+1.6
K76-15	+0.5
K76-16	+0.8
K76-17	+0.5
K76-18	+2.1
K76-19	+1.8
K76-20	+1.7
K76-21	+2.3
K76-22	+2.3
K76-23	+2.5
K76-24	−2.7
K76-25	+1.9
K76-26	−9.6
K76-30	+1.0
K76-31	+2.1
K76-50	+0.4
K76-51	+1.9
K76-52	−1.0
K76-53	+1.9
K76-55	−30.3
K76-58	−23.3
K76-59	−41.5
K76-61	+1.5
K76-62	+2.6
K76-65	−2.5
K76-66	+1.7
K76-67	−1.6
MIT-1	+2.8
MIT-2	−27.5
MIT-3	−4.2
MIT-4	+2.5
MIT-5	+1.2

The K samples were collected by T. B. Kellogg in 1976. The MIT samples were collected by G. H. Denton in 1975.

An example of isotope enrichment was found for our permafrost cores stored at the Antarctic Core Library at Florida State University, Tallahassee. Core 12 was sampled about 3 years apart in January 1975 and February 1978. During storage, the outer portion of the ice evaporated, and only the inner portion still had the solid consistency of the 1975 core. When compared with the expected $\delta^{18}O$ values, obtained by interpolation of the 1975 results, the 1978 values were enriched, on average, by 11‰ (Table 3). Two explanations are possi-

TABLE 3.   The Influence of Core Lab Storage on $\delta^{18}O$ Ratios

Depth, m	$\delta^{18}O$,‰ February 1978	July 1975
46.3	−33.3	−40
47.8	−27.5	−42
48.1	−30.5	−37
59.4	−26.5	−40
66.4	−26.6	−34
132.9	−31.6	−45
133.2	−25.2	−42

ble: (1) enrichment through freeze drying or (2) addition to the core of water vapor that is of local origin and thus higher in ^{18}O. In view of the water loss, we suspect enrichment through freeze drying to be a main factor.

The above experience points to the desirability of storage of cores in impermeable containers. The thin plastic foil used in the Core Library was not sufficient to prevent extensive change of permafrost oxygen isotope composition. In addition, when glacier ice is collected in the field, the ice not directly exposed to air should be taken. For instance, it is possible that freeze drying or water vapor additions caused the somewhat higher $\delta^{18}O$ ratios of the three Taylor Glacier samples reported in the previous section.

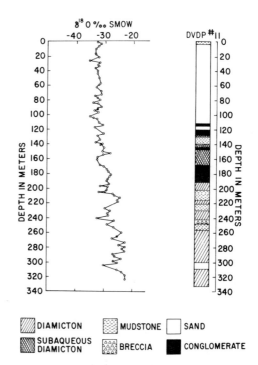

Fig. 5.   The $\delta^{18}O$ values of permafrost relative to the SMOW standard in DVDP core 11.

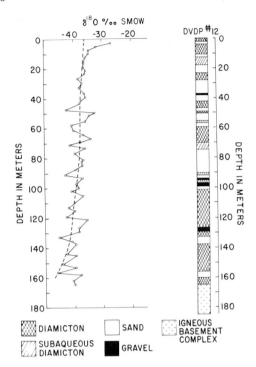

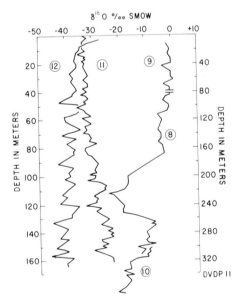

Fig. 6. The δ values of permafrost relative to the SMOW standard in DVDP core 12. The dashed line is a calculated profile by assuming freezing from the surface in a closed system.

Fig. 7. The δ^{18}O values of permafrost versus depth in cores 8-10 (New Harbor), core 11 (Commonwealth Glacier), and core 12 (Lake Hoare). Cores 8-10 and core 12 depths are given on the left scale.

The permafrost samples used in our study, of which the δ^{18}O values are given in Figures 5-7, were taken within a few months of core arrival in the Tallahassee Antarctic Core Library. Also a few months elapsed between core retrieval and arrival in the United States. It is possible that during this interval of shipping and storage the oxygen isotope ratios were slightly altered. As the overall change in 3 years of storage is 11‰, the change induced prior to sampling may have been about 1‰.

The McMurdo Ice Shelf and glacier samples for which ^{18}O ratios are reported here were stored for a few months in closed plastic containers and are not influenced by the above-mentioned isotope exchange.

Permafrost Oxygen

Oxygen isotope composition of permafrost waters is given versus depth for cores 11 and 12 in Figures 5 and 6. The δ^{18}O curves are compared in Figure 7 with the permafrost δ^{18}O curve obtained earlier for cores 8, 9, and 10. The latter curve was published previously [*Stuiver et al.*, 1976]. The interpretation of the permafrost isotope ratios is more complicated than for cores from ice sheets because the accumulation of sediments is not necessarily continuous. In addition, a portion of the ^{18}O record may be missing because temperatures were perhaps not always sufficiently low for permafrost formation. Postdepositional melting, and subsequent replacement by 'younger' waters, also can erase part of the record.

The growth of permafrost is accompanied by the freezing of pore water, either in situ or after migration to form ice lenses. In some soils there is also water expulsion. Ions migrate downward and away from the freezing front; thus salinity is increased in the unfrozen groundwaters below the freezing front. In addition, a slight enrichment in ^{18}O during freezing results in a lowering of ^{18}O content for the subpermafrost waters.

Mackay and Lavkulich [1974], in a study of permafrost formation in sediments exposed after lake drainage, reported extensive ionic and oxygen isotope fractionation. Subpermafrost waters had oxygen isotope ratios about 7‰ lower than the ice permafrost samples, whereas the specific conductance of the subpermafrost waters was about 10 times the conductance of surface waters.

Postdepositional melting of permafrost, and subsequent formation of new permafrost, can be expected for core 12. At this site, glacial Lake Washburn [*Péwé*, 1960] was impounded through glacial damming at the mouth of the Taylor Valley by a grounded Ross Sea ice sheet. There is considerable evidence for a succession of grounded Ross Sea ice sheets in the McMurdo region. These ice sheets were fed by grounded ice that flowed into the sound from an area to the east and southeast that is now covered by the Ross Ice Shelf. The highest lake levels of glacial Lake Washburn in the Taylor Valley were at approximately 300 m above sea level about 20,000 years ago. During subsequent stages of retreat of the Ross Sea ice sheet, the lake level was lowered to an altitude of about 50 m some 12,000 years ago, close to the present level of Lake Fryxell [*Stuiver et al.*, 1981]. The drift deposited by the last ice sheet to occupy McMurdo Sound and the Ross Sea (Ross Sea drift) covers the sites of boreholes 8-10 and 11, but the Ross Sea drift margin is about 1 km east of the borehole 12 site [*Stuiver et al.*, 1981]. Therefore at site 12 a deep, solar-heated lake existed between 20,000 and 12,000 years ago. Permafrost formed prior to the formation of Lake Washburn was melted. Water from the lake appears to have flushed downward through the unfrozen sediments; thus groundwaters that are relatively homogeneous in oxygen isotope composition and salinity were produced.

Freezing from the top, after drainage of the lake, could produce the oxygen isotope record given in Figure 6. The main problems in interpreting the oxygen isotope record are possible horizontal movements of subpermafrost waters, and the extent of vertical diffusion of water underneath the permafrost. Our simple model assumes a closed system in which horizontal flow can be disregarded. Furthermore, we assume sufficiently fast vertical diffusion and mixing for the subpermafrost waters, so that this reservoir is considered well mixed at all times. The last assumption appears justified because the freezing front of the permafrost is moving very slowly downward and it may have taken thousands of years to freeze down to the bottom of core 12. Basement is encountered in this core at 165-m depth.

The fractionation factor for ^{18}O for ice formed at 0°C is 1.0035 [*O'Neil*, 1968]. Thus ice is enriched by 3.5‰ in ^{18}O in comparison to the water from which it is formed. For high-salinity waters the fractionation is less. *Stewart* [1975] derived a fractionation factor of 1.0025 for concentrated seawater at

−10°C. In our calculations a value of 1.0035 was used for the fractionation factor.

Because the ice is enriched in ^{18}O during permafrost formation, the subpermafrost waters will be depleted in ^{18}O. For a well-mixed subpermafrost water reservoir, the fraction f of water remaining differs in oxygen isotope ratio from the original water isotope ratio by δ‰ according to $f^{\alpha-1} - 1 = \delta$, where α is the fractionation factor. The derivation of this formula is similar to the derivation given by *Dansgaard* [1964] for isotope fractionation during Rayleigh distillation processes.

With basement at 165 m (the fraction f is depth above 165 m divided by 165 m), and for a homogeneous water content in the sediments, the calculated fractionation follows the dashed line in Figure 6. Starting with water of −39.5‰, the first ice formed at the surface has a $\delta^{18}O$ value of −36.0‰. Progressive ^{18}O depletion of the subpermafrost waters, and subsequent freezing of these waters, results in ^{18}O depletion by 7-10‰ for the last 10% of the core above basement.

The good agreement between the calculated and the experimental curves shows that the assumption of a homogeneous subpermafrost reservoir was justified. Departure from the calculated curve for the top 5 m may well be caused by the fact that permafrost formation at these shallow depths was much faster; thus there was no time for isotope equilibration in a large body of water. It is also possible that the top meters of permafrost were replaced by heavier water during shallow depth freezing-thawing processes.

Lacustrine carbonates, although sparsely present, are also encountered in core 12. Five samples that were collected at approximately 35, 38, 58, 95, and 142 m deep were precipitated from waters with $\delta^{18}O$ values estimated at −33.6, −36.2, −42.2, −33.7, and −34.3‰, respectively [*Hendy et al.*, 1979]. These values are obtained from the carbonate $\delta^{18}O$ values by assuming (1) a temperature of precipitation of 0°C and (2) no postdepositional reequilibration of oxygen isotopes between carbonate and pore waters, nor postdepositional carbonate addition. There is neither definite proof nor disproof of assumption 2. The carbonates were most likely deposited prior to the last episode of permafrost formation and reflect isotope composition $\delta^{18}O$ fairly close to the isotope ratio of the waters from which at a later stage permafrost was formed (the model calculations use $\delta^{18}O$ = −39.5‰ for the water originally present).

The above model of permafrost formation agrees with the information derived from the core 12 salinity curve [*McGinnis et al.*, 1980]. Pore ice salinity increases continuously with depth from nearly zero at the surface to 70 ppt at the base. During permafrost deepening the salts expelled during ice formation migrate downward and away from the freezing front. McGinnis et al. conclude from the salinity curve that the freezing wave advanced slowly downward with little or no escape of salts, thus indicating static hydraulic conditions (i.e., no groundwater movement or flushing). This picture of permafrost formation is in agreement with the above model derived from the oxygen isotope ratio measurements.

The $\delta^{18}O$ versus depth trend for core 11 is opposite to the trend found for core 12 (Figure 7). Very low pore ice salinities (~3 ppt) are encountered in core 11 to 200-m depth [*McGinnis et al.*, 1980]. Salinity increases below 200 m, and a maximum salinity of 87 ppt is encountered at 255-m depth. Seawater addition, as found in the New Harbor core (see later discussion), is a possibility in core 11 for depths exceeding 200 m.

Five core 11 carbonate samples, from depths of about 70, 77, 88, 93, and 94 m, were analyzed by *Hendy et al.* [1979] for ^{18}O content. These carbonates were precipitated from waters with estimated $\delta^{18}O$ values of -25.0, -30.1, -27.2, -28.9, and $-29.4\%_{o}$, respectively. The average $\delta^{18}O$ value for the permafrost at these depths is $-32.6\%_{o}$. Because ice is enriched in ^{18}O during freezing, the water reservoir for the permafrost should have approximately $\delta^{18}O$ values of $-29\%_{o}$. The waters from which the carbonates were precipitated had an average $\delta^{18}O$ value of $-28.1\%_{o}$. Thus the waters from which permafrost was formed and from which the carbonates were precipitated (assuming isotopic equilibrium) have the same isotopic confirmation.

There are two possible mechanisms that can explain the oxygen isotope distribution in core 11.

1. The top 200 m of permafrost may have been formed from waters entrapped in the sediment at the time of deposition (connate waters). The apparent small increase in $\delta^{18}O$ values with depth would point to a slight cooling toward the present. The permafrost could only have been formed either when sea level was appreciably lower or when the valley floors were tectonically less depressed. A lower sea level is inferred because at a 200-m depth in the core, the sediment is 120 m below present sea level.

Below 200 m the higher ^{18}O content can be attributed to postdepositional mixing with small amounts of seawater. The high salinity could have been originally derived from a saline lake.

2. Another alternative is permafrost formation from the top, in either a closed or open system. Here permafrost melting at an earlier time is inferred. Although the surficial deposits of the Ross Sea drift extend into the valley beyond the borehole 11 site, one could postulate that this site was perhaps covered during a short time interval by an extension of glacial Lake Washburn. For alternative 2*a*, permafrost formation from the top in a closed system is proposed, similar to our interpretation for core 12. Subsequent to, or during, permafrost formation, seawater intruded from below, changing the ^{18}O isotope versus depth distribution from the slope found for core 12 to the one found for core 11. Alternative 2*b* would involve freezing from the top in an open system. The subpermafrost water would be flushed out and replaced by fresh water; thus waters enriched in salts and ^{18}O would be removed. The higher ^{18}O content below about 200 m, and higher salinities, could be attributed to mixing with seawater.

Conclusive evidence is lacking to make a definite choice from the above-mentioned alternatives. If the ice in the permafrost is connate, however, the trend of lower ^{18}O values toward the present may reflect a trend toward colder climatic conditions in Taylor Valley.

The permafrost isotope ratios of the composite New Harbor core (8-10) clearly show alternate seawater-freshwater sequences. The upper 90 m of the core has $\delta^{18}O$ values close to 0$\%_{o}$. Originally, the idea of connate seawater for the entire 90 m was favored [*Stuiver et al.*, 1976], but additional evidence now indicates postdepositional implacement of permafrost marine waters in part of the core. Seven carbonates collected from the core at 45, 55, 57, 63, 73, 74, and 90 m deep were precipitated from waters with isotope ratios of approximately -22.8, -29.5, -34.3, -29.4, -28.3, -28.8, and $-25.7\%_{o}$ [*Hendy et al.*, 1979]. If these freshwater carbonates were formed in situ and not transported from elsewhere, then one has to conclude that the fresh water originally present during carbonate formation between 40- and 90-m depth in the core was subsequently replaced by seawater.

The permafrost waters between 100 and 125 m are most likely influenced by minor addition of seawater because the carbonates at 104.0, 113.0, and

124.9 m deep were precipitated from waters with isotope ratios somewhat lower than the permafrost waters encountered at these depths ($-25\%o$ for the three carbonate samples versus about $-20\%o$ for the permafrost waters).

For parts of the New Harbor core the $\delta^{18}O$ versus depth trends would fit the idea of permafrost formation by freezing from the top in a closed system.

The interpretation of the ^{18}O history of the New Harbor core is complex. This core, located near the shoreline about 1 m above sea level, shows the greatest range in salt content [McGinnis et al., 1980], and alternating saltwater and freshwater lenses are indicated by the salinity profiles. For a more detailed discussion of the complexities of this core, the reader is referred to the papers by Stuiver et al. [1976] and McGinnis et al. [1980].

CHRONOLOGY

Although Pliocene and Miocene faunas are encountered in the lower portions of the cores, the oxygen isotope record often appears to be defined by nonconnate waters deposited much later. For the Lake Hoare core (core 12) the record now preserved in the permafrost is given by waters that were frozen after drainage of glacial Lake Washburn circa 10,000 years ago. For the New Harbor core (8-10), replacement of fresh water by seawater between 40- and 90-m core depth also took place fairly recently, most likely during the Holocene rise in sea level. In the upper portion of the New Harbor core, connate seawaters are still a possibility because the upper 23 m were deposited as Ross Sea drift during the last 5800 years [Stuiver et al., 1976].

For the Commonwealth Glacier core (core 11), permafrost formation from connate waters in the upper 200 m is one hypothesis. If valid, the sedimentary record would span several ice ages. Glacial-interglacial changes in oxygen isotope ratios of Taylor Valley waters stored in the sediments would amount to only $3\%o$ for this part of the core. The long-term overall change toward lower $\delta^{18}O$ values would indicate a cooling trend toward the present.

Acknowledgments. This research was supported by NSF grant DPP76-24403, Office of Polar Programs. D. S. Cassidy, Florida State University, provided samples from the Antarctic Core Library.

REFERENCES

Craig, H., Standard for reporting concentrations of deuterium and oxygen 18 in natural waters, *Science, 133*, 1833, 1961.
Dansgaard, W., Stable isotopes in precipitation, *Tellus, 16*, 436-468, 1964.
Debenham, F., A new mode of transportation by ice: The raised marine muds of South Victoria Land, *Quart. J. Geol. Soc. London, 75*, 51-76, 1920.
Gow, A. J., and S. Epstein, On the use of stable isotopes to trace the origins of ice in a floating ice tongue, *J. Geophys. Res., 77*, 6552-6557, 1972.
Hendy, C. H., T. R. Healy, E. M. Rayner, J. Shaw, and A. T. Wilson, Late Pleistocene glacial chronology of the Taylor Valley, Antarctica, and the global climate, *Quaternary Res., 11*, 172-184, 1979.
Kellogg, T. B., M. Stuiver, D. E. Kellogg, and G. H. Denton, Marine microfossils on the McMurdo Ice Shelf, *Antarct. J., 12*, 82-83, 1977.
Mackay, J. R., and L. M. Lavkulich, Ionic and oxygen isotope fractionation in permafrost growth, *Geol. Surv. Can., 74*(1), 255-256, 1974.
McGinnis, L. D., D. R. Osby, and F. A. Kohout, Paleohydrology inferred from salinity measurements on Dry Valley Drilling Project (DVDP) core in Taylor Valley, Antarctica, in *Antarctic Geoscience*, edited by C. Craddock, University of Wisconsin Press, Madison, in press, 1980.
McKelvey, B. C., Preliminary site reports, DVDP sites 10 and 11, Taylor Valley, *Dry Val. Drilling Proj. Bull., 5*, 16-80, 1975.
Nakai, N., Y. Kioysu, H. Wada, and M. Takimoto, Stable isotope studies of salts and water from dry valleys, Antarctica, I, Origin of salts and water, and the geologic history of Lake Vanda, *Mem. 4*, pp. 30-44, Nat. Inst. Polar Res., Tokyo, 1975.
O'Neil, J. R., Hydrogen and oxygen isotope fractionation between ice and water, *J. Phys. Chem., 72*, 3683-3684, 1968.
Péwé, T. L., Multiple glaciation in the McMurdo Sound region, Antarctica—A progress report, *J. Geol., 68*, 498-514, 1960.
Stewart, M. K., Hydrogen and oxygen isotope studies on the McMurdo Ice Shelf, Antarctica, *N.Z. J. Geol. Geophys., 18*, 49-64, 1975.
Stuiver, M., I. C. Yang, and G. H. Denton, Permafrost oxygen isotope ratios and chronology of three cores from Antarctica, *Nature, 261*, 547-550, 1976.
Stuiver, M., G. H. Denton, and T. J. Hughes, Late Würm and Holocene history of the marine ice sheet in West Antarctica: A working hypothesis, in *The Last Great Ice Sheets*, edited by G. H. Denton and T. J. Hughes, John Wiley, New York, 1981.
Treves, S. B., and B. C. McKelvey, Drilling in Antarctica, September-December 1974, *Dry Val. Drilling Proj. Bull., 5*, 4-10, 1975.

LIMNOLOGICAL STUDIES OF SALINE LAKES
IN THE DRY VALLEYS

Tetsuya Torii

Chiba Institute of Technology, Tsudanuma, Narashino City, Chiba, Japan

Noboru Yamagata

Institute of Public Health, Shirokanedai, Minato-Ku, Tokyo, Japan

The dry valley area is characterized by the presence of a number of lakes and ponds, some of them containing extremely saline water. In this report a general description of the nature of these saline waters is given. The exceptionally high water temperatures found throughout the year in some dry valley lakes have drawn special attention for many investigators interested in the heat source. Solar radiation was suggested and is now generally accepted as the source. In recent studies conducted during the Dry Valley Drilling Project (DVDP), no geothermal activity was found. The possible sources of salts in the saline lakes are (1) geothermal and hydrothermal, (2) trapped seawater, (3) chemical weathering of rocks, (4) sea spray, (5) glacial meltwater, and (6) groundwater discharge. An explanation using a single source has never been successful, and most workers are now inclined to favor multiple sources of salt. The salts in the saline waters of the dry valleys are much more likely to have originated from the accumulation of atmospheric salts than from the alteration of trapped seawater. The saline waters originated as fresh glacial meltwater and other running waters in which the chemical composition gradually approached, in the course of time, that of seawater. Subsequent changes in chemical composition were the same as those of seawater under frigid conditions, as delineated by Thompson and Nelson (1956).

INTRODUCTION

Stephenson first used the term 'oasis' as a name for ice-free areas in polar regions when the British Rymill party (1934-1937), of which he was a member, discovered an ice-free area on Alexander Island (70°45′S) in the Antarctic [*Rymill*, 1938]. Subsequent discoveries and explorations of oases in Antarctica have been made by many others, including a German party in 1938-1939 (Schirmacher Oasis) and an American party in 1946-1947 (Bunger Oasis). At present, 16 oases are known, of which most are in east Antarctica; in west Antarctica, oases have been found only on the Antarctic Peninsula [*Markov et al.*, 1970; *Simonov*, 1971]. The surface areas of oases are variable, ranging from 25 km^2 (Schirmacher Oasis) to 4000 km^2 (McMurdo Oasis).

The climate and the landscape of Antarctic oases are much different from those in surrounding ice-covered areas. The most conspicuous features of the oases are the lakes and ponds. Since the initiation of Antarctic limnological study in limited areas around the permanent coastal stations during the International Geophysical Year (IGY) (1957-1958), investigations of oases have been extended to other coastal regions. Comprehensive reviews of Antarctic limnology have been made from geomorphological, geological, geochemical, and biological points of view by *Nichols* [1966] and *Heywood* [1972]; one attractive subject is the geochemical investigation of the highly saline lakes and pools ubiquitous in the oases.

A history of the discovery of saline waters in the coastal areas of Antarctica begins in 1946-1947, when the Bunger Hills were found by U.S. Opera-

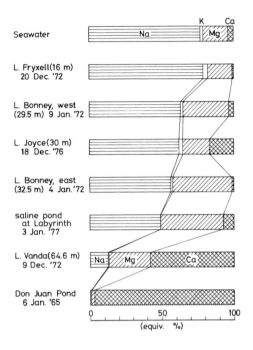

Fig. 1.   Percent composition in equivalent of cations in saline lakes.

tion Highjump [*Byrd*, 1947; *Apfel*, 1948], and follows the discovery of saline lakes in Vestfold Hills by Australian parties [*Law*, 1959; *McLeod*, 1964]. Subsequently, discoveries and surveys were made in several oases and by several parties, including the Bunger [*Glazovskaya*, 1958] and Shirmacher oases [*Simonov and Bonch-Osmolovskaya*, 1968] by Soviet parties, the Syowa oasis by Japanese parties [*Yoshida*, 1970; *Yoshida et al.*, 1973, 1975], and the Alexander oasis by British parties [*Heywood*, 1977]. All of the saline lakes are located in coastal areas. Their common characteristic features are as follows: (1) the lake surface is frequently below sea level; (2) paleolake shorelines and evaporites such as mirabilite (Na_2SO_4 • $10H_2O$), thenardite (Na_2SO_4), calcite ($CaCO_3$), gypsum ($CaSO_4$ • $2H_2O$), halite ($NaCl$), as well as microfossils of marine organisms, are present; and (3) the ratios of the major chemical components to chloride in the waters are close to those of seawater. Such geomorphological and·geochemical findings strongly suggest a relict lake origin of the salts [*McLeod*, 1964; *Watanuki et al.*, 1977].

On the other hand, a number of saline lakes have been discovered far from the coast in the dry valleys of south Victoria Land since the first reconnaissance by *Ball and Nichols* [1960]. The nature of saline waters in the dry valleys is complicated

both in chemical composition (Figure 1) and salinity, and to these salts a simple relict lake origin theory cannot be applied; debates on this problem have continued for a long time. Since the start of the Dry Valley Drilling Project (DVDP) (1971-1972), numerous geophysical and geochemical data from the saline lakes and their surroundings have been accumulated. Observations on unconsolidated sediments and groundwaters encountered during drilling are particularly helpful for understanding the mechanisms of formation of saline lakes.

GENERAL DESCRIPTION OF SALINE LAKES IN THE DRY VALLEYS

The dry valleys, an ice-free area in south Victoria Land, are situated west of McMurdo Sound, which is part of the Ross Sea. This area is divided into three main valleys, all of which are wide, flat, and trending east-west. In order from north to south, the valleys are named Victoria Valley, Wright Valley, and Taylor Valley. In these glacial troughs there are many enclosed drainage basins containing ponds and lakes, both fresh and saline. The most famous saline lakes are Lake Vanda and Don Juan Pond in Wright Valley and lakes Fryxell and Bonney in Taylor Valley. Lake Vanda, 5.6 km long, 1.5 km wide, and 68.8 m at the deepest [*Nelson and Wilson*, 1972], is situated in the lowest part of Wright Valley, approximately 50 km from the Ross Sea. Don Juan Pond is situated in the south fork of the same valley about 9 km west of Lake Vanda. Lake Fryxell, about 5 km long, 2 km wide, and 18 m deep (1974), occupies the center of a wide shallow basin in lower Taylor Valley and is approximately 6.5 km from the coast at New Harbor. Lake Bonney, about 10 km west of Lake Fryxell, is composed of west and east lobes that are connected through a shallow and narrow channel. The west lobe, in front of Taylor Glacier, is 2.6 km long, 0.8 km wide, and 35 m deep, whereas the east lobe is 4.8 km long, 0.8 km wide, and 33 m deep (1974). In addition, there is Lake Joyce in Pearse Valley, west of Taylor Valley, as well as small saline ponds in depressions in glacial troughs and in the Labyrinth, which is in upper Wright Valley and has an elevation of about 950 m above sea level. Some of these small saline ponds occasionally dry up in summer, forming salt beds. No saline lake has yet been discovered in Victoria Valley.

Saline lakes have no outlet and are fed princi-

TABLE 1. Chemical Composition of Saline Lakes in the Dry Valleys

		Lake Bonney			Don Juan Pond	Don Juan Pond	Saline Pond at Labyrinth	Lake Joyce
	Lake Vanda	East Lobe	West Lobe	Lake Fryxell				
Sampling date	Dec. 9, 1972	Jan. 4, 1972	Jan. 9, 1972	Dec. 20, 1972	Jan. 6, 1965	Jan. 9, 1975	Jan. 3, 1977	Dec. 18, 1976
Sampling depth, m	64.6	32.5	29.5	16	surface	surface	surface	30
Temperature, °C	24.3	−2.4	−4.6	1.6	10.4	6.3	1.8	−0.1
pH	5.45	6.51	5.73	7.07	4.6	4.4	7.54	6.59
Specific gravity	1.092	1.203	1.102	1.00	1.386	1.265	1.017	1.00
Na, g/kg	6.11	56.9	32.1	2.98	1.63	9.85	4.13	1.06
K, g/kg	0.59	2.30	1.47	0.203	0.26	0.12	0.03	0.07
Ca, g/kg	24.40	1.22	1.48	0.027	137.1	91.48	0.53	0.25
Mg, g/kg	7.40	21.71	8.34	0.331	1.8	1.2	1.9	0.14
Cl, g/kg	74.28	161.5	78.12	3.710	251.1	182.0	7.71	1.45
SO$_4$, g/kg	0.615	2.94	4.45	0.253	0.00	0.03	2.13	1.08
Br, g/kg	0.020	1.240	0.375	0.011	0.123	0.008	0.009	0.001
HBO$_2$, g/kg	0.012	0.108	0.078	0.009	0.004	0.003	0.000	0.000

pally by glacial meltwater. The annual average air temperature is very low (about −20°C), so that big lakes having a depth of 18-68 m are covered perennially with ice 3-4 m thick. Small and shallow saline ponds, on the other hand, are open in summer and usually frozen in winter, with the exception of Don Juan Pond, which never freezes completely in winter, even at temperatures as low as −51°C (A. M. Bromley, personal communication, 1974).

Although the presence of saline lakes in the dry valleys was first recorded in preliminary reports of reconnaissance in the early stage of the IGY [Ball and Nichols, 1960; McKelvey and Webb, 1961], lake ice coring and sampling of water were conducted for the first time in lakes Vanda and Bonney in December 1960 by Armitage and House [1962] and Angino and Armitage [1963]. Their discoveries included highly saline water in bottom layers, chemical stratification, and abnormal vertical water temperature profiles. These discoveries stimulated a number of scientists to further limnological studies.

Major Ionic Components

The chemical nature of these saline waters has been investigated by many workers [Armitage and House, 1962; Angino and Armitage, 1963; Angino et al., 1962, 1964, 1965; Goldman, 1964; Yamagata et al., 1967; Boswell et al., 1967a, b; Barghoorn and Nichols, 1961; Meyer et al., 1962; Tedrow et al., 1963; Mudrey et al., 1973 (saline ponds)]. Table 1 shows the chemical composition of water in the main saline lakes in the dry valleys [Torii et al., 1975, 1977b, 1979]. All the saline lakes are of the

chloride type and meromictic and all are chemically stratified, having a layer of fresh water underneath an ice cover and deeper water that is increasingly saline with depth. Except in lakes Fryxell and Joyce, the bottom waters have a chlorinity higher than that of seawater. At a depth of 32.5 m, the east lobe of Lake Bonney shows a chlorinity of 161.5‰, more than 8.5 times that of seawater and, with the exception of Don Juan Pond, the highest in the dry valleys. Don Juan Pond showed a maximum chlorinity of 251‰ in 1965.

The predominant components of water in these saline lakes differ considerably from valley to valley as well as from lake to lake. For example, Ca and Cl are predominant in Lake Vanda and Don Juan Pond, Na followed by Mg and Cl in Lake Bonney, and Na and Cl in lakes Fryxell and Joyce. Figure 1 depicts the cationic composition of these saline waters in comparison with that of seawater. Although the cationic composition of Lake Fryxell is quite similar to that of seawater, in anionic composition the lake is very much depleted in sulfate (Table 1). Generally, the depletion of sulfate and sodium ions with respect to seawater is common in saline waters in the dry valleys.

Lake Fryxell, having the composition most similar to that of seawater, also has a Br/Cl ratio (2.96 × 10^{-3}) similar to that of seawater (3.42 × 10^{-3}), but the HBO$_2$/Cl ratio (2.43 × 10^{-3}) is a little larger than that of seawater (0.97 × 10^{-3}). In the east and west lobes of Lake Bonney, the above ratios resemble those of seawater [Angino et al., 1964; Torii et al., 1977a]. In Lake Vanda, by contrast, these ratios are considerably lower than those in seawater [Torii et al., 1977a].

Nutrients

After the pioneering investigations of nutrient matters such as silicate, phosphate, nitrate, and nitrite compounds [*Armitage and House*, 1962; *Angino et al.*, 1963, 1964; *Goldman*, 1964; *Yamagata et al.*, 1967], further comprehensive research was conducted, especially in Lake Bonney [*Parker et al.*, 1973, 1975; *Hoehn et al.*, 1974; *Weand et al.*, 1975; *Torii et al.*, 1975]. This research demonstrated that the concentrations and the distribution patterns of nutrients differ greatly from lake to lake, as in the case of major salt components. One of the most characteristic features of nutrient distribution is the high concentration of nitrogen compounds in and around the lakes. Among all dry valley lakes, the water in the east lobe of Lake Bonney has the highest content of nitrogen compounds [*Torii et al.*, 1975].

Gas chromatography and combined gas chromatography-mass spectrometry revealed the presence of gaseous nitrogen, carbon dioxide, and argon in the botton layer in lakes Vanda, Bonney, and Fryxell, but methane was detected only in Lake Vanda and the west lobe of Lake Bonney [*Waguri*, 1976; *Kaminuma et al.*, 1977].

There is still scanty information on organic compounds in the lake waters [*Parker et al.*, 1974]. The contents of total organic carbon and extractable organic carbon in saline waters showed a wide range of variation depending on the sampling depths and locations. At a depth of 68 m in Lake Vanda, the content of total organic carbon was 63.8 mg/l, which is considerably higher than that of lake waters on other continents. Saturated fatty acids with carbon numbers C_8 through C_{32} and unsaturated fatty acids with carbon numbers C_{16} and C_{18} were detected, but generally the concentration of hydrocarbons was extremely low. However, it is of interest that characteristic hydrocarbons such as branched C_{18} were found in the bottom water. These compounds may be derived from organisms such as algae, bacteria, and fungi [*Matsumoto and Hanya*, 1977; *Matsumoto et al.*, 1979].

Stable Isotopes

The isotopic ratios of hydrogen and oxygen in these saline lakes are nearly the same as those of snow, ice, and glaciers in this region, which suggests a meteoric origin of the lake waters [*Ragotzkie and Friedman*, 1965; *Ambe*, 1966, 1974; *Nakai et al.*, 1975; *Matsubaya et al.*, 1979]. For example, on the standard mean ocean water (SMOW) scale, the values of $\delta^{18}O$ and δD of water in Lake Vanda are very low, ranging from -28.2 to $-32.4‰$ in $\delta^{18}O$ and from -229 to $-251‰$ in δD, whereas those of Onyx River water and glacial meltwaters from the catchment areas range from -26.1 to $-32.4‰$ in $\delta^{18}O$ and from -230 to $-251‰$ in δD [*Matsubaya et al.*, 1979]. These values are considerably lower than those of seawater. This indicates that the present lake water was derived from glacial meltwaters and not from seawater. *Craig* [1966] emphasizes that the present isotopic ratios in these lake waters are not the same as the original ones because of modification by inflow of meltwater and isotopic exchange with atmospheric moisture.

In contrast, the $\delta^{34}S$ values of sulfate ion in water in Lake Vanda are high and increase with depth ($+15$ to $+22.5‰$), reaching $+46‰$ near the bottom. The highest value, $+48.8‰$, is found in the groundwater [*Nakai et al.*, 1975]. However, the $\delta^{34}S$ values of H_2S below a depth of 60 m are low, ranging from $+7.7$ to $+10.5‰$, giving evidence of bacterial activity [*Nakai and Jensen*, 1964].

Jones and Östlund [1971] determined the concentration of ^{14}C in water in Lake Vanda by using $\delta^{14}C$ values. They estimated that the age of the water at a depth of 60 m was 2130 ± 80 years B.P. *Wilson* [1964] calculated from a consideration of the diffusion of salts from the bottom of Lake Vanda that the lake was flooded from low volume about 1200 years ago and cited a similar ^{14}C age of Wellman's confirming this age estimate. On the other hand, remains of algae found on the paleolake shorelines lying up to 55 m above the present surface of Lake Vanda gave ^{14}C ages ranging from 1300 to 2900 years B.P., but a reasonable time-altitude relationship could not be found. It seems likely that the variation in the carbon age of the algal remains is due to fluctuations in lake level during the last 3000 years [*Yoshida et al.*, 1975].

HIGHLIGHTS OF LAKE VANDA

Heat Source

Unusual water temperature profiles have been reported in saline lakes in the dry valleys since the first coring (1960). Lakes Fryxell and Bonney have similar temperature profiles, showing maxima in the mixolimnion (in the east lobe of Lake Bonney,

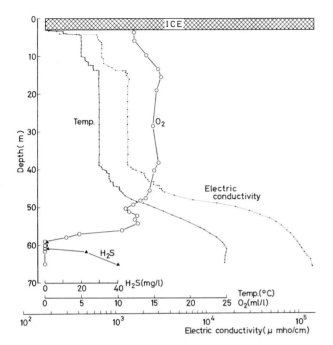

Fig. 2. Vertical distribution of temperature, dissolved oxygen, and electric conductivity in Lake Vanda.

7.3°C in January 1965; and in Lake Fryxell, 2.0°C in January 1965), whereas in Lake Vanda the temperature is 0°C just below the lake ice and rises to 25.1°C (in January 1965) near the bottom; this pattern is present throughout the year (Figure 2) [Torii et al., 1967]. I. Zotikov (personal communication, 1965) recorded a bottom temperature at 25°C in early September 1964. The exceptionally high temperature at the bottom has drawn the special attention of many investigators. In the mixolimnion of Lake Vanda the vertical profile of temperatures in the upper layer has a typical steplike structure in which isothermal layers alternate with sheetlike layers having steep temperature gradients [Hoare, 1966; Shirtcliffe and Calhaem, 1968]. The thick middle layer (19-40 m) is convective, having a uniform temperature of about 7.6°C (in January 1972). Below 40 m in the monimolimnion, the steplike structure appears again, and then the temperature increases steadily from 50 m down to the bottom. This peculiar temperature profile is stable, having varied very little over the last decade [Yusa, 1975, 1977].

In order to explain the abnormal vertical profile of temperatures in lakes Bonney and Fryxell, models involving solar heating and molecular diffusion have been postulated and are now generally accepted [Shirtcliffe, 1964; Shirtcliffe and Benseman, 1964; Hoare et al., 1964, 1965]. However, no reasonable explanations for the peculiar thermal structure of Lake Vanda, which exhibits a maximum water temperature near the bottom, have been postulated until recent years. It has been suggested that there are two possible sources of heat regulating the water temperatures in Lake Vanda; one is solar radiation [Wilson and Wellman, 1962; Ragotzkie and Likens, 1964; Hoare, 1966, 1968] and the other is geothermal or hydrothermal activity underneath the lake [Nichols, 1962; Armitage and House, 1962; Angino and Armitage, 1963; Ragotzkie and Likens, 1964; Angino et al., 1965; Goldman et al., 1967; Kriss and Thomson, 1976].

A few years ago, Yusa [1972, 1975] concluded that the main heat source for Lake Vanda was solar radiation and that the influence of geothermal or hydrothermal activity is insignificant, as suggested previously by Wilson and Wellman [1962]. This conclusion is in accord with the results of DVDP geothermal studies, which revealed no geothermal activity in the dry valleys area [Decker, 1974; Wilson et al., 1974b]. Unusually high water temperature in saline lakes in the polar regions may be explained as a result of a combination of the effects of solar radiation, thermohaline convection of water, and depth of the lake [Yoshida et al., 1975]. The generation and development of thermohaline convection not only promote the formation of steplike structures in the temperature profile of lake water but also increase the magnitude of the apparent thermal diffusivity of the whole lake. The development of thermohaline convection was analyzed fully by Yusa [1977], who found close agreement between his theoretical model and empirical evidence from Lake Vanda. The maximum temperature in Lake Vanda has decreased very slowly during the last decade, and the temperature in the thick convection layer of Lake Vanda also seems to have dropped [Yusa, 1975]. A similar tendency has also been observed at Lake Bonney [Torii et al., 1972]. Yoshida et al. [1975] suggested the following three possible causes: (1) an increase in lake level due to inflow of water, (2) a tendency toward a decrease in insolation, (3) the generation of convection in the past due to a steep temperature gradient. Among these causes, the first may be the most probable because it influences the amount of solar heating in the lake. Since 1969 a general rise in the levels of dry valley lakes has been observed [Chinn, 1976]. A consistent pattern is apparent throughout the region, including lakes Vanda and Bonney.

TABLE 2.   Nutrient Matters in Lake Vanda (Sampling Date: December 9, 1972)

Depth, m	Conductivity, micromho/cm at 18°C	pH	Temperature, °C	Dissolved Oxygen, ml/l	Alkalinity, milliequivalent/l	NO_2-N, μg-atom/l	NO_3-N, μg-atom/l	NH_4-N, μg-atom/l	SiO_2-Si, μg-atom/l	PO_4-P, μg-atom/l
3.9	471	7.11	4.7	12.2	0.64	0.11	2.9	nd	111	0.1
28.6	1.53×10^3	7.40	7.6	14.8	1.31	0.01	1.3	nd	268	0.0
48.4	3.90×10^3	6.72	10.6	13.1	1.49	0.19	2.1	nd	323	nd
55.1	4.88×10^4	6.46	18.1	15.4	2.43	2.27	56.4	nd	911	nd
64.6	1.23×10^5	5.45	24.3	0	4.73	0	0	10.1	1690	nd

nd, not detected.

Nutrients

As shown in Figure 2, the electric conductivity of lake water also increases with depth, and the vertical profile of electric conductivity shows a steplike structure which resembles the pattern of temperature [Hoare, 1966; Shirtcliffe and Calhaem, 1968]. The water is saturated or supersaturated with oxygen down to about 55 m, below which the dissolved oxygen content decreases quickly with depth until 59 m, where an anaerobic state is attained and hydrogen sulfide begins to occur. The concentration of hydrogen sulfide reaches 78.6 mg/l at the bottom (68 m) near the DVDP drilling site [Nakai et al., 1975], but it almost disappears in the groundwaters.

Such chemical conditions strongly suggest the presence of sulfate-reducing bacteria in the water, as anticipated by Barghoorn and Nichols [1961], although Kriss and Thomson [1976] found very little or no evidence for such living bacteria in samples of ooze collected at the bottom of Lake Vanda.

Phosphate-P is entirely absent in Lake Vanda water except in the eastern part, where Onyx River water flows in (Table 2). By contrast, silicate-Si is abundant and reaches a maximum value of 1.7 mg-atom/l near the bottom. Nitrate-N content is 2-3 μg-atom/l in the upper layers, increasing with depth to very high values below 50 m and a maximum value of 56.4 μg-atom/l at 10 m above the bottom. No nitrite-N was detected proximate to the bottom. Almost the same pattern was observed in the vertical distribution of nitrite-N, though its content is much smaller than that of nitrate-N [Torii et al., 1975].

The diminution or even extinction of inorganic nitrogen compounds near the bottom clearly shows anaerobic conditions and the occurrence of a deni-trification reaction producing molecular nitrogen and some ammonia. In fact, there is less than 0.2 ml/l of dissolved oxygen in the bottom water. As in the case of the sulfate-reducing bacteria mentioned above, these observations lead us to conclude that denitrification bacteria are also to be found in the bottom water, despite the extraordinarily high salinity.

According to Johannesson and Gibson [1962] and Morikawa et al. [1975a], the evaporites and soils in the dry valleys are rich in nitrogen compounds. The ubiquity of nitrogen compounds was also pointed out by Claridge and Campbell [1968]. Water samples from Lake Canopus, located 0.4 km south of Lake Vanda, and Don Quixote Pond, in the north fork, contained, respectively, 51 μg-atom/l and 24.8 mg-atom/l of nitrate-N [Torii et al., 1975]. The nitrate and nitrite contents in the DVDP core 4 (Lake Vanda) and the adjacent groundwaters are also very high [Torii et al., 1975]. The hypothesis that nitrate salt found abundantly in this area might have been transported by air [Claridge and Campbell, 1968] appears to be supported by the fact that considerable amounts of nitrogen compounds are also detected in lakes and in meltwater inflowing from the surrounding glaciers. On the other hand, the abundance of nitrogen compounds, silicates, and other major elements in the groundwaters beneath Lake Vanda supports the view that these compounds and elements have come from lake bottom sediments [Torii et al., 1975].

A microflora study in Lake Vanda by Sugiyama et al. [1967] shows that the population and distribution of the fungal strains increase gradually from the upper layer of the lake to the bottom sediment. It was found that the fungal strains could be isolated in greatest numbers from bottom sediments incubated at 25°C.

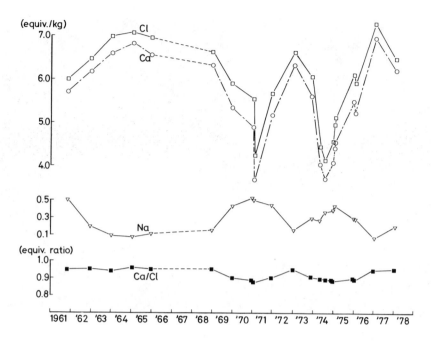

Fig. 3.   Secular changes in the main chemical components of pond water during 1961–1978 in Don Juan Pond.

DON JUAN BASIN

Chemical Characteristics of
Don Juan Pond Water

Don Juan Pond, discovered by *Meyer et al.* [1962], is situated nearly in the center of a closed, flat lacustrine plain surrounded by moraine heaps and talus deposits in the south fork of Wright Valley, about 9 km west of Lake Vanda. The pond is small and very shallow, usually only 10 cm deep, 300 m long, and 100 m wide [*Yamagata et al.*, 1967], and in midsummer is fed mainly by melt-water flowing from a moraine containing ice lenses on the west edge of the basin [*McGinnis et al.*, 1972].

Don Juan Pond water contains an extremely large amount of salt and is nearly a saturated solution of calcium chloride (Table 1). The salinity, chemical composition, and size of pond change considerably from year to year due to changes in the amount of water supplied. The salt concentration is the highest known among saline lakes in Antarctica. *Cartwright and Harris* [1978], on the basis of their hydrogeologic studies, suggested that the pond basin is a discharge zone for a regional groundwater system, which may transport water originating beneath the east Antarctic ice sheet [*McGinnis and Jensen*, 1971].

Measurements of secular changes in the major components of the pond water since summer 1961-1962 [*Meyer et al.*, 1962; *Torii et al.*, 1977b] show that the highest concentration occurred in the 1964-1965 summer season and the lowest in the 1970-1971 summer season. This low concentration was caused by an unusually great inflow of water, which raised the pond surface level to 25 cm. Changes in the amount of water were reflected by changes in the concentrations of calcium and chloride in the pond water, but the ratio of calcium to chloride remained nearly constant (Figure 3). The calcium and sodium concentrations in the pond water seemed to be in inverse correlation, i.e., when the concentration of calcium was low, the concentration of sodium was relatively high. In addition, in an ordinary year, practically no sulfate was detected, except when the water was sufficiently dilute (Table 1). These observations can be explained by the solubilities of sodium chloride and calcium sulfate in a concentrated calcium chloride solution.

The chemical composition of the groundwaters obtained at the DVDP drilling site shows that the chlorinity ratios of the elements in the pond and groundwaters are the same, with the exception only of sulfate. This suggests that the solutes in the pond are probably supplied from sediment and/ or groundwater. In both kinds of water the sulfate concentration approximately corresponds to the

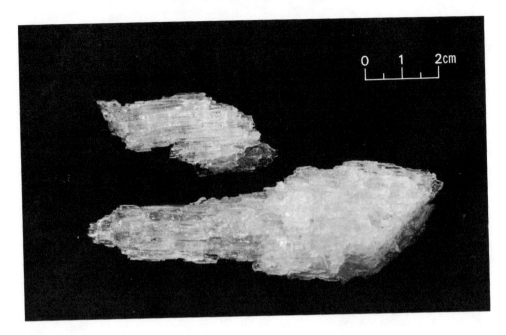

Fig. 4.   Antarcticite.

solubility of calcium sulfate in a concentrated calcium chloride solution [*Torii et al.*, 1977*b*].

The isotopic ratios of hydrogen (δD) and oxygen (δ^{18}O), as well as the salinity of the pond water, are higher in summer than in winter owing to evaporation during the summer season, though groundwater discharges to the pond throughout the year [*Harris and Cartwright*, 1978; *Matsubaya et al.*, 1979].

Antarcticite

Since the discovery of the pond, white crystals occurring in and around the pond have been studied by a number of investigators [*Tedrow et al.*, 1963; *House et al.*, 1966]. Some crystals found in water and at the bottom of the pond were identified as a new mineral, $CaCl_2 \cdot 6H_2O$, to which the name 'antarcticite' (Figure 4) was given, and approved by the Commission on New Minerals and Mineral Names, International Mineralogical Association [*Torii and Ossaka*, 1965]. The crystal is uniaxial in the form of needles, and the optic sign is negative; the refractive indices measured by the immersion method are $n_o = 1.550$ and $n_e = 1.495$. These values are in accord with those of artificial crystals of calcium chloride hexahydrate ($CaCl_2 \cdot 6H_2O$) determined by *Winchell and Winchell* [1964]. A powder X ray pattern of the crystals collected in the pond agrees very well with that of the pure calcium chloride hexahydrate given in the *American Society for Testing Materials* [1963] X ray diffraction data card. The results of chemical analysis of antarcticite are shown in Table 3. In addition to major constituents, magnesium, sodium, potassium, and strontium were detected by emission arc spectral analysis.

Because of its highly hygroscopic nature, calcium chloride hexahydrate should never be found in any part of the globe except this extremely dry, cold place. During the summer, antarcticite could be found only in 1963-1964, 1964-1965, 1965-1966, 1976-1977, and 1977-1978, but even in 1974, when it was not found in the summer season (1973-1974), A. B. Bromley (personal communication, 1974) confirmed the presence of this mineral in July, in the cold winter season.

TABLE 3.   Chemical Composition of Antarcticite

Component	Weight, %	Mole, %	Molar Ratio
Ca	17.5	0.437	1.000
Mg	0.41	0.017	0.039
Na	0.34	0.015	0.034
K	0.008	0.002	0.000
Cl	32.7	0.923	2.11
H_2O	49.2	2.730	6.25
Total	100.1		

When antarcticite was absent, the pond usually was larger and the flat basin around the pond was whiter in color, but in small puddles around the pond, antarcticite was observed. This suggests that the amount of water supply exceeded the loss by evaporation, which might result in the dilution of water in the main pond. Study of the conditions of occurrence of antarcticite in Don Juan Pond shows that the crystallization of the mineral is influenced by both the concentration of calcium chloride and the water temperature [Torii et al., 1970, 1977b].

Salt Distribution in the Don Juan Basin

Powder X ray diffraction studies of the salt deposits in and around the pond showed that all the white crystals in the pond were antarcticite, whereas the majority of the white crystals on the adjacent flat basin were halite with only a small fraction of antarcticite and gypsum. The east slope of the basin was dominated by halite, with chlorite and vermiculite at higher elevations. By contrast, in the upper layers of the DVDP core, sediments were dominated by halite and gypsum. Another remarkable finding in the core was the comparatively large amount of calcite in fractures of the basement rock, especially in the dolerite. The presence of calcite in the basement should be helpful in the attempt to examine the marine origin hypothesis for Don Juan salt [Torii et al., 1977b], because calcite is the first precipitate from seawater that is concentrated under frigid conditions [Thompson and Nelson, 1956].

The soluble salts in sediments of DVDP core 13 (Don Juan Pond), which penetrated 12.7 m of sorted sand and silt, were leached in the laboratory by water and perchloric acid successively. The chemical analyses of both leachates can be summarized as follows: (1) the major component of the water-soluble part is calcium chloride (more than 90%); (2) the upper layers of the sediments contain a large amount of sulfate; and (3) the perchloric acid-soluble part is enriched in potassium and magnesium as compared with the water-soluble part. Because the sediment layer contains much potassium, magnesium, and sulfate as compared with contemporary saline water, we conclude that the sediment layer was formed before the present Don Juan Pond [Torii et al., 1977b].

The vertical distribution of salt in the DVDP core shows that the principal component is calcium chloride and that the total amount of salts in each layer does not change markedly with depth. This suggests that salts were deposited continuously during sedimentation in the Don Juan basin.

Salt Balance in the Don Juan Basin

In order to deal with the salt balance, the Don Juan basin can be divided simply into three components, i.e., (1) the liquid and (2) the solid phases of the pond, and (3) the surrounding flat basin and slopes. The salt balance within the liquid and solid phases can be analyzed using the data in Table 1. In the 1964-1965 summer season the salinity was the highest (specific gravity: 1.386) and the size of the pond was the smallest (105×205 m, 10 cm deep, volume of water: 2150 m^3), whereas in the 1974-1975 summer season, the specific gravity was 1.265 and the size was much larger (134×326 m, 15 cm deep, volume: 6550 m^3). The total amount of NaCl in the liquid phase was 12.4 tons in 1964-1965 and 192 tons in 1974-1975; the difference (179.6 tons) should have been held in the solid phase in 1964-1965. The total $CaCl_2 \cdot 6H_2O$ in the liquid phase was 2240 tons in 1964-1965 and 3840 tons in 1974-1975; the difference (1600 tons) should have been present in the solid phase (as antarcticite) in 1964-1965 [Torii et al., 1977b].

The area of the Don Juan basin is about 400×900 m. If 10% of the deposits is soluble and the specific weight of the solid phase is 2, the total salt content in the basin should amount to 72,000 tons per 1-m thickness and 940,000 tons to a depth of 13 m. This means that more than 95% of salt in the basin is present in the solid phase around and beneath the pond.

Cartwright et al. [1974a] reported that there are at least two, and possibly three higher paleo-shorelines above the present flat basin floor. In order to examine the hypothesis of a seawater origin for the salts, we estimated the volume of the paleolake. If the paleo-shoreline was 20 m above the present basin floor, the depth of seawater was at least 20 m; if the thickness of the sedimentary deposits is included, the total depth was 33 m. Thus the original volume of seawater can be calculated as 1.2×10^7 m^3 (Table 4). The original volume of seawater may also be estimated using the total salt amount of 940,000 tons now present in the basin. If all of the salt came from the evaporation of seawater, the original volume of water was 2.6×10^7 m^3.

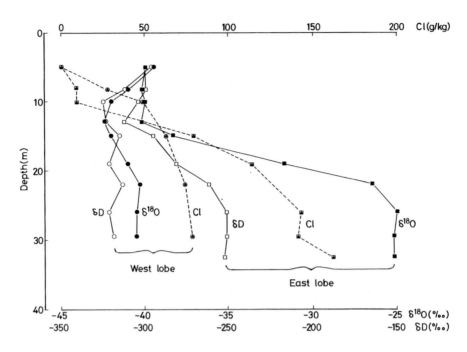

Fig. 5   Vertical distribution of δD, δ¹⁸O, and chlorinity in Lake Bonney.

If the total amounts of each of the major chemical elements K, Mg, SO₄, and Cl present in the basin are used to estimate the original volume of seawater, the estimated volumes range from 1.2 to 2.2 × 10^7 m³. These estimates are of the same order of magnitude as that based on the paleolake shoreline. However, when the estimation is made on the basis of total calcium, a much larger value of almost 37×10^7 m³ is obtained.

FORMATION OF LAKE BONNEY

One of the characteristics of Lake Bonney is the great difference in chemical composition (Table 1) and nutrient profile between waters in the east and west lobes [*Torii et al.*, 1975].

The hydrogen and oxygen isotopic ratios in Antarctic fresh and saline lakes, as well as in glaciers and their meltwaters, have been determined by *Hendy et al.* [1977] and *Matsubaya et al.* [1979], who discussed the origin and evolution of saline lakes using the isotopic data. The vertical profiles of hydrogen and oxygen isotopes show that the ratios underneath the lake ice are almost the same in the east and west lobes, but differ considerably with depth (Figure 5); δ¹⁸O and δD in the east lobe increase remarkably with depth, but in the west lobe they do not. The strongly stratified distribution of isotopic ratios and chlorinity in water in the

east lobe of Lake Bonney is characteristic, and diffusive mixing is occurring between the bottom saline water and overlying fresh water, as in the case of Lake Vanda [*Wilson*, 1964; *Shirtcliffe*, 1964]. *Matsubaya et al.* [1979] concluded that the distribution of isotopic ratios in the monimolimnion in the east lobe indicates that the lobe was originally a shallow saline lake with an open surface layer prior to the inflow of fresh water and the onset of vertical diffusive mixing. In contrast, the monimolimnion of the west lobe was never exposed to the air long enough to become isotopically differentiated.

The isotopic ratios of hydrogen and oxygen in alpine glaciers and their meltwaters in this area range mostly from -210 to $-260‰$ in δD and -26 to $-34‰$ in δ¹⁸O. These values are in accord with those reported for snow and ice from other coastal areas [*Picciotto et al.*, 1960; *Epstein et al.*, 1963;

TABLE 4.   Estimates of the Original Volume of Seawater

Estimation From	Volume, × 10^7 m³
Paleolake shoreline	1.2
Total salt deposit	2.6
Calculation from K, Mg, SO₄, and Cl	1.2–2.2
Calculation from Ca	36.6

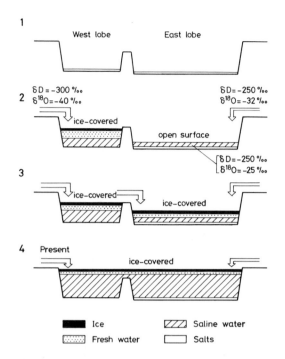

Fig. 6.   Scheme of the formation of Lake Bonney.

Lorius et al., 1969]. However, Taylor Glacier, which flows from the polar plateau, has much lower isotopic ratios ($\delta D = -330\%o$, $\delta^{18}O = -42.5\%o$). Using these data, the following model of formation of Lake Bonney can be postulated.

Initially, dry salt deposits were accumulated in the lower parts of each of the lobes (stage 1 in Figure 6). Then, meltwater with isotopic ratios of approximately $-300\%o$ in δD and $-40\%o$ in $\delta^{18}O$ flowed inland to the west basin, forming a saline lake by dissolving the salt deposits. At the same time a saline lake was formed in the east basin by meltwater flowing from the east and having ratios of approximately $-250\%o$ in δD and $-32\%o$ in $\delta^{18}O$. In the west lobe the amount of inflow was large enough to form a two-layer lake having a saline bottom layer and a fresh upper layer. The fresh water was no doubt covered with ice soon after the formation of the west lobe, so that the water of the west lobe did not become isotopically different from that of the inflow. The east lobe remained a shallow saline lake of uniform concentration because the amount of inflow was not enough to allow the formation of lake ice (stage 2 in Figure 6). During this stage, evaporation of water in the east saline lake resulted in the enrichment of $\delta^{18}O$ in the water. When the west lobe filled up, surface meltwater started to flow into the east lobe through the channel (stage 3 in Figure 6). Finally, the present

stage of Lake Bonney was formed when water levels of the east and west lobes became equal (stage 4 in Figure 6). At present, because of the higher inflow rate of meltwater into the west lobe, surface water still moves from the west to the east. This movement is confirmed by the chemical similarity of surface waters in both lobes. This model suggests that the stratification of the isotopic ratios and the major elements in water in the east lobe started only after stage 3, so that it may be said that the east lobe is younger than the west.

Taylor Glacier meltwaters contributed to Lake Bonney during its formation contained much calcium and sulfate (Table 5), which, upon concentration in the west lobe, precipitated as gypsum. This can be assumed because the present bottom water of the west lobe is saturated with calcium sulfate and because two different kinds of evaporites are found at the lake bottom, gypsum in the west lobe and halite and/or hydrohalite with a small fraction of gypsum in the east lobe [*Yamagata, et al.*, 1967; *Wilson and Hendy*, 1974; *Wilson et al.*, 1974a; *Craig et al.*, 1974, 1975]. As a result, waters relatively enriched in sodium and magnesium form the bottom waters of both lobes. Considerable amounts of calcium and sulfate are still being supplied by Taylor Glacier meltwater, and the formation of gypsum on the bottom of the west lobe seems to be continuing. *Hendy et al.* [1977] estimated that the total amount of sodium chloride in the bottom deposits in the east lobe is at least 4×10^6 tons. The very cold bottom water in the east lobe (Table 1) is thought to be nearly saturated with sodium chloride and continues to dissolve sodium chloride from the sediment.

The channel connecting the two lobes was 5.2 m wide in 1903 [*Scott*, 1905], 30.5 m wide and 6.1 m deep in 1911 [*Taylor*, 1916], 39 m wide and 8.2 m deep in December 1963 [*Hoare et al.*, 1964], and 47.5 m wide in December 1974 (T. Torii, unpublished data, 1974). These observations support the hypothesis that during the last 60 years a large amount of water flowed into the lake, as originally proposed by *Shirtcliffe* [1964] on the basis of a diffusion curve drawn from chemical profile data. By using ^{14}C ages and/or diffusion equation of the bottom waters, *Hendy et al.* [1977] estimated the age of formation of the west lobe to be 15,000 years B.P. and that of the east lobe to be from 1800 to 4600 years B.P., whereas *Matsubaya et al.* [1979] estimated the ages of the lobes to be 6000 and from 1200 to 2600 years B.P., respectively.

TABLE 5. Chemical Composition of Snow and Glacial Meltwaters in Antarctica (mg/kg)

Sample	Na	K	Ca	Mg	Cl	SO$_4$
Snow and ice (35), μg/kg	17–94	0.9–26	1.9–4.0	1.6–12.8	57–140	
Onyx River (8)	5.5–15	0.9–1.8	8.5–21.0	1.4–4.2	7.1–20.1	3.7–14.1
Inflow water from glacier to						
West lobe of Lake Bonney	12.6	1.2	19.0	4.0	12.0	10.7
East lobe of Lake Bonney	25.6	3.8	20.0	6.0	25.0	29.1
Lake Canopus	19.6	1.9	24.2	8.0	20.7	23.6
Lake Miers	<1.0	1.0	15.2	0.6	3.5	2.0
Lake Brownworth	3.9	0.5	1.9	0.6	4.9	6.2
Lake Bull	9.0	3.3	14.4	3.8	21.5	9.7

Numbers in parentheses represent the number of samples. Data from *Boutron et al.* [1972], *Murozumi et al.* [1969], *Torii et al.* [1966, 1975, 1979], and *Yamagata et al.* [1967].

EVOLUTION OF SALINE WATERS

Extensive research on saline lakes in the dry valleys, which has mostly been conducted in recent years, enables us to form general views concerning the limnological nature of these lakes and the surrounding areas. Geochemical studies, including geochronological studies using radioisotopes, also have helped to explain some of the environmental changes associated with different climatic and glacial events in the past. From the geochemical point of view, one of the most interesting and important problems is the riddle of the source of the salts and of the evolution of saline lakes in this region.

As described in detail above, there is a great difference in chemical properties between the waters of the two neighboring valleys. This difference suggests that both the salt source and the process of evolution are complex and that it is not easy to arrive at a reasonable and consistent explanation without making a special effort.

Since the publication of the first report by *Armitage and House* [1962], many investigators have struggled to explain the formation of lakes Vanda and Bonney on the basis of chemical studies. In the earliest stage a special concern was the true source of the salt. Hypotheses concerning the origin of the salts postulated by different investigators are summarized as follows: (1) hydrothermal springs by *Armitage and House* [1962], *Angino and Armitage* [1963], and *Angino et al.* [1962, 1964, 1965]; (2) trapped seawater by *Nichols* [1961, 1963], *Angino et al.* [1964], *Craig* [1966], *Boswell et al.* [1967a, b], *Webb* [1972], and *Morikawa et al.* [1975b]; (3) chemical weathering of rocks by *Meyer et al.* [1962],

Gibson [1962], *Tedrow et al.* [1963], *Angino et al.* [1965], *Jones and Faure* [1967, 1968, 1969], and *Claridge and Campbell* [1977]; (4) atmospheric salt (primarily sea spray) by *Angino et al.* [1962], *Nichols* [1963], *Dort and Dort* [1970], and *Claridge and Campbell* [1977]; (5) glacial meltwater by *Ragotzkie and Friedman* [1965], *Boswell et al.* [1967a, b], and *Mudrey et al.* [1973]; and (6) groundwater discharge by *Cartwright et al.* [1974b, 1975] and *Cartwright and Harris* [1978].

As noted previously, if we assume that the salts in the Don Juan basin come entirely from the evaporation of seawater, we cannot explain the presence of excessive calcium in the basin. This forces us to seek sources of salts other than trapped seawater, such as rock weathering, sea salt spray, and/or glacial meltwater [*Mudrey*, 1974]. For example, if sodium in the saline pond in the Labyrinth is assumed to be meteoric only, the age of the pond can be estimated. Assuming an annual snowfall of 30 g/cm^2/yr and a sodium content in snow of 50 ppb, the annual deposition of sodium is 1.5 μg/cm^2. As the content of sodium in the saline pond is 4.13 g/kg (Table 1), the total amount of sodium in a pond water column 20 cm deep (the present depth of the pond) by 1 cm^2 is 84 mg. Dividing 84 mg by 1.5 μg/yr yields 56,000 years, which is the highest possible estimate of the age of the pond, because factors such as water supply from the surrounding catchment area and the contribution from rock weathering and ion exchange are neglected. The true age should be much less than the above estimate.

Cartwright et al. [1975] emphasize the contribution of groundwater to surface waters in Wright

Valley. The groundwater contains salts and is in most places found only in the active layer of permafrost. The possible existence of a groundwater system at Don Juan Pond is also suggested by studies of a temperature log of DVDP borehole 13, which recorded alternation of high and low temperature gradients [Decker et al., 1975; Bucher and Decker, 1976].

In conclusion, most workers are now inclined to believe that multiple sources of salt account for such a large amount of salts. In order to determine the salt sources, it is necessary to study further the chemical nature of not only the saline waters but also the whole drainage system, including soil and sediments, and the glacial history of the valleys.

Soil and DVDP Cores in Wright Valley

The chemical analysis of three DVDP cores from Wright Valley (Lake Vanda, Don Juan Pond, and North Fork) revealed that halite and calcite were prevalent in the evaporites in the cores, that sulfates such as gypsum and mirabilite were concentrated in the upper layers, and that the principal cationic components in the leachates were mainly calcium and magnesium. The concentration of alkaline earth elements greatly exceeded that of the alkali metals [Morikawa et al., 1975a; Torii et al., 1977b; Murayama et al., 1979].

The vertical distribution of the chemical components in the cores from Wright Valley suggests that climatic changes took place in the past. For example, a peak in the total salt content may reflect a diminished water supply during a cold climatic period, while a minimum may be due to the leaching of salt deposits as a result of a greater water supply during a warm climatic period [Torii et al., 1979]. This tendency can be seen also in the size frequency distribution of sediments in DVDP core 14 (North Fork), as determined by granulometric analysis; the distribution shows two types corresponding to aeolian and lacustrine depositions [Nakao et al., 1979]. Behling [1975] studied in detail a buried soil that has features indicating a more humid and perhaps a warmer climate and suggested that the DVDP cores in Wright Valley might have evidence of similar changes, such as in the proportion and species of clay minerals in fine-grained sediments.

Successive precipitation and dissolution of salts in Lake Vanda has been suggested by a number of investigators using evidence from diffusion calcula-

tions [Wilson, 1964; Shirtcliffe, 1964], from the dating of algal remains on the paleolake shoreline [Yoshida et al., 1975], and from the study of stable isotopes in DVDP core 4 (Lake Vanda) [Nakai et al., 1975]. These observations lead us to consider, in addition to salts from trapped seawater, salts from sea spray and/or rock weathering accumulating over a long period of time under frigid conditions.

We wish to point out that the most important clue to the chemical puzzle of the saline lakes is the chemical composition of snow and ice and of glacial meltwaters feeding the lakes. The concentration of salt in snow and ice is very small (Table 5), but even very small annual accumulations should, in the course of the late Cenozoic, have amounted to a large quantity of salt; this salt should have remained either in concentrated solution or in deposits in glacial troughs. The salts thus left in the valleys should have been leached or deposited according to their different solubilities during the course of water movements. As studies of core samples show, the less soluble sulfate evaporites prevail on the ground surface and in the upper layer of sediments, while the soluble calcium and magnesium compounds prevail in the lower layers [Torii et al., 1979].

Jones and Faure [1967] suggest that the major source of the soil salts may be weathering of rocks. Soil-forming processes in the dry valleys take place in an ahumic system. Saline soils that closely approximate evaporite soils are present in the valleys [Tedrow and Ugolini, 1966], and mineral synthesis is taking place [Ugolini, 1976]. Morikawa et al. [1975a] found that some soil samples collected around Lake Vanda contain water-soluble solids in an amount exceeding 20%, that these solids consist principally of sodium and chloride, and that the most prevalent mineral is halite. Generally, the concentrations of calcium and sulfate are low.

The cationic composition of the leachates from soil around Lake Vanda and from the DVDP cores is quite similar [Murayama et al., 1979]. As described previously, the water leachates are characterized by high contents of alkali metals, whereas the perchloric acid leachates are characterized by high contents of calcium and magnesium; the latter composition resembles that of the bottom waters in Lake Vanda.

Evolutionary Processes of Saline Waters

The deposition of salts from seawater under

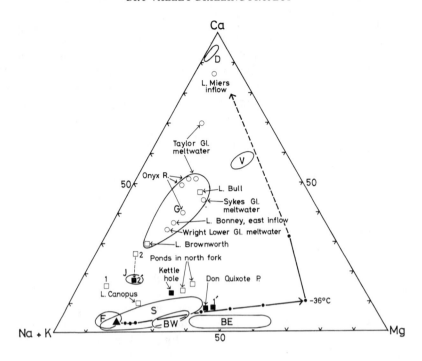

Fig. 7. Chemical composition of different types of waters in the dry valleys (in equivalent percentages). BE, Lake Bonney, east; BW, Lake Bonney, west; D, Don Juan Pond; F, Lake Fryxell; G, glacial meltwater; J, Lake Joyce; S, snow and ice; V, Lake Vanda; open square, freshwater lake or pond; open circle, meltwater; solid square, saline lake or pond; solid triangle, seawater; arrow, changes of seawater under frigid conditions after Thompson and Nelson; 1, fresh pond at Labyrinth; 1', saline pond at Laybrinth; 2, surface water in Lake Joyce; 2', bottom water in Lake Joyce. Original data for waters plotted here are given in Table 5.

frigid conditions is quite different from the deposition processes of salts during the course of evaporation under normal temperature conditions [*Thompson and Nelson*, 1956]. When seawater is concentrated under frigid conditions, calcium carbonate crystallizes first, followed by mirabilite at $-8.2°C$ and sodium chloride dihydrate at $-22.9°C$. The amount of brine remaining at this temperature is only about 2% of the original volume. The magnesium concentration increases until, at $-36°C$, a mixture of magnesium chloride dodecahydrate and potassium chloride begins to crystallize. At a temperature of $-54°C$, the remaining brine consists principally of calcium chloride. Figure 7 is a triangular coordinate diagram showing relative concentrations of sodium plus potassium, magnesium, and calcium in equivalent percentages. The small solid triangle in the diagram represents seawater, and the solid, arrowed line shows the course of evolution of the brine under frigid conditions, as determined by Thompson and Nelson.

The diagrammatic positions of chemical compositions of waters in saline lakes in the dry valleys show that (1) water in Lake Fryxell closely resem-

bles seawater, (2) waters in the west and east lobes of Lake Bonney correspond to brines derived from seawater, respectively, at successive lower temperatures, and (3) waters in Lake Vanda and Don Juan Pond correspond to the residual brine rich in calcium chloride [*Yamagata et al.*, 1967]. Small saline ponds such as Don Quixote Pond in the north fork and 'Kettle Hole' in the south fork of Wright Valley were observed only when the meltwater supply was enough, and at other times were found dry. Water in Lake Bonney that is derived from Taylor Glacier has almost the same composition as that in these ponds, which suggests that these ponds and the lake are earlier in evolutionary sequence than Lake Vanda and Don Juan Pond. And, if the water supply were to be decreased by climatic change, saline water in Lake Bonney should undergo further concentration and, after deposition of halite, potassium chloride, and magnesium chloride dodecahydrate, approach the composition of water in Lake Vanda.

The chemical composition of water in saline pond (1') in the Labyrinth in upper Wright Valley resembles that in the east lobe of Lake Bonney, whereas those in Lake Joyce (2′) in Pearse Valley and in Lake Fryxell resemble each other, except that calcium is richer in the former.

In considering a modification to the simple hypothesis of seawater origin, an important clue is the chemical composition of glacial meltwater (circle G), which has much larger relative concentrations of calcium and sulfate than seawater (Table 5). On the other hand, the similarity of the chemical composition of continental snow and ice (circle S) to that of seawater indicates that a main source of salt in snow and ice is sea spray. Note that the inflow from the Adams and Miers glaciers to Lake Miers is especially rich in calcium. It can be suggested that, in the course of running through glacier beds, meltwater is enriched gradually in calcium and sulfate. Generally, the enrichment in calcium can be explained as a result of the dissolution under low temperatures of more easily soluble components in deposits underneath the ice sheet or in catchment areas.

The compositions of fresh water in the pools in Wright Valley are generally located somewhere between running glacial meltwater (G) and snow (S), as in the case of a fresh pond (1) in the Labyrinth, whereas the composition of water in a saline pond (1′) in the Labyrinth is quite close to that in Don Quixote Pond, which is similar in composition to seawater but richer in magnesium. If the salt in the water in fresh pond (1) were concentrated under frigid conditions, the chemical composition of the pond should approach that of a saline pond (1′) after precipitation of $CaCO_3$ and Na_2SO_4. The chemical composition of water in the lower, saline layers of Lake Joyce (2′) is displaced downward in the diagram relative to that of fresh water in the upper layer (2).

The evolution of saline waters in the dry valleys can be summarized as follows (Figure 7): starting with glacial water and other meltwaters held in ponds or lakes, the composition should approach that of seawater and then move to the right in the diagram, as in the case of seawater under frigid conditions in the Thompson and Nelson model. In a word, later stages of precipitation or concentration of salt should be the same, whether the original water body was meltwater or trapped seawater.

Water in Lake Bonney contains principally sodium and magnesium chlorides, while salts in water in Lake Vanda are composed mainly of calcium chloride. This difference can be explained from not only the chemical but also the glaciogeological

point of view. *Péwé* [1960] described multiple glaciations in Antarctica and multiple advances of Taylor Glacier. Similar advances have been recognized in Wright Valley [*Bull et al.*, 1962; *Nichols*, 1961, 1971]. The earliest advance was the most extensive and is considered to have occurred more than 3.7 million years ago [*Denton et al.*, 1970; *Bull and Webb*, 1973]. In the Neogene, marine waters invaded Wright Valley, converting it into a fjord, and in the latest Pliocene or earliest Pleistocene time, nonglacial and glacio-isostatic uplift brought about a regression of the waters [*Webb*, 1972]. Recent $^{234}U/^{230}Th$ and $^{14}C/^{12}C$ determinations on algal limestones from Taylor Valley by *Hendy* [1978, and unpublished manuscript, 1980] and *Hendy et al.* [1977, 1979] revealed a long history of lakes in the dry valleys, with algal limestones being deposited during the Holocene and each of the previous three interglacial periods.

After the glaciation of this area, copious supplies of meltwater could have washed disseminated efflorescent salts into lakes, where further changes in concentrations and chemical compositions would have been brought about through selective crystallization [*Dort and Dort*, 1970], evaporation, and so forth. A big geomorphological difference between Wright and Taylor valleys is that the former is closed, while the latter is open to McMurdo Sound. In the past, this naturally resulted in different movements of glaciers and water masses in the two valleys. This difference, together with the different solubilities of the compounds of alkali metal and alkaline earth elements, should have affected the distribution of salts following deglaciation of this area.

CONCLUDING REMARKS

Although the Dry Valley Drilling Project very much helped to clarify the geochemistry of unconsolidated sediment near the saline lakes and to produce new and interesting ideas concerning the origin of salt in the region, no quantitative explanation of the balance of salt can yet be achieved. Further studies of sedimentary processes and sediment provenance seem to be necessary. Nevertheless, the geochemical studies of saline lakes in the dry valleys increased our understanding of the entire history of evolution of this unique environment. Chemical and temperature profiles in lakes were interpreted in terms of paleo-environment [*Wilson*, 1964]. Efflorescences of sodium sulfate and other evaporites ubiquitous in the ice-free areas and relatively recent marine sediments in and around the lakes have been investigated in connection with the isolation of the lakes from the sea. Chemical stratification is thought to be an indication of an earlier arid period. Accumulation and differentiation of salt materials in some lakes have proceeded for a very long time. Therefore analysis of salt layers interbedded in or mixed with clastic and biological sediments is an essential prerequisite to elucidating the Quaternary history of the lakes and their surroundings, as well as to the biostratigraphic study of their sediments.

Acknowledgments. The authors are grateful to Y. Miyake, H. Harris, and C. Hendy, who read through the manuscript and gave valuable advice, and to their colleagues, who cooperated and assisted during field work. The authors wish to express their thanks to the Division of Polar Programs of the National Science Foundation, and to the Antarctic Division of the Department of Scientific and Industrial Research, New Zealand, for supporting their research in Antarctica. The Japanese Ministry of Education, Science, and Culture and the Japan Polar Research Association financially supported this work.

REFERENCES

Ambe, M., Deuterium content of water substances in Antarctica, 1, Geochemistry of deuterium in natural water on the East Ongul Island, *Japan. Antarct. Res. Exped. Sci. Rep. 6, Ser. C*, pp.1-13, Nat. Inst. of Polar Res., Tokyo, 1966.

Ambe, M., Deuterium content of water substances in Antarctica, 2, Geochemistry of deuterium of lake waters in Victoria Land, *Antarct. Rec., 48*, 100-109, 1974.

American Society for Testing Materials, X-ray powder data file, *card 1-1220*, Philadelphia, Penn., 1963.

Angino, E. E., and K. B. Armitage, A geochemical study of lakes Bonney and Vanda, Victoria Land, Antarctica, *J. Geol., 71*, 89-95, 1963.

Angino, E. E., K. B. Armitage, and J. C. Tash, Chemical stratification in Lake Fryxell, Victoria Land, Antarctica, *Science, 138*, 34-36, 1962.

Angino, E. E., K. B. Armitage, and J. C. Tash, Nutrient elements in two Antarctic saline lakes, *Bull. Ecol. Soc. Amer., 44*, 38, 1963.

Angino, E. E., K. B. Armitage, and J. C. Tash, Physicochemical limnology of Lake Bonney, Antarctica, *Limnol. Oceanogr., 9*, 207-217, 1964.

Angino, E. E., K .B. Armitage, and J. C. Tash, A chemical and limnological study of Lake Vanda, Victoria Land, Antarctica, *Univ. Kans. Sci. Bull., 65*, 1097-1118, 1965.

Apfel, E. T., Bunger's 'Oasis' Antarctica, *Geol. Soc. Amer. Bull., 59*, 1308–1309, 1948.

Armitage, K. B., and H. B. House, A limnological reconnaissance in the area of McMurdo Sound, Antarctica, *Limnol. Oceanogr., 7*, 36-41, 1962.

Ball, D. G., and R. L. Nichols, Saline lakes and drill-hole brines, McMurdo Sound, Antarctica, *Geol. Soc. Amer. Bull., 71*, 1703-1708, 1960.

Barghoorn, E. S., and R. L. Nichols, Sulfate-reducing bacteria and pyritic sediments in Antarctica, *Science, 134*, 190, 1961.

Behling, R. E., A buried soil in Wright Valley—Evidence for a wetter climate, *Dry Val. Drilling Proj. Bull., 6*, 3, 1975.

Boswell, C. R., R. R. Brooks, and A. T. Wilson, Some trace elements in lakes of McMurdo Oasis, Antarctica, *Geochim. Cosmochim. Acta, 31*, 731-736, 1967a.

Boswell, C. R., R. R. Brooks, and A. T. Wilson, Trace element content of Antarctic lakes, *Nature, 213*, 167-168, 1967b.

Boutron, C., M. Echevin, and C. Lorius, Chemistry of polar snows, Estimation of rates of deposition in Antarctica, *Geochim. Cosmochim. Acta, 36*, 1029-1041, 1972.

Bucher, G. J., and E. R. Decker, Geothermal studies in the McMurdo Sound region, *Antarct. J. U.S., 11*, 88-89, 1976.

Bull, C., and P. N. Webb, Some recent developments in the investigation of the glacial history and glaciology of Antarctica, in *Palaeoecology of Africa and of the Surrounding Islands and Antarctica*, vol. 8, edited by E. M. van Zinderen Bakker, pp. 55-84, A. A. Balkema, Cape Town, 1973.

Bull, C., B. C. McKelvey, and P. N. Webb, Quaternary glaciations in southern Victoria Land, Antarctica, *J. Glaciol., 4*, 63-78, 1962.

Byrd, R. E., Our navy explores Antarctica, *Nat. Geogr., 92*, 497-500, 1947.

Cartwright, K., and H. J. H. Harris, Origin of water in lakes and ponds of the dry valley region, Antarctica, *Dry Val. Drilling Proj. Bull., 8*, 8, 1978.

Cartwright, K., S. B. Treves, and T. Torii, Geology of DVDP 5, Don Juan Pond, Wright Valley, Antarctica, *Dry Val. Drilling Proj. Bull., 3*, 75-91, 1974a.

Cartwright, K., H. Harris, and M. Heidari, Hydrogeological studies in the dry valleys, *Antarct. J. U.S., 9*, 131-133, 1974b.

Cartwright, K., H. Harris, and L. R. Follmer, DVDP hydrogeological studies, *Dry Val. Drilling Proj. Bull., 5*, 134-138, 1975.

Chinn, T. J., Hydrological research report, dry valleys, Antarctica 1974-75, Min. of Works and Develop., Christchurch, 1976.

Claridge, G. G. C., and I. B. Campbell, Origin of nitrate deposits, *Nature, 217*, 428-430, 1968.

Claridge, G. G. C., and I. B. Campbell, The salts in Antarctic soils, their distribution and relationship to soil processes, *Soil Sci., 123*, 377-384, 1977.

Craig, H., Origin of the saline lakes in Victoria Land, Antarctica, *Eos Trans. AGU, 47*, 112-113, 1966.

Craig, J. R., R. D. Fortner, and B. L. Weand, Halite and hydrohalite from Lake Bonney, Taylor Valley, Antarctica, *Geology, 2*, 389-390, 1974.

Craig, J. R., J. F. Light, B. C. Parker, and M. C. Mudrey, Jr., Identification of hydrohalite, *Antarct. J. U.S., 10*, 178-179, 1975.

Decker, E. R., Preliminary geothermal studies of the Dry Valley Drilling Project holes at McMurdo Station, Lake Vanda, Lake Vida, and New Harbor, Antarctica, *Dry Val. Drilling Proj. Bull., 4*, 22-23, 1974.

Decker, E. R., K. H. Baker, and H. Harris, Geothermal studies in the dry valleys and on Ross Island, *Antarct. J. U.S., 10*, 176, 1975.

Denton, G. H., R. L. Armstrong, and M. Stuiver, Late Cenozoic glaciation in Antarctica: The record in the McMurdo Sound region, *Antarct. J. U.S., 5*, 15-21, 1970.

Dort, W., Jr., and D. S. Dort, Low temperature origin of sodium sulfate deposits, particularly in Antarctica, in *Third Symposium on Salt*, edited by J. L. Rau and L. F. Dellwig, pp. 181-203, Northern Ohio Geological Society, Inc., Cleveland, 1970.

Epstein, S., R. P. Sharp, and I. Goddard, Oxygen-isotope ratios in Antarctic snow, firn, and ice, *J. Geol.*, *71*, 698-720, 1963.

Gibson, G. W., Geological investigations in southern Victoria Land, Antarctica, 8, Evaporite salts in the Victoria Valley region, *N.Z. J. Geol. Geophys.*, *5*, 361-374, 1962.

Glazovskaya, M. A., Weathering and primary soil formation in Antarctica (in Russian), *Sci. Pap. 1*, pp. 63-76, Inst. of Moscow Univ., Fac. of Geogr., Moscow, 1958.

Goldman, C. R., Primary productivity studies in Antarctic lakes, in *Biologie Antarctique: Proceedings of the First Symposium Organized by SCAR*, edited by R. Carrick et al., pp. 291-299, Hermann, Paris, 1964.

Goldman, C. R., D. T. Mason, and J. E. Hobbie, Two Antarctic desert lakes, *Limnol. Oceanogr.*, *12*, 295-310, 1967.

Harris, H. J. H., and K. Cartwright, Hydrogeology and geochemistry of Don Juan Pond, *Dry Val. Drilling Proj. Bull.*, *8*, 21,1978.

Hendy, C. H., The chronology of glacial sediments in the Taylor Valley, *Dry Val. Drilling Proj. Bull.*, *Suppl.*, *8*, 1978.

Hendy, C. H., Isotopic measurements on carbonates from the DVDP core 12, Taylor Valley, Antarctica, unpublished manuscript, 1980.

Hendy, C. H., A. T. Wilson, K. B. Popplewell, and D. A. House, Dating of geochemical events in Lake Bonney, Antarctica, and their relation to glacial and climate changes, *N.Z. J. Geol. Geophys.*, *20*, 1103-1122, 1977.

Hendy, C. H., T. R. Healy, E. M. Rayner, J. Shaw, and A. T. Wilson, Late Pleistocene glacial chronology of the Taylor Valley, Antarctica, and the global climate, *Quaternary Res.*, *11*, 172-184, 1979.

Heywood, R. B., Antarctic Limnology: A review, *Brit. Antarct. Surv. Bull.*, *29*, 35-65, 1972.

Heywood, R. B., A limnological survey of the Ablation Point area, Alexander Island, Antarctica, in *Scientific Research in Antarctica*, pp. 39-54, Royal Society of London, London, 1977.

Hoare, R. A., Problems of heat tansfer in Lake Vanda, A density-stratified Antarctic lake, *Nature*, *210*, 787-789, 1966.

Hoare, R. A., Thermohaline convection in Lake Vanda, Antarctica, *J. Geophys. Res.*, *73*, 607-612, 1968.

Hoare, R. A., K. B. Popplewell, D. A. House, R. A. Henderson, W. M. Prebble, and A. T. Wilson, Lake Bonney, Taylor Valley, Antarctica: A natural solar energy trap, *Nature*, *202*, 886-888, 1964.

Hoare, R. A., K. B. Popplewell, D. A. House, R. A. Henderson, W. M. Prebble, and A. T. Wilson, Solar heating of Lake Fryxell, A permanently ice-covered Antarctic lake, *J. Geophys. Res.*, *70*, 1555-1558, 1965.

Hoehn, R. C., B. C. Parker, and R. A. Paterson, Toward an ecological model of Lake Bonney, *Antarct. J. U.S.*, *9*, 297-300, 1974.

House, D. A., R. A. Hoare, K. B. Popplewell, R. A. Henderson, W. M. Prebble, and A. T. Wilson, Chemistry in the Antarctic, Proceedings of the California Association of Chemistry Teachers, *J. Chem. Educ.*, *43*, 502-505, 1966.

Johannesson, J. K., and G. W. Gibson, Nitrate and iodate in Antarctic salt deposits, *Nature*, *194*, 567-568, 1962.

Jones, L. M., and G. Faure, Origin of the salts in Lake Vanda,

Wright Valley, Southern Victoria Land, Antarctica, *Earth Planet. Sci. Lett.*, *3*, 101-106, 1967.

Jones, L. M., and G. Faure, Origin of the salts in Taylor Valley, *Antarct. J. U.S.*, *3*, 177-178, 1968.

Jones, L. M., and G. Faure, The isotope composition of strontium and cation concentrations of Lake Vanda and Lake Bonney in southern Victoria Land, Antarctica, *Rep. 4*, pp. 1-82, Lab. for Isotope Geol. and Geochem., Inst. of Polar Stud. and Dep. of Geol., Ohio State Univ., Columbus, 1969.

Jones, L. M., and H. G. Ostlund, Carbon-14 age and tritium content of Lake Vanda, Wright Valley, *Antarct. J. U.S.*, *6*, 200-201, 1971.

Kaminuma, K., T. Torii, K. Yanai, G. Matsumoto, and Y. Tanaka, Activities of Japanese party in McMurdo Sound area 1976-1977 (in Japanese), *Antarct. Rec.*, *60*, 132-146, 1977.

Kriss, A. E., and R. B. Thomson, Origin of warm water in Lake Vanda, *Sov. Antarct. Exped. Inform. Bull.*, *8*, 678–682, 1976.

Law, P. G., The Vestfold Hills, *Austral. Nat. Antarct. Res. Exped. Rep.*, *Ser. A*, *1*, 5-50, 1959.

Lorius, C., L. Merlivat, and R. Hagemann, Variation in the mean deuterium content of precipitations in Antarctica, *J. Geophys. Res.*, *74*, 7027-7031, 1969.

Markov, K. K., V. I. Bardin, V. L. Lebedev, A. I. Orlov, and I. A. Suetova, *The Geography of Antarctica* (in Russian), pp. 260-286, Israel Program for Scientific Translations, Jerusalem, 1970.

Matsubaya, O., H. Sakai, T. Torii, H. Burton, and K. Kerry, Antarctic saline lakes—stable isotopic ratios, chemical compositions and evolution, *Geochim. Cosmochim. Acta*, *43*, 7-25, 1979.

Matsumoto, G., and T. Hanya, Organic carbons and fatty acids in Antarctic saline lakes, *Antarct. Rec.*, *58*, 81-88, 1977.

Matsumoto, G., T. Torii, and T. Hanya, Distribution of organic constituents in lake waters and sediments of the McMurdo Sound region in the Antarctic, in *Proceedings of the Seminar III on Dry Valley Drilling Project 1978*, *Mem. Spec. Issue 13*, edited by T. Nagata, pp. 103-120, National Institute of Polar Research, Tokyo, 1979.

McGinnis, L. D., and T. E. Jensen, Permafrost-hydrogeologic regimen in two ice-free valleys, Antarctica, from electrical depth sounding, *Quaternary Res.*, *1*, 389-409, 1971.

McGinnis, L. D., K. Nakao, and C. C. Clark, Geophysical identification of frozen and unfrozen ground, Antarctica, *Dry Val. Drilling Proj. Bull.*, *1*, 30-60, 1972.

McKelvey, B. C., and P. N. Webb, Geological reconnaissance in Victoria Land, Antarctica, *Nature*, *189*, 545-547, 1961.

McLeod, I. R., The saline lakes of the Vestfold Hills, Princess Elizabeth Land, in *Antarctic Geology*, edited by R. J. Adie, pp. 65-72, North Holland, Amsterdam, 1964.

Meyer, G. H., M. B. Morrow, O. Wyss, T. E. Berg, and J. L. Littlepage, Antarctica: The microbiology of an unfrozen saline pond, *Science*, *138*, 1103-1104, 1962.

Morikawa, H., I. Minato, J. Ossaka, and T. Hayashi, The distribution of secondary minerals and evaporites at Lake Vanda, Victoria Land, Antarctica, in *Geochemical and Geophysical Studies of Dry Valleys, Victoria Land, Antarctica*, *Mem. Spec. Issue 4*, edited by T. Torii, pp. 45-59, National Institute of Polar Research, Tokyo, 1975a.

Morikawa, H., I. Minato, and J. Ossaka, Origin of magnesium and potassium ions in Lake Vanda, Antarctica, *Nature*, *254*, 583-584, 1975b.

Mudrey, M. G., Jr., A model for chemical evolution of surface waters in Wright Valley, Antarctica, *Dry Val. Drilling Proj. Bull., 4,* 41-42, 1974.

Mudrey, M. G., Jr., N. F. Shimp, C. W. Keighin, G. L. Oberts, and L. D. McGinnis, Chemical evolution of water in Don Juan Pond, Antarctica, *Antarct. J. U.S., 8,* 164-166, 1973.

Murayama, H., S. Nakaya, S. Murata, T. Torii, and K. Watanuki, Interpretation of salt deposition in Wright Valley, Antarctica: Chemical analysis of DVDP 14 core, in *Proceedings of the Seminar III on Dry Valley Drilling Project 1978, Mem. Spec. Issue 13,* edited by T. Nagata, pp. 60-72, National Institute of Polar Research, Tokyo, 1979.

Murozumi, M., T. J. Chow, and C. Patterson, Chemical concentrations of pollutant lead aerosols, terrestrial dusts and sea salts in Greenland and Antarctic snow strata, *Geochim. Cosmochim. Acta, 33,* 1247-1294, 1969.

Nakai, N., and M. L. Jensen, The kinetic isotope effect in the bacterial reduction and oxidation of sulfur, *Geochim. Cosmochim. Acta, 28,* 1893-1912, 1964.

Nakai, N., Y. Kiyosu, H. Wada, and M. Takimoto, Stable isotope studies of salts and water from dry valleys, Antarctica, I, Origin of Salts and water, and the geologic history of Lake Vanda, in *Geochemical and Geophysical Studies of Dry Valleys, Victoria Land in Antarctica, Mem. Spec. Issue 4,* edited by T. Torii, pp. 30-44, National Institute of Polar Research, Tokyo, 1975.

Nakao, T., T. Torii, and K. Tanizawa, Paleohydrology of Lake Vanda in Wright Valley, Antarctica, inferred from granulometric analysis of DVDP 14 core, in *Proceedings of the Seminar III on Dry Valley Drilling Project 1978, Mem. Spec. Issue 13,* edited by T. Nagata, pp. 73-83, National Institute of Polar Research, Tokyo, 1979.

Nelson C. S., and A. T. Wilson, Bathymetry and bottom sediments of Lake Vanda, Antarctica, *Antarct. J. U.S., 7,* 97-99, 1972.

Nichols, R. L., Multiple glaciation in the Wright Valley, McMurdo Sound, Antarctica, paper presented at the Tenth Pacific Science Congress, Honolulu, Hawaii, 1961.

Nichols, R. L., Geology of Lake Vanda, Wright Valley, south Victoria Land, Antarctica, in *Antarctic Research, Geophys. Monogr.,* vol. 7, edited by H. Wexler et al., pp. 47-52, AGU, Washington, D. C., 1962.

Nichols, R. L., Geologic features demonstrating aridity of McMurdo Sound area, *Amer. J. Sci., 261,* 20-31, 1963.

Nichols, R. L. Geomorphology of Antarctica, in *Antarctic Soils and Soil Forming Processes, Antarct. Res. Ser.,* vol. 8 edited by J. C. F. Tedrow, pp. 1-46, AGU, Washington, D. C., 1966.

Nichols, R. L., Glacial geology of the Wright Valley, McMurdo Sound, Research in the Antarctic, *Publ. 93,* pp. 293-340, Amer. Ass. for the Advance. of Sci., Washington, D. C., 1971.

Parker, B. C., R. C. Hoehn, and R. A. Paterson, Ecological model for Lake Bonney, southern Victoria Land, Antarctica, *Antarct. J. U.S., 8,* 214-216, 1973.

Parker, B. C., J. T. Whitehurst, and R. C. Hoehn, Observations of in situ concentrations and production of organic matter in an Antarctic meromictic lake, *Va. J. Sci., 25,* 136-140, 1974.

Parker, B. C., R. A. Paterson, A. E. Linkins, and R. C. Hoehn, Lake Bonney ecosystem, *Antarct. J. U.S., 10,* 137-138, 1975.

Péwé, T. L., Multiple glaciation in the McMurdo Sound region, Antarctica: A progress report, *J. Geol., 68,* 498-514, 1960.

Picciotto, E., X. De Maere, and I. Friedman, Isotopic composition and temperature of formation of Antarctic snows, *Nature, 187,* 857-859, 1960.

Ragotzkie, R. A., and I. Friedman, Low deuterium content of Lake Vanda, Antarctica, *Science, 148,* 1226-1227, 1965.

Ragotzkie, R. A., and G. E. Likens, The heat balance of two Antarctic lakes, *Limnol. Oceanogr., 9,* 412-425, 1964.

Rymill, J., *Southern Lights, The Official Account of the British Graham Land Expedition 1934-1937,* pp. 194-195, Chatto and Windus, London, 1938.

Scott, R. F., *The Voyage of the 'Discovery,'* vol. 2, p. 290, Smith, Elder, and Co., London, 1905.

Shirtcliffe, T. G. L., Lake Bonney, Antarctica: Cause of the elevated temperatures, *J. Geophys. Res., 69,* 5257-5268, 1964.

Shirtcliffe, T. G. L., and R. F. Benseman, A sunheated Antarctic lake, *J. Geophys. Res., 69,* 3355-3359, 1964.

Shirtcliffe, T. G. L., and I. M. Calhaem, Measurements of temperature and electrical conductivity in Lake Vanda, Victoria Land, Antarctica, *N.Z. J. Geol. Geophys., 11,* 976-981, 1968.

Simonov, I. M., *The Oases of Eastern Antarctica* (in Russian), Hydrometeorological Publishing Office, Leningrad, 1971.

Simonov, I. M., and K. S. Bonch-Osmolovskaya, Chemistry of water, ice, and snow in the Schirmacher Ponds, *Sov. Antarct. Exped. Inform. Bull., 7,* 260-264, 1968.

Sugiyama, J., Y. Sugiyama, H. Iizuka, and T. Torii, Report of the Japanese summer parties in dry valleys, Victoria Land, 1963-65, IV, Mycological studies of the Antarctic fungi, II, Mycoflora of Lake Vanda, an ice-free lake, *Antarct. Rec., 28,* 23-32, 1967.

Taylor, T. G., *With Scott: The Silver Lining,* p. 136, Smith, Elder and Co., London, 1916.

Tedrow, J. C. F., and F. C. Ugolini, Antarctic soils, in *Antarctic Soils and Soil Forming Processes, Antarct. Res. Ser.,* vol. 8, edited by J. C. F. Tedrow, pp. 161-177, AGU, Washington, D. C., 1966.

Tedrow, J. C. F., F. C. Ugolini, and H. Janetschek, An Antarctic saline lake, *N.Z. J. Sci., 6,* 150-156, 1963.

Thompson, T. G., and K. H. Nelson, Concentration of brines and deposition of salts from sea water under frigid conditions, *Amer. J. Sci., 254,* 227-238, 1956.

Torri, T., and J. Ossaka, Antarcticite: A new mineral, calcium chloride hexahydrate, discovered in Antarctica, *Science, 149,* 975-977, 1965.

Torii, T., S. Murata, Y. Yoshida, J. Ossaka, and N. Yamagata, Report of the Japanese summer parties in dry valleys, Victoria Land 1963-1965, I, On the evaporites found in Miers Valley, Victoria Land, Antarctica, *Antarct. Rec., 27,* 1-12, 1966.

Torii, T., N. Yamagata, and T. Cho, Report of the Japanese summer parties in dry valleys, Victoria Land, 1963-1965, II, General description and water temperature data for the lakes, *Antarct. Rec., 28,* 1-14, 1967.

Torii, T., S. Murata, J. Ossaka, and N. Yamagata, Report of the Japanese summer parties in dry valleys, Victoria Land, 1963-65, VIII, Occurrence of antarcticite in Don Juan Pond— Sequential change and the conditions of crystallization, *Antarct. Rec., 37,* 26-32, 1970.

Torii, T., Y. Yusa, K. Nakao, and T. Hashimoto, Report of the Japanese summer parties in dry valleys, Victoria Land, X, A preliminary report of the geophysical and geochemical studies at Lake Vanda and in the adjacent dry valleys in 1971–1972 (in Japanese) *Antarct. Rec., 45,* 76–88, 1972.

Torii, T., N. Yamagata, S. Nakaya, S. Murata, T. Hashimoto, O. Matsubaya, and H. Sakai, Geochemical aspects of the Mc-

Murdo saline lakes with special emphasis on the distribution of nutrient matters, in *Geochemical and Geophysical Studies of Dry Valleys, Victoria Land, Antarctica, Mem. Spec. Issue 4*, edited by T. Torii, pp. 5–29, National Institute of Polar Research, Tokyo, 1975.

Torii, T., N. Yamagata, S. Nakaya, and S. Murata, Chemical characteristics of Antarctic saline lakes (in Japanese), *Antarct. Rec.*, *58*, 9–19, 1977a.

Torii, T., N. Yamagata, J. Ossaka, and S. Murata, Salt balance in the Don Juan basin, *Antarct. Rec.*, *58*, 116–130, 1977b.

Torii, T., N. Yamagata, J. Ossaka, and S. Murata, A view on the formation of saline waters in the dry valleys, in *Proceedings of the Seminar III on Dry Valley Drilling Project 1978, Mem. Spec. Issue 13*, edited by T. Nagata, pp. 22–33, National Institute of Polar Research, Tokyo, 1979.

Ugolini, F. G., Weathering and mineral synthesis in antarctic soils, *Antarct. J. U.S.*, *11*, 248–249, 1976.

Waguri, O., Isolation of microorganisms from salt lakes in the dry valley, Antarctica, and their living environment, *Antarct. Rec.*, *57*, 80–96, 1976.

Watanuki, K., T. Torii, H. Murayama, J. Hirabayashi, M. Sano, and T. Abiko, Geochemical features of Antarctic lakes, *Antarct. Rec.*, *59*, 18–25, 1977.

Weand, B. L., R. D. Fortner, R. C. Hoehn, and B. C. Parker, Subterranean flow into Lake Bonney, *Antarct. J. U.S.*, *10*, 15–19, 1975.

Webb, P. N., Wright fjord, Pliocene marine invasion of an antarctic dry valley, *Antarct. J. U.S.*, *7*, 227–234, 1972.

Wilson, A. T., Evidence from chemical diffusion of a climatic change in the McMurdo dry valleys 1,200 years ago, *Nature*, *201*, 176–177, 1964.

Wilson, A. T., and C. H. Hendy, McMurdo dry valley lakes sediments—A record of Cenozoic climatic events, *Dry Val. Drilling Proj. Bull.*, *4*, 72, 1974.

Wilson, A. T., and H. W. Wellman, Lake Vanda: An Antarctic lake, *Nature*, *196*, 1171–1173, 1962.

Wilson, A. T., C. H. Hendy, T. R. Healy, J. W. Gumbley, A. B. Field, and C. P. Reynolds, Dry valley lake sediments: A record of Cenozoic climatic events, *Antarct. J. U.S.*, *9*, 134–135, 1974a.

Wilson, A. T., R. Holdsworth, and C. H. Hendy, Lake Vanda: Source of heating, *Antarct. J. U.S.*, *9*, 137–138, 1974b.

Winchell, A. N., and H. Winchell, *The Microscopical Characters of Artificial Inorganic Solid Substances: Optical Properties of Artificial Minerals*, Academic, New York, 1964.

Yamagata, N., T. Torii, and S. Murata, Report of the Japanese summer parties in dry valleys, Victoria Land, 1963–1965, V, Chemical composition of lake waters, *Antarct. Rec.*, *29*, 53–75, 1967.

Yoshida, Y., Raised beaches and saline lakes on the Prince Olav Coast, Antarctica (in Japanese), *Modern Geogr.*, 93–118, 1970.

Yoshida, Y., T. Torii, and N. Yamagata, Antarctic saline lakes, in *Proceedings of Symposium on Hydrogeochemistry and Biogeochemistry*, vol. 1, pp. 652–660, *Hydrogeochemistry*, Clark Company, Washington, D. C., 1973.

Yoshida, Y., T. Torii, Y. Yusa, S. Nakaya, and K. Moriwaki, A limnological study of some lakes in the Antarctic, in *Quaternary Studies*, edited by R. P. Suggate and M. M. Cresswell, pp. 311–320, Royal Society of New Zealand, Wellington, 1975.

Yusa, Y., The re-evaluation of heat balance in Lake Vanda, Victoria Land, Antarctica, *Contrib. Geophys. Inst. Kyoto Univ.*, *12*, 87–100, 1972.

Yusa, Y., On the water temperature in Lake Vanda, Victoria Land, Antarctica, in *Geochemical and Geophysical Studies of Dry Valleys, Victoria Land, Antarctica, Mem. Spec. Issue 4*, edited by T. Torii, pp. 75–89, National Institute of Polar Research, Tokyo, 1975.

Yusa, Y., A study on thermosolutal convection in saline lakes, *Mem. Fac. Sci. Kyoto Univ., Ser. Phys. Astrophys. Geophys. Chem.*, *35*, 149–183, 1977.

HYDROLOGY OF THE DON JUAN BASIN, WRIGHT VALLEY, ANTARCTICA

H. J. H. HARRIS AND K. CARTWRIGHT

Illinois State Geological Survey, Champaign, Illinois 61820

Don Juan Pond is an intermittent, chemically unique brine pond situated in a closed basin in the south fork of Wright Valley, Antarctica. The floor of the basin is a discharge zone for groundwater brines confined in an underlying dolerite aquifer. Although the flux of groundwater is quite small, groundwater probably provided about 70% of the water entering the pond in the austral summer of 1975-1976. Intermittent, freshwater streams enter the western end of the pond late in the austral summer; the melting of ice in near-surface frozen ground gives rise to the streams. In relation to groundwater discharge and stream flow, precipitation is an insignificant source of water for the pond. Evaporation and sublimation are, with aerosols, the sole means whereby water leaves the basin; during the austral summer of 1975-1976, evaporation exceeded inflow, and the volume of the pond decreased by 60%. The pond appears to be in a precarious hydrologic equilibrium; the cessation of either streamflow or groundwater flow would cause the pond to go dry. The major element chemistry of the groundwater is like that of the pond, indicating that the dissolved salts in both have a common origin. Evaporation effects both a relative concentration of salts in the pond and the precipitation of salts on the floor of the basin. Laboratory data indicate that the chemistries of the pond and the groundwater are sensitive to changes in temperature. The chemistry of the groundwater may reflect both the rock type of the aquifer and the mean annual air temperature. Groundwater currently is transporting salt into the basin. The history of salt accumulation and the history of groundwater flow are related and are obscure. Winds may be removing salts from the basin.

INTRODUCTION

In 1961, *Meyer et al.* [1962] discovered a unique brine pond in southern Victoria Land, Antarctica, in one of the ice-free ('dry') valleys lying between the East Antarctic Ice Cap and the Ross Sea. Don Juan Pond, situated in a topographically closed basin in the south fork of Wright Valley (Figure 1), is chemically unique among surface waters in Antarctica and apparently among surface waters worldwide. Water in the pond is, at times, about 40 wt % salt, and more than 90% of the salt is calcium chloride. The salinity of the brine is so great that the pond remains at least partly unfrozen even at the height of the Antarctic winter, when air temperatures drop below $-50°C$.

Previous investigations in the Don Juan basin have been intended primarily to elucidate the geochemical history of the pond. *Torii and Yamagata* [this volume] include a review of these investigations in their excellent discussion of the chemistry

of surface waters in the Dry Valley region. Although there is no consensus on the origin of the salts in the Don Juan basin, all investigators are apparently agreed that the chemistry of the pond is, in some sense, a reflection of the remarkably cold and arid climate of the region.

In preparation for the Dry Valley Drilling Project (DVDP)—a major drilling program designed to study the geologic history, geophysics, and geochemistry of the region—*McGinnis and Jensen* [1971] and *McGinnis et al.* [1973] conducted geophysical surveys in the Don Juan basin. The surveys indicated that the rocks underlying the floor of the basin were unfrozen, suggesting the presence of a reservoir of and possible conduit for liquid groundwater. Together with the work of Torii and associates, these surveys inspired the siting of two DVDP boreholes in the basin.

This report describes in detail hydrologic research carried out on the floor of the Don Juan basin in conjunction with DVDP during the three

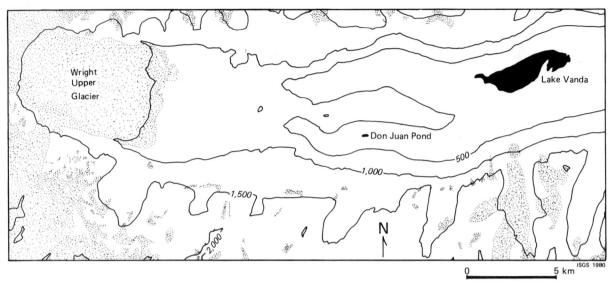

Fig. 1.   Topography of western Wright Valley. Stippled areas are ice or snow covered. Contour interval, 500 m.
Adapted from U. S. Geological Survey (unpublished map).

austral summers between 1973 and 1976; a companion paper [*Cartwright and Harris*, this volume] gives a general discussion of the hydrogeology of the Dry Valley region. Although the subject has been touched upon in a few geochemical studies, there have been no previous detailed examinations of the hydrology of the basin. The emphasis of this report is upon the physical characteristics of currently active hydrologic systems, with particular attention given to hydrogeology. In a previous, brief presentation of part of this research [*Harris et al.*, 1979] it was shown that the chemical behavior of Don Juan Pond is closely involved with the hydrology of the basin. Although geochemical issues are not treated in detail here, we reemphasize the geochemical significance of the findings of this research. Groundwaters currently are transporting dissolved salts into Don Juan Pond; if the present is indeed the key to the past, groundwaters must be taken into account in any discussion of the origin of the salts in the basin.

GENERAL PHYSICAL DESCRIPTION

The floor of the Don Juan basin lies in the axis of and is the lowest point (approximately 117 m above mean sea level) in the south fork of Wright Valley. The floor, 800 m long and 350 m wide, is exceedingly flat and is bounded on both the north and the south by a valley wall rising 1000 m or more in a horizontal distance of less than 1 km. The lower slopes of both walls are thinly veneered with scree;

extensive outcrops of bedrock begin on the south wall at about 150 m and on the north wall at about 500 m above the floor.

To the east and west the topography is much less precipitous (Figure 2). Eastward, the axis of the south fork rises through a series of relatively small depressions to end about 200 m above and 5 km distant from the floor of the basin. This eastern end of the south fork overlooks central Wright Valley and Lake Vanda, the latter about 4 km to the northeast and 220 m below. The eastern half of the fork, almost 1.5 km wide, is floored with unconsolidated sediments of unknown, but probably small, thickness; linear outcroppings of rock found in the axis of the fork and within 300 m of the floor of the Don Juan basin have been described as bedrock ledges [*Cartwright et al.*, 1974]. To the west the floor of the basin is bordered by the toe of a rock glacier that extends westward for about 3 km. The head of the rock glacier is almost 500 m above its toe and abuts the cliffs and scree slopes that mark the beginning of the south fork. Farther west lies the rugged scabland landscape of the Labyrinth.

At elevations below 1000 m, bedrock cropping out in the south fork is either Ferrar Dolerite, which occurs primarily in regionally extensive sills up to a few hundred meters thick, or granitic rocks of the basement complex [*McKelvey and Webb*, 1962; *Smithson et al.*, 1970, Figure 2]. Wright Valley was carved at least in part by glaciers spilling eastward from the East Antarctic Icecap to the

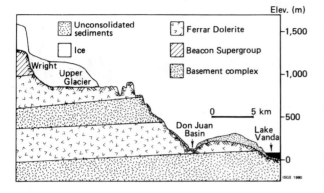

Fig. 2. Generalized geologic cross section of western Wright Valley from the Wright Upper Glacier through Don Juan Pond and the south fork to Lake Vanda. Vertical exaggeration, ×10. Compiled from *Barrett et al.* [1972], *McKelvey and Webb* [1962], U. S. Geological Survey (unpublished map), and our own observations.

Ross Sea; there is evidence for at least four such episodes during the past 4 m.y. Alpine glaciers originating nearby in the bordering mountains have also entered the valley. Much of the unconsolidated sediment mantling the floor of the south fork is undoubtedly of glacial origin, but there is also considerable evidence of mass movements and of fluvial and aeolian activity. In addition, the character of deposits found east of Lake Vanda suggests to some workers that there has been at least one marine invasion of Wright Valley. Detailed discussion of the Late Cenozoic history of the valley is given by *Calkin and Bull* [1974].

As in the rest of the Dry Valley region, surface water is extremely scarce in the south fork, particularly at lower elevations. Ice fields and perennial snowbanks occur in the mountains to the south but are nonexistent at elevations below about 1000 m. A thin (10-20 mm) dusting of light snow accumulates in places protected from the wind during the austral winter; most of this sublimates early in the summer, but some shaded locations retain snow until late summer. Summer snowfalls at lower elevations are rare and sublimate rapidly.

Don Juan Pond has in recent years been the largest body of water in the south fork. Webb and McKelvey (P. N. Webb, personal communication, 1978) traversed the basin in 1957-1958 and did not find the pond, which indicates that it is intermittent. Torii and associates [*Torii and Yamagata*, this volume] noted that the pond changed in size from year to year between 1961 and 1978. When present, the pond occupies the north central part of the floor of the basin; that part of the floor not occupied by the pond is encrusted with salt. *Cartwright et al.* [1974] noted that at least two benches are cut into the sediments above the salt pan, which may indicate stable, higher water levels in the past. The highest of these benches is 21 m above the salt pan.

East of Don Juan Pond there are between 20 and 30 much smaller ponds, many of them intermittent and all of them at elevations below about 400 m; all those present early in the austral summer are frozen. As discussed in our companion paper [*Cartwright and Harris*, this volume], some of these ponds are sustained by shallow groundwater. About 2.5 km west of Don Juan Pond there are two ponds that with the exception of narrow, peripheral moats remain frozen throughout the summer. A broken and ill-defined channel cuts the surface of the rock glacier for at least part of the interval between these ponds and the floor of the Don Juan basin. However, we never found liquid water in the channel. Two other short, relatively well-defined channels cut the toe of the rock glacier, enter the floor of the basin in alluvial fans, and meander across the western third of the salt pan to Don Juan Pond. At times late in the austral summer, intermittent streams, apparently fed by melting ice in the rock glacier, occupy these two channels. South of and on a bench about 600 m above Don Juan Pond, there is a single, small intermittent pond.

METHODS

Measurements and Samples

The floor of the Don Juan basin was mapped topographically, both to produce a general reference map and to allow detailed calculations; the locations of the instruments and sampling locations discussed below are shown on the map (Figure 11). In mapping, particular attention was given to the bathymetry of Don Juan Pond, in which measurements of elevation were taken at about 1200 positions. Over 1100 of these data points were contoured by computer at 1-cm intervals and at a scale of 1 : 300. The volume and surface area of the pond were also calculated by computer for each of 12 different pond surface elevations. A Stevens water level recorder (type F, model 68) with a 1 : 1 pulley, a 40-cm float, and an 8-day clock was installed on a platform in the pond; the float was housed in a stilling box. Measurements of water temperature and water samples for chemical analysis were taken periodically at a shaded location near the recorder.

Evaporation rates were measured by using two pans filled with water from the pond. Periodically, water levels in the pans were measured, the residual pan waters were sampled and discarded, and the pans were refilled.

DVDP boreholes 5 and 13 were drilled on the floor of the basin and core was recovered from both holes by DVDP personnel. The porosity and hydraulic conductivity of sediments from borehole 5 were measured in the laboratory by using standard gravimetric techniques and both constant-head and falling-head permeameters, respectively (see, for example, *Lambe* [1951] for a description of equipment and procedures). Borehole 13, which remained open and cased subsequent to drilling, was fitted with a Stevens water level recorder (type F, model 68) with a 1 : 1 pulley, a 20-cm float, and an 8-day clock. Water samples for chemical analysis were taken periodically at different depths in the borehole, and measurements of the temperature of the samples were made. A thermal log of the borehole was made [*Decker and Bucher*, 1980].

Piezometers, consisting of ½-in.-diameter iron pipes perforated and fitted with a steel tip at one end, were driven at 10 locations in the salt pan to depths ranging from 0.5 to 2.2 m. Water levels in the piezometers were measured periodically.

Barometric pressure and temperature were measured with an aneroid barograph and a metallic coil thermograph, respectively, both equipped with 7-day clocks and both manufactured by Taylor Instruments. These instruments were housed in a shaded, ventilated enclosure at ground surface in the southwest corner of the basin.

Analyses of water samples for Na, K, Ca, and Mg were made by using atomic emission and absorption spectrophotometry (Perkin-Elmer Model 306). Chloride concentrations were determined by titration with mercuric nitrate, and sulfate concentrations were determined gravimetrically, by addition of barium chloride.

Conventions

Unless otherwise noted, any equations given are part of the standard language of hydrogeology; complete discussions of these equations may be found in any standard hydrogeologic text [e. g.,

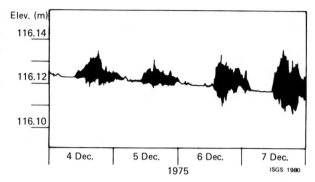

Fig. 3.   Detail of hydrograph of Don Juan Pond. Disturbances late in the day are seiches.

Bear, 1972; *Domenico*, 1972; *Freeze and Cherry*, 1979]. The chemical and physical properties of the calcium chloride brines found in the basin obviously are quite unlike those of fresh water and are significant in the hydrology of the basin. When no measurements of our own are available, we assume that the properties of these brines are like those reported in standard references [e.g., *Robinson and Stokes*, 1959; *Weast*, 1974] for pure solutions of calcium chloride. In partial justification of this assumption, we note that the salts in the brines are, in fact, primarily calcium chloride and that the relation between the density and the content of dissolved solids of the brines is like that of a pure solution of calcium chloride. The elevations of geomorphic features, water levels, and instruments are reported in relation to mean sea level. This convention is adopted for convenience only and does not indicate that all measurements were made in relation to sea level. Most measurements were made in relation to ground surface elevation at DVDP 5, which is 116.7 m above mean sea level (T. J. Chinn, personal communication, 1976). The discussion of groundwater in cold regions poses special nomenclatural problems. In this paper, 'water' is defined primarily as a substance; the use of the word says nothing about the state of the substance. The only exception is in the use of the phrase 'water table,' which always refers to liquid water.

SURFACE WATERS

Don Juan Pond

Don Juan Pond was examined and sampled intermittently between November 28 and December

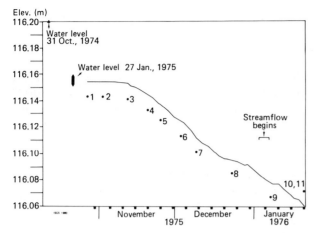

Fig. 4.   Hydrograph of Don Juan Pond, October 27 through January 20, with seiches removed. Numbered dots indicate dates of samples for chemical analysis.

stable until about November 12 and then decreased. The western edge of the pond was covered with ice about 5 cm thick in late October and early November 1974; the pond was otherwise free of ice.

A continuous hydrograph of the pond was obtained between October 27, 1975, and January 20, 1976. In detail, the entire hydrograph is much like that portion shown in Figure 3. The stilling box damped out the effects of individual waves; the relatively large fluctuations occurring in daily cycles, between about 1200 and 0100, are the record of wind-generated seiches. The cyclic pattern is evidence of the persistent, primarily westerly diurnal winds at lower elevations in the western dry valleys [*Bull*, 1966; *Thompson et al.*, 1971]. Seiches of the magnitude shown in Figure 3 were most common, but temporary, wind-generated changes in water level of up to 7 cm were recorded during 12-hour periods on November 15, November 26, and January 11. As the salt pan is exceedingly flat, changes of this magnitude are sufficient to shift the western and eastern edges of the pond about 40 m from their undisturbed positions. *Heine* [1971] observed seiches of almost ± 10 cm in Lake Vanda in 1970.

3, 1973, between October 27, 1974, and January 29, 1975, and between October 27, 1975, and January 21, 1976. During the first of these periods the size of the pond decreased; during the second its size remained stable until early November, decreased until late January, and increased slightly at the end of January; during the last its size remained

TABLE 1.   Surface Area, Volume, and Volume Change of Don Juan Pond

Elevation of Pond Surface, m	Date	Surface Area, m² × 1000	Volume, m³	Volume Change, m³	Rate of Depth Change, mm/day
116.154	Oct. 27, 1975	59.5	7780		
				0	0
116.154	Nov. 7, 1975	59.5	7780		
				− 240	− 0.7
116.150	Nov. 13, 1975	58.5	7540		
				− 570	− 1.0
116.140	Nov. 23, 1975	56.5	6970		
				− 560	− 1.6
116.130	Nov. 29, 1975	55.0	6410		
				− 540	− 1.5
116.120	Dec. 6, 1975	52.5	5870		
				− 510	− 2.2
116.110	Dec. 10, 1975	50.5	5360		
				− 500	− 1.5
116.100	Dec. 17, 1975	48.5	4860		
				− 470	− 0.8
116.090	Dec. 29, 1975	46.5	4390		
				− 460	− 1.8
116,080	Jan. 4, 1976	44.5	3930		
				− 440	− 1.0
116.070	Jan. 14, 1976	42.5	3490		
				− 410	− 1.8
116.060	Jan. 20, 1976	40.5	3080		
− 94 mm (change in water level)	84.4 (days elapsed)			− 4700 m³ (total change)	− 1.1 mm/day (weighted mean rate)

Fractions of days used in calculations but not shown in table.

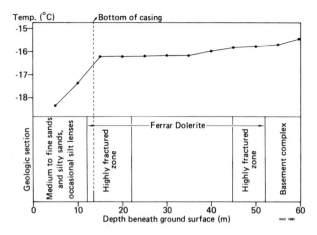

Fig. 5. Geologic and geothermal logs of DVDP borehole 13, Don Juan basin. Geologic log from *Mudrey et al.* [1975]; geothermal log (December 23, 1975) from *Decker and Bucher* [1980].

The complete hydrograph of Don Juan Pond is shown at reduced scale and without seiches in Figure 4; two earlier, individual measurements of water level (November 1, 1974, and January 29, 1975) are also shown. Computer-calculated surface areas and volumes are given in Table 1 for 11 different water levels. The size of the pond decreased during both austral summers but apparently changed relatively little during the winter of 1975. The decrease during the summer of 1975-1976 was particularly dramatic, in that the pond lost about 60% of its volume in 68 of the 84 days of record.

Streams

Intermittent, freshwater streams flowing from the toe of the rock glacier to the salt pan were observed and sampled in January 1975 and January 1976. The sources for the streams appeared to be melting ice contained in the rock glacier. A stream cutting the center of the toe was most active, but some flow was also observed in a channel entering the southwest corner of the salt pan; the courses of the streams were not mapped. Throughout both periods of observation, ice could be found in the parts of these channels on and adjacent to the rock glacier.

Streamflow during 1975 begun prior to January 6 and continued intermittently through at least January 30. In 1976, flow began between January 3 and 6 and continued intermittently through at least January 21. The volume of flow was highly variable. Although no direct measurements of flow

rates were taken, the volume discharged to the salt pan during 1974-1975 appeared to be much greater—perhaps several times greater—than that discharged during 1975-1976. Diurnal variations in flow rate were particularly apparent during 1974-1975; peak flows occurred late in the afternoon and early in the evening, whereas low flows occurred in the early morning. There were also noticeable variations in flow rate from day to day. As with other streams in the region (see, for example, *Chinn* [1976]), all of these variations were undoubtedly a function of air temperature and insolation. Estimates of the total volume of stream flow are given in a subsequent section of this paper.

GROUNDWATERS

Confined Groundwater

Observations made at DVDP borehole 13 are the primary source of information about groundwater in the sediments and bedrock beneath the Don Juan basin. Descriptions of drilling operations and complete geologic logs are given for DVDP 13 by *Mudrey et al.* [1975] and for DVDP 5 by *Cartwright et al.* [1974]; a summary geologic log and geothermal log for DVDP 13 are shown in Figure 5.

DVDP 5 was finished at a depth of about 3.5 m and, with the exception of two thin (0.01 and 0.003 m thick) organic zones lying about 1 m below ground surface, encountered a sedimentary sequence like that cored by DVDP 13. The rocks

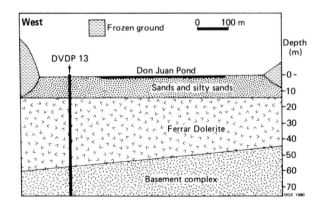

Fig. 6. Generalized geologic section (east-west) across the floor of the Don Juan basin. Compiled from *Decker and Bucher* [1980], *McGinnis and Jensen* [1971], *McGinnis et al.* [1973], *Mudrey et al.* [1975], and our own observations.

penetrated by both boreholes were saturated with highly saline, unfrozen water. *McGinnis et al.* [1973] had previously conducted electrical resistivity and seismic surveys in the Don Juan basin and concluded that the rocks underlying the floor of the entire basin were saturated with unfrozen, saline groundwater. The distribution of frozen ground and the location of the surface of the basement complex shown in Figure 6 are taken from these surveys.

Observations made during the drilling of DVDP 5 (December 1-3, 1973) indicated an upward gradient in groundwater potential; the magnitude of the gradient—about 0.13—suggested that it would be prudent to abandon and close the hole. During the drilling of DVDP 13 (January 8-13, 1975), upward gradients were again encountered, and the borehole sometimes overflowed. On five different occasions subsequent to drilling and prior to January 30, 1975, water samples and measurements of water level were taken in the borehole, which was cased from about 13.5 m below to 0.62 m above ground surface. On all of these occasions the borehole was open to its bottom (75 m deep); water levels ranged from about 0.2 to about 0.5 m above ground surface.

On October 27, 1975, at the end of the austral winter, the upper portion of the casing was found to be plugged with a slush of ice and liquid water; by December 3, 1975, the slush was broken up and extracted from the casing. Although a few crystals of ice were seen floating in the well in December, the borehole was otherwise free of ice throughout its length on six sampling and logging runs during the field season, which ended on January 21, 1976.

A continuous record of water levels in DVDP 13 was obtained between December 5, 1975, and January 11, 1976 (Figure 7a). The most peculiar events exhibited in the hydrograph are the numerous large, exceedingly rapid fluctuations, which appear in the figure as 'spikes'; during at least 11 of these events the well overflowed. These fluctuations, which are to our knowledge unlike any other natural water pressure fluctuations reported from aquifers in other parts of the earth, are analyzed at length elsewhere [*Harris and Cartwright*, 1980]; we argue that they were caused by stress changes effected by small, abrupt movements in the rock glacier bordering the Don Juan basin.

Other, less rapid fluctuations are also apparent. For example, DVDP 13 exhibited a barometric response, as shown by comparison of Figures 7a and

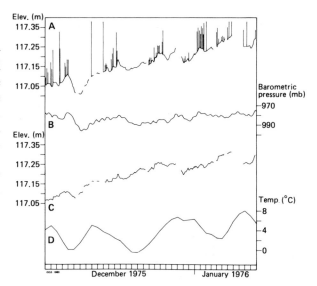

Fig. 7. (a) Hydrograph of DVDP borehole 13, December 5 through January 11. The numerous vertical lines are rapid fluctuations in water level. (b) Barometric pressure (inverted to show relation to Figure 7a) on the floor of the Don Juan basin. (c) Hydrograph of DVDP borehole 13, with fluctuations caused by changes in barometric pressure removed. Scaled for a barometric efficiency of 60%. (d) Mean daily air temperature at the soil surface, floor of the Don Juan basin.

7b. A barometric response usually indicates that the aquifer tapped by the well is confined, meaning that the potential of fluid in the aquifer is not immediately affected by changes in the potential of fluid at the overlying water table. The barometric efficiency (BE) of the aquifer is defined as the ratio of the change in water pressure in the well to the change in atmospheric pressure. In theory, the barometric efficiency is also given by the relation

$$BE = \frac{1}{1 + (C_m/nC_w)}$$

where n and C_m are the porosity and compressibility of the aquifer, respectively, and C_w is the compressibility of the fluid. Barometric efficiency is shown as a function of n and C_m in Figure 8.

As determined from three of the largest barometric responses recorded, the barometric efficiency of the aquifer tapped by DVDP 13 ranges from 50 to 60% during compression (increasing barometric pressure) and from 70 to 80% during expansion. These values are characteristic of bedrock and not of sediments, which commonly have compressibilities greater than 10^{-9} m²/N (Figure 8) and therefore very small barometric efficiencies. The lower limit of compressibility for jointed rock is about 10^{-10} m²/N; jointed rock with a barometric

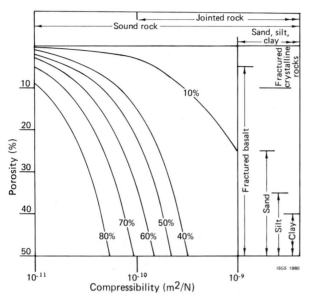

Fig. 8. Barometric efficiency, as a function of porosity and compressibility, for a porous medium saturated with pure water. Ranges of typical porosities and compressibilities taken from *Freeze and Cherry* [1979]. Note that typical compressibilities of sediments extend off scale to the right.

efficiency of 50% or above must therefore have a porosity in excess of about 20%. This suggests that the porosity of at least the highly fractured zones of the Ferrar Dolerite is quite high. Porosities higher than 20% are not uncommon in fractured basalt. *McGinnis et al.* [1973] were unable to detect the boundary between the dolerite and the overlying sediments using geophysical techniques, probably because of the high porosity of the dolerite.

Using a computer, we attempted to filter the barometric responses from the hydrograph of DVDP 13: after removal of the spikes from the hydrograph, both the hydrograph and barograph were digitized and multiplied by appropriate scale factors, and the barograph was then 'subtracted' from the hydrograph. The result of one such attempt, scaled for a barometric efficiency of 60%, is shown in Figure 7c. As the original records are at different temporal scales that are in places difficult to match exactly, some of the fluctuations of limited duration (i.e., of the order of a day or less in length) remaining in the filtered hydrograph probably result from inefficient filtration. Some of these short-term fluctuations may also be associated with the extremely rapid fluctuations mentioned above. Over the entire length of the hydrograph a gradual increase of about 25 cm is apparent. With the possible exception of this gradual increase, none of the changes in water level ob-

served in DVDP 13 bears any apparent relation to the mean air temperature at the soil surface (Figure 7d).

During the austral summer of 1978-1979, L. D. McGinnis (personal communication, 1979) visited and examined DVDP 13. The casing was again plugged with ice. The ice was broken up on December 9, 1978, whereupon the well began to flow at about 4l/mm (6.7 × 10^{-2} l/s). The well remained artesian until it was capped on December 12, 1978.

Water Table Groundwater

During the field seasons of 1973-1974, 1974-1975, and 1975-1976 a water table was found to occur in the sediments underlying the salt pan surrounding Don Juan Pond; piezometers driven to depths of up to 2.5 m did not encounter frozen ground. Beyond the boundary of the salt pan, by contrast, sediments saturated with liquid groundwater were very uncommon, and ice could usually be found at depths less than 0.5 m.

Liquid groundwater occurring intermittently near ground surface and overlying frozen ground is a significant source of water for the small ponds lying east of Don Juan Pond in the south fork, as discussed in our companion paper [*Cartwright and Harris*, this volume]. 'Shallow' groundwater of this kind also feeds the intermittent, freshwater streams entering Don Juan Pond from the west, as discussed above. We were unable, however, to find any other significant discharge of shallow groundwaters to the floor of the Don Juan basin. To the east, no shallow, liquid groundwaters could be found within about 700 m of the edge of the salt pan. The surface of the talus slope south of the salt pan shows long, vertical tongues or ribbons of discoloration (Figure 9). Early in the field season these tongues are darkened and 'damp' in appearance; late in the season they are often white with efflorescences of salt. In other parts of the south fork the unfrozen sediments underlying many such features contain, at times during the austral summer, a zone of partial saturation underlain by a thin zone of saturation. On several occasions, however, examination of the unfrozen sediments underlying these tongues revealed no saturated ground, suggesting that any saturated zone associated with them is vanishingly thin. The damp appearance of the tongues probably is caused by hygroscopic moisture. North of the salt pan the slope is in shade throughout much of the austral summer

Fig. 9.   Oblique aerial photograph of Don Juan Pond and surroundings, looking south, January 21, 1976. Floor of basin is encrusted with salt. Salt in western end of pond has been redissolved by plume of fresh stream water; salt in eastern end has not. Groundwater discharging at eastern end of pond (arrow) has dissolved part of salt crust on floor of basin. Note tongues on south wall of basin.

and had no features indicating the presence of liquid groundwater.

Records of water levels in piezometers situated in the salt pan are shown in Figure 10. In general, the behavior of these water levels mimicked that of the surface of Don Juan Pond: decline during the two austral summers and relatively little apparent change during the intervening winter. Piezometers 4 and 9 are notable exceptions, in that water levels in both increased at least 9 cm during the winter. Piezometer 9 was close to the edge of the pond: the water level in this piezometer may have been affected at times by wind-generated changes in pond level. Piezometers 1, 2, and 4 all were close to the channels of intermittent streams; water levels in these piezometers may have been affected by streamflow late in the austral summers.

The position of the water table beneath the salt pan on October 27, 1975, and January 20, 1976, is shown in Figures 11a and 11b. In making these maps we assumed that (1) the water level in any piezometer was equivalent to the level of the water table around that piezometer, (2) the position of the edge of the pond was equivalent to the position of the water table, and (3) the elevation of the water table at any point could be calculated by linear interpolation between two points at which its elevation was known.

In general, the water table sloped gently toward Don Juan Pond. The one exception was a tongue of relatively high water table beneath the southwestern corner of the salt pan. In some areas, the interpolated position of the water table lay above the surface of the salt pan. These areas, shown in a stippled pattern, generally occurred within about 40 m of the edge of the pond. These areas were, in effect, those for which the piezometer data predicted the occurrence of zones of groundwater discharge at the ground surface; they were, in fact, zones of groundwater discharge during the field season of 1975-1976. This suggests that the assumptions employed in making the water table maps were reasonable assumptions. The discharge zone at the southwest corner of the pond was particularly apparent early in the season; numerous small pools and slow moving streams of liquid water were present. The discharge zone at the eastern end of the pond, which we did not notice while in the field at the end of the season, is apparent in aerial photographs taken at that time (Figure 9). Photographs of the basin taken in a previous year (U.S. Navy, unpublished photographs, 1964) show similar features.

Discharge zones other than those shown on the maps were probably present, particularly in the western third of the salt pan along the stream

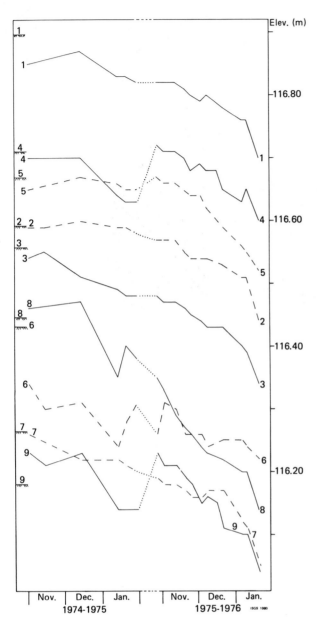

Fig. 10. Water levels in nine piezometers on the floor of the Don Juan basin, November 1, 1974, through January 20, 1976. No measurements were taken during winter 1975 (dotted portion of lines). Ground surface elevations at piezometers are indicated on the left side of the diagram. Locations of piezometers shown in Figure 11.

channels connecting the pond and rock glacier. The topography of the channels was not mapped in detail, but in places they are low enough in elevation to have intersected the water table. Small accumulations of sediment in the shape of shallow cones (up to 6 cm in diameter, 1 cm in height, and sometimes with a central vertical passage up to a few

millimeters in diameter) were found scattered on the floor of the salt pan south and west of the pond early in the field seasons of 1974-1975 and 1975-1976. These features may have resulted from earlier episodes of groundwater discharge and associated higher water tables, perhaps in the preceding austral winters.

Groundwater Flux

The flux of liquid groundwater in the sediments underlying the floor of the basin was calculated using Darcy's law,

$$Q = Ki$$

where Q is the volume of flow per unit area per unit time, K is the hydraulic conductivity, and i is the gradient in potential energy per unit weight (total head) of the fluid. The hydraulic conductivity is given by

$$K = k\gamma/\mu$$

where γ is the weight per unit volume, μ the dynamic viscosity of the fluid, and k the intrinsic permeability of the sediments at saturation.

As indicated in Figure 12, sediments taken from DVDP 5 had permeabilities ranging between 1×10^{-9} and 5×10^{-10} cm^2, values typical of silts and silty sands. The weight per unit volume of brines taken from DVDP 13 was about 1.17 g/cm^3, and the viscosity of calcium chloride brines of this density is about 2 cP at 20°C. Although the viscosity of these brines at lower temperatures was not measured, the viscosity of pure water increases by a factor of 2 with a drop in temperature from 20° to −10°C. These values suggest that the materials encountered by the DVDP boreholes had hydraulic conductivities of the order of 10^{-5} cm/s. In our calculations we assumed that all materials underlying the floor of the basin had conductivities of 1×10^{-5} cm/s (8.64×10^{-3} m/day).

In determining the flux between the dolerite and the floor of the basin we calculated the gradient in total head between the dolerite just below the cased part of DVDP 13 and two other 'points': namely, the bottom of Don Juan Pond and piezometer 4 (Table 2). Gradients were calculated for five successive periods of time using the average total heads in each period. As the density of the fluid in the cased part of DVDP 13 was the same as that of fluid just below the cased part, the total head in the dolerite was taken directly from the hydrograph of DVDP 13 without correction for variations in density. It is, however, not known

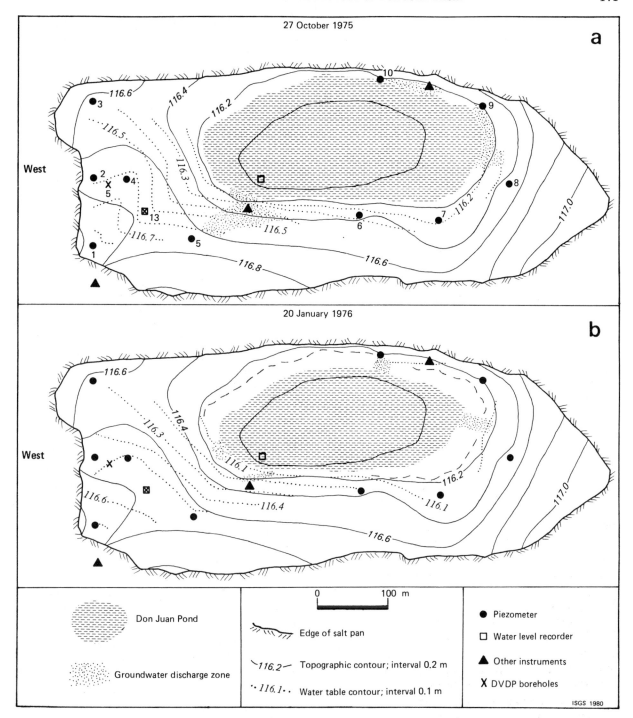

Fig. 11.   Floor of the Don Juan basin; water table elevation and extent of Don Juan Pond are shown.

whether the density of the fluid in piezometer 4 differed significantly from that of fluid surrounding the ports of the piezometer; we assumed that the densities were the same. We further assumed that all of the loss in head between the dolerite and the floor of the basin occurred in a vertical direction only. This is equivalent to assuming that there was little horizontal gradient in head in the dolerite, an assumption which is perhaps reasonable in view of the fractured condition and apparently high porosity of the dolerite.

As shown in Table 2, the fluxes—all upward

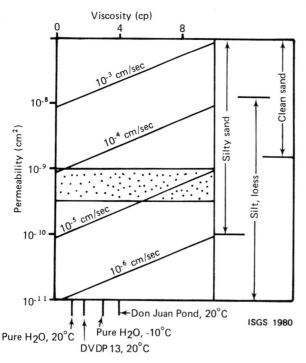

Fig. 12. Hydraulic conductivity (sloping lines) as a function of permeability and viscosity for a fluid having a specific gravity of 1.17. Range of permeabilities of sediments taken from DVDP borehole 5 shown by stippled area. Typical viscosities taken from *Weast* [1974]; ranges of typical permeabilities taken from *Freeze and Cherry* [1979].

from the dolerite to the basin—were very small, of the order of 5×10^{-4} m/day (0.18 m/y). There were small increases in the fluxes between December 6 and January 10. If fluxes like those calculated by using piezometer 4 occurred over the entire area of the salt pan (approximately 1.9×10^5 m^2), then the rate of groundwater discharge to the entire pan was about 0.7 l/s (60 m^3/day). In sediments having a porosity of 30%, this is equivalent to a rise of about 1 mm/day in the level of the water table. If fluxes like those calculated by using the hydrograph of the Don Juan Pond occurred over the entire area of the pond (approximately 5×10^4 m^2), then the rate of groundwater discharge to the entire pond was about 0.4 l/s (35 m^3/day).

As noted previously, the surface of the water table around Don Juan Pond sloped gently toward the pond, indicating that there was some horizontal movement of near-surface groundwater toward the pond. As determined from the water table maps (Figure 11), the gradient in the surface of the water table was about 0.005. The length of the perimeter of the pond was about 1000 m; if the thickness of the zone through which the horizontal flux

occurred is taken to be 1 m, then groundwater entered the pond from the water table at a rate of the order of 6×10^{-3} l/s (0.5 m^3/day).

ATMOSPHERIC WATERS

Precipitation

Snowfall in the Don Juan basin was exceedingly sparse during the field seasons of 1974-1975 and 1975-1976. To our knowledge the only falls to reach the floor of the basin occurred at the end of January and in late December 1975. On both occasions, less than 10 mm of light snow accumulated; most sublimated within 24 hours of its precipitation. Various workers [*Ugolini and Bull*, 1965; *Bull*, 1966; *Thompson et al.*, 1971] have published data indicating that mean annual precipitation (always as snow) in the interior of the Dry valley region ranges between 5 and 10 g/cm^2 (50 and 100 mm). It is our impression that from 1974 through 1976, snowfall in the Don Juan Basin was at least at the lower end of and perhaps below this range.

Evaporation and Sublimation

The rates of change during 1975-1976 of water levels in evaporation pans situated on the floor of the basin are given in Table 3. These rates are an index of the rates of evaporation from Don Juan Pond, but they must be carefully interpreted. Strictly speaking, they are not evaporation rates, since the pans (and the pond) collected small amounts of snow. Furthermore, there is some indication in the hydrograph of Don Juan Pond that losses of water from the pond were relatively large during windy periods. Although increased wind speed increases evaporation, it is also possible— and, we believe, likely—that some water was lost from the pond as aerosols during windy periods. We note in passing that salts would be carried from the basin in any such aerosols, perhaps affecting the chemistry of soils and waters in the rest of the south fork. (Winds might also carry dry salts from the salt pan.) Finally, the high salinity of the waters and presence of salt crusts on the floor of the basin suggest the possibility that during periods of high relative humidity, some hygroscopic water enters the floor of the basin from the atmosphere. Hygroscopic moistening of salt crusts has been noted elsewhere in Wright Valley [*Ugolini*, 1963; *Everett*, 1971] and probably accounts

TABLE 2. Groundwater Flux to the Floor of the Don Juan Basin From the Ferrar Dolerite, Calculated by Using the Gradient in Head Between DVDP Borehole 13 and Two Points on the Floor

	Average Total Head, DVDP 13, m	Average Total Head, Pond, m	Average Total Head, Piezometer 4, m	Head Gradient, Pond	Head Gradient, Piezometer 4	Flux, Pond, 10^{-4} m/day	Flux, Piezometer 4, 10^{-4} m/day
Dec. 6, 1975	117.10	116.12	116.68	0.077	0.034	6.65	2.94
Dec. 10, 1975	117.16	116.11	116.68	0.082	0.039	7.08	3.37
Dec. 17, 1975	117.23	116.10	116.65	0.089	0.047	7.69	4.06
Dec. 29, 1975	117.25	116.09	116.63	0.091	0.050	7.86	4.32
Jan. 4, 1976	117.29	116.08	116.63	0.095	0.053	8.21	4.56
Weighted mean						7.57	3.18

Hydraulic conductivity taken to be 10^{-5} cm/s.

TABLE 3. Evaporation From Two Pans on the Floor of the Don Juan Basin, the Pans Being Filled With Pond Water

	Decrease in Water Level, Pan 1, mm	Loss Rate, Pan 1, mm/day	Decrease in Water Level, Pan 2, mm	Loss Rate, Pan 2, mm/day	Corrected Rate, Pan 1, mm/day	Corrected Rate, Pan 2, mm/day	Average Rate, mm/day
Oct. 28, 1975	4.5	0.9	...	...	0.9	...	0.9
Nov. 2, 1975	6.0	0.6	...	...	0.6	...	0.6
Nov. 12, 1975	8.0	1.0	8.0	1.0	1.1	1.1	1.1
Nov. 20, 1975	8.5	1.7	7.5	1.5	1.9	1.8	1.8
Nov. 25, 1975	13.5	1.5	9.5	1.1*	1.9	*	1.9
Dec. 3, 1975	6.5	2.2	5.5	1.8	2.3	2.7	2.5
Dec. 6, 1975	6.0	2.0*	5.5	1.8*	*	*	*
Dec. 9, 1975	10.0	0.8*	8.0	0.7*	*	*	*
Dec. 21, 1975	18.5	1.5*	14.5	1.2*	*	*	*
Jan. 2, 1976	13.0	1.8*	10.5	1.2*	*	*	*
Jan. 11, 1976							

* Heavy precipitation of salt in pan.

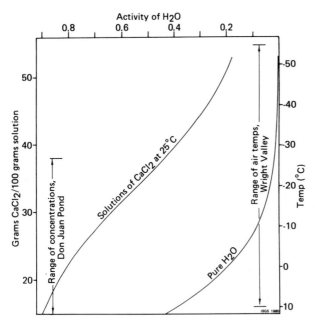

Fig. 13.   Activity of water at temperatures ranging from about 10°C to −53°C [*Weast*, 1974] and in solutions of CaCl₂ at 25°C [*Robinson and Stokes*, 1959]. The work of *Voznesenskaya* [1977] suggests that the curve for CaCl₂ at 0°C lies slightly below the 25°C curve.

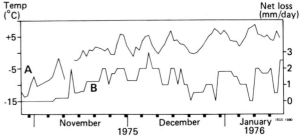

Fig. 14.   (*a*) Mean daily air temperature at the soil surface, floor of Don Juan basin and (*b*) net rate of water level decline in Don Juan Pond, October 27 through January 20.

for the sometimes damp appearance of the 'tongues' on the south wall of the basin.

In theory, the rate of evaporation (or sublimation) from a body of water is given by an equation of the form

$$E = G(a - h)$$

where E is the evaporative flux, G is a 'conductivity' coefficient, a is the activity of water at the surface of the body, and h is the activity of water vapor (in effect, the relative humidity) in the atmosphere overlying the body [e.g., *Petterssen*, 1958]. The wide range of physical conditions in and chemical compositions of the waters at the floor of the basin has a pronounced effect upon the activities of those waters (Figure 13) and therefore upon the rates at which they evaporate and sublimate.

In order to make the loss rates of the pan waters representative of those in the pond, the pans were filled with pond water at the beginning of each period of measurement. During all of these periods, however, the salinity of the pan waters increased more rapidly than that of the pond, both because the pans were shallower than the pond and because the pond received additional, relatively fresh water, while the pans did not. This difference was pronounced after late November, when evaporation rates increased markedly; in most instances

the pans contained a slush of precipitated salt at the end of a period of measurement. Corrections for the difference in salinity between pond and pan waters were made for the periods prior to heavy precipitation in the pans (i.e., prior to December 6) and are given in Table 3. All other factors being equal then, the evaporation rates of pan waters must have been below, and sometimes markedly below, those of the pond water; the difference in rates must have been least in the early part of the field season.

The effect of weather upon evaporation rates is suggested by comparison of Figures 14a and 14b; prior to the onset of streamflow to Don Juan Pond, there is a noticeable correspondence between loss rates in the pond (as determined from the hydrograph) and mean air temperatures at the soil surface. Changes in the temperature of the pond, in the relative humidity, and in the wind speed all would affect evaporation from the pond; all were undoubtedly correlated with air temperature at the soil surface. In addition, there were probably large daily changes in evaporation rate, given the normal large daily excursion of air temperature in the basin (Figure 15).

Other workers have examined summer losses from the predominantly ice-covered surface of

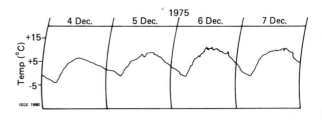

Fig. 15.   Air temperature at the soil surface, floor of Don Juan basin.

TABLE 4. Summary Fluid Balance of Don Juan Pond, October 28, 1975, Through January 11, 1976 (See Tables 5–7)

Volume Evaporated	Volume of Groundwater	Volume of Streamflow	Change in Volume
7751 m³	2525 m³	432 m³	− 4794 m³
92 m³/day	30 m³/day		− 57 m³/day
1.8 mm/day	0.6 mm/day		− 1.1 mm/day

Lake Vanda; the rates reported range between 1.9 and 3.5 mm/day [*Ragotzkie and Likens*, 1964; *Yusa*, 1972; *Chinn*, 1976]. *Everett* [1971] reported that summer evaporation rates measured in eastern Wright Valley using an evaporimeter ranged from 0.5 mm/hr (12 mm/day) for 10 hours up to 1.0 mm/hr (24 mm/day) for 8 hours, the latter on days with conditions 'very favorable' for evaporation.

FLUID BALANCE

The pieces of data reported above may be combined to obtain a more complete picture of the fluid balance for Don Juan Pond and surroundings during the field season of 1975-1976. Balances are given in Table 4 and depicted in Figure 16; the calculation of these balances is explained below.

For the period from October 28 through December 6, during most of which DVDP 13 was plugged with ice, we calculated the difference between volume changes in and volumes evaporated from the pond and assumed that the difference was supplied by groundwater (Table 5). For the period from December 6 through January 4, during which DVDP 13 was open but the evaporation pans were precipitating salt, the procedure was reversed, and volume changes and groundwater discharge were added to give evaporation rates (Table 6).

Fluid balance for the pond between January 4 and 20 was difficult to determine because of the influx of stream water. Additional difficulties were that there was heavy precipitation of salt in the evaporation pans and that no hydrograph of DVDP 13 was taken between January 11 and January 20. The behavior of water levels in the piezometers (Figure 10) suggests that evaporation rates between January 4 and 20 were at least as great as any in the preceding month. Furthermore, mean air temperatures (Figure 14) were as high as at any time during the field season. For these reasons we conservatively assumed that evaporation from the pond between January 4 and 20 averaged 3

mm/day. Water levels in DVDP 13 on January 19 and 20 were similar to those just prior to January 11 (i.e., 0.10 to 0.15 m below the top of the casing), so we assumed that groundwater discharge to the pond between January 11 and 20 was the same as that just prior to January 11 (i.e., 36 m³/day). These assumptions were used, together with the previously calculated changes in volume, to calculate streamflow to the pond (Table 7).

The decline of the water table beneath the salt pan indicates that water evaporated from the water table through the unsaturated zone during 1975-1976. Water levels in those piezometers not affected by streamflow and not adjacent to the pond (numbers 1, 3, 5, 6, 7, 8) dropped an average of about 0.13 m between October 27 and January 20 (Figure 10). In sediments having a porosity of 30%, this is equivalent to a loss of about 0.4 mm of water per day. To this must be added the flux of groundwater into the sediments (about 0.3 mm/day, see Table 2), giving a weighted mean evaporation rate of about 0.7 mm/day from the water table. This is less than half the weighted mean of the evaporation rate for Don Juan Pond (1.8 mm/day) during the same period. For the entire salt pan (about 1.9×10^5 m² in area) this amounts to an evaporative loss of about 130 m³/day.

The sensitivity to measurement error of the calculations given above requires comment. Consider, for example, the data reported in Table 5. For the period between October 28 and November 2, 1975, the water level in the evaporation pans decreased by more than 4.5 mm, whereas the water level recorded in the hydrograph of Don Juan Pond did not change. Assuming, for the sake of argument, that no individual measurement of water level was

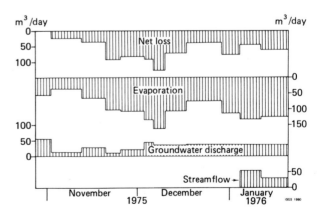

Fig. 16. Fluid balance of Don Juan Pond, October 27 through January 20. (See Tables 4 through 7.)

TABLE 5. Groundwater Discharge to Don Juan Pond, Calculated by Using Evaporation Rates and Changes in Volume

	Evaporation From Pan, mm/day	Surface Area of Pond, 10^3 m^2	Daily Evaporation From Pond, m^3/day	Total Volume Evaporated From Pond, m^3	Total Volume Change in Pond, m^3	Total Groundwater Discharge to Pond, m^3	Daily Groundwater Discharge, m^3/day
Oct. 28, 1975	0.9	59.5	54	270	0	270	54
Nov. 2, 1975	0.6	59.0	35	350	-230	120	12
Nov. 12, 1975	1.1	58.2	64	510	-290	220	28
Nov. 20, 1975	1.8	56.8	102	510	-460	50	10
Nov. 25, 1975	1.9	54.9	104	830	-660	170	21
Dec. 3, 1975	2.5	53.2	133	400	-270	130	43
Dec. 6, 1975							
Total or weighted mean	1.3	56.9	74	2870	-1910	960	25

TABLE 6. Evaporation From Don Juan Pond, Calculated by Using Groundwater Flux and Changes in Volume

	Groundwater Flux to Pond, 10^{-4} m/day	Surface Area of Pond, 10^3 m^2	Daily Groundwater Discharge, m^3/day	Total Groundwater Discharge, m^3	Total Volume Change in Pond, m^3	Total Volume Evaporated From Pond, m^3	Daily Evaporation From Pond, m^3/day	Daily Evaporation From Pond, mm/day
Dec. 6, 1975	6.65	51.3	34	136	−510	646	162	3.2
Dec. 10, 1975	7.08	49.5	35	425	−500	745	106	2.2
Dec. 17, 1975	7.69	47.5	36	432	−470	902	75	1.6
Dec. 29, 1975	7.86	45.5	36	216	−460*	676	113	2.5
Jan. 4, 1976	8.21	43.5	36	216				
Jan. 10, 1976			36					
Total or weighted mean (to Jan. 4)	7.43	48.4	36	1029	−1940	2969	102	2.2

*Volume change may have been affected by streamflow on January 3.

TABLE 7. Streamflow to Don Juan Pond, Calculated by Using Changes in Volume, Estimated Groundwater Flux, and Estimated Evaporation Rates

	Daily Evaporation From Pond, mm/day	Surface Area of Pond, 10^3 m^2	Daily Evaporation From Pond, m^3/day	Total Volume Evaporated From Pond, m^3	Daily Groundwater Discharge, m^3/day	Total Groundwater Discharge, m^3	Total Volume Change in Pond, m^3	Total Streamflow, m^3	Daily Streamflow, m^3/day
Jan. 4, 1976	3.0	43.8	131	920	36	252	−308	360	51
Jan. 11, 1976	3.0	41.8	125	1129	36	324	−542	263	30
Jan. 20, 1976									
Total or weighted mean	3.0	42.7	128	2049	36	576	−850	623	39

TABLE 8.   Chemical Analyses of Water Samples Taken From Don Juan Pond, DVDP Borehole 13, and the Stream Feeding the
Pond, All on January 20, 1976

	Na	K	Ca	Mg	Cl$^-$	SO$_4^{2-}$	pH	Specific Gravity at 20°C	NaCl, %	CaCl$_2$, %
Don Juan Pond (sample 11)	4700	213	165,000	2500	320,000 (299,000)	10	4.1	1.355	0.9	34
DVDP 13 (15 m)	6900	91	72,000	1000	140,000 (138,000)	100	6.4	1.164	1.5	17
DVDP 13 (58 m)	7500	105	77,500	1100	158,000 (149,000)	100	6.7	1.182	1.6	18
Stream	42	2	28	22	140	60	7.7	...	...	...

Unless otherwise indicated, all analyses are in milligrams per liter. NaCl and CaCl$_2$ columns are in weight percent of solution.
Procedure for calculating weight percents: all Na and Ca from analyses was used; amount of Cl$^-$ required to balance Na$^+$ and Ca^{2+}
(shown in parentheses, in milligrams per liter, in Cl$^-$ column) was added; remainder of solution was taken to be H$_2$O (see Figure 17).

more accurate than ± 1 mm, the level of the pond
might have changed by as much as ± 2 mm and the
pan level by as little as 2.5 mm or as much as 6.5
mm. In the most extreme cases then the pond (59.5
$\times 10^3$ m^2 in area at the time, Table 1) could have
gained as little as 0.5 mm (about 6 m^3/day) or as
much as 8.5 mm (about 100 m^3/day) of groundwater. Although errors of this kind are undoubtedly embodied in the values given in Table 5, we
believe the errors in these values to be, on average, less extreme.

As suggested in Table 2, the calculations of
groundwater flux made using Darcy's law are comparatively insensitive to errors in the measurement of water levels. A total error of about 0.5 m
(i.e., approximately the difference between the
water level in DVDP 13 and that in piezometer 4)
would change the calculated flux only by a factor of
about 2. More significant are the uncertainties involved in determining the hydraulic conductivity of
the fluids and sediments (Figure 12). If, for example, the hydraulic conductivity falls between $5 \times
10^{-6}$ and 5×10^{-5} cm/s and the potential gradient
between 0.03 and 0.10, the flux could be as little as
1.3×10^{-4} m/day (about 6 m^3/day to the entire
pond) or as much as 4.3×10^{-3} m/day (about 216
m^3/day to the entire pond). Furthermore, the volume calculations made by using these fluxes (for
example, Table 6) embody the assumptions that (1)
any spatial variation in the hydraulic conductivity
is insignificant, and (2) the fluxes occurred over the
entire area of the pond (or salt pan). Strictly
speaking then, the volume calculations that we report are informed estimates; we believe, however,
that they accurately reflect the orders of magnitude of the fluxes.

CHEMISTRY OF WATERS

The major element chemistry of ground and surface waters occurring on and beneath the floor of
the Don Juan basin during the field season of 1975-
1976 is briefly discussed below. A more complete
discussion of the chemical behavior of Don Juan
Pond during 1975-1976 and, to a lesser degree,
during earlier years is given by *Harris et al.* [1979];
Torii and Yamagata [this volume] provide a good
review of other geochemical research in the basin.

Chemical analyses of waters taken on January
20, 1976, from DVDP 13, Don Juan Pond, and the
stream feeding the pond are reported in Table 8.
Chloride, calcium, and sodium constituted more
than 99% of the salt dissolved in the pond and deep
groundwater; this is also true of all other analyses
of these waters of which we are aware. The salinity of these solutions is so great that they are not
simple aqueous solutions; it is sometimes appropriate to treat water as just another component of the
solutions.

In a triangular diagram for which H$_2$O, CaCl$_2$,
and NaCl are the components, analyses of Don
Juan Pond waters and deep groundwaters fall in
the region indicated in Figure 17*a*; detail within
the region is shown in Figure 17*b*. As the latter
figure is of considerable use in explaining the
chemistry of these waters, it is described here in
some length. The region is broken into fields by
what are, in effect, phase boundaries of the minerals antarcticite [*Torii and Ossaka*, 1965], halite,
hydrohalite [*Craig et al.*, 1974], and ice; the fields
are crossed by isotherms. The positions of both the
phase boundaries and the isotherms are taken from
published laboratory data [*Washburn*, 1928; *Linke*,

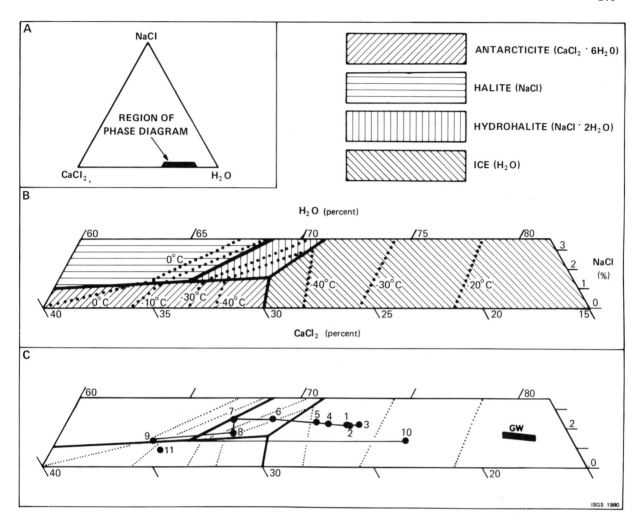

Fig. 17. (a) Component diagram for H₂O, CaCl₂, NaCl; chemical analyses of saline waters from the Don Juan basin fall within the region indicated (see Table 8). (b) Detail within the region shown in Figure 17a. Phase fields are patterned areas. Fields are cut by isotherms; all but 0° isotherm are in negative degrees celsius. (c) Phase diagram of Figure 17b with Don Juan Pond (dots) and groundwater (GW) analyses plotted. Pond analyses are numbered; numbers correspond to those in Figure 4. All analyses are of samples taken between October 27, 1975, and January 20, 1976. Modified from *Harris et al.* [1979].

1958; *Makarov and Shcharkova*, 1969; *Weast*, 1974]. Each phase field and the isotherms within it indicate the range of solution compositions and temperatures at which a particular mineral in contact with the solution should be stable. For example, a solution having a composition of 20% CaCl₂ and 2% NaCl would be free of ice at temperatures above about −20°C. Lowering of the temperature would cause the solution to 'precipitate' ice, simultaneously forcing the composition of the solution away from the H₂O vertex of the component triangle, increasing the concentrations of CaCl₂ and NaCl in the solution, and lowering its freezing point. At a temperature of about −40°C the com-

position of the solution would have moved to the boundary between hydrohalite and ice; further lowering of the temperature would cause the simultaneous precipitation of hydrohalite and ice, and the composition of the solution would move toward the 'triple point' for antarcticite, hydrohalite, and ice. At temperatures below −50°C (isotherm not shown for lack of sufficient data) these three minerals would precipitate from the solution simultaneously.

Analyses of 32 samples of saline waters taken from the floor of Don Juan basin in the field season of 1975-1976 are plotted in Figure 17c. The composition of the pond near the water level recorder

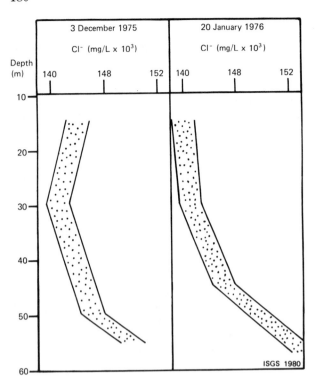

Fig. 18.   Chloride concentrations in DVDP borehole 13, December 3, 1975, and January 20, 1976. Stippled band indicates range of analytical accuracy.

varied widely during the season, primarily as a result first of evaporation (samples 1 through 9) and then of the influx of fresh stream water (samples 9 and 10). Sample 11, taken at the eastern end of the pond on January 20, was sufficiently distant from the streams to remain relatively saline. This incomplete mixture of stream and pond waters is revealed in Figure 9, which shows that salts precipitated from the pond prior to January 20 were, in much of the western end of the pond, redissolved by January 20.

The chemistry of the groundwaters taken from DVDP 13 varied little, either spatially or temporally; analyses of 21 samples taken at different times (December 3, December 5, December 20, January 7, January 20) and depths all plot within the bar at the right-hand end of Figure 17c. Most of the variation in the chemistry of the groundwaters was spatial; the concentrations of chloride (Figure 18), calcium, and sodium all were higher in the basement complex than in the dolerite.

The chemistry of Don Juan Pond is apparently unique among brine ponds and lakes worldwide. Brines in which the predominant cation is calcium are very unusual; we are aware of no other surface

water brines in which calcium is so overwhelmingly predominant [e.g., *Eugster and Hardie*, 1978]. However, the 'type connate' water of *White* [1965] has salinity and major ion chemistry virtually identical to those of groundwater found in the dolerite aquifer beneath Don Juan Pond. This type connate water was taken from a depth of about 900 m in a Precambrian basalt in Michigan, in the north central United States.

DISCUSSION

The evidence of groundwater discharge to the floor of the Don Juan basin during 1975-1976 is, we believe, substantial. The difference between the potential of fluid in the dolerite and that of fluid in the pond, the volume balance calculations, the visible springs: all indicate that discharge occurred. The behavior of water levels in DVDP 5 and DVDP 13 during the austral summers of 1973-1974, 1974-1975, and 1978-1979 indicates that discharge occurred in those summers as well; and there is photographic evidence suggesting that there were springs on the floor of the basin in at least one earlier summer, that of 1963-1964. There is, however, no direct evidence of discharge at other times. As groundwater is potentially significant in the hydrologic and geochemical history of the basin, it is instructive to examine the consequences of the hypothesis that discharges like those we observed have been persistent.

Consider first a simple model of the monthly hydrologic behavior of Don Juan Pond during a 'typical' recent year. The model incorporates the following features: (1) groundwater discharging to the pond at 30 m^3/day throughout the year, (2) net losses of water to the atmosphere like those reported above and like those observed year round at Lake Vanda [*Chinn*, 1976], and (3) streamflow varying from year to year and (to simplify the model) occurring only in the month of January. Monthly and cumulative fluxes and volume changes for the model, with streamflow excluded, are given in Table 9.

Three hydrographs of Don Juan Pond, all spanning a period of 16 months, are shown in Figure 19; two of these are model hydrographs constructed by using the values given in Table 9, and the third is the actual hydrograph of the pond for October 1974 through January 1976. One of the model hydrographs excludes all streamflow, whereas the other includes streamflow sufficient to make the

TABLE 9.   Model Fluid Balance of Don Juan Pond in a 'Typical' Year, But Without Streamflow

	Cumulative Volume Change, m³	Daily Evaporation, mm/day	Daily Evaporation, m³/day	Total Evaporation, m³	Cumulative Evaporation, m³	Cumulative Groundwater Gain, m³	Volume Change, m³
January		2.5	125	3875			−2945
	−2945				3,875	930	
February		1.4	70	1960			−1120
	−4065				5,835	1,770	
March		0.5	25	775			+155
	−3910				6,610	2,700	
April		0.5	25	750			+150
	−3760				7,360	3,600	
May		0.5	25	775			+155
	−3605				8,135	4,530	
June		0.5	25	750			+150
	−3455				8,885	4,530	
July		0.5	25	775			+155
	−3300				9,660	6,360	
August		0.5	25	775			+155
	−3145				10,435	7,290	
September		0.5	25	750			+150
	−2995				11,185	8,190	
October		0.5	25	775			+155
	−2840				11,960	9,120	
November		1.4	70	2100			−1200
	−4040				14,060	10,020	
December		2.0	100	3100			−2170
	−6210				17,160	10,950	
January							
Total or weighted mean	120 mm/yr	0.94 mm/day	47 m³/day		340 mm/yr	220 mm/yr	

Calculated assuming constant (30 m³/day) groundwater discharge, evaporation (net loss) rates like those observed at Lake Vanda [e.g., *Chinn*, 1976] and like those reported in Table 3, 6, and 7, and pond area of 5×10^4 m² (see Figure 19).

model follow the actual hydrograph of the pond as closely as possible.

The model suggests that under current conditions the pond is in a precarious hydrologic equilibrium. Without streamflow the yearly water deficit of the pond would be more than 6000 m³; without groundwater discharge, the deficit would be almost 11,000 m³. If either of these inflows were to cease, the pond would go dry within a year. Furthermore, the model suggests that the continued existence of the pond is predicated upon streamflow larger than those we observed.

Consider next the chemistry of the waters in the dolerite and the pond. As shown in Figure 17, the salts in Don Juan Pond clearly bear a genetic relation to the salts in the groundwater. Furthermore, the salinity and major ion chemistry of the groundwater in the dolerite are almost identical to those of groundwater taken from a basalt in the United States. Figure 5 indicates that the temperature of the groundwater in DVDP 13 is consistently

slightly above the regional mean annual air temperature of −18°C; Figure 17 reveals that the chemistry of the groundwater appears to be delicately equilibrated to this temperature. In other words, any decrease in the temperature of the liquid groundwater would cause ice to precipitate, forcing a simultaneous change in the chemistry of the liquid. In fact, it may well be that ice occurs in the upper parts of the geologic section at times during the year, as winter air temperatures are below −50°C.

Consider finally the salinity of the discharging groundwater. If the salinity of this groundwater is like that of the groundwater in DVDP 13 (i.e., about 225 kg salt/m³ solution), then, given a discharge of 30 m³/day, approximately 2500 metric tons (T) of salt must accumulate at or near ground surface every year. If the discharge to the entire floor of the basin is of the order of 100 m³/day, then some 8200 T of salt must accumulate yearly. In comparison to the amount of salt that apparently is

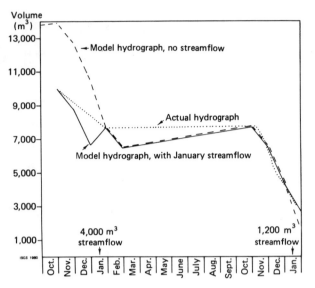

Fig. 19. Model and actual hydrographs of Don Juan Pond over a 16-month period (October through January). First model hydrograph excludes streamflow; second includes January streamflow of 4000 m^3 and 1200 m^3. Actual hydrograph from November 1974 through January 1976; no measurements were taken between January and November 1975.

actually present in the basin, these quantities are surprisingly large. For example, if the density of the dry salt is taken to be 2 T/m^3, then a crust only 1 cm thick covering the entire salt pan contains 4800 T of salt. *Torii et al.* [1977] estimated that each meter of sediment beneath the salt pan contains about 72,000 T of salt; if so, the entire sedimentary sequence contains about 940,000 T. A flow of 100 m^3/day would therefore require only about 100 years to import a quantity of salt equal to that now present in the sediments.

Taken at face value then, these calculations indicate that the discharge of deep groundwater to and accumulation of salt in the Don Juan basin are comparatively recent phenomena, having begun of the order of 10^2 to 10^3 years ago. This length of time seems to us extraordinarily short, particularly in view of the recent climatic history of the region. We can offer no simple explanation of how such a discharge might have been initiated and sustained, especially in a region of deep and ubiquitous frozen ground. The encroachment of frozen ground upon liquid groundwaters which had been in prolonged contact with the dolerite is perhaps the best of the complex explanations that occur to us, but mechanisms for causing the resulting discharge to develop solely in or to become limited to the Don Juan basin seem cumbersome and unlikely. There

are two obvious alternatives remaining. First, perhaps the volume of groundwater discharge has varied widely over time. The mechanism that might effect such variation is, again, obscure. Second, perhaps salt brought into the basin by groundwater has been carried out of the basin by some other means. Wind transport—the refuge of the perplexed Antarctic geochemist—is one possible mechanism.

SUMMARY

Don Juan Pond is the hydrologic centerpiece of the Don Juan basin, the most visible and easily studied piece of the hydrologic system active in the basin. The pond is, however, a passive actor; its behavior is largely determined by and reflective of the behavior of the other pieces of the system.

Intermittent freshwater streams flowing into the western end of the pond are sources of significant volumes of water, but only late in the austral summer. Strictly speaking, the streams are groundwater discharges, in that they are supplied primarily by meltwater derived from frozen ground. There are no outflowing streams; the basin is topographically closed.

Unfrozen, saline groundwater discharges from a deep, confined, basaltic aquifer to the pond and the floor of the basin and probably provides a large percentage of the annual water supply of the pond. Groundwater in the sediments at the surface of the basin floor is unfrozen and unconfined and is recharged from the deep aquifer and from streamflow. Most of this 'water table' groundwater passes through the unsaturated zone and evaporates; a very small percentage moves laterally, discharging in Don Juan Pond or in springs feeding the pond.

Evaporation and sublimation are the predominant facts of hydrologic life in the basin, as in most of the Dry Valley region, and are, with aerosols, the sole means whereby water is lost from pond and basin. Evaporation far exceeds precipitation; without persistent streamflow and groundwater discharge, Don Juan Pond would rapidly go dry.

The salts in Don Juan Pond and in the groundwater discharging to the pond are alike and must have a common origin. The groundwater is similar in major element chemistry and salinity to deep groundwater taken from basalt in the United States, suggesting that the salts originated in the weathering of bedrock. The chemistry of the groundwater also appears to be delicately equilibrated to mean annual air temperature. The history of salt accumulation and the history of groundwater flows in the basin are related but obscure and must be deciphered together.

Acknowledgments. Much of the field work upon which this research is based was made possible by the untiring, highly competent, and diverse efforts of Michael Chapman-Smith of the University of Auckland, New Zealand. David Gross and Leon Follmer, both of the Illinois State Geological Survey, provided valuable assistance and observations in Antarctica, as did Gerry Bucher, Ed Decker, and Keith Baker of the University of Wyoming. Robert Ayers, of Virginia Polytechnic Institute, David Read, of Northern Illinois University, and Larry Wigginton, of the University of Texas at Austin, all assisted in mapping. The men and women of New Zealand's Vanda Station, of U.S. Navy Squadron VXE-6, of the U.S. National Science Foundation, and of Holmes & Narver, Inc., all rendered vital support and comfort in Antarctica. Chemical analyses were performed by the Analytical Chemistry Section of the Illinois State Geological Survey under the direction of R. Ruch. Financial support was provided in part by National Science Foundation grants GV-40436, OPP-73-05917, and DPP 76-23045 through the University of Illinois, Urbana-Champaign.

REFERENCES

Barrett, P. J., G. W. Grindley, and P. N. Webb, The Beacon Supergroup of East Antarctica, in *Antarctic Geology and Geophysics.*, edited by R. J. Adie, pp. 319-32, Universitetsforlaget, Oslo, 1972.

Bear, J., *Dynamics of Fluids in Porous Media*, 764 pp., Elsevier, New York, 1972.

Bull, C., Climatological observations in ice-free areas of Southern Victoria Land, Antarctica, in *Studies in Antarctic Meteorology, Antarct. Res. Ser.*, vol. 9, edited by M. J. Rubin, pp. 177-194, AGU, Washington, D. C. 1966.

Calkin, P. E., and C. Bull, The glacial history of the ice-free area, Southern Victoria Land, Antarctica, *Polar Rec., 17*(107), 129-137, 1974.

Cartwright, K., and H. J. H. Harris, Hydrogeology of the dry valley region, Antarctica, this volume.

Cartwright, K., S. B. Treves, and T. Torii, Geology of DVDP 5, Don Juan Pond, Wright Valley, Antarctica, *Dry Val. Drilling Proj. Bull., 3*, 75-91, 1974.

Chinn, T. J., Hydrological research report, Dry Valleys, Antarctica, 1974-75, N.Z. Min. of Works and Develop., Christchurch, 1976.

Craig, J. R., R. D. Fortner, and B. L. Weand, Halite and hydrohalite from Lake Bonney, Taylor Valley, Antarctica, *Geology, 2*(8), 389-390, 1974.

Decker, E. R., and G. J. Bucher, Preliminary geothermal studies in the Ross Island–Dry Valley Region, in *Antarctic Geoscience*, edited by C. Craddock, University of Wisconsin Press, Madison, in press, 1980.

Domenico, P. A., *Concepts and Models in Groundwater Hydrology*, 405 pp., McGraw-Hill, New York, 1972.

Eugster, H. P., and L. A. Hardie, Saline lakes, in *Lakes: Chemistry, Geology, Physics*, edited by A. Lerman, pp. 237-293, Springer, New York, 1978.

Everett, K. R., Soils of the Meserve Glacier area, Wright Val-

ley, Southern Victoria Land, Antarctica, *Soil Sci., 112*(6), 425-538, 1971.

Freeze, R. A., and J. A. Cherry, *Groundwater*, 604 pp., Prentice-Hall, Englewood Cliffs, N.J., 1979.

Harris, H. J. H., and K. Cartwright, Pressure fluctuations in an Antarctic aquifer: The freight-train response to a moving rock glacier, in *Antarctic Geoscience*, edited by C. Craddock, University of Wisconsin Press, Madison, in press, 1980.

Harris, H. J. H., K. Cartwright, and T. Torii, Dynamic chemical equilibrium in a polar desert pond: A sensitive index of meteorological cycles, *Science, 204*(4390), 301-303, 1979. (Also see *Science, 204*(4396), 909, 1979.)

Heine, A. J., Seiche observations at Lake Vanda, Victoria Land, Antarctica, *N. Z. J. Geol. Geophys., 14*(3), 597-599, 1971.

Lambe, T. W., *Soil Testing for Engineers*, 165 pp., John Wiley, New York, 1951.

Linke, W. F. (Ed.), *Solubilities, Inorganic and Metal-Organic Compounds*, vol. I, 4th ed., D. Van Nostrand, Princeton, N. J., 1958.

Makarov, S. Z., and E. F. Shcharkova, Solubility isotherms (10°) for ternary systems consisting of calcium and sodium hypochlorites and chlorides (in Russian), *Zh. Neorg. Khim., 14*(11), 3096-3099, 1969.

McGinnis, L. D., and T. E. Jensen, Permafrost-hydrogeologic regimen in two ice-free valleys, Antarctica, from electrical depth soundings, *Quaternary Res., 1*(3), 390-409, 1971.

McGinnis, L. D., K. Nakao, and C. C. Clark, Geophysical identification of frozen and unfrozen ground, Antarctica, in *Permafrost: 2nd International Conference, North American Contribution*, pp. 136-146, National Academy of Science, Washington, D. C. , 1973.

McKelvey, B. C., and P. N. Webb, Geological investigations in Southern Victoria Land, Antarctica, 3, Geology of Wright Valley, *N. Z. J. Geol. Geophys., 5*(1), 143-162, 1962.

Meyer, G. H., M. B. Morrow, O. Wyss, T. E. Berg, and J. L. Littlepage, Antarctica: The microbiology of an unfrozen saline pond, *Science, 138*(3545), 1103-1104, 1962.

Mudrey, M. G., T. Torii, and H. Harris, Geology of DVDP 13, Don Juan Pond, Wright Valley, Antarctica, *Dry Val. Drilling Proj. Bull., 5*, 78-99, 1975.

Petterssen, S., *Introduction to Meteorology*, 327 pp., McGraw-Hill, New York, 1958.

Ragotzkie, R. A., and G. E. Likens, The heat balance of two Antarctic lakes, *Limnol. Oceanogr., 9*(3), 412-425, 1964.

Robinson, R. A., and R. H. Stokes, *Electrolyte Solutions*, 2nd ed., 559 pp., Butterworths, London, 1959.

Smithson, S. B., P. R. Fikkan, and D. J. Toogood, Early geologic events in the ice-free valleys, Antarctica, *Geol. Soc. Amer. Bull., 81*(1), 207-210, 1970.

Thompson, D. C., R. M. F. Craig, and A. M. Bromley, Climate and surface heat balance in an Antarctic Dry Valley, *N. Z. J. Sci., 14*(2), 245-251, 1971.

Torii, T., and J. Ossaka, Antarcticite: A new mineral, calcium chloride hexahydrate, discovered in Antarctica, *Science, 149*(3687), 975-977, 1965.

Torii, T., and N. Yamagata, Limnological studies of saline lakes in the dry valleys, this volume.

Torii, T., N. Yamagata, J. Ossaka, and S. Murata, Salt balance in the Don Juan basin, *Antarct. Rec., 58*, 116-130, 1977.

Ugolini, F. C., Soil investigations in the lower Wright Valley, Antarctica, in *Proceedings, International Permafrost Con-*

ference, pp. 55-61, National Academy of Science, Washington, D. C., 1963.

Ugolini, F. C., and C. Bull, Soil development and glacial events in Antarctica, *Quaternaria*, *VII*, 251-269, 1965.

Voznesenskaya, I. E., Effect of electrolyte nature on the character of concentration and temperature dependences of water activity (in Russian), *Zh. Fiz. Khim.*, *51*(10), 2695-2696, 1977.

Washburn, E. W. (Ed.), *International Critical Tables*, vol. IV, McGraw-Hill, New York, 1928.

Weast, R. C. (Ed.), *Handbook of Chemistry and Physics*, 55th ed., Chemical Rubber Company, Cleveland, Ohio, 1974.

White, D. E., Saline waters of sedimentary rocks, in *A Symposium, Fluids in Subsurface Environments, Mem. 4*, edited by A. Young and J. E. Galley, pp. 342-366, American Association of Petroleum Geologists, Tulsa, Okla., 1965.

Yusa, Y., The re-evaluation of heat balance in Lake Vanda, Victoria Land, *Contrib. 12*, pp. 87-100, Geophys. Inst., Kyoto Univ., Kyoto, Japan, 1972.

A REVIEW OF THE GEOCHEMISTRY AND LAKE PHYSICS OF THE ANTARCTIC DRY AREAS

A. T. WILSON[1]

Antarctic Research Unit, University of Waikato, New Zealand

Although much of the Antarctic continent is covered with ice and snow, there are some small areas which are ice free. This paper reviews the lake physics and geochemistry of these dry areas. The Antarctic dry areas represent a climatic region not found elsewhere in the world. Because of the extreme cold and aridity and the fact that these areas are truly rainless, many features and phenomena exist that are not found on the other continents. The major source of the salts in the soil and lakes appears to be from the sea via snow. A relative humidity mechanism for salt separation is proposed in order to explain the distribution of salts in the soils, glaciers, streams, and lakes of Antarctic dry areas. This mechanism leaves the least deliquescent salts in the soil and delivers the more deliquescent salts to the groundwater which flows along the surface of the ice-cemented layer and hence to the saline lakes. While in the groundwater system, trace elements such as strontium may be leached from the soils. Climatic conditions are such that perennially ice-covered freshwater lakes can exist, and the conditions for the existence of these are defined. Solar heating of density-stratified lakes is common and in some cases very spectacular, the bottom of Lake Vanda being 46°C above the mean temperature of the region.

INTRODUCTION

The extent of glaciation in most parts of the world is controlled largely by summer temperature [*Charlesworth*, 1957]. Antarctica is a polar desert, and the extent of ice and snow cover is controlled by the precipitation/evaporation balance (see, for example, *Wilson* [1967]). The extremes of aridity and the fact that it is truly a rainless desert have produced many features and phenomena not found in other parts of the world. Although much of the Antarctic continent is covered with ice and snow, there are some small areas which are ice free. One of the largest and best studied of these is the 'McMurdo Oasis' area. The names 'dry valley' or 'oasis' are not completely satisfactory terms, since a similar physical situation and many of the same phenomena exist generally in all the more arid areas of the Antarctic, some of which are neither oases nor dry valleys. For example, the areas on the side of all the valleys south of the 'McMurdo dry valleys' exhibit some of the same phenomena as the McMurdo dry valleys. They have not been called dry valleys because they have a glacier passing

through them, yet in many other respects they are similar. A specific term such as ice-free area, dry area, or perhaps 'lapissec' should probably be used.

PHYSICAL SETTING OF THE DRY AREAS

The first question usually asked is, Why are the dry areas not ice covered like the rest of the continent? To answer this question, it is necessary to consider the precipitation/evaporation balance [*Wilson*, 1967]. In this cold and arid region (mean annual temperatures of the McMurdo Oasis region are only approximately −20°C) a very small fraction of the snow that falls ever melts, and almost all of it is lost directly by sublimation. The best way of understanding the precipitation/evaporation balance is to consider it in terms of the single parameter net precipitation. This is defined as being equal to the total precipitation less the total evaporation. If this value is positive for any area in this region, the land surface will be covered with ice and snow. If this value is negative (i.e., sublimation is greater than precipitation), the area will be ice free (a so-called 'dry area'), that is, unless ice can flow into the area from a region of positive net precipitation. The imaginary line which di-

[1] Now at Duval Corporation, 4715 East Fort Lowell Road, Tucson, Arizona 85712.

vides these two regions is called the 'snow line' and is the point at which precipitation equals sublimation [*Charlesworth*, 1957]. For a given region the net precipitation increases as altitude is increased. As one moves from the sea inland, the snow line rises, presumably because total precipitation decreases. As one descends below the snow line, the environment becomes more and more arid. Parts of the McMurdo dry valleys are so dry that $CaCl_2 \cdot 6H_2O$ crystallizes from ice-free saline ponds (for example, Don Juan Pond [*Torii and Ossaka*, 1965; *House et al.*, 1966]). The fact that such a deliquescent salt can exist demonstrates that this area is among the most arid on the earth. One could map the relative aridity of this region by drawing lines parallel to the snow line corresponding to, say, 80%, 70%, 60%, 50%, and 40% relative humidities. These lines not only control such physical phenomena as the depth to the ice-cemented layer and length of time the ice in an ice-cored moraine will survive but also have important ecological consequences in this region so inhospitable to living things. An example is the growth of lichen. It is well known that fungi can only grow at relative humidities of about 80%; thus lichens can only inhabit a strip between the 80% relative humidity line and the snow line. The dry areas are therefore those areas which lie below the snow line and into which ice from above the snow line cannot flow.

In the Ross Dependency area of Antarctica one can draw an east-west line along the edge of the Ross Ice Shelf. The mountains to the north of this line are generally covered with ice and snow. The area to the south of this line has many dry areas, presumably because of the lowered precipitation caused by the presence of the Ross Ice Shelf. In the special case in which a glacier cannot pass down a valley the valley is called a dry valley.

If one digs into the loose soil of a dry area in this cold and arid region, one passes through loose dry debris and can suddenly strike a very hard and impenetrable layer which on examination proves to be debris as above but cemented with ice crystals. This phenomenon is equivalent in some respects to the water table of the more temperate regions of the world. The depth of the top of this ice-cemented layer seems to be related to distance below snow line and is probably a balance between distillation of water downward during brief periods of snow cover and evaporation outward during periods of low relative humidity at the surface. Certainly, as one goes to higher altitudes and hence

nearer to the snow line the depth to the top of the ice-cemented layer decreases. During summer, cold saline water can be found moving downslope along the surface of the ice-cemented layer.

The dry valley areas of Antarctica frequently consist of a number of enclosed drainage basins. The lowest part of many of these is occupied by a lake, many of which are saline but some of which are relatively fresh.

If we consider the evaporation/precipitation balance of the entire drainage system, it can be seen that the net excess precipitation of a snow field will flow below the snow line as a glacier. If the surface area of the glacier is insufficient to balance the sublimation/precipitation budget, the glacier will advance further and further below the snow line toward a situation where the total positive net precipitation above the snow line is balanced by the total negative net precipitation below the snow line. Usually, the glacier has pushed sufficiently far below the snow line so that some summer melting takes place. In such cases for a few days during the hottest part of the summer a stream flows away from the glacier snout and feeds a lake which occupies the lowest point of that particular enclosed drainage basin, or flows into that rather special saline lake called the sea. The size of the lake is determined by that area needed to balance the evaporation/precipitation equation for that particular drainage area. If there is a net precipitation increase to the area, the lake levels will rise, and if there is a decrease in precipitation, the lake levels will fall.

The susceptibility of the lake levels to change depends on what fraction of the total evaporation of the system is accounted for by the lake surface. If this is small, a very small increase in net precipitation to the area will lead to an increase in lake area. Thus some lakes are much more susceptible to fluctuation of surface area than others. What is usually measured is lake level. This not only is a function of lake area but also depends on the shape of the depression occupied by the lake. For a flat basin such as that occupied by Lake Fryxell, a large increase in area can be achieved with little increase in depth, whereas in a steep U-shaped trough such as that occupied by Lake Bonney, the reverse is true. Thus these Antarctic lakes are very sensitive indicators of changes in net precipitation, which in turn is related to glacial advances and retreats.

In some cases, dried algae can be found in old

lake levels. For example, the upper level of Lake Vanda has been dated at 3000 ± 50 years [*Wilson*, 1967], and G. Denton (private communication, 1977) has dated the upper lake levels in the Taylor Valley. These dates give the time when the lake dropped from the level in question.

There can be a contribution of water to lakes and ponds from groundwater flow. This is discussed below and is reviewed in a paper by *Harris and Cartwright* [this volume].

Under certain conditions, Antarctic lakes deposit aragonite and gypsum. This material is particularly suitable for U/Th dating and has been used extensively by *Hendy et al.* [1979] for dating Quaternary climatic events. Besides being very suitable for dating, the $^{18}O/^{16}O$ isotope ratio of this material enables one to determine the source of water supplying the ancient lake that laid down the deposit. This is possible because the three possible sources of water—Polar Plateau ice, local glacial ice, or the ocean water—have very different $^{18}O/^{16}O$ ratios.

ORIGIN OF CHEMICAL STRATIFICATION

Many of the lakes that occupy the lowest part of the various enclosed drainage basins are chemically stratified, and these chemical concentration gradients contain paleoclimatic information if we can understand the system sufficiently well to interpret them.

The paleoclimatic data in the lakes are particularly welcome in the Antarctic because the more usual methods of dating climatic events are not often applicable.

Turning to the problem of the origin of chemical gradients in the saline lakes of the McMurdo Oasis, an extreme example is Lake Vanda. The lake is 66 m deep, the lower 20 m being strongly density stratified with the specific gravity rising from 1.005 at a depth of 46 m to 1.10 at the bottom. It is interesting to speculate on the origin of the salt concentration gradient in the lake. *Wilson* [1964] suggested that at some period in the past the climate was such that the Onyx River did not supply appreciable water to Lake Vanda. Under these conditions the lake level would have dropped until only a few feet of concentrated calcium chloride remained. When the climate changed, the Onyx would flow during the summer, and fresh water would have flowed on top of this strong salt solution. Since that time the salts have been diffusing

upward. If such a model is assumed, it is possible to calculate the time in the past when this climatic change occurred. *Wilson* [1964] calculated that Vanda rose relatively suddenly 1200 years ago. This same approach has been used to calculate diffusion ages for Lake Bonney [*Shirtcliffe*, 1964; *Hendy et al.*, 1977]. However, as discussed below, saline groundwater inflow can also provide a mechanism for providing chemical stratification [*Wilson*, 1979].

THE GEOCHEMISTRY OF ANTARCTIC DRY AREAS

When snow falls on a snow field or land surface, it contains small quantities of salts, the principal cations being sodium and calcium and the principal anion being chloride with smaller amounts of Mg^{2+}, K^+, SO_4^{2-}, and other materials. The chemical composition of atmospheric precipitation has been the subject of much study (see, for example, *Wilson* [1959, 1960] and *Junge* [1963]). Although the chemicals in precipitation are generally believed to come from the ocean, the actual composition is not merely diluted seawater but has been modified by separation processes at the ocean surface. The actual composition varies as one moves away from the sea, owing it is believed, to the existence of two types of aerosol particles in the atmosphere. The first type is large, falls out rapidly, and has the composition of bulk seawater. Particles of the second type are very much smaller and have a composition very different from seawater. They have, for example, higher K^+/Na^+ and Ca^{2+}/Mg^{2+} ratios than bulk seawater. Both types of aerosols arise from the bursting of bubbles on the surface of the ocean [*Wilson*, 1959, 1960].

The salt and lake geochemistry in the McMurdo dry area has been studied extensively by New Zealand, American, and Japanese scientists (see review by *Torii and Yamagata* [this volume]). Despite extensive studies over almost 20 years, the geochemistry of this region is far from being understood, and there is much apparently conflicting evidence. Even the origin of the salts in the lakes is in dispute. Since chloride is a minor constituent of Antarctic rocks and very little chemical weathering takes place, one might expect the chlorides in the lakes (which are mostly chlorides of Ca, Mg, and Na) to have originated from the sea. However, some lakes such as Lake Vanda are essentially calcium chloride brines, a fact that is difficult to explain by normal geochemical models. Further,

Jones and Faure [1967] have presented isotopic evidence to suggest that most of the strontium in Lake Vanda does not have an oceanic origin. Other problems are that adjacent lakes frequently have very different chemical compositions which often do not seem to match those of their inflow streams. Furthermore, different layers of the same lake sometimes have different ratios of their ionic constituents. The bottoms of some chemically stratified lakes, which on the basis of other evidence must have been out of contact with the atmosphere for hundreds if not thousands of years, contain measurable amounts of tritium (half-life of 12.3 years) in their bottom waters. Glaciers which would be expected to have the same composition as the regional snowfall, because polar glaciers do very little cutting, frequently produce meltwater streams with very different chemical compositions from the snow which falls on their névés.

The following is a description of phenomena which can only occur in polar deserts and presents evidence in support of a hypothesis which could explain all the apparently anomalous data.

Over very long periods of time, small quantities of snow fall on the land surface and on sublimation leave salts behind. These salts include Na^+, K^+, Ca^{2+}, Mg^{2+}, SO_4^{2-}, and Cl^-. It is proposed that fluctuations in relative humidity, due to the occasional snowfall and the summer-to-winter change, cause some of the more deliquescent salts to percolate slowly downward through the soil. The more deliquescent will reach the ice-cemented layer. At high altitudes, for example, on the top of the Asgard Range and Olympus Range and in the Alatna Valley, only the least deliquescent, for example, sodium sulphate, can remain in the soil; the remaining salts, $CaCl_2$, $MgCl_2$, and much of $NaCl$, eventually reach the ice-cemented layer, where during the late summer they move downslope along the surface of the ice-cemented layer. In more arid situations at lower altitudes, the $NaCl$ (77% equilibrium relative humidity at 0°C) cannot reach the ice-cemented layer. On the sides of the McMurdo dry valleys, for example, above Don Juan Pond and at similar altitudes along the sides of the Wright Valley, large quantities of $NaCl$ can be found in the 'soils' and under boulders. At higher altitudes in the same region, almost pure Na_2SO_4 and $Na_2SO_4 \cdot 2H_2O$ can be found in the same position (equilibrium relative humidity of sodium sulphate is 98% at 0°C). The calcium chloride (equilibrium relative humidity is 44% at 0°C) and magnesium chloride (equilibrium relative humidity is 38% at 0°C) eventually move to the lowest part of the drainage system—in the case of the Wright Valley this is Lake Vanda and Don Juan Pond. Thus it is proposed that we have huge geochemical separation systems in the Wright Valley and other Antarctic dry areas, the salt fractionation being controlled by relative humidity and the deliquescent properties of the possible salts that can form from the anions and cations left by evaporated snow. Very saline groundwater can be detected flowing along the surface of the ice-cemented layer in the late summer. For example, groundwater flowing along the top of the ice-cemented layer in the south fork of the Wright Valley in late January 1974 had the following composition: 61,400 ppm Cl^-, 11 ppm SO_4^{2-}, 18,100 ppm Ca^{2+}, 7,000 ppm Mg^{2+}, 9,900 ppm Na^+, and 635 ppm K^+ (sample collected by A.T.W. and analyzed by A. Field). The groundwater is high in Ca^{2+} with respect to Na^+ and low in SO_4^{2+} with respect to Cl^- and has the chemical composition necessary to explain the origin of the calcium chloride type lakes discussed above.

In the above discussion, relative deliquescence is proposed as the primary mechanism for the separation of water soluble salts in the Antarctic dry areas. For completeness, freezing must also be considered. Eutectic temperature would at first seem to provide an alternative parameter that could regulate the composition of the liquid phase, and such a mechanism will probably play some part in the initial separations. Evidence that relative deliquescence is the dominating mechanism is provided by the 'calcium chloride' type lakes and the 'groundwater' (see data above) which are not enriched in K^+ with respect to Na^+. This would be expected on the relative deliquescent model proposed (equilibrium relative humidity of 0°C of $NaCl$ is 77% and of KCl is 88%), whereas on a eutectic temperature model the reverse would be expected. When brines of the general composition expected in Antarctic soils are frozen, the K^+/Na^+ ratio remains constant until -8.2°C, at which point sodium sulphate decahydrate forms. The K^+/Na^+ ratio continues to increase with falling temperatures until -22.9°C, at which point sodium chloride dihydrate forms. The ratio further increases until at -36°C, sylvite (KCl) separates, as does the dodecahydrate of magnesium chloride [*Thompson and Nelson*, 1956]. On a eutectic freezing model, one would have expected potassium to follow magnesium and be enriched in the calcium chloride type lakes and ponds. In fact, the reverse is true, leading one to conclude that it is the relative deliquescent mechanism rather than a freezing or a solubility mechanism which controls the separation of soluble salts in polar deserts.

GROUNDWATER CHEMISTRY

The mean annual temperature of the McMurdo Dry Valley area is in the region of $-20°C$. Any liquid aqueous phase as deep groundwater at temperatures below $0°C$ must be saline. Saline groundwater has been found in boreholes drilled in the Dry Valley Drilling Project (DVDP) program (see review by *Harris and Cartwright* [this volume]). Some of these are calcium magnesium chloride brines. The source of these brines could be from the ocean, or relic [*Harris and Cartwright*, this volume], or they may enter the deep groundwater system from the surface groundwater which moves along the surface of the ice-cemented layer as described in this paper. As discussed above, brines produced from freezing seawater have a different chemical composition from those produced by the relative humidity separation process, particularly with respect to Na^+/K^+ ratios. However, since groundwater can leach ions from the material through which it percolates, the interpretation may be difficult.

Harris et al. [1979] have described the interesting case of a calcium chloride lake, Don Juan Pond, in the upper Wright Valley, where they claim that deep groundwater is supplying the saline water. Clearly, cold saline water flowing down the surface of the ice-cemented layer can be stored in voids and aquifers and overflow into surface lakes and ponds.

GLACIER CHEMISTRY

The snow that falls on the snow fields that feed the glaciers of the McMurdo region contains a fraction of a part per million of inorganic salts. It can be seen that if we measure, say, the sodium or chloride content of the snow in the névé or a glacier and also the sodium or chloride content of the stream leaving the snout of the glacier, it is a simple calculation to determine how much water has been lost from the glacier by sublimation, provided always that the ionic ratios of the stream water are similar to the ionic ratios of the snow in the névé. This is not always the case (see below). Such measurements indicate that only a fraction of a percent of the water that falls as snow in névés survives to flow as a stream from the snout of the glacier.

The groundwater described above can also flow into the sides of glaciers and may explain the origin of the clear brown saline 'amber ice' often found in the basal layers of the local glaciers in the Antarctic dry areas (see, for example, *Holdsworth* [1969]). It also provides an explanation for the fact that some glacial meltwater streams have different ionic ratios than the snow falling on their névés.

LAKE CHEMISTRY

The streams from the glaciers described above eventually end in a lake (or the sea). Since chloride is a very minor constituent of rock and since the glaciers do almost no cutting and carry very little moraine, it might be concluded that the chloride in the lakes should be the result of the concentration of large quantities of snow. All the lakes in the Antarctic dry areas in practice have salts entering from both groundwater flow and meltwater from snow and glacier ice. Even glacial meltwater streams seem to have a significant contribution from groundwater salt, particularly early in the season. The fact that the composition of inflow streams varies throughout the summer and does not match the composition of the lake surface waters is an anomaly which is readily explained when the importance of the salt contribution by groundwater is realized. The interfingering of high-Ca^{2+} groundwater to its appropriate density level can produce layers of very different Ca^{2+}/Na^+ in a lake. See, for example, the data of *Bell* [1967] for Lake Miers. Layers of high Ca^{2+}/Na^+ ratio presumably contributed by groundwater can be seen at 17 m below the surface and at the bottom.

Samples collected from the bottoms of Lake Bonney (east lobe) and Lake Vanda contain 14.9 ± 1.1 and 6.4 ± 0.4 tritium units, respectively. Surface snow in this area contains about 100 tritium units. This was discovered when the author was attempting to provide the New Zealand Institute of Nuclear Science Tritium Dating Laboratory with tritium-free water for calibration purposes. It is generally believed that the bottom of the east lobe of Lake Bonney has been out of contact with the atmosphere for at least 60 years [*Hendy et al.*, 1977] and the bottom of Lake Vanda for 1200 years [*Wilson*, 1964]. As the half-life of tritium is 12.3 years, there should be essentially no tritium present in the bottom waters of either lake. The author confirmed that the 'age' of the bottom waters of Lake Vanda is at least 1000 years by radiocarbon measurements. If the brine flowing along the top of the ice-cemented layer enters a density-stratified lake, it will interfinger to its appropriate density layer; and it is proposed that this is the mechanism whereby tritium has entered an apparently isolated geochemical system. Note that the inflow of saline groundwater into Antarctic lakes as proposed here would introduce small amounts of carbonate which had been in at least partial equilibrium with the atmosphere. This introduces complications for those attempting to use radiocarbon dating techniques to determine the times at which the lake levels, and hence climate, changed. The interfingering of saline groundwater into Antarctic lakes also provides an additional mechanism for the formation of chemical stratification over and above that proposed by *Wilson* [1964]. The use of diffusion calculations to determine the times at which lake levels changed must be made with caution, with particular attention being paid to how all the ionic ratios vary with depth.

It is proposed that the salt content of the lakes in the Antarctic dry areas is derived from three sources: (1) seawater, (2) evaporated snow or glacier ice meltwater, these salts in turn having been ultimately largely derived from seawater spray [*Wilson*, 1959], and (3) groundwater which in turn is a product of a deliquescent separation process on salts derived from evaporated snow water (described above) together with materials that have been leached from the material through which the groundwater has passed. Lakes which have derived their salts directly from seawater flooding are characterized by high Mg^{2+}/Ca^{2+} ratios. Leaving these aside, we have lakes which range between two principal chemical compositions (together with some of multiple origin): (1) 'Sodium chloride lakes' are those in which the principal anion is chloride and the principal cation sodium. These have a composition which would suggest that they have been derived from the concentration of large quantities of glacial meltwater. This type of lake has a Ca^{2+}/Na^+ ratio of ~0.1. (2) 'Calcium chloride lakes' are those in which the principal cations are calcium and magnesium and the principal anion chloride, the Ca^{2+}/Na^+ ratio being >1. These lakes are generally relatively low in sulphate. The chemical composition suggests that these lakes have been largely derived from groundwater which has been produced in the manner described above.

TABLE 1. Ca^{2+}/Na^+ Ratios for Antarctic Waters

Body of Water	Ca^{2+}/Na^+ Ratio
Inland snow, e.g., ice from upper Taylor Glacier	0.13
Groundwater, e.g., from surface of frozen water table east of Don Juan Pond	1.8
Lake Canopus	0.1
Lake Fryxell	0.4
Lake Hoare	0.09
Lake Joyce	0.03
Lake Miers (bottom)	5.0
Lake Vanda	
Upper water (13 m)	1.5
Lower water (61 m)	5.4
Don Juan Pond	10.2
Amber ice at base of Meserve Glacier [*Holdsworth*, 1969]	1.4
Seawater (for comparison)	0.04

Examples of lakes with salts of different origin are found in the lakes and ponds of the central Wright Valley: Lake Canopus and the upper waters of Lake Vanda have compositions of the sodium chloride type. However, the adjacent Don Juan Pond and the bottom waters of Lake Vanda have a very different composition, having a much greater proportion of calcium and magnesium chloride (see Table 1).

The rather startling implication of this hypothesis is that much of the million tons of chloride contained in Lake Vanda has entered from groundwater flow. The amounts of Na^+ and SO_4^{2-} needed to bring the overall composition back to that of snow water are to be found in the soils on the lower sides of the valley (as NaCl) and high up on the Asgard and Olympus ranges (as Na_2SO_4 and $Na_2SO_4 \cdot 2H_2O$). Such a system would have taken a very long time to develop and implies a great age for the Wright Valley geochemical system, a conclusion supported by other evidence (see, for example, *Wilson* [1973]).

LAKE PHYSICS

The heat balance of the numerous lakes in this cold and arid region presents some interesting problems. Perhaps the most important is why they exist at all. One might, at first thought, expect freshwater lakes to be completely frozen in a region where the mean annual temperature is −20°C. The lakes fall into two types.

Type I: Perennially ice-covered lakes. Many of the lakes contain water on which floats a permanent ice cover ranging from 4 m to 6.5 m in thickness. Each summer, water flows into the lakes and under the ice. This inflow water must be equal to the total evaporation from the lake surface or the level will change with time. We will call this class of lake 'perennially ice-covered lakes.'

Type II: Ice block lakes.   These 'lakes' are really solid blocks of ice with flat tops that are frozen to the bottom. The yearly inflow water which makes up for evaporation losses flows on top of the ice and freezes in early winter.

The problems are the following: (1) Why should some lakes be of one sort and some of another? (2) What controls the ice thickness on the lakes that are not frozen to the bottom? (3) How can liquid water survive in lakes in a region whose mean annual temperature is −20°C? To understand these problems, let us first consider the perennially ice-covered lakes (i.e., type I above). Each winter, ice freezes onto the bottom of the floating ice. The amount that freezes must exactly match the total ice lost by evaporation during the whole year and melting during late summer (i.e., total annual ablation). The amount that freezes is determined by the winter temperatures, the thermal conductivity of the ice, and the ice thickness. It therefore follows that the ice will be thinner where the ablation is greatest and will increase in thickness as the net ablation decreases, approaching infinite thickness as the snow line is approached. Perennially ice-covered lakes, while common in the more arid areas of the Antarctic, are to the author's knowledge unknown in other parts of the world. Their existence requires a region that is (1) sufficiently warm during the heights of the summer for some liquid water to flow into the lake, to replace evaporation, (2) insufficiently warm during the summer for the surface ice to melt out completely, and (3) sufficiently arid for significant water to ablate off the surface ice cover during the year so that when it is replaced by new ice being frozen on the bottom of the ice during winter the latent heat of freezing produced will match that conducted through the ice cover during winter. Clearly, the resulting equilibrium thickness must not exceed the depth of the lake.

Thus it can be seen why perennially ice-covered lakes are common in the Taylor Valley at low altitude, for example, Lakes Fryxell, Hoare, Chad, Bonney, and Joyce, whereas the lakes in the Victoria Valley, which is at a much greater altitude and consequently closer to the snow line, are all frozen to their base, for example, Lakes Vaska, Vida, and Webb.

The above discussion explains how a freshwater lake, such as Lake Hoare or Lake Miers, does not freeze to the bottom and can exist in a region with a mean annual temperature of −20°C. Essentially, such lakes use the latent heat of fusion of water to make up for the heat that escapes through their surfaces in winter. Such lakes can have bottom waters as warm as 4.1°C, the maximum density of water. This can either be produced by solar heating described below or be due to warm (between 0°C and 4.1°C) inflow water fingering to its appropriate density level. In strong sunlight, stream water can be heated to several degrees above freezing.

SOLAR HEATING OF LAKES IN THE ANTARCTIC DRY AREAS

Some of the lakes have temperatures considerably higher than 4°C, and an alternative source of heating must be found. In Lake Vanda, for example, the bottom waters are at 26°C. It was the attempt to find the source of this heating that led *Wilson and Wellman* [1962] to suggest a solar heating mechanism for Antarctic lakes. Solar heating was also proposed for Lake Bonney [*Hoare et al.*, 1964; *Shirtcliffe*, 1964] and Lake Fryxell [*Hoare et al.*, 1965]. Solar heating of Antarctic lakes was not accepted initially by other workers, who advocated high geothermal gradients and/or hot springs as sources of heating [*Nichols*, 1962; *Armitage and House*, 1962; *Angino and Armitage*, 1963]. During the DVDP program a hole was drilled in the deepest part of Lake Vanda. Heat measurements in the hole [*Wilson et al.*, 1974] showed that heat is presently being conducted downward through the bottom of the lake. This is strong evidence against the high geothermal gradient argument. In fact, Antarctic dry area lakes are particularly favorable for solar-heated lakes. Their surface is usually snow free for most of the summer, and the ice which grows onto the bottom of the floating ice cover every winter consists of very clear ice (6% solar energy transmission through 4 m of ice on Lake Vanda). The high light transmission of this ice is due to the fact that for hundreds of years the ice has been growing onto the bottom of the ice cover and ablating from the top. The ice now contains large crystals several square centimeters in cross section and 4 m long with the C axis aligned vertically. These crystals act as light pipes. The light energy from the sun passes through the ice and into the water, where it is absorbed. Because of the very strong density gradient (produced by dissolved salts and probably the result of past climate changes) in the bottom of these lakes, convection is suppressed so that heat can only escape by conduction. Since water is a very poor conductor of heat, the solar energy which is being absorbed from the sun every summer heats the lake water. This can sometimes be to very high temperatures. For example, the bottom waters of Lake Vanda are 26°C in a region where the mean temperature is −20°C. Lake Vanda is by far the most spectacular naturally solar-heated lake in the world. However, in recent years the Israelis [Tarbour, 1961] have constructed artificial solar-heated lakes 2 m deep in which they have obtained temperatures of 90°C. There is considerable current interest in using such solar ponds for trapping solar energy for industrial purposes and domestic heating.

CONCLUSION

The dry areas of Antarctica provide a climatic zone which can be considered an example of an extreme desert situation. This is because of the extreme aridity and complete absence of flash flooding which is an important feature of the other deserts of the world. As a result we find features such as perennially ice-covered lakes and phenomena such as unique geochemical separation systems not found on other continents.

Acknowledgments. Much of the work reported here was made possible by grants from the New Zealand University Grants Committee and the cooperation of the New Zealand Antarctic Division, Department of Scientific and Industrial Research, U.S. Navy, and the U.S. Antarctic Research Program.

REFERENCES

Angino, E. E., and K. B. Armitage, A geochemical study of Lakes Bonney and Vanda, Victoria Land, Antarctica, *J. Geol.*, *71*, 89-95, 1963.
Armitage, K. B., and H. B. House, A limnological reconnaissance in the area of McMurdo Sound, Antarctica, *Limnol. Oceanogr.*, *7*, 36-41, 1962.
Bell, R. A. I., Lake Miers, South Victoria Land, Antarctica, *N. Z. J. Geol. Geophys.*, *10*, 540-556, 1967.
Charlesworth, J. K., *The Quaternary Era*, vol. 1, 591 pp., Edward Arnold, London, 1957.
Harris, H. J. H., and K. Cartwright, Hydrology of the Don Juan Basin, Wright Valley, Antarctica, this volume.
Harris, H. J. H., K. Cartwright, and T. Torii, Dynamic chemical equilibrium in a polar desert pond: A sensitive index of meteorological cycles, *Science*, *204*, 301-303, 1979.
Hendy, C. H., A. T. Wilson, K. B. Popplewell, and D. A. House, Dating of geochemical events in Lake Bonney, Ant-

arctica, and their relation to glacial and climate changes, *N. Z. J. Geol. Geophys.*, *20*, 1103-1122, 1977.
Hendy, C. H., T. R. Healy, E. M. Rayner, J. Shaw, and A. T. Wilson, Late Pleistocene glacial chronology of the Taylor Valley, Antarctica, *Quaternary Res.*, *11*, 172-184, 1979.
Hoare, R. A., K. B. Popplewell, D. A. House, R. A. Henderson, W. M. Prebble, and A. T. Wilson, Lake Bonney, Taylor Valley, Antarctica: A natural solar energy trap, *Nature*, *202*, 886-888, 1964.
Hoare, R. A., K. B. Popplewell, D. A. House, R. A. Henderson, W. M. Prebble, and A. T. Wilson, Solar heating of Lake Fryxell, a permanently ice-covered Antarctic lake, *J. Geophys. Res.*, *70*, 1555-1558, 1965.
Holdsworth, G., Mode of flow of Meserve Glacier, Wright Valley, Antarctica, Ph.D. thesis, Ohio State Univ., Columbus, 1969.
House, D. A., R. A. Hoare, K. B. Popplewell, R. A. Henderson, W. M. Prebble, and A. T. Wilson, Chemistry in the Antarctic, Proceedings of the California Association of Chemistry Teachers, *J. Chem. Educ.*, *43*, 502-505, 1966.
Jones, L. M., and G. Faure, Origin of the salts in Lake Vanda, Wright Valley, Southern Victoria Land, Antarctica, *Earth Planet. Sci. Lett.*, *3*, 101-106, 1967.
Junge, C. E., *Air Chemistry and Radioactivity*, Int. Geophys. Ser., vol. 4., Academic, New York, 1963.
Nichols, R. L., Geology of Lake Vanda, Wright Valley, South Victoria Land, Antarctica, *Antarctic Research, The Matthew Fontaine Maury Memorial Symposium, Geophys. Monogr. Ser.*, vol. 7, pp. 47-52, AGU, Washington, D. C., 1962.
Shirtcliffe, T. G. L., Lake Bonney, Antarctica: Cause of the elevated temperatures, *J. Geophys. Res.*, *69*, 5257-5268, 1964.
Tarbour, H., Large area solar collectors (solar ponds) for power production, U. N. Conference on New Sources of Energy, *Conf. 35/S/47*, United Nations Organ., Geneva, 1961.
Thompson, T. G., and K. H. Nelson, Concentration of brines and deposition of salts from sea water under frigid conditions, *Amer. J. Sci.*, *254*, 227-238, 1956.
Torii, T., and J. Ossaka, Antarcticite: A new mineral, calcium chloride hexahydrate, discovered in Antarctica, *Science*, *149*, 975-977, 1965.
Torii, T., and N. Yamagata, Limnological studies of saline lakes in the dry valleys, this volume.
Wilson, A. T., The surface of the ocean as a source of airborne nitrogenous material and other plant nutrients, *Nature*, *184*, 99-100, 1959.
Wilson, A. T., Sodium/potassium ratio in rainwater, *Nature*, *186*, 705-706, 1960.
Wilson, A. T., Evidence from chemical diffusion of a climate change in the McMurdo dry valleys, 1200 years ago, *Nature*, *201*, 176-177, 1964.
Wilson, A. T., The lakes of the McMurdo Dry Valleys, *Tuatara*, *15*(3), 152-164, 1967.
Wilson, A. T., The great antiquity of some Antarctic landforms—Evidence for an Eocene temperate glaciation in the McMurdo region, in *Palaeoecology of Africa*, vol. 8, pp. 23-35, edited by E. M. van Zinderen Bakker, A. A. Balkema, Cape Town, 1973.
Wilson, A. T., Geochemical problems of the Antarctic dry areas, *Nature*, *280*, 205-208, 1979.
Wilson, A. T., and H. W. Wellman, Lake Vanda: An Antarctic lake, *Nature*, *196*, 1171-1173, 1962.
Wilson, A. T., R. Holdsworth, and C. H. Hendy, Lake Vanda: Source of heating, *Antarct. J. U.S.*, *9*, 137-138, 1974.

HYDROGEOLOGY OF THE DRY VALLEY REGION, ANTARCTICA

K. Cartwright and H. J. H. Harris

Illinois State Geological Survey, Champaign, Illinois 61820

The polar desert climate of the dry valley region of southern Victoria Land, Antarctica, severely restricts the occurrence and movement of all liquid water. The significance of liquid groundwater in the annual hydrologic cycle of the region is particularly limited. However, under appropriate thermal and chemical conditions, liquid groundwaters and groundwater flow systems do occur; these systems are locally significant in the transport of water and solutes. Three varieties of groundwater and of flow systems are identified. Shallow flow systems, which generally occur in the active layer, are widespread at lower elevations; they are the primary source of water for numerous small, intermittent ponds; they have a significant effect on the distribution of soluble salts in surficial materials and soils. There is apparently very little movement of groundwater in frozen ground; limited evidence of movement at two locations is presented. Deep, liquid groundwaters are found penetrating or lying entirely beneath frozen ground in Taylor and Wright valleys. The discharge of these deep groundwaters significantly affects the mass balance and chemistry of Don Juan Pond; the chemistry of bottom waters in lakes Vanda and Bonney may also be affected.

INTRODUCTION

Antarctica is a vast polar desert almost entirely covered with ice. Small ice- and snow-free regions compose less than 5% of the area of the continent. The 'dry' valleys of southern Victoria Land, with a combined area of some 4000 km², together are one of the largest ice-free regions in Antarctica (Figure 1). Primarily as a result of the vastness, remoteness, and inhospitable climate of the continent, little is known of its hydrology. The hydrology of the dry valley region is perhaps most fully understood, largely because of an excellent program of research sponsored by New Zealand. Hydrologic research has focused upon surface waters, both frozen and unfrozen, and the meteorologic factors affecting them. This paper describes the hydrology of the dry valley region from a different perspective, the perspective of the hydrogeologist.

The research reported here was a part of the Dry Valley Drilling Project (DVDP), a major drilling program sponsored by Japan, New Zealand, and the United States to study the geologic history, geophysics, and geochemistry of the region.

All field work was conducted during three consecutive austral summers between 1973 and 1976. Detailed observations were made at DVDP boreholes and in western Wright Valley at elevations below 1000 m; a companion paper (in this volume of DVDP papers) discusses in detail the hydrogeology of the Don Juan basin. Reconnaissance observations were made in Taylor, Wright, and Victoria valleys, primarily at elevations below 1000 m. We present here a general discussion of the hydrogeology of the region, a synthesis of all our observations.

The emphasis of this report is upon the characteristics of hydrogeologic systems that now occur in the region. As these systems are most often observed in connection with ponds, lakes, and bodies of ice or snow, surface water hydrology also is discussed here. The diverse chemistry of soluble salts found in the surface waters and regolith of the region has provoked considerable research by other workers. The great age and polar desert climate of the dry valleys have engendered a tendency to regard the region as a chemical tomb, in which the original character and distribution of accumula-

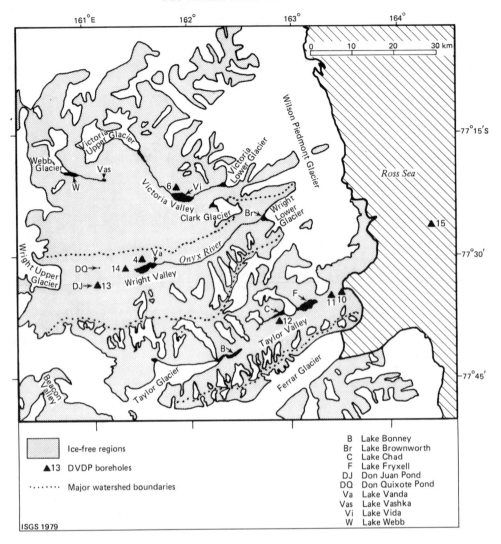

Fig. 1.   The dry valley region of southern Victoria Land, Antarctica, showing locations of DVDP boreholes and other features of interest.

tions of soluble salts are little altered. Our primary interest here is neither to explain the origin of these salts nor to unravel their complex hydrologic history. We do wish to foster among geochemists and soil scientists an awareness that groundwater flow systems are currently transporting salts in much, if not all, of the dry valley geologic section and that many of these systems are probably not in, or cannot long remain in, an unchanging (steady state) condition. The significance of these facts is that, in any study of the origin and distribution of soluble salts, attention must be paid both to active groundwater flows and to the likelihood that flows once occurred in places where they are no longer active.

CLIMATIC SETTING

Modern climate and the modern effects of recent paleoclimate are among the most important influences upon hydrogeologic systems in the dry valleys. The mean annual air temperature at lower elevations in central Wright Valley is approximately −20°C. Temperatures range from midwinter lows of about −55°C to highs of up to 10°C for a few hours of each of several days in late summer [Bull, 1966; Thompson et al., 1971b]. Air temperatures at higher elevations are much colder; even during the summer, it is unlikely that 0°C is ever exceeded at elevations above 1000 m. Mean annual precipitation is between 5 and 10 g/cm² [Thompson et al.,

1971*b*], whereas potential annual sublimation at lower elevations is of the order of at least 50 g/cm² [*Ragotzkie and Likens*, 1964; *Yusa*, 1972; *Anderton and Fenwick*, 1976, *Chinn*, 1976]. Potential sublimation decreases with decreasing temperature, so that snow and ice are much more likely to persist high in the surrounding mountain ranges than in the valleys themselves. Other factors of geography, such as proximity to the open ocean or the polar ice cap, affect the relative amounts of precipitation and sublimation. Paleoclimates equally severe have prevailed for thousands of years. As a result, permafrost (ground in which the temperature is perennially below 0°C) is obliquitous and deep; in geothermal studies at DVDP boreholes, *Decker and Bucher* [1980] estimated that the range of permafrost thickness in the region is from 240 to 970 m.

The general hydrologic effect of this extremely cold and xeric climate is to restrict severely the occurrence and movement of liquid water. Most of the surface water in the valleys occurs as glacial ice. The glaciers are cold based and frozen to their beds [*Holdsworth and Bull*, 1970; *McSaveney*, 1973; *Holdsworth*, 1974]. It has long been noted, however, that under the proper chemical and thermal conditions, unfrozen surface waters do occur locally; the landlocked, highly saline lakes of the region are well-known examples. Fresh, liquid surface water is abundant at some locations in summer, when there are 24 hours of daylight. Solar radiation during summer may heat some soil surfaces to relatively high temperatures (20°C), causing the margins of some glaciers and snowfields to melt.

PREVIOUS WORK

Although groundwaters in the dry valleys become liquid under the proper conditions, until recently they have received only passing mention in the literature of the Antarctic. However, published research into the hydrogeology of the Arctic is relatively extensive, as are laboratory studies of water in frozen ground. Many of the findings of these works are directly applicable in Antarctica; those of particular relevance here are briefly recounted below. For readers interested in specific references and more detailed accounts, the best single source is the two-volume publication of the proceedings of the Second International Conference on Permafrost [*National Academy of Sciences*, 1973, 1978]. Included in the publication are excellent reviews with extensive bibliographies by

Anderson and Morgenstern [1973], *Gold and Lachenbruch* [1973], and *Williams and van Everdingen* [1973]; geochemical papers of especial interest are those by *Anisimova* [1978], *Kononova* [1978], and *Yas'ko* [1978].

The Occurrence and Behavior of Groundwater in Frozen Ground

The properties of frozen and freezing ground give rise to hydrogeologic phenomena that are unique to and common in cold regions. Frozen ground is defined here as any porous earth material in which a 'high' percentage of the pore water is ice. Even at temperatures as low as −50°C, some small percentage of the pore water remains unfrozen. There are theoretical and empirical reasons to suppose that the unfrozen water forms a thin film bordering the pore walls and that the film in any pore is connected to films in adjacent pores, so that continuous, albeit tortuous, liquid pathways pervade all frozen ground. In consequence, frozen ground is permeable to liquid groundwater. The hydraulic conductivity of saturated porous materials can decrease by many orders of magnitude as freezing progresses, owing primarily to the decreasing thickness of the films and to the increasing tortuosity of the flow paths. This phenomenon is of great significance in the hydrogeology of cold regions and is often used (as in this paper) as a basis of field classification. In effect, unfrozen zones are the aquifers of permafrost terrain, whereas frozen zones often are confining layers. Frozen ground is obviously very sensitive to changes in thermochemical conditions. Variations of salinity and temperature may therefore produce large changes in hydraulic conductivity, with consequent changes in the characteristics of any flow systems involved. These effects are readily seen, for example, in the active layer (surficial materials in which the temperature exceeds 0°C at some time during the year), as discussed below in the section, Shallow Environments. This behavior complicates the study of modern groundwaters; it must also complicate paleohydrologic and predictive studies.

In addition to those causes of groundwater motion that are usual to all near-surface environments, there are others of particular significance in freezing and frozen, saturated porous media. Pore water expands during the phase change from liquid to ice; the expansion must be accommodated within existing pore volume, by an increase in pore volume, or by a combination of the two. The first and third alternatives entail an increase in the potential of any remaining pore fluid, which causes the fluid to move away from the freezing zone to regions of lower potential. The second and third alternatives entail expansion and disruption of the medium, as evidenced in the morphologies of frost heaves and closed-system pingos. Chemical potential and temperature gradients can also be significant causes of mass transport within frozen and freezing ground. In some instances, the fluid flux caused by these gradients is toward, rather than away from, the freezing zone.

The chemistry of pore fluids is affected by freezing. As ice can hold few ions in solid solution, most salts dissolved in freezing water are excluded from the solid phase and concentrated in the remaining liquid, selectively precipitated, or both. Concentration of solutes, with concomitant depression of freezing point, is in part responsible for the persistence of liquid water in frozen ground. If fluid motion occurs during freezing, some percentage of the salt originally present in any pore is forced from that pore. This process has been adduced to explain large-scale spatial variations in the solution chemistry of pore waters found in cold regions. Because other factors, including interaction with the medium, can affect the pore water, the interpretation of solution salinity and chemistry can be exceedingly difficult. *Ugolini et al.* [this volume], in a study of DVDP cores, discuss these matters at some length and provide specific citations of previous work.

Studies of Groundwater in Antarctica

DVDP enabled the first purely hydrogeologic research in Antarctica, initiated by the exploratory geophysics of *McGinnis and Jenson* [1971] and *McGinnis et al.* [1973]. Most previous research having hydrogeologic import was concerned primarily with soil-forming processes characteristic of the polar desert environment. All the studies cited below were conducted in the dry valleys. The most detailed work published is that of Ugolini and coworkers [*Ugolini*, 1966; *Ugolini and Bull*, 1965; *Ugolini and Grier*, 1969; *Ugolini and Anderson*, 1972], who focused upon the vertical movement of water and ions in unsaturated soils. The paper of *Ugolini and Grier* [1969] is noteworthy as the first recorded observation in any natural setting of the migration of ions through an unsaturated frozen soil. *Black and Berg* [1963], *Berg and Black* [1966], and Behling and others [*Behling and Calkin*, 1970;

Behling, 1971; *Behling et al.*, 1974] also discussed 'unsaturated flow.' *Nichols* [1963] and *Ugolini and Bull* [1965] briefly discussed meltwater motion in the active layer. A number of workers reported the occurrence of saline, fluid water in saturated ground on slopes adjacent to small saline ponds and inferred downslope movement of the water [*Ball and Nichols*, 1960; *Gibson*, 1962; *Campbell and Claridge*, 1969]. Tedrow and coworkers [*Tedrow and Ugolini*, 1966; *Tedrow et al.*, 1963] and *Campbell and Claridge* [1969] also concluded that salts were moving from high to low ground, a conclusion that implies active saturated flows. Various workers [*Angino and Armitage*, 1962, 1963; *Armitage and House*, 1962; *Angino et al.*, 1964, 1965] believed that the high salinity and temperature of bottom waters in lakes Bonney and Vanda might result from the discharge of warm groundwater, an explanation that more recently has become unpopular. *Oberts* [1973], on the basis of chemical analyses of surface waters, suggested that groundwaters were discharging to some ponds and lakes. *Angino et al.* [1964] and *Weand et al.* [1975] suggested that both temporal changes in bottom water chemistry and calculations of mass balance indicated that groundwater entered Lake Bonney at depth. Solifluction, which apparently entails fluid flow in saturated ground, has been both postulated and observed [*Nichols*, 1963; *Selby*, 1971; *Black*, 1973; *Miagkov*, 1976].

Gibson [1962] studied layered deposits of gypsum ($CaSO_4 \cdot 2H_2O$) occurring in moraine around Lake Vashka, in Victoria Valley, and concluded that groundwater had been important in their formation. Gibson hypothesized that (1) saline groundwater moved downslope along the 'impermeable' surface of frozen ground during warm periods, (2) freezing caused movement of the groundwater to be slow and intermittent, and (3) both evaporation and freezing effected the deposition of gypsum from the groundwater.

In preparation for DVDP, McGinnis and others [*McGinnis and Jensen*, 1971; *McGinnis et al.*, 1973] conducted geophysical studies throughout the region. Using seismic refraction profiles and electrical depth soundings, they concluded that (1) 'confining permafrost' (frozen ground) was present only as a thin surface layer at lower elevations over much of Taylor and Wright valleys; (2) some unfrozen saline ponds and evaporite basins were underlain by sediments containing unfrozen saline groundwater; and (3) none of the large lakes in either valley was underlain by frozen ground. They also postulated that (1) groundwater flow systems occurred in these unfrozen materials, (2) groundwater discharged to some of the ponds and evaporite basins, and (3) groundwater played a 'significant role in . . . replenishment of the large lakes.' As a result of investigations at DVDP boreholes, it now appears that frozen ground is far more thick and extensive than McGinnis and others believed and that liquid groundwater is consequently far less abundant. Nevertheless, McGinnis and coworkers were correct in suggesting that hydrogeologic systems like those found in arctic permafrost regions must also exist in the Antarctic. The research reported here is the direct result of that suggestion.

TYPES OF HYDROGEOLOGIC ENVIRONMENT

For the purposes of this report, we identify three physically distinct types of hydrogeologic environment. As the criterion of distinction, we employ the occurrence and distribution of frozen ground; discussion is limited to water-saturated materials. The three types are (1) 'shallow, unfrozen,' or simply 'shallow,' characterized by unfrozen materials that overlie frozen ground; (2) 'frozen'; and (3) 'deep, unfrozen,' or simply 'deep,' characterized by unfrozen materials that underlie, interlayer with, or penetrate frozen ground. This scheme of classification is similar to that used by *Cederstrom et al.* [1953] in the Arctic. A classification so simple could be difficult to apply in gradational or intermediate cases (e.g., 'partly frozen' materials) and in detailed studies of particular flow systems. It is, however, appropriate for a general description of groundwater in the dry valleys. In principle, different parts of a single groundwater flow system could simultaneously lie within all three environments; that is, a given mass of water could follow a flow path passing through all three. In practice, reconnaissance cannot identify flow paths of this kind. For this reason, we frequently employ the scheme to describe flow systems (e.g., a shallow flow system) as well as environments, thereby designating the one environment in which the system occurs. The notable and common exception is found in some discharge zones; for example, many shallow flow systems have discharge zones in lakes that overlie deep, unfrozen ground. We note that, in consequence of the sensitivity of frozen ground to changing thermochemical conditions, the environments we describe may be temporally in-

termittent or even in periodic alternation with one another.

The nomenclature employed to describe permafrost terrains is often not consistent from text to text. We emphasize that permafrost and active layer are defined in this paper on the basis of temperature alone; frozen ground and frost table are defined by the presence of ice. This distinction is particularly important in the dry valleys, where unfrozen saline groundwaters are found at temperatures well below 0°C. We also note that, as a practical matter, it is difficult to distinguish in the field between saturated and unsaturated frozen ground. We assume that the frozen ground adjacent to (or recently adjacent to) ground saturated with liquid water is saturated frozen ground.

At various points in the remainder of this report we discuss chemical analyses of dry valley waters. To facilitate discussion, the analyses are individually numbered and are grouped in five tables (Tables 1, 3, 5, 6, and 7), each of which pertains only to waters from a particular type of environment or water body. Figure 3 shows the mole percentages of major cations in each and every analysis, allowing ready comparison of waters (1) of different types, (2) taken at one time from different locations in a single flow system, and (3) taken from a single location at different times. In referring to this figure, the reader should bear in mind that it brings together analyses of waters having diverse salinities, ranging from those that are virtually distilled to brines.

Shallow Environments

Shallow environments are widespread at lower elevations throughout Taylor Valley and in the eastern end and north and south forks of Wright Valley; they are relatively uncommon in Victoria Valley. The base of any shallow environment is usually no deeper than the base of the active layer. Shallow environments very rarely are as much as a meter thick. They are subject to the most severe and variable temperatures felt in any of the three environments. The temperature in the active layer may vary from −50°C in winter to 10°C in summer; the diurnal range in summer may exceed 10°C [*Ugolini*, 1966; *Thompson et al.*, 1971a]. As a consequence, shallow environments are very likely to be intermittent. During winter, it is probable that almost all are replaced by frozen environments; even in summer, many exist for only a few days or

Fig. 2a

Fig. 2.   Three oblique photographs, taken at different times and from different perspectives, of an unnamed shallow groundwater pond and its surroundings in the axis of the south fork, Wright Valley. Figure 2a was taken on October 30, 1975, looking northeast. Shallow groundwater pond (white oval near center of photo), about 40 m long, is frozen to its bottom. Crescent of dark ground contiguous to upper edge of pond and lobate patch 's' to left of pond are saline ground. The deep shade characteristic of the north wall of the south fork is at the top of the picture. Position from which Figure 2b was taken is indicated by 'b.'

Fig. 2b.   November 16, 1975, looking southeast. Note human standing near right edge of pond. Pond ice is bordered by a narrow (<1 m wide) moat of liquid water; wide, dark band bordering lower edge of pond and dark patches above human are damp but unsaturated ground overlying shallow hydrogeologic environments.

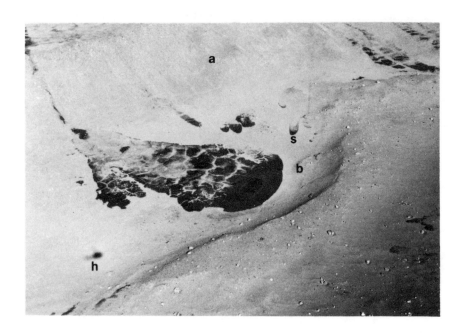

Fig. 2c.   January 21, 1976, aerial, looking southwest. The pond, completely free of ice, is surrounded by the darkened ground indicative of underlying shallow environments. The pond is at the lowest point in a flow system that receives at least part of its recharge from a flat upland, some of which appears in the upper left corner of the picture. The flow system occurs in a region of polygonal patterned ground; polygons extend to either side of the flow system, but these are difficult to see in the photograph. Numerous smaller, separate patches of damp ground are evident; the lobate patch at 's' is coterminous with the lobate patch of saline ground shown in Figure 2a. The effects of topography are evident in the position, shapes, and orientations of all of the shallow environments. Positions from which Figures 2a and 2b were taken are indicated by 'a' and 'b,' respectively; 'h' indicates shadow of helicopter.

hours. Only those containing saline water are persistent.

All shallow flow systems are recharged by meltwater from one or more of four sources: (1) surface water (glaciers, perennial snowfields, and ponds), (2) snowfall, (3) ice in frozen ground, and (4) buried massive ice. As all of these sources are highly responsive to fluctuations in temperature, shallow groundwater flows intermittently and only in summer. The polar desert climate therefore restricts shallow flows in two ways: by limiting both the occurrence of and recharge to shallow environments.

Topographic factors are crucial in the genesis and evolution of shallow environments and flows.

1.   Mesoclimate is strongly affected by topography. Other influences being equal, shallow environments are much less likely to develop at a poorly insolated site than at a well-insolated one; shaded slopes tend to remain frozen.

2.   The fluid potential driving any shallow flow is provided entirely by the difference in elevation between fluid masses in different parts of the system.

Although zones of recharge and discharge can develop at many points along the length of a flow system, recharge zones are always found at the topographically highest point, and discharge zones at the lowest.

3.   As the surface of frozen ground (the frost table) tends to be a subdued replica of topography, the shapes, extents, and distribution of shallow environments are strongly influenced by topography. For example, in areas of poorly dissected topography, such as that often characteristic of recent glacial deposits, many small shallow environments may be found in close proximity, but separated by low 'divides' or 'watersheds' of frozen ground.

Discharge from shallow flow systems takes place by evaporation and/or capillary transport into the overlying unsaturated zone, in seeps or springs, and by flow directly into the margins of ponds and lakes. Where the upper boundary of a shallow environment lies near ground surface, discharge through the unsaturated zone often gives the surface a characteristic darkened, 'damp' appearance

TABLE 1. Chemical Characteristics of Selected Shallow Groundwaters

Analysis	Location	Concentration, mg/l						pH	Date	Valley
		Na	K	Ca	Mg	Cl	SO			
1	VXE-6 flow system, upper part	75	3	50	18	170	65	8.3	Dec. 27, 1975	Wright, south fork
2	VXE-6 flow system, lower part	141	6	117	36	360	80	9.6	Dec. 27, 1975	Wright, south fork
3	South fork axis flow system, upper part	1,200	33	1,700	418	5,600	540	6.7	Dec. 27, 1975	Wright, south fork
4	South fork axis flow system, lower part	2,300	57	4,100	1,000	12,400	690	6.7	Dec. 27, 1975	Wright, south fork
5	South fork axis flow system, upper part	970	29	1,400	339	4,600	420	7.0	Jan. 7, 1976	Wright, south fork
6	South fork axis flow system, lower part	1,700	46	2,900	711	9,000	600	6.7	Jan. 7, 1976	Wright, south fork
7	Springs feeding west end of Don Juan Pond	42	2	28	22	140	60	7.7	Jan. 20, 1976	Wright, south fork
8	North shore, Lake Fryxell	107	8	136	128	240	740	7.6	Jan. 2, 1974	Taylor, eastern

(Figure 2). Such discharges frequently are associated with crusts and efflorescences of water-soluble salts, particularly at the lower ends of flow systems. Solifluction on a small scale is common where shallow groundwater discharges into ponds or lakes, particularly in fine-grained sediments. Polygonal patterned ground is almost always present where there are shallow environments, although the converse is not true. The connection is probably not coincidental; the work of *Black and Berg* [1963] and *Berg and Black* [1966] indicates that the formation of mixed- and ice-wedge polygons is favored by the presence of liquid water in the active layer.

Although the chemistry of shallow groundwater is variable, both spatially and temporally, the topographically lower parts of any shallow flow system tend to be more saline than the upper (Table 1). This pattern probably develops because the recharge water is fresh and because evaporation occurs over the entire length of the flow system and is greatest at the lower end, where the water table often is relatively close to ground surface. In some cases, water may also acquire or precipitate solutes as it traverses the flow system, so that the relative concentrations of solutes are altered. For example, in those flow systems we examined in detail, the lower parts were, by comparison with the upper, enriched in calcium relative to sodium and in chlorine relative to sulfate (Figure 3, flow

systems A and B; Table 1). The chemistry of the recharge waters, evaporation, freezing, variations in temperature, and interaction with surficial materials all must affect the chemical evolution of shallow groundwaters; detailed consideration of these factors is beyond the scope of this paper. The study of the evaporites and precipitates common in the surficial materials of the region is obviously an important part of any such detailed consideration. Recent studies of these soluble salts are those by *Nishiyama and Kurasawa* [1975], *Jones and Faure* [1978], and *Nishiyama* [1978]. The papers of *Jones and Faure* [1978] and *Dort and Dort* [1970] include good reviews and extensive bibliographies.

The temporal development of shallow environments and flows can be very complex. In general, the polar desert climate favors the development of an active layer that is not saturated with water [*Ugolini*, 1966; *Ugolini and Bull*, 1965]. In most places, the potential annual discharge (evaporation and sublimation) from the active layer far exceeds the annual recharge to it. Beneath the active layer, the temperature is always below 0°C, so that only saline pore waters are liquid. In consequence of these conditions, liquid groundwaters, although widespread, are found in only a small proportion of all surficial materials. We estimate that shallow environments occur over an area of approximately 100 km^2, which is less than 1% of the region. In most places, even during summer, moisture is lost

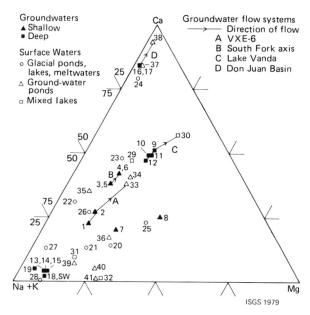

Fig. 3. Triangular component diagram for major dissolved cations, in mole percent, with selected dry valley waters plotted. The number of each sample is included to allow reference to Tables 1, 3, 5, 6, and 7.

from the frost table; the losses occur only by sublimation and vapor phase transport through an overlying, unsaturated active layer [*Everett*, 1971]. This situation has prompted the suggestion that depth to the frost table may be, with correction for mesoclimatic effects, an index of the relative ages of landforms [*Ugolini*, 1966]. Implicit in this suggestion is the assumption that, once the overlying active layer is unsaturated, net loss of moisture from the frost table proceeds at a rate that is more or less uniform over a wide area. Older landforms, presumably exposed to such losses for relatively long and uninterrupted periods of time, should consequently have deeper frost tables than nearby, younger landforms. *Linkletter et al.* [1973] apparently found a good correlation between landform age and depth to the frost table at sites in Beacon Valley, whereas *Porter* [1969], *Everett and Behling* [1970], and *Everett* [1971] found no such consistent correlation at other sites in the region.

Given the prevalence of these conditions, it is surprising that shallow flows recharged from ice in frozen ground should occur at all. In fact, such flows do occur without apparent connection to sources of water lying above the surface of the ground. The work of *Anderton and Fenwick* [1976], who noted widespread 'melting permafrost' in Wright Valley during a period of 'abnormally high temperatures' in 1973-1974, implies that unu-

sual weather may initiate such flows. Although this may be true in some cases, aerial photographs of Wright Valley taken in 1964 [*U.S. Navy*, 1964] show the darkened ground characteristic of shallow environments in many of the same places where shallow environments recharged from frozen ground were found in 1974 and 1975. This indicates that many of the flows are persistent features that are not caused simply by unique meteorologic conditions. The possible explanations for such flows are (1) climatic changes that are progressively deepening the base of the active layer, (2) progressive deflation or other disturbance of the ground surface that also affects the position of the base of the active layer, (3) local anomalies in the accumulation or effects of winter snowfall that are not apparent by the beginning of the summer, and (4) obscure connection to surface sources lying as much as several thousand meters distant from the shallow flow. We favor, but cannot establish, the last of these explanations.

The meltwater streams of the region typically exhibit pulses or surges of flow, which are related to variations in the rate and volume of melting at the source of the stream. It is likely that shallow groundwater flows are similarly erratic. The travel time of any groundwater surge must be a function of both the slopes and the hydraulic conductivities of the shallow environment; it also must be a function of heat exchange with the surroundings. For example, meltwater issuing from the margin of an alpine glacier on a warm day might percolate into the soil and travel only a short distance downslope before freezing on a subsequent, cooler day. The groundwater would thus be stored within the active layer and on a slope, in position to thaw and move during the next warm period. In this way, the travel time of a given mass of groundwater might be greatly prolonged.

Frozen Environments

Of the hydrogeologic environments penetrated in the DVDP boreholes, frozen environments were the most common (Table 2); only one of the holes sited on frozen ground (DVDP 10) encountered deep liquid groundwater. *Decker and Bucher* [1980] found that these frozen environments had temperatures ranging from about −24°C to about −5°C, that all exhibited vertical temperature gradients, and that the lowest temperatures were invariably near ground surface. They also found that, with one exception discussed below, there was little temporal variation in temperature at any

DRY VALLEY DRILLING PROJECT

TABLE 2. Hydrogeologic Characteristics of DVDP Boreholes

Name	Hole	Evidence of Groundwater Flow	Types of Hydro-geologic Environment	Ground Surface*	Hole Depth*	Hole Bottom*	0°C*†	Frozen/Unfrozen Contact*	Sediment/Basement Contact*
Lake Hoare	12	no	frozen	75	185	−110	−285	not penetrated	−91
Commonwealth Glacier	11	no	frozen/deep? (drilling fluid loss at −248 m)	80	328	−248	−325	−248?	not penetrated
New Harbor	10	no	frozen/deep	2	185	−183	−236 to −308	−183	not penetrated
McMurdo Sound	15	no	deep	−122 (sea surface at 0)	65	−187	−180	all unfrozen	not penetrated
Don Juan Pond	13	yes (continuous water level records, geothermal)	deep	118	75	43		all unfrozen	105
North Fork	14	? (displacement of DFA left in hole)	frozen	68	78	−10	−282 to −292	not penetrated	40
Lake Vanda	4, 4A	yes (water level records)	deep	15 (Vanda surface at 83)	18	−3		all unfrozen	2
Lake Vida	6	yes (geothermal, displacement of DFA left in hole)	frozen	349	306	43	−451 to −621	not penetrated	338

* With the exception of depth, all are elevations (MSL); all in meters.
† Temperature measured or calculated by *Decker and Bucher* [1980].

given point in the frozen ground sections. These observations strongly suggest either that there currently is no groundwater flux in most of the frozen ground in the dry valleys or that the flux is exceedingly slow, too slow to have a detectable effect on heat distribution.

The single notable exception to the rule is found in DVDP 6, situated on the northern shore of Lake Vida and drilled entirely in frozen ground. *Decker and Bucher* [1980] report that temperatures taken in DVDP 6 have unusual gradients and show some temporal variation. Decker and Bucher suggest that groundwater flows may occur as deep as 260 m below ground surface. Some of the drilling fluid (Diesel Fuel Arctic, or DFA) left in the completed hole in 1973 was found in subsequent summers to have been displaced, indicating that groundwater is slowly entering the hole. It is particularly surprising that groundwater motion should occur here because DVDP 6 is drilled primarily in the massive Vida granite and is the coldest of all DVDP boreholes, having temperatures ranging from about −24°C to about −16°C. Although temperature profiles give no evidence of it, a small volume of drilling fluid was also apparently displaced in DVDP 14, situated near the western end of Lake Vanda and also drilled entirely in frozen ground. Beyond these fragments of information, nothing is known about the physical behavior and chemistry of groundwater flowing in frozen ground in the dry valleys.

A number of workers have extracted groundwaters and soluble salts from frozen DVDP cores for analysis [*McGinnis et al.*, this volume; *Murayama et al.*, 1978; *Nakao et al.*, 1978; *Stuiver et al.*, this volume; *Ugolini et al.*, this volume]. The degree to which dissolved salts were segregated between liquid and solid phases in the groundwater at the time of coring is not known. The salinity and chemistry of the bulk pore waters are quite variable. Within a single frozen core, the salinity of bulk pore water may range from that of fresh water to that of seawater. Since currently active groundwater flows cannot explain the distribution of salts in these frozen cores, the water and salts must be relict. The possible origins of the solutions are various; for example, flows like that evidenced in DVDP 6, or like those described in the following section, might have affected the solutions in the past. These difficulties are discussed at length by *Ugolini et al.* [this volume], who interpret the solution chemistry of cores from boreholes 8, 9, and 10.

TABLE 3. Chemical Characteristics of Selected Deep Groundwaters and Seawater

Analysis	Location	Depth in Bore-hole, m	Na	K	Ca	Mg	Cl	SO	pH	Specific Gravity at 20°C	Date
9	DVDP 4A, Lake Vanda	70.9-71.6	7,110	826	26,800	7,900	80,000	800	6.3	1.092	Nov. 1973
10	DVDP 4A, Lake Vanda	72.2	9,940	1,010	31,800	9,300	103,800	560	6.3	1.119	Nov. 1973
11	DVDP 4A, Lake Vanda	75.7-76.7	11,150	1,185	37,200	11,300	123,300	400	6.3	1.136	Nov. 1973
12	DVDP 4A, Lake Vanda	79.7-80.6	12,510	1,390	37,400	11,300	129,700	280	6.6	1.146	Nov. 1973
13	DVDP 10, New Harbor	41	21,600	560	1,820	2,300	57,200	3,200		1.062	Nov. 13, 1974
14	DVDP 10, New Harbor	91	18,400	440	1,370	2,000	43,200	3,100		1.047	Nov. 13, 1974
15	DVDP 10, New Harbor	152	16,400	390	1,240	2,000	41,300	4,600		1.044	Nov. 13, 1974
16	DVDP 13, Don Juan basin	15	6,900	91	72,000	1,000	140,000	100	6.4	1.164	Jan. 20, 1976
17	DVDP 13, Don Juan basin	58	7,500	105	77,500	1,100	158,000	100	6.7	1.182	Jan. 20, 1976
18	DVDP 15, Ross Sea	184	10,400	286	358	1,210	19,200	2,700			Nov. 21, 1975
19*	Ice Cone, Taylor Glacier, ppm		328,000	6,100	31,500	17,500	536,400	66,500			Nov. 5, 1962
Sea-water†	Seawater, ppm		10,556	380	400	1,272	18,980	2,649	8.3		

Concentrations in mg/l unless otherwise indicated.

* *Black et al.* [1965].

† *Krauskopf* [1967].

Deep Environments

Liquid groundwaters must occur beneath frozen ground everywhere in the region, but in only one case (DVDP 10) was frozen ground penetrated and liquid water encountered. *Decker and Bucher* [1980] estimated the depth to the 0°C isotherm (the base of permafrost) in most DVDP boreholes (Table 2); the depth to the base of permafrost is the maximum possible depth of frozen ground. At DVDP 10, on the shore of McMurdo Sound, groundwater at a temperature of about −4°C invaded the borehole at an elevation of about −185 m and rose to about −40 m while rapidly freezing. Subsequent to freezing, the bulk water (a slush of ice and liquid) had approximately the ionic ratios of seawater, but was up to twice as saline (Figure 3, Table 3). At DVDP 11, 2.5 km inland, drilling fluid was lost to the sediments near the bottom of the hole, at an elevation of about −250 m; Decker and Bucher measured a temperature of about −4°C near this point. Similar losses occurred at DVDP 10 prior to groundwater invasion, suggesting that the bottom of DVDP 11 was close to the interface between frozen ground and liquid groundwater

when drilling ended. At DVDP 15, 15 km offshore in McMurdo Sound, none of the core was frozen when obtained. *Decker and Bucher* [1980] measured temperatures ranging from about −2°C at the sea floor to just above 0°C at the bottom of the hole. Liquid water taken from the bottom of the hole, at an elevation of about −185 m, had the composition of seawater (Table 3, Figure 3) and probably was a mixture of Ross Sea bottom water and pore water. Owing to disturbance during drilling and to the coarse texture and originally unfrozen condition of the core, it is doubtful whether the chemistry of waters sampled in the laboratory from the artificially frozen core represents the chemistry of pore waters in situ.

The observations at DVDP 10, 11, and 15 strongly suggest that deep liquid groundwaters at the eastern end of Taylor Valley are in hydraulic connection with the Ross Sea. Combined with data from DVDP 12, these observations also suggest that surface waters in the eastern half of the valley (lakes Chad and Fryxell) are isolated from the deep liquid groundwaters by a thick layer of frozen ground. The invasion of DVDP 10 by groundwater and loss of drilling fluid at DVDP 11 do not indi-

TABLE 4.   Hydrogeologic Data for DVDP Borehole 4A, Lake Vanda

	Depth in Borehole, m				
	68.0	72.4	75.8	80.0	82.0
Length of water column in borehole, m	64.1	64.0	67.6	72.1	74.9?
Average density of water in borehole at 20°C, g/cm	1.010	1.117	1.134	1.144	1.164?
Pressure at end of borehold, kg/cm	6.47	7.16	7.68	8.26	8.73?
Density of pore water at 20°C, g/cm	1.085	1.117	1.134	1.144	1.164?
Pressure head, m	59.6	64.0	67.6	72.1	74.9?
Elevation head, m (datum is hole bottom)	14.0	9.6	6.2	2.0	0
Total head, m (potential)	73.6	73.6	73.8	74.1	74.9?

	Interval in Borehole, m			
	68.0-72.4	72.4-75.8	75.8-80.0	80.0-82.0
Change in total head, m	0.0	0.2	0.3	0.8?
Gradient in total head	0.00	0.06	0.07	0.40?
Permeability,* cm	10^{-6} (sand and gravel)	10^{-11} (silts)	10^{-11} (silts)	10^{-13} (fractured crystalline bedrock)
Hydraulic conductivity, cm/s	5×10^{-2}	5×10^{-7}	5×10^{-7}	5×10^{-9}
Flux, cm/yr	0.00	0.94	1.10	0.06?

Question mark indicates that numerical value of measurement (and, consequently, of result of calculation employing measurement) is uncertain.

* Bear [1972].

cate that, under natural conditions, groundwater is moving in the unfrozen sediments encountered by these boreholes. Such phenomena are common in boreholes passing from a confining layer (in this case, frozen ground) into an aquifer. Accurate measurements of natural fluid potential are possible only if potential within the aquifer is in equilibrium with that in the borehole. As no accurate determination of fluid potentials could be made, it is not known if the deep groundwaters at the eastern end of Taylor Valley are moving.

Deep liquid groundwaters penetrating frozen ground are found at Don Juan Pond (DVDP 5 and 13) and Lake Vanda (DVDP 4 and 4A), in western and central Wright Valley, respectively. They may also occur at Don Quixote Pond, in western Wright Valley, and at Lake Bonney, in western Taylor Valley, as discussed below. The hydrogeology of the Don Juan basin is one subject of a lengthy companion paper [Harris and Cartwright, this volume], as well as discussions published elsewhere [Cartwright and Harris, 1976; Harris and Cartwright, 1980] and is mentioned only briefly here.

Don Juan basin.   Liquid groundwater is found in all rock units beneath the Don Juan basin to a depth of at least 75 m; the water temperature through most of this interval is about −16°C. The

section penetrated by DVDP 13 comprises 13 m of sediment overlying about 40 m of Ferrar dolerite sill and 20 m of crystalline basement complex. The groundwater is a concentrated (200,000 ppm) calcium chloride solution similar to, but less concentrated than, that found in Don Juan Pond. The water in the basement complex is slightly more saline than that in the overlying dolerite sill (Table 3). The groundwater is confined and flows upward, discharging in the basin. Geothermal logs of DVDP 13 show the effects of fluid motion [Decker and Bucher, 1980]. Base level fluid potential in the aquifer changed slightly during 80 days of continuous record in the summer of 1975-1976, indicating that the highly variable meteorologic conditions had at most a delayed and muted effect on discharge. Calculations of the mass balance of Don Juan Pond and measurements of fluid potential suggest that between 10,000 and 30,000 m³ of deep groundwater enter the basin annually; this is approximately equal to a flux of between 0.03 and 0.1 m³ per year per square meter of basin surface. We suggest that this flux has persisted for at least 20 years and probably for much longer; if so, at least 200,000 m³ of deep groundwater have entered the basin.

Lake Vanda.   Liquid groundwater is found in all rock units beneath the center of Lake Vanda to

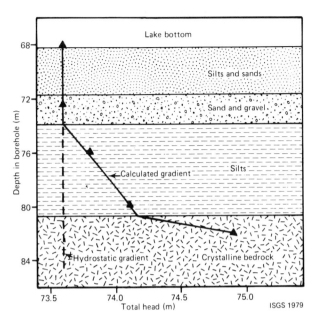

Fig. 4. Calculated fluid potentials (total heads) and potential gradients in DVDP borehole 4A, Lake Vanda, as determined from measurements made during pauses in drilling. Refer to Table 4.

a depth of at least 18 m. The groundwater temperature through this interval exceeds 20°C, as in the bottom waters of the lake [*Bydder and Holdsworth*, 1977; *Decker and Bucher*, 1980]. The section penetrated by DVDP 4A consists of about 12 m of sediment, including a sand and gravel unit some 4 m beneath the lake bottom, overlying about 5 m of crystalline basement rocks. The groundwater is a Ca >> Mg > Na − Cl brine similar in composition to, but slightly more concentrated than, the bottom waters of the lake (Tables 3 and 6, Figure 3).

Fluid level measurements made during pauses in drilling, when combined with density measurements, allow the calculation of fluid potentials. It is apparent (Table 4, Figure 4) that the gradient in potential is different in different parts of the section and is in part a function of material type (permeability), and that groundwater flows upward into the sand and gravel unit. In a preliminary study [*Cartwright*, 1974; *Cartwright et al.*, 1974] we concluded that water flowed from the lake into the underlying sediments; we now consider this preliminary conclusion invalid. As the potential is the same at the lake bottom and in the sand and gravel, the sand and gravel must be connected to the lake by a zone of high permeability not intersected by DVDP 4A. In other words, the ground-

water flowing into the sand and gravel does not pass through the overlying silts and sands in the vicinity of the borehole but moves laterally, eventually discharging to the lake.

Using values of permeability characteristic of the rock types cored in DVDP 4A, the rate of groundwater discharge may be calculated (Table 4); the calculated flux is very small, about 0.01 m³ per year per square meter of lake bottom. As groundwater levels were measured intermittently over a few days only, nothing is known of the temporal behavior of fluid potential. If levels like those we measured are persistent, and if groundwater discharges only in the deepest third (2 km² in area) of the lake, some 20,000 m³ of deep groundwater enter Lake Vanda each year. This volume is too small to be detected by the methods of mass balance determination now in use at the lake.

Temperatures beneath Lake Vanda were measured with thermistors during pauses in the drilling of DVDP 4 [*Decker and Bucher*, 1980] and subsequent to drilling with two thermocouples [*Bydder and Holdsworth*, 1977], one left on the lake bottom and one left in the rock units beneath the lake. Using the thermocouple measurements, Bydder and Holdsworth concluded that heat was being conducted downward and out of the lake, whereas Decker and Bucher believed that the significance of the borehole measurements was ambiguous. As discussed above, our data evidence groundwater motion, suggesting that there must be convective, as well as conductive, transport of heat in some of the rock units beneath the lake. We note that, in cases of combined convective and conductive transport, the net flow of heat can be in a direction opposite to the direction of groundwater flow [*Bredehoeft and Papadopulos*, 1965]. This situation is most likely to prevail where the rate of goundwater motion is very slow, as it apparently is at Lake Vanda.

TYPES OF LAKES AND PONDS

Using modern sources of water as a basis of classification, we identify three types of lakes and ponds in the dry valleys: 'glacial' lakes and ponds, supplied by meltwaters from glaciers and snowfields; 'groundwater' lakes and ponds; and 'mixed' lakes and ponds, supplied by glacial meltwaters and groundwaters. The names of the first two categories indicate the primary source of waters currently entering the lake or pond. Mixed lakes and ponds are supplied primarily by glacial meltwaters

TABLE 5.   Chemical Characteristics of Selected Glacial Ponds, Lakes and Meltwaters

Analysis	Location	Na	K	Ca	Mg	Cl	SO	pH	Date	Valley
20	Wright Upper Glacier	4.3	5.4	3.2	3.7	4	10		Jan. 16, 1976	Wright, western
21	Pond fed by Wright Upper Glacier	6.6	0.7	2.3	2.1	8	9		Jan. 16, 1976	Wright, western
22*	Onyx River, ppm	4.8	1.1	4.8	0.6				1969–1970	Wright, central
23†	Stream feeding western Lake Vida	2.45	0.5	6	1.1	<1	8		Dec. 19, 1971	Victoria
24	Lake Vida	11	0.7	88	2.4	160	<1		Dec. 17, 1973	Victoria
25†	Commonwealth Glacier	12.6	3.8	14	13	<1	7		Dec. 25, 1971	Taylor, eastern
26	Wales Stream	69	5	57	16	200	42		Nov. 22, 1974	Taylor, eastern
27	Lake Fryxell, surface water	13	2	4	1	31	<1	10.1	Dec. 27, 1973	Taylor, eastern
28‡	Lake Fryxell, bottom water (12 m)	2050	187	33	229	2740	460	7.0	1962	Taylor, eastern

Concentrations in mg/l unless otherwise indicated.
* *Jones* [1973].
† *Oberts* [1973].
‡ *Angino et al.* [1962].

but also receive both deep and shallow groundwaters. Groundwater lakes and ponds are of two kinds: those supplied only by shallow flows and those supplied by both shallow and deep flows. Precipitation also contributes water directly to every water body in the region. As noted by *Nichols* [1963], many of these are typical desert water bodies, in that they occupy basins from which there are no outflowing streams; both evaporation and sublimation remove large quantities of water from all surface waters in the region. As with all classifications, ours is artificial; the archetypal water body does not exist. For example, some shallow groundwater enters most lakes, even those we choose to call glacial. Nevertheless, the classification is useful as a general descriptive tool. Detailed descriptions of the hydrology of some dry valley surface waters are available from the New Zealand Ministry of Works and Development. New Zealand has sponsored a systematic program of hydrologic and meteorologic research in the dry valley region since 1969.

The largest lakes in the region are lakes Vanda, Bonney, Vida, and Fryxell. The chemistries and physical behaviors of lakes Vanda, Bonney, and Fryxell have been subjects of intensive investigation for about 2 decades and are not discussed in detail here. Good discussions and bibliographies are given by *Dort and Dort* [1970], *Yusa* [1977], *Jones and Faure* [1978], *Matsubaya et al.* [1979], and *Torii and Yamagata* [this volume].

Glacial Lakes and Ponds

Glacial lakes are the most common of the three types in the dry valleys and are similar to proglacial lakes found in many parts of the world. Characteristically, they are situated adjacent to the glacier supplying them, although some are fed by intermittent meltwater streams flowing overland. Lake volume is controlled by meltwater supply, evaporation, and sublimation. In some proglacial lakes (e.g., Lake Brownworth), meltwater supply greatly exceeds evaporation and sublimation during the summer, giving rise to outflowing streams (e.g., the Onyx River). In some cases, lake level may be affected by intrusion of the glacier into the lake basin. As the water entering glacial lakes is virtually distilled, they also tend to be fresh, although the bottom waters of some of the larger lakes (.e.g, Lake Fryxell) are saline (Table 5, Figure 3). These saline waters may be, in part, the product of long-continued evaporation, entrapment of seawater, or both. Waters and solutes in lakes with outflowing streams have relatively short residence times; such lakes are uniformly fresh. All glacial lakes and ponds have a thick ice cover; many of the shallower, fresher lakes are frozen to the bottom. We suggest that all are underlain by frozen ground.

Large glacial lakes are associated with the Canada, Commonwealth, and Suess glaciers in Taylor Valley (lakes Chad and Fryxell), with the Wright

TABLE 6. Chemical Characteristics of Mixed Lakes

Analysis	Location	Concentration, ppm						pH	Specific Gravity at 20°C	Date
		Na	K	Ca	Mg	Cl	SO			
29*	Lake Vanda, surface water	6	2	17	4	<28	0	7.5	1.000	Jan. 1, 1962
30*	Lake Vanda, bottom water (66 m)	6761	766	24,254	7,684	75,870	770	6.1		Jan. 1, 1962
31†	Lake Bonney, east lobe, surface water	303	22	70	75	688	168	8.5	1.000	Dec. 16, 1961
32†	Lake Bonney, east lobe, bottom water (32 m)	51,400	2,840	1,650	24,200	162,000	3,320	6.8	1.196	Nov. 3, 1961

* Angino et al. [1965].
† Angino et al. [1964].

Lower Glacier in Wright Valley (Lake Brownworth), and with the Webb (lakes Webb and Vashka), Victoria Upper, and Clark glaciebs in the Victoria Valley system. Smaller glacial lakes and ponds are associated with many of the alpine glaciers entering the valleys. Lake Vida, a very large water body near the eastern end of Victoria Valley, is little studied. *Calkin and Bull* [1967] presented convincing evidence that the lake is frozen to the bottom in at least one place in its western half and that it has been frozen throughout its history. However, *McGinnis et al.* [1973], on the basis of geophysical evidence, suggested that fluid, saline bottom waters might be present at the center of the lake. The principal inflows to Lake Vida are meltwater streams originating at the Victoria Lower and Victoria Upper glaciers. Shallow groundwater environments are not common around the lake. Groundwater flowing in frozen ground occurs near Lake Vida, as discussed above, but there is no evidence that this water discharges to the lake.

Mixed Lakes and Ponds

Of the four largest lakes in the region, only Bonney and Vanda are mixed lakes; they are the only surface waters of this type in the dry valleys. Both occupy large, deep, closed basins in the axis of a major valley. With the exception of narrow peripheral moats formed during the summer, both lakes are perennially covered with ice several meters thick. They apparently have lengthy, complex histories, unusual thermal regimes, and saline bottom waters that may be relict. The great preponderance of water currently entering these lakes is glacial meltwater. Both are discharge zones for shallow groundwaters and probably for deep groundwaters as well, although shallow environments are much less common around Lake Vanda than around Lake Bonney.

Lake Bonney. Lake Bonney is in the western end of Taylor Valley; the Taylor Glacier intrudes into the western end of the lake and probably is its primary source of water. Small meltwater streams originating at nearby alpine glaciers may contribute as much as 25% of the annual supply from surface sources [*Weand et al.*, 1975]. Determinations of the total volume of water entering the lake are made uncertain by the position of the terminus of Taylor Glacier; meltwater discharge from the snout of the glacier is difficult to measure, and if the snout were to recede or advance farther into the lake basin, it would displace water and change the level of the lake. Shallow groundwater flows, most recharged by alpine glacial meltwaters, are common on both shores and provide an unknown but probably small quantity of water. The deep waters of the lake are Na >> Mg − Cl brines (Table 6, Figure 3). Water temperature usually is between about −4°C and 0°C at the bottom and is as high as 7°C about 15 m below the lake surface [*Angino et al.*, 1964].

Evidence of deep groundwater discharge to Lake Bonney is indirect. *Angino et al.* [1964] noted a marked decrease in the salinity and an increase in the temperature of bottom waters at one location in the eastern half of the lake during the austral summer of 1961-1962; they also estimated the mass balance of the lake. They concluded that relatively fresh, warm waters had discharged to some

TABLE 7. Chemical Characteristics of Selected Groundwater Ponds

Analysis	Location	Concentration, mg/l						pH	Specific Gravity at 20°C	Date
		Na	K	Ca	Mg	Cl	SO			
		Wright Valley, South Fork								
33	VXE-6 Pond	2,200	88	3,700	1,200	11,800	71	7.6		Dec. 27, 1975
34	VXE-6 Pond	2,900	112	5,200	1,700	16,000	890	7.3		Jan. 7, 1976
35*	Rock Glacier Pond, west of Don Juan Pond	7	1	8	1	4	20			Dec. 13, 1971
36	Rock Glacier Pond, west of Don Juan Pond	19	1	10	9	40	42			Jan. 27, 1975
37	Don Juan Pond	12,200	105	132,000	1,900	260,000	84	4.7	1.289	Dec. 3, 1975
38	Don Juan Pond	4,700	213	165,000	2,500	320,000	10	4.1	1.355	Jan. 20, 1979
39	Bluff Pond, south of Don Juan Pond	460	11.1	74	114	860	380	8.6		Dec. 19, 1975
		Wright Valley, North Fork								
40*	Don Quixote 4	6239	193	760	2,570	9,790	4,690			Dec. 16, 1971
41	Don Quixote 4	33,600	1,100	720	14,800	87,500	4,050			Dec. 14, 1974

* *Oberts* [1973].

deeper parts of the lake. *Weand et al.* [1975] estimated but did not precisely measure the volume of surface water entering the lake during the austral summer of 1973-1974, measured the concurrent change in lake level, and concluded that almost 2 × 10⁶ m³ of groundwater had discharged to the lake. This is equivalent to a flow of about 0.6 m³ per square meter of lake surface area, if the discharge is assumed to have occurred over the entire 3 × 10⁶ m² of surface. They also concluded that increases in oxygen and sulfate concentrations and decreases in principal cation concentrations in bottom waters at the eastern end of the lake indicated the intermittent inflow of relatively fresh, oxygenated groundwater during the austral summer. Several workers [*Hamilton et al.*, 1962; *Black et al.*, 1965; *Black and Bowser*, 1967; *Black*, 1969; *Stephens and Siegel*, 1969] noted the occurrence of salt deposits and associated intermittent discharges of saline water at the terminus of the Taylor Glacier (Table 3, analysis 19). The discharges, which may be groundwater springs, were apparently observed and studied in detail by H. Keys of Victoria University, New Zealand, but we are not aware of any published reports of his work.

Lake Vanda. Lake Vanda, near the center of Wright Valley, is the largest lake in the region. The mass balance of Lake Vanda has in recent years been carefully studied under the auspices of the Antarctic Research Program of New Zealand [e.g., *Hawes*, 1972; *Anderton and Fenwick*, 1976; *Chinn*, 1976]. These studies show that the Onyx River, which enters the eastern end of the lake and

is the principal inflow, supplies of the order of 10⁶ m³ of fresh water every year, although inflow varies by a factor of 4 or 5 from year to year. The only other visible sources of water are a few small, shallow groundwater flows on the northern shore. The deep waters of Lake Vanda are highly saline, unusually warm, and show vertical gradients in temperature and salinity; both the highest temperature (about 24°C) and the greatest salinity (in excess of 100,000 ppm) are found near the bottom (Table 6, Figure 3). These phenomena, discovered by *Angino and Armitage* [1962] and *Armitage and House* [1962], have inspired considerable debate. Some researchers suggested that the influx at depth of heat, saline waters, or both were possible causes of the elevated temperature and salinity. The weight of recent opinion, after *Wilson and Wellman* [1962] and *Wilson* [1964], is that the elevated temperature is caused by solar heating and that the salinity profile is best explained as the result of the diffusion of salts from an ancestral evaporite brine into younger fresh water 'deposited' above the brine [*Yusa*, 1977; *Matsubaya et al.*, 1979]. As discussed above, our data indicate that deep, saline groundwater was discharging at the bottom of Lake Vanda for at least a few days in 1973. If the flux we observed is temporally persistent, approximately 1 cm of highly saline water is added at the bottom of Lake Vanda every year. This addition obviously would not preclude upward diffusion of solutes through the lake; it might increase the rate of diffusion. Convective transport of heat by groundwater probably affects the heat flux between the bottom waters and the underlying rock units.

Groundwater Ponds

By comparison with most other surface waters in the region, groundwater ponds are very small; some are also intermittent or situated near no obvious source of water. In consequence, they have received comparatively little attention. We recognize two types of groundwater ponds: those supplied solely by shallow flow systems and those supplied by both shallow and deep flow systems. Don Juan Pond and Don Quixote Pond 4 are the only two ponds of the latter type in the dry valleys.

Shallow groundwater ponds. Ponds supplied entirely by shallow groundwater flows are widespread in the region; they are rarely more than a few centimeters deep and a few hundred square meters in area. Aerial photographs [*U.S. Navy,* 1964] show that many of these ponds and associated shallow groundwaters were present in 1964. All those we examined in detail were underlain by frozen ground within less than a meter of the ground surface. Many of them are positioned topographically beneath persistent accumulations of ice or snow; meltwaters from the accumulations probably are perennial sources of summer recharge to the flow systems feeding the ponds. In some cases, the distrinction between 'groundwater' pond and 'glacial' pond becomes indistinct as melting increases. Recharge exceeds the capacity of the hydrogeologic environment, so that stream flow and groundwater flow occur simultaneously. Like the flows feeding them, most ponds supplied entirely by shallow groundwater are intermittent; once groundwater discharge ceases, they are dried up by evaporation and sublimation.

The chemistry of the ponds (Table 7, Figure 3) is highly variable, both spatially and temporally, and is strongly affected by the chemistry of the groundwater supplies and by evaporation; the age of each pond and the effects of temperature changes probably are additional factors of importance. In general, any of these ponds is more saline, and probably is isotopically heavier, than the groundwater supplying it. No detailed measurements of the mass balance of shallow groundwater ponds have been made. However, we can offer a conservative estimate of the mass balance of VXE-6 Pond, a groundwater pond situated in the south fork of Wright Valley. During three consecutive field seasons (1973-1976), some water was present

in the pond at all times, indicating that evaporation and sublimation approximately equaled groundwater discharge. The area of the pond is about 100 m^2; if evaporation rates here are like those elsewhere in Wright Valley, the pond loses at least 50 m^3 of water per year. The southern half of the VXE-6 watershed comprises about 0.1 km^2, indicating that the frozen water table in the watershed would, without replenishment, lose an average of at least 0.5 mm of water per year to the pond.

Don Quixote Pond. Don Quixote Pond is the name applied to any one of a string of five small ponds lying in the axis of the north fork of Wright Valley; by convention, the westernmost pond is Don Quixote 1 and the remaining ponds are numbered in sequence to the east. Four of the ponds are present in aerial photographs taken in 1964. All of the ponds are supplied by shallow groundwater, all probably are intermittent, and all except Don Quixote 4 are usually ice covered and contain fresh water. *Clark* [1972] gave geophysical evidence indicating that saline ground was present beneath pond 5, and *Oberts* [1973] suggested that saline ground might underlie some or all of the freshwater ponds. Chemical analyses to substantiate the suggestion have not been made.

Don Quixote 4, which is 20 m in elevation above the surface of Lake Vanda, lies in a depression that is partly floored with very soft mud, the surface of which is encrusted with soluble salts when the pond is dry; the muddy zone has an area of about 100 m^2. Highly saline, liquid groundwater is found in the muddy zone throughout the austral summer. The water is a Na > Mg − Cl brine (Table 7, Figure 3). The depth to the water table increases during the summer. Relatively fresh, shallow groundwater enters the pond from the west; the flow system probably is recharged in part from the ponds lying to the west. Unlike the other shallow groundwater ponds we investigated, Don Quixote 4 is not underlain by frozen ground within a meter of the ground surface. Like Don Juan Pond (discussed below), it is situated at the toe of an active rock glacier or solifluction lobe; as with Don Juan Pond, this may not be coincidental in that liquid groundwater is often associated with such mass movements. The very high salinity and apparent absence of frozen ground suggest that Don Quixote 4 may overlie a deep, unfrozen hydrogeologic environment. If so, it is peculiar that highly saline water is not found in Don Quixote Pond 5, which is situated in a deeper depression only a few hundred meters to the east.

The volume of groundwater discharge to Don Quixote Pond 4 may be estimated. If the high salinity of the water and the depth of the water table reduce evaporation to one tenth of the regional value, approximately 5 m^3 of water (or 0.05 m^3 per square meter of discharge area) enter the pond every year.

Don Juan Pond. Don Juan Pond is discussed at length in other papers in this series [*Harris and Cartwright*, this volume; *Torii and Yamagata*, this volume] and is the subject of papers published elsewhere [e.g., *Harris et al.*, 1979]; complete references are included in our companion paper. The pond, situated in a large closed basin in the south fork of Wright Valley, has persisted since at least 1961; P. N. Webb (personal communication, 1978) reported that it did not exist in the austral summer of 1957-1958, although evaporites encrusted the floor of the basin. With the possible exception of Don Quixote Pond 4, Don Juan Pond is the only groundwater pond in the dry valleys that is known to receive deep groundwater discharge. Shallow groundwater springs also feed the pond from the west (Table 1, analysis 7). The shallow flow systems probably are fed by recharge zones farther west in the south fork; one likely source of recharge is a freshwater pond situated near the western end of the rock glacier that enters the western end of the Don Juan basin (Table 7, analysis 35). Other shallow flow systems supply a negligible amount of water. In response to variations in evaporation rate and shallow groundwater flux, the size of the pond varies greatly; during the summer of 1975-1976 the volume decreased from 8 × 10^3 m^3 to 3 × 10^3 m^3 and the surface area from 6 × 10^4 m^2 to 4 × 10^4 m^2. Without persistent groundwater discharges, the pond would be dry within a year. The chemistry of the pond is highly responsive to fluid balance and to temperature and is consequently highly variable; in general, the water is a calcium chloride brine that is similar to, but more concentrated than, the deep groundwater (Table 7).

DISCUSSION

Several important questions about the characteristics of groundwaters flowing in deep and frozen hydrogeologic environments remain unanswered. There are three questions of particular interest. What are the sources of the groundwaters and what are the causes of flow? How is the chemistry of the groundwaters best explained?

Why are discharge zones for deep flow systems found at Don Juan Pond, Lake Vanda, and perhaps Lake Bonney and Don Quixote Pond, but apparently nowhere else? These questions probably are not directly answerable without more extensive drilling and data collection. However, it is apparent that (1) all of them are to some degree historical questions, (2) as historical questions, they are intimately related to one another, and (3) the number of plausible answers to them may be reduced by informed speculation about the history of each flow system. Although the detailed discussion of these issues is beyond the scope of this paper, they are considered briefly below.

Hydrogeologic Speculations

There are, in general, three possible causes of the deep groundwater flows we observed: recharge from surface sources topographically above the discharge zones of the flow systems; encroachment of frozen ground into deep, unfrozen hydrogeologic environments; and, as explained in more detail below, hydraulic connection to regions of relatively high fluid potential lying entirely beneath frozen ground. In view of both the climate of the region and the regionally pervasive frozen ground, recharge from surface sources seems a highly unlikely cause. However, the encroachment of frozen ground into unfrozen zones must be considered a plausible explanation until further exploration or analysis of the mechanics of the process shows otherwise; the process might also account in part for the peculiar chemistry of the groundwaters. Two possible causes of relatively high fluid potential in deep dry valley groundwaters were suggested or implied in the works of others. As noted previously, several researchers believed that lakes Bonney and Vanda might be discharge zones for thermal groundwaters, implying that high temperatures at depth were causing expansion and consequent upward motion of the fluids. More recently, in an extension of a thesis put forward by *McGinnis* [1968] for North American Pleistocene ice sheets, *McGinnis and Jensen* [1971] and *Cartwright and Harris* [1976] suggested that water might now be moving from beneath the continental Antarctic ice sheet to discharge in the dry valley region; the driving fluid potential in such flow systems would be provided by the enormous weight of the ice sheet. This hypothesis was made more attractive by the work of *Drewry* [1980], who gave geophysical evidence indicating that very large bodies of liquid water occur immediately beneath the ice sheet west of the region.

Most recent explanations of the chemistry of briny dry valley surface waters invoke processes occurring above ground. The brines are thought to derive from ponded seawater, fresh meltwater that may contain weathering products and precipitated aerosols, or some mixture of the two. Concentration and selective precipitation of dissolved ions during evaporation and/or freezing, perhaps followed by selective removal of salts, are thought to explain the high salinity and many details of the chemistry. However, we now know that the picture is complicated in some places by the presence of groundwater brines; at Lake Vanda, as at Don Juan Pond, the solutes in the surface water are virtually identical to and obviously have the same origin as the solutes in the groundwater. Although it is possible that the sedimentary sequences at both places were deposited in brines, the brines in the crystalline basement rocks obviously are not syndepositional. Those who believe that the solutes derive from surface sources must therefore invent mechanisms to emplace them in the crystalline bedrock. Solute diffusion into fresh pore water seems an unlikely mechanism, particularly at Don Juan Pond, where brines are found at least 60 m below the bedrock surface; at both Lake Vanda and Don Juan Pond, the deepest groundwaters are more saline than those higher in the section. Infiltration (convective motion) of brines from the surface into the bedrock is a possible mechanism, but then the history becomes very cumbersome: at a minimum, it involves ponding of waters, evaporation and/or freezing to produce brines, segregation and removal of selected salts, downward motion of the brines, and upward motion of the brines.

At least in the cases of Don Juan Pond and Lake Vanda, it is obviously simpler to believe that the solutes in the saline surface waters derive from discharges of saline groundwaters, with some modification at the surface by evaporation, freezing, or mixture with fresh surface waters. These very processes may be seen in action at Don Juan Pond; hypothetical incursions of seawater, segregation and disappearance of selected salts, and reversing directions of groundwater motion are not required. Brines found in crystalline bedrock in other parts of the earth are usually thought to be the product of prolonged interaction between bedrock and groundwater; the oldest and most saline groundwaters in any flow system usually are found at the discharge end of the system. The chemistry of the

groundwater is thought to be in large part a function of the mineralogy of the rocks with which the water has been in contact; many varieties of solution chemistry may be explained, at least hypothetically, by varieties of rock type. In fact, a brine virtually identical in salinity and major ion chemistry to that found in the dolerite aquifer beneath Don Juan Pond has been found at a depth of about 900 m in a Precambrian basalt in Michigan, in the north-central United States [White, 1965]; to our knowledge, no one has posited that this brine is an evaporite. Explanations invoking reactions with bedrock are no more hypothetical than those that invoke elaborate but unobserved sequences of precipitation, and the problems with 'missing salts' that have plagued the proponents of precipitation are avoided.

A Summary Hydrology

In a description of the hydrologic regime of the modern dry valleys there are three primary concerns: the mechanisms that store and transport water, the volumes of water involved in each mechanism, and the chemistries of the waters. Very large volumes of water are stored as pore ice in thick frozen ground and as pore fluid beneath the frozen ground; only an exceedingly small percentage of this water actively participates in the yearly hydrologic cycle above ground. Glaciers and snowfields store most of the surface water in the valleys, but lakes and ponds are the largest repositories of liquid water. On a regional ($\sim$4000 km^2) scale, atmospheric mechanisms (sublimation, precipitation, evaporation) are overwhelmingly predominant in the transport of water and may involve of the order of 10^9 m^3 annually. Meltwater streams, active only in summer, probably transport of the order of 10^7 m^3 every year; they are major hydrologic factors locally and play a predominant role in the mass balance of some lakes and ponds.

The volume of water transported by groundwater flow systems is not significant in the surface water hydrology of the region as a whole; groundwater is, however, an important contributor to some lakes and ponds. Although precise measurements are not available, we estimate that of the order of 10^5 m^3 of liquid water from shallow flow systems discharge at the ground surface very year. Approximately 10^5 m^3 of deep groundwater discharge at the surface annually. Deep flow systems lying entirely beneath frozen ground may also connect the Ross Sea and fluids at the base of the polar icecap.

Glacial meltwaters contain only very small amounts of dissolved solids. Groundwaters in shallow systems have highly variable chemistry and locally significant effects on the distribution of salts at the ground surface. Deep groundwaters are highly saline and significantly affect the chemistry of some surface waters.

Acknowledgments. Much of the field work upon which this research is based was made possible by the untiring, highly competent, and diverse efforts of Michael Chapman-Smith of the University of Auckland, New Zealand. David Gross and Leon Follmer, both of the Illinois State Geological Survey, provided valuable assistance and observations in Antarctica, as did Gerry Bucher, Ed Decker, and Keith Baker of the University of Wyoming. The men and women of U.S. Navy squadron VXE-6, of the U.S. National Science Foundation, of New Zealand's Vanda station, and of Holmes and Narver, Inc., all rendered vital support and comfort in Antarctica. Chemical analyses were performed by the Analytical Chemistry Section of the Illinois State Geological Survey under the direction of R. Ruch. Financial support was provided in part by National Science Foundation grants GV-40436, OPP 73-05917, and DPP 76-23045 through the University of Illinois, Urbana-Champaign.

REFERENCES

Anderson, D. M., and N. R. Morgenstern, Physics, chemistry and mechanics of frozen ground: A review, in *Permafrost, 2nd International Conference, North American Contribution*, pp. 257-288, National Academy of Sciences, Washington, D. C., 1973.

Anderton, P. W., and J. K. Fenwick, Hydrological research, Dry Valleys, Antarctica, 1973-74, *Annu. Rep. 37*, N.Z. Min. of Works and Develop., Christchurch, 1976.

Angino, E. E., and K. B. Armitage, Geochemical study of lakes Bonney and Vanda, Antarctica, *Geol. Soc. Amer. Spec. Pap., 68*, 129, 1962.

Angino, E. E., and K. B. Armitage, A geochemical study of lakes Bonney and Vanda, Victoria Land, Antarctica, *J. Geol., 71*(1), 89-95, 1963.

Angino, E. E., K. B. Armitage, and J. C. Tash, Chemical stratification in Lake Fryxell, Victoria Land, Antarctica, *Science, 138*(3536), 34-36, 1962.

Angino, E. E., K. B. Armitage, and J. C. Tash, Physicochemical limnology of Lake Bonney, Antarctica, *Limnol. Oceanogr., 9*(2), 207-217, 1964.

Angino, E. E., K. B. Armitage, and J. C. Tash, A chemical and limnological study of Lake Vanda, Victoria Land, Antarctica, *Univ. Kans. Sci. Bull., 45*(10), 1097-1118, 1965.

Anisimova, N. P., Cryogenous metamorphization of chemical composition of subsurface water (exemplified in Central Yakutia), in *Permafrost, 2nd International Conference, USSR Contribution*, pp. 365-370, National Academy of Sciences, Washington, D. C. , 1978.

Armitage, K. B., and H. B. House, A limnological reconnaissance in the area of McMurdo Sound, Antarctica, *Limnol. Oceanogr., 7*(1), 36-41, 1962.

Ball, D. G., and R. L. Nichols, Saline lakes and drill-hole brines, McMurdo Sound, Antarctica, *Geol. Soc. Amer. Bull., 71*(11), 1703-1707, 1960.

Bear, J., *Dynamics of Fluids in Porous Media*, 764 pp., Elsevier, New York, 1972.

Behling, R. E., Rate of chemical weathering in a cold desert environment, *Abstr. Programs, 1971 Meetings Geol. Soc. Amer., 3*, 501-502, 1971.

Behling, R. E., and P. E. Calkin, Wright Valley soil studies, *Antarct. J. U.S., 5*(4), 102-103, 1970.

Behling, R. E., J. P. Reger, and P. E. Calkin, Soil and glacial history studies in Wright Valley (revisited), *Antarct. J. U.S., 9*(4), 148-149, 1974.

Berg, T. E., and R. F. Black, Preliminary measurements of growth of nonsorted polygons, Victoria Land, Antarctica, in *Antarctic Soils and Soil Forming Processes, Antarctic Res. Ser.*, vol. 8, edited by J. C. F. Tedrow, pp. 61-108, AGU, Washington, D. C., 1966.

Black, R. F., Saline discharges from Taylor Glacier, Victoria Land, Antarctica, *Antarct. J. U.S., 4*(3), 89-90, 1969.

Black, R. F., Cryomorphic processes and microrelief features, Victoria Land, Antarctica, in *Research in Polar and Alpine Geomorphology*, edited by B. D. Fahey and R. D. Thompson, pp. 11-24, GeoAbstracts, Ltd., Norwich, England, 1973.

Black, R. F., and T. E. Berg, Hydrothermal regimen of patterned ground, Victoria Land, Antarctica, *Int. Ass. Sci. Hydrol. Publ., 61*, 121-127, 1963.

Black, R. F., and C. J. Bowser, Salts and associated phenomena of the termini of the Hobbs and Taylor Glaciers, Victoria Land, Antarctica, *Int. Ass. Sci. Hydrol. Publ., 79*, 226-232, 1967.

Black, R. F., M. L. Jackson, and T. E. Berg, Saline discharge from Taylor Glacier, Victoria Land, Antarctica, *J. Geol., 73*(1), 175-181, 1965.

Bredehoeft, J. D., and I. S. Papadopulos, Rates of vertical groundwater movement estimated from the earth's thermal profile, *Water Resour. Res., 1*(2), 325-328, 1965.

Bull, C., Climatological observations in ice-free areas of southern Victoria Land, Antarctica, in *Studies in Antarctic Meteorology, Antarctic Res. Ser.*, vol. 9, edited by M. J. Rubin, pp. 177-194, AGU, Washington, D. C. 1966.

Bydder, E. L., and R. Holdsworth, Lake Vanda (Antarctica) revisited, *N.Z. J. Geol. Geophys., 20*(6), 1027-1032, 1977.

Calkin, P. E., and C. Bull., Lake Vida, Victoria Valley, Antarctica, *J. Glaciol., 6*(48), 833-36, 1967.

Campbell, I. B., and G. G. C. Claridge, A classification of frigic soils—The zonal soils of the Antarctic continent, *Soil Sci., 107*, 75-85, 1969.

Cartwright, K., Hydrogeology studies in the dry valleys from the DVDP boreholes, *Dry Val. Drilling Proj. Bull., 4*, 19, 1974.

Cartwright, K., and H. J. H. Harris, Ground water at Don Juan Pond, Wright Valley, southern Victoria Land, Antarctica: A probable origin at the base of the east Antarctic icecap, Abstracts, 1976 Meetings, p. 804, Geol. Soc. Amer., Boulder, Colo., 1966.

Cartwright, K., S. B. Treves, and T. Torii, Geology of DVDP 4, Lake Vanda, Wright Valley, Antarctica, *Dry Val. Drilling Proj. Bull., 3*, 49-74, 1974.

Cederstrom, D. J., P. M. Johnston, and S. Subitsky, Occurrence and development of groundwater in permafrost regions, *U.S. Geol. Surv. Circ., 375*, 30 pp., 1953.

Chinn, T. J., *Hydrological Research Report, Dry Valleys, Ant-*

arctica, 1974-75, N.Z. Min. of Works and Develop., Christchurch, 1976.

Clark, C. C., Geophysical studies of permafrost in the dry valleys, M.S. thesis, 97 pp., Northern Ill. Univ., DeKalb, 1972.

Decker, E. R., and G. J. Bucher, Preliminary geothermal studies in the Ross Island-Dry Valley region, in *Antarctic Geoscience*, edited by C. Craddock, University of Wisconsin Press, Madison, in press, 1980.

Dort, W., Jr., and D. S. Dort, Sodium sulfate deposits in Antarctica, *Mod. Geol., 1*(2), 97-117, 1970.

Drewry, D. J., Geophysical investigations of ice sheet and bedrock inland of McMurdo Sound, Antarctica, in *Antarctic Geoscience*, edited by C. Craddock, University of Wisconsin Press, Madison, in press, 1980.

Everett, K. R., Soils of the Meserve Glacier area, Wright Valley, southern Victoria Land, Antarctica, *Soil Sci., 112*(6), 425-438, 1971.

Everett, K. R., and R. E. Behling, Chemical and physical characteristics of Meserve Glacier morainal soils, Wright Valley, Antarctica: An index of relative age? *Int. Ass. Sci. Hydrol. Publ., 86*, 459-460, 1970.

Gibson, G. W., Geological investigations in southern Victoria Land, Antarctica, 8, Evaporite salts in the Victoria Valley region, *N.Z. J. Geol. Geophys., 5*(3), 361-374, 1962.

Gold, L. W., and A. H. Lachenbruch, Thermal conditions in permafrost: A review of North American literature, in *Permafrost, 2nd International Conference, North American Contribution*, pp. 3-25, National Academy of Sciences, Washington, D. C., 1973.

Hamilton, W., I. C. Frost, and P. T. Hayes, Saline features of a small ice platform in Taylor Valley, Antarctica, *U.S. Geol. Surv. Prof. Pap., 450B*, 73-76, 1962.

Harris, H. J. H., and K. Cartwright, Pressure fluctuations in an Antarctic aquifer: The freight-train response to a moving rock glacier, in *Antarctic Geoscience*, edited by C. Craddock, University of Wisconsin Press, Madison, in press, 1980.

Harris, H. J. H., and K. Cartwright, Hydrology of the Don Juan basin, Wright Valley, Antarctica, this volume.

Harris, H. J. H., K. Cartwright, and T. Torii, Dynamic chemical equilibrium in a polar desert pond: A sensitive index of meteorological cycles, *Science, 204*, 301-303, 1979. (See also Erratum, *Science, 204*, 909, 1979.)

Hawes, J., *Report on the 1971-71 Hydrological-Glaciological Programme, Southern Victoria Land—Dry Valleys Region*, 69 pp., Antarctic Division, New Zealand Division of Scientific and Industrial Research, Wellington, 1972.

Holdsworth, G., Meserve Glacier, Wright Valley, Antarctica, Part I, Basal processes, *Rep. 37*, 104 pp., Inst. Polar Stud., Ohio State Univ., Columbus, 1974.

Holdsworth, G., and C. Bull, The flow of cold ice: Investigations of Meserve Glacier, Antarctica, *Int. Ass. Soc. Hydrol. Publ., 86*, 204-216, 1970.

Jones, L. M., Principal cation concentrations for a length profile of the Onyx River, Wright Valley, *Antarct. J. U.S., 8*(5), 274-276, 1973.

Jones, L. M., and G. Faure, A study of strontium isotopes in lakes and surficial deposits of the ice-free valleys, southern Victoria Land, Antarctica, *Chem. Geol., 22*, 107-120, 1978.

Kononova, R. S., Hydrochemical zoning of subsurface water as one of the indicators of paleopermafrost conditions, in *Permafrost, 2nd International Conference, USSR Contribution*, pp. 422-424, National Academy of Sciences, Washington, D. C., 1978.

Krauskopf, K. B., *Introduction to Geochemistry*, 721 pp., McGraw-Hill, New York, 1967.

Linkletter, G., J. Bockheim, and F. C. Ugolini, Soils and glacial deposits in the Beacon Valley, southern Victoria Land, Antarctica, *N.Z. J. Geol. Geophys., 16*(1), 90-108, 1973.

Matsubaya, O., H. Sakai, T. Torii, H. Burton, and K. Terry, Antarctic saline lakes—Stable isotopic ratios, chemical compositions and evolution, *Geochim. Cosmochim. Acta, 43*(1), 7-25, 1979.

McGinnis, L. D., Glaciation as a possible cause of mineral deposition, *Econ. Geol., 63*, 390-400, 1968.

McGinnis, L. D., and T. E. Jensen, Permafrost-hydrogeological regimen in two ice-free valleys, Antarctica, from electrical depth-sounding, *Quaternary Res., 1*(3), 389-409, 1971.

McGinnis, L. D., K. Nakao, and C. C. Clark, Geophysical identification of frozen and unfrozen ground, Antarctica, in *Permafrost, 2nd International Conference, North American Contribution*, pp. 136-146, National Academy of Sciences, Washington, D. C., 1973.

McGinnis, L. D., J. S. Stuckless, D. R. Osby, and P. R. Kyle, Gamma-ray, salinity, and electric logs of DVDP boreholes, this volume.

McSaveney, M. J., Recession of Meserve Glacier, Wright Valley, between 1966 and 1972, *Antarct. J. U.S., 8*(6), 346-347, 1973.

Miagkov, S., Phototheodolite resurvey in the dry valleys, *Antarct. J. U.S., 11*(2), 96-97, 1976.

Murayama, H., S. Nakaya, S. Murata, T. Torii, and K. Watanuki, Interpretation of salt deposition in Wright Valley, Antarctica: Chemical analysis of DVDP core, *Dry Val. Drilling Proj. Bull., 8*, 62-63, 1978.

Nakao, K., T. Torii, and K. Tanizawa, Interpretation of salt deposition in Wright Valley, Antarctica: Granulometric analysis of DVDP 14 core, *Dry Val. Drilling Proj. Bull., 8*, 68, 1978.

National Academy of Sciences, *Permafrost, 2nd International Conference, North American Contribution*, 783 pp., National Academy of Sciences, Washington, D. C., 1973.

National Academy of Sciences, *Permafrost, 2nd International Conference, USSR Contribution*, 866 pp., National Academy of Sciences, Washington, D. C., 1978.

Nichols, R. L., Geologic features demonstrating aridity of the McMurdo Sound area, Antarctica, *Amer. J. Sci., 261*(1), 20-31, 1963.

Nishiyama, T., Distribution and origin of evaporite minerals from dry valleys, Victoria Land, Antarctica, *Dry Val. Drilling Proj. Bull., 8*, 71, 1978.

Nishiyama, T., and H. Kurasawa, Distribution of evaporite minerals from Taylor Valley, Victoria Land, Antarctica, *Dry Val. Drilling Proj. Bull., 6*, 22, 1975.

Oberts, G. L., The chemistry and hydrogeology of dry valley lakes, Antarctica, M.S. thesis, 55 pp., Northern Ill. Univ., DeKalb, 1973.

Porter, S. C., Weathering and soil-forming processes in the Antarctic dry valleys, *Antarct. J. U.S., 4*(4), 136, 1969.

Ragotzkie, R. A., and G. E. Likens, The heat balance of two Antarctic lakes, *Limnol. Oceanogr., 9*(3), 412-425, 1964.

Selby, M. J., Some solifluction surfaces and terraces in the ice-free valleys of Victoria Land, Antarctica, *N.Z. J. Geol. Geophys., 14*(3), 469-476, 1971.

Stephens, G. C., and F. R. Siegel, Calcium salts from Taylor Glacier, southern Victoria Land, *Antarct. J. U.S., 4*(4), 133, 1969.

Stuiver, M., I. C. Yang, and C. H. Hendy, Permafrost oxygen isotope ratios of Antarctic DVDP cores, *Dry Val. Drilling Proj. Bull.*, *8*, 90-91, 1978.

Tedrow, J. C. F., and F. C. Ugolini, Antarctic soils, in *Antarctic Soils and Soil Forming Processes*, *Antarct. Res. Ser.*, vol. 8, edited by J. C. F. Tedrow, pp. 161-177, AGU, Washington, D. C., 1966.

Tedrow, J. C. F., F. C. Ugolini, and H. Janetschek, An Antarctic saline lake, *N.Z. J. Soil Sci.*, *6*(1), 150-156, 1963.

Thompson, D. C., A. M. Bromley, and R. M. F. Craig, Ground temperatures in an Antarctic dry valley, *N.Z. J. Geol. Geophys.*, *14*(3), 477-483, 1971a.

Thompson, D. C., R. M. F. Craig, and A. M. Bromley, Climate and surface heat balance in an Antarctic dry valley, *N.Z. J. Sci.*, *14*(2), 245-251, 1971b.

Torii, T., and N. Yamagata, Limnological studies of saline lakes in the dry valleys, this volume.

Ugolini, F. C., Soil investigations in the lower Wright Valley, Antarctica, Proceedings, 1st International Conference on Permafrost, *Publ. 1287*, pp. 55-61, Nat. Acad. of Sci., Washington, D. C., 1966.

Ugolini, F. C., and D. M. Anderson, Ionic migration in frozen Antarctic soil, *Antarct. J. U.S.*, *7*(4), 112-113, 1972.

Ugolini, F. C., and C. Bull, Soil development and glacial events in Antarctica, *Quaternaria*, *7*, 251-269, 1965.

Ugolini, F. C., and C. C. Grier, Biological weathering in Antarctica, *Antarct. J. U.S.*, *4*(4), 110-111, 1969.

Ugolini, F. C., W. Deutsch, and H. J. H. Harris, Chemistry and clay mineralogy of selected cores from the Antarctic Dry Valley Drilling Project, this volume.

U.S. Navy, Aerial photography of western Wright Valley, *TMA 1310*, scale 1:15000, U.S. Geol. Surv., Reston, Va., 1964.

Weand, B. L., R. D. Fortner, R. C. Hoehn, and B. C. Parker, Subterranean flow into Lake Bonney, *Antarct. J. U.S.*, *10*(1), 15-19, 1975.

White, D. E., Saline waters of sedimentary rocks, A Symposium, *Fluids in Subsurface Environments*, Mem. *4*, pp. 342-366, Amer. Ass. of Petrol. Geol., Tulsa, Okla., 1965.

Williams, J. R., and R. O. van Everdingen, Ground-water investigations in permafrost regions of North America: A review, in *Permafrost, 2nd International Conference, North American Contributions*, pp. 435-446, National Academy of Sciences, Washington, D. C., 1973.

Wilson, A. T., Evidence from chemical diffusion of a climatic change in the McMurdo dry valleys 1200 years ago, *Nature*, *201*, 176-177, 1964.

Wilson, A. T., and H. W. Wellman, Lake Vanda: An Antarctic lake, *Nature*, *196*, 1171-1173, 1962.

Yas'ko, F. G., Subsurface water regime in areas of occurrence of cryogenous processes and hydrochemical zoning of fossil ice, in *Permafrost, 2nd International Conference, USSR Contribution*, pp. 438-441, National Academy of Sciences, Washington, D. C., 1978.

Yusa, Y., The re-evaluation of heat balance in Lake Vanda, Victoria Land, Antarctica, *Contrib. 12*, pp. 87-100, Geophys. Inst., Kyoto Univ., Kyoto, Japan, 1972.

Yusa, Y., A study on thermosolutal convection in saline lakes, Article 10, *Mem. Fac. Sci. Kyoto Univ.*, *35*(1), 149-183, 1977.

SOME TRACE ELEMENT RELATIONSHIPS IN THE CENOZOIC VOLCANIC ROCKS FROM ROSS ISLAND AND VICINITY, ANTARCTICA

S. S. Goldich,[1,5] J. S. Stuckless,[2] N. H. Suhr,[3] J. B. Bodkin,[3] and R. C. Wamser[4,6]

The variability in chemical composition of the basanitoid (basanite) rocks of Ross Island and vicinity, Antarctica, results from a number of causes. A large part reflects crystal-liquid fractionation processes which produced the silica-undersaturated rock series—basanitoid, trachybasalt, trachyandesite, phonolite. Some of the variability in the basanitoids that is recognizable in the scatter of data points in variation diagrams, however, is directly the result of contamination with xenocrysts of olivine, clinopyroxene, and spinel derived by attrition of mafic and ultramafic xenoliths. Cr-Ni-Co relationships are useful in demonstrating contamination with xenolithic material that may be as much as 10-20 wt %. Ni-MgO and Cr-MgO relationships suggest that reactions took place between the xenoliths and the basanitoid and trachybasalt magmas and that the metals were added to the liquids by diffusion. The large lithophile cations (K, Rb, Ba, and Sr) behaved as incompatible elements, and a similar behavior is noted for La and Zr. Large variations in the concentrations of these elements and in ratios indicate differences in the parental basanitoid magmas. These differences may reflect different degrees of partial melting of garnet peridotite mantle and also heterogeneity in composition of the mantle.

INTRODUCTION

This paper supplements an earlier study [*Goldich et al.*, 1975] of the geochemistry of the Cenozoic volcanic rocks of Ross Island and vicinity. Since then the Dry Valley Drilling Project (DVDP) recovered core of volcanic rocks from three drill holes at McMurdo Station, Ross Island. Of the 17 core samples from DVDP 2 analyzed for this study, 7 are classed as basanitoid, 2 as trachybasalt, and 8 as trachyandesite. The samples intermediate in composition between basanitoid and phonolite are a significant contribution of the drilling program, and the bimodal distribution that characterized the surface samples of the earlier study has been largely eliminated.

DVDP stimulated considerable geochemical research on the volcanic rocks of Ross Island and vicinity. Papers published by *Sun and Hanson* [1975*a*, *b*, 1976], *Kurasawa* [1975], *Kyle and Rankin* [1976], *Stuckless and Ericksen* [1976], and *Kyle et al.* [1979] have been drawn on in the present paper. In addition, two papers in this volume contain the basic analytical data and background material for this contribution. The paper by *Stuckless et al.* [this volume] presents a *Q* mode factor analysis of the chemical data, and a paper by *Weiblen et al.* [this volume] provides data on the occurrence and composition of clinopyroxene in the basalts and the associated mafic and ultramafic nodules.

The alkali basalts of Ross Island and vicinity are variable in composition. This is evident in the scatter of data points in variation diagrams [*Goldich et al.*, 1975; *Stuckless et al.*, this volume] and introduces complications in defining the lines of liquid descent and the differentiation processes that controlled the evolution of the rock series: basanitoid →trachybasalt →trachyandesite →phonolite. *Sun and Hanson* [1975*a*, *b*] recognized some of the probable causes for the chemical differences in the basanitoids:

[1]Department of Geology, Northern Illinois University, DeKalb, Illinois 60115.

[2]U.S. Geological Survey, Federal Center, Denver, Colorado 80225.

[3]Mineral Constitution Laboratories, Pennsylvania State University, University Park, Pennsylvania 16802.

[4]Department of Geology, Northern Illinois University, DeKalb, Illinois 60115.

[5]Now at Department of Geology, Colorado School of Mines and U.S. Geological Survey, Federal Center, Denver, Colorado 80225.

[6]Now at 22554 Market Square Lane, Katy, Texas 77450.

1. The analyzed rocks represent more or less fractionated magma.

2. The rocks are porphyritic, and the phenocrysts may be cumulates. They also recognized that the megacrysts may represent some xenocrystic material.

3. Differences in the mantle source rocks were also considered by *Sun and Hanson* [1975b, p. 85]: variations in the mineralogy of the mantle, variations in the chemistry of the mantle, and variations in the percent of partial melting.

Sun and Hanson also showed that (1) none of the basanitoid samples that they analyzed qualify as a primary magma, although a number approach such a liquid in composition, and (2) a number of lines of liquid descent are indicated, and these probably were appreciably influenced, if not controlled, by differences in composition of the parental basaltic magmas, and by geologic processes related to the pressure or depth of crystallization.

The Q mode factor analysis [*Stuckless et al.*, this volume] brings out many of the points mentioned above. Again it emphasizes that the general differentiation scheme for the evolution of the phonolites and intermediary rock types from basanitoid magma is well established but that the evolution of these rocks was accomplished in distinct stages at different stratigraphic levels and that the lines of liquid descent are complex in detail. The analysis also demonstrates that some of the basanitoid samples, particularly those cored in DVDP 2, contain xenocrystic material that contaminated the magma.

The surface samples [*Goldich et al.*, 1975] were all collected by S. B. Treves. He sampled two cycles of basanitoid and related rocks at a number of localities in the Ross Island area, and there is little doubt that the basanitoid magmas were erupted at different times. K-Ar age determinations [*Armstrong*, 1978] indicate that the volcanic activity, which continues today at Mount Erebus, spans at least 4 m.y.

The mafic and ultramafic nodules referred to in this paper have been described by *Stuckless and Ericksen* [1976]. These include samples of dunite, harzburgite, wehrlite, and plagioclase-pyroxene granulites. The nodules in the basanitoids are of sufficient size so that they can be avoided in the selection of material for analysis. Contamination, however, resulted from the xenocrysts derived by attrition of the nodules and by reaction between the xenocrystic material and magma. Xenocrystic

clinopyroxene in the basanitoids is described by *Weiblen et al.* [this volume]. Some can be recognized in thin sections; others require chemical data to differentiate them from phenocrysts or groundmass clinopyroxene. Some of the phenocrysts are zoned and have cores of xenocrystic clinopyroxene.

The extent of the contamination of the basanitoids by interactions with xenolithic material has not been fully appreciated. Cr, Ni, and Co, because of their high concentrations in the xenoliths, are useful in probing the probable extent of contamination, whereas, Rb, Sr, Ba, the REE, and Pb and Sr isotopes are relatively insensitive because of their low abundances in the nodules. This paper is concerned primarily with the interrelationships between Cr, Ni, and Co and their relationships to some major and minor constituents and trace elements, such as Rb, Ba, Sr, La, and Zr. The latter suggest possible original differences in the basanitoid magmas and are briefly considered.

ANALYTICAL METHODS

Thirty chemical analyses were made for the present study. Of these, 17 are core samples of flows from DVDP 2, three are surface flow samples, and 10 are samples of mafic and ultramafic inclusions in surface flows. Nine of the drill core samples were analyzed (S.S.G.) by conventional methods at Northern Illinois University (NIU). The remaining samples were analyzed by a combination of methods; SiO_2, Al_2O_3, TiO_2, total Fe, MnO, MgO, CaO, Na_2O, K_2O, Rb, Sr, Cu, and Zn were determined (N.H.S.) by atomic absorption spectrophotometry at Pennsylvania State University (PSU), and FeO, H_2O+, H_2O-, and CO_2 (S.S.G.) by conventional methods at NIU. P_2O_5, F, Cl, and S were determined (J.B.B.) at PSU. A colorimetric method for P_2O_5 and an ion-selective electrode method for F, both utilizing the lithium metaborate fusion for dissolution, were devised during the course of this investigation [*Bodkin*, 1976, 1977]. Some of the P_2O_5 determinations also were made gravimetrically (S.S.G.) at NIU.

Cr, Ni, Co, V, Yb, La, Ce, Zr, and Be were determined (N.H.S.) spectrographically at PSU; however, high concentrations of Cr and Ni were repeated (N.H.S.) by atomic absorption. Determinations of Rb and Sr on 13 samples were made (S.S.G. and J.L. Wooden) by isotope dilution at NIU. The Rb and Sr determinations, which were made previously by atomic absorption and emission spectrography, respectively, were repeated (J.S.S.) by X ray fluorescence in the U.S. Geological Survey laboratory in Denver. In relation to the isotope dilution results the X ray fluorescence (XRF) values for Rb and Sr are accurate to 5% of the amounts reported.

All the Ba determinations were made (R.C.W.) by X ray fluorescence at NIU, using the samples analyzed by *Sun and Hanson* [1976] as reference material. In relation to the isotope dilution results the XRF determinations of Ba at levels of 300 ppm or greater are accurate to 10%, but at concentrations below 100 ppm the errors range to 30%. The La determinations by emission spectrography have errors ranging to 25% relative to the isotope dilution results reported by *Sun and Hanson* [1976]. Zr was determined (J.S.S.) on all samples by X ray fluorescence, and because the results are more consistent than the spectrographic data, they are used in preference to the latter.

The analyses of the mafic and ultramafic nodules were made by a combination of methods. FeO, Na$_2$O, and K$_2$O on these samples were determined (S.S.G.) at NIU; all other determinations were made (N.H.S.) at PSU. The isotope dilution values for Rb and Sr were made by *Stuckless and Ericksen* [1976] at NIU.

TERMINOLOGY

The volcanic rocks of Ross Island and vicinity generally are so fine grained that modal classification is impracticable, although the phenocrysts in the porphyritic varieties are useful. The rocks are classified largely on the basis of the chemical composition as basanitoid (basanite), trachybasalt, trachyandesite, and phonolite. The trachybasalts are further divided into low-Si and high-Si varieties. They are essentially equivalent to the hawaiite of *Coombs and Wilkinson* [1969], and the trachyandesites to mugearite, grading with increasing SiO$_2$ to benmoreite. Our preference is to avoid locality names.

DISCUSSION

Analytical Data

All the chemical data, including major, minor, and trace elements, from the present and earlier investigation [*Goldich et al.*, 1975], are tabulated in the accompanying paper by *Stuckless et al.* [this

volume]. The sample numbers as given in the latter paper are retained here.

Averages of the bulk chemical analyses for the principal rock groups are given in Table 1 of this paper. These averages include not only a larger number of samples but also the trachyandesite group, not included in the earlier averages by *Goldich et al.* [1975], and also some values for F, Cl, and S. Averages for the trace elements are given in Table 2.

Diagrams

A number of diagrams are used to illustrate some of the relationships between trace elements, major, and minor constituents. These diagrams commonly are interpretive of the scatter of the data points; hence the lines showing trends have been fitted visually. In computing averages and in plotting the diagrams the isotope-dilution data that are available have been used.

Parental Basanitoid Magma

The basanitoid magmas, generated in the mantle, may have undergone some fractionation by separation of crystals during ascent to the surface, and it may not be advisable to refer to the composition of any of the basanitoids as primitive. Some of the analyzed samples, however, are more suitable than others in indicating a probable composition for the basanitoid magma from which other rock types were derived by crystal-liquid fractionation. To facilitate the present discussion, the least fractionated rock compositions are referred to as possible parental magmas. Recognition of the parental composition is, of course, necessary to evaluate probable causes for the variations in composition of the basanitoid samples as well as to establish the fractionation trends that produced the rock series basanitoid→phonolite.

As a first approach, an average composition has been calculated from the 20 analyses of basanitoid (Table 1, C-1). If the samples were truly representative of all phases of the basanitoid, the average composition should approach a possible parental composition with the fractionated samples compensated for by samples with cumulus phenocrysts. The average, however, may be heavily weighted by the seven core samples from DVDP 2. Four samples from a thick flow are chemically similar, and an average for these samples is given in

TABLE 1. Average Composition of Some Principal Rock Types

	Basanitoid			Trachybasalt		Trachyandesite	Phonolite	
	C-1, 20 Samples, All DVDP Samples	C-2, 4 Samples, DVDP 2	C-3, 2 Samples, DVDP 2	C-4, 5 Samples, Low Si	C-5, 5 Samples, High Si	C-6, 8 Samples	C-7, 4 Samples, Anorthoclase	C-8, 6 Samples, Mafic
SiO_2	42.83	41.65	42.30	44.66	48.98	50.67	55.78	56.15
Al_2O_3	13.24	13.08	12.75	16.26	18.10	18.42	19.28	20.47
TiO_2	3.57	4.16	3.82	3.41	2.53	2.11	1.09	0.64
Fe_2O_3	3.99	3.28	3.24*	3.68†	3.21	3.60	1.86	3.10
FeO	8.09	8.43	8.33*	8.47†	5.80	3.91	4.02	1.58
MnO	0.19	0.18	0.18	0.25	0.21	0.19	0.21	0.19
MgO	11.07	12.03	11.69	4.26	3.06	2.91	1.21	0.71
CaO	11.05	11.34	11.18	9.41	7.33	6.19	3.30	2.55
Na_2O	3.35	3.11	3.12	4.95	6.02	7.11	7.91	8.85
K_2O	1.35	1.45	1.44	2.04	3.06	3.70	4.38	4.68
P_2O_5	0.74	0.81	0.76	1.55	0.89	0.53	0.46	0.18
H_2O+	0.22	0.09	0.17	0.18	0.30	0.23	0.19	0.31
H_2O-	0.13	0.06	0.11	0.09	0.11	0.18	0.09	0.16
CO_2	0.03	0.02	0.09	0.03	0.03	0.07	0.02	0.03
F	0.10	0.10	0.11	0.14	0.14	0.11	0.15	0.09
Cl	0.09	0.09	...	0.14	...	...	0.15	0.34
S	0.02	0.02	...	0.02	...	...	0.05	0.01
Total	100.06	99.90	99.29	99.54	99.77	99.93	100.15	100.04
$<O \equiv$ F, Cl, S	0.07	0.07	0.05	0.10	0.06	0.04	0.12	0.12
Total	99.99	99.83	99.24	99.44	99.71	99.89	100.03	99.92
FeO_t	11.68	11.38	11.25	11.78	8.69	7.15	5.69	4.37

Averages computed from individual analyses of *Stuckless et al.* [this volume]. Values are weight percent.
* Fe_2O_3 and FeO in C-3 adjusted to $FeO/FeO_t = 0.74$; original values determined were 6.99 and 4.96, respectively.
† Fe_2O_3 and FeO in C-4 adjusted to $FeO/FeO_t = 0.72$, intermediate between C-1 and C-2; original values determined were 7.50 and 5.03, respectively.

TABLE 2. Average Trace Element Contents and Ratios for Principal Rock Types

		Basanitoid		Trachybasalt		Trachyandesite	Phonolite	
	C-1, 20 Samples, All DVDP Samples	C-2, 4 Samples, DVDP 2	C-3, 2 Samples, DVDP 2	C-4, 5 Samples, Low Si	C-5, 5 Samples, High Si	C-6, 8 Samples	C-7, 4 Samples, Anorthoclase	C-8, 5 Samples,* Mafic
				Parts Per Million				
Rb	29.4	31.3	29.6	51	69	97	94	123
Sr	905	850	870	1493	1304	1050	900	838
Ba	390	395	390	648	700	744	1100	922
Cr	500	498	420	~10	~10	~45	<5	<5
Ni	230	290	250	~10	~7	~20	<10	<10
Co	57	63	70	27	~13	~20	<10	<10
Cu	51	46	58	19	23	24	25	23
Zn	93	74	76	114	100	106	110	105
Be	2.8	2.8	2.7	3.5	4.5	6.6	6.1	6.5
V	278	315	260	166	132	156	17	~25
Y	35	45	45	45	40	37	57	36
La	48	42	39	89	100	134	125	129
Ce	<100	93	100	165	162	214	240	220
Zr	304	327	339	487	580	815	916	798
				Weight Ratio				
K/Rb	381	385	404	332	368	317	387	316
K/Sr	12.4	14.2	13.7	11.3	19.5	29.3	40.4	46.4
K/Ba	29	31	31	26	36	41	33	42
Rb/Sr	0.032	0.037	0.034	0.034	0.053	0.092	0.104	0.147
Ca/Zr	87	95	92	45	40	42	26	21
Ti/Zr	70	76	68	42	26	16	7.1	4.8
TiO_2/P_2O_5	4.8	5.1	5.0	2.2	2.8	4.0	2.4	3.6
F/P_2O_5	0.14	0.12	0.14	0.09	0.16	0.21	0.33	0.50
Fe/Zn	976	1195	1150	803	675	525	402	324
Fe/V	326	281	336	552	511	356	2600	~1360

*Sample 26 was omitted.

TABLE 3.   Atomic and Weight Ratios for Composite and Individual Samples of
Basanitoid

Sample*	Atomic Ratio Mg/Mg + Fe²⁺	Weight Ratio		
		Ni/MgO × 1000	Ni/Cr	Cr/Fe × 1000
4	0.740	2.36	0.35	10.4
C-2	0.718	2.41	0.58	5.62
C-3	0.714	2.14	0.59	4.80
C-1	0.709	2.08	0.46	5.51
5	0.634	1.33	0.52	2.09

*Sample 4, surface sample approaching a parental magma. Sample C-2, composite
of core samples 30, 31, 32, and 33. Sample C-3, composite of core samples 34 and 36.
Sample C-1, composite of 20 samples. Sample 5, surface sample of a fractionated basanitoid.

Table 1 as C-2. Two samples from a relatively thin flow also are similar, and the average composition is given in C-3. In the latter average the ratio of FeO/Fe₂O₃ has been adjusted to the ratio in C-2, because the flows are oxidized.

The three composite averages (Table 1) are similar as are also the averages for the trace elements (Table 2). There are, however, some significant differences between the core averages (C-2, C-3) and the averall average (C-1), as, for example in TiO₂, Na₂O, and K₂O. The core samples appear to be relatively enriched in TiO₂ and K₂O and depleted in Na₂O. Analytically, these differences are real, and they suggest differences in the primitive liquids generated in the mantle.

Green [1971] suggested the use of the atomic ratio Mg/Mg + Fe²⁺ as an index of crystal fractionation of basaltic magma. Values for this ratio for the composite analyses are given in Table 3. The ratios for two selected surface samples are also included. Sample 4 most nearly approaches the parental basanitoid liquid derived by Q mode factor analysis [Stuckless et al., this volume], whereas sample 5 is clearly a fractionated basanitoid with Mg/Mg + Fe²⁺ considerably lower than the other samples (Table 3). Sample 4 and the composite analyses yield a progressive series in which the ratio falls from 0.74 (sample 4) to 0.71 for the overall composite (C-1). Sun and Hanson [1975b] have discussed some of the limitations to the use of the ratio Mg/Mg + Fe²⁺ and also of Ni content, and the need for caution is illustrated by the weight ratios included in Table 3. Values for the ratio Ni/MgO parallel the atomic ratios Mg/Mg + Fe²⁺; however, the values for the Ni/Cr ratio are all similar, except for the low ratio for sample 4. Similarly, the Cr/Fe ratio for sample 4 is markedly high, reflecting the high Cr content (870 ppm).

Cr-Ni-Co Relationships

The plot of Ni versus MgO (Figure 1) shows considerable scatter of the data points, but explanations are suggested for three lines drawn for sets of data. The curve is the generalized differentiation trend. The straight line with a high slope is a mixing line defined by the core samples. These are characterized by a wide range in Ni (240-440 ppm) and a sharply restricted range in MgO (11.65-12.18%). The second straight line, defined by samples 6, 11, and 9, simulates a mixing line, but these samples, unlike the cores, are from different areas and probably of different ages.

The fractionation curve may be extended to samples 4 and 8; however, an alternative is an extension to the core samples. The latter interpretation is based on the intersection of the projection of the line for the core samples with the fractionation curve at approximately 11.8% MgO and 230 ppm of Ni. Regardless of which interpretation is used, the mixing line indicates an addition of a Ni-rich phase to the basanitoid core samples. The second straight line intersects the fractionation curve in the field of the trachybasalts and suggest that a Ni-rich phase was added to trachybasalt magma. Samples 4 and 8 might be explained by cumulus olivine, but the restricted range for MgO in the core samples poses a problem requiring addition of a nickel sulfide or of Ni by diffusion from xenocrystic Ni-rich olivine.

The plot of Cr versus MgO (Figure 2) contrasts sharply with the Ni-MgO relationships (Figure 1). The core samples have a restricted range in Cr (410-550 ppm) compared to the surface samples. Samples 4 and 8 have similar Ni contents (300 ppm) but Cr concentrations of 870 and 570 ppm, respectively. A wide range in Cr (400-870 ppm) characterizes the samples in a restricted range of MgO

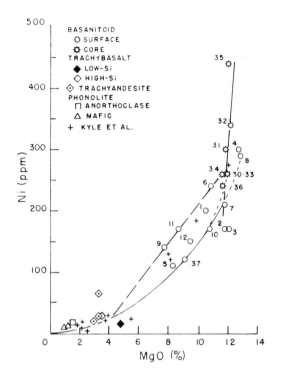

Fig. 1. Plot of Ni versus MgO. The curve is the fractionation trend with interpretations at the upper end in dashed lines. The straight lines are mixing lines. Crosses are data from *Kyle and Rankin* [1976] and *Kyle et al.* [1979].

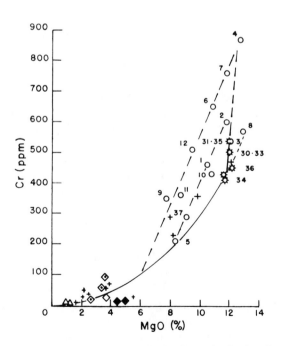

Fig. 2. Plot of Cr versus MgO. Curve is the fractionation trend; straight lines are possible mixing lines.

(11.6-12.8%), an admixture of Cr-rich phases is indicated as in Figure 1, and the alignment of samples along straight lines in Figure 2 suggests mixing lines.

The Ni- and Cr-rich nodules and xenocrysts in the basaltic rocks are obvious contamination sources, and in Figure 3 the variations of Ni with Cr in all samples of basanitoid are shown together with 10 samples of peridotite and plagioclase-pyroxene granulites. The Ni and Cr determinations have analytical errors of approximately 10%, and a greater scatter of the data points is anticipated in Figure 3 than in Figures 1 and 2 in which MgO, with errors of 1-3% is the abscissa. The plot, however, shows trends that support interpretations derived from Figures 1 and 2.

In Figure 3 the core samples cluster above the surface samples, except for sample 8. An additional core sample from DVDP 2 [*Kyle and Rankin*, 1976] falls within the cluster at approximately the average Ni (277 ppm) and Cr (472 ppm) contents of the six core samples in C-2 and C-3 (Table 2). Sample 4 (300 ppm, Ni) falls well to the right because of its high Cr content. The patterns developed in Figures 1, 2, and 3 suggest the following:

1. The parental basanitoid magmas contained approximately 400-550 ppm of Cr and 200-300 ppm of Ni.

2. Samples with lower concentrations may represent fractionated basanitoids; those with higher concentrations are contaminated.

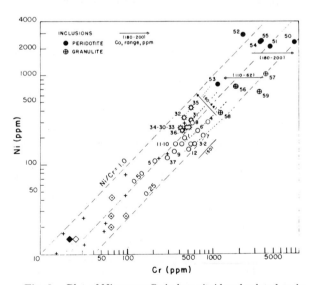

Fig. 3. Plot of Ni versus Cr in basanitoid and related rocks and in inclusions. Dotted lines from basanitoids into the field of inclusions are possible mixing lines.

3. The samples along the curves in the log Ni versus log Cr plot (Figure 3) fall on straight lines in an arithmetic plot. The curves, extrapolated into the field of the mafic and ultramafic inclusions, point to the inclusions as the source of Cr and Ni contamination. In samples with Ni/Cr ratios greater than 0.5, peridotite is the suggested source, and in samples with a lesser ratio, granulite is suggested. The three propositions above probably are oversimplifications but serve as a basis of discussion in which they are considered in reverse order.

In the peridotitic nodules (Figure 3), Cr ranges from 1160 to 8900 ppm, with the largest concentrations in the dunite samples (50 and 51) which contain chromite, with the exception of sample 53 (wehrlite), which contains 1160 ppm of Cr and 785 ppm of Ni; the remaining samples (52, 54, and 55, harzburgite) all have large contents of Ni (2100-2800 ppm). The peridotitic nodules, except for sample 53, have large contents of Co (180-200 ppm, Figure 3). In contrast, the plagioclase-pyroxene nodules range from 1230 to 4170 ppm in Cr, 380 to 1020 ppm in Ni, and 44 to 72 ppm in Co. Thus in large part the high concentration of Ni can be associated with olivine and that of Cr with clinopyroxene.

Weiblen et al. [this volume] report numerous electron microprobe analyses of clinopyroxene from the nodules and from the basanitoids. A clinopyroxene from a harzburgite nodule contains 6570 ppm of Cr and 1570 ppm of Ni. The average Cr content of four samples of clinopyroxene from granulites is 3640 ppm with a range from 2400 to 6160 ppm. Four samples of xenocrystic clinopyroxene from basanitoids range from 1780 to 6500 ppm, averaging 3800 ppm of Cr. In contrast, four samples of groundmass clinopyroxene from basanitoids range from 340 to 620 ppm, averaging 430 ppm of Cr.

If xenocrystic clinopyroxene is a principal source of Cr contamination, a positive correlation between Cr and CaO is expected. Of the basanitoid samples with Cr contents greater than 500 ppm, three (samples 4, 6, and 12) have exceptionally high contents of CaO (Figure 4a). Rough calculations indicate that an addition of 10-20% of clinopyroxene would explain both the high Cr and CaO content of sample 4. Samples 6 and 12, as indicated in SiO_2 content (Figure 4a), are fractionated basanitoids, and much larger additions of CaO are needed. Calcic plagioclase from granulite would be an obvious

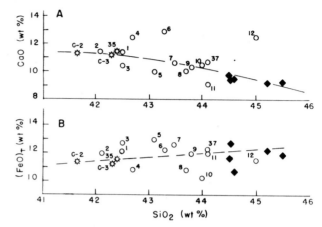

Fig. 4. Plot of (a) CaO and (b) FeO, versus SiO_2 for basanitoid and low-Si trachybasalt samples.

source of CaO, except for the lack of a corresponding enrichment in Al_2O_3 in these samples. The relatively large content of Cr in some of the basanitoid samples (for example, sample 7, 760 ppm of Cr) cannot be explained by addition of xenocrystic clinopyroxene, and chromite is a more likely contaminating phase.

A plot of total iron as FeO (FeO_t) versus SiO_2 (Figure 4b) illustrates the enrichment in iron that results from the crystallization of olivine and clinopyroxene. Cr therefore should show a negative correlation with FeO_t as is seen in Figure 5a. Samples 2, 3, 4, 6, and 7 are clearly aberrant, with Cr contents much too large for the contents of FeO_t. Sample 10 is unusual, with a low FeO_t (see Figure 4b). Samples 8 and 12, although they fall on the extension of the Cr-FeO_t line (Figure 5a), are enriched in Cr.

Although the spectrographic determinations of Co are not precise, a linear relationship between Co and FeO_t is indicated (Figure 5b). The core samples and surface sample 37 appear to be strongly enriched in Co. The peridotite nodules are characterized by relatively large consentrations of Co (Figure 3), and the core samples that appear to be appreciably enriched in Ni are also enriched in Co. Thus the relationships between Cr, Ni, and Co and between the trace elements and major constituents support the contention that the basanitoids are contaminated.

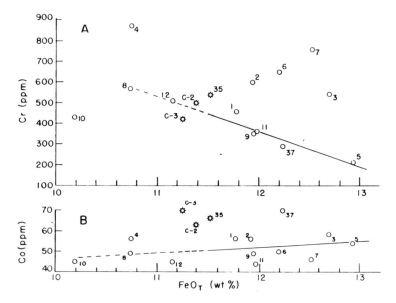

Fig. 5.   Plot of (a) Cr and (b) Co versus FeO, for basanitoid samples.

Rb-Ba-Sr Relationships

The fractionation of basaltic magmas to produce rocks with a compositional range of the basanitoids is dominated by the crystallization of olivine, pyroxene, and spinel. The large cations K, Rb, Ba, and Sr are incompatible with the crystal structures and do not enter these minerals in significant amounts. They are concentrated together in the residual liquids and should show linear relationships, but as will be noted in plots of K versus Rb, Ba, and Sr (Figures 6a, 6b, and 6c), there is considerable scatter in the data points representing the basanitoid samples.

Most of the basanitoid samples plot, within analytical error, along a line with a K/Rb ratio of 350 (Figure 6a). Sample 2 (K/Rb = 214) falls well below the line, and samples 5, 9, and 12 fall along a line with a ratio of approximately 500. Seven of the samples (1, 2, 3, 4, 6, 7, and 12) contain less than 1% of K (0.52-0.97%), and in these the Rb concentrations range from 11 to 35 ppm, a threefold increase compared to less than a twofold increase in K.

Eight of ten samples of trachybasalt have K/Rb ratios approaching 350 (Figure 6a), but the trachyandesites define a separate trend. Similarly, the anorthoclase and mafic phonolites define a third trend. An alternate path shown in Figure 6a suggests that the trachyandesites were derived from fractionated basanitoid liquid via the low-Si trachybasalts by separation of kaersutite, whereas the anorthoclase phonolites were derived via the high-Si trachybasalts, with kaersutite a minor phase, as suggested by Sun and Hanson [1976].

The plot of K versus Ba (Figure 6b) shows a relatively good fit of the basanitoid samples to a line with a K/Ba ratio of 27. Samples 1 and 2, with a ratio of approximately 18, fall well below the line. In the samples with less than 1% of K, the Ba range is from 240 to 390 ppm. The overall pattern in Figure 6b contrasts sharply with the plot of K versus Rb (Figure 6a). In the high-Si trachybasalts and trachyandesites, Ba shows a restricted range (660-860 ppm) in samples with a wide range in K (2.2-3.4%). In these samples, Ba can be correlated with the Ca/K ratio rather than with K content.

Ba is strongly partitioned into anorthoclase, and the anorthoclase phonolites contain 1000 ppm of Ba. Sun and Hanson [1976, p. 146] reported concentrations of 3615 ppm and 227 ppm of Ba in anorthoclase and kaersutite, respectively, from phonolite sample 28. The pattern in Figure 6b then can be related to the appearance of anorthoclase and kaersutite. Declining temperature is a controlling factor, and the highly fractionated phonolite (sample 26) contains approximately 15 ppm of Ba.

The plot of K versus Sr (Figure 6c) has some of the features of Figures 6a and 6b. The basanitoid samples show considerable scatter, and in the sam-

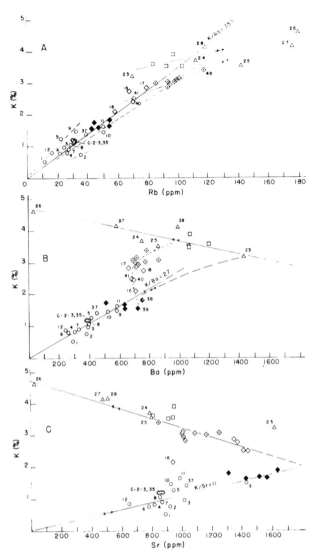

Fig. 6.   Plot of K versus (a) Rb, (b) Ba, and (c) Sr for volcanic
rocks of Ross Island and vicinity.

fails. This failure is most clearly apparent in the trachyandesites and in the phonolites, in which the correlation of La with K is negative and the K/La ratio decreases with increasing La. The wide range in La content (20-60 ppm) in the basanitoid samples containing less than 1% of K conforms to the range noted for Rb, Ba, and Sr.

The variations of Zr with K (Figure 7b) resemble those of La. Again, there is considerable scatter in the data points for the basanitoids, with a range from 190 to 320 ppm of Zr in samples with less than 1% of K. Zr, like La, behaves more strictly as an incompatible cation than does K. Most of the basanitoid samples fall on a line with a K/Zr ratio of approximately 37. The trachyandesite and anorthoclase phonolite samples fall on or near this line, and kaersutite and anorthoclase probably are controlling mineral phases.

The alignment of the core composites C-2 and C-3 and core sample 35 with the surface basanitoid samples (Figure 7b) suggests that the core samples may be somewhat fractionated basanitoid, and this interpretation finds some support in the plots of

ples with less than 1% of K, they range in Sr from 650 to 1020 ppm. The low-Si trachybasalts and fractionated basanitoid sample 9, with high concentrations of Sr (1430-1640 ppm), help to define a line with a K/Sr ratio of approximately 11.

La-Zr Relationships

Plots of K versus La (Figure 7a) and Zr (Figure 7b) show considerable scatter in the basanitoids. The ratios K/La and K/Zr range from 100 to over 400 and from 17 to 50, respectively. In general, La increases with increasing K, and both behave as incompatible cations, but in detail this relationship

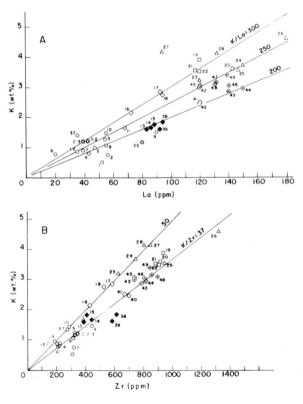

Fig. 7.   Plot of K versus (a) La and (b) Zr.

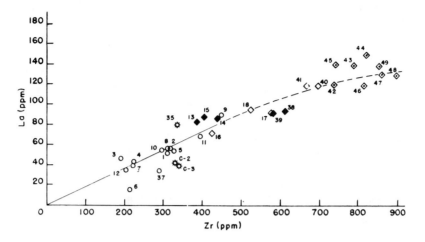

Fig. 8.   Plot of La versus Zr for basanitoid, trachybasalt, and trachyandesite samples.

Figure 6, but not in the K versus La plot of Figure 7a. A plot of La versus Zr (Figure 8) brings out these relationships. Most of the basanitoid samples fit a line with a La/Zr ratio of 0.19 within analytical error. In the trachybasalts and trachyandesites the trend is modified to a curve as the ratio La/Zr drops off because of the crystallization of anorthoclase and kaersutite. Samples 3 and 35 fall well above the line, and composites C-2 and C-3, together with samples 6 and 37, plot below the line.

The plot of Ti versus Zr (Figure 9) separates the principal rock groups and shows some apparent trends. There are, however, a number of exceptions or conflicts in the data that in part may result from analytical error but also may be explained as fractionation trends or the results of contamination. On the basis of Ti/Zr ratios greater than 75, samples 3, 4, 6, 7, and 12 would appear to be less fractionated than samples 1, 37, C-2, and C-3. The surface samples, however, are contaminated (see Figures 2 and 5a), and on the basis of Ni-MgO relationships (Figure 1), samples 3 and 12 are fractionated basanitoid.

Fractionated basanitoid samples fall in the field of the low-Si trachybasalts. The high-Si trachybasalts, with the exception of sample 16, have lower Ti/Zr ratios of approximately 25. Fractionation in the phonolites, with a decrease in the ratio (CaO + MgO)/(Na$_2$O + K$_2$O), can be correlated with decreasing Ti/Zr ratios.

SOME IMPLICATIONS OF THE DATA

Kuno [1960] was among the first to recognize alkali olivine basalt as a primary magma type, which he assigned to deep sources in the mantle, and *Gast* [1968] was first to present a geochemical model for the derivation of alkali olivine basalt by a small degree of melting of garnet peridotite mantle. Although much work has been done and many papers have been written since then, many of the questions raised by *Green and Ringwood* [1967] concerning the relative concentrations of major, minor, and trace elements have not been resolved. The concentration of the light rare earth elements (LREE), in part, can be explained by the retention of the heavy rare earths in garnet in the residuum, but as recently noted by *Hanson* [1977], the mechanism is inadequate to explain the LREE enrich-

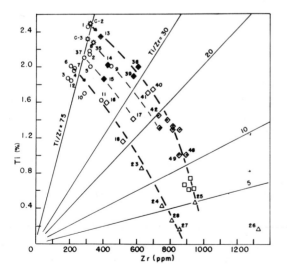

Fig. 9.   Plot of Ti versus Zr.

ment. Discussion of the many problems related to the origin of alkali olivine basalt, and of the basaltic rocks in general, is beyond the scope of this paper; however, a few implications of the data for these problems are presented.

The variability in composition of the basanitoid rocks of Ross Island and vicinity that is apparent in the scatter of data points in variation diagrams results from a combination of causes. Some of the variability is directly the result of contamination derived from inclusions of peridotite and plagioclase-pyroxene granulite. Xenocrysts of olivine and clinopyroxene have been identified in the basanitoids, but some contamination also resulted from reactions between xenocrysts and the magma. Nickel, for example, probably was added to the magma by diffusion from xenocrystic olivine at temperatures just below the melting point of the olivine. Subsequently, some of the Ni-depleted xenocrysts were separated, either by settling or by differential crystal-liquid movement. Such a process would explain the enrichment in Ni without a corresponding enrichment of MgO. An alternate explanation is nickel sulfide as the contaminant.

Large amounts of Cr also were introduced into the basanitoid and trachybasalt magmas from sources such as clinopyroxene, chromite, and possibly chromian hercynite. It follows that additions of Cr, Ni, and Co should be expected, but the proportions can vary considerably depending on the contaminant phases that were involved. Additions of olivine and clinopyroxene will reduce the apparent percentages of constituents such as Al_2O_3, K_2O, Na_2O, and the large lithophile trace elements, Rb, Ba, Sr, and so forth.

Some of the variability in composition of the basanitoids also can be assigned to differences that were generated at the time of partial melting of garnet peridotite mantle as was suggested by *Sun and Hanson* [1975b, 1976]. These differences may reflect heterogeneity in the mantle source, varying degrees of partial melting, mixing of magma from undepleted and partially depleted mantle sources, or some combination of conditions. Testing and discriminating between the possibilities are complicated by a lack of precision in some of the data, but some of the results are suggestive, if not definitive.

There is a large range in concentrations of the incompatible light-ion lithophile (LIL) elements in the 20 samples classed as basanitoids. K ranges from 0.52 to 1.63%, Rb from 11 to 50 ppm, Ba from 240 to 580 ppm, and Sr from 650 to 1430 ppm. La shows a similar high range from approximately 15 to 90 ppm, and Zr from 193 to 448 ppm. The ranges can be explained, in large part, as the result of fractionation of the parental basanitoid magmas. In part, the range can also be explained by varying degrees of contamination that diluted the LIL elements.

If the K content is used as an approximate index of fractionation and the seven samples of less fractionated basanitoid that contain less than 1% (0.52-0.97%) are considered, a wide range in LIL elements still exists: Rb, 11-35 ppm; Ba, 240-390 ppm; and Sr, 650-1015 ppm. These samples also show a wide range in ratios: K/Rb, 214-540; K/Ba, 18-36; and K/Sr, 6-13). The range in concentrations and in ratios cannot be explained by a contamination-dilution mechanism (see Figures 6a, 6b, and 6c). Differences in composition of the parental basanitoid magmas are indicated.

Differences in parental magmas resulting from varying degrees of partial melting appear likely. In addition, variability in the garnet peridotite source is indicated by $^{87}Sr/^{86}Sr$ variations found by E. L. Garner (Table 4). Garner's value for core sample 32 confirms the earlier measurement by *Stuckless and Ericksen* [1976], but the lower value of 0.7030_2 for sample 4 is outside of analytical error and indicates some inhomogeneity of the source region of the basanitoid magmas. This was not apparent in the work of Stuckless and Ericksen because sample 4 was included in the 11-sample basanitoid composite. Stuckless and Ericksen found little variation in samples of the trachybasalt and phonolite—all gave a ratio of 0.7032. Only the trachyte from Mount Cis (sample 29, *Goldich et al.* [1975]), which contains inclusions of metasedimentary rock, has a high ratio of 0.7049. Stuckless and Ericksen also measured the $^{87}Sr/^{86}Sr$ ratios of 16 samples of mafic and ultramafic nodules, and the range in the ratio for these samples is given in Table 4. Sr occurs only in small amounts in the peridotite nodules, and in the granulite nodules the $^{87}Sr/^{86}Sr$ ratio increases with the Sr content. Granulite samples with more than 400 ppm of Sr have a ratio of 0.7031. Because of the low Sr concentrations in the nodules, a significant reduction in the $^{87}Sr/^{86}Sr$ ratio of the basanitoids by contamination with Sr derived from the nodules is not possible.

The nodules are useful in the interpretation of the movements of the basanitoid magmas. The enrichment in K, Rb, Ba, Sr, and so forth requires an undepleted mantle source. The relative enrichment

TABLE 4.   Determinations of $^{87}Sr/^{86}Sr$ on Basanitoid and Nodule Samples

Sample*	Parts Per Million		$^{87}Sr/^{86}Sr$†	Analyst (Reference)
	Rb	Sr		
			Basanitoid	
4	26	817	0.7030_2	E. L. Garner (this paper)*
32	30	846	0.7032_0	E. L. Garner (this paper)
			0.7032_1	J. S. Stuckless and R. L. Ericksen [*Stuckless and Ericksen*, 1976]
1-11	30	993	0.7032	C. E. Hedge [*Stuckless and Ericksen*, 1976]
			0.7031_8	J. S. Stuckless and R. L. Ericksen [*Stuckless and Ericksen*, 1976]
12	16	650	0.7032_2	J. S. Stuckless and R. L. Ericksen [*Stuckless and Ericksen*, 1976]
37	30	1031	0.7031_8	J. S. Stuckless and R. L. Ericksen [*Stuckless and Ericksen*, 1976]
			Nodules	
Peridotite (6 samples)	0.17-0.56	3.0-75	0.7027-0.7059	J. S. Stuckless and R. L. Ericksen [*Stuckless and Ericksen*, 1976]
Granulite (10 samples)	0.94-6.6	27-479	0.7027-0.7038	J. S. Stuckless and R. L. Ericksen [*Stuckless and Ericksen*, 1976]

* Sample 4, surface sample, White Island. Sample 32, core sample, DVDP 2, 126 m below drill collar. Sample 1-11, composite of samples 1-11. Sample 12, surface sample, Mount Melania, Black Island. Sample 37, surface sample, Crater Hill, McMurdo Station.

† Normalized to $^{86}Sr/^{88}Sr = 0.1194$.

in the LREE favors garnet peridotite mantle sources at depth. Possibly, portions of the mantle were separated and moved upward as diapirs, or some process such as shear deformation acted to separate small amounts of interstitial fluid formed by a low degree of partial melting and to concentrate volumes of magma large enough to move upward. If this first magma staging occurred above the melting zone but at depths corresponding to pressures of the order of 27 kbar, garnet would be the liquidus phase [*Green and Ringwood*, 1967, p. 131]. Crystallization of garnet would result in further enrichment in LREE.

Neither garnet nor garnet peridotite nodules have been found in the basanitoids, and we are not able to say whether magma, following the first staging at depth, moved directly to the surface or whether there was a second staging at higher levels during which the basanitoid liquids were derived by fractionation of a more primitive liquid, for example, by olivine fractionation of a high-MgO

magma. The 'high-MgO versus low-MgO primary magma controversy' has been recently reactivated by a paper by *Hart and Davis* [1978] and discussions by *Clarke and O'Hara* [1979], *Hart and Davis* [1979], and *Elthon and Ridley* [1979].

The basanitoid magmas, once formed, moved rapidly to the surface entraining a variety of nodules, and it does not appear likely that the composition of the liquid could have been appreciably changed by crystal settling during this rapid ascent. The trachybasalt magmas were derived by differentiation of basanitoid magmas during stillstands in the upper mantle or lower crust. The trachybasalts also contain numerous xenoliths or nodules. Granulite nodules are common, but peridotite appears to be lacking, and possibly this indicates that the xenoliths in the trachybasalts were derived from higher levels than the peridotitic nodules of the basanitoids. The inclusions of dunite and peridotite in the basanitoids probably were derived from layered gabbroic rocks. For the present we have few data that are useful in placing constraints on the sources and depths from which the nodules were derived.

The xenoliths are not cognate in the sense that they represent residual material from which the primitive magmas were derived, nor in the sense of their being early crystallization phases that were broken up and carried along in the magma. The $^{87}Sr/^{86}Sr$ values, as low as 0.7027 for samples of both peridotite and granulite nodules, indicate that they were derived from Rb-depleted sources. Some of the megacrysts in the basaltic rocks may be cognate [*Stuckless and Ericksen,* 1976], but a larger part are xenocrysts derived by attrition of xenoliths. A large part of the variability in chemical composition of the basanitoids must be assigned to contamination with xenolithic material. Some chemical reactions occurred between the xenocrysts and the magma, but such reactions are difficult to assess in the differentiates.

In conclusion, we wish to emphasize the complexity of the data and the difficulty of eliminating probable mechanisms in explaining the variability in the chemical data. Thus we cannot differentiate between (1) different degrees of melting of a relatively homogeneous source and (2) melting of undepleted and of partially depleted mantle. We have not considered the possibility of contamination in the LIL elements to explain, for example, the variability in K. There are no relicts, and patterns in the LIL elements similar to those of Cr, Ni, Ca, and Mg have not been recognized. The present paper is qualitative but a necessary forerunner to more quantitative assessments.

Acknowledgments. It is our pleasure to acknowledge and thank E. L. Garner of the National Bureau of Standards for the $^{87}Sr/^{86}Sr$ determinations. J. L. Wooden assisted with the mass spectrometric determinations of Rb and Sr at Northern Illinois University and R. J. Pottorf with the bulk chemistry. C. E. Hedge and R. F. Marvin, U.S. Geological Survey, Denver, and G. N. Hanson, State University of New York at Stony Brook, provided constructive criticism to improve the manuscript. We thank S. B. Treves, University of Nebraska-Lincoln, for helpful discussions. This study was supported by the National Science Foundation Office of Polar Programs.

REFERENCES

Armstrong, R. L., K-Ar dating: McMurdo volcanics and Dry Valley glacial history, Victoria Land, Antarctica, *N.Z. J. Geol. Geophys., 21,* 685-698, 1978.

Bodkin, J. B., Colorimetric determination of phosphorus in silicates following fusion with lithium metaborate, *Analyst, 101,* 44-48, 1976.

Bodkin, J. B., Determination of fluorine in silcates by use of an ion-selective electrode following fusion with lithium metaborate, *Analyst, 102,* 409-413, 1977.

Clarke, D. B., and M. J. O'Hara, Nickel, and the existence of high-MgO liquids in nature, *Earth Planet. Sci. Lett., 44,* 153-158, 1979.

Coombs, D. S., and J. F. G. Wilkinson, Lineages and fractionation trends in undersaturated volcanic rocks from the East Otago Province (New Zealand) and related rocks, *J. Petrol., 10,* 440-501, 1969.

Elthon, D., and W. I. Ridley, Comments on: The partitioning of nickel between olivine and silicate melt, by S. R. Hart and K. E. Davis, *Earth Planet. Sci. Lett., 44,* 162-164, 1979.

Gast, P. W., Trace element fractionation and the origin of tholeiite and alkaline magma types, *Geochim. Cosmochim. Acta, 32,* 1057-1086, 1968.

Goldich, S. S., S. B. Treves, N. H. Suhr, and J. S. Stuckless, Geochemistry of the Cenozoic volcanic rocks of Ross Island and vicinity, Antarctica, *J. Geol., 83,* 415-435, 1975.

Green, D. H., Compositions of basaltic magmas as indicators of conditions of origin: Application to oceanic volcanism, *Phil. Trans. Roy. Soc. London, 268,* 707-725, 1971.

Green, D. H., and A. E. Ringwood, The genesis of basaltic magmas, *Contrib. Mineral. Petrol., 15,* 103-190, 1967.

Hanson, G. N., Geochemical evolution of the suboceanic mantle, *J. Geol. Soc. London, 134,* 235-253, 1977.

Hart, S. R., and K. E. Davis, Nickel partitioning between olivine and silicate melt, *Earth Planet. Sci. Lett., 40,* 203-219, 1978.

Hart, S. R., and K. E. Davis, Reply to D. B. Clarke and M. J. O'Hara, 'Nickel, and the existence of high-MgO liquids in nature,' *Earth Planet. Sci. Lett., 44,* 159-161, 1979.

Kuno, H., High-alumina basalt, *J. Petrol., 1,* 121-145, 1960.

Kurasawa, H., *Geochemical and Geophysical Studies of Dry Valleys, Victoria Land in Antarctica, Mem. Spec. Issue 4,* edited by T. Torii, pp. 67-74, National Institute of Polar Research, Tokyo, 1975.

Kyle, P. R., and P. C. Rankin, Rare earth element geochemistry of Late Cenozoic alkaline lavas of the McMurdo Volcanic Group, *Geochim. Cosmochim. Acta, 40,* 1497-1507, 1976.

Kyle, P. R., J. Adams, and P. C. Rankin, Geology and petrology of the McMurdo Volcanic Group at Rainbow Ridge, Brown Peninsula, Antarctica, *Geol. Soc. Amer. Bull.,* part 1, *90,* 676-688, 1979.

Stuckless, J. S., and R. L. Ericksen, Strontium isotopic geochemistry of the volcanic rocks and associated megacrysts and inclusions from Ross Island and vicinity, Antarctica, *Contrib. Mineral. Petrol., 58,* 111-126, 1976.

Stuckless, J. S., A. T. Miesch, S. S. Goldich, and P. W. Weiblen, A Q mode factor model for the petrogenesis of the volcanic rocks from Ross Island and vicinity, Antarctica, this volume.

Sun, S. S., and G. N. Hanson, Evolution of the mantle: Geochemical evidence from alkali basalt, *Geology, 3,* 297-302, 1975a.

Sun, S. S., and G. N. Hanson, Origin of the Ross Island basanitoids and limitations upon the heterogeneity of mantle sources for alkali basalts and nephelinites, *Contrib. Mineral. Petrol., 52,* 77-106, 1975b.

Sun, S. S., and G. N. Hanson, Rare earth element evidence for differentiation of McMurdo volcanics, Ross Island, Antarctica, *Contrib. Mineral. Petrol., 54,* 139-155, 1976.

Weiblen, P. W., J. S. Stuckless, W. C. Hunter, K. J. Schulz, and M. G. Mudrey, Jr., Correlation of clinopyroxene compositions with environment of formation based on data from Ross Island volcanic rocks, this volume.

CORRELATION OF CLINOPYROXENE COMPOSITIONS WITH ENVIRONMENT OF FORMATION BASED ON DATA FROM ROSS ISLAND VOLCANIC ROCKS

P. W. Weiblen,[1] J. S. Stuckless,[2] W. C. Hunter,[3] K. J. Schulz,[4] and M. G. Mudrey, Jr.[5]

Clinopyroxenes from the alkali basalts and associated mafic and ultramafic nodules from Ross Island, Antarctica, may be classified on the basis of occurrence and textures as xenocrysts, megacrysts, phenocrysts, or groundmass pyroxenes. This classification is not entirely unambiguous because of fragmentation and reaction of xenocrysts. The Cr content of unreacted xenocrysts is distinctly higher (>0.50 wt % Cr_2O_3) than other pyroxene types (<0.10). Other elemental variations, notably Ti, Si, Mn, and Na, indicate a range of reaction between host magmas and xenocrysts. The megacryst, phenocryst, and groundmass clinopyroxenes have a narrow range of compositions compared to tholeiitic lava suites. A reversal in the Fe/(Fe + Mg) ratio in zoned phenocrysts may be related to oxidation of alkalic magma during ascent or to complicated local reactions between pyroxene xenocrysts and magma. More detailed experimental investigation of pyroxene equilibria in alkalic systems is needed before definitive conclusions can be drawn from the pyroxene data.

INTRODUCTION

Data are presented and interpreted in this paper on a variety of clinopyroxenes from a suite of alkali basalts and associated mafic and ultramafic nodules from Ross Island and vicinity, Antarctica. The samples were collected from outcrop and drill core as a part of the Dry Valley Drilling Project (DVDP).

Ross Island is part of the Ross Sea petrologic province [*Hamilton*, 1972], which extends as a discontinuous belt that is roughly parallel to the trend of the Transantarctic Mountains [*Goldich et al.*, 1975]. The petrologic province is characterized by alkalic, nepheline-normative lavas and pyroclastic rocks that range in age from Tertiary to Recent [*Armstrong*, 1978].

At most localities on Ross Island, the volcanic sequences consists of basanitoid flows, trachybasalt flows and cones, and phonolite domes and flows [*Treves*, 1967]. Commonly, earlier flows in the sequence are penetrated and overlain by late basanitoid cinder cones. Knowledge of the areal extent of the various rock types is limited by exposures which are restricted to a 1- to 2-km fringe at the edge of the island and some isolated monadnocks. The snow and ice cover constitutes about 95% of the island's surface. On the basis of the exposures and the pyroclastic platform penetrated during drilling at McMurdo Station the volcanic pile at Ross Island is inferred to be 3.5 to 4 km thick.

The volcanic rocks overlie a basement complex which, on the basis of exposures in the Dry Valley Region 80 km to the west, includes a pre-Devonian group of granites, gneisses, and metasedimentary rocks; Devonian to Jurassic continental sandstones of the Beacon Supergroup; a Jurassic series of tholeiitic sills (Ferrar Dolerite); and locally, Quaternary glacial deposits. *Smithson* [1972] has estimated the crustal thickness at Ross Island to be about 30 km on the basis of gravity data. With the exception of the Ferrar Dolerite the above men-

[1] Department of Geology and Geophysics, University of Minnesota, Minneapolis, Minnesota 55455.
[2] U. S. Geological Survey, Denver Federal Center, Denver, Colorado 80225.
[3] 9220 Clarewood, Houston, Texas 77036.
[4] Department of Earth and Planetary Sciences, Washington University, St. Louis, Missouri 63130.
[5] Geological and Natural History Survey of Wisconsin, Madison, Wisconsin 53706.

TABLE 1. Textural Data for Clinopyroxenes and Classification by Mode of Occurrence

Classification	Size	Shape	Color in Plane Light	Cleavage	Textural Occurrence	Source(s)
Xenocrysts	variable, generally <1 cm	anhedral, rounded and corroded	dark to olive green*	well to poorly developed	isolated fragments, usually with reaction rims, cores of phenocrysts	mafic and ultramafic nodules, crystallized from a melt in equilibrium with nodules at an early stage of magmatic history
Megacrysts	2–7 cm	anhedral to subhedral	brown to purplish brown			disaggregated xenoliths; nucleated on xenoliths; very large phenocrysts
Phenocrysts	0.2–1.5 mm	subhedral to euthedral	brown to purple, may have green or clear cores, a clear zone, and brown to yellowish-brown rim†	poorly developed or lacking	isolated crystals or grown on the sides of ultramafic nodules	primary crystallization from the melt
				poorly developed or lacking	isolated crystals or glomeroporphyritic clusters with olivine	
Groundmass	<0.2 mm	subhedral prismatic	brown to yellowish-brown‡	apparently absent	intergranular	primary crystallization from the melt and fragments of above classes

*There appear to be two green colors: dark green like CC 792 (Table 2a, 1) and a more olive green, which is also characteristic of many phenocryst cores.

†Green cores are anhedral and often internally corroded. Clear cores are generally euhedral. Rims are almost always brown, rarely clear. When three zones exist, the intermediate one is clear.

‡True color difficult to ascertain owing to small crystal size. Many crystals may appear transparent and colorless owing to their extreme thinness (<10 μm).

TABLE 2a. Clinopyroxene Compositions From the Alkali Basalts and Associated Mafic and Ultramafic Nodules From Ross Island, Antarctica: Analyses 1-9, Xenocrysts

Anal. # Lab No.	1 CC 792	2 MS 6-1	3 MS 4-1	4 BV 171	5 CC 821	6 MS 2115	7 MS 2113	8 MS 1-6	9 MS 9-4
SiO_2	55.30	54.30	53.20	55.90	54.60	51.90	50.50	50.20	52.80
Al_2O_3	1.99	1.04	2.12	0.44	3.13	3.68	3.36	4.40	2.57
FeO	2.72	2.93	6.50	4.86	3.07	3.36	3.40	4.89	3.64
MgO	18.90	21.10	16.20	17.00	16.10	19.00	18.90	17.40	17.30
CaO	21.10	19.60	20.50	18.90	20.50	22.40	21.90	21.40	21.40
Na_2O	0.23	0.52	0.66	0.71	0.19	0.62	0.63	0.49	0.23
TiO_2	0.02	0.19	0.23	0.22	0.08	0.42	0.74	0.52	0.63
MnO	0.18	0.33	0.31	0.22	0.10	0.22	0.34	0.17	0.08
Cr_2O_3	0.62	0.39	0.26	0.95	0.96	0.90	0.38	0.50	0.35
TOTAL	101.06	100.40	99.98	99.20	98.73	102.50	100.15	99.97	99.00
Si	1.971	1.949	1.950	2.043	1.993	1.858	1.845	1.845	1.933
Al	0.084	0.044	0.092	0.019	0.135	0.155	0.145	0.191	0.111
Fe	0.081	0.088	0.199	0.149	0.094	0.101	0.104	0.150	0.111
Mg	1.004	1.129	0.885	0.926	0.876	1.014	1.030	0.953	0.944
Ca	0.806	0.754	0.805	0.740	0.802	0.859	0.858	0.843	0.839
Na	0.016	0.036	0.047	0.050	0.013	0.043	0.045	0.035	0.016
Ti	0.001	0.005	0.006	0.006	0.002	0.011	0.020	0.014	0.017
Mn	0.005	0.010	0.010	0.007	0.003	0.007	0.011	0.005	0.002
Cr	0.018	0.011	0.008	0.028	0.028	0.026	0.011	0.015	0.010
TOTAL	3.986	4.026	4.002	3.967	3.945	4.074	4.068	4.050	3.985
Al_4	0.0289	0.0440	0.0501		0.0074	0.1421	0.1448	0.1554	0.0671
Al_6	0.0547		0.0416	0.0190	0.1273	0.0132		0.0352	0.0439
Fe++	0.081	0.029	0.164	0.149	0.094			0.039	0.111
Fe+++		0.059	0.035			0.101	0.104	0.112	
FER		2.01	0.21					2.89	
EN	53.10	57.29	46.85	51.03	49.45	51.37	51.71	48.98	49.82
FS	4.29	4.46	10.55	8.19	5.29	5.10	5.22	7.72	5.88
WO	42.61	38.25	42.61	40.78	45.26	43.53	43.07	43.30	44.30
FM	0.075	0.072	0.184	0.138	0.097	0.090	0.092	0.136	0.106
FMO	0.126	0.122	0.286	0.222	0.160	0.150	0.152	0.219	0.174

Analyses were made at 20 kV and 0.015×10^{-6} A sample current. Values are averages of at least three replicate, 20-s analyses corrected for background. Si, Al, Fe, Mg, Ca, Na data were reduced by using pyroxene standards close in composition to the pyroxene analyzed. An ilmenite standard with 3.15% MnO was used for Ti and Mn and a chromite standard for Cr. The following counting rates (per second) for analysis 8 are typical for other analyses. Si ~ 1000, Al ~ 100, Fe ~ 500, Mg ~ 150, Ca ~ 2000, Na ~ 5, Ti ~ 100, Mn ~ 50, Cr ~ 20. The data for Na, Mn, and, to a lesser extent, Ti and Cr have large errors, but the data define petrogenetically significant trends (Figures 4 and 5). Analyses 1–51 were made during the period 1972–1976, and most do not meet the stoichiometric requirements of superior analyses [Papike et al., 1974]. Analyses 52–56 are selected analyses of a zoned phenocryst that have good stoichiometry. Al^{IV}: tetrahedral Al calculated from $2 - Si$. Al^{VI}: octahedral Al calculated from Al (total) $- Al^{IV}$ (zero if Al^{IV} is greater than Al total). Fe^{3+}: ferric iron estimated from the charge balance ($Fe^{3+} + Cr + Al^{VI} + 2Ti = Na + Al^{IV}$). Fe^{2+}: estimated from Fe total $- Fe^{3+}$ (zero if Fe^{3+} is greater than Fe total). Fer: Fe^{3+}/Fe^{2+} (atomic). EN, FS, WO: enstatite, ferrosilite, wollastonite, respectively (calculated from Mg, Fe, and Ca cations). FM: Fe/(Fe + Mg) (atomic). FMO: Fe/(Fe + Mg) (weight percent). Lab no. refers to sample numbers used by Stuckless and Ericksen [1976]. The numbers are keyed to general sample localities. Letter prefixes refer to geographical localities: CB, Cape Bird; CC, Cape Crozier; HMC, Half Moon Crater; MS, McMurdo Station; DV, DVDP drill hole 2; BV, 'Brandau Vent,' and ME, Mount Erebus.

TABLE 2*b*. Clinopyroxene Compositions From the Alkali Basalts and Associated Mafic and Ultramafic Nodules From Ross Island, Antarctica: Analyses 10-17, Uncertain Occurrences and Megacrysts

Anal. # Lab No.	10 CC 6861	11 CC 6911	12 CC 6911	13 CC 6922	14 CB 6111	15 CB 5511	16 MS 481	17 CC 661
SiO_2	49.20	51.20	50.10	49.20	44.10	48.30	47.00	45.50
Al_2O_3	8.27	6.10	7.62	8.76	9.75	8.40	8.09	8.81
FeO	6.09	4.61	6.34	6.22	8.65	7.61	9.05	8.03
MgO	10.40	16.80	14.40	15.10	12.00	11.80	11.30	12.20
CaO	20.50	19.00	19.20	19.40	21.30	22.00	21.20	22.20
Na_2O	0.21	0.50	0.50	0.50	0.61	0.42	0.67	0.53
TiO_2	1.76	0.71	1.62	2.11	2.83	2.17	2.53	2.24
MnO	0.10	0.10	0.16	0.10	0.38	0.27	0.25	0.15
Cr_2O_3		0.05	0.08					0.09
TOTAL	96.53	99.07	100.02	101.39	99.62	100.97	100.09	99.75
Si	1.853	1.864	1.824	1.769	1.661	1.771	1.761	1.714
Al	0.367	0.262	0.327	0.371	0.433	0.363	0.357	0.391
Fe	0.192	0.140	0.193	0.187	0.272	0.233	0.284	0.253
Mg	0.584	0.912	0.781	0.809	0.674	0.645	0.631	0.685
Ca	0.827	0.741	0.749	0.747	0.860	0.864	0.851	0.896
Na	0.015	0.035	0.035	0.035	0.045	0.030	0.049	0.039
Ti	0.050	0.019	0.044	0.057	0.080	0.060	0.071	0.063
Mn	0.003	0.003	0.005	0.003	0.012	0.008	0.008	0.005
Cr		0.001	0.002					0.008
TOTAL	3.893	3.978	3.961	3.979	4.037	3.975	4.013	4.049
Al_4	0.1466	0.1363	0.1762	0.2311	0.3390	0.2289	0.2386	0.2860
Al_6	0.2207	0.1255	0.1508	0.1402	0.0939	0.1342	0.1188	0.1052
Fe^{++}	0.192	0.135	0.193	0.175	0.143	0.228	0.258	0.163
Fe^{+++}		0.006		0.012	0.129	0.005	0.026	0.090
FER		0.04		0.07	0.90	0.02	0.10	0.55
EN	36.42	50.84	45.34	46.41	37.31	37.01	35.74	37.35
FS	11.97	7.83	11.20	10.73	15.09	13.39	16.06	13.79
WO	51.61	41.33	43.46	42.86	47.60	49.60	48.20	48.86
FM	0.247	0.133	0.198	0.188	0.288	0.266	0.310	0.270
FMO	0.369	0.215	0.306	0.292	0.419	0.392	0.445	0.397

See Table 2*a* footnote.

tioned basement rocks, as well as a diverse suite of mafic granulites and ultramafic nodules, are common as xenoliths in the alkali basalts at many of the localities sampled [*Stuckless and Ericksen*, 1976].

Clinopyroxenes occur as xenocrysts, megacrysts [*Irving*, 1974], ubiquitous phenocrysts, as important groundmass constituents in the alkali basalts, and as a major phase in associated mafic and ultramafic nodules (Table 1). This broad range of occurrences combined with the extensive volume of recent experimental data on clinopyroxenes [e.g., *Green and Ringwood*, 1967; *Green*, 1973; *Huebner and Ross*, 1975; *Lofgren et al.*, 1975; *Grove and Bence*, 1977] makes this phase a significant petrogenetic indicator in alkali basalts.

TABLE 2c. Clinopyroxene Compositions From the Alkali Basalts and Associated Mafic and Ultramafic Nodules From Ross Island, Antarctica: Analyses 18-26, Phenocrysts and Related Xenocrysts of Questionable Origin

Anal. # Lab No.	18 MS 3-2	19 MS 123	20 MS 6-1	21 MS 6-1	22 MS 11A	23 MS 122	24 DV 49-31	25 DV 49-22	26 DV 49-26
SiO_2	51.20	50.90	40.60	50.60	50.60	52.00	46.60	48.90	47.40
Al_2O_3	4.61	5.45	8.30	4.56	5.31	5.80	8.29	8.19	6.61
FeO	10.60	14.60	16.10	14.30	14.40	14.60	8.12	8.19	11.00
MgO	9.38	6.76	11.60	9.48	7.13	6.63	11.80	11.40	10.00
CaO	20.60	19.20	21.10	19.20	19.50	19.70	22.50	20.80	21.30
Na_2O	1.76	0.75	0.81	0.52	0.57	0.50	0.74	0.66	0.13
TiO_2	1.38	0.87	1.91	1.16	1.20	1.24	2.42	2.45	1.89
MnO	0.45	0.73	0.73	0.68	0.32	0.32	0.12	0.19	0.30
Cr_2O_3	0.01								0.02
TOTAL	99.99	99.26	101.15	100.50	99.03	100.79	100.59	100.78	98.65
Si	1.924	1.946	1.587	1.912	1.936	1.949	1.739	1.801	1.816
Al	0.204	0.246	0.382	0.203	0.240	0.256	0.365	0.356	0.298
Fe	0.333	0.467	0.526	0.452	0.461	0.458	0.253	0.252	0.352
Mg	0.526	0.385	0.676	0.534	0.407	0.370	0.656	0.626	0.571
Ca	0.830	0.786	0.884	0.778	0.800	0.791	0.900	0.821	0.874
Na	0.128	0.056	0.061	0.038	0.042	0.036	0.054	0.047	0.010
Ti	0.039	0.025	0.056	0.033	0.035	0.035	0.068	0.068	0.054
Mn	0.014	0.024	0.024	0.022	0.010	0.010	0.004	0.006	0.010
Cr									0.001
TOTAL	3.999	3.934	4.197	3.972	3.930	3.906	4.038	3.977	3.986
Al_4	0.0756	0.0542	0.3824	0.0876	0.0636	0.0510	0.2613	0.1989	0.1844
Al_6	0.1286	0.1914		0.1156	0.1760	0.2053	0.1034	0.1567	0.1141
Fe++	0.333	0.467	0.195	0.452	0.461	0.458	0.178	0.252	0.352
Fe+++			0.332				0.076		
FER			1.70				0.43		
EN	31.13	23.51	32.40	30.28	24.40	22.88	36.28	36.84	31.76
FS	19.73	28.49	25.23	25.63	27.64	28.26	14.00	14.85	19.60
WO	49.14	48.00	42.37	44.09	47.96	48.86	49.72	48.31	48.63
FM	0.388	0.548	0.438	0.458	0.531	0.553	0.279	0.287	0.382
FMO	0.531	0.684	0.581	0.601	0.669	0.688	0.408	0.418	0.524

See Table 2a footnote.

ANALYTICAL METHODS

Data reported in this study were obtained on pyroxenes selected after detailed study of thin sections by Hunter and Stuckless. Modal analyses and textural descriptions of samples were made by *Hunter* [1974]. Bulk chemical analyses and ferrous-ferric iron determinations were made using conventional analytical techniques as described by *Goldich et al.* [1975]. Electron microprobe analyses were made on the model 400 MAC electron microprobe at the University of Minnesota. Data reductions are based on natural mineral standards as described by *Grant and Weiblen* [1971]. The analyses were made over a period of time (1972-1977) and are of varying quality as noted in Tables 2a-2f.

TABLE 2d. Clinopyroxene Compositions From the Alkali Basalts and Associated Mafic and Ultramafic Nodules From Ross Island, Antarctica: Analyses 27-36, Phenocrysts

Anal. # Lab No.	27 DV 49-24	28 Dv 49-42	29 CC 9613	30 HMC 3522	31 HMC 3511	32 HMC 3531	33 HMC 3121	34 DV 32-61	35 DV 32-73	36 MS 11B
SiO_2	48.50	48.10	49.10	50.80	50.40	50.70	48.40	48.50	47.80	48.90
Al_2O_3	8.52	7.76	6.19	5.52	6.01	8.36	8.63	6.45	8.40	6.77
FeO	8.22	7.88	5.89	9.50	13.40	12.00	16..10	7.60	8.63	10.10
MgO	11.80	13.30	13.80	8.03	7.74	5.23	4.33	12.30	9.81	8.44
CaO	21.40	20.20	20.40	22.30	19.10	20.20	20.40	21.90	22.08	19.40
Na_2O	0.32	0.52	0.50	0.50	0.50	0.50	0.50	0.50	0.50	0.60
TiO_2	2.49	2.65	1.38	1.60	1.72	2.04	2.59	2.14	2.60	1.81
MnO			0.57	0.27	0.29	0.32	0.38	0.13	0.25	0.20
Cr_2O_3		0.56	0.45					0.06		
TOTAL	101.25	100.97	98.29	98.52	99.16	99.35	101.33	99.58	100.07	96.22
Si	1.780	1.776	1.846	1.926	1.914	1.907	1.832	1.814	1.786	1.892
Al	0.369	0.338	0.274	0.247	0.269	0.371	0.385	0.284	0.370	0.309
Fe	0.252	0.243	0.185	0.301	0.426	0.378	0.510	0.238	0.270	0.327
Mg	0.645	0.732	0.774	0.454	0.438	0.293	0.244	0.686	0.546	0.487
Ca	0.841	0.799	0.822	0.906	0.777	0.814	0.827	0.878	0.884	0.804
Na	0.023	0.037	0.036	0.037	0.037	0.036	0.037	0.036	0.036	0.036
Ti	0.069	0.074	0.039	0.046	0.049	0.058	0.074	0.060	0.073	0.053
Mn			0.018	0.009	0.009	0.010	0.012	0.004	0.008	0.007
Cr		0.017	0.014					0.002		
TOTAL	3.979	4.016	4.009	3.924	3.920	3.868	3.920	4.003	3.974	3.923
Al_4	0.2204	0.2237	0.1535	0.0744	0.0855	0.0926	0.1683	0.1855	0.2138	0.1078
Al_6	0.1482	0.1141	0.1209	0.1722	0.1836	0.2782	0.2167	0.0989	0.1563	0.2010
Fe^{++}	0.252	0.243	0.185	0.301	0.426	0.378	0.510	0.237	0.270	0.327
Fe^{+++}								0.001		
FER										
EN	37.11	41.25	43.44	27.32	26.70	19.75	15.45	38.07	32.14	30.09
FS	14.50	13.71	10.40	18.14	25.94	25.42	32.23	13.20	15.86	20.20
WO	48.38	45.04	46.16	54.54	47.36	54.83	52.32	48.73	52.00	49.71
FM	0.281	0.249	0.193	0.399	0.493	0.563	0.676	0.257	0.330	0.402
FMO	0.411	0.372	0.299	0.542	0.634	0.696	0.788	0.382	0.468	0.545

See Table 2a footnote.

RESULTS

Petrographic Description and Classification of Pyroxene Types

We developed a working classification for the clinopyroxenes from the volcanic rocks of the Ross Island area based on the petrographic parameters of size, shape, color, cleavage, and textural occurrence. This permitted classification of the clinopyroxenes as xenocrysts, megacrysts, phenocrysts, and groundmass pyroxenes (Table 1). The differences in physical properties obviously reflect distinctive conditions of formation of each of the types.

TABLE 2e. Clinopyroxene Compositions From the Alkali Basalts and Associated Mafic and Ultramafic Nodules From Ross Island, Antarctica: Analyses 37-46, Phenocrysts

Anal. # Lab No.	37 DV 49-23	38 MS 103	39 DV 49-32	40 MS 11C	41 MS 4-2	42 Ms 105	43 DV 49-33	44 MS 101	45 DV 32-72	46 DV 32-62
SiO_2	49.10	46.10	47.10	45.10	45.20	48.30	46.00	48.10	48.10	45.70
Al_2O_3	7.09	9.90	8.21	7.83	9.48	6.62	8.20	8.02	6.96	8.27
FeO	5.51	8.10	5.40	8.36	6.56	7.30	6.20	8.28	7.67	8.03
MgO	14.40	11.55	14.40	9.10	11.40	13.15	13.20	12.10	11.80	10.95
CaO	22.00	20.40	22.70	20.30	22.20	20.40	22.60	20.10	21.76	22.50
Na_2O	0.61	0.82	0.27	0.81	0.64	0.45	0.34	0.32	0.50	0.50
TiO_2	1.97	3.87	2.24	3.41	3.57	3.49	3.84	2.62	2.80	3.42
MnO	0.05	0.25	0.11	0.13	0.24	0.18	0.18	0.08	0.26	0.23
Cr_2O_3	0.47									
TOTAL	101.20	100.99	100.43	95.04	99.29	99.89	100.56	99.62	99.85	99.60
Si	1.797	1.703	1.736	1.776	1.699	1.792	1.705	1.790	1.795	1.725
Al	0.306	0.431	0.357	0.363	0.420	0.290	0.358	0.352	0.306	0.368
Fe	0.169	0.250	0.166	0.275	0.206	0.227	0.192	0.258	0.239	0.253
Mg	0.785	0.636	0.791	0.534	0.639	0.727	0.730	0.671	0.657	0.616
Ca	0.863	0.808	0.897	0.856	0.894	0.811	0.898	0.802	0.870	0.910
Na	0.043	0.059	0.019	0.062	0.047	0.032	0.024	0.023	0.036	0.037
Ti	0.054	0.108	0.062	0.101	0.101	0.097	0.107	0.073	0.079	0.097
Mn	0.002	0.008	0.003	0.004	0.008	0.006	0.006	0.003	0.008	0.007
Cr	0.014									
TOTAL	4.032	4.003	4.033	3.972	4.013	3.982	4.021	3.972	3.991	4.013
Al_4	0.2035	0.2966	0.2636	0.2242	0.3010	0.2079	0.2946	0.2097	0.2046	0.2754
Al_6	0.1024	0.1346	0.0933	0.1393	0.1191	0.0817	0.0638	0.1422	0.1017	0.0925
Fe++	0.146	0.245	0.101	0.275	0.179	0.227	0.151	0.258	0.239	0.228
Fe+++	0.022	0.006	0.065		0.027		0.041			0.025
FER	0.15	0.02	0.65		0.15		0.27			0.11
EN	43.24	37.55	42.67	32.06	36.73	41.21	40.09	38.79	37.17	34.62
FS	9.28	14.77	8.98	16.53	11.86	12.84	10.57	14.89	13.56	14.24
WO	47.48	47.67	48.35	51.41	51.41	45.95	49.34	46.32	49.27	51.13
FM	0.177	0.282	0.174	0.340	0.244	0.237	0.209	0.277	0.267	0.291
FMO	0.277	0.412	0.273	0.479	0.365	0.357	0.320	0.406	0.394	0.423

See Table 2a footnote.

The general properties of the different types overlap, and this highlights the essential difficulty in using clinopyroxenes as petrogenetic indicators: A given pyroxene in thin section or hand specimen cannot be unambiguously assigned to a given type, and thus compositional data cannot be straightforwardly construed to record conditions of formation of a specific pyroxene type. In fact, as discussed below, we must first use compositional data to substantiate further the petrographic classification. In view of this we must accept the fact that our petrogenetic deductions and those of others [Bacon and Carmichael, 1973] are presently somewhat circular in nature.

TABLE 2f. Clinopyroxene Compositions From the Alkali Basalts and Associated Mafic and Ultramafic Nodules From Ross Island, Antarctica: Analyses 47-56, Groundmass Clinopyroxenes and Selected Analyses of a Zoned Phenocryst

Anal. #	47	48	49	50	51	52	53	54	55	56
Lab No.	DV 49 GM	DV 49 M6	MS 126	DV 32-32	Me 10-1	DV 49-M1	DV 49 M2	DV 49 M3	DV 49 M4	DV 49 M5
SiO_2	43.90	42.50	48.00	46.80	50.34	47.70	47.90	49.00	47.50	47.90
Al_2O_3	11.20	11.65	4.75	8.72	2.08	9.50	9.80	8.30	8.75	9.75
FeO	6.50	7.15	8.60	8.00	10.74	9.20	8.58	4.30	5.91	4.90
MgO	9.15	10.18	13.40	13.00	12.93	9.10	9.15	13.30	12.10	11.50
CaO	21.10	21.20	19.90	19.60	20.80	20.50	21.00	21.10	20.80	20.70
Na_2O	1.40	1.18	0.40	0.71	0.15	1.91	1.95	1.30	1.44	1.97
TiO_2	5.50	5.62	1.72	2.71	0.92	2.15	2.40	2.65	3.20	4.10
MnO	0.18	0.11	0.09	0.20	0.72	0.18	0.19	0.09	0.11	0.12
Cr_2O_3	0.06	0.09	0.05	0.05	0.05	0.02	0.02	0.62	0.16	0.02
TOTAL	98.99	99.68	96.91	99.79	98.73	100.26	100.99	100.66	99.97	100.96
Si	1.656	1.602	1.847	1.744	1.923	1.781	1.773	1.792	1.759	1.746
Al	0.498	0.518	0.215	0.383	0.094	0.418	0.428	0.358	0.382	0.419
Fe	0.205	0.225	0.277	0.249	0.343	0.287	0.266	0.132	0.183	0.149
Mg	0.514	0.572	0.769	0.722	0.736	0.507	0.505	0.725	0.668	0.625
Ca	0.853	0.856	0.820	0.783	0.852	0.820	0.833	0.827	0.825	0.808
Na	0.102	0.086	0.030	0.051	0.011	0.138	0.140	0.092	0.103	0.139
Ti	0.156	0.159	0.050	0.076	0.026	0.060	0.067	0.073	0.089	0.112
Mn	0.006	0.004	0.003	0.006	0.023	0.006	0.006	0.003	0.003	0.004
Cr	0.002	0.003	0.002	0.001	0.002	0.001	0.001	0.018	0.005	0.001
TOTAL	3.992	4.026	4.012	4.016	4.011	4.019	4.017	4.020	4.018	4.003
Al_4	0.3442	0.3979	0.1532	0.2562	0.0767	0.2186	0.2271	0.2077	0.2412	0.2544
Al_6	0.1539	0.1199	0.0623	0.1269	0.0170	0.1996	0.2005	0.1503	0.1407	0.1645
Fe^{++}	0.205	0.183	0.257	0.222	0.327	0.251	0.233	0.132	0.162	0.146
Fe^{+++}		0.043	0.020	0.027	0.016	0.036	0.032		0.021	0.004
FER		0.23	0.08	0.12	0.05	0.14	0.14		0.13	0.03
EN	32.72	34.59	41.19	41.17	38.13	31.38	31.49	43.07	39.85	39.48
FS	13.04	13.63	14.83	14.21	17.77	17.80	16.57	7.81	10.92	9.44
WO	54.24	51.78	43.97	44.62	44.09	50.82	51.95	49.12	49.23	51.08
FM	0.285	0.283	0.265	0.257	0.318	0.362	0.345	0.154	0.215	0.193
FMO	0.415	0.412	0.391	0.381	0.454	0.503	0.484	0.244	0.328	0.299

See Table 2a footnote.

Xenocrysts

Known sources for xenocrystic clinopyroxenes are ultramafic nodules, mafic granulites, and possibly, local equivalents of tholeiitic intrusive rocks, such as the Ferrar Dolerite.

The mineralogic composition of ultramafic nodules found on Ross Island, which can be the source of some xenocrysts, include harzburgites, lherzolites, websterite, and wehrlite. Clinopyroxenes in the harzburgites and lherzolites are light to dark green in hand specimen and clear to pale green in thin section in plane-polarized light, and they exhibit poorly to well-developed cleavage. Clinopyroxenes in the websterites and most of the wehrlites are black in hand specimen and brown to

purplish brown in plane-polarized light, and they
exhibit only poorly developed or no cleavage. In
thin section those clinopyroxene xenocrysts which
are found less than 20 μm away from the contact
with an ultramafic nodule are, in general, optically
similar to those within the nodules, except for evidence of reaction with the melt. Such crystals (presumed to have been derived from the nodules) are
generally anhedral in shape, although the shape
may be modified by clear and/or brown overgrowths. Thus among clinopyroxenes which have
been classified as xenocrysts, those with green
cores and reaction rims are presumed to be fragments from ultramafic nodules which have equilibrated to varying degrees with alkali basalt
magma.

Clinopyroxenes from the mafic granulites (all-
magnesium gneisses of R. B. Forbes (written communication, 1978)) can be distinguished from those
in the ultramafic nodules by their darker to olive
green color in thin section.

Clinopyroxene xenocrysts presumed to have
been derived from plagioclase-pyroxene granulites
on the basis of color and shape are corroded and
have erose internal structures or are intergrown
with opaque phases.

No obvious inclusions of the Ferrar Dolerites or
other tholeiitic, intrusive rocks were found in the
Ross Island samples. However, one clinopyroxene
grain classified as a xenocryst on the basis of its
size and shape has a composition (Table 2a, 3) characteristic of the Ferrar Dolerites or some other
tholeiitic material [Himmelberg and Ford, 1976].

Megacrysts

Clinopyroxenes classified as megacrysts (>2 cm)
are optically homogeneous, are anhedral to subhedral in shape, and generally lack cleavage. They
are black in hand specimen and clear to greenish-
brown in plane-polarized light. Poikilitic inclusions,
similar to those in the phenocrysts described below, are common.

Phenocrysts

A group of euhedral to subhedral clinopyroxene
grains which on casual observation appear to be
phenocrysts fall in the size range of 0.2 to 1.5 mm
and are almost always zoned. Zoning is generally
concentric, although hourglass structures were
found in crossed polarized light. Cores are light to
olive green or clear, intermediate zones are clear,

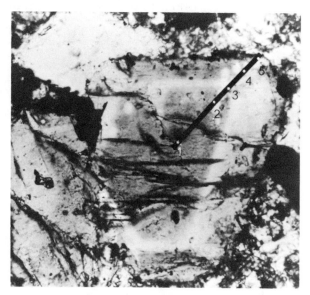

Fig. 1. Zoned phenocryst in trachybasalt (DV-49). Three distinct zones are optically distinguishable: A green core, a clear intermediate zone, and a brown rim. Absence of cleavage and exsolution is characteristic of all phenocrysts and ground-mass clinopyroxenes. Location of electron microprobe traverse (dark line) (see Figure 3) and analyses (1-5) (see Table 2f, 52-56) are indicated. Distance between points 1 and 4 is 160 μm.

and rims are brown to yellowish brown (Figure 1).
Cleavage is poorly developed or lacking. Poikilitic
inclusions of opaque oxides, olivine, and apatite are
common. On the basis of the above observations,
clinopyroxenes with an euhedral shape, light green
coloredcores, and complex zoning have been classified as phenocrysts. The fact that the cores of
some xenocrysts (anhedral grains with reaction
rims) have the same green color as the cores of
some phenocrysts (euhedral grains with complex
zoning) makes the classification of xenocrysts and
phenocrysts ambiguous. Some cores of phenocrysts
may be remnant xenocrysts; on the other hand, in
our classification scheme, some anhedral green
pyroxenes with reaction rims may be fragments of
broken phenocrysts.

Groundmass Pyroxenes

Groundmass pyroxenes (<0.2 mm) occur as subhedral prisms that lack apparent cleavage. In
plane-polarized light the crystals are brown to yel-
lowish-brown, although on thick edges they appear
to be clear. A few small crystals that exhibit var-

TÁBLE 3. Host Rock Types and Textural Descriptions of Analyzed Clinopyroxenes

Number*	Host†	Textural Description and Data on Location of Analyses (Tables 2a–2f)
		Xenocrysts
1	B	Green xenocrystic megacryst.
2	MG	Nodule is isotopically distinct from enclosing basanitoid.
3	B	Green color, only weakly reacted, near a mafic granulite inclusion.
4	B	Green xenocrystic megacryst.
5	UM	Nodule is isotopically distinct from enclosing basanitoid.
6	MG	Nodule is isotopically indistinct from enclosing basanitoid and contains glass.
7	MG	Nodule is isotopically indistinct from enclosing basanitoid and contains glass.
8	MG	Nodule is isotopically indistinct from enclosing basanitoid and contains glass.
9	MG	Nodule is isotopically indistinct from enclosing basanitoid and contains glass.
		Uncertain Occurrences
10	UM	Nodule contains glass.
11	UM	Nodule is isotopically distinct from enclosing basanitoid.
		Megacrysts
12	UM-B	Megacryst on side of nodule 11, analysis near contact with nodule.
13	UM-B	Megacryst on side of nodule 11, analysis near contact with basanitoid.
		Megacrysts 12 and 13 are isotopically indistinct from basanitoid.
14	TB	Megacryst isotopically indistinct from enclosing trachybasalt.
15	TB	Megacryst.
16	B	Megacryst.
17	B	Megacryst.
		Phenocrysts and Related Xenocrysts of Questionable Origin
18	B	Green, anhedral xenocrystic core.
19	B	Green, anhedral xenocrystic core.
20	B	Green, anhedral xenocrystic core.
21	B	Green, anhedral xenocrystic core.
22	B	Subhedral phenocryst core possibly xenocrystic.
23	B	Subhedral phenocryst core possibly xenocrystic.
24	B	Subhedral phenocryst core possibly xenocrystic.
25	B	Subhedral phenocryst core possibly xenocrystic.
26	B	Strongly poikilitic with opaque oxides.
		Phenocrysts
27	B	Phenocryst core.
28	B	From a glomeroporphyritic segregate of olivine and clinopyroxene.
29	B	From a glomeroporphyritic segregate of olivine and clinopyroxene.
30	TB	Core.
31	TB	Core.
32	TB	Core.
33	TB	Core.
34	TB	Core.
35	TB	Core.
36	B	Intermediate zone.
37	B	Intermediate zone.
38	B	Intermediate zone.
39	B	Intermediate zone.
40	B	Rim.
41	B	Rim.
42	B	Rim.
43	B	Rim.
44	B	Rim.
45	TB	Rim.
46	TB	Rim.
		Groundmass Clinopyroxenes
47	B	Groundmass.
48	B	Groundmass.
49	B	Groundmass.
50	TB	Groundmass.
51	P	Clinopyroxene within an anorthoclase crystal.
		Selected Analyses of a Zoned Phenocryst
52	B	Core.
53	B	Core.
54	B	Intermediate zone.
55	B	Intermediate zone.
56	B	Rim.

*Corresponding analysis in Tables 2a–2f.
†Host: B, basanitoid; TB, trachybasalt; P, phonolite; UM, ultramafic nodule; MG, mafic granulite nodule.
‡Possibly phenocrysts adjacent to nodules.

ious shades of green could be fragments of xenocrysts or phenocrysts.

Despite the wide range of textures described
above, exsolution in the clinopyroxenes is conspicuous by its absence. This feature is best developed
in clinopyroxenes from layered and hypabyssal intrusions which form in intermediate depth regimes. Thus on the basis of the textural data the
clinopyroxenes reported on here apparently record
two distinctly different environments of formation:
deep-seated and subaerial, with a bimodal range of
cooling rates: slow with complete equilibration and
fast with quenched compositional gradients. However, this conclusion should be evaluated in light of
recent experimental data on clinopyroxene textures [Lofgren et al., 1975; Grove and Bence, 1977;
Grove and Randsepp, 1978, and references
therein].

Clinopyroxene Compositions

Microprobe analyses of representative clinopyroxenes of the petrographic types described
above are presented in Tables 2a-2f. The analyzed
xenocrysts, megacrysts, phenocrysts, and groundmass clinopyroxenes are from chemically diverse
suites of rocks (Table 3), and the phenocrysts are
strongly zoned. Thus we believe we have analyzed
the overall spectrum of clinopyroxene compositions
that are to be found in the alkali basalts of the Ross
Island area.

A remarkable general feature of the analyzed
compositions of all the types of clinopyroxenes described above is the rather restricted range of major element compositions (Figure 2). This is surprising in view of the diverse sources of the
clinopyroxenes and the fact that they have crystallized in or equilibrated with magmas ranging in
bulk chemistry from basanitoid through trachybasalt (Table 2a, footnotes; Goldich et al. [1975]). In
addition to the restricted range of compositions
there are two other distinctive features of the compositional data: (1) The compositions of xenocrysts
and clinopyroxenes in associated ultramafic nodules do not overlap with the compositions of the
other clinopyroxenes (Figure 2a; Table 2a, 1-9). (2)
There is a decrease in the Fe/(Fe + Mg) ratio in
the intermediate zones and rims compared to cores
of zoned phenocrysts and in some groundmass
clinopyroxenes from more differentiated lavas
(Figures 2a-2e; Tables 2c-2f, 27-56). The best developed reversal in the Fe/(Fe + Mg) trend is be-

tween phenocryst cores and intermediate zones
(Figures 2e and 3; Table 2f, 53-54).

The above two compositional characteristics contrast markedly with the regular, extended compositional trends of clinopyroxenes found in intrusive
tholeiitic rocks [Deer et al., 1966] and the broad
complex variations found for clinopyroxenes in lunar rocks [Bence and Papike, 1972]. The tholeiitic
trend provides a record of melt crystal equilibria
characterized by relatively uniform cooling rates
and with disequilibria restricted primarily to the
effects of gravity segregation of minerals in a
closed system and subsolidus exsolution. The
broad range of pyroxene compositions found in lunar basalts reflects the special conditions of fast
cooling [Lofgren et al., 1975; Grove and Bence,
1977] and very low partial pressure of oxygen
[Sato, 1976] under which the lunar pyroxenes
formed. The restricted and 'reversed' trend of the
clinopyroxenes from the alkali basalts and associated xenoliths of the Ross Island area appears to
be characteristic of alkali basalts in general [Huck-
enholz, 1973; Binns et al., 1970]. The problem remains to deduce the significance of the trend for
models of alkali basalt petrogenesis.

INTERPRETATION

Attempts have been made to use clinopyroxene
compositional data to place constraints on temperature and pressure and to delimit modes of magma
generation, emplacement, and cooling history [Ba-
con and Carmichael, 1973]. Such efforts are all
predicated, however, on the assumption that the
different pyroxene types can be reliably distinguished and each of the textural types and their
compositional variations assigned to the appropriate stages of any proposed petrogenetic model.

Several factors limit what can be done with the
data in this direction at present: (1) The classification of pyroxenes (Table 1) is not entirely unambiguous. (2) The present experimental data on the effect of T, P, f_{O_2}, and cooling rate on pyroxene
equilibria [Grove and Bence, 1977] are not directly
applicable to alkali basalt bulk compositions. In
view of this, we wish here only to emphasize observed correlations between textures and compositions. Our interpretation of factors responsible for
these correlations is eclectic and largely intuitive.
In a broad sense the correlations are a reflection of
the systematics of crystal-chemical, melt composition, T, P, f_{O_2}, and cooling history.

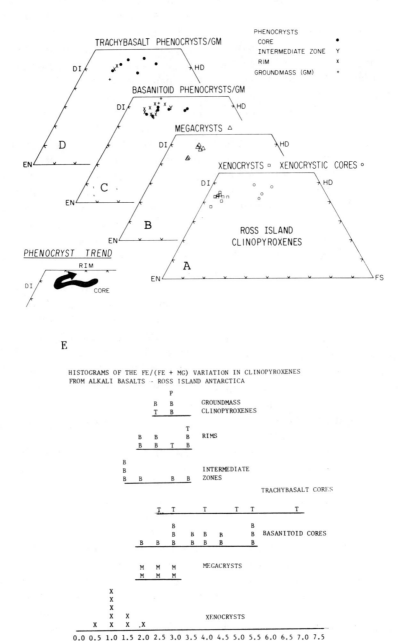

Fig. 2.   Compositional variations in clinopyroxenes. Inset outlines the general trend of the reversed zoning in phenocrysts. (*a*) Clinopyroxene compositions of xenocrysts. Over 80% of the components are represented on the quadrilateral (Table 2*a*, 1-9). (*b*) Clinopyroxene compositions of megacrysts (Table 2*b*, 12-17). (*c*) Clinopyroxene compositions from primitive (low silica) basanitoid lavas (Tables 2*c*-2*f*, 24-29, 37, 39, 43, 47) and from differentiated (high silica) basanitoid lavas (Tables 2*c*-2*f*, 18-23, 36, 38, 40-42, 44, 49, 52-56). (*d*) Clinopyroxene compositions from trachybasalts (Tables 2*b*-2*f*, 14-15, 30-35, 45-46, and 50). (*e*) Histogram of atomic Fe/(Fe + Mg) ratios of clinopyroxenes. The data define a reversal in the Fe/(Fe + Mg) ratio for zoned phenocrysts from basanitoid through phonolite. Data plotted are the FM values in Tables 2*a*-2*f*.

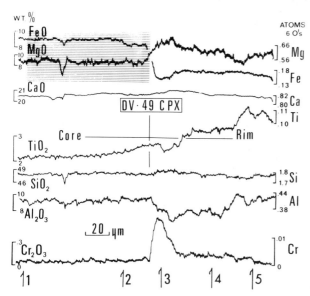

Fig. 3. Electron microprobe traverse across zoned clino-pyroxene phenocryst DV-49 (Figure 2). Ratemeter traces were made at 0.3 s/μm and calibrated with five-point analyses as indicated (Table 2f, 52-56). Depending on the level of accuracy assumed from the Mg-Fe data reversal, more than one major reversal in the Fe/(Fe + Mg) ratio may be inferred. Zoning is similar to that found in megacrysts in a lamprophyre dike in east Greenland by *Brooks and Rucklidge* [1973].

Correlations Between Textures and Compositions

The data in Tables 2a-2f and 3 suggest certain correlations between the petrographic types described above and compositions:

1. As mentioned above, most clinopyroxenes from known xenocrystic source rocks (Table 2a, 1-9) are chemically distinct from the clinopyroxenes classified as megacrysts, phenocrysts, or groundmass clinopyroxenes (Figure 2e). With the exception of the clinopyroxenes from some wehrlites (Table 2b, 10 and 11), crystals from the xenocrystic sources are lower in TiO_2, Al_2O_3, and FeO than the other clinopyroxenes. Xenocrysts contain a maximum of 0.7 TiO_2, 4.41 Al_2O_3, and 4.6 FeO (average weight percent), whereas the other clinopyroxenes have average minimums of 1.2 TiO_2, 4.6 Al_2O_3, and 9.6 FeO. The most diagnostic element for distinguishing xenocrysts from the other clinopyroxene types appears, however, to be Cr_2O_3 which is greater than 0.5 wt % in all clinopyroxenes analyzed in nodules and below the limits of detection (<0.05 wt %) for the phenocryst cores and rims. The only exception to this in our data is the composition of the intermediate zones of phenocrysts, which contain up to 0.62 wt % Cr_2O_3 (Table 2f, 54).

2. Chemical data for the megacrysts suggest that most of the megacrysts crystallized at the same time as the phenocryst cores of the basanitoids (Table 2b, 12-17). Two xenocrystic megacrysts (Table 2a, 1 and 4) are obvious exceptions. They are green in color, low in Al_2O_3 and TiO_2, and high in Cr_2O_3, and thus although they are large crystals, they are xenocrysts on the basis of composition. Also, they were collected from localities where ultramafic inclusions are abundant. Megacryst CC 6922 (Table 2b, 13) is also chemically different from the phenocryst cores. However, this megacryst was attached to the outside of a wehrlite nodule and may have grown in a melt that was MgO rich owing to reaction between the basanitoid liquid and the nodule. Partial wet chemical analyses for two of the megacrysts (CB 61 and CB 55) are surprisingly low in ferrous iron (4.15 wt % and 4.29 wt %, respectively (S. S. Goldich, unpublished data, 1975) and suggest large amounts of oxidized iron (see Table 2b, 14 and 15, for electron microprobe analyses).

3. Phenocrystic clinopyroxenes from the basanitoids exhibit a correlation of optical zoning with composition: light to olivine green cores are iron rich, clear zones are lower in iron, and brownish rims are intermediate in iron content (Tables 2c-2e, 22-46).

A typical zoned phenocryst from the basanitoids is shown in Figure 2, and microprobe traverses across the crystal are shown in Figure 3. Five analyses along the traverse are given in Table 2f, 53-56. It is interesting to note that Cr_2O_3 is detectable in only the intermediate (clear) zone of the phenocrysts, reaching a maximum of 0.62% in the traverse shown in Figure 3. Therefore crystals that started to nucleate during the growth of the clear zone may appear to be chemically similar to the xenocrysts. Tw of the phenocryst core analyses (Table 2d, 28 and 29) are high in Cr_2O_3 and may represent crystals which nucleated while the intermediate zones were crystallizing. Both of these clinopyroxenes poikilitically enclose olivine and have the optical appearance of cognate phases.

The reversal in zoning within the phenocrysts is mirrored by a change in composition of phenocryst cores in the differentiated sequence of basanitoid through trachybasalt. The core compositions exhibit an iron depletion, followed by iron enrichment as a function of increasing silica content of the host rocks (Figures 2a-2d) [*Goldich et al.*, 1975].

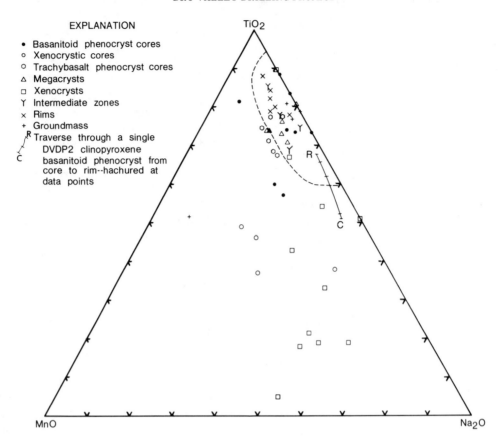

Fig. 4.   Ternary plot of the Ross Island area clinopyroxene data in the system MnO-TiO₂-Na₂O. The field for clinopyroxenes from within plate alkali basalts [*Nisbet and Pearce*, 1977] is shown by dashed line. Traverse through a single clinopyroxene phenocryst from a DVDP 2 basanitoid is shown by the solid line with hachures at data points. A
generalized upward trend in the diagram is given by unreacted xenocrysts, reacted xenocrysts, phenocryst cores
through rims, and groundmass clinopyroxenes.

4.   The groundmass clinopyroxenes form a continuation of the chemical trend of phenocrysts as
exhibited by their iron-rich compositions (Table 2*f*,
47-50; Figure 2*c*).

*Factors Responsible for Correlations
Between Textures and Compositions*

The details of coupled substitutions and other
facets of crystal-chemical control on clinopyroxene
compositions are complicated by lack of data on
ferric iron [*Neumann*, 1976; *Edwards*, 1976, and
references therein]. As discussed below in connection with the reversal in the Fe/(Fe + Mg) trend,
our data do not permit an accurate estimate of ferric iron content. In view of this we have not attempted to evaluate the details of coupled substitutions   suggested   by   studies   of   similar

clinopyroxenes from other rocks [*Tracy and Robinson*, 1977].

There is a substantial data base on bulk compositions of the volcanic rocks from Ross Island
[*Goldich et al.*, 1975]. Serious evaluation of the relevance of these data to explaining the variations in
clinopyroxene compositions must await the constraints of appropriate experimental studies. In
what follows we explore only some general aspects
of bulk compositional controls on clinopyroxene
compositions.

A recent statistical study by *Nisbet and Pearce*
[1977] using a method of discriminant analysis by
Pearce [1976] has demonstrated that there are
some special compositional variations of clinopyroxene with tectonic setting and thus bulk composition. Si, Al, Ti, Na, and Mn variations distinguish at least 90% of the clinopyroxene data for
'within plate' alkali basalts from clinopyroxene

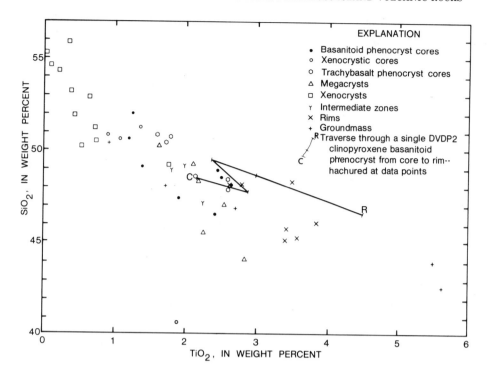

Fig. 5.   Variation of TiO$_2$ with SiO$_2$ for clinopyroxenes from volcanic rocks and xenoliths from Ross Island area. A traverse through a single clinopyroxene phenocryst, DV-49 (Table 2f, 52-56) is shown by the solid line.

data for basalts from other settings (for example, tholeiitic basalts from oceanic islands or continental rifts or basalts from the ocean floor or island arc settings). The bulk of the clinopyroxene data used by Nisbet and Pearce, as well as our Ross Island data, plot in a restricted field on the MnO-TiO$_2$-Na$_2$O diagram (Figure 4). *Nisbet and Pearce* [1977] offer no explanation for the data that plot outside their chosen field for within plate alkali basalts. We find, however, that our data define systematic trends on the MnO-TiO$_2$-Na$_2$O and SiO$_2$-TiO$_2$ plots (Figures 4 and 5). Furthermore, these trends can be interpreted in light of the mode of clinopyroxene genesis as inferred from our petrographic classification. The Ross Island data indicate an increase in Ti relative to Mn and Na (Figure 4) and an increase in Ti with decreasing Si (Figure 6) for the sequence: unreacted xenocrysts, reacted xenocrysts, megacrysts, phenocryst cores, intermediate zones of phenocrysts, phenocryst rims, and groundmass clinopyroxenes. Although there are large errors in the data (see Table 2a, footnotes), they may also be construed to define a crude decrease in the Na/Mn ratio for the above sequence. The following may be pertinent to any explanation for the relationship between the chemical trends and our textural classification.

1.   Xenocrysts must have equilibrated initially with unspecified bulk compositions that were relatively higher in Si and lower in Ti and Na than the alkalic magmas. Thus varying degrees of equilibration between xenocrysts and magma would produce that part of the vertical trend (as observed in Figure 4) for xenocrysts to reacted xenocrysts. Our interpretation of this reaction-controlled trend is supported by a few Sr isotopic analyses of nodules: nodules that contain lower Ti clinopyroxenes are isotopically distinct from their host lavas, whereas those that contain higher Ti clinopyroxenes are isotopically indistinguishable from their host lavas and are inferred to have reacted with the melt [*Stuckless and Ericksen*, 1976].

2.   Cooling rates affect the amounts of nonquadrilateral components (for example, Al^{IV} and Ti^{VI}) in clinopyroxene [*Usselman et al.*, 1975]. The trends in Figures 4 and 5 for zones of phenocrysts through groundmass clinopyroxene could be produced by increased cooling rates during magma ascent and eruption.

3.   The pressure dependence on partitioning of Ti between melt and clinopyroxene is uncertain and is complicated by cooling rate effects [*Grove and Bence*, 1977], but a decrease the Al/Ti ratio with decreasing pressure of crystallization is sug-

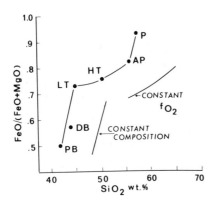

Fig. 6. Plot of FeO/(FeO + MgO) versus SiO_2 in weight percent. Bulk compositions of the alkali basalts from Ross Island [*Goldich et al.*, 1975] may be construed to define a change in crystallization from constant composition in primitive basanitoid (PB), differentiated basanitoid (DB), and low-silica trachybasalt (LT) to constant f_{O_2} in high-silica trachybasalt (HT), anorthoclase phonolite (AP), and phonolite (P). Schematic variation for experimental trends in the FeO, Fe_2O_3, MgO, and SiO_2 systems and tholeiitic basalt are indicated [from *Osborn*, 1959].

gested by limited experimental data [*Akella and Boyd*, 1973]. The decrease in the Al/Ti ratio (atomic) of about 20% from cores to rims of phenocrysts (Figure 3) could have originated in a changing depth regime from deep to shallow. Therefore the decrease in Al/Ti ratio in this sequence could be attributed to pressure and/or cooling rate effects.

The above explanations for the Si, Na, Mn, and Ti variations in the clinopyroxenes in alkali basalts must await experimental verification, but we tentatively conclude that accurate analyses for Si, Ti, Mn, and Na can provide valid checks on the textural classification of pyroxenes in alkali basalts.

Reversed Compositional Trends

We have suggested previously that the compositional variations observed in clinopyroxene phenocrysts in the Ross Island lavas (Figures 2 and 3) record early crystallization at depth in a basanitoid magma, continued crystallization during ascent with increased oxidation causing a compositional reversal, and final crystallization under near-surface conditions [*Weiblen et al.*, 1974]. Oxidation has been proposed by others [*Brooks and Rucklidge*, 1973] to account for similar reversals in the Fe/(Fe + Mg) variation in clinopyroxenes in alkalic rocks. The proposed change in oxidation should

be reflected in the Fe^{3+}/Fe^{2+} ratio in the zoned clinopyroxenes. We have made an attempt to estimate these ratios from the electron microprobe data (Tables 2a-2f). However, there is no systematic variation in the calculated Fe^{3+}/Fe^{2+} ratios for our data (Tables 2a-2f). This is probably due to the relatively large errors in the Si, Al, and Na analyses. By selecting analyses of a zoned phenocryst (Figures 1 and 3) it is possible to define a trend which would indicate oxidation in the intermediate zone, but we have not been able to reproduce the trend with our most careful analyses (Table 2f, 52-56). Also, analyses of intermediate zones of other phenocrysts do not require ferric iron to balance their stoichiometry (Tables 2d and 2e, 36-39).

The bulk compositional data on the lavas [*Goldich et al.*, 1975] indicate an increase in the Fe^{3+}/Fe^{2+} ratio from an average of 0.340 for basanitoids to 0.66 for trachybasalts. Furthermore, the silica versus Fe/(Fe + Mg) variation for the lavas may be interpreted to indicate a change in crystallization conditions from closed system, constant composition to constant f_{O_2} from basanitoid to trachybasalt compositions (Figure 6). The bulk compositional data therefore support the idea (but do not provide direct evidence) that oxidation occurred during the crystallization of the clinopyroxenes in these lavas.

Other explanations for the apparent reversal in clinopyroxene zoning may lie in magma mixing processes [*Brooks and Printzlaw*, 1978] or in the nature of the reaction of xenoliths with basanitoid magma with regard to the latter. The Fe-Mg compositional reversal (Figure 2) and the Si-Ti reversal (Figure 5) between cores and intermediate zones of phenocrysts could be related to a complex origin for phenocryst cores: They could be an unidentified class of xenocryst fragments. Such relic xenocrysts (Table 2c, 18-26) would have had a higher Fe/(Fe + Mg) ratio and experienced a greater degree of reaction with the melt than xenocrysts recognized by petrographic criteria (Table 1 and Table 2a, 1-9). Instead of postulating an increase in f_{O_2} during crystallization, alternatively we suggest that initial crystallization may have occured at relatively high f_{O_2}. During this period, megacrysts, intermediate zones of phenocrysts around reacted xenocrysts, and rims around xenocrysts may have formed.

Wet chemical analyses of megacrysts (CB 61 and CB 55) and of a coarse-grained pyroxene separate from the thick basanitoid flow from DVDP 2 reveal relatively high ferric-ferrous iron ratios of approximately 1 : 1 for both samples at levels of approximately 4% ferric iron [*Goldich et al.*, 1975]. These data suggest early crystallization under conditions of high f_{O_2}. Subsequent crystallization under conditions of constant and/or lower f_{O_2} would have produced the normal (increasing) Fe/(Fe + Mg) trends in the pyroxenes (Figure 2).

Trace element data on the lavas may be construed to indicate extensive reaction of xenocrystic material with basanitoid magma: basanitoid lavas with xenocrysts exhibit considerable chemical and isotopic variability [*Sun and Hanson*, 1975, 1976]. In particular, the thick basanitoid flow from DVDP 2 contains abundant xenocrystic clinopyroxene and has the most radiogenic lead of the analyzed basanitoids [*Sun and Hanson*, 1975].

The difficulty with postulating a xenocryst origin for the reversed zoning lies in the fact that there is no compositional trend between recognized xenocrysts (Table 2*a*, 1-9) and phenocryst cores (Tables 2*c* and 2*d*, 18-35); we must make the ad hoc assumption that a special type of xenocryst now represented only by reacted cores of phenocrysts (Table 2*c*, 18-26) was involved.

A definitive explanation for the unique trend in the Fe/(Fe + Mg) ratio of clinopyroxenes in alkali rocks must await refined electron microprobe analyses and experimental data on phase equilibria and cooling rates.

REFERENCES

Akella, J., and F. R. Boyd, Partitioning of Ti and Al between coexisting silicates, oxides, and liquids, *Proc. Lunar Sci. Conf. 4th*, 1049-1059, 1973.

Armstrong, R. L., K-Ar dating: Late Cenozoic Volcanic Group and Dry Valley glacial history, Victoria Land, Antarctica, *N. Z. J. Geol. Geophys.*, *21*(6), 685-698, 1978.

Bacon, C. R., and I. S. E. Carmichael, Stages in the *P-T* path of ascending basalt magma: An example from San Quintin, Baja California, *Contrib. Mineral. Petrol.*, *41*, 1-22, 1973.

Bence, A. E., and J. J. Papike, Pyroxenes as recorders of lunar basalt petrogenesis: Chemical trends due to crystal-liquid interaction, *Proc. Lunar Sci. Conf. 3rd*, 431-469, 1972.

Binns, R. A., M. B. Duggan, and J. F. G. Wilkinson, High pressure megacrysts in altaline lavas from northeastern New South Wales, *Amer. J. Sci.*, *269*, 132-168, 1970.

Brooks, C. K., and I. Printzlau, Magma mixing in mafic alkaline volcanic rocks: The evidence from relict phenocryst phases and other inclusions, *J. Volcanol. Geotherm. Res.*, *4*, 315-331, 1978.

Brooks, C. K., and J. C. Rucklidge, A tertiary lamprophyric dike with high-pressure xenoliths and megacrysts from Wiedemanns Fjord, East Greenland, *Contrib. Mineral. Petrol.*, *42*, 197-212, 1973.

Deer, W. A., R. A. Howie, and J. Zussman, *An Introduction to the Rock Forming Minerals*, p. 528, Longmans, Green, Toronto, Ont., 1966.

Edwards, A. C., A comparison of the methods for calculating Fe^{3+} contents of clinopyroxenes from microprobe analysis, *Neues Jahrb. Mineral. Monatsh.*, *11*, 508-512, 1976.

Goldich, S. S., S. B. Treves, N. H. Suhr, and J. S. Stuckless, Geochemistry of the Cenozoic rocks of Ross Island and vicinity, Antarctica, *J. Geol.*, *83*, 415-435, 1975.

Grant, J. A., and P. W. Weiblen, Retrograde zoning in garnet near the second sillimanite isograd, *Amer. J. Sci.*, *270*, 281, 1971.

Green, D. H., Experimental melting studies on a model upper mantle composition at high pressure under water-saturated and water-undersaturated conditions, *Earth Planet. Sci. Lett.*, *19*, 37-53, 1973.

Green, D. H., and A. E. Ringwood, The genesis of basaltic magmas, *Contrib. Mineral. Petrol.*, *15*, 103-190, 1967.

Grove, T. L., and A. E. Bence, Experimental study of pyroxene-liquid interaction in quartz-normative basalt 15597, *Proc. Lunar Sci. Conf. 8th*, 1549-1579, 1977.

Grove, T. L., and M. Raudsepp, Effects of kinetics on the crystallization of quartz normative basalt 15597: An experimental study, *Proc. Lunar Sci. Conf. 9th*, 585-599, 1978.

Himmelberg, G. R., and A. B. Ford, Pyroxenes of the Dufek Intrusion, Antarctica, *J. Petrology*, *17*(2), 219-243, 1976.

Hamilton, W., The Hallett volcanic province, Antarctica, *U. S. Geol. Surv. Prof. Pap.*, *456-C*, 62 pp., 1972.

Huckenholz, H. G., The origin of Fassaitic augite in the alkali basalt suite of the Hocheifel Area, Western Germany, *Mineral. Petrol.*, *40*, 315-326, 1973.

Huebner, J. S., and M. Ross, Estimation of the minimum temperature for co-existence of orthopyroxene, pigeonite augite and its application to prediction of temperatures of crystallization of lunar pyroxene, in *Lunar Science VI*, part II, pp. 689-691, Lunar Science Institute, Houston, Tex., 1975.

Hunter, W. C., Petrography and petrology of a basanitoid flow from Hut Point Peninsula, Antarctica, M.S. thesis, 91 pp., N. Ill. Univ., DeKalb, 1974.

Irving, A. J., Pyroxene-rich ultramafic xenoliths in the newer basalts of Victoria, Australia, *Neues Jahrb. Mineral. Abh.*, *120*, 147-167, 1974.

Lofgren, G. E., C. H. Donaldson, and T. M. Usselman, Geology, petrology and crystallization of Apollo 15 quartz normative basalts, *Proc. Lunar Sci. Conf. 6th*, 79-99, 1975.

Neumann, E. R., Two refinements for the calculation of structural formulae for pyroxenes and amphiboles, *Norsk Geol. Tidsskr.*, *56*, 1-6, 1976.

Nisbet, E. G., and J. A. Pearce, Clinopyroxene composition in mafic lavas from different tectonic settings, *Contrib. Mineral. Petrol.*, *63*, 149-160, 1977.

Osborn, E. F., Role of oxygen pressure in the crystallization and differentiation of basaltic magma, *Amer. J. Sci.*, *257*, 609-647, 1959.

Papike, J. J., K. L. Cameron, and K. Baldwin, Amphiboles and pyroxenes: Characterization of other than quadrilateral components and estimates of ferric iron from microprobe data, *Geol. Soc. Amer. Abstr. Programs*, *6*, 1053-1054, 1974.

Pearce, J. A., Statistical analysis of major element patterns in basalts, *J. Petrology*, *17*, 15-43, 1976.

Sato, M., Oxygen fugacity and other thermochemical parameters of Apollo 17 on high-Ti basalt and their implications on the reduction mechanism, *Proc. Lunar Sci. Conf. 7th*, 1323-1344, 1976.

Smithson, S. S., Gravity interpretations in the Transantarctic Mountains near McMurdo Sound, Antarctica, *Geol. Soc. Amer. Bull.*, *83*, 3437-3442, 1972.

Stuckless, J. S., and R. L. Ericksen, Strontium isotopic geochemistry of the volcanic rocks and associated megacrysts and inclusions from Ross Island and vicinity, Antarctica, *Contrib. Mineral. Petrol.*, *58*, 111-126, 1976.

Sun, S. S., and G. N. Hanson, Origin of Ross Island basanitoids and limitations upon the heterogeneity of mantle sources for alkali basalts and nephelinites, *Contrib. Mineral. Petrol.*, *52*, 77-106, 1975.

Sun, S. S., and G. N. Hanson, Rare earth element evidence for differentiation of the McMurdo Volcanics, Ross Island, Antarctica, *Contrib. Mineral. Petrol.*, *54*, 139-155, 1976.

Tracy, R. J., and P. Robinson, Zoned titanium augite in alkali olivine basalt from Tahiti and the nature of titanium substitutions in augite, *Amer. Mineral.*, *62*, 634-645, 1977.

Treves, S. B., Volcanic rocks from the Ross Island, Márguerite Bay, and Mount Weaver areas, Antarctica, *Jap. Antarct. Res. Exped. Sci. Rep.*, *1*, 136-149, 1967.

Usselman, T. M., et al., Experimentally reproduced textures and mineral chemistries of high-titanium mare basalts, *Proc. Lunar Sci. Conf. 6th*, 997-1020, 1975.

Weiblen, P. W., K. J. Schulz, J. S. Stuckless, W. C. Hunter, and M. G. Mudrey, Clinopyroxenes in alkali basalts from the Ross Island area, Antarctica: Clues to stages of magma crystallization, paper presented at the 1st Dry Valley Drilling Project Seminar, Natl. Sci. Found., Seattle, Wash., 1974.

U-Pb ZIRCON AGES AND PETROGENETIC IMPLICATIONS FOR TWO BASEMENT UNITS FROM VICTORIA VALLEY, ANTARCTICA

ROBERT D. VOCKE, JR., AND GILBERT N. HANSON

Department of Earth and Space Sciences, State University of New York, Stony Brook, New York 11794

U-Pb ages for zircons have been determined for granite and gneiss recovered from Dry Valley Drilling Project (DVDP) drillhole 6 in Victoria Valley, Antarctica. These samples have been correlated with the Vida granite and the Olympus granite-gneiss, respectively. U-Pb ages for zircons from the gneiss are very discordant, defining upper and lower concordia intercepts of 2555 and 467 m.y. The $^{207}Pb/^{206}Pb$ age on the most discordant zircons suggests the gneiss contains a component that is at least 950 m.y. old and may be Archean in age. Cathodoluminescence and microprobe studies of the gneiss zircons indicate a low U core and a higher U rim which truncates luminescent banding in the core. If the cores of the zircons are primarily detrital, this age may date the provenance of these zircons. If, on the other hand, the zircon cores were formed in situ, this older age may date an event that affected the gneiss. Based on the high initial $^{87}Sr/^{86}Sr$ ratio, rare earth elements (REE), and other geochemical considerations, the granite is suggested to be derived by partial melting of metasedimentary rocks such as those found in the Asgard formation or the Olympus granite-gneiss. Because they are strongly discordant, the U-Pb ages for zircon from the granite only support, but do not define, an age of about 470 m.y. for the time of intrusion of the Vida granite at the end of the Ross orogeny.

INTRODUCTION

Structural and stratigraphic relationships within the basement complex of the dry valleys in southern Victoria Land, Antarctica, suggest the possibility of rocks of Precambrian age [*Webb and McKelvey*, 1959; *McKelvey and Webb*, 1962]. This is not surprising, as Precambrian ages have been found in other parts of Antarctica [*Grew*, 1978]; however, radiometric studies in the dry valleys have only given well-documented ages as old as the early Cambrian or late Precambrian. The majority of the ages date the Ross orogeny at 450-500 m.y. ago [*Stuckless and Ericksen*, 1975; *Faure and Jones*, 1974; *Deutsch and Grögler*, 1966; *Deutsch and Webb*, 1964; *Angino et al.*, 1962; *Goldich et al.*, 1958]. In an attempt to better define the time of the Ross orogeny and to detect components that might predate the Ross orogeny, this investigation has used U-Pb dating of zircons from core samples of gneiss and granite recovered from Dry Valley Drilling Project (DVDP) drillhole 6.

The stratigraphy in the dry valleys of the Trans-Antarctic Mountains, Antarctica, is known primarily from the studies of *Webb and McKelvey* [1959] and *McKelvey and Webb* [1962]. The basement rocks consist of the pre-Ordovician Asgard formation and the younger, lower Paleozoic intrusions, the Olympus granite-gneiss, Dais granite, and Vida granite (Figure 1). The Asgard formation consists of a thick metasedimentary sequence of variably deformed marble, hornfels, and schist. The contacts of the Olympus granite-gneiss and Dais granite with the Asgard formation are often gradational. The strike of the compositional bedding within the metasedimentary rocks parallels the foliation in the Olympus granite-gneiss, the oldest intrusion [*McKelvey and Webb*, 1962]. The parallel foliation and the presence of aligned inclusions of the metasedimentary rocks within the granite-gneiss allow the possibility that the Olympus granite-gneiss represents higher grade or granitized Asgard sediments. *Smithson et al.* [1970] noted that while the Asgard formation has at least three different phases of deformation, the Olympus granite-gneiss contains evidence of only the latest period of folding. This latest period of deformation was associated with the Ross orogeny, which resulted in widespread disruption and resetting of radiometric systems.

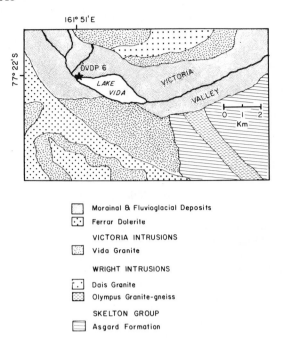

Fig. 1. Generalized geologic map of the Victoria Valley region around Lake Vida, southern Victoria Land, Antarctica, showing the location of DVDP drillhole 6 (modified after *Webb and McKelvey* [1959]).

The postkinematic Vida granite cuts the Asgard formation, the Olympus granite-gneiss, and Dais granite [*McKelvey and Webb*, 1962; *Smithson et al.*, 1970]. The rocks younger than the Vida granite consist of the overlying mid-Paleozoic to mid-Mesozoic Beacon sandstone, the intrusive Jurassic Ferrar dolerites, a thin covering of Pleistocene diamictites, and locally, the alkaline McMurdo volcanics of Quaternary age.

Radiometric age determinations in the dry valleys have been mainly by the Rb-Sr method (Table 1). Mineral, as well as whole rock, ages for the Vida granite are generally concordant at around 470 m.y. However, mineral and whole rock ages for the Olympus granite-gneiss range from 235 to 972 m.y., suggesting that its geologic history is more complex.

The samples studied in this investigation were obtained from DVDP drillhole 6, located on the extreme southwest shore of Lake Vida, in Victoria Valley, southern Victoria Land, Antarctica (Figure 1). The nearest outcrops of bedrock are approximately 2 km to the north and 1 km to the south and consist predominantly of Vida granite intruded by Ferrar dolerites. The nearest outcrop of Olympus granite-gneiss is approximately 5 km to the north within a large exposure of Asgard forma-

tion. The drill core, as described by *Kurasawa et al.* [1974], consists, starting from the top, of approximately 10 m of sands and pebbles of Quaternary age, 146 m of strongly foliated biotite-hornblende gneiss, followed by 150 m of massive biotite granite. The foliation within the gneiss becomes progressively weaker as the biotite granite is approached, suggesting contact effects. *Kurasawa et al.* [1974] correlated the biotite granite with the Vida granite and the gneiss with the Olympus granite-gneiss and suggested that the gneiss forms a large raft in the Vida granite. The specimens of gneiss (ANGN-1) and granite (ANGR-1) used in this study came, respectively, from the zones of the biotite-hornblende gneiss and the deeper, coarse-grained biotite granite and were also used in the study of the fission track ages reported by *Stuckless* [1975] and *Vocke et al.* [1978].

ANALYTICAL DATA

Samples of the granite and the gneiss were processed through heavy liquids and a magnetic separator, with visual removal of foreign fragments, to concentrate the zircons. The zircons were sieved to give a range in size fractions, with individual analyzed fractions ranging from 1 to 14 mg. Two aliquots of zircons from the granite were analyzed. ANGR-1C was hand picked to consist of clear euhedral zircons, whereas ANGR-1D consisted of a mixed population of clear and cloudy zircon grains.

The chemical procedure of *Krogh* [1973] was used for the digestion of the zircons and the separation of the U and Pb fractions. A mixed, isotopically enriched ^{208}Pb and ^{235}U spike was used for the determination of abundances of U and Pb by isotope dilution analysis [*Grauert et al.*, 1973]. The Pb isotope ratios were measured using the single filament, silica gel-phosphoric acid method of *Cameron et al.* [1969]. The U isotope ratios were measured using the triple-filament method of *Garner et al.* [1971]. The total Pb blank was less than 0.3 ng, while the U blank was much less than 1 ng. Precision for ^{207}Pb/^{206}Pb and ^{208}Pb/^{206}Pb ratios was about 0.1%, while that for ^{206}Pb/^{204}Pb ratios greater than 1000 was about 0.3%. Mass fractionation corrections for Pb and U were determined from repeated runs of National Bureau of Standards equal atom Pb standard and U standard U-500, and applied to the data.

The whole rock sample (ANGR-1) was prepared by crushing the fresh rock by hand and splitting it

TABLE 1a. Summary of Radiometric Age Determinations on the Vida Granite and the Olympus Granite-Gneiss

	Rb-Sr		U-Pb Zircon Discordia		Fission Track	
	87/86 Sr$_i$	Age	Lower Intercept	Upper Intercept	Zircon	Apatite
		Vida Granite				
Deutsch and Webb [1964]	0.709*	B 472 ± 15				
		F 462 ± 80				
Faure and Jones [1974]	0.7104 ± 0.0008	WRI 471 ± 44				
Stuckless and Erickson [1975]	0.7098 ± 0.00075	WRI 475 ± 14				
Vocke et al. [1978]					500 ± 83	48 ± 12
R. Vocke and G. Hanson (this study)			0	447 ± 34		
		Olympus Granite-Gneiss				
Deutsch and Webb [1964]	0.709*	B 474 ± 15				
		F 916 ± 80				
		WR 972 ± 80				
Deutsch and Grögler [1966]			200 ± 193	638 ± 92		
Faure and Jones [1974]	0.7109 ± 0.0007	WRI 485 ± 43				
Stuckless and Erickson [1975]	0.7082 ± 0.00272	B-WR 235 ± 31				
Vocke et al. [1978]					180 ± 24	123 ± 16
R. Vocke and G. Hanson (this study)			462 ± 6	2554 ± 330		

All ages are in millions of years and were recalculated with the decay constants for ^{238}U, ^{235}U, and ^{87}Rb of *Steiger and Jäger* [1977]. B, biotite; F, feldspar; WR, whole rock; WRI, whole rock isochron.
*Assumed initial for model age calculation.

TABLE 1b. Undifferentiated Basement Metasediments and Gneisses

	K-Ar Age,* m.y.	Rock Type and Locality
Goldich et al. [1958]	B 520	Paragneiss in Wright Valley
Angino et al. [1962]	B 500	Marble from Marble Point
	B 425	Gneiss from Gneiss Point

*Ages were recalculated with the decay constants for ^{238}U, ^{235}U, and ^{87}Rb of *Steiger and Jäger* [1977]; B, biotite.

to approximately 30 g. This was then ground to −200 mesh size. The rare earth elements (REE) and Ba concentrations were determined by isotope dilution on a 0.3-g aliquot following the procedure of J. G. Arth given by *Hanson* [1976]. The Ba determination was carried out directly on the decomposed sample. Analytical precision for the REE is about 1% except for Gd and La for which it is better than 3% and 5%, respectively. Ba precision is about 1% by comparison with repeated determinations of BCR-1. All trace element and isotopic analyses were carried out on a fully automated 12-in. radius, 90° sector, NBS design mass spectrometer. The major element chemistry was determined on an aliquot of the whole rock powder by atomic absorption by N. Suhr, Pennsylvania State University.

The analytical results are presented in Tables 2 and 3, and the U-Pb ages for the zircons are plotted on a concordia diagram in Figure 2 (the uncertainties are 2 σ). The gneiss zircon U-Pb data (ANGN-1) are very discordant, approaching a

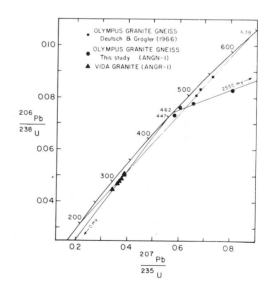

Fig. 2. ^{206}Pb/^{238}U versus ^{207}Pb/^{235}U concordia diagram showing the data points for different zircon fractions from the Vida granite and the Olympus granite-gneiss. Also shown are the zircon data of *Deutsch and Grögler* [1966].

TABLE 2.   U-Pb Analytical Data

Sample No.	Sieve Fraction (Mesh)	U Content, ppm	Pb Content, ppm	Atomic Ratios Observed			Atomic Ratios Calculated			Apparent Ages, m.y.		
				$^{206}Pb/^{204}Pb$	$^{207}Pb/^{206}Pb$	$^{208}Pb/^{206}Pb$	$^{206}Pb/^{238}U$	$^{207}Pb/^{235}U$	$^{207}Pb/^{206}Pb$	$^{206}Pb/^{238}U$	$^{207}Pb/^{235}U$	$^{207}Pb/^{206}Pb$
Vida Granite												
ANGR-1C	80/140	1119	60.11	1582	0.06490	0.1526	0.05062	0.3864	0.05536	318	332	419
	140/200	976.4	52.57	1564	0.06553	0.1650	0.05023	0.3870	0.05589	316	332	441
	270/325	1037	50.97	874.1	0.07297	0.1793	0.04458	0.3425	0.05572	281	299	433
ANGR-1D	80/140	1681	85.28	1468	0.06630	0.1557	0.04763	0.3680	0.05603	300	318	446
	140/200	1257	65.45	1452	0.06638	0.1605	0.04868	0.3758	0.05599	306	324	444
	270/325	1150	58.35	1163	0.06896	0.1696	0.04681	0.3614	0.05600	295	313	445
	<325	1176	61.32	1306	0.06773	0.1714	0.04818	0.3733	0.05619	303	322	452
ANGR-1AP				19.76	0.7905	2.004						
Olympus Granite Gneiss												
ANGN-1	>80	421.8	46.87	229.1	0.1350	0.2892	0.08282	0.8086	0.07081	513	602	947
	80/140	359.5	28.41	4089	0.06465	0.1098	0.07789	0.6552	0.06101	484	512	633
	140/200	287.8	36.58	114.9	0.1876	0.4160	0.07629	0.6031	0.05734	474	479	497
	275/325	366.4	26.96	4980	0.06059	0.1055	0.07310	0.5803	0.05758	455	465	507

All ANGR and ANGN data are for zircon analyses, while ANGR-1AP data are for apatite analyses. The isotopic composition of common Pb used for the corrections was $^{206}Pb/^{204}Pb$: 19.23; $^{207}Pb/^{204}Pb$: 16.15; $^{208}Pb/^{204}Pb$: 40.00. $\lambda_{238U} = 1.55125 \times 10^{-10}/y$, $\lambda_{235U} = 9.8485 \times 10^{-10}/y$.

TABLE 3.   The Concentration of the Major and Minor Elements (in Weight %), Sr, Rb, Ba, and REE Values (in ppm) for the Vida Granite (ANGR-1) Listed Together With the Chondrite Normalizing Values for Ba and the REE

Element	Abundance	Normalizing Value [*Sun and Hanson*, 1975]
SiO_2	74.1	
TiO_2	0.17	
Al_2O_3	14.10	
FeO	1.77	
MgO	0.20	
MnO	0.04	
CaO	1.37	
Na_2O	3.42	
K_2O	4.74	
Σ	99.91	
Sr	149	
Rb	207	
Ba	794	3.51
La	31.0	0.315
Ce	59.8	0.813
Nd	21.7	0.597
Sm	3.72	0.192
Eu	0.747	0.0722
Gd	2.88	0.259
Dy	2.37	0.325
Er	1.27	0.213
Yb	1.22	0.200

The Rb and Sr values are from *Stuckless and Erickson* [1975].

lower concordia intersection of 462 ± 6 m.y. with a projected upper intersection for a least-squares fit chord of 2555 ± 330 m.y. and show a relatively large scatter about the chord. The $^{207}Pb/^{206}Pb$ age for the most discordant zircon (ANGN-1 > 80) is 947 m.y., which suggests that a component of the zircons is at least Proterozoic in age. The U content of the gneiss zircons is low, averaging 360 ppm.

The six zircon fractions for the granite also give quite discordant ages with only a small spread in the data, leading to a very large uncertainty in the upper and lower concordia intercept ages. A least-squares fitted line through the data has a negative intercept for $^{206}Pb/^{238}U$ of −0.000979 and an upper intersection with the concordia at 433 m.y. If the lower intercept is constrained to 0 m.y., the chord through the data gives an upper intercept of 447 ± 34 m.y., consistent with the granite being the Vida granite which intruded during the Ross orogeny. The U content of the zircons is high and relatively constant at approximately 1200 ppm. An apatite (ANGR-1AP) from the granite was also analyzed; however, it has essentially common lead (Table 2).

In an attempt to understand some of the factors controlling zircon paragenesis and possible causes of zircon discordance, cathodoluminescence employing a Nuclide Luminoscope attached to a Leitz Microscope has been used to reveal the internal morphology of zircon crystals. Cathodoluminescence is electron excited optical fluorescence in which activator ions such as REE in zircon emit visible light under cold electron bombardment. The internal structure is revealed by contrasts in the luminescing and nonluminescing portions of the grain which are a result of variations in the concentration of the activator ions. Based on microprobe analyses for U, the luminescing portion generally correlates with regions of high U abundance.

The zircons from the gneiss (ANGN-1) show two different growth phases (Figure 3a).

1. A core, which makes up 80% of the zircon, consisting of fine alternating luminescent and nonluminescent bands, subhedral to anhedral in shape. Rarely, a nonfluorescent zone is seen at the center.

2. A strongly fluorescent rim which occasionally shows internal banding. The rim zone often truncates the banding in the core.

These morphologies imply a core with a variable U concentration that may have undergone some modification in shape due to high-grade metamorphism or mechanical abrasion. This core is then surrounded by a rim of slightly higher U abundance that probably formed in situ.

The granite zircons have three distinct zones (Figure 3b).

1. A low luminescent core that is anhedral in shape with no discernible internal banding.

2. A large, strongly luminescent zone surrounding the core that preserves only occasional banding and is subhedral to anhedral in shape.

3. A finely and multiply banded euhedral rim, consisting of alternating luminescent and nonluminescent zones. These bands often show evidence of chemical resorption and discontinuous growth.

Generally, a given zircon does not have all three zones. However, the final zone is almost always present, and makes up less than 20% of the zircon. As with the gneiss zircons, the luminescing regions are generally correlated with regions of higher U abundance. While the overall high U content of the zircon (1200 ppm) suggests that it formed in situ [Köppel and Sommerauer, 1974], the cathodoluminescence observations suggest that this high U content may be an overprint of U-rich rims on U-poor inherited zircon cores.

OLYMPUS GRANITE-GNEISS

The gneiss analyzed in this study was classified as a granodioritic biotite gneiss by Kurasawa et al. [1974] and correlated on the basis of color index with the Olympus granite-gneiss. Locally, the gneiss from the drillcore contains areas rich in hornblende. The radiometric ages determined on the Olympus granite-gneiss, listed in Table 1, range from greater than 950 m.y. to 123 m.y., implying an involved history.

The low U abundance (about 360 ppm) and scatter about the chord for zircons from the gneiss dated in this study (Figure 2) are characteristic of zircon populations with a significant inherited detrital component [Köppel and Sommerauer, 1974]. The scatter may also reflect partial Pb loss since the Ross orogeny. Considering these data and the cathodoluminescence information, if the zircons are detrital, the upper intercept age of 2555 m.y. may date their provenance. However, if the zircon cores were formed in situ during a high-grade metamorphism, the older age may date an event that affected the gneiss prior to 950 m.y. ago, perhaps even in the Archean. The lower intersection of the chord with the concordia at 462 m.y. may represent the growth of the rim or partial Pb loss in the older core during the regional metamorphism associated with the Ross orogeny or the contact metamorphism accompanying the intrusion of the Vida granite [see, for example, Davis et al., 1968; Anderson, 1965].

The previously determined 638 m.y. zircon U-Pb age [Deutsch and Grögler, 1966] on the Olympus granite-gneiss, also plotted in Figure 2, is in apparent conflict with the ages from this study. However, the high U concentration (1060 ppm) together with the collinearity and near concordance of these zircons suggest they have undergone a significantly different history, probably reflecting in situ formation around 640 m.y. ago. Due to the superficial similarity of orthogneiss and paragneiss after tectonic reworking, this age may date the emplacement of a younger intrusive phase into the Olympus granite-gneiss. The other possibility is that the gneiss in DVDP drillhole 6 may not be the Olympus granite-gneiss but rather granitized Asgard formation.

The 480 m.y. Rb-Sr whole rock isochron age suggests that either a component was added to the gneiss at the time of the Ross orogeny or that Sr in the Olympus granite-gneiss was regionally homog-

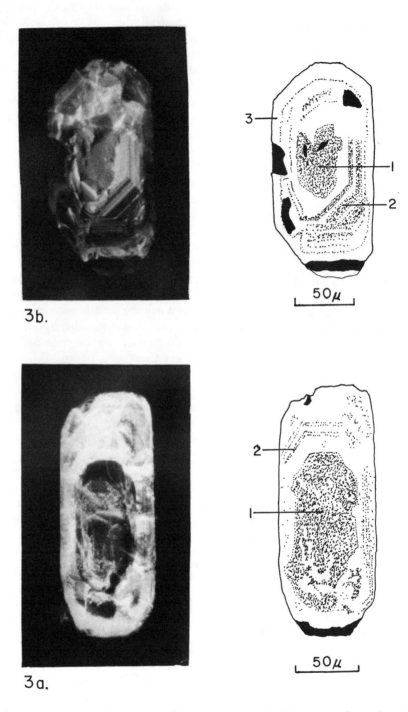

Fig. 3. Cathodoluminescent photographs of zircon grains from the Olympus granite-gneiss and the Vida granite.
(a) Olympus granite-gneiss zircon showing (1) the finely banded, subhedral core rimmed by a (2) generally unbanded
but strongly luminescing zone. (b) Vida granite zircon showing (1) low luminescent anhedral core, (2) strongly luminescing zone with occasional internal banding, and (3) a finely and multiply banded euhedral rim.

enized at the time of the Ross orogeny. The 480 m.y. biotite age suggests that some parts of the gneiss underwent little subsequent thermal history. While the 200 m.y. age from zircon fission tracks and a biotite-whole rock Rb-Sr determination has been interpreted as resulting from a localized thermal event [*Stuckless*, 1975], an equally plausible explanation would interpret this as an uplift age when the gneiss rose to depths of less than 15 km and rock temperatures became less than 350°C. The apatite fission track age of 123 m.y. could then imply subsequent uplift to depths of 2-3 km by that time. It is evident then that, based on the available radiometric data, the Olympus granite-gneiss has had an involved history and may have components whose ages are widely varying. The gneiss may include a component as old as Archean but at least 950 m.y. old, a component about 640 m.y. old, and possibly components added during the Ross orogeny. In any case, the Ross orogeny severely affected the gneiss and any events since the Ross orogeny would appear to be minor and of local extent.

VIDA GRANITE

The Vida granite is quartz monzonitic in composition with normative Ab/An = 4.5 and less than 2.5% ferromagnesian component. The high aluminum/alkali and low Na_2O/K_2O ratios and the presence of biotite as the main ferromagnesian phase give the Vida granite the characteristics of S-type granites which *Chappel and White* [1974] suggest are derived by partial melting of sedimentary rocks. The high initial Sr ratio for the granite of 0.710 (Table 1) implies a parent that had a relatively high Rb/Sr ratio for a significant period of time. The chondrite normalized pattern for Ba and the REE is shown in Figure 4. While the light rare earth elements show a progressive linear increase in abundance with decreasing atomic number from 11 times chondrites for Gd to 98 times chondrites for La, the heavy rare earth elements show a concave upward pattern with Yb at 5.8 times chondrites. There is a significant negative Eu anomaly (Eu/Eu* = 0.703) and a marked Ba enrichment. This pattern is characteristic of granitic rocks of this composition and is similar to those reported by *Arth and Hanson* [1975] and *Buma et al.* [1971] which were interpreted to be the result of partial melting of greywacke at crustal depths. Possible parents for the Vida granite could be similar to metamorphosed greywacke present in the Asgard

formation or Olympus granite-gneiss [*McKelvey and Webb*, 1962].

As shown by *Nance and Taylor* [1976], the shape of the REE patterns for Proterozoic and younger greywackes, the REE abundances, as well as the major element chemistry [*Pettijohn*, 1975] vary only slightly (Figure 4). The mineralogical composition of possible residues was calculated for different degrees of melting using both major and trace elements, the greywacke representing the parent and the Vida granite, the derived melt. The equations and approaches used for this type of modeling are given by *Hanson* [1978, 1980]. The most consistent fit of the data is for 30% melting leaving a residue of quartz + plagioclase + amphibole + biotite at upper amphibolite grade of metamorphism (Figure 4). While melting of greywacke leaving an amphibolite grade residue gives the best fit, melting producing a granulite grade residue cannot be eliminated [*Hanson*, 1980].

Previous radiometric age determinations using Rb-Sr mineral model ages and whole rock isochrons and zircon fission track ages (Table 1) are all concordant, suggesting that the Vida granite has not undergone any major disturbance since its emplacement at around 470 m.y. Because fission track ages for zircon are generally interpreted as giving the time at which the temperature became less than 350°C [*Nishida and Takashima*, 1975], it

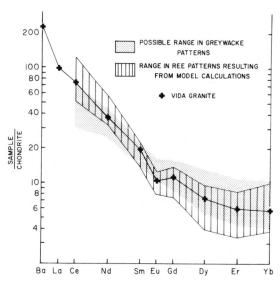

Fig. 4.   Chondrite normalized rare earth element plot of the Vida granite compared with the possible ranges in pattern for the greywacke parent and the melt resulting from 30% melting leaving a residue of quartz + plagioclase + biotite + amphibole.

is likely that this portion of the Vida granite has not been heated for any significant time above 350°C since it was emplaced.

In view of these earlier data, the very strong discordance of the U-Pb ages for the zircons determined in this study (Figure 2) is quite surprising because it suggests a major, recent disruption of the U-Pb systematics. The cause of the zircon discordance is problematic. While the zircon internal morphology (Figure 3b) suggests a multistage history for these minerals, the general correlation of higher U abundance with the rim growths implies that the 450 m.y. age represents in situ zircon growth over inherited cores. *Stuckless* [1975] suggested that the 50 m.y. apatite fission track age for the granite represents the time when it was displaced upward along the vertical shear zone observed in DVDP drillhole 6. This possible uplift may have caused zircon discordance due to a dilatancy effect [*Goldich and Mudrey*, 1972]. Using 50 m.y. as the lower intercept of a chord fitted through the data yields an upper concordia intercept of 477 ± 6 m.y., in agreement with the previously determined whole rock Rb-Sr isochron ages of 471 and 475 m.y. for the Vida granite.

Acknowledgments. The assistance of W. J. Holzwarth in the microprobe study is gratefully acknowledged. Financial support for this study was provided by a grant from the Office of Polar Programs NSF GV-040762 and National Science Foundation grant EAR76-13354 (Geochemistry).

REFERENCES

Anderson, D. H., Uranium-thorium-lead ages of zircons and model lead ages of feldspars from the Saganaga, Snowbank and Grants Range granites of northeastern Minnesota, Ph.D. thesis, 128 pp., Univ. of Minn., Minneapolis, 1965.

Angino, E. E., M. D. Turner, and D. J. Zeller, Reconnaissance geology of lower Taylor Valley, Victoria Land, Antarctica, *Geol. Soc. Amer. Bull.*, *73*, 1553-1562, 1962.

Arth, J. G., and G. N. Hanson, Geochemistry and origin of the early Precambrian crust of northeastern Minnesota, *Geochim. Cosmochim. Acta*, *39*, 325-362, 1975.

Buma, G., F. A. Frey, and D. R. Wones, New England granites: Trace element evidence regarding their origin and differentiation, *Contrib. Mineral. Petrol.*, *31*, 300-320, 1971.

Cameron, A. E., D. H. Smith, and R. L. Walker, Mass spectrometry of nanogram-size samples of lead, *Anal. Chem.*, *41*, 525-526, 1969.

Chappell, B. W., and A. J. R. White, Two contrasting granite types, *Pacific Geol.*, *8*, 173-174, 1974.

Davis, G. L., S. R. Hart, and G. R. Tilton, Some effects of contact metamorphism on zircon ages, *Earth Planet. Sci. Lett.*, *5*, 27-34, 1968.

Deutsch, S., and N. Grögler, Isotopic age of Olympus Granitegneiss (Victoria Land-Antarctica), *Earth Planet. Sci. Lett.*, *1*, 82-84, 1966.

Deutsch, S., and P. N. Webb, Sr/Rb dating on basement rocks from Victoria Land: Evidence for a 1000 million year old event, in *Antarctica Geology*, edited by R. J. Adie, pp. 557-569, John Wiley, New York, 1964.

Faure, G., and L. M. Jones, Isotopic composition of strontium and geologic history of the basement rocks of Wright Valley, southern Victoria Land, Antarctica, *N.Z. J. Geol. Geophys.*, *17*, 611-627, 1974.

Garner, E. L., L. A. Machlan, and W. R. Shields, Standard reference materials: Uranium isotopic standard reference materials, *Nat. Bur. Stand. Spec. Publ.*, *260-27*, 150 pp., 1971.

Goldich, S. S., and M. G. Mudrey, Dilatancy model for explaining uranium-lead zircon ages, in *Ocherki Sovremennoy Geokhimii i Analithicheskoy Khimmi*, pp. 415-418, Izd-Vo Nanka, Moscow, 1972.

Goldich, S. S., A. O. Nier, and A. L. Washburn, Ar^{40}/K^{40} age of gneiss from McMurdo Sound, Antarctica, *Eos Trans. AGU*, *39*, 956-958, 1958.

Grauert, B., R. Hänny, and G. Soptrajanova, Age and origin of detrital zircons from the prePermian basements of the Bohemian Massif and the Alps, *Contrib. Mineral. Petrol.*, *40*, 105-130, 1973.

Grew, E. S., Precambrian basement at Molodezhnaya Station, East Antarctica, *Geol. Soc. Amer. Bull.*, *89*, 801-813, 1978.

Hanson, G. N., Rare earth element analyses by isotope dilution, *Nat. Bur. Stand. Spec. Publ.*, *422*, 937-949, 1976.

Hanson, G. N., The application of trace elements to the petrogenesis of igneous rocks of granitic composition, *Earth Planet. Sci. Lett.*, *38*, 26-43, 1978.

Hanson, G. N., Rare earth elements in petrogenetic studies of igneous systems, *Annu. Rev. Earth Planet. Sci.*, *8*, 372-406, 1980.

Köppel, V., and J. Sommerauer, Trace elements and the behavior of the U-Pb system in inherited and newly formed zircons, *Contrib. Mineral. Petrol.*, *43*, 71-82, 1974.

Krogh, T. E., A low-contamination method for hydrothermal decomposition of zircon and extraction of U and Pb for isotopic age determinations, *Geochim. Cosmochim. Acta*, *37*, 485-494, 1973.

Kurasawa, H., Y. Yoshida, and M. G. Mudrey, Jr., Geologic log of the Lake Vida core-DVDP6, *Dry Val. Drilling Proj. Bull.*, *3*, 94-108, 1974.

McKelvey, B. C., and P. N. Webb, Geological investigation in southern Victoria Land, Antarctica, 3, Geology of Wright Valley, *N.Z. J. Geol. Geophys.*, *5*, 143-162, 1962.

Nance, W. B., and S. R. Taylor, Rare earth element patterns and crustal evolution, I, Australian post-Archean sedimentary rocks, *Geochim. Cosmochim. Acta*, *40*, 1539-1551, 1976.

Nishida, T., and Y. Takashima, Annealing of fission tracks in zircons, *Earth Planet. Sci. Lett.*, *27*, 257-264, 1975.

Pettijohn, F. J., *Sedimentary Rocks*, 3rd ed., 628 pp., Harper and Row, New York, 1975.

Smithson, S. B., P. R. Fikkan, and D. J. Toogood, Early geologic events in the ice-free valleys, Antarctica, *Geol. Soc. Amer. Bull.*, *81*, 207-210, 1970.

Steiger, R. H., and E. Jäger, Subcommission on geochronology, Convention on the use of decay constants in Geo- and Cosmochronology, *Earth Planet. Sci. Lett.*, *36*, 359-362, 1977.

Stuckless, J. S., Geochronology of core samples recovered from DVDP6, Lake Vida, Antarctica, *Dry Val. Drilling Proj. Bull.*, *6*, 27, 1975.

Stuckless, J. S., and R. L. Erickson, Rb-Sr ages of basement rocks recovered from borehole DVDP6, southern Victoria Land, Anarctica, *Antarct. J. U.S.*, *10*, 302-307, 1975.

Sun, S. S., and G. N. Hanson, Origin of Ross Island basanitoids and limitations upon the heterogeneity of mantle sources for alkali basalts and nephelinites, *Contrib. Mineral. Petrol.*, *52*, 77-106, 1975.

Vocke, R. D., Jr., G. N. Hanson, and J. S. Stuckless, Ages for the Vida granite and Olympus granite-gneiss, Victoria Valley, southern Victoria Land, *Antarct. J. U.S.*, *13*, 15-17, 1978.

Webb, P. N., and B. C. McKelvey, Geological investigations in southern Victoria Land, Antarctica, I, Geology of Victoria dry valley, *N.Z. J. Geol. Geophys.*, *2*, 120-136, 1959.

A Q-MODE FACTOR MODEL FOR THE PETROGENESIS OF THE VOLCANIC ROCKS FROM ROSS ISLAND AND VICINITY, ANTARCTICA

J. S. STUCKLESS, A. T. MIESCH, AND S. S. GOLDICH

U. S. Geological Survey, Denver Federal Center, Denver, Colorado 80225

P. W. WEIBLEN

Department of Geology and Geophysics, University of Minnesota
Minneapolis, Minnesota 55455

Major and minor elemental data examined by extended Q-mode factor analysis show that the petrogenetic evolution of the volcanic rocks from Ross Island and vicinity occurred in distinct stages that were complex in detail. The chemical analyses for 24 oxides in 49 samples can be modeled closely by a three-dimensional system that consists of a starting liquid and two solidus assemblages. This fact suggests that all of the samples are related to a common parent magma. However, the three-dimensional system shows a large compositional gap between the basanitoids and trachybasalts. Furthermore, attempts to match possible solidus compositions to actual minerals failed to yield assemblages similar to those observed in the samples. Modeling was therefore attempted by dividing the samples into two related groups: (1) basanitoids and (2) trachybasalts and phonolites. The basanitoid sequence can be accounted for by separation of two similar solidus assemblages, which are dominated by titaniferous clinopyroxene with lesser amounts of olivine, iron-titanium oxide, and spinel or ilmenite. These assemblages approximate those reported for experimentally determined, high-pressure solidus assemblages of basanite magma. They also approximate the assemblages observed in the basanitoid samples. The inferred depth of crystallization of 60 to 90 km is consistent with the abundance of mantle-derived xenoliths found in the basanitoids. The basanitoid model indicates that the thick flow from drill hole DVDP 2 had xenolithic olivine and clinopyroxene added, which agrees with inferences drawn from both mineralogic and isotopic data. The model for the trachybasalts and phonolites suggests that they were derived from the basanitoids but that differentiation occurred at lower pressures. The postulated solidus assemblages of plagioclase, titaniferous clinopyroxene, kaersutite iron-titanium oxide, and apatite for the trachybasalts are consistent with crystallization under conditions of lower-crustal pressure. Reaction of the magma with wall rocks similar to the inclusion suite found with the trachybasalts would account for the more radiogenic lead in the trachybasalts. The dominance of anorthoclase in the calculated solidus for the phonolites and the absence of all but upper-crustal xenoliths suggest that these rocks were differentiated at very low pressure.

INTRODUCTION

The volcanic rocks of Ross Island and vicinity lie near the midpoint of a discontinuous, 2000-km-long line of continental, intraplate alkalic extrusives which extends northward from Mount Weaver to the Balleny Islands [*Goldich et al.*, 1975]. The volcanic lineament, which is roughly parallel to the trend of the Transantarctic Mountains, was referred to by *Hamilton* [1972] as the Ross Sea petrologic province. Major subdivisions of the petrologic province are the McMurdo, Hallett, and Balleny volcanic provinces.

Hamilton described the exposed volcanic rocks in the McMurdo volcanic province as dominated by subaerial flows, and those in the Hallett province as mostly formed by subglacial eruptions. Drilling as part of the Dry Valley Drilling Project (DVDP) has demonstrated that Ross Island is built on a large platform of subaqueous or subglacial hyaloclastite [*Treves*, 1978].

Throughout the Ross Sea petrologic province, and at Mount Weaver, the volcanic sequences are dominated by alkali olivine basalts [*Treves*, 1967; *Cole and Ewart*, 1968; *Hamilton*, 1972], which, for the most part, can be classified as basanitoids because they contain more than 5% normative nepheline but no modal nepheline [*MacDonald and Katsura*, 1964]. The dominance of basanitoid in the McMurdo volcanic province is exemplified by the materials recovered from DVDP 3 which are basaltic from a depth of 64 to 381 m. The lowermost 214 m are basaltic hyaloclastite [*Treves*, 1978], and this unit presumably underlies the entire island with thicknesses in excess of 300 m.

Within the McMurdo volcanic province, basal basanitoids are overlain by a series of rocks of intermediate composition (nepheline hawaiites, nepheline mugearites, and nepheline benmoreites of *Kyle and Rankin* [1976]). These are collectively referred to as trachybasalts in this paper. Subequal in abundance and stratigraphically superior to the trachybasalts are phonolites. Many of the phonolites contain phenocrysts of anorthoclase (as much as 30% by volume [*Goldich et al.*, 1975]), but phonolites that lack anorthoclase are common.

As a result of the Dry Valley Drilling Project, several recent papers have included petrogenetic models as explanations for the sequence of volcanic rocks from the Ross Island area [*Stuckless et al.*, 1974; *Goldich et al.*, 1975; *Kurasawa*, 1975; *Sun and Hanson*, 1975, 1976; *Stuckless and Ericksen*, 1976; *Kyle and Rankin*, 1976]. Most of these models propose a fairly simple evolution starting with derivation of basanitoid deep within the mantle and differentiation by crystal fractionation along one or two lines of liquid descent to yield trachybasalts and phonolites. *Sun and Hanson* [1975, 1976] concluded from rare earth and lead isotopic data that several unrelated basanitoid magmas were generated from a heterogeneous mantle. *Kurasawa* [1975] and *Stuckless and Ericksen* [1976] concluded from strontium isotopic data that some of the late differentiates were contaminated with crustal materials.

In this paper we examine major and minor elemental data for volcanic rocks, xenoliths, and minerals using an extended form of Q-mode factor analysis [*Miesch*, 1976a, b]. Isotopic, petrographic, and field data are used as constraints and tests for the model. The advantage of Q-mode factor analysis is that it allows quantitative testing of petrologic models. Furthermore, the testing of several variables can be done simultaneously.

SAMPLES AND ANALYTICAL METHODS

Samples

Samples 1-29 are all surface samples reported by *Goldich et al.* [1975]. Of the 30 new chemical analyses (samples 30-59), 8 are basanitoids, of which 6 are cores from DVDP 2, and 12 are trachybasalts, of which 6 are cores from DVDP 2. The remaining 8 samples out of this group and the 10 samples of ultramafic inclusions are surface samples (see appendix). The inclusions range in composition from dunite to plagioclase-pyroxene granulites.

Chemical Procedure

The chemical procedures are essentially the same as those used in the earlier study by *Goldich et al.* [1975]. All determinations of FeO, H_2O, and CO_2 were made at Northern Illinois University (NIU) by conventional methods. Four basanitoid and four trachybasalt samples were analyzed (NIU) by conventional methods in which Na_2O and K_2O were determined by flame spectrophotometry. Alkalis in the inclusions were determined similarly (NIU). The major constituents in 12 samples of basanitoids and trachybasalts and in the 10 samples of inclusions were determined in the Mineral Constitution Laboratory at the Pennsylvania State University, except for P_2O_5, which was determined by colorimetry, and F, which was determined by selective ion electrode [*Bodkin*, 1977]. The trace elements were determined by a variety of methods, including isotope dilution and mass spectrometry, X ray fluorescence, atomic absorption, and emission spectrography. The trace elements are considered in greater detail in the accompanying paper by Goldich et al.

The CIPW normative mineral compositions in Table 1 were computed by using the program GNAP [Bowen, 1971].

Computer Techniques

Q-mode factor analysis is a natural extension of the commonly used petrologic methods that graphically display the variations of one element as a function of another (e.g., a plot of SiO_2 versus MgO). Graphical examinations of the simultaneous variations of several elements (e.g., an AMF plot must implicitly assume a known relationship between certain elements (e.g., in the AMF diagram Na_2O and K_2O are added on the implicit assumption that they behave in an identical fashion). Q-mode analysis replaces assumed relationships with mathematically determined, best fit relationships. The method allows simultaneous examination of a large number of variables that are related to each other in a large number of samples.

The principal method used in analysis of the chemical data was an extended form of Q-factor analysis as described by Miesch [1976a, b]. The method requires that all chemical abundance data for each sample sum precisely to a constant (usually 100%). For this reason, all data (except for F) were expressed as oxide percentages, and each analysis was then adjusted to 100%. In order to give each oxide an equal chance to influence the outcome of the factor analysis the data for each oxide were then scaled to range from zero to one, following the suggestion of Imbrie [1963, p. 29]. Without some kind of scaling, the major oxides, such as SiO_2 and Al_2O_3, would be overwhelmingly dominant over such minor oxides as Rb_2O and BeO. The scaled data were regarded as N vectors in n-dimensional space (where N is the number of samples and n is the number of oxides). As is customary in most Q-mode factor analyses, all the conceptual vectors were then adjusted to unit length by dividing each scaled analysis by the square root of the sum of squares (that is, the data matrix was row-normalized). The advantage of the extended form of Q-mode factor analysis over conventional methods is that some of the factor analysis results, derived from these scaled and row-normalized data, can be converted back to the same percentage of concentration units as the original data and therefore are more easily interpreted

in terms of rock and mineral compositions. This allows selection of reference axes for the vector systems on petrographic rather than purely mathematical grounds, so that the factor solutions can be a great deal more compatible both with the observed mineral compositions of the rocks and with isotopic data and field observations. The factored matrix was equivalent to the major product of the scaled and row-normalized data matrix, although the minor product was actually used in accordance with the shortcut procedure described by Klovan and Imbrie [1971]. The major product is equal to a matrix of coefficients of proportional similarity (cosine theta) among the samples [Imbrie and Purdy, 1962].

The vectors in the n-dimensional vector system were projected into subspaces of 2 to 10 dimensions by using the conventional principal components methods [see Harman, 1967]. After each projection the compositions represented by the vectors in their new positions were determined and compared with the original data for the corresponding samples. The comparisons were generally poor for Fe_2O_3, CO_2, H_2O, and CuO, and it seems evident that the abundances of these constitutents in the 49 samples have been caused in large part by processes other than those that controlled the other oxides. Substantial proportions of the variation in CO_2, H_2O, and the oxidation state of iron may have been caused by secondary processes, and an important part of the variation in CuO may be due to analytical error. For all subsequent computations the data for CO_2, H_2O, and CuO were omitted, and Fe_2O_3 was recomputed and combined with FeO.

The compositions represented by the projected vectors will be referred to as the recomputed data. The means of each set of recomputed data, each set representing a projection of groups of samples into 2 to 10 dimensions, are essentially the same as the corresponding means for the original 49 samples (Table 2). The variances, however, are smaller by an amount that varies with each oxide and with the dimensions of the subspace into which the vectors were projected. The ratios of the variances in the recomputed data to those in the original data are given in Table 3. These ratios are equal to the squares of the correlation coefficients (coefficients of determination) between the original and recomputed data. The ratios, or coefficients of determination, are measures of the proportions of the vari-

DRY VALLEY DRILLING PROJECT

TABLE 1. Chemical Data and Normative Mineralogy for Volcanic Rocks From Ross Island and Vicinity, Antarctica

Oxide	1	2	3	4	5	6	7	8	9	10	11	12	13	14	15
SiO_2	41.30	42.10	42.50	42.70	43.10	43.30	43.47	43.70	43.80	44.00	44.10	45.00	44.50	44.60	45.50
Al_2O_3	12.90	13.00	13.30	12.80	14.30	12.80	12.85	11.90	14.50	14.20	14.60	13.30	16.60	16.90	17.00
Fe_2O_3	4.87	3.37	3.00	3.10	4.89	3.19	3.18	1.89	4.56	3.12	3.57	5.30	6.09	11.80	3.81
FeO	7.39	8.90	10.00	7.96	8.55	9.33	9.67	9.04	7.85	7.35	8.77	6.38	6.01	0.09	8.41
MgO	10.20	11.80	12.10	12.70	8.30	10.90	11.78	12.90	7.75	10.80	8.71	9.50	4.78	3.78	3.74
CaO	11.00	11.50	10.40	12.50	10.00	12.90	10.60	10.00	10.30	10.50	9.07	12.50	9.74	9.44	9.30
Na_2O	3.59	3.70	2.85	2.85	4.31	2.62	2.93	3.52	4.11	4.38	4.45	2.61	1.92	5.37	5.07
K_2O	0.63	0.90	1.17	0.96	1.51	0.93	1.09	1.28	1.78	1.77	1.96	1.03	0.26	2.01	2.13
H_2O	1.91	0.49	0.16	0.17	0.66	0.10	0.07	0.20	0.43	0.30	0.19	0.53	3.91	0.20	0.36
TiO_2	4.08	3.57	3.12	3.27	3.32	3.42	3.28	3.64	3.47	2.83	2.70	3.07	1.53	3.46	3.11
P_2O_5	0.82	0.81	0.68	0.63	0.63	0.55	0.58	0.82	0.98	0.65	0.71	0.54	0.25	1.39	1.56
MnO	0.20	0.19	0.20	0.19	0.21	0.19	0.19	0.19	0.23	0.19	0.22	0.19	0.03	0.22	0.24
CO_2	0.82	0.02	0.03	0.00	0.02	0.03	0.09	0.01	0.03	0.02	0.03	0.08		0.02	0.01
F	0.11	0.11	0.09	0.05	0.10	0.07	0.09	0.11	0.12	0.10	0.10	0.07	0.17	0.13	0.14
BaO	0.0324	0.0424	0.0435	0.0335	0.0458	0.0290	0.0363	0.0435	0.0648	0.0569	0.0564	0.0268	0.0703	0.0703	0.0567
SrO	0.1050	0.1088	0.1203	0.0966	0.1118	0.0925	0.1022	0.1022	0.1691	0.1093	0.1168	0.0769	0.1691	0.1794	0.1549
Rb_2O	0.0007	0.0038	0.0026	0.0028	0.0024	0.0023	0.0030	0.0031	0.0035	0.0055	0.0055	0.0017	0.0046	0.0059	0.0048
Cr_2O_3	0.0672	0.0877	0.0789	0.1271	0.0307	0.0950	0.1111	0.0833	0.0512	0.0628	0.0526	0.0745	0.0003	0.0003	0.0003
NiO	0.0255	0.0216	0.0216	0.0382	0.0140	0.0305	0.0267	0.0369	0.0178	0.0216	0.0216	0.0191	0.0018	0.0004	0.0004
CoO	0.0071	0.0071	0.0074	0.0069	0.0069	0.0064	0.0058	0.0062	0.0061	0.0057	0.0056	0.0057	0.0041	0.0033	0.0028
CuO	0.0051	0.0071	0.0046	0.0070	0.0063	0.0080	0.0071	0.0053	0.0040	0.0055	0.0079	0.0106	0.0043	0.0031	0.0021
ZnO	0.0137	0.0105	0.0105	0.0149	0.0190	0.0098	0.0139	0.0098	0.0141	0.0097	0.0129	0.0157	0.0136	0.0138	0.0143
BeO	0.0007	0.0008	0.0006	0.0007	0.0007	0.0006	0.0006	0.0009	0.0010	0.0008	0.0009	0.0006	0.0010	0.0009	0.0009
V_2O_3	0.0412	0.0427	0.0368	0.0412	0.0353	0.0456	0.0412	0.0382	0.0382	0.0309	0.0279	0.0441	0.0279	0.0177	0.0191
Y_2O_3	0.0039	0.0044	0.0039	0.0032	0.0039	0.0037	0.0039	0.0027	0.0062	0.0039	0.0030	0.0029	0.0047	0.0042	0.0052
La_2O_3	0.0061	0.0066	0.0055	0.0050	0.0063	0.0012	0.0056	0.0069	0.0106	0.0062	0.0082	0.0041	0.0094	0.0101	0.0094
Ce_2O_3	0.0047	0.0047	0.0047	0.0047	0.0047	0.0047	0.0047	0.0047	0.0141	0.0047	0.0047	0.0047	0.0047	0.0117	0.0129
ZrO_2	0.0419	0.0435	0.0261	0.0315	0.0442	0.0283	0.0301	0.0301	0.0605	0.0393	0.0524	0.0278	0.0527	0.0593	0.0551
Total	100.17	100.35	99.96	100.29	100.23	100.68	100.26	99.57	100.37	100.57	99.55	100.41	100.50	99.79	100.71
−O	0.05	0.05	0.04	0.02	0.04	0.03	0.04	0.05	0.05	0.04	0.04	0.03	0.07	0.05	0.06
Total	100.12	100.30	99.92	100.27	100.19	100.65	100.22	99.52	100.32	100.53	99.51	100.38	100.43	99.74	100.65
Normative mineral															
Q	0.00	0.00	0.00	0.00	0.00	0.00	0.00	0.00	0.00	0.00	0.00	0.00	0.00	0.00	0.00
Z	0.06	0.06	0.04	0.05	0.07	0.04	0.04	0.05	0.09	0.06	0.08	0.04	0.08	0.09	0.08
OR	3.80	5.31	6.95	5.68	8.99	5.48	6.44	7.63	10.56	10.46	11.69	6.11	11.36	11.97	12.58
AB	16.76	5.31	7.92	3.44	10.99	5.99	10.85	10.06	12.95	4.97	10.80	15.86	23.24	29.04	20.60
AN	17.57	16.19	20.13	19.32	15.32	20.34	18.70	13.00	15.92	13.86	14.20	21.60	20.12	16.19	17.33
LC	0.00	0.00	0.00	0.00	0.00	0.00	0.00	0.00	0.00	0.00	0.00	0.00	0.00	0.00	0.00
NE	7.71	14.07	8.84	11.21	13.94	8.72	7.56	10.83	11.91	17.38	14.73	3.41	7.37	9.07	12.07
AC	0.00	0.00	0.00	0.00	0.00	0.00	0.00	0.00	0.00	0.00	0.00	0.00	0.00	0.00	0.00
NS	0.00	0.00	0.00	0.00	0.00	0.00	0.00	0.00	0.00	0.00	0.00	0.00	0.00	0.00	0.00
WO	11.33	14.70	11.23	16.16	12.58	16.51	12.25	13.10	11.96	14.05	10.92	15.22	7.49	4.54	7.78
EN	8.94	10.53	7.59	11.96	8.76	11.35	8.40	9.27	8.54	10.24	7.16	12.14	6.47	3.93	4.15
FS	1.12	2.86	2.79	2.63	2.78	3.84	2.87	2.70	2.36	2.51	3.00	1.33	0.01	0.00	3.39
FO	11.90	13.20	15.90	13.81	8.45	11.00	14.68	16.22	7.60	11.67	10.32	8.12	3.82	3.90	3.62

	16	17	18	19	20	21	22	23	24	25	26	27	28	29	30
FA	1.64	3.95	6.44	3.35	2.96	4.10	5.52	5.22	2.31	3.15	4.77	0.98	0.01	0.00	3.26
MT	7.20	4.88	4.37	4.50	7.14	4.61	4.61	2.76	6.64	4.52	5.22	7.71	8.84	0.00	5.52
CM	0.10	0.13	0.12	0.19	0.05	0.14	0.16	0.12	0.08	0.09	0.08	0.11	0.00	0.00	0.00
HM	0.00	0.00	0.00	0.00	0.00	0.00	0.00	0.00	0.00	0.00	0.00	0.00	0.00	11.89	0.00
IL	7.91	6.77	5.95	6.22	6.35	6.47	6.23	6.97	6.62	5.37	5.17	5.85	7.43	0.67	5.90
PF	0.00	0.00	0.00	0.00	0.00	0.00	0.00	0.00	0.00	0.00	0.00	0.00	0.00	5.34	0.00
AP	1.93	1.92	1.62	1.49	1.50	1.30	1.37	1.96	2.33	1.54	1.70	1.28	3.63	3.32	3.69
FR	0.03	0.08	0.06	0.00	0.09	0.04	0.08	0.08	0.07	0.09	0.08	0.04	0.07	0.01	0.00
CC	1.90	0.05	0.07	0.00	0.05	0.07	0.20	0.02	0.07	0.05	0.07	0.18	0.07	0.05	0.02
Oxide															
SiO_2	49.50	50.00	50.36	55.10	55.71	56.10	56.20	54.00	55.00	55.90	56.80	57.30	57.90	58.10	41.41
Al_2O_3	17.20	18.70	19.59	19.50	18.14	19.57	19.90	19.80	19.60	20.50	21.70	21.60	19.60	15.90	13.04
Fe_2O_3	2.98	2.85	2.82	1.91	1.83	1.78	1.91	2.51	3.79	2.65	2.80	1.70	5.16	2.72	2.65
FeO	6.38	5.42	5.16	4.12	4.90	3.46	3.59	3.83	2.04	1.58	0.64	0.80	0.58	5.06	9.07
MgO	2.95	2.58	2.14	1.13	1.37	1.19	1.16	1.39	1.17	0.87	0.07	0.29	0.44	0.90	12.01
CaO	7.95	6.37	6.46	3.30	3.35	3.24	3.32	4.33	3.65	2.98	0.90	1.26	2.17	2.40	11.33
Na_2O	5.02	6.35	6.57	8.03	7.92	7.80	7.89	7.18	7.76	9.04	10.95	10.00	8.18	5.28	3.15
K_2O	2.57	3.43	3.32	4.68	4.35	4.26	4.24	3.85	4.45	4.26	5.56	5.00	4.97	4.98	1.48
H_2O	0.87	0.36	0.25	0.39	0.11	0.44	0.19	0.59	0.81	0.39	0.20	0.80	0.04	2.19	0.14
TiO_2	2.64	2.36	1.93	1.04	1.22	1.10	1.01	1.42	0.70	0.76	0.27	0.26	0.44	1.00	4.34
P_2O_5	1.08	0.99	0.85	0.42	0.49	0.45	0.46	0.55	0.23	0.17	0.01	0.03	0.11	0.22	0.82
MnO	0.21	0.20	0.20	0.21	0.25	0.19	0.19	0.22	0.12	0.20	0.20	0.17	0.04	0.23	0.18
CO_2	0.02	0.03	0.07	0.01	0.03	0.07	0.02	0.22	0.06	0.01	0.01	0.03	0.05	0.06	0.01
F	0.15	0.14	0.17	0.12	0.18	0.13	0.17	0.15	0.07	0.11	0.11	0.05		0.10	0.10
BaO	0.0793	0.0737	0.0849	0.1195	0.1340	0.1195	0.1195	0.1597	0.0837	0.0960	0.0017	0.0648	0.1105	0.0581	0.0435
SrO	0.1110	0.1596	0.1620	0.1120	0.0957	0.1113	0.1073	0.1916	0.0931	0.0946	0.0031	0.0563	0.0601	0.0088	0.1015
Rb_2O	0.0063	0.0086	0.0073	0.0106	0.0091	0.0101	0.0113	0.0075	0.0122	0.0155	0.0197	0.0192	0.0129	0.0138	0.0043
Cr_2O_3	0.0003	0.0003	0.0003	0.0003	0.0003	0.0003	0.0003	0.0003	0.0015	0.0015	0.0003	0.0003	0.0003	0.0003	0.0731
NiO	0.0004	0.0004	0.0004	0.0004	0.0017	0.0006	0.0004	0.0004	0.0006	0.0015	0.0032	0.0006	0.0006	0.0006	0.0331
CoO	0.0006	0.0019	0.0006	0.0006	0.0006	0.0006	0.0006	0.0006	0.0006	0.0006	0.0006	0.0006	0.0006	0.0006	0.0086
CuO	0.0045	0.0040	0.0015	0.0049	0.0030	0.0024	0.0025	0.0010	0.0038	0.0043	0.0030	0.0024	0.0035	0.0035	0.0056
ZnO	0.0118	0.0113	0.0111	0.0147	0.0168	0.0123	0.0122	0.0108	0.0154	0.0134	0.0149	0.0122	0.0217	0.0217	0.0091
BeO	0.0009	0.0012	0.0011	0.0017	0.0017	0.0016	0.0017	0.0012	0.0017	0.0021	0.0024	0.0026	0.0015	0.0018	0.0007
V_2O_3	0.0191	0.0106	0.0057	0.0007	0.0047	0.0024	0.0024	0.0024	0.0063	0.0071	0.0007	0.0007	0.0007	0.0007	0.0397
Y_2O_3	0.0047	0.0055	0.0047	0.0047	0.0084	0.0058	0.0070	0.0058	0.0047	0.0039	0.0055	0.0050	0.0033	0.0094	0.0058
La_2O_3	0.0084	0.0108	0.0110	0.0141	0.0164	0.0129	0.0141	0.0141	0.0152	0.0164	0.0211	0.0141	0.0176	0.0176	0.0043
Ce_2O_3	0.0117	0.0152	0.0152	0.0258	0.0281	0.0234	0.0269	0.0246	0.0305	0.0293	0.0351	0.0234	0.0281	0.0340	0.0117
ZrO_2	0.0577	0.0788	0.0708	0.1268	0.1244	0.1201	0.1236	0.0847	0.1016	0.1282	0.1797	0.1155	0.1093	0.1445	0.0435
Total	99.86	100.16	100.27	100.39	100.29	100.20	100.67	100.35	99.82	99.83	100.51	99.60	100.25	99.45	100.11
−O	0.06	0.06	0.07	0.05	0.08	0.05	0.07	0.06	0.03	0.05	0.05	0.02	0.02	0.04	0.04
Total	99.80	100.10	100.20	100.34	100.21	100.15	100.60	100.29	99.79	99.78	100.46	99.58	100.23	99.41	100.07
Normative mineral															
Q	0.00	0.00	0.00	0.00	0.00	0.00	0.00	0.00	0.00	0.00	0.00	0.00	0.00	1.15	0.00
Z	0.09	0.12	0.11	0.19	0.19	0.18	0.18	0.13	0.15	0.19	0.27	0.17	0.16	0.22	0.06
OR	15.39	20.38	19.68	27.74	25.75	25.31	25.02	22.90	26.63	25.39	32.80	29.96	29.37	30.31	8.77
AB	31.60	27.06	27.38	32.89	37.40	40.27	39.41	37.05	36.42	38.04	24.06	36.80	43.14	46.02	1.47

TABLE 1. (continued)

Oxide	16	17	18	19	20	21	22	23	24	25	26	27	28	29	30
AN	17.03	12.46	14.20	3.35	1.10	5.84	6.35	10.50	5.58	2.80	0.00	0.00	2.08	5.13	17.12
LC	0.00	0.00	0.00	0.00	0.00	0.00	0.00	0.00	0.00	0.00	0.00	0.00	0.00	0.00	0.00
NE	6.19	14.61	15.38	19.11	16.10	14.14	14.70	13.06	16.29	21.19	30.59	25.81	14.13	0.00	13.68
AC	0.00	0.00	0.00	0.00	0.00	0.00	0.00	0.00	0.00	0.00	8.09	1.21	0.00	0.00	0.00
NS	0.00	0.00	0.00	0.00	0.00	0.00	0.00	0.00	0.00	0.00	0.65	0.00	0.00	0.00	0.00
WO	6.43	5.17	4.76	4.12	4.76	2.71	2.61	2.87	4.44	4.34	1.48	2.39	3.18	1.99	14.07
EN	3.60	3.01	2.54	1.58	1.77	1.25	1.14	1.49	2.95	2.19	0.17	0.73	1.10	2.31	10.18
FS	2.57	1.91	2.07	2.61	3.08	1.44	1.46	1.30	0.00	0.00	1.10	0.30	0.00	6.00	2.60
FO	2.74	2.42	1.97	0.87	1.16	1.22	1.22	1.40	0.00	0.00	0.00	0.00	0.00	0.00	13.88
FA	2.15	1.69	1.76	1.60	2.22	1.55	1.73	1.35	0.00	0.00	0.00	0.00	0.00	0.00	3.90
MT	4.38	4.15	4.10	2.78	2.66	2.60	2.76	3.66	5.00	3.58	0.00	1.89	1.28	4.06	3.85
CM	0.00	0.00	0.00	0.00	0.00	0.00	0.00	0.00	0.00	0.00	0.00	0.00	0.00	0.00	0.11
HM	0.00	0.00	0.00	0.00	0.00	0.00	0.00	0.00	0.00	0.21	0.00	0.00	4.28	0.00	0.00
IL	5.08	4.51	3.68	1.98	2.32	2.10	1.92	2.71	1.35	1.46	0.51	0.50	0.84	1.96	8.27
PF	0.00	0.00	0.00	0.00	0.00	0.00	0.00	0.00	0.00	0.00	0.00	0.00	0.00	0.00	0.00
AP	2.59	2.36	2.02	1.00	1.16	1.07	1.09	1.31	0.55	0.41	0.02	0.07	0.26	0.54	0.95
FR	0.11	0.11	0.19	0.17	0.28	0.19	0.26	0.21	0.10	0.20	0.22	0.10	0.08	0.17	0.06
CC	0.05	0.07	0.16	0.02	0.07	0.16	0.05	0.07	0.14	0.02	0.02	0.07	0.09	0.14	0.02

Oxide	31	32	33	34	35	36	37	38	39	40	41	42	43	44	45
SiO_2	41.65	41.66	41.89	42.20	42.40	42.40	44.10	45.20	44.50	47.24	47.80	48.80	50.20	50.40	50.50
Al_2O_3	13.14	12.96	13.19	12.80	12.80	12.70	13.60	15.40	15.40	17.93	17.10	17.70	17.70	18.00	17.90
Fe_2O_3	3.77	2.79	3.92	7.24	3.52	6.73	3.96	4.97	10.83	4.07	3.34	7.16	2.76	4.57	3.22
FeO	8.05	8.99	7.61	4.82	8.35	5.10	8.67	7.68	2.97	5.49	6.53	1.80	5.27	3.32	4.50
MgO	11.91	12.18	12.01	11.71	12.08	11.65	9.08	4.16	4.86	3.74	3.86	3.11	3.52	3.27	3.28
CaO	11.32	11.31	11.38	11.22	11.44	11.13	10.70	9.20	9.38	7.91	7.98	6.84	6.90	6.53	6.54
Na_2O	3.16	3.08	3.05	3.13	3.14	3.11	3.29	4.98	4.99	6.18	6.00	6.42	6.85	6.87	7.15
K_2O	1.45	1.45	1.44	1.47	1.43	1.40	1.70	2.20	1.95	2.98	3.02	3.64	3.47	3.55	3.68
H_2O	0.14	0.14	0.17	0.26	0.16	0.30	0.28	0.20	0.31	0.21	0.32	1.41	0.16	0.31	0.21
TiO_2	4.08	4.11	4.12	3.81	3.80	3.82	3.50	3.42	3.16	2.91	2.83	2.42	2.32	2.20	2.18
P_2O_5	0.82	0.79	0.80	0.76	0.78	0.76	0.82	1.77	1.51	0.76	0.77	0.56	0.61	0.51	0.56
MnO	0.18	0.18	0.18	0.18	0.17	0.18	0.19	0.26	0.26	0.24	0.18	0.22	0.20	0.19	0.20
CO_2	0.01	0.01	0.04	0.12	0.01	0.06	0.04	0.02	0.08	0.03	0.03	0.03	0.03	0.15	0.18
F	0.10	0.10	0.10	0.11	0.11	0.12	0.15	0.14	0.14	0.12	0.13	0.11	0.11	0.11	0.11
BaO	0.0413	0.0424	0.0447	0.0447	0.0447	0.0424	0.0502	0.0837	0.0804	0.0770	0.0759	0.0782	0.0793	0.0804	0.0782
SrO	0.1012	0.1054	0.0992	0.1031	0.1012	0.1026	0.1218	0.1939	0.1880	0.1715	0.1673	0.1377	0.1211	0.1193	0.1191
Rb_2O	0.0036	0.0039	0.0044	0.0034	0.0034	0.0032	0.0043	0.0072	0.0056	0.0082	0.0077	0.0093	0.0102	0.0103	0.0106
Cr_2O_3	0.0789	0.0658	0.0731	0.0599	0.0789	0.0628	0.0424	0.0029	0.0032	0.0048	0.0003	0.0003	0.0140	0.0096	0.0094
NiO	0.0382	0.0433	0.0331	0.0331	0.0560	0.0305	0.0153	0.0013	0.0013	0.0006	0.0006	0.0006	0.0034	0.0056	0.0034
CoO	0.0086	0.0076	0.0071	0.0086	0.0084	0.0092	0.0089	0.0032	0.0039	0.0024	0.0028	0.0006	0.0006	0.0006	0.0006
CuO	0.0056	0.0063	0.0056	0.0075	0.0068	0.0069	0.0050	0.0013	0.0013	0.0013	0.0031	0.0030	0.0044	0.0039	0.0029
ZnO	0.0093	0.0095	0.0091	0.0095	0.0106	0.0095	0.0107	0.0149	0.0144	0.0131	0.0149	0.0137	0.0143	0.0137	0.0137
BeO	0.0008	0.0008	0.0008	0.0007	0.0011	0.0008	0.0007	0.0011	0.0011	0.0012	0.0018	0.0018	0.0020	0.0020	0.0021
V_2O_3	0.0471	0.0515	0.0471	0.0382	0.0471	0.0382	0.0441	0.0265	0.0309	0.0279	0.0338	0.0279	0.0279	0.0309	0.0265

	46	47	48	49	50	51	52	53	54	55	56	57	58	59
Y_2O_3	0.0061	0.0057	0.0051	0.0058	0.0050	0.0062	0.0051	0.0074	0.0074	0.0061	0.0043	0.0074	0.0053	0.0043
La_2O_3	0.0047	0.0047	0.0043	0.0047	0.0094	0.0045	0.0041	0.0110	0.0108	0.0129	0.0141	0.0141	0.0176	0.0164
Ce_2O_3	0.0059	0.0117	0.0059	0.0117	0.0047	0.0117	0.0047	0.0223	0.0211	0.0234	0.0199	0.0234	0.0281	0.0246
ZrC_2	0.0446	0.0436	0.0451	0.0455	0.0453	0.0461	0.0389	0.0829	0.0792	0.0943	0.0904	0.0997	0.1110	0.1002
Total	100.17	100.15	100.28	100.20	100.61	99.83	100.43	100.05	100.78	100.25	100.32	100.63	100.41	100.62
−O	0.04	0.04	0.04	0.05	0.05	0.05	0.06	0.06	0.06	0.05	0.05	0.05	0.05	0.05
Total	100.13	100.11	100.24	100.15	100.56	99.78	100.37	99.99	100.72	100.20	100.27	100.58	100.36	100.57
Normative mineral														
Q	0.00	0.00	0.00	0.00	0.00	0.00	0.00	0.00	0.00	0.00	0.00	0.00	0.00	0.00
Z	0.07	0.07	0.07	0.07	0.07	0.07	0.06	0.12	0.12	0.14	0.13	0.15	0.17	0.15
OR	8.59	8.59	8.52	8.71	8.44	8.33	10.06	13.07	11.51	17.67	17.91	20.50	21.02	21.72
AB	3.51	2.12	4.59	8.47	4.13	9.81	12.44	22.85	22.79	17.91	18.34	21.00	24.40	21.85
AN	17.43	17.30	18.07	16.59	16.58	16.68	17.35	13.24	13.85	12.42	10.85	7.30	7.81	5.87
LC	0.00	0.00	0.00	0.00	0.00	0.00	0.00	0.00	0.00	0.00	0.00	0.00	0.00	0.00
NE	12.62	13.01	11.51	9.81	12.13	9.05	8.37	10.57	10.51	18.72	17.67	20.01	18.34	20.90
AC	0.00	0.00	0.00	0.00	0.00	0.00	0.00	0.00	0.00	0.00	0.00	0.00	0.00	0.00
NS	0.00	0.00	0.00	0.00	0.00	0.00	0.00	0.00	0.00	0.00	0.00	0.00	0.00	0.00
WO	13.91	14.02	13.72	13.90	14.49	13.88	12.41	8.79	9.35	8.99	9.74	9.39	8.37	8.96
EN	10.59	10.14	10.69	12.01	10.76	12.00	8.63	5.60	8.08	6.52	6.07	6.05	7.23	6.34
FS	1.88	2.60	1.52	0.00	2.32	0.00	2.75	2.61	0.00	1.65	3.09	2.71	0.00	1.84
FO	13.41	13.50	13.50	12.09	13.51	12.07	9.82	3.37	2.81	1.98	2.51	1.90	0.65	1.28
FA	2.62	4.01	2.12	0.00	3.21	0.00	3.45	1.73	0.00	0.55	1.41	0.94	0.00	0.41
MT	5.48	4.06	5.69	5.11	5.09	6.00	5.75	7.24	1.26	5.92	4.86	4.00	4.96	4.66
CM	0.12	0.10	0.11	0.09	0.12	0.09	0.06	0.00	0.00	0.01	0.00	0.02	0.01	0.01
HM	0.00	0.00	0.00	0.00	0.00	0.00	0.00	0.00	9.95	0.00	0.00	0.00	0.00	0.00
IL	7.77	7.83	7.84	3.74	7.20	2.64	6.66	6.53	6.00	5.54	5.37	4.40	4.19	4.14
PF	0.00	0.00	0.00	7.26	0.00	7.31	0.00	0.00	0.00	0.00	0.00	0.00	1.16	0.00
AP	1.95	1.88	1.90	1.81	1.84	1.81	1.95	4.21	3.57	1.81	1.83	1.44	1.21	1.33
FR	0.06	0.06	0.06	0.09	0.08	0.11	0.16	0.00	0.01	0.11	0.13	0.11	0.13	0.12
CC	0.02	0.02	0.09	0.27	0.02	0.14	0.09	0.05	0.18	0.07	0.07	0.07	0.34	0.41

Oxide	46	47	48	49	50	51	52	53	54	55	56	57	58	59
SiO_2	50.61	50.79	51.60	52.46	39.70	40.00	42.70	42.70	44.20	45.20	46.50	47.50	48.10	49.20
Al_2O_3	19.32	18.68	18.60	19.48	0.85	0.60	1.13	5.63	1.21	0.95	6.23	2.15	12.00	3.32
Fe_2O_3	2.37	2.73	2.98	2.99	1.94	0.84	4.03	7.57	2.89	0.83	1.44	0.88	1.23	2.55
FeO	4.87	2.92	4.32	2.86	8.43	9.33	5.83	4.44	5.73	6.79	6.92	5.60	3.14	2.86
MgO	2.64	2.92	2.27	2.29	46.70	46.90	41.50	24.10	43.00	43.50	24.40	30.90	16.70	23.30
CaO	6.05	6.13	5.57	4.93	0.50	0.61	3.15	11.50	1.34	0.92	12.80	11.90	16.60	17.00
Na_2O	7.15	7.08	7.60	7.74	0.04	0.04	0.22	0.52	0.05	0.04	0.69	0.34	0.90	0.40
K_2O	3.66	3.79	3.74	4.09	0.02	0.01	0.02	0.02	0.02	0.01	0.06	0.05	0.08	0.12
H_2O	0.15	0.22	0.20	0.39	0.00	0.00	0.00	0.00	0.00	0.03	0.00	0.00	0.00	0.00
TiO_2	2.15	2.13	1.77	1.70	0.09	0.05	0.15	1.53	0.03	0.00	0.31	0.29	0.31	0.31
P_2O_5	0.46	0.55	0.56	0.42	0.00	0.00	0.00	0.00	0.00	0.00	0.00	0.00	0.00	0.00
MnO	0.20	0.19	0.18	0.17	0.16	0.15	0.15	0.17	0.12	0.12	0.15	0.12	0.08	0.10
CO_2	0.02	0.07	0.04	0.03	0.00	0.00	0.00	0.00	0.00	0.00	0.00	0.00	0.00	0.00
F	0.11	0.11	0.10	0.10	0.00	0.00	0.00	0.00	0.00	0.00	0.00	0.00	0.00	0.00

TABLE 1. (continued)

Oxide	46	47	48	49	50	51	52	53	54	55	56	57	58	59
BaO	0.0860	0.0793	0.0960	0.0871	0.0000	0.0000	0.0000	0.0000	0.0000	0.0000	0.0012	0.0016	0.0019	0.0025
SrO	0.1291	0.1194	0.1471	0.0992	0.0004	0.0002	0.0051	0.0089	0.0007	0.0004	0.0142	0.0035	0.0215	0.0032
Rb_2O	0.0102	0.0110	0.0108	0.0124	0.0001	0.0000	0.0001	0.0001	0.0001	0.0000	0.0001	0.0001	0.0001	0.0005
Cr_2O_3	0.0037	0.0095	0.0003	0.0051	0.9879	0.5174	0.2514	0.1286	0.3565	0.3727	0.2046	0.4633	0.1368	0.3873
NiO	0.0006	0.0025	0.0006	0.0006	0.3003	0.2698	0.3602	0.1000	0.3003	0.3100	0.0900	0.1298	0.0331	0.0900
CoO	0.0022	0.0023	0.0006	0.0013	0.0254	0.0267	0.0229	0.0140	0.0242	0.0203	0.0092	0.0084	0.0056	0.0074
CuO	0.0225	0.0019	0.0023	0.0025	0.0029	0.0003	0.0058	0.0054	0.0062	0.0003	0.0025	0.0006	0.0013	0.0238
ZnO	0.0121	0.0117	0.0149	0.0117	0.0093	0.0075	0.0081	0.0081	0.0062	0.0056	0.0067	0.0042	0.0021	0.0031
BeO	0.0014	0.0015	0.0022	0.0017	0.0003	0.0003	0.0003	0.0003	0.0003	0.0003	0.0001	0.0001	0.0001	0.0002
V_2O_3	0.0177	0.0191	0.0191	0.0147	0.0082	0.0034	0.0071	0.0382	0.0091	0.0063	0.0177	0.0154	0.0177	0.0221
Y_2O_3	0.0030	0.0028	0.0056	0.0028	0.0000	0.0000	0.0000	0.0000	0.0000	0.0000	0.0003	0.0003	0.0003	0.0003
La_2O_3	0.0141	0.0141	0.0152	0.0164	0.0000	0.0000	0.0000	0.0000	0.0000	0.0000	0.0008	0.0008	0.0008	0.0008
Ce_2O_3	0.0211	0.0211	0.258	0.0211	0.0000	0.0000	0.0000	0.0000	0.0000	0.0000	0.0047	0.0047	0.0047	0.0047
ZrO_2	0.1098	0.1163	0.1214	0.1154	0.0014	0.0108	0.0019	0.0073	0.0008	0.0015	0.0007	0.0007	0.0024	0.0032
Total	100.19	100.11	99.99	100.07	99.76	99.36	99.54	98.49	99.28	99.10	99.85	99.46	99.36	99.70
–O	0.05	0.05	0.04	0.04	0.00	0.00	0.00	0.00	0.00	0.00	0.00	0.00	0.00	0.00
Total	100.14	100.06	99.95	100.03	99.76	99.36	99.54	98.49	99.28	99.10	99.85	99.46	99.36	99.70
Normative mineral														
Q	0.00	0.00	0.00	0.00	0.00	0.00	0.00	0.00	0.00	0.00	0.00	0.00	0.00	0.00
Z	0.16	0.17	0.18	0.17	0.00	0.02	0.00	0.01	0.00	0.00	0.00	0.00	0.00	0.00
OR	21.69	22.49	22.22	24.31	0.12	0.05	0.12	0.12	0.12	0.06	0.36	0.30	0.48	0.71
AB	22.32	22.89	26.15	27.38	0.34	0.00	1.87	4.47	0.43	0.34	4.65	2.89	6.29	3.40
An	9.84	8.03	5.62	6.37	2.09	1.44	2.05	13.18	3.04	2.41	13.75	4.22	28.66	6.93
LC	0.00	0.00	0.00	0.00	0.00	0.00	0.00	0.00	0.00	0.00	0.00	0.00	0.00	0.00
NE	20.78	20.19	20.86	20.86	0.00	0.18	0.00	0.00	0.00	0.00	0.65	0.00	0.75	0.00
AC	0.00	0.00	0.00	0.00	0.00	0.01	0.00	0.00	0.00	0.00	0.00	0.00	0.00	0.00
NS	0.00	0.00	0.00	0.00	0.17	0.00	0.00	0.00	0.00	0.92	0.00	0.00	0.00	0.00
WO	7.00	7.58	7.53	6.18	1.48	0.67	5.70	18.70	1.53	25.59	20.83	23.03	22.66	32.45
EN	4.20	5.20	4.69	5.14	0.17	0.52	15.30	22.11	24.63	2.88	15.77	22.26	18.14	28.33
FS	2.43	1.78	2.39	0.26	80.70	0.07	1.16	0.00	1.96	55.70	2.92	2.72	1.89	1.29
FO	1.68	1.48	0.70	0.42	10.48	82.05	62.08	27.24	58.36	7.27	31.62	37.06	16.63	20.96
FA	1.07	0.56	0.39	0.02	2.82	12.90	5.18	0.00	5.12	1.21	6.46	4.99	1.91	1.05
MT	3.45	3.97	4.34	4.36	1.46	1.23	5.87	10.71	4.22	0.55	2.09	1.26	1.80	3.71
CM	0.01	0.01	0.00	0.01	0.00	0.77	0.37	0.19	0.53	0.00	0.30	0.69	0.20	0.57
HM	0.00	0.00	0.00	0.00	0.17	0.00	0.00	0.30	0.00	0.06	0.00	0.00	0.00	0.00
IL	4.10	4.06	3.38	3.25	0.00	0.10	0.29	2.95	0.06	0.00	0.59	0.55	0.59	0.59
PF	0.00	0.00	0.00	0.00	0.00	0.00	0.00	0.00	0.00	0.00	0.00	0.00	0.00	0.00
AP	1.09	1.31	1.33	1.00	0.00	0.00	0.00	0.00	0.00	0.00	0.00	0.00	0.00	0.00
FR	0.14	0.13	0.10	0.13	0.00	0.00	0.00	0.00	0.00	0.00	0.00	0.00	0.00	0.00
CC	0.05	0.16	0.09	0.14	0.00	0.00	0.00	0.00	0.00	0.00	0.00	0.00	0.00	0.00

Numbers 1–59 are sample numbers.

TABLE 2. Means and Variances of Chemical Constituents in Volcanic Rocks From Ross Island and Vicinity, Antarctica

Constituent	All 49 Samples		Basanitoid Group (21 Samples)*		Trachybasalt-Phonolite Group (29 Samples)*	
	Mean	Variance	Mean	Variance	Mean	Variance
SiO_2	48.25	30.7155	43.16	1.1614	51.81	19.7667
Al_2O_3	16.42	8.5350	13.47	0.9969	18.57	2.6337
FeO	9.21	8.6322	11.69	0.4859	7.50	6.9920
MgO	5.95	20.4529	10.77	4.0943	2.43	1.9161
CaO	7.91	11.6455	11.05	0.8634	5.70	6.9787
Na_2O	5.52	4.7534	3.42	0.3637	7.00	2.3582
K_2O	2.72	1.9647	1.38	0.1270	3.67	1.0098
TiO_2	2.61	1.3726	3.60	0.2081	1.94	1.0552
P_2O_5	0.70	0.1429	0.78	0.0425	0.68	0.2346
MnO	0.20	0.000697	0.20	0.000381	0.21	0.000937
F	0.11	0.000891	0.10	0.000655	0.12	0.001033
BaO	0.0687	0.000945	0.0448	0.000117	0.0862	0.000792
SrO	0.1176	0.001619	0.1109	0.000479	0.1243	0.002453
Rb_2O	0.0074	0.000020	0.0034	0.000001	0.0102	0.000014
Cr_2O_3	0.0313	0.001438	0.0698	0.000719	. . .	. . .
NiO	0.0126	0.000235	0.0277	0.000140	. . .	. . .
CoO	0.0038	0.000010	0.0071	0.000002	. . .	. . .
ZnO	0.0129	0.000007	0.0118	0.000008	0.0138	0.000005
BeO	0.0013	<0.000001	0.0008	<0.000001	0.0016	<0.000001
V_2O_3	0.0256	0.000259	0.0404	0.000038	. . .	. . .
Y_2O_3	0.0051	0.000002	0.0046	0.000001	0.0055	0.000003
La_2O_3	0.0107	0.000024	0.0059	0.000005	0.0141	0.000009
Ce_2O_3	0.0162	0.000096	0.0066	0.000011	0.0228	0.000048
ZrO_2	0.0774	0.001465	0.0413	0.000084	0.1027	0.000830

*Sample number 13 is included in both groups.

ances in the original data that could be accounted for by factor analysis models with 2 to 10 end members.

It is clear from Table 3 that with the exceptions of MnO, ZnO, and Y_2O_3, most of the variances in the original data are preserved after projection of the vectors into three-dimensional space. Thus a model consisting of a starting magma and two solidus compositions could be developed that could account for 60 to 98% of the variability in 21 of the chemical variables. However, a three-end-member model would require participation of the same two solidus compositions throughout the differentiation sequence, and this seems unlikely on petrographic grounds. Consequently, the 49 samples were divided into two groups which became apparent upon examination of the three-dimensional vector system. The two groups consisted of (1) 20 basanitoid samples plus trachybasalt sample 13, and (2) 29 samples of trachybasalts and phonolites. Trachybasalt sample 13 was included in group 1 and group 2 to provide a bridge between the two series.

Factor analysis of each of the two groups followed the same procedures that were used for the total group of 49 samples and was performed on the same scaled and row-normalized data. Although the variances in the total data set are about the same for each oxide, they differ among oxides within each group, and the variances for any one oxide differ between groups. The effect of these differences is to endow various oxides with different degrees of influence on the factor analysis outcomes, although the influence is in proportion to the fraction of the total variance of the oxide that is in the group being analyzed rather than in proportion to the abundance of the oxide in the rock samples. Most of the variance in SiO_2, for example, is in the trachybasalt-phonolite group (Table 2), whereas most of the variance in MgO is in the basanitoids. Consequently, SiO_2 had more influence than MgO in analysis of the trachybasalts and phonolites, and MgO had more influence than SiO_2 in analysis of the basanitoids. Because of the scaling, however, each oxide had about the same influence on the factor analysis results as a whole.

Each subset of vectors, one set representing the basanitoid samples and the other representing the trachybasalt-phonolite samples, was initially projected into three dimensions. The petrogenetic

TABLE 3. Proportions of Variances That Can be Accounted for by Factor Models With 2 to 10 End Members

Constituent	Number of End Members								
	2	3	4	5	6	7	8	9	10
SiO_2	0.92	0.94	0.95	0.95	0.98	0.98	0.98	0.98	0.99
Al_2O_3	0.92	0.93	0.97	0.97	0.98	0.98	0.98	0.98	0.98
FeO	0.83	0.92	0.96	0.97	0.97	0.97	0.97	0.98	0.98
MgO	0.92	0.98	0.98	0.99	0.99	0.99	0.99	0.99	0.99
CaO	0.95	0.97	0.97	0.97	0.99	0.99	0.99	0.99	0.99
Na_2O	0.93	0.93	0.95	0.95	0.96	0.96	0.96	0.97	0.97
K_2O'	0.96	0.98	0.98	0.98	0.98	0.98	0.98	0.98	0.99
TiO_2	0.85	0.91	0.91	0.92	0.97	0.97	0.97	0.97	0.97
P_2O_5	0.21	0.83	0.84	0.85	0.88	0.89	0.91	0.91	0.92
MnO	0.01	0.45	0.65	0.68	0.72	0.73	0.73	0.90	0.98
F	0.02	0.65	0.65	0.83	0.82	0.86	0.97	0.98	0.99
BaO	0.42	0.60	0.61	0.67	0.68	0.91	0.91	0.92	0.96
SrO	0.01	0.66	0.74	0.74	0.81	0.91	0.92	0.92	0.94
Rb_2O	0.87	0.92	0.92	0.92	0.94	0.96	0.97	0.98	0.98
Cr_2O_3	0.74	0.92	0.92	0.92	0.94	0.95	0.96	0.97	0.97
NiO	0.68	0.86	0.86	0.92	0.93	0.93	0.93	0.93	0.94
CoO	0.85	0.89	0.90	0.92	0.93	0.94	0.96	0.97	0.99
ZnO	0.17	0.17	0.54	0.89	0.90	0.90	0.93	0.96	0.97
BeO	0.77	0.83	0.82	0.82	0.94	0.94	0.96	0.96	0.96
V_2O_3	0.85	0.86	0.87	0.87	0.97	0.97	0.98	0.98	0.98
Y_2O_3	0.08	0.14	0.69	0.90	0.94	0.96	0.96	0.96	0.98
La_2O_3	0.85	0.86	0.87	0.88	0.91	0.91	0.92	0.93	0.97
Ce_2O_3	0.83	0.84	0.89	0.89	0.93	0.95	0.95	0.95	0.97
ZrO_2	0.87	0.89	0.91	0.92	0.95	0.95	0.95	0.95	0.97

models were then developed by finding the positions of vectors (reference axes) that might represent the compositions of materials that were involved in the genesis of the representative lavas. In development of the model for the trachybasalts and phonolites it was necessary to expand the vector system to four dimensions because the mathematical solutions could not be matched to the observed mineral assemblages. Finally, estimates of the mixing (and unmixing) proportions were derived from the coordinates (loadings) of the sample vectors with respect to these selected reference axes.

The recomputed data can be represented precisely in an m-dimensional vector system. Thus the recomputed data for each sample can be expressed as a combination of m-end-member compositions. The end member compositions can be represented by any m selected vectors in the m-dimensional space, so long as each of the m selected vectors is independent of the others (that is, none is a linear combination of the others). The mixing proportions for each sample are derived from the coordinates of the sample vector with respect to the vectors that represent the end member compositions. The modeling procedure therefore consists of selecting end member vectors that represent compositions of magmas, minerals, or other materials that may

have been involved in the petrogenetic processes that formed the lavas, and then examining the plausibility of the derived mixing (or unmixing) proportions. Materials that may have been involved in the petrogenetic processes are those that are either observed in the lavas or judged as possibly involved on theoretical grounds. The plausibility of derived mixing proportions, and especially of their signs, which indicate addition or subtraction of the end member, is judged against the petrography and field occurrence of the sample and other independent data such as REE (rare earth elements) abundances and isotopic determinations.

DISCUSSION

General Geochemistry

The volcanic sequence of Ross Island is geochemically similar to alkalic sequences in both oceanic and continental rift environments. On an AMF diagram (Figure 1) the sequence exhibits a small iron enrichment in the basanitoids, followed by an alkali enrichment in the trachybasalts and phonolites. The trend is generally similar to that exhibited by alkalic suites from Hawaii, Tristan Island, Gough Island, and Tahiti (see *Irvine and Baragar* [1971] for summary), but there are two discontinui-

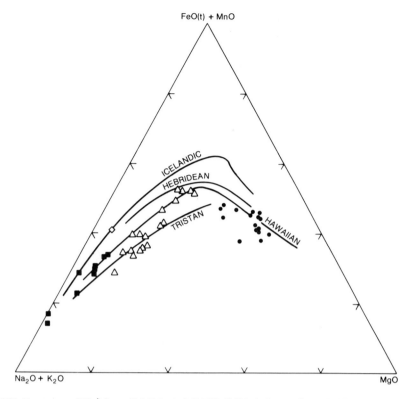

Fig. 1. AMF diagram modified from *Goldich et al.* [1975]. Solid circles are basanitoids, triangles are trachybasalts, solid squares are phonolites, and the diamond is a trachyte. Oxide data used for the plot are in weight percent. FeO(*t*) is total iron calculated as ferrous iron.

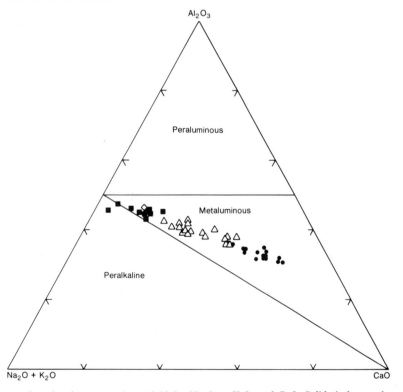

Fig. 2. Ternary plot of molar proportions of Al_2O_3, $Na_2O + K_2O$, and CaO. Solid circles are basanitoids, triangles are trachybasalts, solid squares are phonolites, and the diamond is a trachyte.

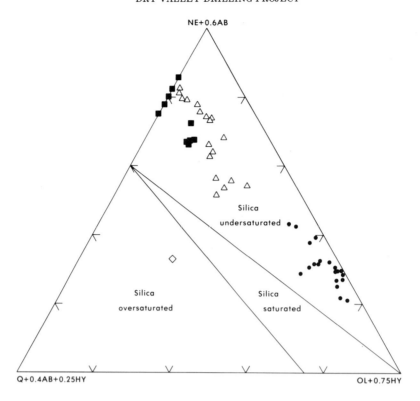

Fig. 3.   Ternary plot using molar proportions of the normative minerals nepheline (NE), albite (AB), quartz (Q), hypersthene (HY), and olivine (OL). Solid circles are basanitoids, triangles are trachybasalts, solid squares are phonolites, and the diamond is a trachyte.

ties: a pronounced compositional break between the basanitoid and trachybasalt trends and a shift away from the MgO corner between the trachybasalt and phonolite trends.

The ternary plot of molar values of Al_2O_3, Na_2O + K_2O, and CaO shows that all but two of the analyzed samples are metaluminous (Figure 2). The ternary plot of normative minerals (modified from *Yoder and Tilley* [1962]) shows that all but one sample is silica undersaturated (Figure 3). *Stuckless and Ericksen* [1976] concluded, on the basis of Sr isotopic data, that this sample was contaminated with crustal materials. Figure 3 also shows a pronounced compositional gap between the basanitoids and trachybasalts.

Silica variation diagrams (Figure 4) for major element concentrations generally suggest simple differentiation trends with scatter about a liquidus line of descent accounted for by differential crystal entrapment. Only the MgO curve shows a major compositional break; this could be accounted for by the disappearance of olivine as a solidus phase during differentiation [*Goldich et al.*, 1975]. Trends for some of the minor and trace element data are more complex. P_2O_5, Sr, and Ba appear to have been

added during the transition from basanitoid to trachybasalt. Sr and Ba appear to have been added again during the transitions from trachybasalt to phonolite, whereas Rb may have been lost.

Q-Mode Models

The configuration of the three-dimensional system of vectors that represents the 49 samples and 24 variables is shown by the stereogram in Figure 5. All the vectors that plot between vectors 4 and 9 represent basanitoid lavas that range from about 41 to 45% silica. Vectors that plot between vectors 13 and 27 represent trachybasalts and phonolites, which range from about 45 to 58% silica.

The trend discontinuities noted in some of the silica variation diagrams (Figure 4) are again discernible on the stereogram for all 49 samples (Figure 5). The large gap between vectors 9 and 13 in Figure 5 represents the major compositional discontinuities shown by Mg, Ni, Sr, Ba, and P at the change from basanitoid to trachybasalt. The small separation between the trachybasalt trend (vectors 13 and 29) and the phonolite trend (vectors 23 and 27) represents the subtle compositional discontinui-

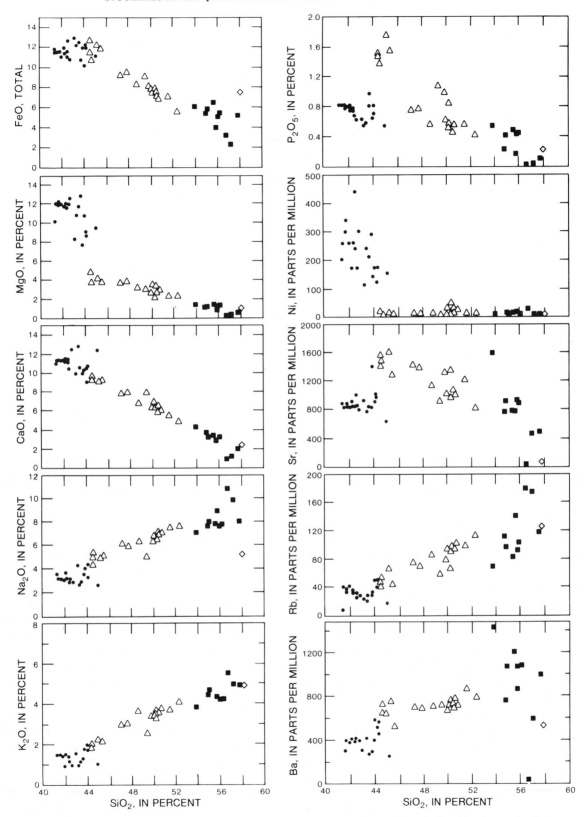

Fig. 4.   Silica variation diagrams for the data given in Table 1. Solid circles are basanitoids, triangles are trachybasalts, solid squares are phonolites, and the diamond is a trachyte.

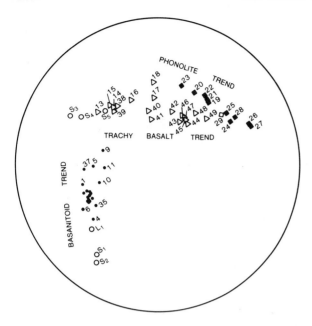

Fig. 5.   Stereogram showing the configuration of 49 samples projected by Q mode factor analysis for 24-dimensional space. In construction of the stereogram (see also Figures 6 and 7), points were projected vertically from the upper hemisphere onto the plane of the diagram. Solid cirlces are basanitoids, triangles are trachybasalts, solid squares are phonolites, and the diamond is a trachyte.

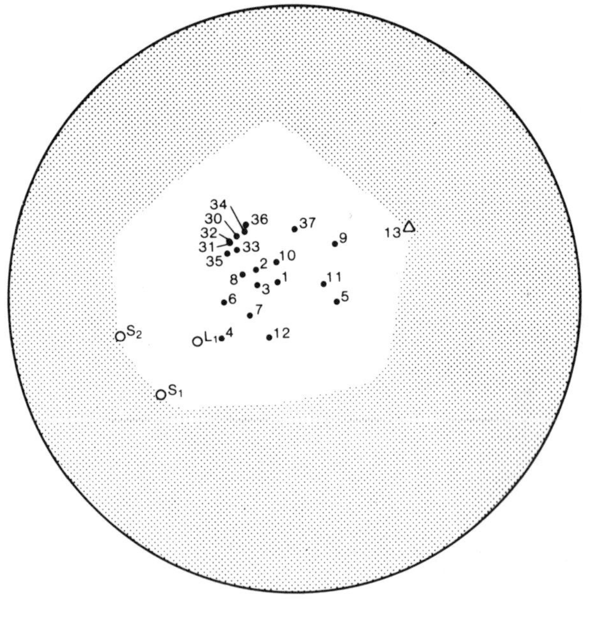

Fig. 6.   Stereogram showing the configuration of vectors representing 20 basanitoids (solid cirlces) and 1 trachybasalt (triangle). Points L_1 and 13 represent the initial and final liquids in this series. Points S_1 and S_2 represent the two solidus compositions that are added to or subtracted from liquid L_1 to match each of the sample vectors. The stippled area represents the locus of points for which one or more oxide values are negative.

ties shown by Sr, Rb, and Ba at the change from trachybasalt to phonolite.

The fact that the 49 sample compositions can be represented so well in only three dimensions (Table 3) is evidence for a common magmatic parentage. However, the sharp angle between the trends for the basanitoid and trachybasalt-phonolite trends shows clearly that the path of differentiation changed abruptly sometime after formation of the basanitoids.

The vector configurations for each of the two groups of samples are shown in Figures 6 and 7. Because the data were scaled in the same manner in the analysis of the basanitoid and trachybasalt-phonolite groups as they were in the analysis of the total 49 samples, apparent differences between the vector configurations for the two smaller groups (Figures 6 and 7) and the vector configuration of the larger group (Figure 5) result solely from the differences in projection from 24- to 3-dimensional space. The vector configuration in Figure 5 describes the chemical relations among all 49 samples, but the configurations in Figures 6 and 7 more accurately describe relations among samples within each group.

The three-dimensional projections for the two groups (Figures 6 and 7) represent the original data adequately and are therefore used to develop the petrogenetic model. Choice of a starting liquid for the basanitoid group was somewhat arbitrary, but several constraints limited the choice. The vector for the starting liquid must lie within the positive space (Figure 6) and should be located near a primitive lava. *Irving and Green* [1976] have suggested that such lavas will have 'Mg values' (Mg/(Mg + Fe^{++}) between 66 and 75). Of the basanitoids, all but samples 34 and 36 (which are strongly oxidized and stained red) and samples 5, 9, 11, and 37 are primitive by this criterion. Thus Mg values constrain the starting liquid to the left side on Figure 6.

Ni contents can be used to further constrain the possible starting liquid. Very high values suggest contamination by xenoliths [*Irving and Green*, 1976], whereas low values suggest the separation of olivine [*Hedge*, 1971]. The vector represented by sample 4 is intermediate in Ni contents. Furthermore, it is from an early sequence of lavas [*Goldich et al.*, 1975] and has a low silica content (Table 1). Therefore the vector that is to represent the initial

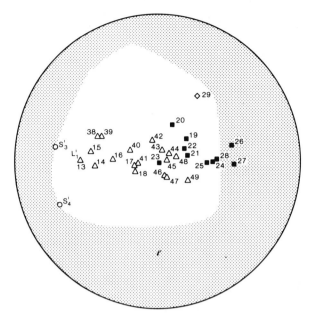

Fig. 7. Stereogram showing the configuration of vectors representing 18 trachybasalts (triangles), 10 phonolites (solid squares), and 1 trachyte (diamond). Points S_3' and S_4' represent the two solidus compositions that are added to or subtracted from liquid L_1'. The stippled area represents the locus of points for which one or more oxide values are negative.

TABLE 4.   Compositions of End Members for the Basanitoid Model

Constituent	End Member		
	L_1	S_1	S_2
SiO_2	42.95	42.95	41.25
Al_2O_3	11.92	10.77	9.64
FeO	11.60	11.59	11.06
MgO	13.65	15.71	18.34
CaO	12.41	13.48	13.98
Na_2O	2.54	1.87	1.36
K_2O	0.75	0.22	0.26
TiO_2	3.24	2.86	3.53
P_2O_5	0.38	0.04	0.08
MnO	0.17	0.16	0.13
F	0.05	0.00	0.01
BaO	0.0227	0.0046	0.0034
SrO	0.0704	0.0383	0.0285
Rb_2O	0.0020	0.0007	0.0014
Cr_2O_3	0.1248	0.1694	0.1780
NiO	0.0390	0.0465	0.0630
CoO	0.0070	0.0066	0.0089
Zno	0.0129	0.0145	0.0089
BeO	0.0006	0.0004	0.0005
V_2O_3	0.0462	0.0502	0.0570
Y_2O_3	0.0029	0.0014	0.0032
La_2O_3	0.0033	0.0014	0.0000
Ce_2O_3	0.0037	0.0009	0.0046
ZrO_2	0.0241	0.0096	0.0125

S_1 is interpreted as an assemblage of approximately 76% pyroxene (Table 5, specimen 35-Cpx), 8% olivine (Table 5, specimen RI-82), 9% spinel (Table 5, specimen HP-2A1), and 7% iron-titanium oxide (Table 5, specimen Ox-3).

S_2 is interpreted as an assemblage of approximately 70% pyroxene, 15% olivine, 8% spinel, 4% iron-titanium oxide, and 3% ideal ilmenite. Refer to Table 5 for mineral compositions. Specimen numbers are given for S_1 above.

starting liquid should lie near vector 4 and be displaced from it in a left or lower-left direction (Figure 6). The exact position is dependent, in part, on the choice of solidus compositions.

The compositions of solidus materials that separated from the initial liquid must be represented by vectors toward the lower-left part of Figure 6. Otherwise, the solidus compositions would contain more SiO_2 and less MgO than the starting liquid, and the observed compositional variations could be accounted for only by additions of solidus.

The vectors representing the solidus compositions should be as far distant from the vector representing the starting liquid as possible, in order to explain the compositional variation with as little differentiation as possible. One solidus composition that seems petrologically plausible was found in the plane of sample vectors 6 and 8 (S_1 in Figure 6). The vector represents a composition that includes no F, and if any farther removed from vector L_1 or from the sample vectors, the F content would be negative. A second plausible solidus composition was found in the plane of vectors 4 and 12 (S_2 in Figure 6). This composition contains no La, and if the vector were any farther removed from the sample vectors, the La value would be negative.

The vector representing the starting magma composition was taken as the vector at the intersection of the planes through sample vectors 6 and 8, and sample vectors 4 and 12 (vector L_1 in Figure 6). The compositions represented by vectors L_1, S_1, and S_2 are given in Table 4. Table 4 also gives possible mineral compositions represented by vectors S_1 and S_2. These were derived by the least squares method of Bryan et al. [1969], using the mineral compositions in Table 5.

Each of the sample compositions shown in Figure 6, except for samples 30 to 36 from DVDP 2, can be approximated by separating or adding solidus materials S_1 and S_2 from liquid L_1 in varying proportions. The required mixing proportions are given in Table 6. For example, the composition of sample 4 can be approximated by subtracting 0.2620 parts of composition S_2 from 1.2620 parts of the initial magma (L_1), and sample 7 can be approximated by subtracting 0.6879 parts of a mixture of about equal parts of the two solidus compositions

TABLE 5. Compositions of Selected Minerals From Volcanic Rocks of Ross Island and Vicinity, Antarctica

Constituent	Pyroxene		Olivine, RI-82	Spinel, HP-2A1	Iron-Titanium Oxide		Horn-blende, RI-50	Apatite, Ap-1	Plagio-clase, Dv-32-P8	Anortho-clase, Anor-1
	35-Cpx	RI-55			Ox-1	Ox-5				
SiO_2	50.46	47.87	40.92	0.51	0.00	0.00	38.90	0.16	55.85	65.46
Al_2O_3	6.36	8.33	0.00	63.32	4.18	0.56	15.07	0.44	28.29	21.32
FeO	6.05	7.54	8.36	18.42	78.21	68.64	11.40	0.28	0.63	0.19
MgO	14.18	11.70	50.68	17.55	2.32	1.91	11.91	0.29	0.00	0.10
CaO	20.96	21.81	0.03	0.20	0.00	0.05	12.63	53.30	11.28	2.00
Na_2O	0.51	0.59	0.00	0.00	0.00	0.00	2.50	0.12	3.59	8.21
K_2O	0.05	0.01	0.00	0.00	0.00	0.00	1.57	0.02	0.36	2.60
TiO_2	1.42	2.15	0.00	0.00	15.29	28.82	6.02	0.05	0.00	0.05
P_2O_5	0.00	0.00	0.00	0.00	0.00	0.00	0.00	45.35	0.00	0.07

Analyst: P. W. Weiblen. All analyses were made by electron microprobe and adjusted to sum 100%.
35-Cpx, RI-55, etc. are specimen numbers.

(0.3458 parts S_1 and 0.3421 parts S_2). The generation of sample 13, the initial sample in the trachybasalt-phonolite series, requires 3.7 parts initial magma; inasmuch as 2.7 parts must be separated to arrive at the approximate composition of sample 13, differentiation of approximately 73% of the magma seems evident. The only alternative would be to suggest that the initial magma was intermediate in composition to the samples themselves and that some of the samples are partly cumulate in nature.

With the selection of end member compositions given in Table 4, samples 30 to 36 (from DVDP 2) can be approximated only by adding materials similar in composition to S_2 and subtracting material of composition S_1 (Table 6). Material of composition S_2 may have been subtracted as well, but the net effect must be one of addition. *Weiblen et al.* [1978] have noted that basanitoids from DVDP 2 contain abundant xenocrystic clinopyroxene and some xenocrystic olivine. Hence additions of material similar to composition S_2 (Table 4) seem petrologically reasonable.

The observed composition of sample 13 and the approximate composition produced by subtraction of solidus compositions S_1 and S_2 from liquid L_1 are given in Table 7, along with the composition represented by vector 13 in Figure 7, which shows the trachybasalt-phonolite trend. The three compositions are sufficiently close to suggest that a liquidus composition produced by the basanitoid model could have been the starting composition for the trachyte-phonolite trend and that sample 13 might be from a lava that represents the end of one series and the beginning of the other.

Two solidus compositions for the trachybasalt-phonolite model (S_3' and S_4' in Figure 7) were

found in the same manner as was described for the basanitoid model. Vector S_3' is in the plane of sample vectors 13 and 14, and vector S_4' is in the plane of sample vectors 13 and 15. Vector 13 is taken to represent the composition of the magma at the beginning of this series. The resulting liquidus composition, and the two solidus compositions, constitute a three-end-member model that can account for most of the variability in the trachybasalt-phonolite series entirely by crystal fractionation. That is, subtraction of the two solidus materials

TABLE 6. Mixing Proportions for the Basanitoid Model

Sample No.	End Member			Extent of Magmatic Differentiation, as a Percent of Liquid L_1
	L_1	S_1	S_2	
4	1.2620	0.0000	−0.2620	20.8
6	1.5649	−0.5649	−0.0000	36.1
12	1.6290	0.0000	−0.6290	38.6
7	1.6879	−0.3458	−0.3421	40.8
3	2.0113	−0.8188	−0.1925	50.3
8	1.9605	−0.9605	0.0000	49.0
32	2.1641	−1.5801	0.4160	53.8
31	2.1381	−1.5274	0.3893	53.2
30	2.2415	−1.6464	0.4049	55.4
33	2.1274	−1.3939	0.2665	53.0
35	2.0315	−1.3875	0.3560	50.8
34	2.3520	−1.7110	0.3590	57.5
36	2.4219	−1.8120	0.3900	58.7
1	2.1729	−0.8245	−0.3484	54.0
2	2.1203	−1.0395	−0.0808	52.8
10	2.3099	−1.1006	−0.2093	56.7
11	2.4640	−0.7168	−0.7477	59.4
5	2.4099	−0.4416	−0.9683	58.5
37	2.7442	−1.6374	−0.1067	63.6
9	2.8958	−1.3111	−0.5848	65.5
13	3.7019	−1.5862	−1.1157	73.0

See Table 4 for end member compositions.

TABLE 7. Composition of Sample 13 and Compositions Represented by Vector 13 in the Basanitoid and Trachyte-Phonolite Models

Constituent	Composition (From Table 1)	Basanitoid Model	Trachyte-Phonolite Model
SiO_2	44.69	44.83	43.99
Al_2O_3	16.67	16.29	16.37
FeO	11.54	12.21	12.02
MgO	4.80	5.14	4.72
CaO	9.78	8.95	10.48
Na_2O	4.37	4.93	4.49
K_2O	1.93	2.13	1.82
TiO_2	3.93	3.50	3.79
P_2O_5	1.54	1.24	1.55
MnO	0.25	0.25	0.24
F	0.17	0.17	0.16
BaO	0.0706	0.0731	0.0815
SrO	0.1698	0.1682	0.2015
Rb_2O	0.0046	0.0048	0.0037
ZnO	0.0136	0.0147	0.0127
BeO	0.0010	0.0010	0.0008
Y_2O_3	0.0047	0.0051	0.0053
La_2O_3	0.0094	0.0101	0.0092
Ce_2O_3	0.0047	0.0072	0.0114
ZrO_2	0.0529	0.0600	0.0543

from a liquid that is similar in composition to sample 13 can approximate each of the sample compositions. However, this model does not account for the observed variations in Eu anomalies, which range from positive to negative [Sun and Hanson, 1976]. Furthermore, the three-end-member model does not account for the compositional variation in Ba and Sr as well as it does for oxides that exhibit no change in trend with increasing SiO_2 (Figure 4).

Eu, Ba, and Sr are all strongly partitioned into anorthoclase [Sun and Hanson, 1976], and anorthoclase is a prevalent phenocryst in most of the analyzed phonolites. Thus a fourth end member was sought that contained this phase. This was done by projecting the 29 sample vectors for the trachybasalts and phonolites from the original 20-dimensional space (data for Cr_2O_3, NiO, CoO, and V_2O_3 were not used because most concentrations were at the detection limit) into four-dimensional space. The compositions represented by vectors S_3' and S_4' were projected into the same four-dimensional space and renamed S_3 and S_4, respectively. The vector representing sample 13 was relabeled L_2. A vector to represent the third solidus composition, S_5, was sought by using computer programs EQMAIN and EQSCAN [Miesch, 1976a]. The compositions represented by vectors L_2, S_3, S_4, and S_5 are given in Table 8. Table 8 also gives possible mineral compositions for the three

solidus compositions. Solidus compositions S_3 and S_4 can be interpreted as varying amounts of titaniferous clinopyroxene, iron-titanium oxide, kaersutite, plagioclase, and apatite, whereas solidus S_5 can be interpreted as containing more than 50% anorthoclase.

Table 9 gives the required mixing proportions based on the selected liquidus composition L_2 and the three solidus compositions S_3, S_4, and S_5. Solidus compositions S_3 and S_4 are subtracted from liquid L_2 to approximate the composition for all samples. Solidus composition S_5 is subtracted from all but nine samples, six of which are phonolites and three of which are high-silica trachybasalts. Four of these nine samples were analyzed for REE [Sun and Hanson, 1976], and three have positive Eu anomalies suggesting feldspar addition. Sample 25 shows a slight negative Eu anomaly, which disagrees with predictions of the model, but Table 9 indicates that solidus S_5 was added in only very mi-

TABLE 8. Compositions of End Members for the Trachyte-Phonolite Model

	End Member			
Constituent	L_2	S_3	S_4	S_5
SiO_2	43.99	40.13	39.41	51.20
Al_2O_3	16.37	14.35	17.36	19.09
FeO	12.02	15.25	12.21	8.74
MgO	4.72	5.99	5.76	1.85
CaO	10.48	12.95	12.94	6.62
Na_2O	4.49	2.75	4.32	5.26
K_2O	1.82	0.86	0.87	2.50
TiO_2	3.79	4.81	4.60	2.32
P_2O_5	1.55	2.08	1.78	1.45
MnO	0.24	0.28	0.21	0.26
F	0.16	0.19	0.16	0.26
BaO	0.0815	0.0759	0.0814	0.1796
SrO	0.2015	0.2324	0.2589	0.2281
Rb_2O	0.0037	0.0001	0.0013	0.0002
ZnO	0.0127	0.0141	0.0079	0.0082
BeO	0.0008	0.0005	0.0005	0.0000
Y_2O_3	0.0053	0.0068	0.0011	0.0070
La_2O_3	0.0092	0.0070	0.0062	0.0046
Ce_2O_3	0.0114	0.0079	0.0001	0.0069
ZrO_2	0.0543	0.0351	0.0171	0.0391

S_3 is interpreted as an assemblage of approximately 32% plagioclase (Table 5, specimen DV-32-P8), 26% pyroxene (Table 5, specimen RI-55), 26% hornblende (Table 5, specimen RI-50), 13% iron-titanium oxide (Table 5, specimen Ox-3), and 3% apatite (Table 5, specimen Ap-1).

S_4 is interpreted as an assemblage of approximately 36% plagioclase, 2% pyroxene, 49% hornblende, 8% iron-titanium oxide, and 5% apatite. Refer to Table 5 for mineral compositions. Specimen numbers are as given for S_3 above.

S_5 is interpreted as an assemblage of approximately 56% anorthoclase (Table 5, specimen Anor-1), 17% plagioclase (Table 5, specimen DV-32-P8), 15% hornblende (Table 5, specimen RI-50), 9% iron-titaniume oxide (Table 5, specimen Ox-1), and 3% apatite (Table 5, specimen Ap-1).

TABLE 9. Mixing Proportions for the Trachyte-Phonolite Model

Sample No.	L₂	S₃	S₄	S₅	Extent of Magmatic Differentiation, as a Percent of Liquid L₁ (Table 6)
13	1.0000	0.0000	0.0000	0.0000	73.0
39	1.6204	−0.0236	−0.5006	−0.0962	76.9
15	1.4292	−0.1473	−0.2011	−0.0808	75.8
14	1.5226	−0.4072	−0.0015	−0.1139	76.3
38	1.6820	−0.0833	−0.4991	−0.0996	77.2
16	1.6627	−0.6029	−0.2076	0.1478	77.1
41	2.5978	−1.1168	−0.2669	−0.2141	81.1
40	2.3964	−0.8601	−0.4461	−0.0902	80.4
17	2.2009	−1.0547	−0.2534	0.1073	79.6
18	2.0288	−1.1188	−0.1608	0.2508	78.9
42	2.9993	−1.1387	−0.7293	−0.1313	82.5
43	3.2149	−1.3806	−0.6231	−0.2112	83.1
45	3.2788	−1.5753	−0.4848	−0.2187	83.3
44	3.3683	−1.5518	−0.6008	−0.2157	83.5
46	2.9856	−1.6641	−0.2458	−0.0757	82.4
47	3.0731	−1.7206	−0.2390	−0.1135	82.7
48	3.4321	−1.6766	−0.5973	−0.1582	83.7
49	3.4794	−2.0817	−0.2960	−0.1017	83.8
24	4.2702	−2.3921	−0.7265	−0.1516	85.7
29	4.7614	−1.4658	−2.1132	−0.1824	86.6
23	2.4420	−1.4117	−0.4171	0.3869	80.6
20	3.1380	−1.2590	−1.1754	0.2964	82.9
22	3.1949	−1.7098	−0.8027	0.3176	83.0
21	3.2754	−1.8266	−0.6903	0.2416	83.3
19	3.4769	−1.6659	−0.9803	0.1693	83.8
25	3.9781	−2.2997	−0.6795	0.0012	85.0
28	4.3151	−2.4604	−0.8036	−0.0511	85.7
26	5.2029	−2.7407	−1.2322	−0.2299	87.3
27	4.9486	−2.9650	−0.8267	−0.1569	86.9

See Table 4 for end member compositions.

nor amounts. Ba attains its maximum value in the five phonolite samples for which the model calls for the addition of S_5. Eight other samples (15, 24, 27, 28, 29, 41, 47, and 49) were analyzed for REE [*Sun and Hanson*, 1976]. All eight samples have negative Eu anomalies, and all eight samples require subtraction of S_5 according to the model (Table 9).

The recomputed data for each sample, produced by mixing the end member compositions in Tables 4 and 8 in the proportions indicated in Tables 6 and 9, closely approximate the original compositional data. The recomputed analyses are shown in Figure 8. A comparison of Figures 8 and 4 shows that the original trends and much of the original scatter are preserved by the differentiation model. Some of the scatter is lost because factor analysis emphasizes similarities in the data and minimizes minor and apparently random dissimilarities. This is readily seen in the FeO, MgO, and CaO trends for

the basanitoids (Figures 4 and 8). The recomputed data also preserve the alkalic AMF trend, and the metaluminous and silica-undersaturated characteristics of the original data.

The degree to which the data recomputed from the models agree with the original analytical data can be seen from Table 10. The model residuals were derived by subtracting the recomputed data from the original data. The means of the residuals were essentially zero for all oxides. The standard deviations of the residuals are given in Table 10, where they can be compared with those of the original data. The final column of Table 10 shows the ratios of the variances in the recomputed data to those in the original data and, as mentioned previously, may be interpreted as the squares of correlation coefficients between the original and recomputed data. The poorest correlation, for V_2O_3 in the basanitoid model, is 0.70. Most of those for the major oxides are close to unity.

The fact that the models do not account for the original data perfectly indicates that the models do not represent all of the processes that caused compositional variations. However, the fact that the models account for more than 70% of the variance in almost all the oxides (and more than 90% for the major element oxides) suggests that they include the major processes that operated during magma petrogenesis. The fact that the major element oxides are better accounted for than the minor and trace element oxides may be due to the omission of partitioning coefficients, which would make the amount of a trace element removed with the solidus dependent upon the concentration of that element in the liquid. The major oxides may also be better accounted for because smaller proportions of their variances are caused by analytical error.

The proposed model does not quantitatively consider the effects of wall rock contamination. This mechanism may well account for some of the variance between original and recomputed data. Also, because of the control of partitioning coefficients for trace elements between liquid and solid phases, partial melting of wall rocks may have a more pronounced effect on trace elements than on major elements.

Lead and strontium isotopic data place some constraints on the petrogenic model and permit some evaluation of the question of wall rock contamination. *Kurasawa* [1975] and *Stuckless and Ericksen* [1976] reported that the strontium iso-

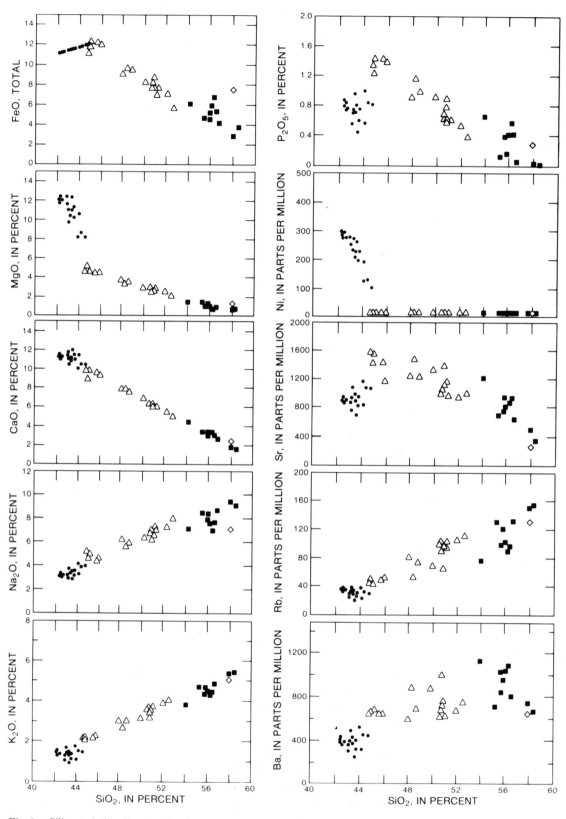

Fig. 8. Silica variation diagrams for sample compositions computed from the Q-mode factor analysis model. Solid circles are basanitoids, triangles are trachybasalts, solid squares are phonolites, and the diamond is a trachyte.

TABLE 10. Standard Deviations of Chemical Variables and of Residuals From the Factor Models

| Constituent | Standard Deviation (49 Samples) | Standard Deviation of Model Residuals | | | Proportion of Total Variance Accounted for by Both Models |
		Basanitoid Model	Trachyte-Phonolite Model	Both Models	
SiO_2	5.54	0.70	0.83	0.79	0.98
Al_2O_3	2.92	0.36	0.51	0.46	0.98
FeO	2.94	0.64	0.39	0.52	0.97
MgO	4.52	0.65	0.36	0.51	0.99
CaO	3.41	0.65	0.35	0.51	0.98
Na_2O	2.18	0.37	0.58	0.51	0.94
K_2O	1.40	0.24	0.16	0.20	0.98
TiO_2	1.17	0.33	0.17	0.25	0.95
P_2O_5	0.38	0.11	0.13	0.12	0.90
MnO	0.026	0.007	0.017	0.013	0.74
F	0.030	0.010	0.016	0.013	0.81
BaO	0.0307	0.0051	0.0194	0.0154	0.75
SrO	0.0402	0.0108	0.0228	0.0188	0.78
Rb_2O	0.0045	0.0010	0.0012	0.0011	0.94
Cr_2O_3	0.0379	0.0102	...	...	0.85*
NiO	0.0153	0.0061	...	...	0.72*
CoO	0.0031	0.0008	...	...	0.61*
ZnO	0.0027	0.0014	0.0009	0.0011	0.82
BeO	0.0006	0.0001	0.0002	0.0002	0.87
V_2O_3	0.0161	0.0043	...	...	0.49*
Y_2O_3	0.0016	0.0006	0.0007	0.0007	0.82
La_2O_3	0.0049	0.0017	0.0014	0.0016	0.90
Ce_2O_3	0.0098	0.0026	0.0031	0.0029	0.91
ZrO_2	0.0383	0.0058	0.0123	0.0102	0.93

*Basanitoid model only.

topes are somewhat more radiogenic in the more siliceous lavas, which suggests that these rocks may be slightly contaminated with crustal strontium. *Sun and Hanson* [1975] reported large variations in lead isotopic ratios which they attributed to derivation of magma from several source materials that were chemically similar but physically separated for about 1500 m.y.

Our *Q*-mode factor analysis does not eliminate the possible petrogenesis by several parallel paths (i.e., similar but separate magmas that evolved independently), but we suggest that a polybaric fractional crystallization mechanism with some contamination offers a better explanation for both the chemical and the isotopic data. Lead isotopic data of *Sun and Hanson* [1975] show that the basanitoids from DVDP 2 are enriched in radiogenic lead relative to other analyzed basanitoids. These same flows contain xenocrystic clinopyroxene [*Weiblen et al.*, 1978], and many contain xenocrystic olivine (P. W. Weiblen and J. S. Stuckless, unpublished data, 1974). Furthermore, these lavas plot in a distinct field that is separated from the trend of the rest of the basanitoids on the stereograms (Figures 5 and 6), and this separation could be accounted for by small additions of foreign material.

A comparison of lead isotopic data for the basanitoids and more evolved rocks yields further evidence of contamination. At each volcanic center the evolved rocks are more radiogenic than the early basanitoids. The trachybasalts from DVDP 2 are distinctly more radiogenic than the basanitoids, but isotopically indistinguishable from a phonolite collected near the drill hole [*Sun and Hanson*, 1975]. A comparison of basanitoids and phonolites at both Cape Bird and Cape Crozier shows the same pattern.

The solidus compositions derived by *Q*-mode factor analysis suggest three different pressures of crystallization. The initial solidus compositions, which are dominated by titaniferous-clinopyroxene with lesser amounts of olivine, iron-titanium oxide, and spinel or ilmenite (S_1 and S_2, Figures 5 and 6), are similar to those obtained at pressures of 20 to 30 kbar or in experimental studies of water-undersaturated alkali-basalt magmas and mantle compositions [*Green*, 1970, 1973]. Higher pressures are precluded by the lack of REE evidence for garnet separation [*Sun and Hanson*, 1976]. Solidus compositions S_3 and S_4, which are interpreted as plagioclase, titaniferous-clinopyroxene, kaersutite, iron-titanium oxide, and apatite, represent crystallizations at much lower pressure, because plagioclase would not be a stable phase at pressures greater than about 11 kbar [*Green*, 1970; *Green and Ringwood*, 1970]. The dominance of anorthoclase in solidus composition S_5 indicates crystallization at very low pressure, because crystallization of a volatile-rich magma at pressures greater than about 5 kbar would yield both plagioclase and potassium feldspar [*Tuttle and Bowen*, 1958].

Data for xenolithic inclusions suggest eruption from three different depths, which are consistent with the three inferred pressures of crystallization. The three different suites of inclusions are (1) ultramafic, which are probably mantle-derived, (2) mafic granulites, which are probably derived from the lower crust, and (3) a variety of igneous, metamorphic, and sedimentary rocks similar to those that crop out in the dry valleys. Depths of derivation for these three suites of inclusions can be obtained from the gravity model of *Smithson* [1972]. This model places the crust-mantle boundary at a depth of about 27 km and the upper-lower crust boundary at about 19 km.

The basanitoids contain inclusions from all three suites, with ultramafic nodules as the dominant type. It is therefore likely that they were erupted rapidly from the mantle. The trachybasalts contain only mafic granulites and upper-crustal xenoliths. The lack of ultramafic nodules suggests that they were erupted from a lower crustal magma chamber for which the total pressure was less than 10 kbar. The phonolites contain only upper-crustal xenoliths, which is consistent with the inferred low-pressure crystal fractionation for this group of rocks. The three separate magma chambers suggest a staging of magma similar to that proposed on geochemical and geophysical grounds for the lavas of Kilauea and Mauna Loa [*Wright*, 1971].

The three separate depths of magma chambers provide three possible sources of extraneous lead and strontium isotopes. The high Sr content for all samples, except number 29 (Table 1), would make it difficult to detect additions of Sr. However, Pb contents are generally much less than 15 ppm [*Goldich et al.*, 1975], and therefore the isotopic composition of lead would be more susceptible to contamination effects. The compositional gap between the basanitoids and L_2 (Figure 5) may indicate a long residence in the intermediate depth magma chamber with complete separation of crystals that formed as S_1 and S_2. During this period of time, wall rock assimilation may have been exten-

sive enough to affect the isotopic composition of lead.

CONCLUSIONS

Major and minor elemental data examined by *Q*-mode factor analysis and constraints imposed by isotopic, petrographic, and field data show that the petrogenetic evolution of the volcanic rocks of Ross Island and vicinity was rather simple in gross aspects, but that it was quite complex in detail. The volcanic rocks can be related to one another by crystal fractionation from a single parent magma, but fractionation occurred at three distinct pressures with minor and variable degrees of wall rock contamination.

Comparison of the composition of the initial magma (Table 4) with experimentally determined compositions [e.g., *Green*, 1970] suggests that the initial magma was derived by a low degree of partial melting within the mantle. Most of the basanitoid series can then be accounted for by the separation of two similar solidus assemblages, which are dominated by clinopyroxene with lesser amounts of olivine and oxide (Table 4), that suggest crystallization of 20 to 30 kbar [*Green*, 1970, 1973]. The origin of basanitoids from DVDP 2 is similar except that some solidus material similar to composition S_2 (Table 4) must be added. Lead isotopic data [*Sun and Hanson*, 1975] and microprobe analyses of clinopyroxenes and olivines [*Weiblen et al.*, 1978] for these lavas suggest that the added material was xenolithic.

The trachybasalt series can be accounted for by the separation of two similar solidus assemblages (S_3 and S_4, Table 8) that are dominated by plagioclase and kaersutite. A third solidus phase, which is dominated by anorthoclase (S_5, Table 8), was added or separated to minor and variable degrees. The dominance of plagioclase in solidus assemblages S_3 and S_4 suggests a crystallization pressure or less than 11 kbar. The pronounced compositional gap between the basanitoid and trachybasalt series (Figure 5) may be due to a complete separation of phenocrysts that formed as S_1 and S_2 during the time that the magma was trapped at the lower pressure.

The phonolite series is closely related to the trachybasalt series, but the anorthoclase-dominated solidus (S_5, Table 8) has a generally stronger effect on the phonolitic compositions. As a result, those phonolites to which S_5 has been added are high in Ba and Sr and have positive Eu anomalies. Conversely, those from which S_5 has been removed are low in Ba and Sr and have negative Eu anomalies. The participation of anorthoclase as a solidus phase indicates that at least some crystal fractionation occurred at very low pressure (<5 kbar [*Tuttle and Bowen*, 1958]).

Mathematical tests of the crystal fractionation model (Table 10) show that more than 70% of the observed variation for most of the oxides can be accounted for by differentiation of a single-parent magma. It seems likely that some of the unexplained variance is due to minor wall rock contamination. This same mechanism provides a reasonable explanation for the observed spread in Pb isotopic data. Xenoliths are abundant in some of the volcanic rocks, and the types found within each group of host rocks are consistent with the depths (inferred from crystallization pressures) of magma chambers: mantle for the basanitoids, lower crust for the trachybasalts, and upper crust for the phonolites. A similar staging or pooling of magma at more than one depth has been proposed for the Hawaiian volcanics [*Wright*, 1971], and it seems likely that such a process is common to many differentiated sequences.

APPENDIX: SAMPLE LOCATIONS AND DESCRIPTIONS

Samples 1-29
Samples described by *Goldich et al.* [1975]: 1-12 are basanitoids, 13-18 are trachybasalts, 19-28 are phonolites, and 29 is a trachyte. Rare earth elements (REE) have been reported for samples 7, 8, 10, 11, 13, 15, 18, 20, 21, 24, 25, 28, and 29 by *Sun and Hanson* [1976]. Pb isotopic data for samples 3, 5, 7, 10, 11, 15, 19, 21, 23, 24, 25, 28, and 29 are given by *Sun and Hanson* [1975]. Sr isotopic data for samples 12, 20, 23, 24, 27, and 29 are given by *Stuckless and Ericksen* [1976].

Sample 30
Basanitoid flow from DVDP 2, 135.3-m depth. Flow is intergranular and glomeroporphyritic with phenocrysts of olivine (11%), titanaugite (9%), opaques (1%), and a trace of plagioclase and xenolithic fragments. Pb isotopic data are given by *Sun and Hanson* [1975, sample DV-52].

Sample 31
Basanitoid flow from DVDP 2, 109.2-m depth. Flow is intergranular and glomeroporphyritic with phenocrysts of olivine (14%), titanaugite (8%), opaques (3%), and a trace of xenolithic fragments. Pb isotopic data are given by *Sun*

and Hanson [1975]; REE data are given by *Sun and Hanson* [1976, sample DV-43].

Sample 32
Basanitoid flow of *Stuckless and Ericksen* [1976, sample DV-49].

Sample 33
Basanitoid flow from DVDP 2, 118.4-m depth. Flow is intergranular and glomeroporphyritic with phenocrysts of olivine (14%), titanaugite (8%), opaques (4%), and a trace of xenolithic fragments.

Sample 34
Basanitoid flow from DVDP 2, 149.8-m depth. Sample is intergranular and stained red with iddingsite after olivine. Phenocrysts are olivine (12%), titanaugite (10%), and traces of opaques and xenolithic fragments.

Sample 35
Basanitoid flow from DVDP 2, 98.8-m depth. Flow is intergranular and porphyritic with phenocrysts of olivine, titanaugite, and opaques.

Sample 36
Basanitoid flow from DVDP 2, 156.2-m depth. Flow is intergranular and stained red with iddingsite after olivine. Phenocrysts are olivine (9%), titanaugite (10%), opaques (1%), and traces of plagioclase and xenolithic fragments.

Sample 37
Basanitoid flow MS-41 of *Stuckless and Ericksen* [1976].

Sample 38
Trachybasalt from Half Moon Crater. Sample is porphyritic and pilotaxitic to intergranular with phenocrysts of titanaugite, kaersutite, olivine, and partly resorbed plagioclase. Apatite and opaque oxides are poikilitic in titanaugite. Xenolithic fragments of olivine-plagioclase granulite occur with the sample.

Sample 39
Trachybasalt HMC-35 of *Stuckless and Ericksen* [1976].

Sample 40
Trachybasalt DV-32 of *Stuckless and Ericksen* [1976]. Pb-isotopic and REE data are reported by *Sun and Hanson* [1975, 1976], respectively.

Sample 41
Trachybasalt flow from DVDP 2, 71.6-m depth. Flow is hyalopilitic with microphenocrysts of titanaugite, plagioclase, kaersutite, and traces of opaque oxides.

Sample 42
Trachybasalt flow from DVDP 2, 46.9-m depth. Flow is hyalopilitic with phenocrysts of plagio-

clase, titanaugite, kaersutite, and trace amounts of olivine and opaque oxides.

Sample 43

Trachybasalt flow from DVDP 2, 28.3-m depth. Flow is intergranular with phenocrysts of kaersutite; microphenocrysts include kaersutite, plagioclase, titanaugite, and trace amounts of olivine and opaque oxides.

Sample 44

Trachybasalt flow from DVDP 2, 19.2-m depth. Flow is hyalopilitic with phenocrysts of plagioclase, kaersutite, and minor olivine.

Sample 45

Trachybasalt flow from DVDP 2, 20.8-m depth. Flow is hyalopilitic with phenocrysts of plagioclase, kaersutite, and minor olivine.

Sample 46

Trachybasalt DV-17 of *Stuckless and Ericksen* [1976].

Sample 47

Trachybasalt DV-8 of *Stuckless and Ericksen* [1976]. Pb-isotopic and REE data are given by *Sun and Hanson* [1975, 1976], respectively.

Sample 48

Trachybasalt flow from DVDP 2, 58.0-m depth. Vesicular trachybasalt with phenocrysts of titanaugite and plagioclase.

Sample 49

Trachybasalt flow DV-5 of *Stuckless and Ericksen* [1976]. Pb-isotopic and REE data are given by *Sun and Hanson* [1975, 1976], respectively.

Sample 50

Friable dunite with opaque bands collected at McMurdo Station, near DVDP 1. Host rock is basanitoid.

Sample 51

Cataclastic dunite collected at McMurdo Station near DVDP 1. Host rock is basanitoid.

Samples 52-59

Samples are described by *Stuckless and Ericksen* [1976]: 52(CC-73) is a harzburgite; 53(CC-69) is a wehrlite; 54(CC-76) is a mixed wehrlite and harzburgite; 55(CC-82) is a lherzolitic harzburgite; and 56(MS-1), 57(MS-21), 58(MS-9), and 59(MS-6) are plagioclase-pyroxene granulites.

Acknowledgments. We are grateful for helpful suggestions from reviewers K. R. Ludwig and W. B. Hamilton and from one anonymous reviewer. Transportation and logistical support were provided by the Naval Task Force 43, U.S. Naval Support Force, Antarctica, Antarctic Development Squadron Six, and the U.S. Coast Guard. The work was supported by a National Science Foundation grant (GV-3695) from the Office of Polar Programs while J. S. Stuckless and S. S. Goldich were at Northern Illinois University.

REFERENCES

Bodkin, J. B., Determination of fluorine in silicates by use of an ion-selective electrode following fusion with lithium metaborate, *Analyst London, 102*, 409-413, 1977.

Bowen, R. W., Graphic normative analysis program, *U.S. Geol. Surv. Comput. Contrib., 13*, 80 pp., 1971.

Bryan, W. B., L. W. Finger, and F. Chayes, Estimating proportions in petrographic mixing equations by least-squares approximation, *Science, 163*, 926-927, 1969.

Cole, J. W., and A. Ewart, Contributions to the volcanic geology of the Black Island, Brown Peninsula and Cape Bird Areas, McMurdo Sound Antarctica, *N.Z. J. Geol. Geophys., 11*(4), 793-828, 1968.

Goldich, S. S., S. B. Treves, N. H. Suhr, and J. S. Stuckless, Geochemistry of the Cenozoic volcanic rocks of Ross Island and vicinity, Antarctica, *J. Geol., 83*, 415-435, 1975.

Green, D. H., A review of experimental evidence on the origin of basaltic and nephelitic magmas, *Phys. Earth Planet. Interiors, 3*, 221-235, 1970.

Green, D. H., Conditions of melting of basanite magma from garnet peridotite, *Earth Planet. Sci. Lett., 17*(2), 456-465, 1973.

Green, D. H., and A. E. Ringwood, Mineralogy of peridotitic compositions under upper mantle conditions, *Phys. Earth Planet. Interiors, 3*, 359-371, 1970.

Hamilton, W., The Hallett volcanic province, Antarctica, *U.S. Geol. Surv. Prof. Pap., 456-C*, 62 pp., 1972.

Harman, H. H., *Modern Factor Analysis*, 2nd ed., rev., 474 pp., University of Chicago Press, Chicago, Ill., 1967.

Hedge, C. E., Nickel in high-alumina basalts, *Geochim. Cosmochim. Acta, 35*, 522-524, 1971.

Imbrie, J., Factor and vector analysis programs for analyzing geologic data, *Tech. Rep. 6, ONR Task 389-135*, 83 pp., U.S. Office of Nav. Res., Geogr. Br., 1963.

Imbrie, J., and E. G. Purdy, Classification of modern Bahamian carbonate sediments, *Mem. Amer. Ass. Petrol. Geol., 1*, 253-272, 1962.

Irvine, T. N., and W. R. A. Baragar, A guide to the chemical classification of common volcanic rocks, *Can. J. Earth Sci., 8*, 523-548, 1971.

Irving, A. J., and D. H. Green, Geochemistry and petrogenesis of the Newer Basalts of Victoria and south Australia, *J. Geol. Soc. Aus., 23*, 45-66, 1976.

Klovan, J. E., and J. Imbrie, An algorithm and Fortran-IV program for large-scale *Q*-mode factor analysis and calculation of factor scores, *J. Int. Ass. Math. Geol., 3*, 61-77, 1971.

Kurasawa, H., Strontium isotopic studies of the Rosso Island volcanics, Antarctica, in *Geochemical and Geophysical Studies of Dry Valleys, Victoria Land in Antarctica, Mem. Spec. Issue 4*, edited by T. Torii, pp. 67-74, National Institute of Polar Research, Tokyo, 1975.

Kyle, P. R., and P. C. Rankin, Rare earth element geochemistry of late Cenozoic alkaline lavas of the McMurdo volcanic group, Antarctica, *Geochim. Cosmochim. Acta, 40*, 1497-1507, 1976.

MacDonald, G. A., and T. Katsura, Chemical composition of Hawaiian lavas, *J. Pet., 1*, 172-177, 1964.

Miesch, A. T., Interactive computer programs for petrologic modeling with extended *Q*-mode factor analysis, *Comput. Geosci., 2*, 439-492, 1976a.

Miesch, A. T., *Q*-mode factor analysis of geochemical and petrologic data matrices with constant row-sums, *U.S. Geol. Surv. Prof. Pap., 574-G*, 47 pp., 1976b.

Smithson, S. S., Gravity interpretation in the Transantarctic Mountains near McMurdo Sound, Antarctica, *Geol. Soc. Amer. Bull.*, *83*, 3437-3442, 1972.

Stuckless, J. S., and R. L. Ericksen, Strontium isotopic geochemistry of the volcanic rocks and associated megacrysts and inclusions from Ross Island and vicinity, Antarctica, *Contrib. Mineral. Petrol.*, *58*, 111-126, 1976.

Stuckless, J. S., P . W. Weiblen, and K. J. Schulz, Magma evolution for the alkalic rocks of the Ross Island Area, Antarctica (abstract), *Eos Trans. AGU*, *55*, 474, 1974.

Sun, S. S., and G. N. Hanson, Origin of Ross Island basanitoids and limitations upon the heterogeneity of mantle sources for alkali basalts and nephelinites, *Contrib. Mineral. Petrol.*, *52*, 77-106, 1975.

Sun, S. S., and G. N. Hanson, Rare earth element evidence for differentiation of McMurdo volcanics, Ross Island, Antarctica, *Contrib. Mineral. Petrol.*, *54*, 139-155, 1976.

Treves, S. B., Volcanic rocks from the Ross Island, Marguerite Bay and Mt. Weaver areas, Antarctica, *Jap. Antarct. Res.* *Exped. Sci. Rep. 1*, pp. 136-149, Nat. Inst. of Polar Res., Tokyo, 1967.

Treves, S. B., Hyaloclastite of DVDP-3, Hut Point Peninsula, Ross Island, Antarctica (abstract), *Dry Val. Drilling Proj. Bull.*, *8*, 103, 1978.

Tuttle, O. F., and N. L. Bowen, Origin of granites in the light of experimental studies in the system $NaAlSi_3O_8$–$KAlSi_3O_8$–SiO_2–H_2O, *Geol. Soc. Amer. Mem.*, *74*, 153, 1958.

Weiblen, P. W., J. S. Stuckless, W. C. Hunter, K. J. Schulz, and M. G. Mudrey, Correlation of clinopyroxene composition with environment of formation based on data from Ross Island volcanic rocks, *Dry Val. Drilling Proj. Bull.*, *8*, 108-115, 1978.

Wright, T. L., Chemistry of Kilauea and Mauna Loa lava in space and time, *U.S. Geol. Survey Prof. Pap.*, *735*, 40 pp., 1971.

Yoder, H. S., and C. E. Tilley, Origin of basalt magmas—An experimental study of natural and synthetic rock systems, *J. Pet.*, *3*, 342-532, 1962.

SEDIMENTOLOGY AND PETROLOGY OF CORE FROM DVDP 15, WESTERN MCMURDO SOUND

P. J. Barrett

Antarctic Research Centre and Department of Geology, Victoria University of Wellington
Wellington, New Zealand

S. B. Treves

Department of Geology, University of Nebraska, Lincoln, Nebraska 68504

The first drilling into the floor of McMurdo Sound (DVDP 15) took place in November 1975 16 km east of Marble Point through 122 m of water. The purpose was to core the Cenozoic glaciogenic sequence estimated to be 300 m thick. The 2-m-thick annual ice was used as a drilling platform until November 21, when the appearance of cracks in the ice caused the hole to be terminated. The drill penetrated 65 m below the sea floor, recovering a total of 34 m of core. Over 278 kg of wash samples were also collected. The core is mainly sand made up chiefly of glassy and crystalline basaltic grains, with persistent quartz, feldspar, and granitic fragments. The sources were the late Cenozoic McMurdo Volcanics to the south and the basement complex, Beacon and Ferrar rocks in the adjacent Transantarctic Mountains. The top 12 m of core (unit 1) has pebbles scattered throughout and has been much disturbed by drilling. A richly fossiliferous interval at the base of the unit contains bryozoa, foraminifera, and ostracods that indicate a Recent age. The texture of the sediment suggests deposition from wind and melting ice. Unit 2, from 13 to 65 m, consists largely of basaltic sand, like unit 1, but differs in its lack of pebbles and the well-developed stratification. The stratification and texture suggest that it accumulated largely from sediment transported by wind and was only slightly modified by bottom currents. The basaltic character of the sand and the slight lithification suggest a Plio-Pleistocene age. Increase in deformation of the stratification toward the top of the unit is attributed to movement of grounded ice. A detailed description of the core is given in the appendix.

SITE DATA

The following are data for the core site DVDP 15.

First core taken	November 6, 1975
Last core taken	November 21, 1975
Position	77°26'14.19"S, 164°22'49.39"E
Water depth	122m
Tidal range	1.4 m
Total penetration	64.6 m
Number of cores	45
Total section cored	61.6 m
Total section recovered	31.8 m
Percentage core recovered	52%
Number of wash samples	37
Weight of sediment recovered	278 kg
Oldest sediment cored	
Depth below sea floor	64.6 m
Lithology	muddy basaltic sand
Age	?Plio-Pleistocene

BACKGROUND AND OBJECTIVES

Most of our knowledge of the history of the East Antarctic Ice Sheet has come from glacial landforms and deposits along the Transantarctic Mountains and particularly in the dry valley region. However, the terrestrial record is confused because of its patchy distribution, and its interpretation has been made difficult by the lack of fossils for correlation. There has been some success in working out a history for the last few million years [*Denton et al.*, 1971; *Armstrong and Stuiver*, 1971; *Bull and Webb*, 1972], and from cores in Taylor Valley [*McKelvey*, 1977; *Webb and Wrenn*, 1977; *Brady*, 1977], but the earlier glacial history remains something of a mystery.

Evidence of a long history of Antarctic glaciation was recently obtained by the *Glomar Challenger* from three drill holes (Deep Sea Drilling Project

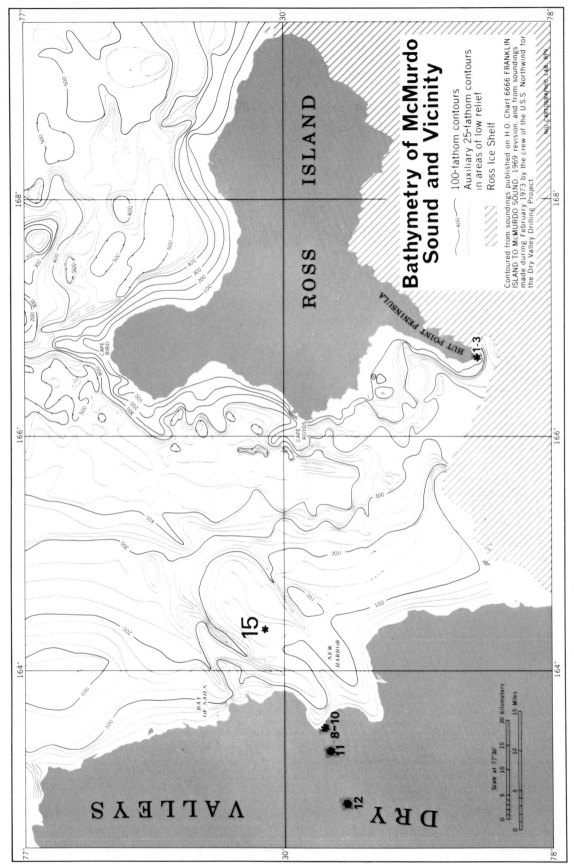

Fig. 1. Bathymetric map of McMurdo Sound [after *McGinnis*, 1973], showing the location of DVDP drill sites.

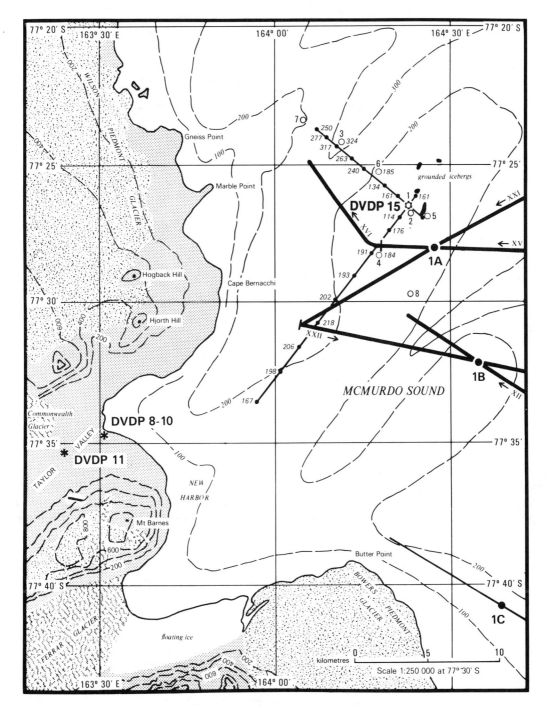

Fig. 2. Map of the New Harbor area, showing DVDP drill sites (stars), bottom sample sites (circles), and soundings from the sea ice (in meters). Land contours and coastline are from H. O. Chart 6666 Franklin Island to McMurdo Sound, 1969 revision; submarine contours are from *McGinnis* [1973] and *Northey et al.* [1975]. The heavy lines are seismic profile tracks of *Northey et al.* [1975]; the three drill sites that they proposed are also shown (1A, 1B, 1C).

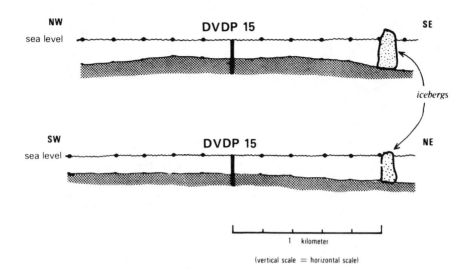

Fig. 3.   Bathymetry around DVDP 15, measured at 200-m intervals along the two perpendicular lines shown in
Figure 2.

(DSDP) sites 270-272) in the south central Ross Sea 400 km east of McMurdo Sound. Those cores revealed a thick sequence of marine glacial mud and mudstone underlain by thin shallow marine and estuarine sand resting in turn on marble breccia and a marble basement [*Hayes et al.*, 1975; *Barrett*, 1975]. Paleontologic, radiometric, and paleomagnetic dating gave a Late Oligocene (25 m.y.) age for the base of the glacial sequence, older than most workers had expected.

Continuous seismic profiling [*Northey et al.*, 1975] showed that McMurdo Sound is underlain by a thick subhorizontal sedimentary sequence. The seismic records clearly show the submarine extensions of Taylor and Ferrar valleys cutting into the sequence, suggesting that most of the sequence is

TABLE 1.   Interpretation of the Sedimentary Sequence Beneath DVDP 15

Reflector	V, km/s	Thickness at DVDP 15 (Water Depth 122 m)	Interpreted Lithology	Age
Sea floor				
	2.1	100 m	pebbly muddy sand	Plio-Pleistocene
A				
	2.1	200 m	pebbly mudstone	Miocene
C				
	2.5		preglacial sand-stone and mudstone	early Cenozoic or older
D*				
	3.2			

*Reflector not recognized in profiles from DVDP 15.

TABLE 2.   Grain Size of Sea Floor Sediment Samples From Around DVDP 15

Sample Site/Number	Water Depth, m	Percent Gravel	Percent Sand	Percent Mud	Comments
1A	122	16	80	4	dark, moderately sorted sand
1B		7	92	1	with variable amount of fine
1C		50	48	2	gravel (pebbles up to 30 mm
1D		20	79	1	long)
1E		20	80	trace	
1F		41	55	4	
1G		17	82	1	
2A	124	trace	70	30	yellowish green muddy fine to medium sand, with abundant organic material including bryozoa and ophiuroids
3C	324	. . .	92	8	moderately sorted medium sand, a few sponge spicules
4A	184	2	69	29	dark, poorly sorted muddy coarse sand with sponge spicules up to 10 cm long
5A	205	1	84	15	dark, poorly sorted fine sand with
5B		trace	39	61	pebbles up to 30 mm in 5A
6A	185	. . .	32	68	dark sandy mud; thin mollusc shells form 1% of sample
7C		trace	88	12	dark, moderately sorted medium sand
8	173	trace	86	14	dark, moderately sorted medium sand
G863	495	9	51	40	gravelly muddy very fine sand
G865	366	trace	67	33	muddy fine sand
G866	291	20	49	31	gravelly muddy fine sand
G867	312	20	63	17	gravelly muddy medium sand

Samples from sites 1–7 were collected with an Eckman grab; for site 8 an Orange Peel grab was used. Sites are shown in Figure 2. Samples prefixed by G have been described previously from the area [Glasby et al., 1975]. Many samples contained a large amount (up to 20 wt %) of biogenic material; this has not been included in the figures.

relatively old (pre-Pliocene) and that coring through it could well provide the best evidence thus far on the initiation and early history of the East Antarctic Ice Sheet. DVDP 15 was the first attempt to core the sequence.

RESULTS OF SITE SURVEYS

Surveys of the area around DVDP 15 fall into two main categories: (1) surveys to provide information for locating the drill site (bathymetric and seismic surveys) and (2) surveys of the present oceanographic and sedimentary environment (by bottom sampling and current measurements). The drill site itself was precisely located (see the section on site data) by using tellurometer and theodolite by John Williams and Bill Wicks, New Zealand Lands and Survey Department, Wellington. In addition, vertical deflection of the sea ice around the drill site was monitored to ensure the safety of the drilling operation (S. B. Treves, cited by *Barrett et al.* [1976]).

Bathymetric Surveys

A bathymetric map of McMurdo Sound (Figure 1) was prepared by *McGinnis* [1973] from existing data and from three north-south profiles on the western side of the sound. McGinnis noted that the western part of the sound represented a broad upland into which were cut valleys whose form suggested that they resulted from erosion by expanded Ferrar and Taylor glaciers. It was evident from this bathymetry that the only sites more than 5 km offshore that could be drilled with the equipment available were on the upland along the western side of the sound. The eastern part of the sound has much deeper water accentuated by a depression west and north of Ross Island attributed to isostatic loading of the sea floor. Subsequent seismic profiling data have provided further evidence for this interpretation [*Northey et al.*, 1975]. Further bathymetric data were collected by Northey et al. during the profiling survey and are shown in part in their Figure 3.

TABLE 3. Composition of Sand Size Grains From Bottom Sediments Around DVDP 15

| | Basaltic Grains, % | | | | | | Grains Mainly From Basement Rocks, % | | | | | |
| | Lithic Fragments | | Crystals | | | | | | | | | |
Site	Glassy	Crystalline	Olivine	Clino-pyroxene	Plagioclase	Total	Plagioclase	K Feldspar	Quartz	Other	Grains Identified	Unknown
2(1)	62	3	7	8	5	85	9	1	3	2	192	11
2(2)	64	6	3	6	7	86	8	2	3	1	238	7
3	62	4	2	8	4	80	7	2	8	3	201	8
4	65	6	3	3	2	79	10	3	5	3	198	7
6	77	5	1	2	3	88	6	x*	4	2	212	10
7	59	4	4	5	3	75	1	10	2	2	216	7
8	76	2	2	2	3	85	10	10	2	3	199	9

Sites are located in Figure 2. Nonbasaltic grains are mainly of granitic and metasedimentary origin but include a few rounded quartz from the Devonian sandstone and rare grains of Jurassic dolerite. Fresh plagioclase has been included with the basaltic grains, and altered plagioclase with the nonbasaltic grains.
*Indicates trace amount.

Figure 2 shows a synthesis of the available bathymetry in the region of the drill site. The bathmyetry is tied to depths that we have measured along two surveyed lines intersecting at the drill site (DVDP 15). These depths were measured through holes in the sea ice by using weighted line and a meter wheel and are believed to be within 2 m of the true depth. The contours are based on data from both *Northey et al.* [1975] and *McGinnis* [1973], remembering that navigational limitations may have resulted in errors of location as much as 2 km. More detailed bathymetry within 2 km of the drill site is shown in Figure 3.

Seismic Surveys

Continuous seismic profiling to determine sediment thickness and structure in McMurdo Sound was first attempted in January 1973. Despite problems with equipment the results were sufficient to indicate at least 135 m of sediment of pre-Pliocene age on the western side of the sound [*Sissons and Northey*, 1974; *Barrett et al.*, 1974]. A similar but more extensive survey was carried out the following season [*Northey et al.*, 1975]. The results show that the sound is underlain by a thick sequence of strata that dip gently east-northeast beneath Ross Island. In only one area covered by the survey, in the southwest corner of the sound, was the sea floor shallow enough (<250 m) for drilling. Five sites were selected, with the following constraints: (1) The first drill hole was to be sited in water no deeper than 150 m. If this hole were successful, a second hole could be drilled in water up to 250 m deep. (2) All sites were picked to lie on seismic lines so that the reflectors identified in the survey data, and which extend right across the sound, could more easily be related to rock types in the drill hole. The submarine ridge between the extension of Taylor and Ferrar valleys was chosen for the prime site because it was close to the mouth of the Taylor Valley, where DVDP had successfully cored a thick sequence of sediments that could be in part correlative, and because there were places on the ridge that filled the requirements of water depth and seismic data. Site 1A (Figure 2) was selected in the only part of the ridge that lay in less than 150 m of water; it was also at the intersection of two seismic tracks.

The sequence beneath the sea floor at DVDP 15 has been interpreted (Table 1) from sonic velocities obtained from three sonobuoy runs further out in

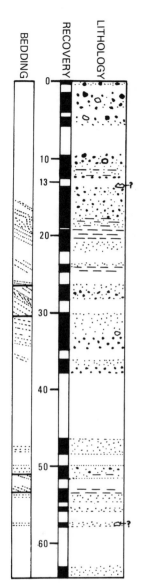

UNIT 1

0-13 m

Fine to coarse sand, olive gray to olive black, moderately to very poorly sorted, with grains mostly basaltic in origin. Scattered pebbles up to 70 mm long of basalt, dolerite, granitic and metasedimentary rocks. Core soft and much disturbed by drilling.

UNIT 2

13-65+ m

Fine to medium sand, olive black, moderately to poorly sorted, with grains mostly basaltic in origin. Lacks pebbles. Most core shows plane parallel stratification on a scale 5 to 50 mm. Thin beds of olive gray silt are common from 19 to 25 m and from 50 to 53 m. Recovered core is firm and shows little drilling disturbance, though intervals with poor recovery may represent softer sand that has washed away.

Fig. 4.   Summary of lithologies and structures in DVDP 15 core.

McMurdo Sound [*Northey et al.*, 1975]. The interpretation is based on velocities obtained from the thick glacial sequence at DSDP site 270 [*Hayes et al.*, 1975] and from pebbles of Eocene sandstone from Black Island [*Barrett and Froggatt*, 1978].

*Texture and Mineralogy of
Sea Floor Sediments*

The nature of the sea floor in the area of DVDP 15 was not known until it was sampled after the ice hole was cut at the site. However, some idea was obtained from four grab samples taken in deeper water about 15 km to the southwest. All four samples are muddy sand with granules and pebbles forming as much as 20% of the sample [*Glasby et al.*, 1975]. Sorting is extremely poor (Folk standard deviation from 2.9 to 5.5 phi), though three of the four samples show a clear mode in the fine and very fine sand grades. The fourth, G867, has a broad mode that includes the medium sand grade also. Glasby et al. noted that the sand had a high proportion of basaltic grains, presumably carried by wind or ice from the late Cenozoic volcanic piles to the south.

288   DRY VALLEY DRILLING PROJECT

SIZE FREQUENCY CURVES FOR DVDP 15

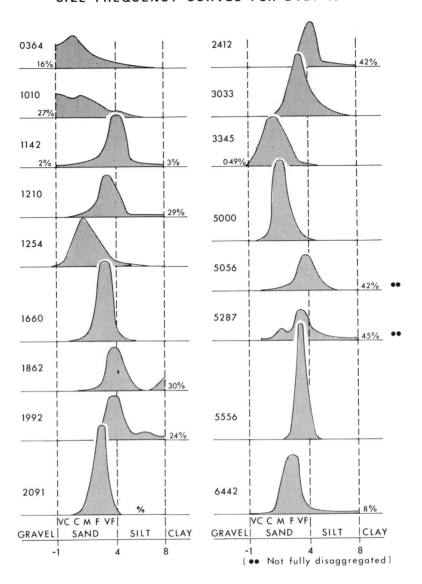

Fig. 5a. Grain size frequency curves, samples from DVDP 15. Four-digit numbers are centimeters subbottom.

Seven bottom samples were taken from the ice hole at the drill site before the rig was set up, and a further eight samples from seven sites were taken through seal holes within 5 km of the site. Proportions of gravel, sand, and mud (Table 2) indicate that the texture of sediment varies considerably not only from site to site but also for different samples from one station. Further work is being carried out to explain the texture of these sediments. It is clear, however, that they are quite different from the marine till that characterizes the late Cenozoic out in the Ross Sea [Barrett, 1975].

The sand on the sea floor at and around the drill site (Table 3) is dominated by material from the late Cenozoic McMurdo Volcanic Group [Smith, 1954; Goldich et al., 1975]. About 75% of the grains are of glassy basalt, and a further 10% are crystals of clinopyroxene and lesser olivine. Most of the remaining 15% is feldspar, some which is fresh and similar in appearance to that in the basaltic fragments. However, a small proportion are sericitized and are more likely to come from the granitic basement rocks of the dry valley region. Other grains from pre-Cenozoic rocks on the 'mainland' include

SIZE FREQUENCY CURVES FOR
SEDIMENT IN MCMURDO SOUND REGION

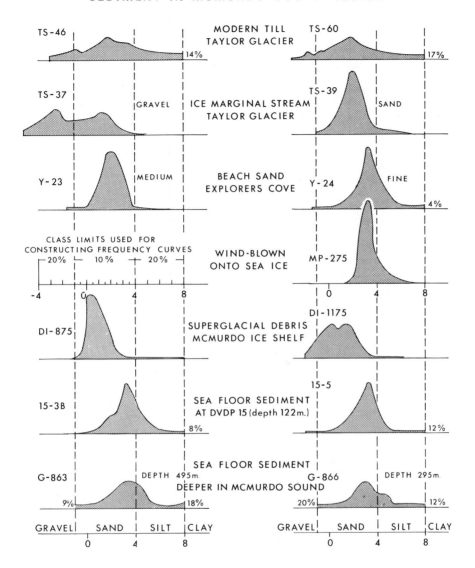

Fig. 5b. Grain size frequency curves, samples from different modern sedimentary environments in Taylor Valley
and McMurdo Sound.

green hornblende, brown biotite, and rounded quartz grains, the latter probably reworked from Devonian sandstone. Together with the feldspar they form less than 5% of the samples, a remarkably low proportion considering the proximity of the area to the coast of south Victoria Land (Figure 1). A feature of the basaltic fragments is the range in degree of rounding from very angular to well rounded. The well-rounded material must have been blown or washed around for some time

prior to deposition. Most of the gravel fraction is of granule and fine pebble grade and forms a significant component of the sediment only at the drill site. The bulk of the material appears to be basaltic.

The biogenic component of grab samples from the drill site was examined by John Oliver [Barrett et al., 1976] and found to be similar to the 'deep shelf mixed assemblage' of Bullivant [1967]. Current velocities were measured at the site over a 24-

TABLE 4. Grain Size Data (as Cumulative Percent) for Samples From DVDP 15

Sample Number	Class Limits, phi units													
	< −1.0	0.0	1.0	1.5	2.0	2.5	3.0	3.5	4.0	5.0	6.0	7.0	8.0	Rest
Unit 1														
0364	16.3	37.1	62.2	72.0	78.1	82.6	87.1	91.2	93.6	N.D.	N.D.	N.D.	N.D.	100.0
1010	27.3	45.4	61.4	70.1	77.9	84.0	88.9	92.0	94.7	N.D.	N.D.	N.D.	N.D.	100.0
1142	1.9	3.9	7.5	9.9	12.5	15.4	20.9	30.1	50.7	88.6	93.3	95.4	96.8	100.0
Unit 2														
1210A	0.1	0.4	1.6	3.3	5.9	10.4	21.3	38.3	53.0	65.3	67.8	69.5	71.4	100.0
1210B				1.9	6.0	12.0	27.3	47.1	61.0	70.5	73.1	75.2	77.2	100.0
1254	0.8	3.5	25.6	45.7	63.2	76.9	86.7	92.2	95.0	N.D.	N.D.	N.D.	N.D.	100.0
1660			2.3	1.4	5.2	17.1	47.5	79.6	94.5	N.D.	N.D.	N.D.	N.D.	100.0
1862			0.1	0.6	1.2	1.9	4.4	16.2	33.4	58.3	64.1	64.2	70.2	100.0
1992					0.1	0.9	5.1	20.5	38.6	60.2	65.5	72.9	76.5	100.0
2091		0.6	0.6	4.0	16.5	43.2	78.7	95.7	99.3	N.D.	N.D.	N.D.	N.D.	100.0
2412							0.2	2.8	16.1	47.0	52.9	55.8	58.0	100.0
3033			0.2	0.9	3.7	13.2	31.8	56.5	73.9	N.D.	N.D.	N.D.	N.D.	100.0
3345	0.4	10.8	43.0	63.0	79.2	91.1	97.2	99.0	99.5	N.D.	N.D.	N.D.	N.D.	100.0
5000		0.3	3.9	30.0	62.1	83.3	93.7	97.7	98.9	N.D.	N.D.	N.D.	N.D.	100.0
5056*			0.1	0.2	0.6	1.5	4.5	13.8	27.6	46.9	54.6	57.8	57.8	100.0
5287*			0.5	2.6	6.8	10.1	15.7	27.7	39.2	47.2	51.3	53.2	55.0	100.0
5556						1.2	19.8	65.8	92.0	N.D.	N.D.	N.D.	N.D.	100.0
6442	0.3	0.5	1.2	5.4	18.4	40.8	64.8	76.6	80.5	84.4	87.0	89.4	92.1	100.0

Sample numbers also indicate centimeters below sea floor. N.D. is not determined.
*Not well disaggregated.

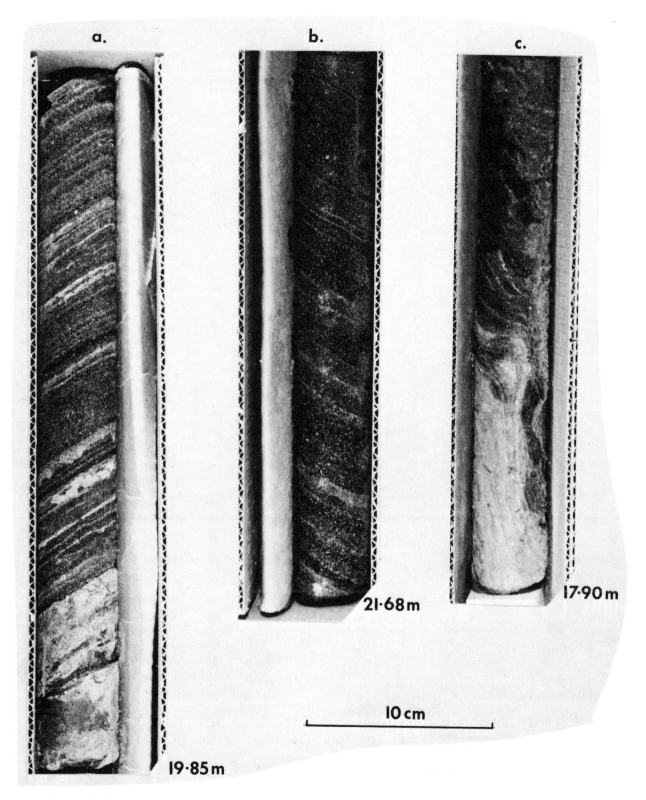

Fig. 6.   Sedimentary features from unit 2, DVDP 15. (*a*) Silt beds (light-colored between 19 and 20 m). (*b*) Graded beds averaging 10 mm thick at 21 m. (*c*) Deformed stratification at 17.9 m.

TABLE 5.   Grain Composition (in Percent)

Wash Number	Depth, m	Crystals										
		Olivine	Augite	Kaersutite	Opaques	Hornblende	Biotite	Quartz	K Feldspar	Plagioclase	Total
1	0.0-0.3	1	6	...	...	...	1	13	1	5	27
3	1.5-2.1	2	20	...	...	...	...	9	2	3	36
4	4.0-4.6	...	10	...	2	1	...	23	3	5	44
6	4.6-5.1	...	9	1	7	...	...	22	4	3	46
7	5.1-5.3	...	13	...	1	...	1	28	6	5	54
8	5.3-5.4	...	13	...	3	...	1	23	8	2	51
11	7.0-7.9	...	12	...	3	...	1	22	4	5	47
12	7.9-8.9	...	10	1	2	...	1	24	5	4	47
13	-9.8	...	12	...	5	...	1	17	7	4	48
14	7.9-12.6	1	11	1	3	...	...	13	7	2	38
15	-12.5	3	15	...	7	...	...	12	6	5	48
16	12.7-13.8	2	11	...	5	...	...	16	5	4	43
(Run 12)	14.7	2	7	...	1	...	...	18	3	4	35
17	13.8-15.0	...	10	...	1	1	1	12	11	8	42
18	11.6-19.0	1	12	1	5	...	...	11	6	2	45
19	19.0-22.1	...	13	...	1	...	...	12	8	4	38
(Run 19)	-27.6	2	10	2	4	...	...	19	4	2	42
20	27.2-30.8	2	15	...	7	...	1	13	6	5	49
21	33.4-37.3	...	17	...	4	1	...	12	4	4	42
22	-40.3	2	6	1	5	...	...	15	5	3	37
23	39.5-41.8	2	9	...	2	...	2	11	2	6	34
24A	32.3-33.2	...	12	...	3	...	...	16	7	6	44
24B	33.2-39.4	...	13	2	2	...	1	6	1	6	31
25	39.4-45.1	...	9	1	...	...	...	13	1	7	31
26	45.1-48.2	...	8	...	...	...	...	5	x	5	19
27	45.6-48.5	...	11	...	1	...	...	5	1	4	22
28	51.6-54.6	...	16	...	x*	...	...	5	1	2	25
29	54.6-56.0	1	15	...	...	...	...	5	2	5	29
31	36.3-39.3	1	13	2	...	...	1	14	2	11	42
32	39.3-45.4	...	8	...	...	...	...	13	1	2	25
(Run 37)	57.5	x	4	...	1	...	...	16	3	5	29
34	51.5-57.6	1	20	1	...	...	...	5	...	3	29
(Run 41)	60.1-63.6	1	11	1	...	...	...	13	2	5	33
35	57.6-61.7	...	15	1	...	...	...	8	1	2	27
36	61.7-62.8	...	26	1	...	...	...	4	...	6	37
(Run 42)	63.6-64.4	...	8	...	1	1	1	14	1	5	31
37	62.8-64.4	...	11	1	...	...	1	18	1	3	35

*Indicates trace amount.

hour period and did not exceed 0.12 m s^{-1} [Barrett et al., 1976, Table 5].

SEDIMENTOLOGY OF DVDP 15 CORE

DVDP penetrated only 65 m of the thick sedimentary sequence in western McMurdo Sound before drilling was terminated by deteriorating ice conditions and a pocket of methane. Nevertheless, 52% of the cored interval was recovered, along with 278 kg of wash material. The degree of core deformation was difficult to assess in places, although it was clear that some of the core had been highly disturbed by drilling. Our assessment of the condition of the core has been recorded in the detailed core log in the appendix, which also includes a brief discussion of the procedures used to collect core and wash samples. The 65 m cored is mainly olive gray to olive black basaltic sand and is divided into two units (Figure 4).

Unit 1

Unit 1 is mainly poorly or very poorly sorted fine to coarse gravelly sand with a few beds up to 0.6 m thick of very fine sandy silt. No stratification finer for Wash Samples From DVDP 15

	Basalt Fragments							Other Lithic Fragments						Grains Counted
Aphyric	Kaersutite	Pyroxene	Plagioclase	Olivine-Pyroxene	Glassy	Tuff and Breccia	Total	Trachyte	Dolerite	Granitic Rock	Gneiss	Shell Fragments	Total	
8	1	8	11	2	33	...	63	...	...	9	...	1	73	156
13	...	4	5	2	35	...	59	1	1	3	...	...	64	219
78	...	2	67	1	37	...	53	...	1	2	...	...	56	161
9	...	4	3	...	35	1	52	1	...	1	...	...	54	164
7	...	1	3	1	33	...	45	...	...	1	...	...	46	163
15	...	...		...	28	...	48	2	...	...	...	...	49	172
12	...	5	2		31	1	51	...	...	2	...	...	53	161
6	..	4	5	...	37	...	52	...	...	1	...	...	53	163
9	...	2	7	2	32	...	52	...	...	2	...	...	54	175
6	...	6	10	...	26	2	50	2	...	10	...	...	62	100
10	...	1	3	...	35	...	50	...	...	2	...	1	52	111
11	...	2	5	1	34	...	53	...	...	4	...	...	57	107
20	...	...	3	...	37	...	60	...	...	5	...	...	65	129
8	...	10	5	...	30	1	54	...	...	3	...	1	58	104
12	...	5	5	...	38	...	60	1	1	2	...	...	64	141
9	...	3	8	1	32	...	53	...	...	9	...	...	62	130
11	...	2	6	...	37	...	56	...	...	2	...	...	58	124
5	...	10	3	...	28	...	46	...	...	4	...	1	51	126
13	...	2	3	3	29	1	51	...	2	4	...	1	58	100
7	...	5	9	...	35	...	56	...	...	7	...	...	63	130
12	...	2	5	...	37	...	56	...	1	9	...	...	66	125
9	...	5	6	1	30	...	51	...	...	5	...	...	56	121
27	...	4	4	6	26	...	67	1	1	1	...	...	69	196
14	...	6	6	...	40	1	57	...	1	1	...	1	69	199
27	...	3	3	4	44	...	81	...	...	...	...	x	...	81
20	x	8	1	1	48		78	...	...	...	...	...	78	403
17	...	5	4	1	47	...	74	1	...	...	...	...	75	255
24	...	10	...	4	32	...	70	...	1	x	...	...	71	247
15	...	2	2	...	37	...	56	1	...	...	...	1	58	188
32	...	6	2	5	29	...	74	1	1	...	...	...	75	167
22	...	3	...	3	40	...	68	...	...	3	...	...	71	206
16	...	5	4	2	43	...	70	...	...	1	...	...	71	130
14	...	2	3	2	43	...	64	1	...	2	...	...	67	180
18	...	6	6	2	39	...	71	...	...	2	...	...	73	195
11	...	13	4	1	32	...	61	1	...	2	...	...	63	156
18	...	1	5	4	39	...	67	...	...	2	...	...	69	147
29	...	5	8	5	14	...	61	1	1	2	...	...	65	148

than a few centimeters could be seen, though it may have been lost by core deformation during drilling. The gravel component is scattered throughout the core and is mainly basaltic in character, though 10–20% of the pebbles are of granite. Some dolerite was also found.

Grain size distributions were determined for three samples from unit 1 (Figure 5a; Table 4), two of coarse gravelly sand and one of sandy coarse silt with very little clay. The gravelly sand samples are similar in texture to a number of local deposits, including some of the grab samples around the drill site [Barrett et al., 1976], stream gravel in Taylor Valley [Powell, 1976] (samples TS 37, TS 38), surface debris on the McMurdo Ice Shelf, and what appears to be ablation till on The Strand Moraines. The textures are quite out of character with the present stable low-energy sedimentary environment, which is below wave base and has barely detectable currents [Barrett et al., 1976]. They are therefore taken to be relict or inherited and are believed to result mainly from ice rafting of sediment already partly sorted by surface streams. The gravelly sand, though poorly sorted, is still much

TABLE 6. Grain Composition (in Percent) for Samples From the Edge of the McMurdo Ice Shelf and from The Strand Moraines

| | McMurdo Ice Shelf | | | | The Strand Moraines | |
	Sample 1	Sample 2	Sample 3a*	Sample 36†	Sample 1	Sample 2
Mineral Grains						
Olivine	6	6	. . .	5	6	5
Augite	6	4	2	3	6	5
Kaersutite	. . .	. . .	. . .	. . .	. . .	TR
Opaque minerals	1	1	. . .	. . .	2	. . .
Hornblende	. . .	. . .	. . .	. . .	1	1
Biotite	. . .	. . .	. . .	. . .	. . .	. . .
Muscovite	. . .	. . .	. . .	. . .	TR	. . .
Quartz	TR	. . .	. . .	1	10	2
K feldspar	TR	TR	. . .	. . .	7	6
Plagioclase	9	1	. . .	1	12	11
Calcite	. . .	. . .	. . .	. . .	. . .	1
Total, %	22	12	2	10	45	31
Rock Fragments						
Basaltic fragments						
Aphyric	13	16	29	19	3	6
Kaersutite	3	2	. . .	TR	. . .	. . .
Pyroxene	5	5	. . .	5	1	2
Olivine-pyroxene	4	25	25	24	7	8
Glassy	49	39	39	39	32	39
Tuff and breccia	. . .	. . .	. . .	. . .	. . .	. . .
Total, %	74	86	93	87	43	55
Other lithic fragments						
Trachyte	3	2	5	2	1	3
Dolerite	1	1	. . .	1	1	2
Granitic rock	TR	. . .	. . .	. . .	9	8
Sandstone	. . .	. . .	. . .	. . .	1	1
Gneiss	. . .	. . .	. . .	. . .	. . .	. . .
Shell fragments	. . .	. . .	. . .	. . .	. . .	. . .
Total, %	4	3	5	3	12	14
Grand total (grains counted)	500	500	82	500	500	300

*Grain size 3-5 mm.
†Grain size average 0.2 mm.

better sorted (owing largely to the lack of silt and clay) than local basal glacial debris [*Powell*, 1976] (samples TS 42, TS 46).

Unit 2

Unit 2 is dominated by basaltic material like unit 1 but tends to be finer grained and better sorted. Also it lacks the pebbles of unit 1; three pebbles were seen in unit 2, but only one, a granitic fragment 15 mm long, was in place in the core. A further difference is in the preservation of stratifica-

tion throughout most of the core of unit 2, presumably due to incipient cementation.

No trend in grain size or sorting is evident in unit 2. The sediment typically ranges between moderately sorted fine sand and poorly sorted medium sand, though lenses and thin beds of coarser and finer sediment do occur. The most notable are the beds of coarse silt up to 5 cm thick (Figure 6a), which are common between 18 and 25 m and between 50 and 53 m. Their fractured appearance and lack of internal stratification indicate that they have been quite severely deformed during drilling. The siltstone fragments between 14 and 16 m may represent similar beds more severely disturbed during drilling.

Stratification can be seen in most of the core recovered in unit 2. In scale it ranges from 5 to 50 mm and is plane parallel virtually throughout the unit. The only exceptions to this are three small-scale cross beds from 27 to 28 m subbottom and some soft sediment deformation from 16 to 19 m. The stratification results from grain size variations which in places are quite subtle. Only one clear instance of graded bedding was found (Figure 6b). The bedding over this 10-cm interval averages 12 mm in thickness. The grading is believed to have resulted from a seasonal effect on sedimentation, with the silt layer representing winter deposition. The layers could possibly have been deposited by a rapid mass movement process, but no loading features could be seen. The moderate to poor sorting, the scarcity of small-scale cross-bedding in such a sand-rich sedimentary environment, and the extensive fine parallel stratification suggest that the sediment was deposited mainly from suspension having been transported to the site by wind, with only a small contribution from floating glacier or shelf ice, and winnowed by very gentle bottom currents. The physical conditions were probably similar to those at the site today.

The dip of the stratification in unit 2 decreases with depth. Between 16 and 30 m subbottom the dip is mostly about 30°, though in at least two places it increases to 90° over a few centimeters (Figure 6c). From 30 to 40 m the dip averages 20° and is mostly horizontal in the beds beneath. The nonhorizontal dips are probably not depositional, for the cross-bedding referred to above and between 27 and 28 m lies within a parallel-bedded sequence dipping at 30° and has sigmoidal foresets dipping at 50°, much steeper than the normal angle of rest. The deformation that affected the upper 6 m of the unit and the tilting for some tens of meters further down are believed to have resulted from grounding of an ice sheet, like that proposed by *Denton et al.* [1971] for McMurdo Sound during the last major ice advance in the area.

PETROLOGY

The petrology of the wash samples (Table 5) indicates that they consist of sand-sized, angular and rounded mineral and lithic fragments. Lithic fragments are more abundant than mineral grains.

The lithic fragments consist of volcanic, plutonic, and very minor amounts of metamorphic rocks. The volcanic rocks are primarily basalt and brown basaltic glass and minor amounts of basaltic tuff and breccia and trachyte. Five varieties of basalt were identified on the basis of their microphenocrysts. They are aphyric basalt, augite basalt, plagioclase basalt, and olivine-augite basalt. Volcanic rock fragments occur as rounded and angular fragments. Finer grained samples tend to be more angular and to be richer in aphyric basalt.

Plutonic and metamorphic rock fragments constitute a very small part of the samples. They occur as rounded and angular fragments and are more abundant in the coarser grained samples. fragments. They consist of quartz, orthoclase, plagioclase, biotite, and green hornblende, though not all of the minerals occur in any one grain. It seems likely that a broad variety of granitic rock types may be represented and that the suite may even include some gneiss and granulite. The one gneiss fragment encountered was identified on the basis of foliation shown by biotite. The rarest plutonic rock is dolerite. It consists of plagioclase and pyroxene and is distinctive in thin section. Shell fragments occur in some of the samples and include rounded and angular fragments, though they constitute a very minor part of the sediment.

The other major component of the sediment samples is the mineral fraction. Mineral grains are less common than lithic fragments but also occur as Granitic rocks are the most common plutonic rock angular and rounded grains. Quartz, plagioclase, and augite are common; olivine, hornblende, kaersutite, orthoclase, and opaque minerals are not abundant.

The quartz occurs as single and composite grains. The composite grains are plutonic rock fragments. Some of the single grains are rounded, some show suggestions of overgrowths, and others exhibit undulatory extinction. Plagioclase occurs both as unaltered and altered grains. The altered grains are commonly oligoclase, whereas the unaltered ones are andesine or labradorite. Only rarely are the plagioclase grains rounded. Augite is commonly pink and has a large 2V. Clear, diopsidic grains and pale green grains are rare. Olivine is clear and only rarely altered. It has a large 2V. Hornblende is pleochroic in pale greens and generally angular. Kaersutite is dark reddish brown and shows parallel or near-parallel extinction. The orthoclase is almost always altered and occurs as blocky subangular grains.

The ultimate source of the grains that constitute grains and pale green grains are rare. Olivine is clear and only rarely altered. It has a large 2V. Hornblende is pleochroic in pale greens and generally angular. Kaersutite is dark reddish brown and the sediment of DVDP 15 is clear. The volcanic rocks and the olivine, augite, kaersutite, and some of the more calcic plagioclase are derived from the alkaline volcanic suite of the McMurdo area [*Kyle*, 1976; *Goldich et al.*, 1975; *Kyle and Treves*, 1973, 1974; *Treves and Kyle*, 1973; *Cole et al.*, 1971; *Cole and Ewart*, 1968; *Smith*, 1954]. It is not possible to assign sources within the province to any of the rocks because they are all quite common throughout the area and lack distinctive characteristics.

The granitic rocks and the dolerite and the quartz, orthoclase, some of the calcic plagioclase, all of the oligoclase, the hornblende, and the biotite represent materials derived from rocks that constitute the Transantarctic Mountains. These mountains consist of a basement complex that includes granitic igneous and metamorphic rocks [*Gunn and Warren*, 1962] that are overlain by a thick sequence of clastic, sedimentary rocks that constitute the Beacon Supergroup [*Barrett et al.*, 1972]. The Beacon rocks and the basement complex are cut by sills of Ferrar Dolerite.

The basement rocks are the probable source of some of the quartz, oligoclase, biotite, and hornblende. The Beacon sequence is probably the source of some of the rounded quartz grains and perhaps the source of some of the altered feldspars. The Ferrar Dolerite is clearly the source of the dolerite fragments and is probably the source for some of the augite and more calcic plagioclase.

How materials with such divergent and widely spread ultimate sources came to be deposited in the McMurdo Sound at the site of DVDP 15 is not clear. However, the wash samples from DVDP 15 do not resemble the basaltic bottom sediment collected by P. Dayton and his group of divers from the west coast of Ross Island and New Harbor and by A. DeVries from the deeper basins offshore of McMurdo Station [*Treves et al.*, 1975]. They do, however, resemble samples collected from the edge of the McMurdo Ice Shelf and The Strand Moraines (Table 6). The Strand Moraines are believed to be remnants of a larger grounded McMurdo Ice Shelf, and the debris on the surface of the ice shelf

today is probably a mixture of marine sediment and moraine. The morainal materials were derived from outlet glaciers that join the ice shelf after passing through the Transantarctic Mountains, as does the Koettlitz Glacier, and hence are a source of basement rocks, Beacon sedimentary rocks, and Ferrar Dolerite. Volcanic morainal materials are derived in much the same way, since ice streams pass Black and White islands, Minna Bluff, Brown Peninsula, Mount Discovery, and Mount Morning. In addition, the sediment on the McMurdo Ice Shelf today includes material from the floor of the sound that was carried to the base of the floating McMurdo Ice Shelf by freezing of seawater. It is brought to the surface by ablation, in much the same way that marine invertebrates and fish are brought to the surface by ablation of the surface of the ice shelf [*Debenham*, 1920].

In light of the above, it seems that expansion and retreat of the McMurdo Ice Shelf might explain how these materials were transported to the site of DVDP 15. The well-rounded nature of some of the grains, however, and field observations indicate that the wind is very effective in transporting large volumes of sediment from the ice shelf and the Strand Moraines to sea and to the surface of the annual ice each year. The data suggested that release of surface and entrained materials by melting of the McMurdo Ice Shelf and of outlet glaciers and transport by wind were the two main processes responsible for the accumulation of the dominantly basaltic sediment at the site of DVDP 15.

SUMMARY AND CONCLUSIONS

The purpose of drilling in McMurdo Sound was to investigate the history of the East Antarctic Ice Sheet. The first attempt was made on November 1975, when DVDP 15 was drilled to a subbottom depth of 65 m from the annual ice in the western part of McMurdo Sound (77°26'14.19"S, 164°22'49.39"E). The water depth at the site was 122 m.

The prospective drill sites were located, and water depth measured, soon after Win Fly (September 2). The drilling camp was established by two tractor trains, which moved 150,000 lb (70,000 kg) of drilling and camp equipment to the site. The camp was maintained from October 24 to November 29. Drilling was terminated on November 21 because cracks in the ice made it unsafe to continue. All drilling equipment and most of the supplies were returned to McMurdo Station and stored; the remaining supplies were cached at Marble Point.

Site surveys established the bathymetry in detail, indicating that the drill site was located on a flat-topped rise, and showed that the currents were negligible. Grab samples showed that the sea floor consisted primarily of gravelly sand derived largely from the late Cenozoic McMurdo Volcanic Group.

The core from DVDP 15 is mainly olive gray to olive black basaltic sand; 52% of the cored interval was recovered, and 278 kg of wash material was collected. The 65 m penetrated can be divided into two units. Core from unit 1, which is 13 m thick, ranges from fine to coarse sand and is mainly poorly sorted and silty. It is generally unconsolidated and was much disturbed by drilling. Unit 2, which extends down from 13 m subbottom, consists also of basaltic sand but is fine to medium grained and better sorted than unit 1. It also differs by having widespread plane-parallel stratification; rare small-scale cross beds and graded beds were also seen. The stratification is tilted up to 35° and locally plastically deformed in the upper part of the unit, but dips become smaller lower down. This bedding disturbance is attributed to movement of grounded ice at the site prior to the deposition of unit 1.

Petrography of the wash samples shows that between 65 and 80% is basaltic material derived from the late Cenozoic McMurdo Volcanic Group. The remainder is material from the granitic and metamorphic basement sedimentary rocks of the Beacon Supergroup and from the Ferrar Dolerite, the major geological units of the Transantarctic Mountains. The cores and wash samples were carefully examined for fossils, but only wash samples from the base of unit 1 were productive, yielding abundant bryozoans, foraminifera, and ostracods of Recent age. Lithification in unit 2 indicates that it is significantly older, though it could still be as young as late Pleistocene.

The scientific achievements of DVDP 15 are significant. Even though the penetration was not great, it was sufficient to establish that unit 2 represents the layer above reflector A, which *Northey et al.* [1975] have traced across the sound and into the moat around Ross Island. No age-diagnostic fossils have been found, but the thickening of the layer in the moat and the dominance of basaltic debris suggest a Plio-Pleistocene age. The sediments of unit 2 are remarkable for the lack of pebbles and coarser debris that would indicate deposition from melting glacial ice. The texture and composition of the sediment suggest that it is wind transported largely from the area of the McMurdo Ice Shelf to the south and has been slightly modified by bottom currents.

Clearly, wind transport is very important for near-shore sediments in extremely cold climates. The nature of the deformation of unit 2 is another point of interest and is believed to indicate the movement of grounded ice from the last Ross Ice Sheet. Unit 1, with its Recent fossil assemblage at the base, appears to represent deposition in open water by wind and melting ice over the last several thousand years.

APPENDIX

The following are some notes on the drilling, and a detailed log of core samples appears on the next 16 pages, starting with a legend for the logs.

1. Downhole depth was measured in feet and inches from the floor of the drill shack, and 404 feet subtracted to give penetration below the sea floor.

2. For conversion from feet to meters, 1 foot = 0.3048 m.

3. Depth below the rig floor was not adjusted to compensate for tidal movement, and the depths of different runs may therefore be in error by as much as 0.6 m (the tidal range). However, available tidal measurements indicate a lower range for runs after November 12 (run 8 on).

4. The drilling fluid was seawater.

5. The hole was drilled to be vertical. Attitudes of the strata are given as dips, i.e., angle between bedding and the perpendicular to the core axis.

6. The core is described from top to bottom for each run in turn. The core is assumed to come from the lower part of each run if recovery is less than 100% (and the core is in good condition). Depths in meters (e.g., 1.37 m) indicate the stratigraphic position assumed for each significant horizon. Intervals and depths in centimeters indicate the distance below the top of a particular run as the core lies in the box.

7. Core diameter is approximately 69 mm for runs 1 to 12 and 17 and is 40 mm for runs 13 to 28 except for run 17.

REFERENCE

CONDITION OF CORE

Core in good condition, unbroken for tens of cm. Structures preserved.

Core slightly deformed or broken. Stratification partly obliterated.

Core intensely deformed or broken. Stratification on a scale of mm's lost. May have become a slurry during drilling.

Sediment carried up from bottom of hole in suspension.

LITHOLOGY

Silt and clay (mud).

Very fine and fine sand.

Medium to very coarse sand.

Basaltic (●) and non basaltic (○) pebbles.

Claystone fragments.

CONTACTS

Gradational.

Sharp and straight.

Sharp and irregular.

STRUCTURES

Stratification.

Fissility, cleavage.
Fractures.

CONDITION OF CORE | RUN | BOX | DEPTH(m) | LITHOLOGY | STRUCTURES

WASH

RUN 1   0 to 0.30 m

0 to 0.30 m. Collected in a bucket and stored in a plastic bag.
FINE SAND, olive black (5Y 2/1), moderately well sorted, soft to soupy. A few foraminifera.
Basaltic and mafic/quartzofeldspathic grains 70:30.

RUN 2   1.37 to 3.97 m. Covers 2.82 m in core box, but upper 180 cm may be wash material.

1.37 to 1.90 m (0 to 57 cm)
FINE TO MEDIUM SAND, olive black (5Y 2/1), moderately sorted. Basaltic and quartzofeldspathic grains subangular to subrounded.

— gradational contact —

1.90 to 3.10 m (57 to 86 cm)
COARSE SAND, olive black (5Y 2/1), poorly sorted. Subrounded to rounded grains of basalt, quartz and feldspar.   A few pebbles up to 30 mm long of basalt, quartz and feldspar.

— moderately sharp contact —

3.10 to 3.51 m (186 to 231 cm)
SILTY SAND, olive black (5Y 2/1), coarse and extremely poorly sorted, ranging from silt to very coarse sand and granules. Mineralogy as above plus obvious K-feldspar. Coarser lithic fragments (up to 5 mm) are mainly basalt.

— gradational contact —

3.51 to 3.69 m (231 to 249 cm)
SILTY SAND, olive black (5Y 2/1), as for 173 to 214 cm but slightly coarser. Subrounded dolerite pebble 20 mm long at 218 cm.

— gradational contact —

3.69 to 3.94 m (249 to 279 cm)
COARSE SAND, olive black (5Y 2/1), very poorly sorted with common granules. Grains up to 4 mm and angular to subrounded. Composed of quartz, feldspar and basalt.

— moderately sharp contact —

3.94 to 3.97 m (279 to 282 cm)
SILTY SAND, olive black (5Y 2/1), extremely poorly sorted, ranging from silt to fine granules. Mineralogy as above.

RUN 3   4.58 to 4.88 m. Core stored in 5 bags.

15-3-5
SILTY FINE SAND, olive grey (5Y 3/2), poorly sorted. Grains subangular to rounded and mainly basaltic with some quartz and feldspar.

15-3-3 and 15-3-4
FINE TO MEDIUM SAND, olive grey (5Y 3/2), poorly sorted. Roundness and grain composition as above.

15-3-2 and 15-3-1
PEBBLY MEDIUM SAND, olive grey (5Y 3/2), extremely poorly sorted, ranging from silt to pebbles of basalt, granodiorite and dolerite up to 70 mm long. Roundness and grain composition as above.

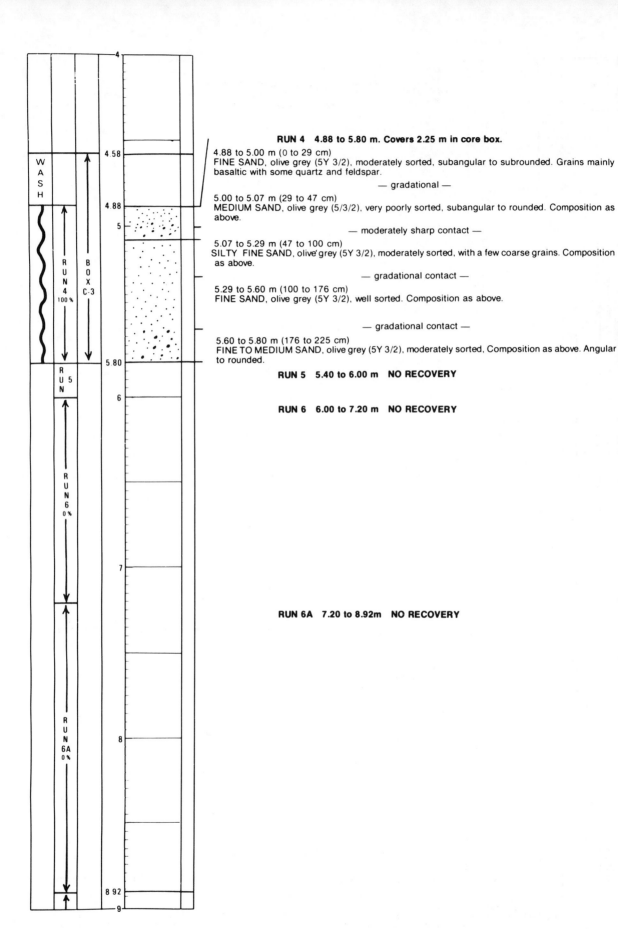

RUN 4   4.88 to 5.80 m. Covers 2.25 m in core box.

4.88 to 5.00 m (0 to 29 cm)
FINE SAND, olive grey (5Y 3/2), moderately sorted, subangular to subrounded. Grains mainly basaltic with some quartz and feldspar.

— gradational —

5.00 to 5.07 m (29 to 47 cm)
MEDIUM SAND, olive grey (5/3/2), very poorly sorted, subangular to rounded. Composition as above.

— moderately sharp contact —

5.07 to 5.29 m (47 to 100 cm)
SILTY FINE SAND, olive grey (5Y 3/2), moderately sorted, with a few coarse grains. Composition as above.

— gradational contact —

5.29 to 5.60 m (100 to 176 cm)
FINE SAND, olive grey (5Y 3/2), well sorted. Composition as above.

— gradational contact —

5.60 to 5.80 m (176 to 225 cm)
FINE TO MEDIUM SAND, olive grey (5Y 3/2), moderately sorted, Composition as above. Angular to rounded.

RUN 5   5.40 to 6.00 m   NO RECOVERY

RUN 6   6.00 to 7.20 m   NO RECOVERY

RUN 6A   7.20 to 8.92m   NO RECOVERY

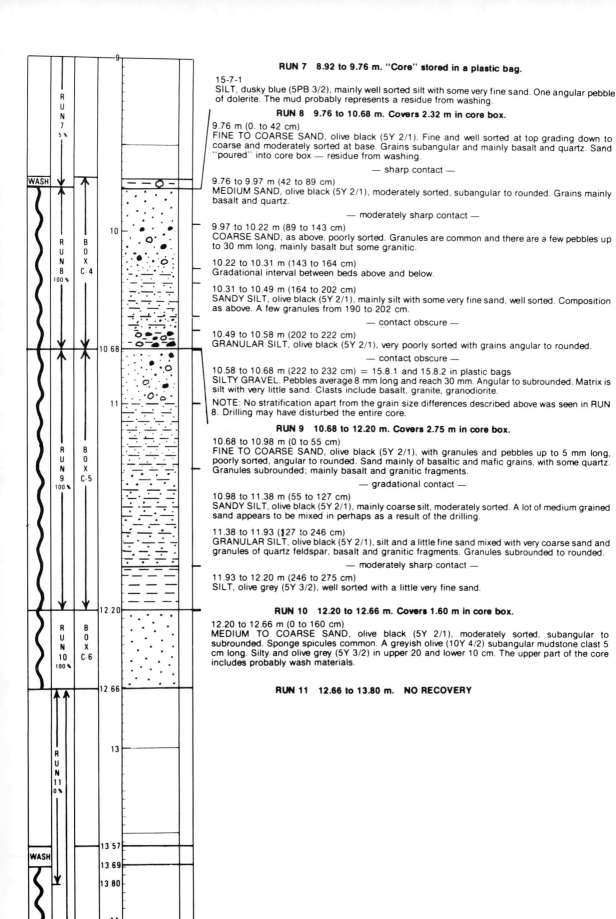

RUN 7   8.92 to 9.76 m. "Core" stored in a plastic bag.

15-7-1
SILT, dusky blue (5PB 3/2), mainly well sorted silt with some very fine sand. One angular pebble of dolerite. The mud probably represents a residue from washing.

RUN 8   9.76 to 10.68 m. Covers 2.32 m in core box.

9.76 m (0. to 42 cm)
FINE TO COARSE SAND, olive black (5Y 2/1). Fine and well sorted at top grading down to coarse and moderately sorted at base. Grains subangular and mainly basalt and quartz. Sand "poured" into core box — residue from washing.

— sharp contact —

9.76 to 9.97 m (42 to 89 cm)
MEDIUM SAND, olive black (5Y 2/1), moderately sorted, subangular to rounded. Grains mainly basalt and quartz.

— moderately sharp contact —

9.97 to 10.22 m (89 to 143 cm)
COARSE SAND, as above, poorly sorted. Granules are common and there are a few pebbles up to 30 mm long, mainly basalt but some granitic.

10.22 to 10.31 m (143 to 164 cm)
Gradational interval between beds above and below.

10.31 to 10.49 m (164 to 202 cm)
SANDY SILT, olive black (5Y 2/1), mainly silt with some very fine sand, well sorted. Composition as above. A few granules from 190 to 202 cm.

— contact obscure —

10.49 to 10.58 m (202 to 222 cm)
GRANULAR SILT, olive black (5Y 2/1), very poorly sorted with grains angular to rounded.

— contact obscure —

10.58 to 10.68 m (222 to 232 cm) = 15.8.1 and 15.8.2 in plastic bags
SILTY GRAVEL. Pebbles average 8 mm long and reach 30 mm. Angular to subrounded. Matrix is silt with very little sand. Clasts include basalt, granite, granodiorite.

NOTE: No stratification apart from the grain size differences described above was seen in RUN 8. Drilling may have disturbed the entire core.

RUN 9   10.68 to 12.20 m. Covers 2.75 m in core box.

10.68 to 10.98 m (0 to 55 cm)
FINE TO COARSE SAND, olive black (5Y 2/1), with granules and pebbles up to 5 mm long, poorly sorted, angular to rounded. Sand mainly of basaltic and mafic grains, with some quartz. Granules subrounded; mainly basalt and granitic fragments.

— gradational contact —

10.98 to 11.38 m (55 to 127 cm)
SANDY SILT, olive black (5Y 2/1), mainly coarse silt, moderately sorted. A lot of medium grained sand appears to be mixed in perhaps as a result of the drilling.

11.38 to 11.93 (127 to 246 cm)
GRANULAR SILT, olive black (5Y 2/1), silt and a little fine sand mixed with very coarse sand and granules of quartz feldspar, basalt and granitic fragments. Granules subrounded to rounded.

— moderately sharp contact —

11.93 to 12.20 m (246 to 275 cm)
SILT, olive grey (5Y 3/2), well sorted with a little very fine sand.

RUN 10   12.20 to 12.66 m. Covers 1.60 m in core box.

12.20 to 12.66 m (0 to 160 cm)
MEDIUM TO COARSE SAND, olive black (5Y 2/1), moderately sorted, subangular to subrounded. Sponge spicules common. A greyish olive (10Y 4/2) subangular mudstone clast 5 cm long. Silty and olive grey (5Y 3/2) in upper 20 and lower 10 cm. The upper part of the core includes probably wash materials.

RUN 11   12.66 to 13.80 m.   NO RECOVERY

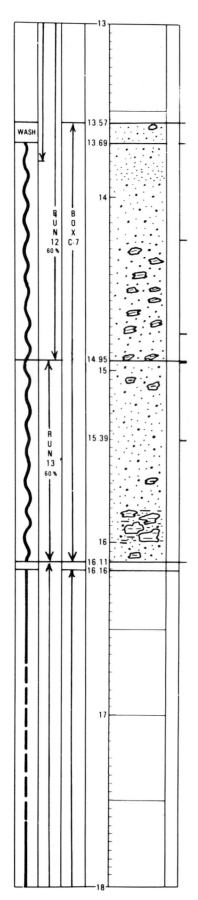

RUN 12   12.66 to 14.95 m.

13.57 to 13.69 m (0 to 12 cm)
FINE TO MEDIUM SAND, olive black (5Y 2/1), moderately well sorted. Structureless and loose. Pebble of medium grained dolerite 50 mm long at top of core. Core probably represents wash material.

— sharp contact —

13.69 to 14.25 m (12 to 68 cm)
FINE TO MEDIUM SAND, olive black (5Y 2/1), moderately well sorted, but finer grained from 24 to 34 cm. No other sign of stratification. Grains angular to subrounded. Basalt and mafic/quartzofeldspathic grains 60:40.

— gradational contact over several cm —

14.25 to 14.78 m (68 to 121 cm)
MEDIUM TO COARSE SAND, olive black (5Y 2/1), poorly sorted. Lacks stratification. Scattered angular fragments of olive grey claystone form as much as 10% of the core. They are mostly 1 to 10 mm long and uniformly spread through the core except for the upper 15 cm of the interval where they show normal size grading. The fragments probably represent discrete layers of claystone fragmented and mixed with the interbedded sand during drilling. Basalt and mafic/quartzofeldspathic grains 60:40.

— gradational contact —

14.78 to 14.95 m (121 to 138 cm)
FINE TO MEDIUM SAND, olive black (5Y 2/1), moderately sorted. The lowest 6 cm is medium-grained and has fragments of the olive grey claystone up to 20 mm long, which are concentrated mainly in the lower 2 cm. One 3-mm-thick lamina of claystone is slightly sheared but still intact 7 cm above the base of the interval.

RUN 13   14.95 to 16.11 m.

15.39 to 15.65 m (0 to 26 cm)
FINE TO MEDIUM SAND, olive black (5Y 2/1), moderately sorted. Unstratified apart from faint coarse lamination from 18 to 26 cm. A few scattered angular claystone fragments up to 20 mm long, presumably broken up and mixed with the sand during drilling. Sand grains angular and subangular except for a few rounded medium to coarse ones. Basalt and mafic/quartzofeldspathic grains 60:40.

— gradational contact —

15.65 to 16.11 m (26 to 72 cm)
FINE TO MEDIUM SAND, olive black (5Y 2/1), moderately sorted. Scattered olive grey claystone fragments from 26 to 42 cm. The interval from 42 to 69 cm is a brecciated sandy mud, probably from the mixing of the mud from the claystone to the interbedded sand.

RUN 14   16.11 to 18.96 m.

16.16 to 16.17 m (0 to 1 cm)
SILT, olive grey (5Y 4/1), rather sheared, dips of 50°.

— sharp contact —

16.17 to 17.76 m (1 to 160 cm)
FINE TO MEDIUM SAND, olive black (5Y 2/1), moderately to poorly sorted, laminated to thinly bedded and dipping mainly at 40°. Vertical dips from 10 to 20 cm and from 145 to 160 cm change gradually to much lower inclinations. The stratification is due mainly to slight grain size variations, but there are also occasional lenses of coarse sand and some intervals have paper thin laminae up to 5 mm long of grey silt scattered through otherwise homogeneous core. Grains are mainly subangular to subrounded, though some larger ones are rounded. Basaltic and mafic/quartzfeldspathic grains 70:30.

— sharp contact horizontal over ²/₃ of the core at 160 cm but plunging down to 170 cm over the remainder —

17.76 to 18.67 m (160 to 251 cm)
COARSE SILT, light olive grey to olive grey (5Y 5/1), with several fine sand intervals up to 2 cm thick. The distinct fine lamination in the upper part of the interval is obscured by drilling brecciation below 195 cm. Stratification dips mostly at 35° except for the interval from 160 to 180 cm, where the dip ranges from 10° to 90°. Scattered clasts of lighter-coloured claystone up to 5 mm across in the well-stratified core. Apart from these, fragments coarser than fine sand are very rare. Moderate effervescence with conc. hydrochloric acid.

— sharp contact —

18.67 to 18.90 m (251 to 274 cm)
FINE SAND, olive black (5Y 2/1), poorly sorted, well laminated and dipping at 25°. No grains coarser than medium sand. Basaltic and mafic/quartzofeldspathic grains 40:60.

— sharp contact —

18.90 to 18.96 m (274 to 280 cm)
SILT, olive grey (5Y 4/1), includes clay and scattered sand grains up to 1 mm long perhaps mixed together by the drilling. Penetrated by a dyke of sand 5 mm wide extending down vertically from the sand above.

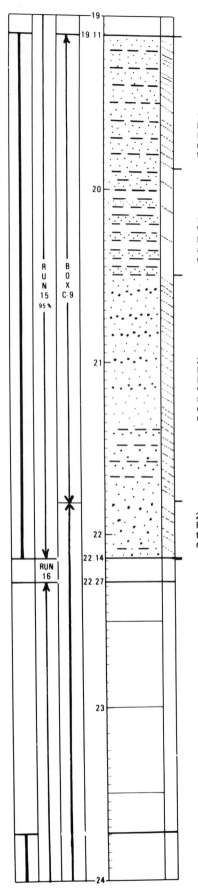

RUN 15 18.96 to 22.14 m NOTE: Bottom 33 cm collected on RUN 16.

19.11 to 19.88 m (0 to 77 cm)
FINE SAND, olive black (5Y 2/1), moderately sorted, well-stratified, ranging from finely laminated to medium-bedded. The most prominent stratification is due to thin silt laminae from 1 mm to 50 m apart. Dip from 25° from 0 to 30 cm, and 30° from 50 to 77 cm. No grains coarser than 0.5 mm. Basaltic and mafic/quartzofeldspathic grains 40:60.

— gradational contact —

19.88 to 20.54 m (77 to 143 cm)
VERY FINE SANDY SILT, olive grey (5Y 4/1), in beds 5 cm thick, separated by thinner beds and laminae of fine, olive black (5Y 2/1), moderately sorted, fine sand dipping at 30°. The silt beds and upper homogeneous but have mainly irregular fractures suggesting drilling brecciation. No grains coarser than 0.2 mm.

— gradational contact —

20.54 to 21.81 m (143 to 270 cm) NOTE: Base of this interval is at 22.14 m
FINE SAND, like that from 0 to 77 cm, olive black (5Y 2/1), moderately sorted, with a few coarse laminae, well stratified. The material also includes fine silt laminae in places clearly capping thin varvoid graded beds. The grading is best seen from 197 to 204 cm where there are 8 cycles of similar thickness. From 227 to 255 cm beds of sandy silt 1 to 2 cm thick are interbedded with coarse sand. The whole interval dips at 30°. The largest grain is 1 mm. Basaltic and mafic/quartzofeldspathic grains 60:40.

— gradational contact —

21.81 to 22.14 m (270 to 303 m). Collected on RUN 16.
FINE TO MEDIUM SAND, olive black (5Y 2/1), poorly sorted. A layer of muddy fine sand 5 mm thick at 294 cm. Indistinct lamination dipping at 30°. No grains coarser than 1 mm. Basaltic and mafic/quartzofeldspathic grains 60:40.

RUN 16 22.14 to 22.27 m. No recovery apart from core drilled on RUN 15 (see above).

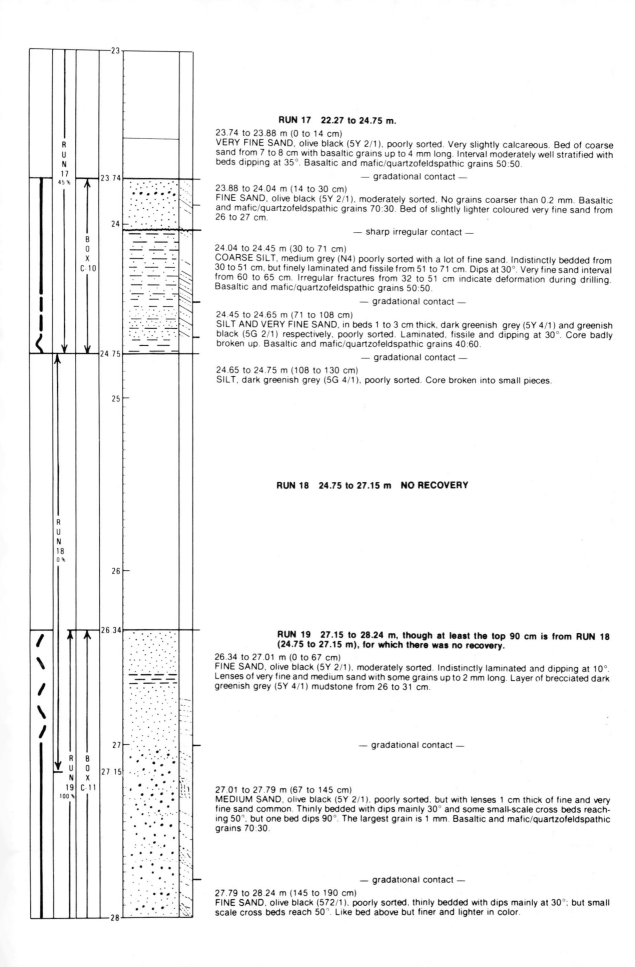

RUN 17   22.27 to 24.75 m.

23.74 to 23.88 m (0 to 14 cm)
VERY FINE SAND, olive black (5Y 2/1), poorly sorted. Very slightly calcareous. Bed of coarse sand from 7 to 8 cm with basaltic grains up to 4 mm long. Interval moderately well stratified with beds dipping at 35°. Basaltic and mafic/quartzofeldspathic grains 50:50.

— gradational contact —

23.88 to 24.04 m (14 to 30 cm)
FINE SAND, olive black (5Y 2/1), moderately sorted, No grains coarser than 0.2 mm. Basaltic and mafic/quartzofeldspathic grains 70:30. Bed of slightly lighter coloured very fine sand from 26 to 27 cm.

— sharp irregular contact —

24.04 to 24.45 m (30 to 71 cm)
COARSE SILT, medium grey (N4) poorly sorted with a lot of fine sand. Indistinctly bedded from 30 to 51 cm, but finely laminated and fissile from 51 to 71 cm. Dips at 30°. Very fine sand interval from 60 to 65 cm. Irregular fractures from 32 to 51 cm indicate deformation during drilling. Basaltic and mafic/quartzofeldspathic grains 50:50.

— gradational contact —

24.45 to 24.65 m (71 to 108 cm)
SILT AND VERY FINE SAND, in beds 1 to 3 cm thick, dark greenish grey (5Y 4/1) and greenish black (5G 2/1) respectively, poorly sorted. Laminated, fissile and dipping at 30°. Core badly broken up. Basaltic and mafic/quartzofeldspathic grains 40:60.

— gradational contact —

24.65 to 24.75 m (108 to 130 cm)
SILT, dark greenish grey (5G 4/1), poorly sorted. Core broken into small pieces.

RUN 18   24.75 to 27.15 m   NO RECOVERY

RUN 19   27.15 to 28.24 m, though at least the top 90 cm is from RUN 18 (24.75 to 27.15 m), for which there was no recovery.

26.34 to 27.01 m (0 to 67 cm)
FINE SAND, olive black (5Y 2/1), moderately sorted. Indistinctly laminated and dipping at 10°. Lenses of very fine and medium sand with some grains up to 2 mm long. Layer of brecciated dark greenish grey (5Y 4/1) mudstone from 26 to 31 cm.

— gradational contact —

27.01 to 27.79 m (67 to 145 cm)
MEDIUM SAND, olive black (5Y 2/1), poorly sorted, but with lenses 1 cm thick of fine and very fine sand common. Thinly bedded with dips mainly 30° and some small-scale cross beds reaching 50°, but one bed dips 90°. The largest grain is 1 mm. Basaltic and mafic/quartzofeldspathic grains 70:30.

— gradational contact —

27.79 to 28.24 m (145 to 190 cm)
FINE SAND, olive black (572/1), poorly sorted, thinly bedded with dips mainly at 30°; but small scale cross beds reach 50°. Like bed above but finer and lighter in color.

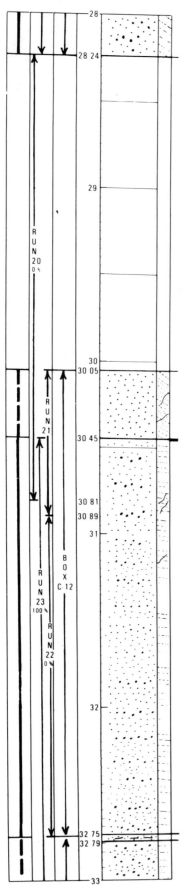

RUN 20   28.24 to 30.81 m   NO RECOVERY

RUN 21   30.81 to 30.89 m 40 cm recovered but located from 30.05 to 30.45 m because of constraint from RUN 23.

30.05 to 30.45 m (0 to 40 cm)
FINE AND VERY FINE SAND, brownish black (5YR 2/1), moderately well sorted. Indistinct thin bedding dipping at 30 to 55° but partly disrupted by fractures at 45° to the core axis (probably acquired during drilling.) Largest grains 1 mm. Basaltic and mafic/quartzofeldspathic grains 60:40.

RUN 22   30.89 to 32.75 m.   NO RECOVERY

RUN 23   32.75 to 33.39 m. 2.94 m was recovered although only 0.64 m was drilled. The lowest core was jammed in the bit, and hence is known to come from 33.39 m. This core is assumed to extend up to 30.45 m, and the core recovered in RUN 21 to come from directly above this.

30.45 to 32.75 m (0 to 230 cm)
VERY FINE, FINE AND MEDIUM SAND, greyish black (N2) to brownish black (5YR 2/1), in beds 5 mm to 8 cm thick, moderately to moderately well sorted. No grains coarser than 2 mm. Well stratified and dipping between 5 and 10°. The small high-angle faults may be due to drilling. Basaltic and mafic/quartzofeldspathic grains 60:40.

— sharp contact —

32.75 to 32.79 m (230 to 234 cm)
MUD, dark greenish grey (5GY 4/1). A breccia (probably due to drilling) of clay fragments in a homogeneous matrix of clay and very fine sand. Smear slide shows rare sponge spicule fragments.

— sharp contact —

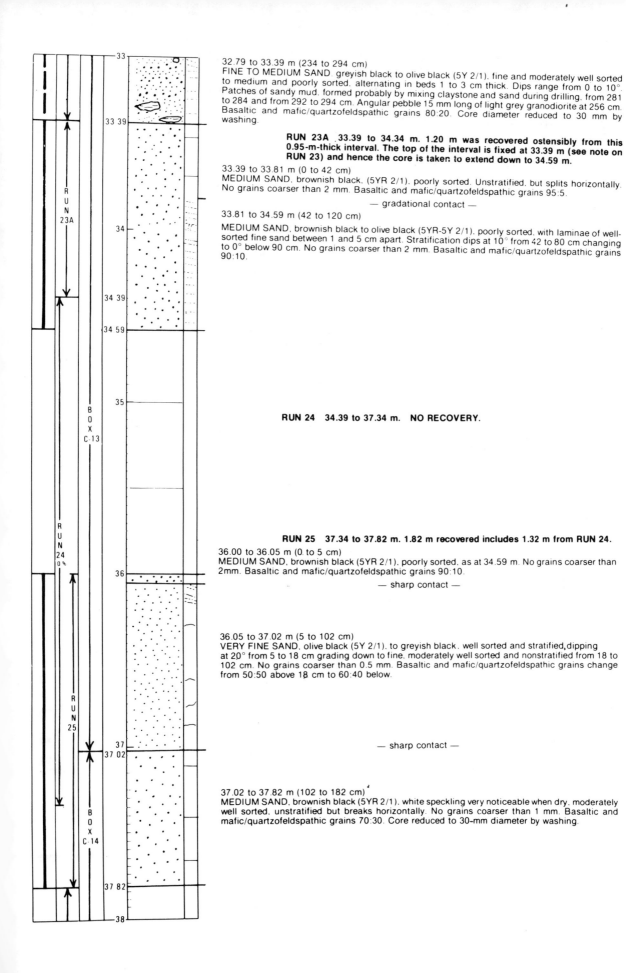

32.79 to 33.39 m (234 to 294 cm)
FINE TO MEDIUM SAND. greyish black to olive black (5Y 2/1). fine and moderately well sorted to medium and poorly sorted. alternating in beds 1 to 3 cm thick. Dips range from 0 to 10°. Patches of sandy mud. formed probably by mixing claystone and sand during drilling. from 281 to 284 and from 292 to 294 cm. Angular pebble 15 mm long of light grey granodiorite at 256 cm. Basaltic and mafic/quartzofeldspathic grains 80:20. Core diameter reduced to 30 mm by washing.

RUN 23A .33.39 to 34.34 m. 1.20 m was recovered ostensibly from this 0.95-m-thick interval. The top of the interval is fixed at 33.39 m (see note on RUN 23) and hence the core is taken to extend down to 34.59 m.

33.39 to 33.81 m (0 to 42 cm)
MEDIUM SAND, brownish black. (5YR 2/1). poorly sorted. Unstratified. but splits horizontally. No grains coarser than 2 mm. Basaltic and mafic/quartzofeldspathic grains 95:5.

— gradational contact —

33.81 to 34.59 m (42 to 120 cm)
MEDIUM SAND, brownish black to olive black (5YR-5Y 2/1). poorly sorted. with laminae of well-sorted fine sand between 1 and 5 cm apart. Stratification dips at 10° from 42 to 80 cm changing to 0° below 90 cm. No grains coarser than 2 mm. Basaltic and mafic/quartzofeldspathic grains 90:10.

RUN 24   34.39 to 37.34 m.   NO RECOVERY.

RUN 25   37.34 to 37.82 m. 1.82 m recovered includes 1.32 m from RUN 24.

36.00 to 36.05 m (0. to 5 cm)
MEDIUM SAND, brownish black (5YR 2/1). poorly sorted. as at 34.59 m. No grains coarser than 2mm. Basaltic and mafic/quartzofeldspathic grains 90:10.

— sharp contact —

36.05 to 37.02 m (5 to 102 cm)
VERY FINE SAND, olive black (5Y 2/1). to greyish black . well sorted and stratified, dipping at 20° from 5 to 18 cm grading down to fine. moderately well sorted and nonstratified from 18 to 102 cm. No grains coarser than 0.5 mm. Basaltic and mafic/quartzofeldspathic grains change from 50:50 above 18 cm to 60:40 below.

— sharp contact —

37.02 to 37.82 m (102 to 182 cm)
MEDIUM SAND, brownish black (5YR 2/1). white speckling very noticeable when dry. moderately well sorted. unstratified but breaks horizontally. No grains coarser than 1 mm. Basaltic and mafic/quartzofeldspathic grains 70:30. Core reduced to 30-mm diameter by washing.

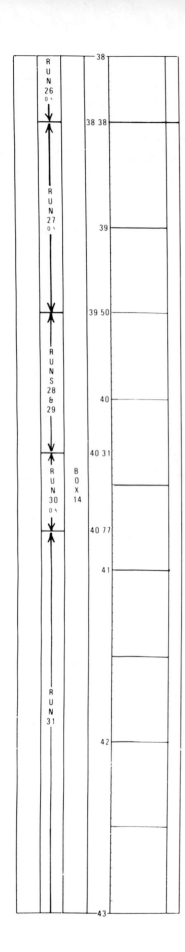

RUN 26   37.82 to 38.38 m.   NO RECOVERY.

RUN 27   38.38 to 39.50 m.   NO RECOVERY.

RUN 28   39.50 to 40.31 m.   NO RECOVERY.
Sand rose in casing to about 30 m subbottom

RUN 29   Washed and then dry drilled to 40.31 m.   NO RECOVERY.

RUN 30   40.30 to 40.77 m.   NO RECOVERY
Sand again rose to about 30 m subbottom.

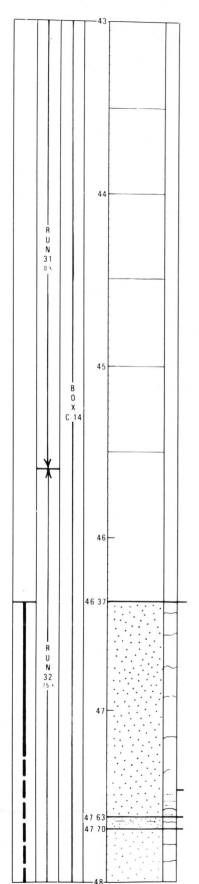

RUN 31 Washed and then dry drilled to 35.38 m.   NO RECOVERY.
Washed down to 45.60 m.

RUN 32   45.60 to 48.50 m.

46.37 to 47.48 m (0 to 111 cm)
FINE SAND, brownish black to olive black (5YR-5Y 2/1), moderately sorted, unstratified except for broad slight colour changes that appear to represent a change in the ratio of basaltic and mafic/quartzofeldspathic grains from 90:10 to 70:30. There is also a gradational change to very fine sand from 45 to 62 cm and from 73 to 84 cm. No grains coarser than 0.5 mm.

— gradational contact —

47.48 to 47.63 m (111 to 126 cm)
FINE SAND, olive black (5Y 2/1), moderately sorted, faintly laminated and slightly fissile. Stratification horizontal

— sharp contact —

47.63 to 47.70 m (126 to 133 cm)
MUDDY VERY FINE SAND, dark greenish grey (5GY 4/1), poorly sorted. Faint horizontal stratification. Possibly mixed by drilling. Basaltic and mafic/quartzofeldspathic grains 30:70.

— sharp contact —

47.70 to 48.50 m (133 to 213 cm)
FINE SAND, greyish black to olive black (N2 to 5Y 2/1) moderately well sorted. No stratification apart from slight changes in grain size over 10-cm intervals. No grains coarser than 0.5 mm. Basaltic and mafic/quartzofeldspathic grains 90:10 to 80:20.

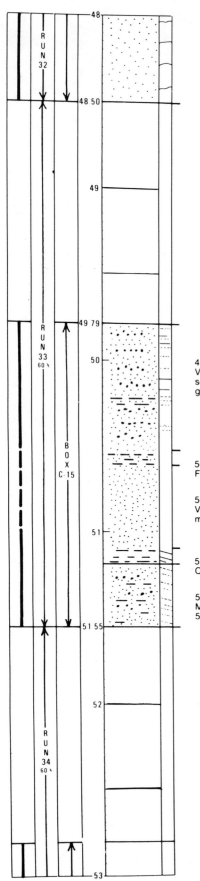

RUN 33   48.50 to 51.55 m

49.79 to 50.52 m (0 to 73 cm)
VERY FINE TO MEDIUM SAND, olive black (5Y 2/1) to greyish black , moderately well sorted. Horizontal stratification evident from color and grain size variation, and the several thin grey mudstone laminae between 44 and 48 cm.

— gradational contact —

50.52 to 50.61 m (73 to 82 cm)
FINE SANDY SILT, dark greenish grey (5GY 4/1), moderately sorted. No clear stratification.

— gradational contact —

50.61 to 51.09 m (82 to 130 cm)
VERY FINE SAND, olive grey (5Y 4/1), moderately sorted. No stratification. Basaltic and mafic/quartzofeldspathic grains 30:70.

— gradational contact —

51.09 to 51.19 m (130 to 140 cm)
COARSE SILT, greyish black , moderately well sorted. Fissile, with cleavage dipping at 15°.

— sharp contact —

51.19 to 51.55 m (140 to 176 cm)
MUDDY VERY FINE, FINE AND MEDIUM SAND, dark greenish grey to olive black (5GY 4/1 to 5Y 2/1), moderately to poorly sorted, finely laminated to very thin-bedded and dipping at 15°.

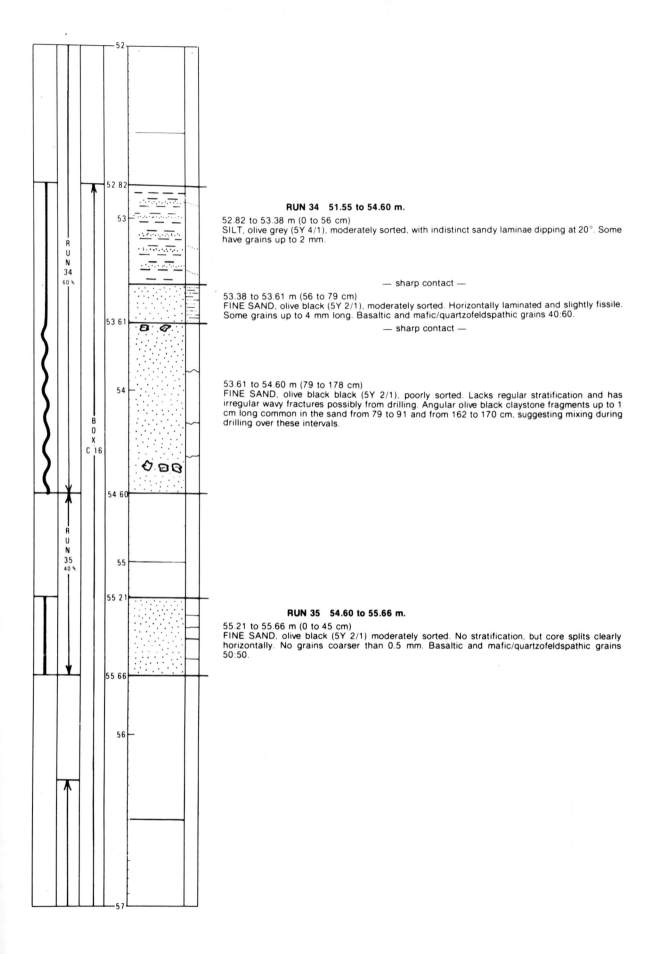

RUN 34   51.55 to 54.60 m.

52.82 to 53.38 m (0 to 56 cm)
SILT, olive grey (5Y 4/1), moderately sorted, with indistinct sandy laminae dipping at 20°. Some have grains up to 2 mm.

— sharp contact —

53.38 to 53.61 m (56 to 79 cm)
FINE SAND, olive black (5Y 2/1), moderately sorted. Horizontally laminated and slightly fissile. Some grains up to 4 mm long. Basaltic and mafic/quartzofeldspathic grains 40:60.

— sharp contact —

53.61 to 54.60 m (79 to 178 cm)
FINE SAND, olive black black (5Y 2/1), poorly sorted. Lacks regular stratification and has irregular wavy fractures possibly from drilling. Angular olive black claystone fragments up to 1 cm long common in the sand from 79 to 91 and from 162 to 170 cm, suggesting mixing during drilling over these intervals.

RUN 35   54.60 to 55.66 m.

55.21 to 55.66 m (0 to 45 cm)
FINE SAND, olive black (5Y 2/1) moderately sorted. No stratification, but core splits clearly horizontally. No grains coarser than 0.5 mm. Basaltic and mafic/quartzofeldspathic grains 50:50.

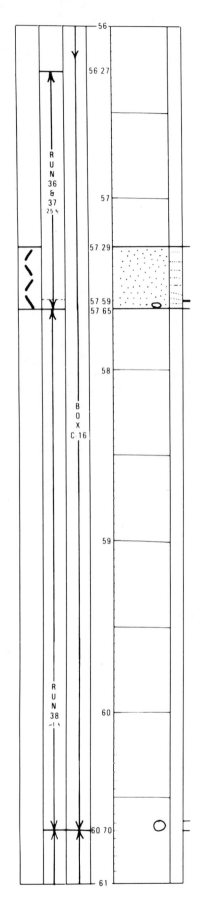

RUN(S) 36 (& 37)   56.27 to 57.65 m. RUN 37 picked up on an additional 6 cm from RUN 36.

57.29 to 57.59 m (0 to 30 cm)
FINE SAND, olive black (5Y 2/1), moderately sorted. No grains coarser than 1 mm. Core broken into pieces 3 to 6 cm long, but several longer pieces show a fine horizontal lamination; in one or two others it dips at 10°.

RUN 37 to 57.65 m.

57.59 to 57.65 m (0 to 6 cm)
SAND, as in RUN 36. One drilled pebble 2 cm long of granitic gneiss.

RUN 38   57.65 to 60.70 m.

60.64 to 60.70 m (0 to 6 cm)
Entire core is a drilled cobble of GNEISS, poorly foliated. quartz-plagioclase-biotite-chlorite.

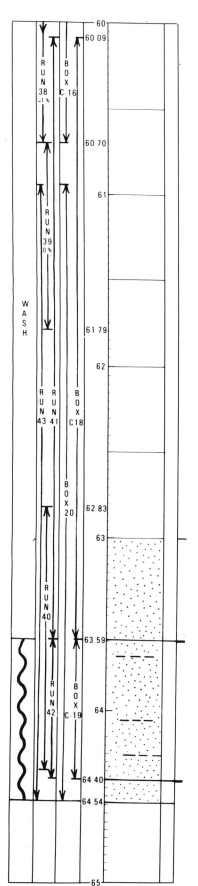

RUN 39   60.70 to 61.79 m.   NO RECOVERY. Lost circulation.
Sand rose in hole to 35 m subbottom. Washed down to 62.38 cm.

RUN 40   62.83 to 64.35 m.   NO RECOVERY
Sand rose in hole to 60 m subbottom.
RUN 41   60.09 to 63.59 m.

63 to 64 m. 31 lb recovered as loose sand and "poured" into a plastic bag.
FINE SAND, olive black (5Y 2/1), moderately well sorted. Basaltic and mafic/quartzofeldspathic grains 50:50.

RUN 42   63.59 to 64.40 m.

63.5 to 64.5 m. 24 lb recovered as loose sand and "poured" into bucket, from which gas samples were taken (O_2 2%, N_2 44%, CO_2 16%, CH_4 38%). Sand stored in plastic bag.
FINE SAND, olive black (5Y 2/1), moderately sorted. Basaltic and mafic/quartzofeldspathic grains 40:60. A few small patches 1 cm across of lighter coloured muddy fine sand. A cluster of angular fragments (broken during drilling) of hard medium grey claystone.

Sand rose in hole to 61.0 m subbottom.

RUN 43   60.96 to 64.54 m.

64 m (0 to 286 cm) Extruded as core into box C-20 but not stratigraphically locatable.
FINE SAND, olive black, (5Y 2/1), moderately sorted. Basaltic and mafic/quartzofeldspathic grains 40:60. No grains coarser than 0.5 mm. Probably wash material.

64.40 to 64.54 m (286 to 300 cm)
FINE SAND, olive black (5Y 2/1), moderately sorted and slightly muddy. Believed to be the only core "in place" in the run.

Sand rose in hole to 57.3 m.

RUN 44   57.3 to 60.3 m.   NO RECOVERY
Sand rose in hole to 51.8 m

RUN 45
Sand from water sampler at 51.8 m

— END OF HOLE —

Acknowledgments. We gratefully acknowledge the logistic support of the U.S. Naval Support Force (Antarctica), Holmes & Narver, the NSF Office of Polar Programs, and the New Zealand Antarctic Research Program. The project was funded by the U.S. National Science Foundation, the Japan National Institute of Polar Research, and New Zealand's Antarctic Division, D.S.I.R., and the New Zealand University Grants Committee. We would like to especially acknowledge the efforts of the New Zealand drilling team and of M. McGale of Longyear Ltd., who worked long and hard to produce the core.

REFERENCES

Barrett, P. J., Textural characteristics of Cenozoic pre-glacial and glacial sediments at Site 270, Ross Sea, Antarctica, in *Initial Reports of the Deep Sea Drilling Project*, vol. 28, edited by D. E. Hayes et al., pp. 757-767, U.S. Government Printing Office, Washington, D. C., 1975.

Barrett, P. J., and P. C. Froggatt, Densities, porosities and seismic velocities of some rocks from Victoria Land, Antarctica, *N.Z. J. Geol. Geophys.*, *21*, 175-187, 1978.

Barrett, P. J., G. W. Grindley, and P. N. Webb, The Beacon Supergroup of East Antarctica, in *Antarctic Geology and Geophysics*, edited by R. J. Adie, pp. 319-332, Universitetsforlaget, Oslo, 1972.

Barrett, P. J., D. A. Christoffel, D. J. Northey, and B. A. Sissons, Submarine extension of Wright Valley into McMurdo Sound *Antarct. J. U.S.*, *8*, 138-140, 1974.

Barrett, P. J., et al., Initial report of DVDP 15, western McMurdo Sound, Antarctica, *Dry Val. Drilling Proj. Bull.*, *7*, 1-100, 1976.

Brady, H. T., Late Neogene history of Taylor and Wright Valleys and McMurdo Sound, derived from diatom biostratigraphy and paleoecology of DVDP cores (abstract), in *Third Symposium on Antarctic Geology and Geophysics, Volume of Abstracts*, Madison, University of Wisconsin Press, 1977.

Bull, C. B., and P. N. Webb, Some recent developments in the investigation of the glacial history and glaciology of Antarctica, in *Quaternary Studies in Antarctica*, edited by E. M. Van Zinderen Bakker, pp. 55-84, A. A. Balkema, Rotterdam, The Netherlands, 1972.

Bullivant, J. S., Ecology of the Ross Sea benthos, in Fauna of the Ross Sea, part 5, General Accounts, Station Lists and Benthic Ecology, *N. Z. Dep. Sci. Ind. Res. Bull.*, *176*, 49-75, 1967.

Cole, J. W., and A. Ewart, Contributions to the volcanic geology of the Black Island, Brown Peninsula, and Cape Bird areas, McMurdo Sound, Antarctica, *N. Z. J. Geol. Geophys.*, *11*, 793-828, 1968.

Cole, J. W., P. R. Kyle, and V. E. Neall, Contributions to Quaternary geology of Cape Crozier, White Island and Hut Point Peninsula, McMurdo Sound region, Antarctica, *N. Z. J. Geol. Geophys.*, *14*, 528-546, 1971.

Debenham, F., A new mode of transportation by ice: The related marine muds of South Victoria Land, *Quart. J. Geol. Soc. London*, *75*, 51-76, 1920.

Denton, G. H., R. L. Armstrong, and M. Stuiver, The Late Cenozoic glacial history of Antarctica, in *Late Cenozoic Glacial Ages*, edited by K. K. Turekian, pp. 267-306, Yale University Press, New Haven, 1971.

Glasby, G. P., P. J. Barrett, J. C. McDougall, and D. G. McKnight, Localized variations in sedimentation characteristics in the Ross Sea and McMurdo Sound regions, Antarctica, *N. Z. J. Geol. Geophys.*, *18*, 605-621, 1975.

Goldich, S. S., S. B. Treves, N. H. Suhr, and J. S. Stuckless, Geochemistry of the Cenozoic volcanic rocks of Ross Island and vicinity, Antarctica, *J. Geol.*, *83*, 415-435, 1975.

Gunn, B. M., and G. Warren, The geology of Victoria Land between The Mawson and Mulock Glaciers, Ross Dependency, Antarctica, *N. Z. Geol. Surv. Bull.* (new series), *71*, 157 pp., 1962.

Hayes, D. E., et al., *Initial Reports of the Deep Sea Drilling Project*, vol. 28, *Antarctica*, 1017 pp., U. S. Government Printing Office, Washington, D. C., 1975.

Kyle, P. R., The geology, mineralogy and geochemistry of the Late Cenozoic McMurdo Volcanic Group, Victoria Land, Antarctica, Ph.D. thesis, 744 pp., Victoria Univ. of Wellington, Wellington, New Zealand, 1976.

Kyle, P. R., and S. B. Treves, Review of the geology of Hut Point Peninsula, Ross Island, Antarctica, *Dry Val. Drilling Proj. Bull.*, *2*, 1-10, 1973.

Kyle, P. R., and S. B. Treves, Geology of DVDP 3, Hut Point Peninsula, Ross Island, Antarctica, *Dry Val. Drilling Proj. Bull.*, *3*, 13-48, 1974.

McGinnis, L. D., McMurdo Sound—A key to the Cenozoic of Antarctica, *Antarct. J. U.S.*, *8*, 166-179, 1973.

McKelvey, B. C., Upper Cenozoic marine and terrestrial glacial sedimentation in eastern Taylor Valley, southern Victoria Land (abstract), in *Third Symposium on Antarctic Geology and Geophysics, Volume of Abstracts*, University of Wisconsin Press, Madison, 1977.

Northey, D. J., C. Brown, D. A. Christoffel, H. K. Wong, and P. J. Barrett, A continuous seismic profiling survey in McMurdo Sound, Antarctica—1975, *Dry Val. Drilling Proj. Bull.*, *5*, 167-179, 1975.

Powell, R. D., Textural characteristics of some glacial sediments in Taylor Valley, Antarctica, M.Sc. thesis, 316 pp., Victoria Univ. of Wellington, Wellington, New Zealand, 1976.

Sissons, B. A., and D. J. Northey, Continuous seismic profiling in McMurdo Sound, *Dry Val. Drilling Proj. Bull.*, *3*, 234-239, 1974.

Smith, W. C., The volcanic rocks of the Ross Archipelago, *Nat. Hist. Rep. Br. Antarct. Terra Nova Exped. 1910* (Geol.), *2*, 1-107, 1954.

Treves, S. B., and P. R. Kyle, Geology of DVDP 1 and 2, Hut Point Peninsula, Ross Island, Antarctica, *Dry Val. Drilling Proj. Bull.*, *2*, 11-82, 1973.

Treves, S. B., R. G. Rinehart, and R. Pochon, Geology and petrography of rocks from the floor of the Ross Sea near Ross Island, *Antarct. J. U.S.*, *9*, 152-154, 1975.

Webb, P. N., and J. H. Wrenn, Late Cenozoic micropaleontology and biostratigraphy of eastern Taylor Valley, Antarctica (abstract), in *Third Symposium on Antarctica Geology and Geophysics, Volume of Abstracts*, University of Wisconsin Press, Madison, 1977.

CHEMISTRY AND CLAY MINERALOGY OF SELECTED CORES FROM THE ANTARCTIC DRY VALLEY DRILLING PROJECT

F. C. Ugolini

College of Forest Resources, University of Washington, Seattle, Washington 98195

W. Deutsch

Department of Geological Sciences, University of Washington, Seattle, Washington 98195

H. J. H. Harris

Illinois State Geological Survey, Urbana, Illinois 61801

Electrical conductivity and ionic composition of solutions extracted from ice-cemented permafrost from cores 8, 9, and 10, New Harbor, show that most of the sediments were deposited in a marine environment and suggest that aggradation of permafrost during exposure of the sediments to sub-aerial conditions caused ionic concentration. Influx of brines capable of moving in permafrost is also suggested. Regions of low conductivity are interpreted as a result either of freshwater episodes or of textural discontinuities. The clay minerals of the above cores, and of core 11, Commonwealth Glacier, and core 12, Lake Leon, show little weathering and complex irregular interstratification of mica, vermiculite, montmorillonite, and chlorite. Clay mineralogy of core 10 can be separated into three major assemblages corresponding to the three major lithologic units.

INTRODUCTION

Because most of the world's ice is contained in Antarctica, the present and past size and extent of the Antarctic ice sheets are of interest from several different viewpoints, including those that encompass problems of climatic change and prediction. The Dry Valley Drilling Project (DVDP) was designed to address these and other problems through examination of the geological and geochemical history of the ice-free areas of southern Victoria Land and other regions of McMurdo Sound, Antarctica. This report presents and discusses the chemical parameters of water extracts from, and the clay mineralogy of, cores from DVDP boreholes 8, 9, and 10, which were drilled in the New Harbor area of southern Victoria Land.

The clay mineralogy, only, is discussed for core 11, Commonwealth Glacier, and core 12, Lake Leon (Figure 1).

The electrical conductivity and ionic composition of solutions extracted from the DVDP cores were determined. Chemical parameters such as these have been used to interpret the postemergence history of marine sediments in the Barrow area [*O'Sullivan*, 1966; *Brown*, 1966; *Sellmann and Brown*, 1973] and the Quaternary stratigraphy of terrestrial deposits near Fairbanks [*Pewe and Sellmann*, 1973; *Sellmann*, 1967, 1968], both in Alaska. *Page* [1978] and *Iskandar et al.* [1978] have more recently examined the salinity and chemical composition of pore water in permafrost in the offshore areas near Prudhoe Bay, Beaufort Sea. *McGinnis et al.* [1980] measured salinity and electrical resistivity of sections of cores from DVDP boreholes 10, 11, and 12, Taylor Valley, Antarctica, and used these data to reconstruct glacial events.

The mineralogy of the clay fraction from selected DVDP core sections was determined both to provide additional information on the lithology of the cores and to shed light on the provenance, weathering, and possible diagenesis of the sediments.

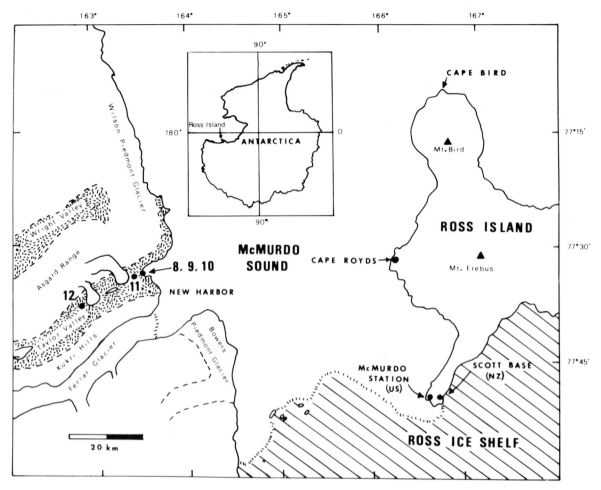

Fig. 1.   The McMurdo Sound region, southern Victoria Land, Antarctica, the New Harbor site of cores 8, 9, and 10, the Commonwealth Glacier site of core 11, and the Lake Leon site of core 12.

Weaver [1958] pointed out that clay minerals constitute more than 50% of the minerals in sedimentary rocks and that more than 95% of all sedimentary rocks contain clay minerals. This statement does not appear controversial, but disagreement exists whether the clay minerals in sediments, specifically in marine sediments, are detrital or diagenetic. *Millot* [1970] concluded that marine clay minerals are mostly detrital but that some of them are subject to diagenesis after sedimentation. It would appear therefore that a systematic analysis of the clay minerals in the cores could provide information on the provenance of the sediments and on the environmental conditions during diagenesis.

This report demonstrates that, in practice, a single interpretation of a particular chemical or mineralogical parameter can be misleading or erro-

neous. An awareness of this difficulty is implied in the work of *O'Sullivan* [1966], who noted that the salinity of sediments in a permafrost region is influenced by many factors, including texture, depth, ice content, and the history of the site. The environment of the sediments of cores 8, 9, and 10 has varied markedly—prior to, during, and subsequent to deposition. In principle, then, several different physical processes, acting singly, simultaneously, and/or at different times, could have left their fingerprints on the sediments and pore fluid. Consequently, some of the variations we have measured could have been caused in a number of different ways, and there is often nothing in the individual parameters themselves to allow an unambiguous historical interpretation. We have met this problem in two ways: (1) by identifying those physical processes that are active, or are thought to be active, in sedimentary sequences deposited in polar regions and (2) by comparing different parameters, including those reported in the literature, in the hope that the comparison will allow a more complete and accurate interpretation.

PROCESSES ACTIVE IN SEDIMENTS OF POLAR REGIONS

In interpreting the chemical parameters of pore fluids, one must remember that both the water and the solutes could be (1) connate, (2) nonconnate, or (3) some mixture of the two. There are three processes that might act, singly or in concert, to change the composition of a pore solution: (1) interaction with sediments, (2) freezing, and (3) groundwater flux.

Much interaction between sediment and solutes in a pore fluid is a function of the cation exchange capacity and weathering of the minerals in the sediment. Sediments in the clay and silt size fractions, often containing minerals having a high exchange capacity, tend to retain the solutes deposited with or passing into them, even under hydrodynamic conditions that might 'flush' a coarser sediment. This kind of effect is a possible explanation for the observations of O'Sullivan [1966], who reported a striking correlation between texture and salinity: silts displaying salinities 3-10 times, and clays 5-10 times, greater than that of adjacent sandy gravels. Nakao et al. [1978] reported data for DVDP 14 that show a similar correlation. The retention of solutes could have a further effect. Several workers have suggested that ions held in clays and silts can act as a barrier to the passage of additional charged solutes. Under such conditions, the ions dissolved in groundwater that passes through fine-grained material are filtered out and retained on the 'upstream' side of the 'charge barrier' [Olsen, 1969; Kharaka and Berry, 1973].

Weathering of the sediments could be another cause of changes in the composition of the pore fluid. The presence of unfrozen water in frozen soils and sediments makes weathering and ionic migration a plausible phenomenon [Anderson, 1967, 1968; Anderson and Morgenstern, 1973; Ugolini and Anderson, 1973; Murrmann, 1973]. Fripiat et al. [1965] have shown that the dissociation of water in films next to the surface of silicates is many times greater than in bulk. The high ionic strength of the film of unfrozen water surrounding the mineral grains leads to corrosion of the silicate

surface. Ions are thus liberated from the solid phase and become available to move in the unfrozen water. Murrmann and Hoekstra [1970] and Murrmann et al. [1968] have shown that although ion diffusion is considerably lower in frozen soils than in unfrozen ones, the diffusion coefficients in the temperature range of $0°$ to $-15°C$ are relatively high, of the order of 1×10^{-7} cm^2 s^{-1} for silt soil at $-3°C$. When the soil is water saturated and in a closed system, conditions existing in ice-rich permafrost, the diffusion coefficient of an ion is virtually independent of total water content and thermal gradients but is dependent on the quantity of unfrozen water, which is controlled by temperature [Murrmann and Hoekstra, 1970; Murrmann et al., 1968; Hoekstra, 1965].

The effects of freezing on pore fluids are complex. It has long been known that dissolved ions are almost entirely excluded from the ice phase as freezing progresses [Thompson and Nelson, 1956; Terwilliger and Dizio, 1970; Hallet, 1976; Anisimova, 1978; Yas'ko, 1978]. Hallet et al. [1978] have recently shown that cryogenic precipitates may originate through rejection of salts during freezing of ionic solutions. In principle, then, the freezing of an originally brackish pore fluid could result in the formation of two adjacent and physically separate 'strata' of solutions, one fresh and frozen, the other, expelled from its original position, highly saline and liquid. In fact, the geometry and rate of advance of the freezing front must have a significant effect on the degree of solute segregation found in a permafrost section. As implied above, material type must also play a significant role in determining how much solute is held in place during freezing. The freezing of pore fluids not only is responsible for ionic exclusion, but also tends to segregate the ions as a function of the lowering of the temperature. The early experiments of Thompson and Nelson [1956] demonstrated that seawater evolves chemically when it is subjected to progressive freezing, owing to a sequential segregation of salts. At a temperature of $-22.9°C$, after calcium carbonate, mirabilite, and sodium chloride dihydrate have crystallized, the original seawater has become a brine enriched in magnesium, potassium, and calcium chloride [Thompson and Nelson, 1956]. As we will later compare our data to those of Stuiver et al. [1976], we also note that hydrogen and oxygen isotopes in water are fractionated between the liquid and solid phases during freezing.

Hydrodynamic systems are obvious mechanisms for the redistribution of water and solutes; they can, in principle, produce very complex effects. For example, connate solutions in bedded sediments might be preferentially removed from the coarser, more permeable layers, owing both to larger groundwater fluxes in those layers and to sediment properties of the kind discussed above. The situation can be further complicated in a permafrost region by the presence of frozen ground which acts as a zone of very low permeability. Prior to the aggradation of permafrost, coarse materials are most likely to be affected by extensive groundwater motion. Subsequent to aggradation, motion can occur only in fluids saline enough to remain unfrozen. This factor favors a reversal of hydrogeologic roles peculiar to permafrost terrains: namely, that fluids are likely to move most rapidly (albeit very slowly) in fine-grained materials, precisely because these materials are most likely to retain saline fluids and unfrozen water [Anderson, 1967, 1968]. A further hydrogeologic complexity is introduced by the increase in volume that occurs during the phase change from liquid water to ice. Even in the absence of the forces that 'normally' give rise to groundwater flow, some flow must occur in a freezing sediment in order to accommodate the expansion. Some of these phenomena are discussed by Cartwright and Harris [this volume], who describe modern hydrogeologic systems in the dry valleys.

MATERIALS AND METHODS

Cores 8, 9, and 10 were obtained at the eastern end of Taylor Valley, southern Victoria Land, Antarctica, at an elevation of 1.9 m, and about 50 m from the southwest shore of New Harbor (Figure 1). Core 8 was drilled to a depth of 157 m; core 9 was obtained to secure a more complete record of the upper 38 m of core 8; core 10 attained a total length of 185.47 m. Because the boreholes were in close proximity and penetrated lithostratigraphic sequences that are virtually identical [Beget, 1977; Porter and Beget, this volume], the three cores are usually treated as a composite in the following discussion. Complete logs of the cores are given in the DVDP bulletins [Chapman-Smith and Luckman, 1974; McKelvey, 1975].

Core 11, the Commonwealth Glacier site, was collected from a borehole approximately 3 km west of New Harbor at the eastern end of Taylor Valley, Antarctica. The site is at an elevation of 80.2 m [McKelvey, 1979]. Core 12, the Lake Leon site, was obtained about 16 km from New Harbor at 75 m in elevation. Core 11 measured 327.96 m in length; core 12, 184.65 m [McKelvey, 1979; Chapman-Smith, 1975].

Measurements of chemical parameters were made on 1 : 5 sediment-water extracts. Ten grams of air-dried sediments were mixed with 50 ml of water; the mixture was agitated, allowed to stand for 10 min, and centrifuged. The conductivity of the decantate was measured with a Yellow Springs model 31 conductivity bridge. Chloride and $SO_4^=$ were determined by autoanalysis (Technicon Industrial method AAII, 99-70W, 1971; and AAII 226-72W, 1971, respectively). Ca^{2+}, Mg^{2+}, Na^+, and K^+ were determined by atomic absorption spectrophotometry (Instrumentation Laboratory 353). For the purpose of evaluating the electrical conductivity readings, a scale was adopted. The lower and upper limits of this scale were taken from the electrical conductivities measured in normal and saline soils [Richards, 1954]. Following this scale, electrical conductivity of 800–2000 μmhos/cm is low, from 2000 to 4000 μmhos/cm is moderately low, from 4000 to 6000 μmhos/cm is moderately high, and above 6000 μmhos/cm is high.

The clay fraction (<2 μm) was obtained from selected samples after salt and iron removal [Mehra and Jackson, 1960]. The deferrated clay fraction was saturated with Mg^{2+} and glycolated. A clay aliquot was saturated with K^+ and heated at 300° and 500°C. X ray diffractograms were obtained with a Picker unit with CuKα radiation and a Ni filter; the unit was operated at 30 kV and 20 mA. A separate aliquot of clay was treated with dimethylsulfoxide (DMSO) to differentiate between kaolinite and chlorite [Abdel-Kader et al., 1978].

RESULTS AND DISCUSSION

Electrical Conductivities

In examining the electrical conductivity values of the fluid extracts (Table 1 and Figure 2), major and minor fluctuations can be observed. The major fluctuations occur from 148 to 183 m and from 44 to 68 m. The central part, from 68 to 148 m, and upper part, from 0 to 44 m, show minor fluctuations. Because conductivity was measured on a 1:5 extract, the values obtained are internally consistent but do not directly measure the original concentration of ions in the interstitial water. However, the

TABLE 1. Electrical Conductivity (EC) for DVDP Cores 8, 9, and 10, New Harbor, Antarctica

Hole 8		Hole 9		Hole 10	
Depth, m	EC, μmhos/cm	Depth, m	EC, μmhos/cm	Depth, m	EC, μmhos/cm
		6	1800		
		9	1800		
		11	1200		
		19	1000		
		21	2700		
		25	1900		
		27	900		
		35	800		
44	890			37	800
46	3100			46	2500
49	960				
52	940				
54	1700				
57	4400				
58	4600				
60	5700				
63	4300				
65	3900				
68	5600			68	1800
71	2600				
74	2300				
78	2400			83	1900
90	3200			87	1200
98	2200				
103	3500			109	2200
112	3000			112	2900
113	2300			113	2500
118	1400			119	1100
123	1900				
126	4800				
131	2500				
133	3500			133	3500
136	1900				
146	2900				
146	2800				
147.9	2500				
148	8400				
150	6400				
154	3400				
155	5500			155	5500
156	7100			158	10000
				168	7900
				171	900
				175	3700
				180	900
				183	8200

All EC are at 25°C.

fluctuations we measured, except for some discrepancies, parallel rather closely the fluctuations in the salinity curve that *McGinnis et al.* [1980] obtained from fluids squeezed from the sediments.

From 148 m down to 183 m, there are widely varying conductivities, with maxima at 148, 150, 156, 158, 168, and 183 m. Lower conductivity values are recorded at 171 and 180 m (Table 1). The salinity curve of *McGinnis et al.* [1980] for

DVDP hole 10 also shows considerable fluctuation below 148 m, which they interpreted as evidence of alternating periods of exposure to freshwater and saltwater environments. They supported the view that the pore water is connate and that the salt enrichment was due to permafrost aggradation and segregation of brines at the base of permafrost. Low salinity values were explained, alternatively, by (1) sea level lowering, (2) crustal uplift, (3) im-

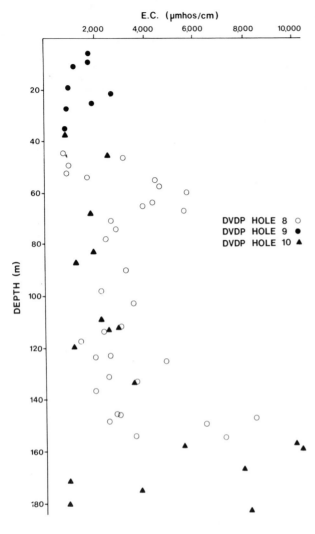

Fig. 2.   Electrical conductivity values for cores 8, 9, and 10 (New Harbor) plotted according to depth (meters).

hand, the low conductivity of the fine sediments (64% finer than 8 φ) [*Powell*, 1978] at 180 m depth must be explained by a freshwater event.

The permafrost oxygen-isotope ratios [*Stuiver et al.*, 1976] have freshwater values between 150 and 180 m. However, only at depths 171 and 180 m could conductivity values indicate freshwater conditions and thus coincide with the isotopic ratios. The high salinities from 148 m downward, ascribed in this report to marine conditions, do not correlate with the oxygen-isotope ratios. Perhaps the salts are connate, but the original interstitial fluids were replaced by fresh water.

Between 68 and 148 m the conductivities were from low to moderately high, with an average value of approximately 2400 μmhos/cm. The electrical conductivity curve between these two depths displays a number of single peaks and valleys, indicating fluctuations in the concentration of dissolved ions. Conductivity values of the order of 4800, 3500, 3500, and 3200 μmhos/cm occur, respectively, at 126, 133, 103, and 90 m. Low values, 1100 and 1200 μmhos/cm, occur at 119 and 87 m. A moderately low value occurs at 146 m (2800 μmhos/cm), and a moderately high (5600 μmhos/cm), and a low (1800 μmhos/cm) value occurs at the same depth (68 m) in, respectively, holes 8 and 10. In general, the high and low conductivities are consistent with the salinity distribution reported by *McGinnis et al.* [1980]. We interpret this interval (68-148 m) as reworked marine sediments [*Webb and Wrenn*, 1980; *Brady*, 1980] that have been affected by fresh water. It is not possible to precisely determine the number of cycles of marine and freshwater environments from our electrical conductivity values. However, values at the 119- and 87-m depths seem to indicate terrestrial environments. Between core depths of 68 and 148 m, *Beget* [1977] distinguished five diamictons, two of which displayed a strong microfabric. The lowest was from 114 to 120 m and the upper from 68 to 74 m. He indentified the lowest as a basal till recording the incursion of the Ross Sea Ice. This interpretation agrees with freshwater isotopic ratios reported by *Stuiver et al.* [1976] and with our low conductivity values. The upper region (68-74 m), which shows a strong microfabric, failed to show freshwater isotopic values but correlated well with the low salinity of *McGinnis et al.* [1980] and with our low conductivity values.

The segment above 68 m has moderately high initial conductivity values, which decrease upward

pounding of fresh water by advances of the Ross Ice Shelf in New Harbor, or (4) a combination of the foregoing. We, too, ascribe the high conductivity values at core depths of 148, 150, 156, 168, and 183 m to brine concentration in the fine-grained marine sediments during freezing. The relatively high values for Na, K, Ca, Mg, and SO$_4$ indicate that repeated freezing and ionic segregation may be involved (Table 2). Although the low conductivity value of 900 μmhos at 171 m coincides with a very coarse sand unit, the low value at 180 m occurs in fine-grained sediments [*Powell and Barret*, 1975; *Powell*, 1978]. Consequently, we infer that the coarse sands at 171 m failed to retain the ions during the aggradation of permafrost. On the other

TABLE 2.   Chemical Characteristics of Selected Samples for DVDP Cores 8, 9, and 10, New Harbor, Antarctica

Core	Depth, m	Ion Content,* ppm					
		Na^+	K^+	Ca^{2+}	Mg^{2+}	Cl^-	SO_4^{2-}
9	21	450	30	21	20	724	25
9	25	380	24	9	12	553	28
10	37	150	14	4	6	250	25
10	46	575	38	5	8	700	155
8	57	970	89	10	15	1426	37
8	60	1170	93	12	25	1812	53
10	68	150	33	4	4	250	45
10	83	350	30	6	11	300	35
10	87	175	21	6	7	450	120
10	109	475	31	10	10	700	30
10	112	640	48	13	15	900	135
10	113	520	54	9	11	1100	140
10	119	225	25	9	5	300	50
10	133	600	47	17	42	1175	135
8	148	520	32	16	10	733	35
10	158	2100	99	85	163	3700	107
10	171	170	10	11	9	278	11
10	180	180	11	4	6	251	9
10	183	1900	84	54	79	2980	86

*In 1 : 5 sediment-water extract.

rather uniformly to low values of 1000-2000 μmhos/ cm above 54 m. Oxygen isotopes [Stuiver et al., 1976] and microfabric data [Beget, 1977] show this to be a marine section. From the surface to a depth of 24.86 m, Brady [1978] reported both marine and nonmarine diatoms. According to Webb and Wrenn [1980], the lowermost part of this interval is a diamicton, possibly deposited by ice rafting, and the upper part is deltaic. A marine delta is sub-aerially exposed during growth and is affected by both influx of fresh water and permafrost aggradation. Combined with the influence of textural variability, these effects may explain the conductivity measurements in this interval. The lowest conductivity recorded coincides with the unit with the largest clasts, and the highest conductivity with fine-grained diamictons [Powell, 1978].

Ionic Concentration and Ionic Ratios

Sodium and Cl are the predominant ions in the 1:5 sediment-water extracts through cores 8, 9, and 10 (Table 2). When Na, K, Ca, and Mg are plotted in the diagram of Torii et al. [1978], the points for cores 8, 9, and 10 tend to concentrate in the left corner of the diagram, close to seawater composition (Figure 3). The points for Taylor Val-

ley and Cape Royds soils also appear to betray the marine influence. The points for the Victoria Valley soils (VV I through VV IV) tend to approach glacier meltwater and the Onyx River composition (Figure 3). Although the 1:5 sediment-water extract tends to dilute the original interstitial fluids and thus favor the solution of Na and K, the chemical characteristics of the extract seem to indicate seawater contamination for all the core depths. A marine origin of the core unit below 150 m is suggested by the microfauna analysis of Webb and Wrenn [1980]. These authors recognized an unconformity at 154 m, and they ascribed the sediments and microfauna below this depth (zone II) to marine deposition in a sublittoral fjord environment extending from early Pliocene to late Miocene. Brady [1978], on the basis of diatoms in core 10, recognized an early Pliocene marine unit from 137 to 183 m. Additional evidence of marine conditions is provided by Beget [1977] and Porter and Beget [1978], who believed that the sediments between 148 and 183 m were intially deposited in a deep-water marine environment, as indicated by rare drop stones and graded beds resembling turbidites.

The ionic profile for the region between 148 and 68 m also betrays a marine influence (Table 2 and Figure 3).

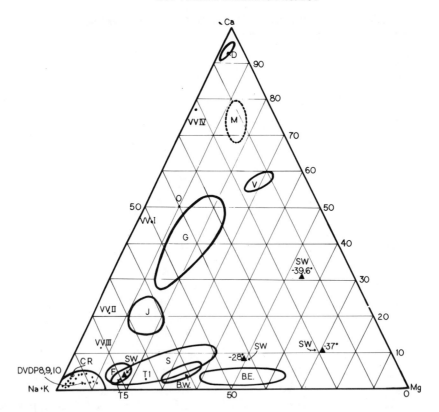

Fig. 3.   Chemical composition of snow and ice, glacier meltwaters, lakes, seawater above and below freezing, soil-water extracts, sediment-water extracts of cores 8, 9, and 10. Dry Valley, Antarctica. Legend: S, snow and ice; M, Lake Miers; J, Lake Joyce; O, Onyx River; F, Lake Fryxell; B.W., Lake Bonney west; B.E., Lake Bonney east; V, Lake Vanda; D, Don Juan Pond; G, glacier meltwaters; VV I to VV IV, soil-water extract, Victoria Valley; T1, soil-water extract, Taylor Valley site 1; T5, soil-water extract, Taylor Valley site 5; CR, soil-water extract, Cape Royds; DVDP 8, 9, and 10, sediment-water extract, cores 8, 9, and 10 to New Harbor; solid triangle, seawater above and below freezing. (Modified from *Torii et al.* [1978].)

The uppermost region (above 68 m) shows a marine influence (Table 2); however, at a depth of 37 m, a low electrical conductivity is recorded. The Na and Cl concentrations are also low, perhaps indicating a terrestrial environment. Especially significant is the Na/Cl ratio, which in core 10 slightly exceeds that of normal seawater (Tables 3 and 4), suggesting that originally the interstitial water was seawater.

The Na/K ratios for the sediments of cores 8, 9, and 10 range from 4.5 to 22.6, compared with a ratio of 28.9 for seawater. Potassium enrichment, resulting in a lower Na/K ratio, may be explained by a number of mechanisms. Those mechanisms related to the postdepositional weathering and diagenesis of the sediments are discussed below, in the clay mineralogy section of this paper. The Na/K ratio might have been influenced by the effect of temperature on the exchangeable cations. With in-

creasing temperature of the sediments, K, Na, and Li are desorbed and diffused into the interstitial water [*Page*, 1978]. As all the samples were treated alike, any artificial temperature effect should have been the same in all samples. The K enrichment could also indicate that the sediments either were transported in a nonmarine environment or were affected by influx of fresh water after deposition. Since freshwater conditions were not common throughout the core, only certain core depths could have been affected by these processes. Core samples at 68, 87, 113, and 119 m, which show low Na/K ratios and moderately low to low conductivities, could have been influenced by freshwater environments. Another attractive explanation of these data is that the K enrichment and the low electrical conductivity could be related to the mechanism of deposition. According to *Beget* [1977], the intervals 114-120 m and 74-68 m coin-

TABLE 3. Ionic Ratios for Selected Depths, DVDP Cores 8, 9, and 10, New Harbor, Antarctica

Core	Depth, m	Na/Cl	Na/Ca	Na/K	Ca/Cl	Cl/SO$_4$	Ca/Mg
9	21	0.62	21.4	15.0	0.03	28.9	1.05
9	25	0.69	42.2	15.8	0.02	19.7	0.75
10	37	0.60	37.5	10.7	0.02	10.0	0.67
10	46	0.82	115.0	15.1	0.01	4.5	0.63
8	57	0.68	97.0	10.9	0.01	38.5	0.67
8	60	0.65	97.5	12.6	0.01	34.2	0.48
10	68	0.60	37.5	4.5	0.02	5.5	1.00
10	83	1.17	58.3	11.7	0.02	8.6	0.55
10	87	0.39	29.2	8.3	0.01	3.7	0.86
10	109	0.68	47.5	15.3	0.01	23.3	1.00
10	112	0.71	49.2	13.3	0.10	6.6	0.87
10	113	0.47	57.8	9.6	0.01	7.8	0.82
10	119	0.75	25.0	9.0	0.03	6.0	1.80
10	133	0.51	35.3	12.8	0.01	8.7	0.40
8	148	0.71	32.5	16.3	0.02	20.9	1.60
10	158	0.57	24.7	21.2	0.02	34.6	0.50
10	171	0.61	15.5	17.0	0.04	25.3	1.20
10	180	0.72	45.0	16.4	0.02	27.9	0.67
10	183	0.64	35.2	22.6	0.02	34.7	0.68
Seawater		0.53	26.0	28.0	0.02	7.20	0.30

cide with basal till and diamictons with strong microfabric, respectively. The grinding and the abrasion of minerals carried in a viscous mass at the bottom of an ice sheet could effectively weather the K minerals, especially mica, and release K. Consequently, the pore fluids could have become enriched in K and, upon freezing, have maintained this chemical signature. Furthermore, this would also be an effective mechanism for degrading the mica and producing the mixed layer assemblages discussed below.

Calcium/magnesium ratios in core 10 show that calcium is enriched with respect to seawater (Tables 3 and 4). This enrichment of Ca may be due to either the cation exchange reactions or the displacement of Ca from the exchange sites due to the influx of K. The Cl/SO$_4^=$ ratios (Tables 3 and 4) vary from 3.75 to 34.7 and, in general, tend to be larger than for seawater (Table 4). The lower part of the core, from 148 to 183 m, appears to be consistently depleted in SO$_4^=$. This depletion may be due to microbial reduction. Enrichment in SO$^{4=}$

TABLE 4. Ionic Ratios for Lakes, Soils, and DVDP Cores 8, 9, and 10, McMurdo Region, Antarctica

	Na/Cl	Na/Ca	Na/K	Ca/Cl	Cl/SO$_4$	Ca/Mg
			Lakes			
Vanda*						
50 m	0.27	0.36	3.90	0.76	17.0	3.90
64 m	0.09	0.23	7.30	0.38	90.0	3.4
Don Juan*	0.06	0.09	73.0	0.50	1300.0	73.0
N. Fork saline pond*	0.64	8.20	32.0	0.08	2.10	0.30
			Soils			
Taylor Valley†						
Site 1	0.45	35.0	6.1	0.014	34.9	0.09
Site 5	0.54	33.3	16.8	0.016	20.4	0.13
Ross Island‡	0.46	10.34	2.41	0.04	6.36	4.3
Beacon Valley‡	1.70		11.93		0.09	
Ross Island‡	0.60	14.2	1.72	0.043	5.20	3.3
			Cores			
DVDP 8, 9, 10 range	0.62–1.17	15.0–115.0	4.5–22.6	0.01–0.03	34.7–3.75	0.4–1.8
Seawater	0.53	26.0	28.0	0.02	7.20	0.30

*Oberts [1973].
†Linkletter [1970].
‡Tedrow and Ugolini [1966].

with respect to Cl⁻ may indicate leaching of Cl⁻ and Na⁺. The region between 114 and 120 m could evidence such a leaching episode. Both the Na/K ratio and the electrical conductivity are low at these depths.

Clay Mineralogy

The clay mineralogy of core 10 consists of a complex association of interstratified montmorillonite-vermiculite-chlorite-mica. Most of the interstratification is irregular; however, regular interstratification leading to the formation of superlattices was also observed. The diffractograms in Figure 4 are from the clay fraction of the 148-m depth. The diffractogram of Mg-saturated clay shows a prominent peak at 14.48 Å and two smaller peaks at 10.04 and 7.10 Å. Upon glycolation, the 14-Å peak resolves itself into a number of diffraction intensities. A much reduced peak at 14.24 Å indicates that a sizable portion of the 14.48 Å (Mg saturation) was montmorillonite, now expanded at high-order spacings. The numerous

small peaks between 17.1 and 24.5 Å reveal the presence of irregularly interstratified montmorillonite-vermiculite with some regular interstratification forming superlattices. The broad mica peak (10.10 Å) indicates irregularly interstratified mica-vermiculite. The 7.10-Å peak may represent the second-order diffraction of vermiculite as well as the presence of kaolinite. The K saturation and heating at 300° and 500°C tend to collapse the expansible minerals. The collapsing of the interstratified montmorillonite-vermiculite induces the reenforcement of the 10-Å peak. Treatment with dimethylsulfoxide [*Abdel-Kader et al.*, 1978] clearly shows the presence of kaolinite at 11.4 Å, mica at 10.16 Å, and a peak for both montmorillonite and vermiculite at 19.11 Å (Figure 4).

In the lower part of the core, from 158 to 182 m, the interstratification is more irregular and complex (Table 5). Diffractograms of Mg-saturated clay show little resolution of the individual minerals and a high degree of irregularities. Interstratified, highly variable montmorillonite-vermiculite is the dominant phase; however,

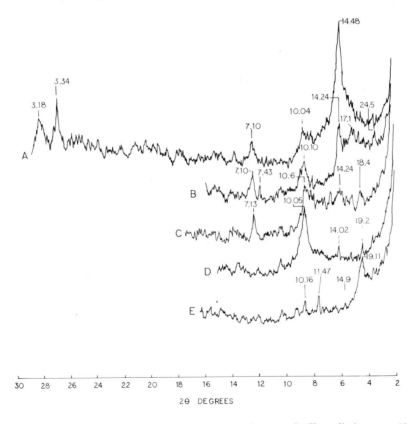

Fig. 4.   X ray diffractograms of clay fraction <2 μm, oriented mount, Cu Kα radiation, core 10, 148-m depth. (*a*) Mg saturated, (*b*) Mg/glycol saturated, (*c*) K saturated ambient, (*d*) K saturated heated at 550°C, and (*e*) Li-DMSO treated.

TABLE 5. Clay Mineralogy by X Ray Diffraction Analysis for Selected Samples From Core 10 (<2-μm Fraction)

Depth, m	Montmorillonite, vermiculite, chlorite, and mica interstratified	Vermiculite	Mica	Chlorite	Kaolinite	Quartz	Feldspar	Amphibole	Pyroxene
6	M					m–t	m–t		
21	M					m	m	m	m
25	M		t	t					
57	M	t	t		t				
60	M	t	t		t	m		m	
71	M					M	m		
90	M					M		m	
118	M					m	M	m	
120	M			t			M	m	
123	M						M		
133	M						M	m	
146	M						M		
148	M				t	m	m		
158	M	m	m	m–t	t	m	m		
171	M	m	m	m	t	M	M	m	
179	M	m	m	m–t		m	m	t	t
182	M	m	m	m–t	t	m	m	m	

M, major; m, minor; t, trace.

noninterstratified vermiculite, mica, and chlorite species, with some kaolinite, appear as minor and trace components together with quartz and feldspar. From 71 to 146 m (Table 5) the core shows a weakly developed interstratification of montmorillonite-vermiculite-chlorite and mica but only a few individual components of these mixed layer complexes were noted. Feldspars are abundant throughout the interval, except at 90 and 71 m, where quartz prevails.

The section of the core from the surface to 60 m (Table 5) shows a complex assemblage of interstratified phyllosilicates similar to that in the underlying unit but containing more unmixed individual components, such as vermiculite, mica, and kaolinite. This upper section is depleted in clay-size quartz and feldspars.

The clay mineral assemblages throughout the core indicate that the sediments experienced a low level of weathering. Mixed layers are formed readily and represent either the initial stage in weathering [Sawheny, 1977] or the mature weathering products of such regions as deserts, where little weathering occurs [Sawheny, 1977; M. L. Jackson, personal communication, 1978]. The mica-vermiculite-montmorillonite interstratification indicates only a partial removal of K from freshly cleaved mica flakes. Removal of K results in structural changes and in reduction in layer charges, which in turn results in the appearance of mixed layers [Norrish, 1973]. If the degradation of mica, via exchange of interlayer K with hydrated cations, had occurred after deposition, it could have caused the K enrichment and the lowering of the Na/K ratios discussed above (Table 3). Ths interpretation, however, implies that diagenetic processes in Antarctic marine environments are different from those reported in the literature. For example, various authors [Millot, 1970; Sawheny, 1977; Fanning and Keramidas, 1977] have suggested that interstratified mica-vermiculite or mica-smectite can form diagenetically in a marine environment through the preferential uptake of K by the expansible 2:1 minerals. Diagenesis of this kind would make it even more difficult to account for the relatively high K content of the water extracts from the DVDP cores. The weathering of mica into interstratified mica-vermiculite-montmorillonite could have occurred under terrestrial conditions, prior to deposition, as commonly observed in the soils in south Victoria Land [Claridge, 1965; Ugolini and Jackson, 1980]. Alternatively, and as mentioned earlier, the degradation of mica could have occurred during glacial transport, by comminution and hydrolysis (Tamm [1924] in Keller [1962]). In any case, the retention of the K in the pore fluid would have depended on the postdepositional history of the sediments. As previously discussed, the possibilities for changing the initial chemical signature of the pore fluid are numerous and mostly unpredictable.

Montmorillonite is rather common as a separate phase in the clay of the soils of Wright and Taylor valleys [*Claridge*, 1965; *Ugolini and Jackson*, 1980]; it has also been found together with iron-rich chlorite in Oligocene to Quaternary turbidites derived from Victoria Land [*Piper and Pe*, 1977]. The origin of montmorillonite in core 10 could be detrital, but it could also have formed in situ in the permafrost, as recently suggested by *Uskov* [1978]. According to this author, hydromica, chlorite, and (more slowly) kaolinite are changed into montmorillonite in the Pleistocene permafrost of Yakutia, U.S.S.R.

In summary, three major units of rather uniform clay mineralogy can be detected in core 10: one below 158 m, one between 60 and 158 m, and one above 60 m. Although showing complex and irregular interstratification, the unit below 158 m displays both noninterstratified individual mineral species and traces of kaolinite and is the most weathered unit in the core. According to *Beget* [1977], the lithological assemblages for this unit indicate a constant source from eastern Taylor Valley. Above 158 m and below approximately 60 m the clay mineralogy, although similar to that of the underlying unit, shows an abundance of feldspar and only a few individual mineral species. This is the least weathered unit. The lithology and texture of the sediments suggest a source southeast of Taylor Valley [*Beget*, 1977]. The clay mineralogy in the uppermost unit has charactertistics of both the lower two units. This is expected since the unit represents a delta that received sediments from the glacial drift reported in the middle unit, which extends up Taylor Valley. In addition, the delta received sediments from the eastern Taylor Valley material, the source of the lowermost core unit.

Clay mineralogy of core 11 consists of a complex association of interstratified montmorillonite-vermiculite-chlorite-mica and noninterstratified vermiculite, mica, chlorite, and kaolinite. This assemblage is comparable to clay mineralogy of core 10. Most of the interstratification is irregular, but also regular interstratification is present. The interstratified montmorillonite-vermiculite-chlorite-mica phase is as abundant as the noninterstratified vermiculite, whereas chlorite, mica, and kaolinite appear as minor to trace components. Feldspars, quartz, and amphibole are also minor components (Table 6). The clay mineral assemblage throughout the core suggests that the sediments have experienced a low level of weathering. Although different lithologies, stratigraphy, microfossils, depositional environment, and salinities are recognized in core 11 [*Webb and Wrenn*, 1976, 1980; *McKelvey*, 1979; *Brady*, 1979a, b; *Powell*, 1979], the clay and the clay-size minerals do not show marked differences or definite units of contrasting mineralogies. In view of the fact that depositional environment has changed across the core, a detrital origin of these minerals seems to be favored. Alternatively, an episode of diagenesis could have left its signature all through the core.

Although core 12 displays a complex interstratification of montmorillonite, vermiculite, chlorite, and mica as does core 11, montmorillonite rather than vermiculite prevails as a separate phase. Mica is a minor component and kaolinite and chlorite are minor to trace constituents. Clay-size quartz and feldspar are minor in amount; amphibole is present as trace (Table 7). According to *Po-*

TABLE 6. Clay Mineralogy by X Ray Diffraction Analysis for Selected Samples From Core 11 (<2μm)

Depth, m	Montmorillonite, Vermiculite, Chlorite, and Mica Interstratified	Vermiculite	Mica	Chlorite	Kaolinite	Quartz	Feldspar	Amphibole
24.2	M	M	t	m	m-t	M	M	m
45.0	M	M	t	m-t	t	m	m	m-t
63.7	M	M	t	m	t	m	m	m
87.1	m	M	t	t	m	m	m	t
106.4	M	M	t	t	m	m	m	m-t
127.5	M	M	m	t	m-t	m	m	m
144.0	m	M	t	m	m-t	m	m	m
166.8	M	m	m	m-t	m	m	m	
306.6	m	M	t	m-t	m	m	M	m

M, major; m, minor; t, trace.

TABLE 7. Clay Mineralogy by X Ray Diffraction for Selected Samples From Core 12 (<2μm)

Depth, m	Montmorillonite, Vermiculite, Chlorite, and Mica Interstratified	Montmorillonite	Vermiculite	Mica	Chlorite	Kaolinite	Quartz	Feldspar	Amphibole
5.3	M	M	m?	m	t	m	m	m	t
26.6	M	M	m?	m-t	t	t	m	m	
47.4									
67.0	M	m		m-t	m	t	m	M	m
105.4	m	M	m	t	m-t	t	m	m	t
126.4	m	M	m	m	t	m	m	m	
162.0	m	M	?	m	m	m	m-t	m	m

M, major; m, minor; t, trace.

well [1979], the sedimentary succession of core 12 indicates a number of Ross Sea ice fluctuations with deposition of till, paratill, gravity flow deposits, and subaqueous traction current deposits. *Brady* [1979*a*, *b*] reports the absence of nonmarine diatoms in core 12. The hydrological history of this core as reconstructed from $\delta^{18}O$ values and salinity data suggests a slow permafrost aggradation with the expulsion of salts and isotopic fractionation [*McGinnis et al.*, 1980; *Stuiver et al.*, this volume]. The limited number of samples analyzed (5) does not warrant a definite conclusion on the origin of clay minerals in core 12. The predominance of montmorillonite in this core indicates, however, either a source or a diagenesis different from cores 11 and 10.

CONCLUSIONS

The electrical conductivity values and the ionic ratios seem to indicate that most of the sediments in cores 8, 9, and 10 were deposited in a marine environment and that aggradation of permafrost during subaerial exposure(s) of the sediments caused ionic concentration and segregation(s). Regions of low conductivity that display more 'terrestriallike' ionic ratios may be interpreted as evidence of freshwater incursions or of the exclusion of salt from coarse sediments during slow downward movement of a freezing front, or of the deposition of sediments containing fresh connate pore water. The clay mineralogy of the entire core (10) does not show distinct weathering horizons or soil formation episodes. The predominance of mixed layer clay minerals and the presence of clay-size feldspars and amphibole suggest that the weathering environment in Antarctica has been, since at least the early Pliocene, of low intensity.

REFERENCES

Abdel-Kader, F. H., M. L. Jackson, and G. B. Lee, Soil kaolinite, vermiculite, and chlorite identification by an improved lithium DMSO X-ray diffraction test, *Soil Sci. Soc. Amer. J.*, *42*, 163-167, 1978.

Anderson, D. M., The interface between ice and silicate surfaces, *J. Colloid Interface Sci.*, *25*, 174-191, 1967.

Anderson, D. M., Undercooling, freezing point depression and ice nucleation of soil water, *Israel J. Chem.*, *6*, 349-355, 1968.

Anderson, D. M., and N. R. Morgenstern, Physics, chemistry, and mechanics of frozen ground: A review, in *Permafrost, 2nd International Conference, North American Contribution*, pp. 257-288, National Academy of Sciences, Washington, D. C., 1973.

Anisimova, N. P., Cryogenous metamorphization of chemical composition of subsurface water (exemplified in central Yakutia), in *Permafrost, 2nd International Conference, U.S.S.R. Contribution*, edited by F. J. Sanger, pp. 365-370, National Academy of Sciences, Washington, D. C., 1978.

Beget, J. E., Stratigraphy and sedimentology of three permafrost cores from New Harbor, south Victoria Land, Antarctica, M.S. thesis, Univ. of Wash., Seattle, 1977.

Brady, H. T., The dating and interpretation of diatom zones in DVDP 10 and 11, *Dry Val. Drilling Proj. Bull.*, *8*, 5-6, 1978.

Brady, H. T., The dating and interpretation of diatom zones in Dry Valley Drilling Project holes 10 and 11, Taylor Valley, south Victoria Land, Antarctica, in *Proceedings of the Seminar III on Dry Valley Drilling Project 1978, Mem. Spec. Issue 13*, edited by T. Nagata, pp. 150-164, National Institute of Polar Research, Tokyo, 1979*a*.

Brady, H. T., A diatom report on DVDP cores 3, 4a, 12, 14, 15, and other related surface sections, in *Proceedings of the Seminar III on Dry Valley Drilling Project 1978, Mem. Spec. Issue 13*, edited by T. Nagata, pp. 165-175, National Institute of Polar Research, Tokyo, 1979*b*.

Brady, H. T., Late Neogene history of Taylor and Wright valleys and McMurdo Sound, derived from diatom biostratigraphy and paleoecology of DVDP cores, in *Antarctic Geoscience*, edited by C. Craddock, University of Wisconsin Press, Madison, in press, 1980.

Brown, J., Ice-wedge chemistry and related frozen ground processes, Barrow, Alaska, Proceedings, 1st International Conference on Permafrost, *Publ. 1287*, pp. 94-98, National Acad. of Sci., Washington, D. C., 1966.

Cartwright, K., and H. J. H. Harris, Hydrogeology of the dry valley region, Antarctica, this volume.

Chapman-Smith, M., Geology of DVDP holes 12 and 14, *Antarct. J. U.S.*, *10*, 170-172, 1975.

Chapman-Smith, M., and P. G. Luckman, Late Cenozoic glacial sequence cored at New Harbor, Victoria Land, Antarctica (DVDP 8 and 9), *Dry Val. Drilling Proj. Bull.*, *3*, 120-147, 1974.

Claridge, G. G. C., The clay mineralogy and chemistry of some soils from the Ross Dependency, Antarctica, *N.Z. J. Geol. Geophys.*, *8*, 186-220, 1965.

Fanning, D. S., and V. S. Keramidas, Micas, in *Minerals in Soil Environments*, edited by J. B. Dixon and J. B. Weed, pp. 195-258, Soil Science Society of America, Madison, Wisc., 1977.

Fripiat, J. J., A. Jelli, G. Poncelet, and J. Andre, Thermodynamic properties of adsorbed water molecules and electrical conductions in montmorillonite and silicas, *J. Phys. Chem.*, *69*, 2185-2197, 1965.

Hallet, B., Deposits formed by subglacial precipitation of $CaCO_3$, *Geol. Soc. Amer. Bull.*, *87*, 1003-1015, 1976.

Hallet, B., R. Lorrain, and R. Souchez, The composition of basal ice from a glacier sliding over limestone, *Geol. Soc. Amer. Bull.*, *89*, 314-320, 1978.

Hoekstra, P., Conductance of frozen bentonite suspensions, *Soil Sci. Soc. Amer. Proc.*, *29*, 519-522, 1965.

Iskandar, I. K., T. E. Osterkamp, and W. D. Harrison, Chemistry of interstitial water from subsea permafrost, Prudhoe Bay, Alaska, Third International Conference on Permafrost, vol. 1, pp. 92–98, Nat. Res. Counc. of Can., Ottawa, 1978.

Keller, W. D., *The Principles of Chemical Weathering*, Lucas Brothers, Columbia, Mo., 1962.

Kharaka, Y. K., and F. A. F. Berry, Simultaneous flow of water and solutes through geological membranes, 1, Experimental investigation, *Geochim. Cosmochim. Acta*, *37*, 2577-2603, 1973.

Linkletter, G. O., Weathering and soil formation in Antarctic dry valleys, Ph.D. thesis, 122 pp., Univ. of Wash., Seattle, 1970.

McGinnis, L. D., D. R. Osby, and F. A. Kohout, Paleohydrology inferred from salinity measurements on dry valley drilling project (DVDP) cores in Taylor Valley, Antarctica, in *Antarctic Geoscience*, edited by C. Craddock, University of Wisconsin Press, Madison, in press, 1980.

McKelvey, B. C., DVDP sites 10 and 11, Taylor Valley, in *Dry Val. Drilling Proj. Bull.*, *5*, 16-60, 1975.

McKelvey, B. C., The Miocene-Pleistocene stratigraphy of eastern Taylor Valley—An interpretation of DVDP cores 10 and 11, in *Proceedings of the Seminar III on Dry Valley Drilling Project 1978*, *Mem. Spec. Issue 13*, edited by T. Nagata, pp. 176-186, National Institute of Polar Research, Tokyo, 1979.

Mehra, P. O., and M. L. Jackson, Iron oxide removal from soils and clay by a dithionite-citrate system with sodium bicarbonate buffer, *Clays Clay Miner.*, *7*, 317-327, 1960.

Millot, G., *Geology of Clays*, translated by W. R. Farrand and H. Paguet, pp. 178-186, Springer, New York, 1970.

Murrmann, R. P., Ionic mobility in permafrost, in *Permafrost, 2nd International Conference, North American Contribution*, pp. 352-358, National Academy of Sciences, Washington, D. C., 1973.

Murrmann, R. P., and P. Hoekstra, Effect of thermal gradient on ionic diffusion in frozen earth materials, 1, Experimental,

Res. Rep. 284, 8 pp., U.S. Cold Regions Res. and Eng. Lab., Hanover, N. H., 1970.

Murrmarn, R. P., P. Hoekstra, and R. C. Biolkowski, Self-diffusion of sodium ions in frozen Wyoming bentonite-water paste, *Soil Sci. Soc. Amer. Proc.*, *32*, 501-506, 1968.

Nakao, K., T. Torii, and K. Tanizawa, Interpretation of salt deposition in Wright Valley, Antarctica, *Dry Val. Drilling Proj. Bull.*, *8*, 68, 1978.

Norrish, K., Factors in the weathering of mica to vermiculite, in *Proceedings of the International Clay Conference*, edited by J. M. Serratosa et al., pp. 417-432, Division de Ciencias, Consejo Superior de Investigationes Cientificas, Madrid, 1973.

Oberts, G. L., The chemistry and hydrogeology of dry lakes, Antarctica, M.S. thesis, 66 pp., N. Ill. Univ., DeKalb, 1973.

Olsen, H. W., Simultaneous fluxes of liquid and charge in saturated kaolinite, *Soil Sci. Soc. Amer. Proc.*, *33*, 338-344, 1969.

O'Sullivan, J. B., Geochemistry of permafrost: Barrow, Alaska, Proceedings, International Conference on Permafrost, *Publ. 1287*, pp. 30-37, Nat. Res. Counc., Nat. Acad. of Sci., Washington, D. C., 1966.

Page, F. W., Geochemistry of subsea permafrost at Prudhoe Bay, Alaska, M.S. thesis, Dartmouth Coll., Hanover, N. H., 1978.

Péwé, T. L., and P. V. Sellmann, Geochemistry of permafrost and Quaternary stratigraphy, in *Permafrost, 2nd International Conference, North American Contribution*, pp. 166-170, National Academy of Sciences, Washington, D. C., 1973.

Piper, D. J. W., and G. G. Pe, Cenozoic clay mineralogy from DSDP holes on the continental margin of the Australia-New Zealand sector of Antarctica, *N.Z. J. Geol. Geophys.*, *20*, 905-917, 1977.

Porter, S. C., and J. E. Beget, Provenance and depositional environments of late Cenozoic sediments in permafrost cores from lower Taylor Valley, Antarctica, *Dry Val. Drilling Proj. Bull.*, *8*, 74-76, 1978.

Porter, S. C., and J. F. Beget, Provenance and depositional environments of late Cenozoic sediments in permafrost cores from lower Taylor Valley, Antarctica, this volume.

Powell, R. D., Sedimentation conditions in Taylor Valley, Antarctica, from DVDP cores, *Dry Val. Drilling Proj. Bull.*, *8*, 77-80, 1978.

Powell, R. D., Conditions of sediment deposition in Taylor Valley, Antarctica from Dry Valley Drilling Project cores 8-12 (extended abstract), in *Proceedings of the Seminar III on Dry Valley Drilling Project 1978*, *Mem. Spec. Issue 13*, edited by T. Nagata, pp. 187-195, National Institute of Polar Research, Tokyo, 1979.

Powell, R. D., and P. J. Barrett, Grain size distribution of sediments cored by the Dry Valley Drilling Project in the eastern part of Taylor Valley, Antarctica, *Dry Val. Drilling Proj. Bull.*, *5*, 150-166, 1975.

Richards, L. A. (Ed.), Diagnosis and improvement of saline and alkali soils, *Agr. Handb. 60*, U.S. Dep. of Agr., Washington, D. C., 1954.

Sawheny, B. L., Interstratification in layer silicates, in *Minerals in Soil Environments*, edited by J. B. Dixon and S. B. Weed, pp. 405-434, Soil Science Society of America, Madison, Wisc., 1977.

Sellmann, P. V., Geology of the USA/CRREL permafrost tunnel, Fairbanks, Alaska, *Tech. Rep. 199*, U.S. Cold Regions Res. and Eng. Lab., Hanover, N. H., 1967.

Sellmann, P. V., Geochemistry and ground ice structures: An aid in interpreting a Pleistocene section, Alaska (abstract), *Geol. Soc. Amer. Spec. Pap.*, *101*, 197-198, 1968.

Sellmann, P. V., and J. Brown, Stratigraphy and diagenesis of perennially frozen sediments in the Barrow, Alaska, region, in *Permafrost, 2nd International Conference, North American Contribution*, pp. 171-181, National Academy of Sciences, Washington, D. C., 1973.

Stuiver, M., I. C. Yang, and G. H. Denton, Permafrost oxygen isotope ratios and chronology of three cores from Antarctica, *Nature*, *261*, 547-550, 1976.

Stuiver, M., I. C. Yang, G. H. Denton, and T. B. Kellogg, Oxygen isotope ratios of antarctic permafrost and glacier ice, this volume.

Tedrow, J. C. F., and F. C. Ugolini, Antarctic soils, in *Antarctic Soils and Soil Forming Processes, Antarctic Res. Ser.*, vol. 8, edited by J. C. F. Tedrow, pp. 161-177, AGU, Washington, D. C., 1966.

Terwilliger, J. P., and S. F. Dizio, Salt rejection phenomena in the freezing of saline solutions, *Chem. Eng. Sci.*, *25*, 1331-1349, 1970.

Thompson, T. G., and K. H. Nelson, Concentration of brines and deposition of salts from sea water under frigid conditions, *Amer. J. Sci.*, *254*, 227-238, 1956.

Torii, T., N. Yamagata, J. Ossaka, and S. Murata, A view of the formation of saline waters in the dry valleys, *Dry Val. Drilling Proj. Bull.*, *8*, 96-101, 1978.

Ugolini, F. C., and D. M. Anderson, Ionic migration and weathering in frozen Antarctic soils, *Soil Sci.*, *115*, 461-470, 1973.

Ugolini, F. C., and M. L. Jackson, Weathering and mineral synthesis in Antarctic soils, in *Antarctic Geoscience*, edited by C. Craddock, University of Wisconsin Press, Madison, in press, 1980.

Uskov, M. N., Clay minerals in Pleistocene permafrost and contemporaneous bottom deposits of central Yakutia (an X-ray study), in *Permafrost, 2nd International Conference, U.S.S.R. Contribution*, edited by F. J. Sanger, pp. 210-212, National Academy of Sciences, Washington, D. C., 1978.

Weaver, C. E., Geologic interpretation of argillaceous sediments, 1, Origin and significance of clay minerals in sedimentary rocks, *Bull. Amer. Ass. Petrol. Geol.*, *42*(2), 254-271, 1958.

Webb, P. N., and J. H. Wrenn, Foraminifers from DVDP holes 8-12, Taylor Valley, *Antarct. J. U.S.*, *11*, 85-86, 1976.

Webb, P. N., and J. H. Wrenn, Late Cenozoic micropaleontology and biostratigraphy of eastern Taylor Valley, Antarctica, in *Antarctic Geoscience*, edited by C. Craddock, University of Wisconsin Press, Madison, in press, 1980.

Yas'ko, F. G., Subsurface water regime in areas of occurrence of cryogenous processes and hydrochemical zoning of fossil ice, in *Permafrost, 2nd International Conference, U.S.S.R. Contribution*, edited by F. J. Sanger, pp. 438-441, National Academy of Sciences, Washington, D. C., 1978.

SEDIMENTATION CONDITIONS IN TAYLOR VALLEY, ANTARCTICA, INFERRED FROM TEXTURAL ANALYSIS OF DVDP CORES

Ross D. Powell[1]

Geology Department, Victoria University of Wellington, Wellington, New Zealand

Six major sediment types have been defined from particle-size analyses of sediment from the Dry Valley Drilling Project (DVDP) cores 8, 9, 10, 11, and 12. Samples of present sedimentary environments in Taylor Valley provide 'baseline' data for interpretation of each sediment type with respect to depositional processes. Type I sediments are poorly sorted gravel and sandy gravel (Folk classification) and are interpreted as lag deposits. Type II samples are poorly to very poorly sorted muddy sandy gravel to gravelly muddy sand which has one mode composed of a wide range of particle sizes. These sediments may have formed from traction currents with high sediment-water density or gravity flows of high viscosity. They could also have originated by winnowing or by loss of fines while dropping through the water column. Samples of type III are well to poorly sorted sands or gravelly sands that lack high proportions of mud; they are the best sorted samples studied. These samples are interpreted as traction-current deposits. Only one size analysis of type IV shows such good sorting that it is similar to temperate and tropical beach sands. Antarctic beach-face deposits are found to be similar to those in the Arctic in that they are not as well sorted and are more platykurtic than temperate and tropical beach deposits. This could be a useful paleo-environmental indicator. Type IV sediment ranges widely from poorly to very poorly sorted silty, clayey sands to very poorly sorted gravelly sandy muds. Sediment of this type was deposited by traction currents with fluctuating flow intensities. Type V samples are mainly very poorly sorted, slightly gravelly sands, muds, or clays. They are interpreted as suspension deposits with included ice-rafted gravel and sand. Type VI sediment contains approximately equal proportions of grain sizes from gravel through clay. They are extremely to very poorly sorted and range from gravelly muddy sands to sandy muds. Several depositional mechanisms can be suggested for these samples, such as till (terrestrial or subaqueous), waterlain till, and gravity-flow deposits. The sorted sediments are thought to be the result of reworking of the syndepositional diamictons. The particle-size analyses show a trend of increase in sorting as size increases or decreases toward 2.5 φ. Two interpretations are proposed for the glacial history of Taylor Valley. The two histories involve similar depositional processes and sedimentary environments; they differ in the timing of glacial events. This problem is created by discrepancies in chronostratigraphy. Thus a preferred glacial history of Taylor Valley cannot be given until the discrepancies are resolved.

INTRODUCTION

Taylor Valley is one of several ice-free valleys cut through the Transantarctic Mountains to the coast of southern Victoria Land. The western portion of the valley is presently occupied by Taylor Glacier, an outlet glacier of the east Antarctic Ice Sheet. The eastern section of the valley inland from McMurdo Sound is filled with glacial sediment. During the course of the Dry Valley Drilling Project (DVDP) the glacial beds were cored to a maximum depth of 328 m. The drill sites of interest here are DVDP holes 8, 9, 10, 11 and 12 (Figure 1).

The primary purpose of this study was to determine conditions of sedimentation for the cored sequences by study of sediment texture and sedimentary structures. Textural characteristics and sedimentary structures in the products of modern sedimentary processes in Taylor Valley were obtained in order to provide 'baseline' data for interpretation of the drill cores.

Particle-size analyses were carried out to complement studies of the stratigraphic sequence [*Chapman-Smith and Luckman*, 1974; *Chapman-Smith*, 1975; *McKelvey*, 1975a, b], chronostratigraphy [*Webb and Wrenn*, 1976, 1980; *Brady*, 1979; *Elston and Bressler*, this volume] and fabric and provenance studies [*Porter and Beget*, this vol-

[1]Now at the Department of Geology and Mineralogy, and Institute of Polar Studies, Ohio State University, Columbus, Ohio 43210.

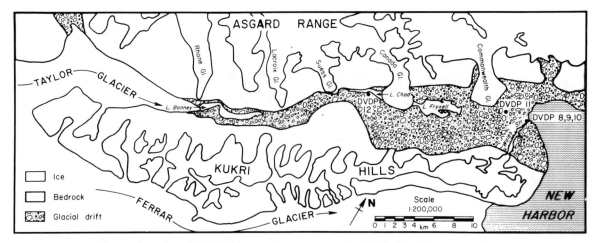

Fig. 1.   Location of DVDP drill sites 8, 9, 10, 11, and 12 in Taylor Valley.

ume; *Purucker et al.*, this volume]. Information from these studies allows different interpretations for the geologic history of the cored sequences. Two alternatives for the glacial history of lower Taylor Valley are presented. However, other interpretations are possible, and until some chronostratigraphic discrepancies are resolved, the most likely history will not be found.

GEOLOGIC SETTING OF THE DRILL CORES

Bedrock is exposed in the Kukri Hills on the south side of the valley and in the Asgard Range on the north side. There are four pre-Cenozoic rock units exposed [*Haskell et al.*, 1965; *Lopatin*, 1972]. The Ross Supergroup (late Precambrian and early Paleozoic age) consists of marble, schist, and quartzite. This sequence is cut by granite gneiss and associated dikes of the Granite Harbor intrusives, which include most of the exposed pre-Cenozoic rocks in lower Taylor Valley. Beacon Supergroup (?Devonian-Triassic) rocks exposed in the valley are predominantly friable quartzose sandstone of the Taylor Group. The lithologically more variable Victoria Group caps the peaks in upper Taylor Valley. Sills of Ferrar Dolerite (Jurassic) intrude the basement complex and the Beacon Supergroup.

Small basaltic scoria cones and flows are located within the valley [*Angino et al.*, 1962; *McCraw*, 1962]. These rocks belong to the McMurdo Volcanic Group (Cenozoic). Superficial glacial drift in Taylor Valley includes basal till, flow till, gravity-flow deposits, lacustrine sediments, and fluvial de-

posits. *Denton et al.* [1969, 1970, 1971] described a tentative glacial history for Taylor Valley encompassing the past 3.5 m.y. which involved fluctuations in ice fronts originating from the east Antarctic Ice Sheet, the Ross Sea, and the local alpine glaciers.

At New Harbor, three DVDP cores were obtained from virtually the same location. DVDP 8 was drilled vertically at 77°34′43″S latitude, 163°31′00″E longitude, and 1.9 m in altitude on a small delta near Wales Stream [*Chapman-Smith and Luckman*, 1974]. DVDP 9 was drilled about the same site but at an angle of 84.25°. DVDP 10 was drilled a short distance away (77°34′43″S latitude, 163°30′42″E longitude) and at an altitude of 2.8 m. This was the longest core of the three, and reached 185-m depth.

McKelvey [1975a, 1980] has divided the cored sediments of DVDP 10 and 11 into five and eight informal lithostratigraphic units, respectively. 'Both cores show the same irregular sedimentary transition, with basal pebble diamictites passing up to sandstones, pebbly sandstones, with minor cobble conglomerates' [*McKelvey*, 1975b, p. 169]. Bedrock was not encountered in either hole. Summary stratigraphic logs for DVDP 10, 11, and 12 are shown in Figure 2.

DVDP 12 (Lake Leon core log) was sited on the southern shore of Lake Hoare (77°38′22″S latitude, 162°51′13″E longitude) at an altitude of 75.1 m. It was drilled through glacial sediment into migmatitic basement and reached a depth of 185 m. *Chapman-Smith* [1975] described the core as largely diamicton (58%), the remaining sediments being graded beds of gravel, sand and mud,

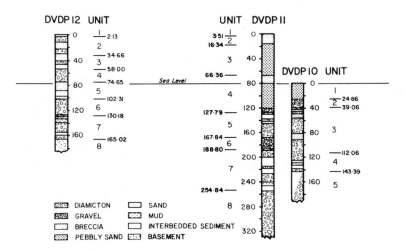

Fig. 2.   Summary geologic logs of DVDP 10, 11, and 12 arranged with sea level as the reference horizon. DVDP 10 and 11 columns are redrawn after *McKelvey* [1980]. All depths are in meters.

'varved' muds and silts, weathered iron-oxide horizons and lenses, and 'carbonate-evaporite' beds.

LABORATORY PROCEDURES

A flow diagram (Figure 3) summarizes laboratory procedures carried out on all samples. Most aspects of this procedure are standard techniques; for more details see *Powell* [1976]. A slab count was carried out to estimate the amount of material coarser than $-3\,\phi$, so that it could be included as a 'weight percent coarser than' in the size analysis and thus make the analyses representative of the total grain population within a unit. To make the slab size equivalent to the sieve size data, a conversion was used [*Adams*, 1974]: sieve mean = 1.00 (slab mean) $+ 0.404\,\phi$.

Reliability checks were made using smear slides and grain mounts to examine the effectiveness of the disaggregation and dispersion procedures. Eight samples have been used to test for reproducibility and accuracy of the analyses.

The computer program used to process the data, the raw data printout, and sample statistics are presented and explained by *Powell* [1979a].

SEDIMENT CLASSIFICATION

Textural Classification

Sediments from DVDP cores are classified texturally on the relative proportions of gravel, sand, silt, and clay. Because of the wide range of sediment textures, the method proposed by *Folk* [1954] of using two ternary diagrams simultaneously is followed for the textural classification. Samples with less than 5% gravel are included on the sand-silt-clay diagram. Inclusion of gravel in such samples with the percentage sand best categorizes the sediment so that the fine end members, silt and clay, do not excessively dominate the coarser end member. From the triangular diagrams (Figure 4), textural names can be assigned to the samples [*Powell and Barrett*, 1975].

Validity of Particle Size for Interpretation of Depositional Processes

Particle-size distributions can be a function of the type of parent material. During transportation and deposition, abrasion as well as physical sorting processes influence the final particle-size distribution. After deposition, particle-size distributions can be altered by material infiltrating down from younger sediment or diagenesis. All of these factors, apart from the sorting processes, are thought to be minimal for sorted samples in this study. *Porter and Beget* [this volume] and *Purucker et al.* [this volume] have shown some variation in volcanogenic components throughout the cores. However, the proportion of volcanic detritus is generally small, and grain mineralogy and clast lithologies vary little throughout the cores [*McKelvey*, 1975a]. Thus the type of parent material changed little with time. Given that different lithologies have different rates of attrition by fractur-

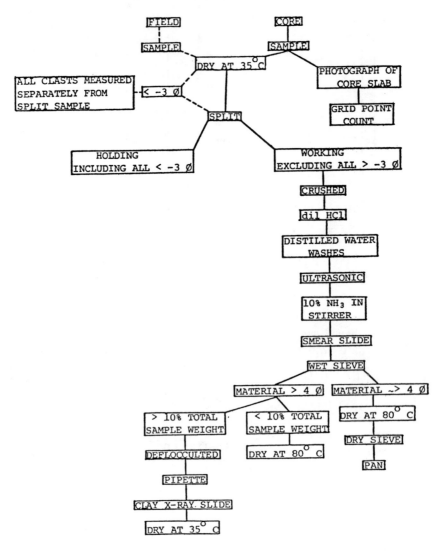

Fig. 3. Flow diagram of procedures by which each sample was treated in the laboratory.

ing and abrasion [*Rogers and Schubert*, 1963], traction-current deposits would not be expected to have uniform mineralogy, as found in the cores, if abrasion was a major factor in producing the particle-size distributions. If permafrost conditions existed after deposition, alteration of the grain size by infiltration would be minimal (e.g., the ice-cored breccias in DVDP 10 and 11). The only recognizable diagenesis is carbonate cementation at the bottom of core 11. Thus a particle-size analysis of a water-sorted sample from the DVDP cores can be used to suggest the process of deposition.

Particle-Size Classification and Interpretation

Some of the sediments deviate widely from a Gaussian distribution using the phi (ϕ) scale with the Wentworth size-grade scale. In this case the assumption of a log-normal distribution of particle sizes is not valid, and standard particle-size statistics, whether calculated graphically or by moments, is not adequate. Moreover, the fine sediment studied was deposited from suspension, probably often in a flocculated state. Modal size is

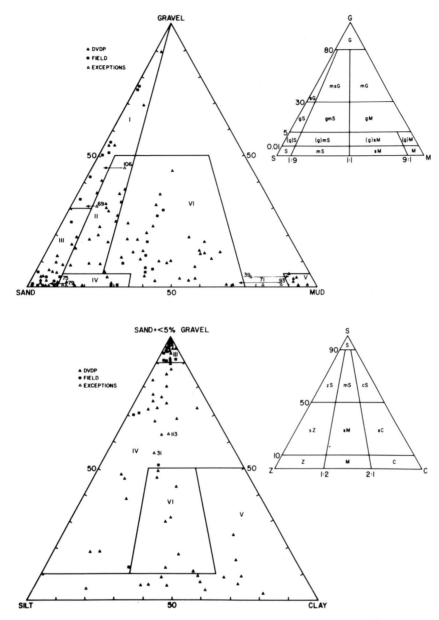

Fig. 4. Textural classification using grain-type end members. The solid straight lines define regions where samples of a particular sediment type (I-VI) preferentially plot. Open symbols with the sample number represent samples that plot outside their sediment type region. (Top) Gravel (coarser than −1 φ), sand (1-4 φ), and mud (finer than 4 φ). The small diagram shows the textural classes defined by *Folk* [1954] where (g)sM = slightly gravelly sandy mud, etc. (Bottom) Sand (−1-4 φ) plus less than 5% gravel, silt (4-8 φ), and clay (finer than 8 φ). The small diagram shows the textural classes defined by *Folk* [1954] where zM = silty mud, etc.

a truer indicator of conditions during suspension deposition than other average measures because the fine tail of the distribution is analyzed in a deflocculated state [*Kranck*, 1975]. Thus the distributions obtained have been grouped using modal

grain size and the shape of the histograms and cumulative frequency curves.

Particle-size analysis data from DVDP cores can be grouped into six natural populations or sediment types (Table 1 and Figure 5), that were de-

TABLE 1. List of Sample Numbers With Corresponding Numbers Used to Represent the Samples in the Text Figures, Grouped Under Their Assigned Types and Subtypes

Sample Number	Number Used in Figures
Type I	
10-14-231	7
8-51-149	35
12-14-38**	91
12-34-95**	101
12-47-129**	106
Z-21	
Z-22	
Type I″	
11-66-60	69
11-69-177	70
11-91-100	78
Y-1	
Y-14	TS-37
TS-38	TS-41
Type II′	
10-4-193	3
10-33-130	17
10-55-61	29
11-35-218	56
11-44-105	63
12-26-72	97
12-50-139	110
Z-23	
Type II″	
11-10-223	46
11-20-120	50
11-34-112	55
11-38-140	60
12-2-125	86
12-9-24**	89
Type III′	
9-1-41	1
10-3-46	2
11-11-55	47
11-30-207	53
11-31-62	54
11-78-236	75
11-93-248	79
12-7-19	88
12-11-29	90
12-23-65	96
12-36-99	102
TG-3	
TG-4	
TG-6	
Y-15	
Y-22	
Y-23	
Y-25	
Y-26	
TS-36	
TS-39	
Type III″	
10-6-167	4
10-29-30	15
10-66-118	38
11-8-125	45

TABLE 1. (continued)

Sample Number	Number Used in Figures
11-23-210	51
11-51-300	66
12-5-13	87
12-20-57**	94
12-36-100	103
12-58-160**	111
Y-2	
TS-40	
Type IV	
10-10-168	5
9-11-31	6
10-16-94	9
8-14-49	11
10-20-119	13
8-29-84	18
10-34-144	19
10-37-165	20
10-52-71	26
10-55-103	30
8-53-157	37
11-1-101	42
11-7-7	43
11-12-196	48
11-37-11	59
11-41-37	61
11-44-67	62
11-63-267	68
11-71-108	71
12-30-82	98
Z-20	
Y-24	
TS-31	
TS-35	
Type V	
10-20-6	12
10-42-88	22
10-46-131	23
10-52-135	28
10-61-205	36
10-67-93	39
10-68-171	40
10-69-73	41
11-13-38	49
11-35-260	57
11-51-122	65
11-77-244	74
11-89-98	77
11-100-40	81
11-118-55	84
Type VI′	
10-31-0	16
10-39-235	21
11-50-135	64
11-72-81/1	72
11-72-81/2	73
12-46-127**	105
12-57-157	112
TG-1	
TG-22/1	
TG-22/2	

TABLE 1.   (continued)

Sample Number	Number Used in Figures
Type VI''	
10-15-100	8
10-51-40	25
10-51.40	10-57-45
31	
8-50-146/1	32
8-50-146/2	33
11-28-66	52
11-80-198	76
12-15-42	92
12-21.59	95
12-33-90/1	99
12-33-90/2	100
12-48-132	107
12-48-134	108
12-60-165	113
TS-42	
TS-46	
TS-60	
TS-61	
TG-62	
Type VI'''	
10-25-165	14
10-52-108	27
11-58-10	67
11-99-49	80
11-107-150/1	82
11-107-150/2	83
11-118-134	85
12-41-114	104
Type VI''''	
10-18-15	10
8-50-147	34
11-7-126	44
12-17-48	93
12-49-136	109

the curve bimodal and platykurtic. The sorting is moderate in both modes.

Sediment of type I has the highest proportion of gravel of the proposed types. It has been transported by strong currents a relatively short distance during which the fine material was being removed. Variations in flow intensity can account for the slight differences in the particle-size distributions. It may also be the remnant of material sorted while being dropped through the water column. Samples with this particle-size distribution occur today in Taylor Valley as wind- and water-winnowed deposits on top of the Tedrow Glacier (Z-21, Z-22); coarse material from the surface of Wales Stream (Y-1); a surface layer below high tide on New Harbor beach (Y-14); a gravel 50 mm

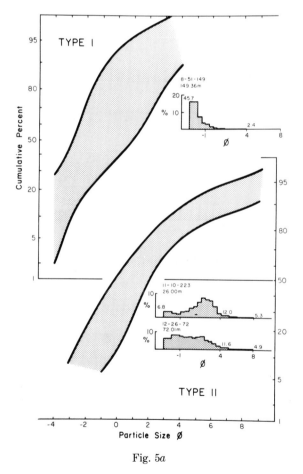

Fig. 5a

Fig. 5.   Particle-size cumulative frequency curve envelopes of the sediment types defined in the text, plotted as weight percent coarser than versus size class (in phi units). A 'typical' sample histogram is presented with each sediment type.

posited under differing sedimentation conditions. Recognition of these populations is based on the number of modes, particle size and frequency percent of the modes and tails, sorting of the modes, and overall sorting of the sediment.

Definition of modes was a problem because of the broad size distributions in the diamictons. The factors used to establish a mode in histograms were (1) where the rise in weight percent occurs over more than a 0.5 φ interval and (2) where the modal class rises more than about 5% by weight above the classes that do not build the mode.

Type I sediments are poorly to very poorly sorted gravels and sandy gravels. They have a pebble-size mode with a sandy tail. Some have an additional, broad medium-sand mode, which makes

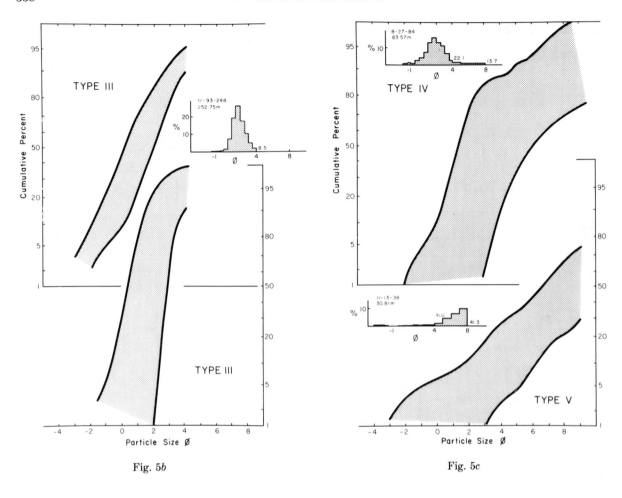

Fig. 5b

Fig. 5c

beneath the surface of the southeast marginal stream of Taylor Glacier (TS-37, TS-38); and a granule layer about 30 mm beneath the same surface (TS-41). Similar processes of deposition are inferred for type I sediment in the DVDP cores.

Type II samples are unimodal, poorly to very poorly sorted muddy, sandy gravels or gravelly muddy sands. The modes are broad, which tends to give a rectangular particle-size distribution. The tails vary in size and extent, but all distributions are positively skewed. One group is quite similar to type I, but the sand mode, which is in the very coarse to coarse sand size, is much broader and the fine tail is much more prominent. Others have a slightly lower percentage of coarser material and a stronger mode in the fine- to medium-sand size. Type II sediments could have formed from a traction current with a high sediment water density or a gravity flow of high viscosity. However, they could also have originated by winnowing of fine material either while at the sediment-water inter-

face or by the loss of fines while being dropped through the water column. The more material in the fine tail, the less likely it is that the whole sediment has been winnowed. Also, the more fine material existing with a coarse first percentile, the more likely is a gravity flow, rather than a traction-current origin. *Hjulstrom* [1939] showed graphically that material of fine-sand size and smaller often is eroded as a mixture. Thus the size distribution of a winnowed gravel-lag would be truncated below the fine-sand size. Mud separates from sand once they are both in transport together [*McCrone*, 1962]. Thus if these samples have been winnowed and still contain material finer than fine sand, though in a small proportion, then the winnowing must have occurred when the entire size range was in transport. Winnowing was caused by currents as the sediment was dropped from ice through a column of water. Type II sediments are therefore gravity-flow or ice-rafted deposits. A present-day example of this type of sediment is a

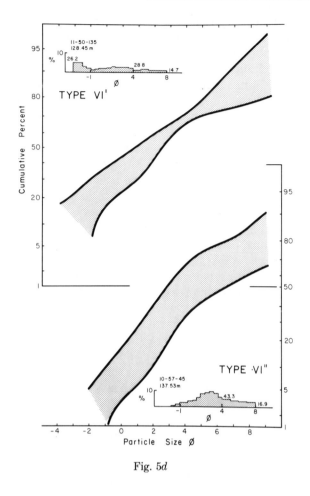

Fig. 5d

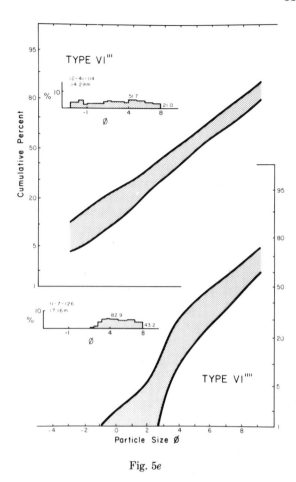

Fig. 5e

diamicton at the top of the most landward ice-cored debris cone on Tedrow Glacier (Z-23).

Samples in type III are well-sorted to poorly sorted sands or gravelly sands that lack high proportions of mud, and are the best-sorted samples studied. The most leptokurtic samples have fine- to medium-sand modes, and both tails are relatively insignificant. Other samples are not as well sorted (moderate to poor sorting) and have a mode in coarse- to medium-sand sizes. Type III sediments are fluvial or at least traction-current deposits. The coarser average particle size indicates a higher flow intensity. Broader modes indicate flow intensity fluctuated about the mean velocity. Samples have an excess of fine or coarse material depending on how widely the current velocity varied about the mean flow intensity. New Harbor beach samples, both intertidal and subtidal have a particle-size distribution similar to those of Wales Stream [Powell, 1976]. Thus fluvial sediment is indistinguishable from deposits on a beach that has ice on its face for most of the time. Similar results have been obtained for an arctic beach with an 8- to 9-month ice cover [Naidu and Mowatt, 1975]. Therefore this may be a criterion to distinguish polar from temperate and tropical beach deposits. Only one sample (11-30-207) from the cores is similar to temperate and tropical beach deposits. Examples of type III sediment in Taylor Valley include New Harbor intertidal beach-face material (Y-15, Y-22, Y-23), 5 m above high tide and below a boulder pavement (Y-25); sand in Wales Stream (Y-2, Y-26); sand in the southeast marginal stream of Taylor Glacier (TS-36, TS-39); a granule layer 200 mm below its surface (TS-40); and supraglacial streams of Taylor Glacier (TG-3, TG-4, TG-6).

Type IV samples plot on both textural triangles because some contain a minor amount of gravel whereas others fall into a general scatter on the diagram for finer-grained sediment. They range from poorly to very poorly sorted silty, clayey sands to very poorly sorted gravelly sandy muds.

These samples have modes that are similar in strength to type III sediment, but they differ from the latter in that the modes are fine to very fine sand and they have a marked fine tail. Thus skewness is positive and all are leptokurtic. Some samples have broader modes (poorly sorted), in which case the modes are either coarser (medium sand) or finer (very fine sand/silt) grained than the better-sorted samples. Some coarser tails are represented by only a few grains. Thus sediments were deposited by traction currents which at times had a lower intensity flow than that which deposited modal grains. The more poorly sorted samples were deposited by flows with a greater range in flow intensity (because of the long tails of the distribution) and a greater variance about the mean flow (because of the broad mode) than the better-sorted sediment. Occasionally, pebble clusters occur in these sediments, which may have been produced by iceberg overturning and dumping or in situ melting of bergs on the sediment surface. Examples of this type from present Taylor Valley come from the fine beach sand (Y-24), mud in an abandoned braid of the southeast marginal stream of Taylor Glacier (TS-31, TS-35), and fine material 5 cm below the surface of Tedrow Glacier (Z-20).

Type V samples are mainly very poorly sorted slightly gravelly sandy muds or clays. Modal particle sizes can be inferred to lie outside the range of analysis in the clay-size fraction. Some samples with a long tail stretching into gravel sizes have a gradual decrease of material as size increases, while others have a mode in silt sizes. The samples have less sand than type IV samples and thus gravel, where it is present, is more significant. The high percentage of clay-size sediment in these deposits implies suspension deposition through a water column. Long gravel tails are produced by erratic material. The sediment is interpreted as iceberg-zone sediment because of the abundance of pelagic material and erratic debris. Burrowing structures and tuffaceous material are associated with this sediment in the cores (e.g., DVDP 11/4) and probably indicate open-water conditions rather than continuous ice cover. When these beds are thin, melting of a grounded iceberg may have produced them, or they may be from an abandoned braid of a braided stream.

Type VI samples contain approximately equal proportions of grain sizes. They are extremely to very poorly sorted and range from gravelly muddy sands to sandy muds. They represent the most poorly sorted of sediment types studied and are generally described in core logs as diamictons. Four subtypes of type VI sediment are recognized. Two contain distinct modes, and two approximate a rectangular size distribution. The first subtype (type VI′) has a higher proportion of pebble-size material than other subtypes. There is a tendency to form a sand mode, but this is not significant with respect to physical conditions during deposition. This subtype grades into the second (type VI″) with a decrease in coarse fraction and a corresponding increase in sand sizes to form a clearly defined mode in medium- to fine-sand sizes. The fine tail is more dominant than the coarse tail. A more equal distribution through sand sizes leads to the most rectangular distributions studied (type VI‴). Sorting is extremely poor with no significant modes. An increase in the dominance of the fine fraction, and a decrease of the coarse- and medium-sand sizes, produces the last subtype (type VI⁗). This has a coarse tail through the sand fraction with a decrease in grain percentage as size increases, but the distribution through the silts is rectangular. Rarely, a sample has a very fine sand mode.

The distinction between different processes of formation for diamictons has been a traditional sedimentologic problem. Comminution (both crushing and abrasion) is the characteristic mechanical process that acts on rock particles in a glacier and thus influences the particle-size distribution of a till. Mineralogy is a dominant factor in this process. Different mineral grains have characteristic terminal modal grain sizes, at which point comminution ceases [*Jarnefors*, 1952; *Davis*, 1958; *Gillberg*, 1965; *Beaumont*, 1971; *Dreimanis and Vagners*, 1971, 1972]. Related to this effect is the length of transport; terminal particle modes are not exhibited by till that has undergone a short distance of glacial transport [*Slatt*, 1972]. Relict modes may be exhibited by the till as a reflection of a mode in the parent material.

Gravity-flow phenomena such as mudflows and debris flows also produce sediment that is very poorly sorted diamicton. The distinction between till, waterlain till, and gravity-flow deposits is virtually impossible to make on particle-size distributions alone, although some have tried [e.g., *Landim and Frakes*, 1968; *Chriss and Frakes*, 1972; *Buller and McManus*, 1973; *Frakes and Crowell*, 1975]. Proper interpretation of these subtypes can only be made with the aid of other sediment data such as unit thickness, associated sediment types, and sedimentary structures.

Some useful criteria from particle-size distributions are as follows.

1. One criterion is the winnowing or loss of fine material in a water column during deposition distinguished by the type of distribution of particles finer than fine sand.

2. Another criterion is the presence of erratic material including medium and coarse sand-size grains [von Huene et al., 1973] within muds, indicating ice rafting.

3. There is the likelihood that if no mode is present, the sediment has a glacial origin. A gravity-flow deposit with this type of size distribution must be flow till because if it originated from several different sediment types, a mode would occur somewhere in its distribution. Furthermore, once the full size range was in suspension in the gravity flow, clays would be more easily lost than sand, creating a sand mode or no fine tail. But a rectangular sand-silt-clay-size distribution could result from winnowing.

4. A high percentage (greater than ~50% of the sample) of silt and clay in a rectangular distribution is more common in subaqueous till than in terrestrial till, unless the latter is derived from fine-grained deposits.

Therefore types VI′ and VI″ can be either glacial or debris-flow deposits; type VI‴ sediments are tills (probably waterlain) or flow tills; and type VI‴′ are waterlain tills.

Field examples of the first subtype are of debris layers at the snout of Taylor Glacier (TG-1, TG-22). Subtype two field samples are derived from material that melted out (and did not flow) from an ice-cored apron at the snout of Taylor Glacier (TS-42, TS-46, TS-60, TS-61, TS-62). There are no field examples of the third and fourth subtypes.

Textural Trends in the Sorted Drift

Powell and Barrett [1975] suggested the bulk of material that is water sorted had been derived from diamictons. The range of sediments is considered to have originated by water reworking, which gives trends similar to those found elsewhere in terrestrial glacial [*Slatt and Hoskin*, 1968] and marine [*Stanley*, 1968] environments. The distribution of grain sizes in a sediment is dependent on the length of time water has acted on the very poorly sorted sediment, and the manner in which the sedi-

ment was deposited. A sorting trend was established from the graphic mean versus inclusive graphic standard deviation plot (Figure 6). Samples of type VI, with a standard deviation of 3.50 φ or more, are not substantially affected by water reworking. A transition zone of the effectiveness of water sorting is represented between 2.75 and 3.50 φ standard deviation, containing type II and some type IV samples. With these samples, water winnowing has occurred but has not yet been fully effective in sorting the type II samples and has mixed subpopulations in the type IV samples due to the fluctuation in flow intensity. Type V samples also result from mixing of subpopulations, but this is caused by ice rather than water. The lag left from type VI sediment after the washing out of the finer material is represented by the unimodal gravel sediments of type I. Material washed out from type VI sediment progresses through type II and then is incorporated in the sorting trend produced with type III and the rest of type IV samples, by deposition from the water-sorting system. The increase in sorting trend occurs with convergence of grain size to about 2.50 φ, which is the easiest material moved by traction [*Inman*, 1949].

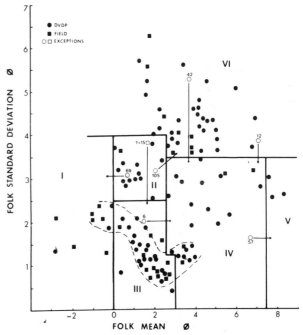

Fig. 6. Graphic mean versus inclusive graphic standard deviation [*Folk and Ward*, 1957], both in phi units. The solid straight lines define the regions where samples of a particular sediment type (I-VI) preferentially plot. Open symbols with the sample number represent samples that plot outside their sediment type region. The region outlined by the dashed lines defines a general trend of increasing sorting.

BEDDING AND STRUCTURAL FEATURES USED FOR INTERPRETATION

General

Evidence of postdepositional sediment readjustment, redeposition, and slumping within the cored sequences is indicated by high angles of bedding dips that change inclination over short lengths in the core, intraformational clasts, boudinage and microfaulting. Redeposition was probably caused by high water contents and/or fast sedimentation rates, or as a result of ice push. Rapid deposition is also indicated by loaded contacts and sand injection features. Intraformational clasts can also be a product of ice rafting [*Ovenshine*, 1970].

The yellowish orange horizons that occur in DVDP 12 were produced either by subaerial oxidation or diagenetic effects under subaqueous conditions similar to those described in arctic environments [*Turner*, 1971].

Diamictons

Thick continuous sequences of diamictons were deposited either from grounded ice, an ice shelf, or an iceberg zone close to an ice front. In shallow water the latter case is a high-energy environment due to calving and meltwater discharge. Thus the diamictons would be associated with coarse-grained lag and traction-current and gravity-flow deposits. Ice shelves must consist of cold ice, below the pressure melting point [*Robin and Adie*, 1964, p. 105], and observations of floating ice tongues in fjord environments suggest that only cold ice can produce these tongues; warm-based glaciers end as tidewater fronts (J. H. Mercer, personal communication, 1979). Release of sediment from the base of a shelf or tongue is controlled by factors other than just cold-ice conditions. Water circulation patterns under the ice and the bottom surface slope of the ice shelf also control basal melting [*Doake*, 1976]. Under favorable conditions, *Carey and Ahmad* [1961] and *Anderson* [1972] have suggested that very little sediment may be released from the base of a shelf or tongue. In present-day conditions, melting dominates at the base of ice shelves, especially near the front [*Robin*, 1975]. Assuming debris is incorporated into ice on land, thick sequences of diamictons may be deposited close to the grounding line of an ice shelf when melting is the dominant sub-ice shelf process, or near the

shelf front when basal freezing occurs near the grounding line.

Subaqueous diamictons can have both a massive or laminated matrix similar to terrestrial tills. Matrix lamination of a waterlain till could be produced by changes in water chemistry (possibly due to seasonal climatic variations) or slight sorting effects by water currents while the sediment is dropping through the water column. A massive waterlain till could be produced by flocculation of fine material, causing it to fall through the water column with sand-size material, which does not allow laminations to be produced.

Sorted Sediment

Thin beds of gravel and breccia can be produced by overturning of icebergs. Debris that has accumulated on the top ablating surface of a berg is dumped as it rolls over. In this situation, bedding can dip at normal angles of repose (≤30°). Thick sequences of gravel may accumulate at a tidewater glacier terminus. Subglacial debris may be sorted and supraglacial debris dumped from calving icebergs, continuously adding coarse debris [*Powell*, 1979b].

Persistent bedding dips up a core at normal angles of repose for sand suggest foreset bedding. Progradation may be implied if the dip decreases upward over tens of meters of core.

CORE INTERPRETATION

DVDP 10

Unit 5 (185.47-148.89 m) is predominantly sandy mud with medium to fine sand, pebble gravel, and breccia. The unit was deposited away from an ice front in an iceberg zone, in which traction currents deposited sediment (type IV). The bergs produced the erratics in the fine-grained sediment (type V) and contributed some of the coarser-grained units by overturning and dumping. Minor redeposition and slumping has occurred. The biostratigraphy implies a long period of time not represented because of erosion in the upper part of the unit.

Unit 4 (143.39-112.06 m) is composed of two beds of massive pebble diamicton separated by interbedded pebble gravel, breccia, sand, and laminated sandy mud. Ice shelf and grounded ice conditions existed during which there was an advance of the ice grounding line or a change to iceberg conditions when the sorted interval was deposited. Samples of both diamictons have type VI'' particle-size distributions and intraformational clasts, so redeposition may have occurred within them.

In unit 3 (112.06-39.06 m), diamictons (type VI'-VI'''') are interbedded with minor pebble gravel (type II) and coarse to fine sand, sometimes with scattered pebbles (mainly type IV, some type III). Deposition was either from an ice shelf or grounded ice. If the whole unit is a paratill, it is mainly waterlain till interbedded with sorted material produced by traction-current winnowing of the former. Some sorted intervals may be thin turbidite pulses and some diamictons, with inclined matrix lamination and intraformational clasts, may be debris-flow or slump deposits. Sandy mud (type V) is rare above 90.49 m, so grounded ice conditions are more likely for sediments above this level. In this upper section, coarser, more continuous sorted deposits occur, and *Porter and Beget* [this volume] have noted a fabric suggestive of possible grounded ice. Furthermore, high-energy structures, such as signs of redeposition and scour infillings, appear to increase in frequency up the unit. However, particle-size distributions of some diamictons indicate deposition of waterlain till (types VI''' and VI''''). Periodic grounding of an ice shelf could explain this upper sequence.

Unit 2 (39.06-24.86 m) is a fining-upward sequence of granule and pebble gravel interbedded with minor medium to coarse sand. These are horizontally bedded lag deposits (type I) interbedded with lower flow intensity, traction-current deposits (type IV). A decrease in flow intensity up the unit is shown by the upward fining of grain size, together with a decrease in abundance of mud clasts. Therefore traction currents weakened with time. This probably happened when the ice grounding line retreated, so that mud clasts had less chance of surviving the longer transport distances. Whether the unit was deposited subaqueously or subaerially was not determinable from these sediments.

Unit 1 (24.86-3.65 m) contains predominantly medium to very coarse sand (mainly type III) interbedded with minor gravel beds (type II). These traction-current deposits show an increase in initial dip (from about 6° to 20° on average) and a decrease in clarity of stratification up the unit. The supplying current strengthened as dip-slope increased with progradation, and the deposits reflect the change of shear by grading between beds rather than by distinct contacts.

DVDP 11

Unit 8 (328.88-254.84 m) is a massive diamicton sequence with rare mud, sand, and gravel beds. Deposition under an ice shelf is indicated by occasional matrix lamination in the thick diamicton, type VI''' particle-size distributions, and thin ice-rafted deposits (type V). Iceberg deposition is likely for some of the thicker sandy mud (type V) intervals. The thin traction-current deposits formed by winnowing of the diamicton. *Webb and Wrenn* [1976, 1980] suggested water depths of 600-900 m for deposition of this sequence.

Unit 7 (254.84-188.80 m) varies lithologically from diamicton interbedded with sandy to silty mud to minor thin bedded gravel, sand (type III), pebbly sand, and breccia. Sandy silty mud (type V) and breccia beds (type I) suggest an iceberg environment. Traction currents intermittently flowed over the sediment surface. Sorted sequences may include thin turbidites. Thicker intervals of diamicton (e.g., 226.40-240.17 m) are probably ice shelf deposits. Thus fluctuations of the ice front took place. The thinner diamicton (types VI' and VI'') intervals may be debris-flow deposits.

The whole of unit 6 (188.80-167.84 m) is poorly sorted, fine to medium pebble gravel (type I). The unit is inferred to have been deposited subaqueously because of its inclusion between units 5 and 7. The gravel was deposited close to an ice grounding line where currents were continuously active and sorted finer material from pebbles and sand as they were deposited through the water column from the ice to form a lag deposit.

Unit 5 (167.84-127.79 m) has a thick diamicton unit at the base followed by a sequence of interbedded gravel, sand (type III), sandy mud (type V), and diamicton. An ice shelf deposited the diamicton sequence. After deposition of unit 6, the grounding line retreated, but traction currents were still present when the lower section of diamicton was deposited. This section has a laminated matrix showing soft sediment deformation and a higher percentage of clasts compared with the upper part. Quieter water conditions allowed a massive matrix and a type VI''' size distribution to form in the upper part. Iceberg conditions existed during deposition of the predominantly traction-current sediment in the upper part of the unit. There are common signs of redeposition and graded turbidite material. Some laminated muds may be geostrophic current deposits.

Unit 4 (127.79-66.36 m) has a rapid lithologic variation. There is a transition from pebble and cobble gravel (type II) at the base, up through a thick sequence of pebbly sand (type IV), to a medium sand (type III) at the top. Water at the site was shallower than previously. The sequence was deposited close to the front of a tidewater glacier and is a subaqueous ice-marginal apron which was occasionally exposed subaerially. The environment is thought to be similar to that described by *Rust and Romanelli* [1975] and *Rust* [1977]. The high-energy environment is suggested by signs of sediment readjustment after deposition, in situ fracturing of a boulder, and some type IV sediment that appears to form turbidite deposits. Thin type V beds are bioturbated, deformed, and faulted, and some are draped over boulders. They may be traction-current deposits, part of turbidite sequences, or they may result from in situ melting of icebergs. Pebble clusters in type IV sediment indicate in situ iceberg melting or iceberg overturning. The upper type III sand is probably a subaerial-intertidal sequence deposited when the ice had less influence over the site. One particle-size analysis indicates an open-water beach environment.

Unit 3 (66.36-16.34 m) is homogeneous pebbly sand (type IV), sand (type III), and minor pebble gravel (type II). Pebbly sand is the predominant lithology, and it grades upward through the unit from fine to very coarse sand. Grounded ice was probably close to this site again when these deposits were produced in a subtidal-tidal-delta/beach environment similar to that in the lower part of the unit 4.

Unit 2 (16.34-3.51 m) is a sequence of uniform well-sorted medium sand. This traction-current sequence shows progradation by indistinct lamination changing up the unit from horizontal to maximum of 7°. There are also two sharp loaded contacts and some sand-injection features. Therefore periods of local erosion were interrupted by periods of rapid deposition.

Unit 1 (3.51-0.46 m) has thin-bedded, laminated silty mud, sand (type IV), and granule gravel. This is a traction-current sequence, perhaps deposited by a braided stream with mud deposition in abandoned braids.

DVDP 12

The Lake Hoare core has not previously been divided into units. The units used here are provisional.

Unit 8 (184.65-165.02 m) is migmatitic basement.

Unit 7 (165.02-130.18 m) contains predominantly massive diamicton (type VI″) with much thinner mud and sand interbeds. Standing-water deposition is indicated by mud units as much as 3.5 m thick. Some of these grade up into diamictons which are therefore waterlain tills. Other diamictons may be subglacial till or subaerial or subaqueous debris-flow deposits. Carbonate occurs within diamicton matrices, and some thin calcareous mud beds resemble an evaporite. Two yellowish orange horizons perhaps indicate subaerial exposure but may also be caused by subaqueous diagenetic effects.

Unit 6 (130.18-102.31 m) is a diamicton sequence. The thickness of the unit and the lack of definite bedding suggest an ice, rather than density-flow, origin. A glacial subaqueous environment is indicated by type VI‴ particle-size distributions. The nature of the diamicton varies up the unit and some redeposition has probably occurred (type VI′ sediment). A yellowish orange horizon occurs just beneath a gradational contact within the sequence. Thus depositional conditions varied, and all of the unit may not be subaqueous glacial sediment.

Unit 5 (102.31-74.64 m) consists largely of traction-deposited fine to medium sands (types III and IV) that vary in thickness and subordinate gravel (type I), mud, and diamicton beds. The muds have been described as 'varved' in the core logs, and there are thin calcareous horizons. With a wide variation in texture, but a relatively high proportion of fine material, this unit is probably a shallow-water deposit. Signs of redeposition and diamictons (type VI″) that may be debris-flow deposits suggest that ice was close to the site and/or sedimentation rates were high.

Unit 4 (74.65-38.00 m) is a diamicton sequence with very minor, thin (<40 cm thick) sand beds (type III). This unit represents an advance of ice to the site following deposition of unit 5. Most sedimentation occurred through a body of standing water. A winnowed waterlain till is present at the bottom of the unit. It has a laminated matrix, a type II particle-size distribution, and a yellowish orange horizon within. Other diamictons are massive type VI″ sediments, so some may be debris-flow deposits. The uppermost diamicton contains a dropped clast, showing that others are original waterlain tills.

Unit 3 (58.00-34.66 m) has rapid alternation of diamicton and either sand or mud. The base of the unit is marked by a calcareous evaporite bed. The diamictons are generally thin (<6 m thick; average 1 m thick) massive, type VI″ sediments that contain mud clasts (rarely evaporite). Two sand beds grade up into diamictons. The diamictons could thus be subaqueous debris-flow deposits from waterlain till. The sands (type III and IV) are massive to bedded and often contain mud clasts. Beds show normal and inverse grading, and some grade with mud and diamicton. Most of the sands are inferred to be part of gravity-flow sequences. Silts and muds are generally massive, often calcareous, and sometimes grade into and out from diamicton units. They are generally suspension deposits, some of which include ice-rafted debris, isolated clasts, and sand and gravel lenses. The unit is composed of subaqueous glacial sediment. The glacial deposits have undergone redeposition by various types of gravity flows and have been reworked by traction-current activity.

Unit 2 (34.66-2.13 m) has sand as its dominant lithology which is interbedded with minor diamicton. Most of the sand is type III traction-current sediment that contains locally eroded mud and evaporite clasts. They are probably proglacial sands deposited by pulsating flows. The diamictons are massive, two have type II size distributions and they are generally thin, although the uppermost is 7 m thick. Whether this unit is subaerial or subaqueous is hard to ascertain. Whichever it is, the water depth decreased over the site after deposition of unit 3 because traction rather than gravity flowage was the dominant depositional mechanism, and there are far fewer mud beds.

Unit 1 (2.3-0.00 m) is recent wind-winnowed glacial drift, a lag deposit caused by deflation.

CONCLUSION

Chronostratigraphies based on microfauna [*Webb and Wrenn*, 1976, 1980], microflora [*Brady*, 1979], and paleomagnetic polarity reversals [*Purucker et al.*, this volume; *Elston and Bressler*, this volume] all differ in their age assignments for sections of DVDP cores 10, 11, and 12 from Taylor Valley. Therefore a preferred glacial history of Taylor Valley cannot be given at the present time, and the problem will continue until discrepancies in chronostratigraphy are resolved. Two possibilities, one based on biostratigraphy, the other on paleomagnetic stratigraphy, for the glacial history of Taylor Valley are presented below. Other alternatives are possible, and these interpretations are intended as examples only.

Using chronologies established by *Webb and Wrenn* [1976, 1980] and *Brady* [1979] as well as fabric and provenance studies by *Porter and Beget* [this volume], a tentative glacial history can be sketched for Taylor Valley from DVDP cores 10 and 11. Based on biostratigraphic zones, unit 5 of DVDP 10 (DVDP 10/5) is correlated with units 7 and 8 of DVDP 11 (DVDP 11/7, 8), being late Miocene(?)-early Pliocene. Pleistocene sediment occurs above these levels in both cores. Correlations for this Pleistocene record are made only on the basis of logical sedimentary facies (and inferred environmental) successions. Figure 7 shows the detailed correlations and inferred environments of deposition.

ASSIGNED AGE	DVDP 11	DEPOSITIONAL ENVIRONMENT	DVDP 10	DEPOSITIONAL ENVIRONMENT
	1, 2, 3	Fluvial/deltaic/beach	1	Fluvial/deltaic/beach
	4	Proglacial (tidewater glacier)	2 to 3.4	Proglacial grounding line
	5.1 to 5.2	Iceberg zone	3.5 to 3.9	Floating ice tongue and/or grounded ice
	5.3 to 5.6, 6	Floating ice tongue and/or grounded ice, proglacial grounding line	4	Proglacial grounding line, grounded ice and/or floating ice tongue
		194.8 m to 202.5 m		~ 137.1 m
	7, 8	Floating ice tongue and iceberg zone	5	Iceberg zone

(Left axis, bottom to top: LATE MIOCENE (?) -PLIOCENE, PLEISTOCENE)

Fig. 7.   Facies correlations between DVDP 10 and DVDP 11 based on biostratigraphy. The solid line is a chronostratigraphic time line given with core depths in meters. Units (e.g., DVDP 10/1) and subunits (e.g., DVDP 10/3.4) are assigned inferred environments of deposition. Units above the time line are correlated using logical sedimentary facies (and inferred environmental) successions.

During the late Miocene(?)-early Pliocene, Taylor Glacier extended down to New Harbor, where it formed a floating ice tongue. The tongue deposited sediment in water depths of 600-900 m near Commonwealth Glacier (DVDP 11), although retreats of its terminus (calving line) created iceberg conditions intermittently over this site. Sediment deposited under iceberg conditions (iceberg zone sediment) is the major contribution to the record for this time at New Harbor beach (DVDP 10).

In Pleistocene time the sediment source was the Ross Ice Shelf (or marine ice sheet), although some could have come from local alpine glaciers. Water depths had decreased to less than 50 m at this time. The oldest late Pleistocene deposits show that a tongue of the Ross Ice Sheet advanced upvalley and deposited proglacial sediment close to its grounding line at DVDP 11. On retreat, floating ice tongue and/or grounded ice deposits accumulated, followed by iceberg zone sediment. Meanwhile proglacial grounding line and/or floating ice tongue sediment was first deposited at DVDP 10 and was followed by floating ice tongue and/or grounded ice sediment.

The Ross Ice Sheet advanced again and deposited proglacial grounding line sediment at New Harbor. From this time onward, shallow-water deposits are preserved in the DVDP cores; a similar but thicker section of proglacial tidewater glacier deposits and fluvial/deltaic/beach facies sediment form the top of DVDP 11 core. This depositional sequence apparently produced the present topographic high at site 11. Ice fluctuations cannot be interpreted for this sequence by itself, and therefore the correlation with DVDP 10 (DVDP 10/1) is uncertain.

Using the chronology established by *Elston and Bressler* [this volume, Figure 3] and magnetic susceptibility zones [*Purucker et al.*, this volume] as sediment provenance indicators, a tentative glacial history can be sketched for Taylor Valley from DVDP cores 10 and 11. Correlations between both cores are based on inferred unconformities, paleomagnetic time lines, and logical sedimentary facies (and inferred environmental) successions. Figure 8 lists the detailed correlations and inferred depositional environments.

During the late Miocene, Taylor Glacier extended down to New Harbor, where it formed a floating ice tongue. The tongue deposited sediment

ASSIGNED AGE	DVDP 11	DEPOSITIONAL ENVIRONMENT	DVDP 10	DEPOSITIONAL ENVIRONMENT
	1	Fluvial/deltaic/beach	1	Fluvial/deltaic/beach
		- - - - 2 m - - - -		- - - - 25 m - - - -
	2, 3, 4.1	Fluvial/deltaic/beach	2, 3.1	Proglacial grounding line
		79 m		41 m
	4.2 to 4.7	Proglacial (tidewater glacier)	3.2 to 3.4	Proglacial grounding line
	4.8 to 4.11	Proglacial grounding line	3.5 to 3.7	Grounded ice
	4.12 to 5.4	Iceberg zone	3.8	Proglacial grounding line
	5.5 upper	Floating ice tongue		?
		161 m		89 m
	?		3.9	Floating ice tongue
	5.5 lower, 6	Grounded ice, proglacial grounding line	4.1	Grounded ice
	7.1 to 7.6	Iceberg zone	4.2 to 4.7	Iceberg zone
		203 m		146 m
	7.7 to 7.9	Iceberg zone	5.1 to 5.6	Iceberg zone
		- - - - 208 m - - - -		- - - - 155 m - - - -
	7.10 to 7.20, 8	Floating ice tongue and iceberg zone	5.4 to 5.22	Iceberg zone

(left margin, reading top to bottom: PLEISTOCENE, ?, PLIOCENE, MIOCENE)

Fig. 8.   Facies correlations between DVDP 10 and DVDP 11 based on paleomagnetic stratigraphy. Solid lines are chronostratigraphic time lines; dotted lines are inferred unconformities [*Elston and Bressler*, this volume], both given with their core depths in meters. Units (e.g., DVDP 10/1) and subunits (e.g., DVDP 10/3.1) are assigned inferred environments of deposition. Units between the chronostratigraphic time lines are correlated using logical sedimentary facies (and inferred environmental) successions. The areas where questions arise are discussed in the text.

in water depths of 600-900 m near Commonwealth Glacier, although retreats of its calving line created intermittent iceberg conditions at the site. Sediment deposited under iceberg conditions is the major contribution to the record for this time at New Harbor beach.

In earliest Pliocene the sediment source was the Ross Ice Shelf (or marine ice sheet), although some could have come from local alpine glaciers. McMurdo Volcanic detritus found in the cores [*Porter and Beget*, this volume; *Purucker et al.*, this volume] has been interpreted to have been carried by ice from McMurdo Sound in a manner described by *Kyle* [this volume]. In earliest Pliocene when the source of ice was changing, iceberg conditions existed at both sites.

Iceberg conditions remained during later early Pliocene time when water depths were shallower than 50 m. Then ice from McMurdo Sound grounded at site 10 (DVDP 10/4.1) and advanced up Taylor Valley, deposited proglacial grounding line sediment at site 11 (DVDP 11/6), and eventually overrode the site. The next event is very difficult to explain sedimentologically and glaciologically. The problem is created by 'the time-transgressive nature of the upper boundary of the zone of high susceptibility (DVDP 10/3.9, middle; DVDP 11/5.5, top) with respect to the polarity zonation' [*Elston and Bressler*, this volume]. The decrease in magnetic susceptibility may be caused by an influx of nonvolcanic material that could have been derived from Taylor Glacier (*Denton et al.* [1971] Taylor IV glaciation?), local alpine glaciers (*Denton et al.* [1971] Alpine III glaciation?), or by the formation of a marine ice sheet in McMurdo Sound that eroded older material containing little or no volcanic detritus. If the first two options are true, then the onset of decreasing magnetic susceptibility should first appear in DVDP 11, whereas it is found first in DVDP 10. Floating ice tongue sediment was deposited at site 10 during this change, and sedimentologically the best correlative of this sediment is the upper part of DVDP 11/5.5 that also appears to be floating ice tongue sediment. Masking of volcanic detritus continued through late Pliocene time, while proglacial grounding line sediment was deposited at New Harbor beach and iceberg zone sediment was deposited near Commonwealth Glacier.

Ice from McMurdo Sound advanced again depositing grounded ice sediment at site 10 and proglacial grounding line sediment at site 11. The present topographic high at site 11 was started at this time, but the greatest proportion of it was built during the latest Pliocene and Pleistocene. Throughout this time the sediment load within the ice was once again high in McMurdo Volcanic detritus and proglacial grounding line sediment was deposited at site 10, while proglacial tidewater glacier sediment (DVDP 11/4.7 to 4.2) and fluvial/deltaic/beach facies sediment (DVDP 11/4.1, 3, 2) were deposited farther up-valley. Finally, a fluvial/deltaic/beach facies was deposited at New Harbor beach while a similar, very thin sequence remains near Commonwealth Glacier. The record in DVDP 10 is very fragmentary compared with that in DVDP 11 for latest Pliocene-Pleistocene time and paleomagnetic stratigraphy is difficult to interpret. Thus the exact timing of ice advances from McMurdo Sound and correlative sediment between sites 10 and 11 cannot be defined.

While the younger sections of the sediment successions in DVDP 10 and 11 were being deposited at the mouth of Taylor Valley, terrestrial and subaqueous glacial, traction-current, and gravity-flow sediments were deposited near Lake Hoare. Unfortunately, events at DVDP 12 cannot be related to those farther down-valley because of the dearth of chronostratigraphic units to aid correlation.

Acknowledgments. This work was carried out at Victoria University of Wellington, New Zealand, as part of an M.Sc. thesis under the invaluable assistance and guidance of Peter Barrett. The author is also grateful for comments made by John Shaw (University of Alberta) on the original thesis, and David Elliot, Ken Stanley, and John Mercer (Ohio State University) on earlier versions of this manuscript. The New Zealand University grants committee provided financial backing for research in Antarctica and the author was also aided by a McKee Trust postgraduate scholarship. The manuscript was typed and final figures prepared by staff at the Institute of Polar Studies and the Department of Geology and Mineralogy, Ohio State University. Gratitude is extended for support in the field to Antarctic Division (N. Z. Department of Scientific and Industrial Research), N. Z. Antarctic Research Program, and the U.S. Navy personnel during the 1974-1975 summer field season. The DVDP coordinators allowed the sampling of the drill cores, and M. Chapman-Smith sampled DVDP 12. The U.S. National Science Foundation and the Institute of Polar Studies (OSU) provided partial financial support for the author to attend the DVDP seminar in Tokyo (1978). The author appreciates receiving preprints from other authors of papers in this volume.

REFERENCES

Adams, J. E., Empirical determination of sieve-size statistics from grain measurement, M.Sc. thesis, 155 pp., Victoria Univ. of Wellington, New Zealand, 1974.

Anderson, J. B., Nearshore glacial-marine deposition from modern sediments of the Weddell Sea, *Nature Phys. Sci.*, *240*, 189-192, 1972.

Angino, E. E., M. D. Turner, and E. J. Zeller, Reconnaissance geology of lower Taylor Valley, Victoria Land, Antarctica, *Geol. Soc. Amer. Bull.*, *73*, 1553-1562, 1962.

Beaumont, P., Break of slope in particle size curves of glacial tills, *Sedimentology*, *16*, 125-128, 1971.

Brady, H. T., The dating and interpretation of diatom zones in Dry Valley Drilling Project holes 10 and 11 Taylor Valley, South Victoria Land, Antarctica, in *Proceedings of the Seminar III on Dry Valley Drilling Project, 1978*, edited by T. Nagata, pp. 150-164, National Institute of Polar Research, Tokyo, 1979.

Buller, A. T., and J. McManus, The quartile-deviation/median diameter relationship of glacial deposits, *Sediment. Geol.*, *10*, 135-146, 1973.

Carey, S. W., and N. Ahmad, Glacial marine sedimentation, in *Proceedings, First International Symposium on Arctic Geology*, vol. 2, pp. 865-894, University of Toronto Press, Toronto, Ont., 1961.

Chapman-Smith, M., Geology of DVDP holes 12 and 14, *Antarct. J. U.S.*, *10*(4), 170-172, 1975.

Chapman-Smith, M., and P. G. Luckman, Late Cenozoic glacial sequence cored at New Harbor, Victoria Land, Antarctica, *Dry Val. Drilling Proj. Bull.*, *3*, 120-147, 1974.

Chriss, T., and L. A. Frakes, Glacial marine sedimentation in the Ross Sea, in *Antarctic Geology and Geophysics*, edited by R. J. Adie, pp. 747-762, Universitetsforlaget, Oslo, 1972.

Davis, S. N., Size distribution of rock types in stream gravel and glacial till, *J. Sediment. Petrol.*, *28*, 87-94, 1958.

Denton, G. H., R. L. Armstrong, and M. Stuiver, Histoire glaciaire et chronologie de la region du Detroit de McMurdo, sud de la Terre Victoria, Antarctide, *Rev. Geogr. Phys. Geol. Dyn.*, *11*, 265-278, 1969.

Denton, G. H., R. L. Armstrong, and M. Stuiver, Late Cenozoic glaciation in Antarctica: The record in the McMurdo Sound region, *Antarct. J. U.S.*, *5*(1), 15-21, 1970.

Denton, G. H., R. L. Armstrong, and M. Stuiver, The Late Cenozoic glacial history of Antarctica, in *Late Cenozoic Glacial Ages*, edited by K. K. Turekian, pp. 267-306, Yale University Press, New Haven, 1971.

Doake, C. S. M., Thermodynamics of the interaction between ice shelves and the sea, *Polar Rec.*, *18*(112), 37-41, 1976.

Dreimanis, A., and U. J. Vagners, Bimodal distribution of rock and mineral fragments in basal tills, in *Till: A Symposium*, edited by R. P. Goldthwait, pp. 237-250, Ohio State University Press, Columbus, 1971.

Dreimanis, A., and U. J. Vagners, The effect of lithology upon the texture of till, in *Research Methods in Pleistocene Geomorphology*, edited by E. Yatsu and A. Falconer, pp. 66-82, Geo Abstracts, Ltd., Norwich, 1972.

Elston, D. P., and S. L. Bressler, Magnetic stratigraphy of DVDP drill cores and Late Cenozoic history of Taylor Valley, Transantarctic Mountains, Antarctica, this volume.

Folk, R. L., The distinction between grain size and mineral composition in sedimentary rock nomenclature, *J. Geol.*, *62*, 344-359, 1954.

Folk, R. L., and W. C. Ward, Brazos River bar: A study in the significance of grain-size parameters, *J. Sediment. Petrol.*, *31*, 3-26, 1957.

Frakes, L. A., and J. C. Crowell, Characteristics of modern glacial marine sediments, in *Gondwana Geology*, edited by K.

S. W. Campbell, pp. 373-380, Australian National University Press, Canberra, 1975.

Gillberg, M., A statistical study of till from Sweden, *Geol. Foren.*, *87*, 84-108, 1965.

Haskell, T. R., J. P. Kennett, W. M. Pebble, G. Smith, and I. A. G. Willis, The geology of the middle and lower Taylor Valley of south Victoria Land, Antarctica, *Trans. Roy. Soc. N.Z.*, *2*(12), 169-186, 1965.

Hjulstrom, F., Transportation of detritus in moving water, in *Recent Marine Sediments, A Symposium*, edited by P. D. Trask, pp. 5-31, American Association of Petrology and Geology, Tulsa, Oklahoma, 1939.

Inman, D. L., Sorting of sediments in the light of fluid mechanics, *J. Sediment. Petrol.*, *19*, 51-70, 1949.

Jarnefors, B., A sediment-petrographic study of glacial till from the Pajala district, North Sweden, *Geol. Foren.*, *74*, 185-214, 1952.

Kranck, K., Sediment deposition from flocculated suspensions, *Sedimentology*, *22*, 111-123, 1975.

Kyle, P. R., Glacial history of the McMurdo Sound area as indicated by the distribution and nature of McMurdo Volcanic Group rocks, this volume.

Landim, P. M. B., and L. A. Frakes, Distinction between tills and other diamictons based on textural characteristics, *J. Sediment. Petrol.*, *38*, 1213-1223, 1968.

Lopatin, B. G., Basement complex of the McMurdo 'Oasis,' south Victoria Land, in *Antarctic Geology and Geophysics*, edited by R. J. Adie, pp. 287-292, Universitetsforlaget, Oslo, 1972.

McCraw, J. D., Volcanic detritus in Taylor Valley, *N.Z. J. Geol. Geophys.*, *5*, 740-745, 1962.

McCrone, A. W., Clarification of the "winnowing" concept in geology, *Geol. Soc. Amer. Bull.*, *73*, 517-578, 1962.

McKelvey, B. C., Preliminary site reports: DVDP sites 10 and 11, Taylor Valley, *Dry Val. Drilling Proj. Bull.*, *5*, 16-50, 1975*a*.

McKelvey, B. C., Stratigraphy of DVDP sites 10 and 11, Taylor Valley, *Antarct. J. U.S.*, *10*, 169-170, 1975*b*.

McKelvey, B. C., Upper Cenozoic marine and terrestrial glacial sedimentation in eastern Taylor Valley, southern Victoria Land, in *Antarctic Geosciences*, edited by C. Craddock, University of Wisconsin Press, Madison, in press, 1980.

Naidu, A. S., and T. C. Mowatt, Depositional environments and sediment of the Colville and adjacent deltas, northern Arctic Alaska, in *Deltas Models for Exploration*, edited by M. S. L. Broussard, pp. 283-309, Houston Geological Society, Houston, Tex., 1975.

Ovenshine, A. T., Observations of iceberg rafting in Glacier Bay, Alaska, and the identification of ancient ice-rafted deposits, *Geol. Soc. Amer. Bull.*, *81*, 891-894, 1970.

Porter, S. C., and J. E. Beget, Provenance and depositional environments of Late Cenozoic sediments in permafrost cores from lower Taylor Valley, Antarctica, this volume.

Powell, R. D., Textural characteristics of some glacial sediments in Taylor Valley, Antarctica, M.Sc. thesis, 316 pp., Victoria Univ., Wellington, New Zealand, 1976.

Powell, R. D., Conditions of sediment deposition in Taylor Valley, Antarctica from Dry Valley Drilling Project cores 8-12 (extended abstract), in *Proceedings, Seminar III Dry Valley Drilling Project, Mem. Spec. Issue 13*, edited by T. Nagata, pp. 187-195, National Institute of Polar Research, Tokyo, 1979*a*.

Powell, R. D., A model for glacial-marine sedimentation by

tidewater glaciers, *Geol. Soc. Amer. Abstr. Programs*, *11*(7), 498, 1979*b*.

Powell, R. D., and P. J. Barrett, Grain-size distribution of sediments cored by the DVDP in the eastern part of Taylor Valley, Antarctica, *Dry Val. Drilling Proj. Bull.*, *5*, 150-166, 1975.

Purucker, M. E., D. P. Elston, and S. L. Bressler, Magnetic stratigraphy of Late Cenozoic glaciogenic sediments from drill cores, Taylor Valley, Transantarctic Mountains, Antarctica, this volume.

Robin, G. de Q., Ice shelves and ice flow, *Nature*, *253*, 168-172, 1975.

Robin, G. de Q., and R. J. Adie, The ice cover, in *Antarctic Research*, edited by R. Priestly et al., pp. 100-117, Butterworths, London, 1964.

Rogers, J. W. W., and C. Schubert, Size distributions of sedimentary populations, *Science*, *141*(3583), 801-802, 1963.

Rust, B. R., Mass flow deposits in a Quaternary succession near Ottawa, Canada: Diagnostic criteria for subaqueous outwash, *Can. J. Earth Sci.*, *14*, 175-184, 1977.

Rust, B. R., and R. Romanelli, Late Quaternary subaqueous outwash deposits near Ottawa, Canada, Glaciofluvial and Glaciolacustrine Sedimentation, *Spec. Publ. 23*, pp. 177-192, Soc. of Econ. Paleontal. and Mineral., Tulsa, Okla., 1975.

Slatt, R. M., Texture and composition of till derived from parent rocks of contrasting textures, southeast Newfoundland, *Sediment. Geol.*, *7*, 283-290, 1972.

Slatt, R. M., and C. M. Hoskin, Water and sediment in the Norris Glacier outwash area, Upper Taku Inlet, southeast Alaska, *J. Sediment. Petrol.*, *38*, 434-456, 1968.

Stanley, D. J., Reworking of glacial sediments in the northwest arm of a fjord-like inlet on the southeast coast of Nova Scotia, *J. Sediment. Petrol.*, *38*, 1224-1241, 1968.

Turner, R. R., The significance of color banding in the upper layers of the Kara Sea sediments, *Oceanogr. Rep. 36*, 36 pp., U.S. Coast Guard, Washington, D. C., 1971.

von Huene, R., E. Larson, and J. Crouch, Preliminary study of ice-rafted erratics as indicators of glacial advances in the Gulf of Alaska, in *Initial Reports of the Deep Sea Drilling Project*, vol. 18, edited by L. D. Kuhn et al., pp. 835-842, U.S. Govt. Printing Office, Washington, D.C., 1973.

Webb, P. N., and J. H. Wrenn, Foraminifera from DVDP holes 8, 9, 10, 11 and 12, Taylor Valley, *Antarct. J. U.S.*, *11*, 85-86, 1976.

Webb, P. N., and J. H. Wrenn, Late Cenozoic micropaleontology and biostratigraphy of eastern Taylor Valley, Antarctica, in *Antarctic Geosciences*, edited by C. Craddock, University of Wisconsin Press, Madison, in press, 1980.

PROVENANCE AND DEPOSITIONAL ENVIRONMENTS OF LATE CENOZOIC SEDIMENTS IN PERMAFROST CORES FROM LOWER TAYLOR VALLEY, ANTARCTICA

STEPHEN C. PORTER AND JAMES E. BEGET

Quaternary Research Center, University of Washington, Seattle 98195

The provenance and depositional environments of frozen late Cenozoic sediments of glacial and nonglacial origin recovered in four drill cores from lower Taylor Valley, Antarctica, were assessed by means of lithic counts, microfabric analyses, and studies of sand grain surface textures, supplemented by isotopic and paleontologic analyses by coinvestigators. Three of the cores, DVDP 8, 9, and 10, were drilled near the shore of New Harbor and intersected 185 m of sediment, whereas the fourth (DVDP 11), drilled about 3 km inland near Commonwealth Glacier, penetrated 328 m of sediment. The uppermost 39 m of section in the New Harbor cores consists of deltaic sediments deposited in a shallow marine environment during the middle Holocene following recession of a grounded ice sheet in the Ross Sea (Ross Sea I). A thick section between 39 and 125 m consists largely of diamictons of inferred glacialmarine origin, but the lowest diamicton unit, between 104 and 125 m, has a well-developed fabric and is interpreted as a possible basal till deposited by grounded Ross Sea ice. A major lithologic change was recognized at about 154 m in cores 8 and 10 and at about 205 m in core 11; above those levels, volcanic clasts of the McMurdo volcanic group that were derived from the region of the Ross Sea were found in all samples, but at lower levels none were detected. This lithic discontinuity coincides approximately with an unconformity recognized on paleontologic and paleomagnetic evidence and may be of regional extent. Diamictons of glacial and glacialmarine origin above the unconformity record incursions of grounded and floating ice into lower Taylor Valley from the east, presumably at times when the West Antarctic Ice Sheet expanded into the Ross Sea region (Ross Sea glaciations). Absence of McMurdo volcanic rocks at lower depths in the cores may mean either (1) that if the sediments had a source in the Ross Sea region, they must have been deposited before the volcanic rocks were extruded, or at least before they were extruded in sufficient volume to form a detectable component of glacially derived rock fragments in the cores, or (2) that those sediments do not record Ross Sea glaciations but rather expansions either of the East Antarctic Ice Sheet (Taylor glaciations) or of local alpine glaciers, in which case Ross Sea glaciations must postdate the initial eruption of the McMurdo volcanics. In contrast to abundant glacially abraded sand grains seen in the upper parts of the cores, many grains examined from marine sediments of early Pliocene or late Miocene age near the bottom of the cores are rounded and lack features suggestive of glacial transport. This may indicate that wind and streams were important local sources of sediment to the marine (fjord) environment at that time.

INTRODUCTION

Geologic investigations in the ice-free ('dry') valleys of the McMurdo Sound region of southern Victoria Land, Antarctica, have disclosed one of the longest records of late Cenozoic glaciation found on any of the continents. Regional mapping and stratigraphic studies, coupled with radiometric dating of volcanic rocks associated with glacial deposits, have shown that evidence of glaciation in the major ice-free valleys extends back nearly 4 m.y. Nonsynchronous expansions of the East Antarctic Ice Sheet, of the West Antarctic Ice Sheet, and of local alpine glaciers have resulted in a complex record of sediments and landforms that provides the basis for a provisional reconstruction of the glacial history [*Denton et al.*, 1970, 1971, 1975; *Calkin and Nichols*, 1972]. As is the case with most terrestrial glacial sequences, the exposed sedimentary record in the ice-free valleys is discontinuous and incomplete, and thus some aspects of the glacial history are obscure and difficult to interpret. As a means of both amplifying and extending the available depositional record, recovery of frozen sediment cores from selected sites in the ice-free valleys was included as an integral part of the Dry Valley Drilling Project (DVDP). The DVDP was organized jointly by the United States, Japan, and New Zealand in 1972 as a multidisciplinary program which has as its primary goal an investigation of the physical, chemical, and biologic environment and of the geologic history of the ice-free valleys of southern Victoria Land by means of deep borings [*McGinnis et al.*, 1972, 1975]. The drilling sites having the greatest potential bearing on the glacial history of the McMurdo Sound region are located in lower Taylor Valley, the southernmost of the three principal ice-free valleys (Figure 1). Four cores (DVDP 8, 9, 10, and 11) recovered at the seaward end of the valley include both marine and nonmarine sediments and are the subject of the present study.

GEOLOGY OF TAYLOR VALLEY AND VICINITY

Taylor Valley, the only one of the major ice-free valleys currently open to the sea, is bordered on the east by New Harbor, a small coastal embayment along the western margin of McMurdo Sound. Taylor Glacier, which drains the East Antarctic Ice Sheet across a bedrock threshold in the Transantarctic Mountains, occupies the western portion of the valley. The largely ice-free eastern part of the valley is some 35 km long and is bordered on the north by the Asgard Range and on the south by the Kukri Hills. Both highlands reach altitudes of 2000 m or more and harbor numerous small alpine glaciers that flow toward the axis of the valley. Two of these glaciers, Canada and Commonwealth, terminate as broad spatulate lobes on the north side of the valley. Three large lakes— Bonney, Hoare, and Fryxell—and many smaller ones occupy closed depressions and locally abut glacier tongues that reach the valley floor. Although bedrock crops out widely along the margins of the valley, the lower elevations are extensively mantled with surficial sediments, largely of glacial origin.

Metasedimentary rocks of the Skelton group (Ross supergroup) and granitic rocks of the Granite Harbor intrusive complex constitute a crystalline basement of Precambrian to early Paleozoic age that crops out widely throughout Taylor Valley [*Haskell et al.*, 1965]. Sedimentary rocks of the late Paleozoic Beacon supergroup and the associated Ferrar dolerite (Jurassic) are restricted to the upper, western part of the valley and do not crop

out within 22 km of New Harbor. Scattered small occurrences of olivine basalt assigned to the McMurdo volcanic group (late Cenozoic) comprise less than a few percent of exposed rocks within the central and western portions of the valley. However, varied lithologies of this volcanic assemblage underlie Ross Island, Black Island, White Island, Brown Peninsula, and portions of the Koettlitz-Blue glacier area east and south of the dry valleys. Gravity, seismic, and fathometer data suggest that rocks of the Ross supergroup comprise the basement beneath western and central McMurdo Sound and that they are overlain to the east by McMurdo volcanics [*McGinnis*, 1973; *Northey et al.*, 1975]. Basement rocks in the central part of the sound are overlain by at least 1 km of sediment, within which lies an angular unconformity possibly corresponding to a Miocene-Pliocene unconformity identified in cores obtained farther north in the Ross Sea [*Hayes et al.*, 1973]. At least the upper part of the sedimentary sequence very likely consists of Pliocene-Quaternary glacial and glacialmarine drift.

A preliminary glacial history of Taylor Valley has been outlined by *Denton et al.* [1969, 1970, 1971]. Advances of Taylor Glacier, resulting from expansions of the East Antarctic Ice Sheet, have led to deposition of drift of Taylor Valley provenance throughout the valley. During the earliest advances (Taylor III, IV, and V), Taylor Glacier reached McMurdo Sound, whereas younger advances were much less extensive. The terminus of Taylor Glacier is now more advanced than it was at any time since before the last glaciation.

Evidence from the McMurdo Sound region indicates that at least four times during the last 1.2 m.y. the Ross Ice Shelf expanded, became a grounded ice sheet, and advanced into the seaward ends of the dry valleys, where it deposited drift containing volcanic erratics from Ross Island. At least two such advances have been documented in the lower part of Taylor Valley (Figure 1; G. H. Denton, personal communications, 1976). Ross Sea I drift, in part cored by ice, extends to the eastern end of Lake Hoare. It mantles most of the valley floor east of Canada Glacier and lies at the surface on slopes surrounding the New Harbor drill site. Older drift of Ross Sea origin, representing one earlier expansion or more of a Ross Sea glacier, crops out beyond and above the limit of the younger drift and extends along the axis of the valley to a point about 7 km west of Canada Glacier. Radiocarbon dates of freshwater algae related to lake

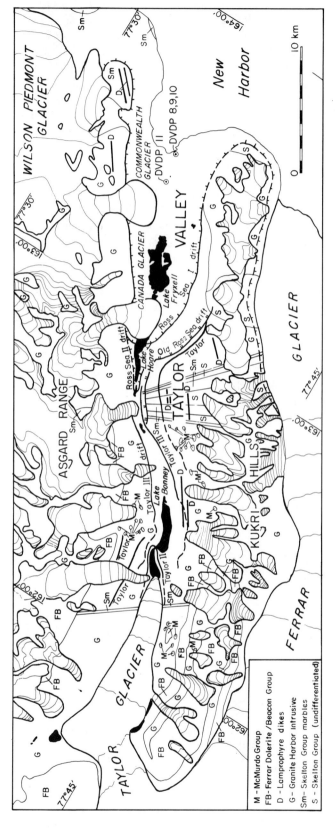

Fig. 1. Geologic map of Taylor Valley and vicinity showing location of DVDP drilling sites at New Harbor and Commonwealth Glacier, limits of Taylor and Ross Sea drift sheets [after *Denton et al.*, 1971; G. H. Denton, personal communication, 1976], and generalized distribution of bedrock [*Haskell et al.*, 1965]. (Topography obtained from U. S. Geological Survey preliminary topographic map of ice-free valleys (1 : 100,000).)

level fluctuations in Taylor Valley indicate that the Ross Sea I advance culminated about 18,000-19,000 years ago and that deglaciation of the area near the New Harbor drill site had occurred by about 8200 years ago [*Denton et al.*, 1971; *Kellogg et al.*, 1978; G. H. Denton, personal communication, 1978]. The ages of the older Ross Sea advances have not been established, but very likely they antedate the last interglaciation [*Denton et al.*, 1970; G. H. Denton, personal communication, 1976]. During each Ross Sea glaciation a large deep lake was impounded in Taylor Valley beyond the encroaching glacier terminus; strandlines marking former lake levels are etched across old drift units along the valley walls.

Three minor fluctuations of local alpine glaciers have been recognized. The two youngest, which occurred since 400,000 years B.P., were not in phase with Ross Sea glaciations, probably reflecting the requirement for an open Ross Sea to provide an adequate source of precipitation for glacier growth.

PERMAFROST CORES

The three holes at New Harbor were drilled at an altitude of 1.9 m about 50 m from the shoreline on the surface of a small raised delta (77°34'33"S latitude, 163°30'42"E longitude, Figure 1) [*Chapman-Smith and Luckman*, 1974; *McKelvey*, 1975; *Stuiver et al.*, 1976a]. DVDP 8 was drilled vertically to a depth of 157.06 m. DVDP 9 was drilled nearby to a depth of 38.34 m at an angle of 84.25° to obtain a continuous record of the upper part of the section, much of which was lost in DVDP 8. DVDP 10 was drilled vertically to a depth of 185.47 m. A fourth hole (DVDP 11) drilled near Commonwealth Glacier (77°35'24.3"S latitude, 163°24'40.3"E longitude) extends from an altitude of 80.2 m to a depth of 327.96 m. The borings were all in permafrost, and none reached bedrock. A preliminary log was made at the drill sites prior to shipment of cores to a storage facility, where more detailed geologic descriptions were made.

Because of the proximity of the two long cores at New Harbor, their gross stratigraphy is similar (Figure 2). Minor stratigraphic differences in the cores may reflect abrupt lateral facies changes, differences in logging technique by personnel from the three countries, or both. DVDP 8 and 10 are divisible into three major units based on dominant sediment type: an upper unit consisting of coarse sand and gravel with interstratified finer layers

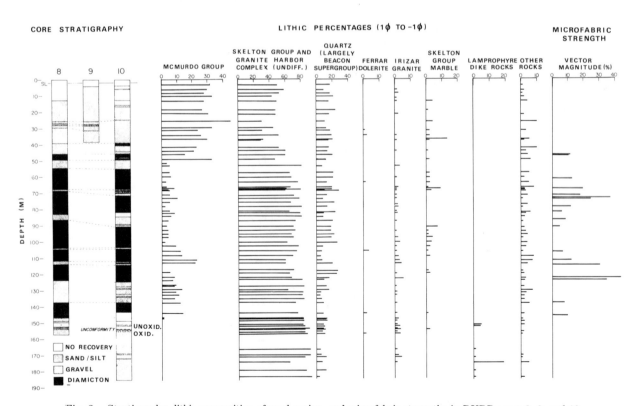

Fig. 2. Stratigraphy, lithic composition of sand grains, and microfabric strengths in DVDP cores 8, 9, and 10.

that extends to a depth of 39 m, a middle unit that is dominated by pebbly diamictons alternating with stratified sand and silt layers and that extends to about 147 m, and a lower unit consisting of interbedded sand and mud layers with some thin pebble gravels.

Preliminary assessment of the frozen sediments suggested that the upper 39 m might be postglacial in view of the location of the drilling site on Holocene deltaic sediments that bear a close resemblance to the sand and gravel in the upper part of the cores. The diamicton sequence was presumed to be glacial in origin and to reflect one or more advances of glacier ice, either floating or grounded, across the drilling site. Because the basal sediments in the two longer cores lack thick diamictons, which are characteristic of the middle unit, they provisionally were considered to be nonglacial and probably largely of marine origin.

The gross stratigraphy of core 11 resembles that of the New Harbor cores. The uppermost 128 m of core penetrated sand and gravel, whereas the remaining sediments consist of diamicton layers that are interbedded with sand and gravel. Unlike cores 8-10, the lowermost recovered sediments are marine diamictons that were presumed to be of glacialmarine origin [*McKelvey*, 1975]. No obvious correlations between this core and those from New Harbor could be made on the basis of preliminary descriptions reported in the initial drill logs.

The primary attention of the present study was directed toward the diamicton layers in order to assess whether they recorded glacialmarine conditions (ice shelf or berg ice) or advances of grounded ice. In addition, the provenance of the diamicton layers as well as the overlying and interstratified sediments was investigated in order to evaluate whether sediment influx might be attributed to expansions of the East Antarctic Ice Sheet (Taylor Glacier), of a grounded Ross Sea ice sheet, or of local alpine glaciers.

LABORATORY ANALYSES

Frozen cores were examined in detail at the Antarctic Core Storage Facility, Florida State University, where DVDP cores are stored at $-20°C$. Selected frozen samples were shipped to the Quaternary Research Center, University of Washington, where they were maintained in frozen condition until they were studied.

Lithic Analysis

Core samples were thawed at room temperature and soaked and agitated by using a Calgon dispersing agent. After air drying, samples were sieved into $>-1\phi$, -1ϕ-1ϕ, and $<1\phi$ fractions. A split of the intermediate fraction consisting of 100-300 grains was placed in a gridded petri dish for mineralogic and petrologic identification and tabulation with a binocular microscope. The entire coarse fraction of each sample was counted in a similar manner. As a means of categorizing clasts and grains in the larger size fractions, representative large clasts and resin-impregnated sand samples were thin sectioned for specific identification with a petrographic microscope.

Microfabric Analysis

Microfabric analyses were made of diamicton units to aid in differentiating basal till, which generally is characterized by strong macrofabric and microfabric [*Ostrey and Deane*, 1963], from glacialmarine drift deposited by floating ice, which normally should be characterized by random orientation of component particles. Frozen diamicton samples were thawed slowly to avoid disturbance. Friable samples were impregnated with a commercial resin by using a method similar to that described by *Evenson* [1971]. Resin-impregnated and naturally indurated samples then were thin sectioned along the horizontal plane.

The orientation of 100 elongate grains having a length-to-width ratio of at least 3 : 2 was measured on each slide. In a few cases, less than 100 grains met the criterion of elongation. For each fabric measurement, grain azimuth data were grouped into 10° class intervals, starting at 355° and ending at 185°. The groupings then were analyzed statistically by using a two-dimensional vector method [*Krumbein*, 1939; *Curray*, 1956]. The calculated azimuth of the resultant vector [*Curray*, 1956] was rotated so as to lie within the first class interval (355°-5°), and a rose diagram was constructed, based on the reoriented data. Next, the strength of the resultant vector was calculated and converted to vector magnitude L, expressed as percent. Finally, the statistical significance of the vector magnitude, p, was determined by using a significance level of 0.05. Values of vector magnitude for each sampled horizon are plotted in Figures 2 and 3.

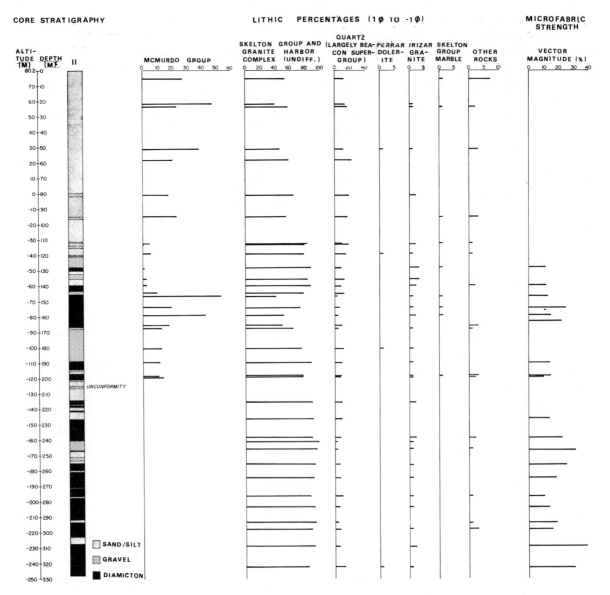

Fig. 3. Stratigraphy, lithic composition of sand grains, and microfabric strengths in DVDP core 11.

Sand Grain Surface Textures

Sand size quartz grains from selected levels in DVDP 8 were prepared for examination with a scanning electron microscope. Analyses of surface textures from resulting photographs were used as a supplementary basis for assessing possible depositional environments and diagenetic effects (Figure 4; *Krinsley and Doornkamp* [1973]).

PROVENANCE OF SEDIMENTS

Evaluation of sediment provenance was based on presence or absence of distinctive indicator lithologies related to source areas in Taylor Valley and in the volcanic archipelago of the Ross Sea. Basaltic and trachytic rocks of the McMurdo volcanic group are common east and south of New Harbor, whereas McMurdo volcanics within Taylor Valley consist of scattered small olivine basalt flows. The presence of large quantities of McMurdo volcanics in the cores therefore probably would reflect incursion of floating or grounded Ross Sea ice into the lower part of the valley. Similarly, presence in the cores of distinctive lithologies common to Taylor Valley but probably absent beneath or marginal to McMurdo Sound would point to a local (Taylor Valley) provenance.

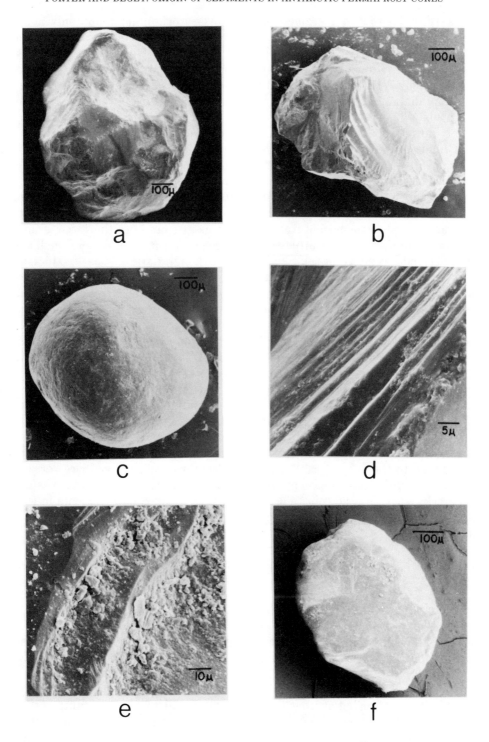

Fig. 4.   Surface features of quartz sand grains from DVDP 8: (*a*) Glacially sculptured grain from 26.00 m showing evidence of subaqueous modification; (*b*) Glacially sculptured grain from glacialmarine diamicton at 95.47 m showing well-developed conchoidal fractures; (*c*) Well-rounded grain of probable eolian origin recovered from glacialmarine drift at 95.47 m; (*d*) Detail of conchoidal fractures on surface of grain from glacialmarine diamicton at 102.95 m; (*e*) Redeposited silica on conchoidal fractures of glacially sculptured grain from within oxidized horizon at 154.07 m; (*f*) Glacially sculptured grain from 171.15 m showing rounded edges attributed to aqueous reworking.

Sand grains were divided into mineralogic and lithologic categories that could be related to rock units in Taylor Valley and adjacent areas (Figures 2 and 3). These include yellowish, pinkish, and greenish quartz probably derived largely from the Beacon supergroup; diabase derived from Ferrar dolerite sills; hornblende crystals and pinkish K feldspar probably derived mainly from the Irizar granite of the Granite Harbor intrusive complex; volcanic rock fragments belonging to the McMurdo volcanic group; marble derived from the Skelton group; clear quartz and white feldspar grains derived largely from the Granite Harbor intrusives but possibly containing a small component from the Beacon supergroup; undifferentiated lithic fragments probably derived largely from the Skelton group; mafic dike rocks, predominantly hornblende lamprophyres; biotite flakes of uncertain origin; and rip-up clasts consisting of consolidated or indurated sediment.

Because sediment coarser than -1ϕ was rare or absent in many core samples, a continuous record of lithologic variation for this size component could not be established, but available data support the interpretation based on the sand size fraction. The principal rock types recognized include granodiorite attributed to the Larsen granodiorite and Theseus granodiorite of the Granite Harbor complex, a pink granite assigned to the Irizar granite, trachytic rocks of the McMurdo volcanic group, mafic and aplitic dike rocks, gneisses and schists of the Skelton group, and other unidentified constituents, including sediment clasts.

New Harbor Cores
(DVDP 8, 9, and 10)

Sand grains attributed to Skelton group and Granite Harbor intrusive complex lithologies are present throughout the cores and collectively constitute between about 30 and 90% of the sediments. Grains attributed to Beacon supergroup sedimentary rocks average between about 10 and 20% of each sample above 125 m but average closer to 10% below that level. Many such grains are well rounded and show surface features consistent with an eolian origin (Figure 4c). Their presence in the cores therefore possibly reflects sediment transport downvalley by persistent winds throughout the period represented by the recovered section.

Rocks of the McMurdo volcanic group, marble from the Skelton group, and rip-up clasts of sedimentary origin are found only above 154 m. Because abundant volcanic detritus could be introduced to New Harbor only from areas east of the drill site, its presence in the cores is interpreted as indicating transport by ice from the region of the Ross Sea. The rip-up clasts, which may have been derived from marine or glacialmarine sediments in McMurdo Sound, are consistent with such an interpretation. McMurdo volcanics average between 25 and 30% of the sand fraction above a depth of 50 m, only about 5% between 50 and 100 m, and about 10% between 100 and 147 m, at which level this component disappears. Other lithologies, including grains attributed to the Irizar granite and Ferrar dolerite, are present only in small amounts (<5%).

Although data for the $> -1\phi$ fraction are less complete, a similar picture emerges. Crystalline rocks of the Skelton group and Granite Harbor complex are present throughout the cores, but the McMurdo volcanics are found only above about 150 m. Lamprophyric dike rocks, which crop out within Taylor Valley, constitute a significant fraction of the clasts below 150 m but were not detected in the middle unit and were found only locally and in small amounts in a gravel layer at 30- to 40-m depth in the upper unit. Mafic dike rocks are present in the sand fraction only below 150 m. Because this lithology is present where McMurdo volcanics are absent, and vice versa, the dike rocks may be the best indicator of a western (Taylor Valley) provenance for sediments in the New Harbor cores.

Commonwealth Glacier Core
(DVDP 11)

As in the New Harbor cores, sand grains attributable to the Granite Harbor complex and Skelton group were observed in DVDP 11 samples and in all but two samples constitute the dominant lithology. The amount of these rock types ranges from 41 to 97%. Minor amounts of quartz (generally <20%) and of Irizar granite (≤3%) are found throughout the core. Small amounts of Ferrar dolerite (1%) were identified in only four widely separated samples. Volcanic rocks of the McMurdo volcanic group and marble of the Skelton group were seen only in the uppermost 200 m of the core. Whereas the latter component occurs only in small amounts (1%), the volcanic rocks locally constitute as much as 53% of the total.

Data for the $> -1\phi$ fraction in DVDP 11 are consistent with those for the sand fraction. Rocks of the McMurdo volcanic group were only encountered above a depth of 200 m. Lamprophyric dike rocks were found in trace amounts only below a depth of 198 m. Ferrar dolerite was more plentiful in this size fraction than in the finer size fraction, being present in 53% of the samples, but with few exceptions it constitutes less than 5% of the total sand fraction.

In summary, rock types that crop out in middle and lower Taylor Valley and possibly underlie at least part of the western section of McMurdo Sound are present throughout the New Harbor and Commonwealth Glacier cores. The absence of rocks of the McMurdo volcanic group below 150 m in DVDP 8 and 10 and below 200 m in DVDP 11 suggests either that the sediments predate eruption of the McMurdo volcanics in and east of New Harbor or that a significant change in provenance takes place close to these levels in the cores.

DEPOSITIONAL ENVIRONMENTS

Assessment of the depositional environments represented by the sediments in the Taylor Valley cores is based on such sedimentary characteristics as particle size distribution, degree of sorting, sedimentary structures, presence or absence of rafted clasts, microfabric strength, and surface morphology of quartz grains. Data provided by co-workers regarding isotopic composition of interstitial ice and fossil content of the sediment also bear on the environmental determinations.

New Harbor Cores

The upper 39 m of section consists largely of unweathered medium to coarse sand and gravel that is well bedded. The coarser sediment tends to be more poorly sorted than the finer-grained units. Silt and clay content is uniformly low and commonly is <15% [*Powell and Barrett*, 1975]. Well-developed cross stratification is evident in the upper parts of the cores, where dips typically range from 15° to 25°; locally, they reach 60°-80° in core 9 [*Chapman-Smith and Luckman*, 1974]. The basal 15 m of the upper unit is dominated by variably sorted pebble and cobble gravel. Some layers are normally graded. The Pecten *Adamussium colbecki* was found in the upper parts of all New Harbor cores; a pronounced concentration occurs in

sandy sediment between about 22 and 24.5 m. In several instances the opposing valves were articulated. In addition, abundant well-preserved and diverse foraminiferal faunas were recovered at 16 m, 23.6 m, and 24.3 m in core 9, and remains of other marine organisms, including ophiuroids, ostracods, and sponges, were found between 13- and 45.5-m depth [*Webb and Wrenn*, 1975].

Collectively, these features suggest that the upper unit of the cores consists predominantly of deltaic sediment deposited in a nearshore marine environment. Such an interpretation is consistent with the $\delta^{18}O$ measurements on interstitial ice which indicate marine conditions in this part of the section [*Stuiver et al.*, 1976b], and with the fact that the drill rig was located on the surface of a landform interpreted as an emerged delta. Ice-rafted clasts are relatively scarce in the upper unit, suggesting that the sediments probably accumulated in a prograding delta in water sufficiently shallow to exclude abundant large bergs. The high concentrations of McMurdo volcanics in the upper unit is also consistent with the inferred environmental conditions, for the low-density vesicular lava fragments tend to be selectively winnowed from surficial glacial drift upslope from the New Harbor site and concentrated by streams entering the sea.

The thick diamictons in the middle section of the core contain Ross Sea lithologies and resulted from one or more expansions of Ross Sea glacier ice across the drill site. The diamictons are as much as 21 m thick and have a texture best described as pebbly silty sand. Most sediment within the middle unit is barren of foraminifera; where specimens have been found, the tests commonly are large, filled with matrix, and damaged [*Webb and Wrenn*, 1975]. Only broken fragments of marine diatoms were found in this unit by *Brady* [1979]. No convincing in situ assemblages have been seen, and the observed bioclastic debris apparently has been reworked from older marine sediments. A predominantly glacial character for the middle unit also is suggested by observations of sand grain morphology. The diamictons contain an abundance of sharp, angular grains displaying irregular concoidal fractures and cleavage plates (Figures 4b and 4d). Although superficially all the diamictons resemble till, the likelihood that the bulk of the sediment is in fact glacialmarine drift deposited from floating ice is supported by isotopic measurements which indicate full or nearly full marine conditions to a depth of 85 m, less saline conditions between 85 and 100 m, fresh or slightly salty conditions between 100 and 125 m, a transition zone between 125 and 130 m, and brackish water between 130 and 155 m [*Stuiver et al.*, 1976*b*]. Most of the diamicton layers display either an extremely weak fabric or no fabric (Figure 2), but at two levels (73 m and 104-125 m), strong fabrics were measured. The lower of these levels corresponds closely to a $\delta^{18}O$ minimum, marking a zone of greatly reduced salinity, but the upper one lies within a zone marked by strongly marine conditions. The coincidence of the lower strong fabric with a zone of interstitial fresh water argues for probable deposition by a grounded ice sheet under conditions where marine water was excluded from the area.

Freshwater conditions might have resulted from a relative sea level lowering of at least 125 m (or about 143 m below the local marine limit, which is at least $+7.5$ m) and/or from the filling of McMurdo Sound with grounded glacier ice. The upper strong fabric, which coincides with a zone of sedimentation that is probably of marine origin, may have resulted either from grounding of an ice shelf, from deposition as a subaqueous debris flow, or possibly from movement of the base of an iceberg across glacialmarine drift on the sea floor, thereby imparting a fabric to the sediment. The much weaker fabric of the other diamicton layers is consistent with deposition of glacialmarine drift from an ice shelf or from floating icebergs.

The lowest unit in cores 8 and 10, extending below a depth of 147 m, consists mostly of sandy mud with interbedded medium to fine sand and pebble gravel. The horizontally thin-bedded sediments tend to be poorly sorted. Locally they display bioturbation and cross stratification, with dips ranging up to 30°. Graded bedding is common, especially above 154 m. Dropstones are dispersed through the sediment but are not as numerous as they are at higher levels in the cores. A conspicuous color change was observed between 154 and 156 m in core 10 and is inferred to mark a zone of oxidation (Figure 2). Abundant but poorly preserved foraminiferal faunas found in the lower unit are regarded by *Webb and Wrenn* [1975] as near-natural assemblages. They consider the sediments between 153 and 168 m to be marine in origin on the basis of the foraminifera. This interpretation is supported by the presence of marine diatoms [*Brady*, 1979]. Many quartz grains below 154 m,

and especially between 154 and 156 m, display diagenetic features resulting from solution and precipitation of silica. Such effects are especially noticeable along concoidal fractures of presumed glacial origin (Figure 4*e*). Most grain edges also show evidence of subsequent rounding, probably in an aqueous environment (Figure 4*f*). Similar surface modification was observed on grains from the upper deltaic unit of the cores (Figure 4*a*). The aqueous and diagenetic overprint suggests that although the quartz grains initially may have received their gross morphology as a result of glacial processes, subsequent transport, deposition, and burial in an aqueous environment have led to their observable surface morphology.

The various features of the lower unit collectively suggest that it is dominantly marine in origin but that nearby glaciers in the Transantarctic Mountains contributed to the sediment influx. A fjord environment would be consistent with the available evidence and with *Webb and Wrenn's* [1975] tentative correlation of foraminifera from the 153- to 155-m level in the cores with the Pecten gravel fauna of adjacent Wright Valley, to which a fjord origin has been ascribed [*Webb*, 1972, 1974]. The top of the oxidation zone at 154 m could represent a significant unconformity, possibly of regional extent, in which case the observed oxidation might represent an interval of subaerial exposure and weathering prior to the accumulation of the diamicton-rich middle unit of the cores.

Commonwealth Glacier Core

The upper 128 m of DVDP 11 consists largely of stratified sand and gravel and locally has dips of 20° or more (Figure 3). Reworked marine diatoms have been found throughout this unit, and in situ nonmarine diatom floras have been identified in the upper 32 m of the core as well as between about 90 and 115 m [*Brady*, 1979, 1980]. The absence of diamictons and of in situ marine fossils in this section suggests that the bulk of the sediment may have been deposited in a nonmarine environment, most likely under lacustrine and/or fluvial conditions. The diatom floras further suggest deposition of at least part of the section in a lake that presumably was dammed by glacier ice in the Ross Sea area, as is inferred from the presence of clasts of the McMurdo volcanic group within this part of the core.

Five diamictons occur between 128 and 205 m

and reflect the incursion of either floating or grounded ice across the drill site. No in situ diatom floras have been found within this part of the section, but reworked marine diatoms of unknown age and foraminifera of Pleistocene age were observed. These fossils, together with the persistent occurrence of rock fragments from the McMurdo volcanic group, imply that the source of much of the sediment in the diamictons was ice from the Ross Sea region. Weak fabrics in the two uppermost diamictons suggest that they may have been deposited by floating ice, but the low percentage of McMurdo volcanic group clasts in these units may mean that the bulk of the sediment came from local alpine glaciers that calved either in a marine embayment or in a proglacial lake impounded beyond a Ross Sea glacier which was the source of the recognized McMurdo volcanic detritus. *Webb and Wrenn* [1980] reported a possible in situ marine littoral foraminifera fauna at a depth of 132.8 m, which suggests that the diamicton immediately above that level may be marine. Although a massive diamicton between depths of about 145 and 167 m has a high component of sediment derived from the McMurdo volcanic group, its microfabric is rather weak, and so it may record incursion of bergs or shelf ice from the Ross Sea. A questionable in situ marine fauna ($>26,800$ ^{14}C years old) was found by *Webb and Wrenn* [1980] in a gravel just below the thick diamicton, indicating possible marine sedimentation prior to the advance of ice that was responsible for the diamicton. Lithology and fabric in the two diamictons between 188 and 201 m suggest deposition by floating ice of Ross Sea origin.

A significant lithologic change is inferred to lie close to a depth of 200 m, for no rocks of the McMurdo volcanic group were found below that level. This is close to the depth where *Webb and Wrenn* [1980] and *Brady* [1980] place a major unconformity on faunal evidence (202 m) and floral evidence (195 m), respectively. *Elston et al.* [1978] reported a major change in magnetic intensity at about the same level (208 m). Fossils below the inferred unconformity are early Pliocene to late Miocene in age. Two diamictons between 188 and 201`m lie above the inferred lithologic change; the lower lies within a diatom zone of early Pliocene age [*Brady*, 1979], whereas the upper lies within an interval that is barren of diatoms [*Brady*, 1979] but contains foraminifera assigned by *Webb and Wrenn* [1980] to the Pleistocene.

Numerous diamictons below the unconformity lack clasts of the McMurdo volcanic group and contain rich in situ diatom floras and foraminifera which indicate that marine water probably was at least 600 to 900 m deep [*Webb and Wrenn*, 1980; *Brady*, 1979]. The strongest fabrics within this core were found at 246, 309, and 323 m and could reflect grounded glacier ice. If the water depth inferred from the fossil evidence is correct, then any grounded ice must have been of at least comparable thickness, implying that major trunk glaciers, presumably fed either from an expanded ice sheet or from local glaciers, reached the drill site during the late Miocene and early Pliocene.

IMPLICATIONS FOR LATE CENOZOIC GLACIAL HISTORY OF ANTARCTICA

Sedimentological analyses of the New Harbor and Commonwealth Glacier cores have shown that the gross units discernible in the recovered sections reflect major changes in depositional environment and sediment provenance that are related to late Cenozoic glacial events in Taylor Valley and vicinity. Rocks of the McMurdo volcanic group from the Ross Sea archipelago are absent below an unconformity at about the 154-m level in DVDP 8 and 10 and at about 205 m in DVDP 11, implying either (1) that those rocks had not yet been extruded or had not been extruded in sufficient volume to form a detectable component of glacially derived exotic rock fragments in the cores or (2) that neither an ice shelf nor a grounded glacier in the Ross Sea region contributed sediment to the New Harbor area prior to the time represented by the unconformity. Although the bulk of Mount Erebus on Ross Island may be less than about 0.7 m.y. old [*Forbes et al.*, 1974], the oldest dated rocks from the island are reportedly about 4.5-5 m.y. old [*Armstrong*, 1981; *Northey et al.*, 1975]; McMurdo volcanic rocks farther south at Mount Morning and at Black Island, both potentially in the path of expanding Ross Sea glacier ice, may be as old as 15.4 and 10.9´ m.y., respectively [*Armstrong*, 1981]. Consequently, it is reasonable to infer that any late Miocene, Pliocene, and Pleistocene glacier ice that entered New Harbor from the southern Ross Sea region could have transported volcanic detritus to the mouth of Taylor Valley. Because volcanic-rich sediments of Ross Sea provenance occur only above the unconformity at 154 m in DVDP 8 and 10 and at 205 m in DVDP 11, Ross Sea glaciations that affected lower Taylor Valley may all be younger than the unconformity. On the basis of diatom zonation, *Brady* [1979] assigned an age of about 4.6 m.y. to this level in the cores, whereas *Elston et al.* [1978], using paleomagnetic results, inferred the age of the unconformity to be about 5 m.y. These ages coincide approximately with the time when volcanic activity began on Ross Island. Therefore one cannot rule out on lithologic evidence from the drill cores the possibility that lower Taylor Valley experienced pre-Erebus Ross Sea glacier advances (>5 m.y. old) which, because of the prevailing direction of ice flow, failed to transport older McMurdo volcanic group rock fragments from the vicinity of Mount Morning and Black Island into the dry valleys. At the same time, however, the lithologic evidence does not preclude the possibility that the mapped Ross Sea drifts in the dry valleys region are younger than about 1.2 m.y., as has been inferred from field evidence and radiometric dates by *Denton et al.* [1971] and *Forbes et al.* [1974].

The sediments near the top of the New Harbor cores are postglacial, as is indicated both by surface mapping and by radiometric dating [*Stuiver et al.*, 1976a], and record the last phase in recession of Ross Sea I ice from lower Taylor Valley. Radiocarbon dates of algae indicate that the maximum Ross Sea I advance probably culminated about 18,000-19,000 years ago [*Kellogg et al.*, 1978; G. H. Denton, personal communication, 1978]; the ice had receded to a position at or close to the drill site by about 8200 years ago. A radiocarbon date of 5820 ± 200 years (QL-191) on *Adamussium* shells from a depth of 23 m in cores 8 and 9 indicates that postglacial deltaic sedimentation took place rapidly after the retreating ice shelf had cleared the New Harbor drill site [*Stuiver et al.*, 1976a, b].

The age of the sediments between depths of about 25 and 154 m in DVDP 8 and 10 is more difficult to assess. Although impoverished Pleistocene foraminiferal faunas occur above 125 m, richer collections have been obtained between 125 and 154 m (P. N. Webb, personal communication, 1976), suggesting that this entire part of the section is of Pleistocene age and therefore less than about 1.8 m.y. old; *Brady* [1979], however, infers an early Pliocene age for sediments below 137 m on diatom evidence. But do all these sediments represent the several Ross Sea drifts mapped in Taylor Valley and elsewhere in the McMurdo Sound region by *Denton et al.* [1969, 1970, 1971]? Clasts of anortho-

clase phonolite derived from Mount Erebus are a common component of Ross Sea moraines in the dry valleys area, but no unequivocal clasts or fragments of this lithology were recognized during the lithic analyses of the lower Taylor Valley cores. The oldest phonolite flows that have been dated on the volcano crop out at Cape Barne and are 0.94 ± 0.05 m.y. old [*Armstrong*, 1981]. The apparent absence of this rock type in the cores may indicate that the volcanic-rich section between about 25 and 154 m possibly predates the phonolite eruptions and that it is more than 1 m.y. old, in which case the diamictons in the middle parts of the New Harbor cores could represent early Ross Sea glaciations that are not represented by sediments currently exposed at the surface in lower Taylor Valley. However, such an interpretation is based on negative evidence. Furthermore, the phonolite was not detected either in the uppermost 25 m of the cores, which are postglacial on the basis both of radiocarbon dating and stratigraphic relationships near the drill site, and therefore should include this lithology among those reworked from Ross Sea I drift on adjacent slopes (Figure 1). Consequently, a definitive age for the diamicton-rich middle unit of the cores cannot be derived on the basis of available lithologic data.

Although a glacial influence is apparent in the sediments of the lower part of the cores, the bulk of the section below 154 m in DVDP 8 and 10 and below about 205 m in DVDP 11 appears to have been deposited under marine conditions and probably in a fjord environment. Paleontologic data from these sediments suggest that a faunal break occurs at about 172-m depth in DVDP 10, below which foraminiferal assemblages are late Miocene (?) in age [*Webb and Wrenn*, 1976; P. N. Webb, personal communication, 1976]; *Brady* [1979], however, interprets diatom evidence to indicate that the lowest sediments are early Pliocene in age. Most sand grains examined from below that level are rounded, and few show the typical conchoidal fractures suggestive of glacial erosion and transport. This may indicate that eolian and fluvial activity contributed more sediment to the marine environment than did glaciers during the earliest part of the Pliocene or the late Miocene, whereas during subsequent intervals the glacial component was predominant.

Although it was anticipated that a more complete depositional record of Quaternary Ross Sea glaciations would be penetrated in the DVDP cores than is exposed at the surface of Taylor Valley, the present assessment of the stratigraphy and chronology suggests in fact that the Quaternary section may have some significant hiatuses and that some of the latest invasions of Ross Sea glacier ice, which are documented in lower Taylor Valley by surface moraines, may not be represented in the dry valley cores.

Acknowledgments. This research was carried out as part of the Dry Valley Drilling Project with the support of the Office of Polar Programs, National Science Foundation (grant OPP 73-05834). Support from the Antarctic Division, New Zealand Department of Scientific and Industrial Research, and the personnel of Scott Base during the 1973-1974 field season is gratefully acknowledged, as is the assistance in core sampling provided by Dennis Cassidy of the Antarctic Core Storage Facility, Florida State University. We thank Yorko Tsukada of the Quaternary Research Center for her help in preparation and photography of SEM samples. We also thank George H. Denton, Minze Stuiver, and Peter N. Webb for useful comments and discussions during the preparation of this paper and for reviewing a preliminary draft of the manuscript.

REFERENCES

Armstrong, R. L., K-Ar dating: Late Cenozoic McMurdo volcanic group and dry valley glacial history, Victoria Land, Antarctica, *N. Z. J. Geol. Geophys.*, in press, 1981.

Brady, H. T., The dating and interpretation of diatom zones in Dry Valley Drilling Project holes 10 and 11, Taylor Valley, South Victoria Land, Antarctica, in *Proceedings of the Seminar III on Dry Valley Drilling Project, 1978, Mem. Spec. Issue 13*, edited by T. Nagata, pp. 150-164, National Institute of Polar Research, Tokyo, 1979.

Brady, H. T., Late Neogene history of Taylor and Wright valleys and McMurdo Sound, derived from diatom biostratigraphy and paleoecology of DVDP cores, in *Antarctic Geoscience*, edited by C. Craddock, University of Wisconsin Press, Madison, in press, 1980.

Calkin, P. E., and R. L. Nichols, Quaternary studies in Antarctica, in *Antarctic Geology and Geophysics*, edited by R. J. Adie, pp. 625-643, Universitetsforlaget, Oslo, 1972.

Chapman-Smith, M., and P. G. Luckman, Late Cenozoic glacial sequence cored at New Harbor, Victoria Land, Antarctica (DVDP 8 and 9), *Dry Val. Drilling Proj. Bull., 3*, 120-118, 1974.

Curray, J. R., The analysis of two-dimensional orientation data, *J. Geol., 64*, 117-131, 1956.

Denton, G. H., R. L. Armstrong, and M. Stuiver, Histoire glaciaire et chronologie de la région du Detroit de McMurdo, Sud de la Terre Victoria, Antarctide: Note préliminaire, *Rev. Geogr. Phys. Geol. Dyn., 11*, 265-278, 1969.

Denton, G. H., R. L. Armstrong, and M. Stuiver, Late Cenozoic glaciation in Antarctica: The record in the McMurdo Sound region, *Antarct. J. U. S., 5*, 15-21, 1970.

Denton, G. H., R. L. Armstrong, and M. Stuiver, The late Cenozoic glacial history of Antarctica, in *The Late Cenozoic Glacial Ages*, edited by K. K. Turekian, pp. 267-306, Yale University Press, New Haven, Conn., 1971.

Denton, G. H., H. W. Borns, Jr., M. G. Grosswald, M. Stuiver, and R. L. Nichols, Glacial history of the Ross Sea, *Antarct. J. U.S., 10*, 160-164, 1975.

Elston, D. P., M. E. Purucker, S. L. Bressler, and H. Spall,

Polarity zonations, magnetic intensities, and the correlation of Miocene and Pliocene DVDP cores, Taylor Valley, Antarctica, *Dry Val. Drilling Proj. Bull., 8*, 12-15, 1978.

Evenson, E. B., The relationship of macro- and microfabric of till and the genesis of glacial landforms in Jefferson County, Wisconsin, in *Till: A Symposium*, edited by R. P. Goldthwait, pp. 345-364, Ohio State University Press, Columbus, 1971.

Forbes, R. B., D. L. Turner, and J. R. Carden, Age of trachyte from Ross Island, Antarctica, *Geology, 2*, 297-298, 1974.

Haskell, T. R., J. P. Kennett, W. M. Prebble, G. Smith, and I. A. G. Willis, The geology of the middle and lower Taylor Valley of South Victoria Land, Antarctica, *Trans. Roy. Soc. N. Z., 2*, 169-186, 1965.

Hayes, D. E., et al., Leg 28 deep-sea drilling in the southern ocean, *Geotimes, 8*(6), 19-24, 1973.

Kellogg, D., M. Stuiver, G. Denton, and T. Kellogg, Freshwater diatoms from perched deltas in Taylor Valley, Antarctica, *Antarct. J. U. S., 13*(4), 26-27, 1978.

Krinsley, D. H., and J. C. Doornkamp, Atlas of Quartz Sand Surface Textures, 91 pp., Cambridge University Press, Cambridge, Mass., 1973.

Krumbein, W. C., Preferred orientation of pebbles in sedimentary deposits, *J. Geol., 37*, 673-706, 1939.

McGinnis, L. D., McMurdo Sound—A key to the Cenozoic of Antarctica, *Antarct. J. U.S., 8*, 166-169, 1973.

McGinnis, L. D., T. Torii, and P. N. Webb, Dry Valley Drilling Project, *Antarct. J. U.S., 7*, 53-56, 1972.

McGinnis, L. D., T. Torii, and R. Clark, Antarctic Dry Valley Drilling Project: Report on seminar 1, *Eos Trans. AGU, 56*, 217-220, 1975.

McKelvey, B. C., DVDP sites 10 and 11, Taylor Valley, *Dry Val. Drilling Proj. Bull., 5*, 16-60, 1975.

Northey, D. J., C. Brown, D. A. Christoffel, H. K. Wong, and P. J. Barrett, A continuous seismic profiling survey in McMurdo Sound, Antarctica, 1975, *Dry Val. Drilling Proj. Bull., 5*, 167-179, 1975.

Ostrey, R. C., and R. E. Deane, Microfabric analyses of till, *Geol. Soc. Amer. Bull., 74*, 165-168, 1963.

Powell, R. D., and P. J. Barrett, Grain size distribution of sediments cored by the Dry Valley Drilling Project in the eastern part of Taylor Valley Antarctica, *Dry Val. Drilling Proj. Bull., 5*, 150-166, 1975.

Stuiver, M., G. H. Denton, and H. W. Borns, Carbon-14 dates of *Adamussium colbecki* (Mollusca) in marine deposits at New Harbor, Taylor Valley, *Antarct. J. U.S., 11*(2), 86-88, 1976a.

Stuiver, M., I. C. Yang, and G. H. Denton, Permafrost oxygen isotope ratios and chronology of three cores from Antarctica, *Nature, 261*, 547-550, 1976b.

Webb, P. N., Wright Fjord, Pliocene marine invasion of an Antarctic valley, *Antarct. J. U.S., 7*, 227-234, 1972.

Webb, P. N., Micropaleontology, paleoecology, and correlation of the Pecten gravels, Wright Valley, Antarctica, and description of *Trochoelphidiella onyxi* n. gen., n. sp., *J. Foraminif. Res., 4*, 184-199, 1974.

Webb, P. N., and J. H. Wrenn, Foraminifera from DVDP holes 8, 9, and 10, Taylor Valley, *Antarct. J. U.S., 10*, 168-169, 1975.

Webb, P. N., and J. H. Wrenn, Foraminifera from DVDP holes 8-12, Taylor Valley, *Antarct. J. U.S., 11*, 85-86, 1976.

Webb, P. N., and J. H. Wrenn, Late Cenozoic micropaleontology and biostratigraphy of eastern Taylor Valley, Antarctica, in *Antarctic Geoscience*, edited by C. Craddock, University of Wisconsin Press, Madison, in press, 1980.

THE TAYLOR FORMATION (HOLOCENE) AND ITS MACROFAUNAS, TAYLOR DRY VALLEY, ANTARCTICA

M. CHAPMAN-SMITH[1]

Antarctic Division, Department of Scientific and Industrial Research
Christchurch, New Zealand

Descriptions are given of macrofaunas that were collected from the New Harbor area at the entrance to the Taylor Valley during the drilling of Dry Valley Drilling Project holes 8 and 9. Small collections by previous workers were reviewed by *Speden* [1962], who ascribed them to fossiliferous sediments of the Taylor Formation. Although the faunas here described comprise the largest macrofossil collections yet made from Cenozoic deposits of the Dry Valley region of Antarctica, they are sparse, with the pectinid *Adamussium colbecki* (Smith, 1902) being the predominant taxon. A number of paleoecological and systematic paleontological observations are presented. Six molluscan species are recorded for the first time from the New Harbor deposits. It is believed that this is the first record of the gastropod taxa *Subonoba contigua* Powell, 1958, *Lorabela plicatula* (Thiele, 1912), and *Philine apertissima* Smith, 1902 having been found fossilized. Faunal elements are analyzed, and a number deduced to be in situ. All species are extant. From consideration of modern habitats it is deduced that the New Harbor faunas lived in a shallow water environment some 5 to 35 m below sea level. Radiocarbon age determinations of valves of *Adamussium colbecki* (Smith, 1902) from the New Harbor deposits are discussed. It is suggested that the lithological sequence in which the faunas are found was deposited in two differing sedimentary regimes. Deposition appears to have taken place during the formation of stranded beach deposits and during shallow water conditions at the head of an ice-covered fjord. *Speden* [1962] erected the Taylor Formation for sediments containing the New Harbor faunas. Consideration of the original description, type locality, type faunas, and correlatives of the Taylor Formation, together with reference to subsequent stratigraphic nomenclature for lithologies outcropping in the lower Taylor Valley, leaves little doubt that this stratigraphic unit requires redefinition. It is concluded that the Taylor Formation has precedence over alternative names and that it is a valid stratigraphic unit which should be redefined by using information obtained by the Dry Valley Drilling Project.

INTRODUCTION

During the 1973/1974 austral summer the Dry Valley Drilling Project (DVDP) completed two shallow holes (DVDP 8 and 9) at New Harbor at the mouth of the Taylor Valley (Figures 1 and 2). Detailed and summary logs of the vertical hole DVDP 9, which intersected intervals of sparsely macrofossiliferous lithologies, were prepared in the Thiel Earth Sciences Laboratory at McMurdo Station [*Chapman-Smith and Luckman*, 1974].

Macrofaunas from the Taylor Valley had been collected previously by three workers: R. E.

Priestley and B. Armytage in 1908 and W. Romanes in 1958 [*Speden*, 1962]. Priestley and Armytage regarded the few specimens they collected from close to the New Harbor shoreline as evidence of past 'raised beaches' [*Shackleton*, 1909; *David and Priestley*, 1914]. Like many later authors [e.g., *Nicols*, 1961a; *Speden*, 1962; *Claridge and Campbell*, 1966], Priestley and Armytage drew attention to the widespread occurrence, throughout the McMurdo Sound region, of 'raised' fossiliferous marine deposits, and much debate has taken place over their origin [*David and Priestley*, 1914; *Chapman*, 1916; *Debenham*, 1920; *Nicols*, 1961a, b]. Correlatives of these deposits have been

---

[1]Now at P. O. Box 28-318, Auckland 5, New Zealand.

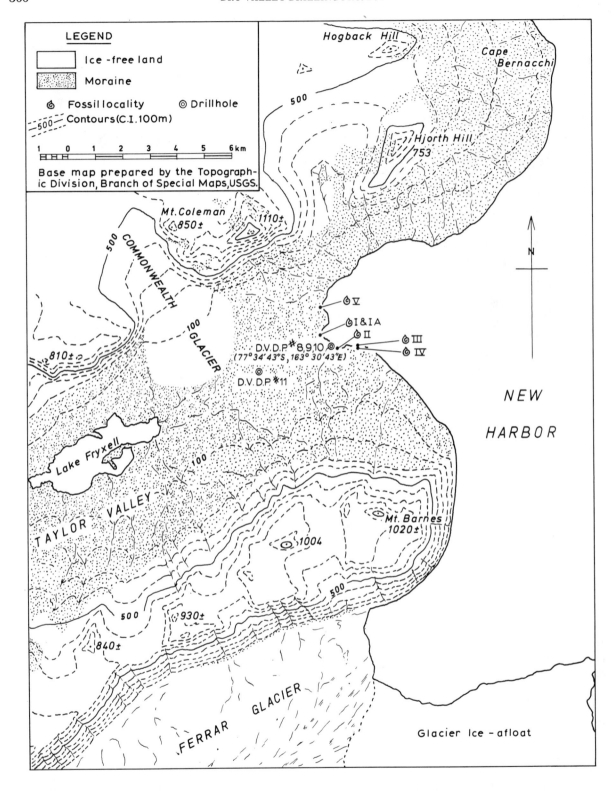

Fig. 1. Locality map of lower Taylor Valley area showing positions of fossil localities and of DVDP drill holes 8, 9, 10, and 11.

Fig. 2.   View south from Mt. Coleman across Taylor Valley to the slopes of Mt. Barnes. The New Harbor site of DVDP holes 8, 9, and 10 is indicated by the arrow. The terminal face of the Commonwealth Glacier is 800 m below the author in the right foreground.

described from Princess Elizabeth Land [*Crespin,* 1960] and from Wilkes Land [*Cameron et al.,* 1959].

Speden [1962] published an extensive review of all the then known fossiliferous Quaternary deposits of the McMurdo Sound region and divided them into two new formations, each with a characteristic faunal assemblage. The older fauna is characterized by the extinct *Chlamys (Zygochlamys) anderssoni* (Hennig) which occurs in cemented tuffaceous sandstones, grits, and conglomerates of the Scallop Hill Formation. The younger Last Interglacial and Postglacial assemblage is characterized by the extant *Adamussium colbecki* (Smith) found amongst unconsolidated marine sediments of the Taylor Formation. Although he was unable to visit New Harbor, Speden designated the type locality for the Taylor Formation as that locality near the south side of the mouth of Taylor Valley where in 1908 Priestley and Armytage had collected *Adamussium colbecki* and other species. Thus the locality is obviously in close proximity to the DVDP's New Harbor site at the mouth of Wales Stream.

As the drill penetrated fossiliferous lithologies in DVDP 8, the author decided it was an opportune time to re-collect and examine macrofossils from the Taylor Formation. A number of previously unrecognized fossil localities were found, and a considerable number of specimens collected. The heights of the fossil localities above high tide mark (HTM) were measured with the aid of a 1.5–m Jacob's staff.

Subsequent intensive investigation of the diamond drill cores recovered by the Dry Valley Drilling Project has revealed considerable information about the micropaleontology and paleoecology of the Lower Taylor Valley deposits [*Brady,* 1977; *Brady and Webb,* 1976; *Webb and Wrenn,* 1976, 1977; *Wrenn and Webb,* 1976; *Wrenn,* 1977], but apart from the present study the only known recent macropaleontological work in the area has been a series of ^{14}C dates of fossil specimens of *Adamussium colbecki* from the New Harbor marine deposits [*Stuiver et al.,* 1976]. Other minor references to the macropaleontology of the lower Taylor Valley area are in the work of *McKelvey* [1975a, b, 1976], who recorded bivalve debris down

Fig. 3.   Fossil locality IV. Superficial lag gravel has been removed showing valves of *Adamussium colbecki*.

to 24.1 m below the surface in the summary log of DVDP 10 at New Harbor, and of *Wrenn and Webb* [1976], who found macrofossil debris consisting of teleost scales and bones, echinoderm spines and ossicles, and molluscan fragments in core samples from DVDP holes 8, 9, 10, 11, and 12.

DESCRIPTION OF FOSSIL LOCALITIES

Priestley and Armytage (cited by *Shackleton* [1909, p. 326] and *David and Priestley* [1914, p. 273]) stated they found 'a raised beach. . .near New Harbor [which] consisted of dark gritty sand containing numerous entire valves of *Adamussium colbecki* up to elevations of about 50 feet [15 m] above sea level, and at a distance of several hundreds of yards [say, 275 m] inland from the present shore' (brackets are by present author). The writer searched the New Harbor coastline from the mouth of the Ferrar fjord beneath Mt. Barnes in the south, northward to below Hjorth Hill and found many fossil localities. However, no significant numbers of fossils were found higher than 7.6 m above HTM or further inland than 75 m from HTM. The localities from which the author made collections include the most fossiliferous seen within several kilometers of the DVDP New Har-

bor drill site, and also one which was the highest above HTM and farthest inland. As stated earlier, Priestley and Armytage's locality is likely to be close to the DVDP drill site, near the mouth of the large Wales Stream which debouches into New Harbor. Here there is a considerable distance (about 110 m) between the high and low tide marks, and the anomalies between the writer's measurements and Priestley and Armytage's statements might be partially explained if the latter took the low tide mark as their datum. It is interesting to note that of the 16 localities from which *Stuiver et al.* [1976] collected *Adamussium colbecki* for ^{14}C dating only two were higher 5.7 m above HTM (at 7.5 m and 8.1 m).

All localities are distinguished by the appearance of fragile brown-red valves of *Adamussium colbecki* protruding through a superficial lag gravel 1-3 cm in thickness (Figure 3). The positions of the fossil localities are shown in Figures 1 and 2.

Localities I and I(a)

These localities are 780 m northwest of DVDP 8 and 17 m south of a small stream. Locality I is 2.0-2.3 m above HTM and 24-28 m inland from HTM. Locality I(a) is 2.9-4.9 m above HTM and 2.8-6.0 m inland from HTM. The locality is distinguished by *Adamussium colbecki* bands and lenses at 2.0-2.3 m, 2.9-3.1 m, 4.4 m, and 4.9 m above HTM.

Locality II

This locality is 250 m southeast of DVDP 8, 0.0-1.2 m above HTM, and extends from HTM inland for 20 m. The locality is approximately 60 m long and contains a lens of *Adamussium colbecki* valves.

Localities III and IV

These localities are 1 km east of DVDP 8. Locality III is 0.0-3.75 m above HTM and from HTM inland from 0.37 m. Locality IV is 7.4-7.6 m above HTM and 70-75 m inland from HTM. Locality IV appears to be the tread of an old beach level. At locality IV the fossiliferous beds are overlain by moraine at about 10 m above HTM.

Locality V

This locality is 1.8 km north-northwest of DVDP 8 in a small bay and extends 10 m either side of the small stream. The locality is 0.0-4.5 m above HTM.

Lithological Description

The fossils occur in lithologies identical to those described by *Chapman-Smith and Luckman* [1974] in DVDP drill holes 8 and 9. They are grayish olive to olive gray unconsolidated medium and coarse sands which are moderately well to well sorted with subrounded to well rounded (average rounded) grains and numerous very thin ice lenses. Grain sizes vary from 0.2 to 20.0 mm (scattered granules) but average 0.3 mm. Grains are composed of mud clasts, quartz, feldspar, biotite, olivine, granite, and weathered basalt. No bedding was discernible at the fossil localities; however, the bands and lenses of *Adamussium colbecki* indicate that the deposits are subhorizontal.

FAUNAL LIST OF NEW HARBOR COLLECTIONS

Table 1 is a faunal list of New Harbor collections. The following points should be noted.

1. Collections from localities I, I(a), II, III, IV, V, DVDP 8 and 9, and New Harbor Campsite are from fossiliferous Holocene beach deposits and were made by the writer. J. H. W. Wrenn helped collect from localities I and II; K. Watanuki and R. G. Smith also assisted at locality II.

2. Specimens recorded from New Harbor Shoreline are recent and collected by the writer and J. H. W. Wrenn.

3. The New Harbor Campsite locality is that used by the Dry Valley Drilling Project as a campsite during the drilling of holes 8 and 9.

4. *Speden* [1962, p. 749] recorded three assemblages from the New Harbor area. They are listed here under their original collectors: Priestley, Armytage, and Romanes but with Speden's locality numbers. The writer believes that many of the specimens found by these earlier workers were recent ones that did not come from the Holocene beach deposits. His main reasons for such a belief are first that the Romanes collection was examined and contained shell material that has not been fossilized and second that the fauna collected by Priestley and Armytage and shown in Figure 1, Plate 89 of *David and Priestley* [1914] is typical in composition and state of preservation with specimens that can be collected from the New Harbor shoreline but unlike that of the fossil deposits (see also discussion section under the Taylor Formation as a stratigraphic unit).

PALEONTOLOGICAL AND PALEOECOLOGICAL NOTES

Phylum **MOLLUSCA**
Class **Bivalvia**
Adamussium colbecki (Smith, 1902)

Pecten colbecki Smith, 1902, p. 212, pl. 30, fig. 11; 1907a, p. 6, pl. 3, figs. 9, 9a—Hedley, 1911, p. 3.

Pecten racovitzai Pelseneer, 1903, p. 27, pl. 8, figs. 101, 102.

Chlamys colbecki Smith, 1915, p. 77.

Adamussium colbecki Thiele, 1936, p. 807.—Soot-Ryen, 1951, p. 16.—Powell, 1958, p. 176; 1960, p. 175.

Smith [1902] erected the species on the basis of a single right valve discovered near Franklin Island. Later he described and figured the previously unknown left valve, but his specimens appear to have been juveniles [*Smith*, 1907a]. *Hedley* [1911] recorded the species living from 4 to 22 m off Cape Royds. *Soot-Ryen* [1951] described the animal, gave a much more detailed shell description includ-

TABLE 1. Faunal List of New Harbor Collections

	Localities											
	I	IA	II	III	IV	V	New Harbor Campsite	DVDP 8, 9	New Harbor Shoreline	Priestley and Armytage, 2*	Priestley and Armytage, 15*	Romanes, 31*
Bivalvia												
Adamussium colbecki (Smith, 1902)	37p, fa	39,p fc	57p, fa	31p, f	61p, fa	24p, f	1f	p, f	92p, f	x		x
Thracia meridionalis Smith, 1885	158p		2	10p	3p	1						
Laternula elliptica (King and Broderip, 1831)	f		f	fc		fa		?p	1f	?		
Lima sp.										x		
Bivalvia indet.		x						x				
Gastropoda												
Subonoba contigua Powell, 1958		2										
Neobuccinum eatoni (Smith, 1875)	3		2f		1f	1						
Lorabela plicatula (Thiele, 1912)			1									
Philine apertissima Smith, 1902	6	1	1		1	sp, 2f						
Echinodermata												
Sterechinus neumayeri (Meissner, 1900)	fc	fa	fc	fc	fa	fc	fa	?	x			?
Odontaster validus Koehler, 1906												x
Psilaster charcoti (Koehler, 1906)									x			
Ophiuroidea—ossicles	10	78	8		1		21					
Annelida												
Spirorbis sp.		3							x			
Serpulidae indet.		1										
Arthropoda												
Pycnogonida	?f								x	x	x	
Ostracoda								x				
Amphipoda									x		x	
Vertebrata												
Teleost—vertebrae, bones					x		x				x	
Pygoscelis adeliae—bones									x			
Cetacea—tooth									x			
Pinnipedia—bones									x			
Pinnipedia—teeth										x		
Foraminifera												
Undifferentiated	x	x	x	x	x	x	x	x				

Numbers, say, 18, are the number of specimens collected; p is presence of paired values; f is fragments, uncommon; fc is fragments, common; fa is fragments, abundant; x is present, undifferentiated.
*Snoden's locality numbers.

ing sculptural variation, and noted that living spec-imens could be obtained from circum-Antarctic wa-ters from 4 to 700 m.

Shells of *Adamussium colbecki* are the dominant element of the New Harbor fossil localities. Bro-ken valves of *A. colbecki* mantle the outcrops, making them visible amidst a thin cover of superfi-cial lag gravel (Figure 3). About 250 fossilized valves were collected. The number of left and right valves is nearly identical, and articulated speci-mens are present at all six main localities and at 24.32 m in the core of DVDP 8 [*Chapman-Smith and Luckman*, 1974], strongly suggesting that the valves are in situ.

The sea ice adjacent to the New Harbor shore-line commonly contains specimens of *A. colbecki* that have their soft parts frozen inside the valves. The shells have been rafted by the ice from their benthic habitat to the surface. This phenomenon is thought to explain adequately the presence of this subtidal benthic species on the upper surface of the ice; specimens become trapped in the ice on the sea floor and pass up through the ice body by ablation on its upper surface and continued freezing at its base [*Debenham*, 1920]. A skua gull was observed to feed by landing beside an *A. colbecki* ablating from the surface of the pack ice, pecking a hole in the upper valve and devouring the contents. In fact, along the New Harbor shoreline, articulated specimens of *A. colbecki* with one valve broken are quite common. Despite this, 31 unbroken, articu-lated recent specimens were collected (Figure 4). At New Harbor today, living *Adamussium colbe-cki* has been observed to have an upper bathymet-ric limit of 4 m and to prefer depths of about 25 m (Oliver, cited by *Stuiver et al.* [1976]).

Thracia meridionalis Smith, 1885

Thracia meridionalis Smith, 1885, p. 68, pl. 6, figs. 4-46; 1907a, p. 1.—Hedley, 1911, p. 3.—Soot-Ryen, 1951, p. 21.—Powell, 1960, p. 184 (full synonymy included).

Of the 173 valves collected, 158 came from fossil locality I. *Smith's* [1885] description accords with the valves collected, but his figures do not. How-ever, *Smith* [1907a] noticed that the shell form is variable. *Soot-Ryen* [1951] compared the sculpture of *T. meridionalis* with that of *T. convexa* and its outline with *T. devexa*. *Hedley* [1911] reported that at Cape Royds the depth range for *T. meridionalis* is 18-146 m. *Powell* [1960] recorded the distribution

of the species as circum-Antarctic and from 18 to 594 m.

The presence of articulated individuals at three fossil localities and the abundance of valves (includ-ing all growth stages) are taken as evidence that the species lived in situ at these localities. About 10 of the valves are pierced by circular conical holes which are thought to have been drilled by a carnivorous gastropod such as *Trophon* or *Neobuc-cinum*. Whereas at least two species of *Trophon* are common in McMurdo Sound, none was found fossil. On the other hand, *Neobuccinum* was col-lected at four of the fossil localities.

Laternula elliptica (King and Broderip, 1831)

Anatina elliptica King and Broderip, 1831, p. 335 (unseen).—Smith, 1902, p. 210, pl. 25, figs. 9, 10; 1907a, p. 1, pl. 3, fig. 3; 1915, p. 78.—Hed-ley, 1911, p. 3—Lamy (for detailed synonymy, see *Smith* [1915]).—Burne, 1920, p. 249, pl. 4, figs. 20-23, 25.
Laternula elliptica Powell, 1960, p. 184 (useful synonymy included).

Abundant broken shells and shell fragments of *Laternula elliptica* are present at the New Harbor localities. Not one unbroken specimen was found. The morphology of the specimens collected at New Harbor fits well with that shown in *Smith's* [1902] figures. However, *Smith* [1907a] noted and figured variations in the form of the species. *Burne* [1920] described and figured in detail the anatomy of *L. elliptica* taken from 9 m off Cape Evans, compar-ing it with *L. truncata*, *L. subrostrata*, and *L. flex-uosa*. *Hedley* [1911] recorded abundant specimens of *L. elliptica* from 11 to 55 m off Cape Royds.

As *Powell* [1965] noted, *Laternula* is a relatively large thin-shelled bivalve which burrows deeply into soft mud, and this insulating factor is probably the main reason why this warm water genus has managed to adjust itself to a cold water habitat. Unbroken specimens of *L. elliptica* were not found at New Harbor, perhaps because it lives there im-mediately subtidal and has a fragile shell that is easily crushed by the pressures of sea ice stranded along the shoreline at low tides.

Class **Gastropoda**
Subonoba contigua Powell, 1958

Subonoba contigua Powell, 1958, p. 184, pl. 1, fig. 8; 1960, p. 136.

Fig. 4.   Shell morphology of *Adamussium colbecki* [*Smith*, 1902]; internal and external views of both valves of a recent specimen collected from the New Harbor shoreline.

Two specimens of *Subonoba contigua* were collected from locality I(a) and constitute the first fossil record of the species. The writer examined three paratypes of *S. contigua* from Commonwealth Bay in the A. W. B. Powell collection at the Auckland Museum. There is little doubt that the five specimens belong to the same species. However, the Rissoidae are a difficult group for systematists, in that modern workers generally accept wider morphological variation within the same species than would have been deemed prudent by scientists in former times. *Powell* [1960] records no less than 20 species of *Subonoba* in Antarctic and sub-Antarctic waters.

Smith [1907b] erected the species of *Rissoia gelida* (subsequently assigned to *Subonoba*) for specimens found at 22-75 m at Winter Quarters Bay. *Thiele* [1912] figured a specimen of *S. gelida* which has similarity to the paratypes of *S. contigua*. *Hedley* [1916b] recorded that 50 specimens of *S. gelida* were dredged from 46 m and another 9 specimens from 82-91 m in Commonwealth Bay. It is interesting to note that *Hedley* [1916b] erected *S. bickertoni* and *S. wilkesiana* as two new species with type localities not only at Commonwealth Bay but also in the same 82- to 91-m interval in which he recorded *S. gelida*. Further confusion seems probable in that the type locality for *S. contigua* is 150 m in Commonwealth Bay.

Thus the following species of *Subonoba* have been recorded by various workers as being present in Commonwealth Bay: *S. bickertoni*, *S. contigua*, *S. deserta*, *S. gelida*, *S. glacialis*, *S. ovata*, and *S. wilkesiana*. The probability of seven species of the same genus coexisting in the same bay and of three of these species being present at the same depth is remote. A review of the systematics of *Subonoba* is obviously long overdue.

Neobuccinum eatoni (Smith, 1875)

Buccinopsis eatoni Smith, 1875, p. 68.
Neobuccinum eatoni Smith, 1887, p. 169; 1902, p. 202; 1907b, p. 1; 1915, p. 72—Hedley, 1911, p. 6.—Powell, 1960, p. 150 (full synonymy included).

Four shells of *Neobuccinum eatoni*, together with fragments of three other specimens and an operculum, were found fossilized at New Harbor. Despite the fact that only one specimen is unbroken, it is thought that *N. eatoni* lived in situ; the

presence of the operculum inside one shell supports this belief. *Powell* [1960] recorded *N. eatoni* as a wide-ranging circum-Antarctic species found living in depths of 6-600 m. The upper limit of this range is consistent with the depths that are the habitats of other members of the thanatocoenose that are deduced to be in situ. Perhaps *Neobuccinum eatoni* fed on *Thracia meridionalis*, for, as previously stated, a number of valves of this bivalve had holes in them similar to those drilled today by carnivorous gastropods like *Neobuccinum*. This is the first record of *N. eatoni* having been found fossilized.

Lorabela plicatula (Thiele, 1912)

Bela plicatula Thiele, 1912, p. 215, pl. 14, fig. 4.—Hedley, 1916b, p. 54.
Lorabela plicatula Powell, 1951, p. 171; 1958, p. 202; 1960, p. 159.

One specimen of *Lorabela plicatula* with its protoconch abraded was collected from locality II. This is the first fossil record of the species. The shell is very similar to that figured by *Thiele* [1912] from the type locality at Gauss Station in the Davis Sea. However, because *Hedley* [1916b] referred to the close similarity between *L. plicatula* and his *L. davisi*, the writer carefully examined the specimens of species of *Lorabela* in the A. W. B. Powell collection at the Auckland Museum. The Powell collection contains a number of specimens of *Lorabela plicatula*, and the fossil shell falls within the range of sculptural variation shown by these specimens.

L. plicatula ranges from Enderby Land to the Ross Sea at depths at 220-640 m [*Powell*, 1960]. This depth range is inconsistent with that of the species deduced to be in situ in the New Harbor fossil localities, and thus it is considered that this specimen of *Lorabela plicatula* was transported into the thanatocoenose from deeper waters, losing its protoconch en route.

Philine apertissima Smith, 1902

Philine apertissima Smith, 1902, p.208, pl. 24, figs. 2, 3.—Hedley, 1911, p. 8.—Powell, 1960, p. 163.

Smith [1902] based this species on specimens taken from 44 m off Cape Adare. *Hedley* [1911] recorded 20 live specimens dredged from depths of 18-37 m off Cape Royds. The writer collected nine shells and fragments of two others from the New Harbor localities. Only four of the shells were unbroken, and all these are smaller than the specimen found by *Smith* [1902] which measured 8 mm × 6.5 mm. It is believed this fragile thin-shelled infaunal gastropod lived in situ at New Harbor, because none of the specimens show abrasions indicative of transport and it is unlikely that its delicate shell could have survived movement from the biocoenose.

Phylum ECHINODERMATA
Sterechinus neumayeri (Meissner, 1900)

Sterechinus neumayeri McKnight, 1976, p. 4 (full snyonymy included).

Two damaged recent specimens of *Sterechinus neumayeri* were collected from the New Harbor shoreline and identified using *Fell's* [1961] colored plate which shows both aboral and adoral aspects. *Bullivant and Dearborn* [1967] concluded that this species is the only sea urchin taken regularly from near McMurdo Station. *McKnight* [1976] listed its range as circum-Antarctic extending north to South Georgia and in depths of 3-640 m. Throughout the fossil localities at New Harbor it was common to find spines, test fragments, and denticles belonging to *S. neumayeri*. Although such fragments are easily transported, its range and common occurrence in the area today suggest that *Sterechinus neumayeri* was part of the biocoenose.

Odontaster validus Koehler

Odontaster validus Shearburn Clark, 1963, p. 35 (full synonymy included).

Speden [1962] recorded that W. Romanes collected *Odontaster*, probably *O. tenuis* Koehler, on a 'beach' near high tide mark near the center of the mouth of Taylor Valley. The writer reviewed the collection, which consists of three specimens, and concluded that they were *O. validus*. No other specimens have been found at New Harbor.

Psilaster charcoti (Koehler, 1906)

Astropecten charcoti Koehler, 1906, p. 4, pl. 3, figs. 20, 21, 31, 32.

TABLE 2. Age Determinations of Fossil Faunas

Fossil Locality	Altitude, m above HTM	Lab Number	^{14}C date, yr B.P.
II	0.0–1.2	Wk 95	5040 ± 70
IV	7.4–7.6	Wk 96	5330 ± 80

Psilaster charcoti Shearburn Clark, 1963, p. 30, pl. 3, figs. 7, 8.

One arm of the asteroid *Psilaster charcoti* was found on the New Harbor shoreline. *Shearburn Clark* [1963] considered *P. charcoti* to live in a bathymetric range of 30-3248 m, and both she and *Bullivant and Dearborn* [1967] believed *P. charcoti* to be a common inhabitant of McMurdo Sound. After identification the specimen was disaggregated with H_2O_2 so that the resistant components could be compared with echinoderm fragments found in the fossil localities. No fragments of *P. charcoti* were found fossilized—perhaps because the normal bathymetric habitat of *P. charcoti* is deeper than that inferred for the biocoenose represented by the New Harbor fossilized deposits.

Phylum ARTHROPODA
Class Cirrepedia
Bathylasma corolliforme (Hoek, 1883)

Balanus corolliformis Hoek, 1883, p. 155, pl. 6, figs. 21, 23, pl. 13, figs. 1-7 (unseen). *Hexelasma corolliforme* Hoek, 1913, p. 245 (unseen).
Hexelasma antarcticum Borradaile, 1916, p. 132.—Speden, 1962, p. 749, fig. 14.
Bathylasma corolliforme Newman and Ross, 1971, pp. 143-149, pls. 15-22, 43A (full synonymy included).

No specimens of *B. corolliforme* were found fossil at New Harbor, and none have been recorded in previous collections from the area. However, *Speden* [1962] recorded *B. corolliforme* or Cirrepede plates from 12 Quarternary fossil localities in the McMurdo Sound region. Although most of the specimens came from exposures of the Scallop Hill Formation, Speden noted that four localities in his Taylor Formation contained *Bathylasma*. As the type locality for the Taylor Formation is at New Harbor, the author is surprised not to find the genus amongst his collections. *Bathylasma* had also been expected in the New Harbor faunas be-

TABLE 3. Analysis of ^{14}C Dates by *Stuiver et al.* [1967] for
          Specimens of *Adamussium colbecki*

Lab Number	Altitude, m	Date yr B.P.	Direction of Younging
QL 138	7.5	5800 ± 70	?
QL 156	2.2	5090 ± 50	↓
QL 155	1.0	5310 ± 60	
QL 158	4.2	5860 ± 70	
QL 159	1.9	5350 ± 70	
QL 137	5.7	6050 ± 70	↓
QL 154	3.3	5630 ± 60	
QL 153	1.4	5200 ± 60	

cause *Newman and Ross* [1971] described how initially they were alerted to the synonymy of *B. corolliforme* and *B. antarcticum* by seeing Speden's figures of plates of *Bathylasma* from the Taylor Formation.

DISCUSSION

Age of Fossil Faunas

Valves of *Adamussium colbecki* from two localities (II and IV) were submitted to the University of Waikato for ^{14}C dating. The age determinations are shown in Table 2.

Stuiver et al. [1976] ^{14}C dated samples of *A. colbecki* from 16 New Harbor localities as well as from a depth of 2.1-22.1 m below the surface from the cores of DVDP 8 and 9. The ^{14}C dates ranged from 4620 ± 60 to 6670 ± 200 years B.P.

Origin of the Marine Faunas

The writer submitted the valves of *Adamussium colbecki* for ^{14}C dating because he was uncertain whether or not the marine fossiliferous sediments at New Harbor had been deposited as a series of beach deposits or as a result of continuous sedimentation in, say, a shallow fjord. In the former case, shells from elevated stranded beach deposits would be older than those in beach deposits close to modern sea level, while in the latter case the fossils in the stratum exposed at the highest altitude would be the youngest. That is to say, if the fossils belonged to a series of stranded beach deposits, they would 'young' downward toward modern sea level, whereas if they belonged to a normal sedimentary sequence they would young upward.

The author's ^{14}C dates indicate a 'younging' downward, and this supports his concept of a series of elevated beach deposits around the mouth of Wales Stream. An analysis of ^{14}C dates given by *Stuiver et al.* [1976] for specimens of *Adamussium colbecki* collected from above sea level in the same area which also supports this hypothesis is given in Table 3. However, the data of Stuiver et al. from the northern coastline of New Harbor (lab numbers QL 162 to QL 165) indicate the direction of younging is the reverse, as does the 6670 ± 200 year B.P. date from the specimens in DVDP 8 and 9. If one postulates that the fossiliferous marine sediments exposed along the New Harbor shoreline have been deposited by a mixture of continuous sedimentation in shallow water near the head of a fjord and sedimentation associated with beach erosion and deposition, one would explain the anomalous ^{14}C dates. Such a mixture of origins would be compatible with the diastems and disconformities already recorded in drill cores from the Taylor Valley [*Chapman-Smith and Luckman*, 1974; *McKelvey*, 1975a, b; *Webb and Wrenn*, 1977].

Paleoecological Evidence From the Macrofaunas

The New Harbor thanatocoenose is a very restricted one in that very few species are present. Only four species of bivalve are present, which corroborates the findings of *Lowry* [1976] that Antarctic waters are impoverished in species of pelecypods. Because the faunas discovered are so meager, it is difficult to draw paleoecological conclusions beyond that of a likely depth range for the biocoenose.

Extant Species Deduced To be in Situ	Current Depth Range
Adamussium colbecki	4–700 m
Thracia meridionalis	18–594 m
Laternula elliptica	11–55 m
Subonoba contigua	?
Neobuccinum eatoni	6-600 m
Philine apertissima	18-37 m
Sterechinus neumayeri	3-640 m

From the shallow water nature of the depth ranges and because the fossiliferous deposits are at least in part associated with beach deposits, it is postulated that the New Harbor biocoenose existed in water depths of 5-35 m. There is micropaleontological evidence to confirm this belief in that *Webb and Wrenn* [1977] recorded very diverse foraminiferal faunas in the upper 24 m of DVDP 8 and

10 which they believed lived in sublittoral (<50 m) waters that were probably blanketed by semipermanent sea ice.

Is the Taylor Formation a Valid Stratigraphic Unit?

Speden [1962, p. 758] defined the Taylor Formation as 'frozen but unconsolidated marine silts, sands, muds, and perhaps gravel containing a varied fauna including *Adamussium colbecki.*' As previously discussed, Speden was unable to visit New Harbor and relied upon the descriptions of *David and Priestley* [1914, p. 272] and *Debenham* [1920, pp. 54-55] for accounts of the locality from which Priestley and Armytage collected their fauna, and which he (Speden) designated as the type locality for the Taylor Formation. *Speden* [1962, p. 761] stated that 'the type locality is the only one that does not overlie ice and is not admixed with moraine.' It seems clear that it is only at the type locality that the faunas are likely to be in situ, fossilized, and older than Recent.

The Dry Valley Drilling Project has demonstrated that the fossiliferous sediments at the type locality extend to well below the surface. The three drill holes at New Harbor, DVDP 8, 9, and 10 terminated at depths of 157.06 m, 38.34 m, and 185.47 m, respectively [*Chapman-Smith and Luckman*, 1974; *McKelvey*, 1975a, b]. Furthermore, *Webb and Wrenn* [1977] correlate sediments encountered in the New Harbor drill holes with lithologies intersected in DVDP 11 at Commonwealth Glacier and with sediments recovered from DVDP 15 in McMurdo Sound. Obviously, these correlatives are totally different from the sediments that Speden ascribed to the Taylor Formation away from its type locality. For instance, *Swithinbank et al.* [1961] and *Gaylord and Robertson* [1975] described some fossiliferous sediments exposed on the surface of the Ross Ice Shelf that are similar to lithologies recorded by Speden as belonging to the Taylor Formation away from its type locality. Clearly, there are now two groups of fossiliferous sediments within the Taylor Formation: first, the sequence at New Harbor which is in excess of 150 m thick at the type locality, and second, a 'carpetbag' group of sediments that are similar in faunal content and in physiographic position in that they are underlain by ice.

In addition, the writer believes it likely that the fauna collected from the type area by Priestley and Armytage is a mixture of fossil and recent specimens. The state of preservation indicated by the specimens in Figure 1, Plate 89 of *David and Priestley* [1914] is inconsistent with that found by the author in his collections. It seems highly probable that at least the crustacean appendage and the seal's tooth are recent specimens collected from the surface of the New Harbor shoreline. Furthermore, *Hedley* [1916a, p. 85] described Priestley's molluscan specimens as 'so fresh and glossy that they appear to have come direct from the sea, rather than to be fossils.'

To make matters worse, several workers in the lower Taylor Valley prior to the Dry Valley Drilling Project erected additional and overlapping stratigraphic units. For instance, *Murrel* [1973] described two members, the 'Hjorth Sand' at New Harbor and the overlying 'Suess Till.' Both members were included in the 'Canada Formation,' which was erected to embrace all deposits in the lower Taylor Valley above the basement complex except the most recent Kukri Drift and perched deltas. Murrell, who had read Speden's paper, failed even to mention the obvious overlap between the Taylor and Canada formations. While not attempting to erect formal lithostratigraphic units, *McCraw* [1967], *Angino et al.* [1962], and *Webb and Neall* [1972] all published important lithological and stratigraphic detail from the lower Taylor Valley.

It is clear that stratigraphic units proposed for deposits in the lower Taylor Valley overlapped even before the information gained from the drilling project, that sediments currently considered to be within the Taylor Formation are inconsistent with those at the type locality, and that a revision of the stratigraphy is required. Cores from drill holes completed by the Dry Valley Drilling Project in the Taylor Valley should be used to redefine the Taylor Formation, a name which should take preference over units such as the Canada Formation of *Murrell* [1973].

Acknowledgments. J. H. W. Wrenn is warmly thanked for assistance in the field. J. S. Oliver and the late J. Rude kindly gave the author some recent specimens from McMurdo Sound. The writer acknowledges with pleasure I. G. Speden of the New Zealand Geological Survey and A. W. B. Powell and W. O. Cernohorsky of the Auckland Institute and Museum for permission to study literature and faunal collections. A. T. Wilson is thanked for arranging ^{14}C dates on valves of *Adamussium colbecki* collected by the author. The author is indebted to the staff of the Geology Department, University of Auckland, for use of research facilities and to J. A. Grant-Mackie in particular for his encouragement and constructive criticism of the manuscript.

REFERENCES

Angino, E. E., M. D. Turner and E. J. Zeller, Reconnaissance geology of Lower Taylor Valley, Victoria Land, Antarctica, *Bull. Geol. Soc. Amer.*, *73*, 1553-1562.

Borradaile, L., Cirrepedia, *Nat. Hist. Rep. Br. Antarct. Terra Nova Exped. 1910* (Zool.), *3*(4), 127-136, 1916.

Brady, H. T., Late Neogene History of Taylor and Wright Valleys and McMurdo Sound, derived from diatom biostratigraphy and paleoecology of D.V.D.P. cores, *Proc. Sci. Comm. Antarct. Res.*, 1977.

Brady, H. T., and R. N. Webb, Microfossils from D.V.D.P. hole 15, Western McMurdo Sound, *Antarct. J. U.S.*, *11*(2), 81-82, 1976.

Bullivant, J. S., and J. H. Dearborn, The fauna of the Ross Sea, 5, *N. Z. Dep. Sci. Ind. Res. Bull.*, *176*, 1-77, 1967.

Burne, R. H., Mollusca, 4, Anatomy of Pelecypoda, *Nat. Hist. Rep. Br. Antarct. Terra Nova Exped. 1910*, *2*(10), 233-256, 1920.

Cameron, R. L., et al., Wilkes Station glaciological data 1957-58, *Rep. 825*, pp. 1-173, Ohio State Univ., Columbus, 1959.

Chapman, F., Report on the foraminifera and ostracoda from elevated deposits on the shores of the Ross Sea, *Rep. Sci. Invest. Br. Antarct. Exped. 1907-09* (Geol.), *2*, 27-52, 1916.

Chapman-Smith, M., and P. G. Luckman, Late Cenozoic glacial sequence cored at New Harbour, Victoria Land, Antarctica (D.V.D.P. #8, 9), *Dry Val. Drilling Proj. Bull.*, *3*, 120-148, 1974.

Claridge, G. G. C. and I. B. Campbell, The raised beaches at Inexpressible Island, Antarctica, *N. Z. J. Sci.*, *9*(4), 889-900, 1966.

Crespin, I., Some recent foraminifera from Vestfold Hills, Antarctica, in Hanazawa Memorial Volume, *Sci. Rep. Tohuku Univ., Spec. Vol. 4*, 19-31, 1960.

David, T. W. E., and R. E. Priestley, Glaciology, physiography, stratigraphy and tectonic geology of south Victoria Land, *Rep. Sci. Invest. Br. Antarct. Exped. 1907-09* (Geol.), *1*, 319 pp, 1914.

Debenham, F., A new mode of transportation by ice: The raised marine muds of South Victoria Land, *Quart, J. Geol. Soc. London*, *75*, 51-76, 1920.

Fell, H. B., The fauna of the Ross Sea, I, Ophiuroidea, *N. Z. Dep. Sci. Ind. Res. Bull.*, *142*, 1-79, 1961.

Gaylord, D. R., and J. D. Robertson, Sediments exposed on the surface of the Ross Ice Shelf, Antarctica, *J. Glaciol.*, *14*(71), 332-333, 1975.

Hedley, C., Mollusca, *Rep. Sci. Invest. Br. Antarct. Exped. 1907-09*, (Biol.), *2*(1), 1-8, 1911.

Hedley, C., Report on Mollusca from elevated marine beds, 'raised beaches' of McMurdo Sound, *Rep. Sci. Invest. Br. Antarct. Exped. 1907-09* (Geol.), *2*, 85-88, 1916a.

Hedley, C., Mollusca, *Australas. Antarct. Exped. 1911-1914*, Ser. C, *4*(1), 1-80, 1916b.

Hoek, *Zool. Chall.*, *8*, 155, 1883 (unseen).

Hoek, *Sigoba Exped.*, *31*, 245, 1913 (unseen).

King, P. P., and W. J. Broderip, Description of the Cirrhipedia conchifera and mollusca in a collection by H. M. S. 'Adventure' and 'Beagle,' *Zool. J.*, *5*, 322-349, 1831 (unseen).

Koehler, R., Echinodermes, *Exped. Antarct. Fr. 1903-05*, 1-28, 1906.

Lowry, J. K., Studies on the macrobenthos of the southern ocean, 1-5, Ph.D. thesis, pp. 1-176, Univ. of Canterbury, New Zealand, 1976.

McCraw, J. D., Soils of Taylor Valley, Victoria Land, Antarctica, with notes on soils from other localities in Victoria Land, *N. Z. J. Geol. Geophys.*, *10*(2), 498-539, 1967.

McKelvey, B. C., Stratigraphy of DVDP sites 10 and 11, Taylor Valley, *Antarct. J. U.S.*, *10*(4), 169-170, 1975a.

McKelvey, B. C., DVDP sites 10 and 11, Taylor Valley, *Dry Val. Drilling Proj. Bull.*, *5*, 16-24, 1975b.

McKelvey, B. C., Cenozoic marine and terrestrial glacial sedimentation in Taylor Valley, *Dry Val. Drilling Proj. Bull.*, *7*, 101-104, 1976.

McKnight, D. G., Echinoids from the Ross Sea and the Balleny Islands, *N. Z. Oceanogr. Inst. Rec.*, *3*(1), 1-6, 1976.

Meissner, 1900.

Murrell, B., Cenozoic stratigraphy in Lower Taylor Valley, Antarctica, *N. Z. J. Geol. Geophys.*, *16*(2), 225-242, 1973.

Newman, W. A., and A. Ross (Eds.), *Antarctic Cirrepedia*, *Antarct. Res. Ser.*, vol. 14, pp. 1-257, AGU, Washington, D.C., 1971.

Nichols, R. L., Coastal geomorphology, McMurdo Sound, Antarctica: Preliminary report, *I.G.Y. Glaciol. Rep.*, *4*, 51-101, 1961a.

Nichols, R. L., Characteristics of beaches formed in polar climates, *I.G.Y. Glaciol. Rep.*, *4*, 103-113, 1961b.

Pelseneer, P. E., *Resultats du Voyage du S. Y. 'Belgica,'* Zoologie Mollusques (Amphineures, Gastropodes et Lamellibranches), pp. 1-85, J.-E. Buschman, Antwerp, Belgium, 1903.

Powell, A. W. B., Antarctic and Subantarctic Mollusca: Pelepoda and Gastropoda, *Discovery Rep.*, *26*, 47-196, 1951.

Powell, A. W. B., Mollusca from the Victoria-Ross quadrants of Antarctica, *Rep. B. A. N. Z. Antarct. Res. Exped.*, Ser. B, *6*(9), 167-215, 1958.

Powell, A. W. B., Antarctic and Subantarctic mollusca, *Rec. Auck. Inst. Mus.*, *5*(3), 117-193, 1960.

Powell, A. W. B., Mollusca of Antarctic and Subantarctic seas, in *Biogeography and Ecology in Antarctica*, edited by S. Oye and J. Van Mieghem, Junk, The Hague, 1965.

Shackleton, E. H., *The Heart of the Antarctic Being the Story of the British Antarctic Expedition 1907-09*, vol. 2, Heinemann, London, 1909.

Shearburn Clark, H. E., The fauna of the Ross Sea, 3, Asteroidea, *N. Z. Dep. Sci. Ind. Res. Bull.*, *151*, 1-84, 1963.

Smith, E. A., Descriptions of some new shells from Kerguelen's Island, *Ann. Mag. Nat. Hist.*, *4*(16), 67-73, 1875.

Smith, E. A., Report on the Lamellibranchiata collected during the voyage of H. M. S. 'Challenger,' *Zool. Chall.*, *13*(35), 1-341, 1885.

Smith, E. A., Mollusca in zoology of the transit of Venus expedition, *Phil. Trans. Roy. Soc. London*, *168*, 167-192, 1887.

Smith, E. A., Mollusca, in *Report on the Collections of Mollusca Made in the Antarctic During the Voyage of the 'Southern Cross,'* part 7, pp. 201-213, William Clowes and Sons, London, 1902.

Smith, E. A., Lamellibranchiata, *Nat. Hist. Rep. Nat. Antarct. Discovery Exped. 1901-1904* (Zool.), *2*, 1907a.

Smith, E. A., Gastropoda, *Nat. Hist. Rep. Nat. Antarct. Discovery Exped. 1901-1904* (Zool.), *2*, 1907b.

Smith, E. A., Mollusca, 1, *Nat. Hist. Rep. Br. Antarct. Terra Nova Exped. 1910* (Zool.), *2*, 61-112, 1915.

Soot-Ryen, T., Antarctic Pelecypods, *Sci. Res. Norw. Antarct. Exped. 1927-29*, *32*, 1-46, 1951.

Speden, I. G., Fossiliferous Quaternary marine deposits in the McMurdo Sound region, Antarctica, *N. Z. J. Geol. Geophys.*, *5*(4), 746-777, 1962.

Stuiver, M., G. H. Denton, and H. W. Borns, Carbon-14 dates of *Adamussium colbecki* (Mollusca) in marine deposits at New Harbour, Taylor Valley, *Antarct. J. U.S.*, *11*(2), 86-88, 1976.

Swithinbank, C. W. M., D. G. Darby, and D. E. Wohlschlag, Faunal remains on an ice shelf, *Science*, *133*(3455), 764-766, 1961.

Thiele, J., Die antarktischen und subantarktischen Muscheln, *Dt. Sudpol. Exped. 1901-03*, *13*, 183-285, 1912.

Thiele, J., *Handbuch der Systematischen Weichtierkunde*, Gustav Fischer, Jena, Germany, 1936.

Webb, P. N., and V. E. Neall, Cretaceous foraminifera from Quaternary deposits in Taylor Valley, Antarctica, in *Antarc-tic Geology and Solid Earth Geophysics*, edited by R. J. Adie, Universitetsforlaget, Oslo, 1972.

Webb, P. N., and J. H. Wrenn, Foraminifera from D.V.D.P. holes 8-12, Taylor Valley, *Antarct. J. U.S.*, *11*(2), 85-86, 1976.

Webb, P. N., and J. H. Wrenn, Late Cenozoic micropaleontology and biostratigraphy of eastern Taylor Valley Antarctica, *Proc. Sci. Comm. Antarct. Res.*, 1977.

Wrenn, J. H., Cenozoic subsurface micropaleontology and geology of eastern Taylor Valley, Antarctica, M.S. thesis, pp. 1-255, N. Ill. Univ., DeKalb, 1977.

Wrenn, J. H., and P. H. Webb, Microfossils from Taylor Valley, *Antarct. J. U.S.*, *11*(2), 82-85, 1976.

THE SIGNIFICANCE OF FOSSIL MARINE AND NONMARINE DIATOMS IN DVDP CORES

HOWARD THOMAS BRADY

School of Biological Sciences, Macquarie University
North Ryde, New South Wales, Australia

Marine and nonmarine diatom fossils are important biostratigraphic indicators in Dry Valley Drilling Project cores. The marine diatoms lived in Miocene and early Pliocene coastal fjords, southern Victoria Land. The nonmarine diatoms lived in lakes within these fjords or on ice shelf surface pools. It is suggested that freshwater lakes or ice shelves have provided a stable environment since the early Pliocene for a nonmarine diatom flora which migrates from ice shelves to terrestrial lakes during warm interglacial periods.

INTRODUCTION

This article summarizes and updates previous publications on the diatom stratigraphy of Dry Valley Drilling Project (DVDP) cores [*Brady*, 1977a, b, c, d, e; 1979a, b], and it presents a new theory on the survival of nonmarine algae in Antarctica during the ice ages. Fossil diatoms have been recovered from DVDP cores 3, 4A, 8, 9, 10, 11, and 15. Since any foraminifera faunas in these sections are long-ranging benthic forms and since silicoflagellates and radiolaria are either rare or absent, diatom stratigraphy is a critical tool for the dating and interpretation of these cores.

The oldest recovered diatom fossils existed in the late Miocene-early Pliocene fjords, as was first suggested by *Webb* [1972] from his analysis of the Pecten deposit of shells and foraminifera in Wright Valley. More recent fossil taxa are from paleo-proglacial lakes and lived during the uplift and consequent marine draining of the dry valleys. These paleo-proglacial lakes in Taylor Valley and McMurdo Sound were sometimes formed or enlarged by ice dams which blocked valley mouths [*Denton et al.*, 1968, 1970] or which then separated Mc-Murdo Sound from the Ross Sea [*Denton and Borns*, 1974; *Brady*, 1977b].

NONMARINE DIATOMS

DVDP 3 (Hut Point, Ross Island)

DVDP 3 is a volcanic sequence of lavas overlying a lower hyaloclastite sequence. The only fossil diatoms recovered from this hole were found in smear slides made by Cameron and Morelli in 1973 from the upper section of the hyaloclastite at 193.55 m downhole. The only species recovered has a close affinity to *Fragilaria bidens* Heiberg. It occurs in five samples. Little interpretation of this interval is possible, but the existence of this nonmarine species is consistent with *Treves'* [1978] suggestion that the hyaloclastites at this level formed in a freshwater environment. This species would then have lived in a freshwater lake created by vulcanism through ice [*Brady*, 1979b]. The K-Ar dates for the upper volcanic sequence have now been published and indicate that the upper lavas were laid down quickly about 1 m.y. ago. [*Treves*, 1979]. These dates place a minimum age on the diatom species in the lower hyaloclastite. It is suggested that they existed during one of the earlier Ross glaciations which filled McMurdo Sound with thick glacial ice [*Denton and Borns*, 1974].

DVDP 4a (Wright Valley)

Fossil nonmarine diatoms were recovered from the uppermost interval (0-3.84 m) in the sediment underlying Lake Vanda in Wright Valley. Since the flora is no different from that in a Pliocene section of DVDP 10 (this article) and since the flora is still living, no age determination is possible. Layers of gypsum within the interval do suggest a complicated lake history.

DVDP 15 (McMurdo Sound)

Nonmarine diatoms were also recovered from McMurdo Sound (DVDP 15) from an interval 141-184 m below present-day sea level [*Brady*, 1977*b*, 1979*b*]. This author has not changed his opinion that these nonmarine diatoms represent a downwasting of a grounded Ross Ice Shelf in McMurdo Sound after a Ross type glaciation as described by *Denton and Borns* [1974]. It is not possible to date this section because the flora ranges from the early Pliocene to the present day. The interval is overlain with sediments which contain a foraminiferal fauna at 134 m below present-day sea level (12 m downhole). This fauna is similar to other Pleistocene benthic faunas described by *Chapman* [1916], *Webb and Wrenn* [1976], and *Wrenn* [1977].

DVDP 10 and 11 (Taylor Valley)

The Dry Valley Drilling Project recovered successions in holes 10 and 11 in Taylor Valley, southern Victoria Land, which show that the major components of the modern terrestrial diatom flora in that region were already present 4 m.y. ago in the Pliocene. Prior to the first drilling of freshwater sediments in Antarctica it was only possible to speculate on the survival of freshwater algae during ice age maxima. Postglacial invasion of terrestrial floras from adjacent continents and Subantarctic islands was regarded as the probable mechanism for post ice age recolonization [*Greene and Longton*, 1970; *Ahmadjian*, 1970; *Rudolph*, 1970].

The sedimentary successions in DVDP 11 which contain nonmarine diatoms lie between 0 and 17.20 m and between 99 and 102 m. Similar successions in DVDP 10 lie between 0 and 5.12 m and between 149.57 and 152.10 m. These successions overlie in situ marine sediments of the Miocene-early Pliocene Taylor Valley fjord [*Webb and Wrenn*, 1976,

1977; *Brady*, 1977*a*]. Some of these overlying sediments may have been transported into Taylor Valley from McMurdo Sound by an expanded Ross Ice Shelf during glacial periods [*Denton et al.*, 1968, 1970, 1974.]. Paleomagnetic studies indicate that the intervals in hole 11, 99-101 m, and hole 10, 149.57-152.10 m, are Pliocene within the Gilbert reversed epoch from 3.97 to 5.0 m.y. [*Elston et al.*, 1977; *Elston and Bressler*, this volume; *Purucker et al.*, this volume].

The succession from 99 to 102 m in hole 11 contains the following nonmarine diatoms: *Hantzschia amphioxys* (Ehrenberg) Grun, *Navicula cryptocephala* Kützing, *Navicula cymatopleura* West and West, *Navicula gibbula* Cleve var. *peraustralis* West and West, *Navicula murrayi* West and West, *Navicula mutica Kützing* f. *cohnii* (Hilse) Grun, *Navicula muticopsis* var. *muticopsis* Van Huerck, *Navicula seminulum* Grun, *Navicula shackletoni* West and West, *Navicula rhynchocephala* var. *amphiceros* (Kützing) Grun, *Melosira distans* (Ehrenberg) Kützing, aff. f. *seriata* Müller, *Nitzschia antarctica* West and West, and *Synedra* sp.

Synedra, *Nitzschia*, and *Melosira* spp. are still attached in colonies. Preservation is excellent. The bedding inclinations of up to 14° within this diatom succession suggest a marginal deltaic environment similar to the environments formed in modern lakes by freshwater streams. Exposed deltaic fans containing rich fossil assemblages of nonmarine diatoms are common in Taylor Valley.

The 149.57- to 152.10-m succession in DVDP 10 contains few nonmarine diatoms together with a well-preserved abundant marine flora which contains *Nitzschia praeinterfrigidaria* McCollum and *Thalassiosira torokina* Brady; these are marker species in the Gilbert epoch. It is possible that this nonmarine flora has been transported to a marine environment or that it is an in situ development on a reworked, glacially transported marine sediment. In any case, the interval is paleomagnetically reversed and is below other reversed intervals; thus the marine fossils and the paleomagnetic interpretations strongly indicate an age between 3.97 and 5.0 m.y.

The fossil diatom flora in DVDP 10 and 11 contains 14 of the 22 species described by *Baker* [1967] from modern Lake Miers, 8 of the 13 species described by *West and West* [1911] from Ross Island and the Ross Ice Shelf, and 4 of the 8 species described by *Fritsch* [1912, 1917] from the same area.

The only diatom which is absent from the fossil flora and which is a dominant species in a locally described nonmarine diatom population is *Tropidonesis laevissima* West and West, which is the dominant species in a pool at Cape Barne, Ross Island [*Fukushima*, 1970]. The dominant species in 20 ice shelf pools and 5 dry valley lake samples examined by this author are contained in these fossil floras. In ice shelf pools, probably because of the adventitious nature of colonization and annual melt, dominant species vary from pool to pool [*Brady*, 1977e]. In lake samples, variations could be ecological or simply due to sampling procedures. However, it is clear that over half the diatom species recorded from Ross Island, the western Ross Ice Shelf, and southern Victoria Land are present in the Pliocene fossil floras of DVDP 10 and 11.

Survival in a Dry Valley Oasis

There is now strong evidence which indicates that sections of Taylor and Wright valleys have not been glaciated since the early Pliocene. High bedrock thresholds to the west have restricted the flow of glaciers from the polar plateau, so that no polar glacier has completely filled Taylor Valley since 3.5 m.y. or Wright Valley since 3.7 m.y. [*Denton et al.*, 1968, 1970]. Indeed, Wright Valley may not have been filled by a polar glacier since the late Miocene [*Brady*, 1979b]. Fission track dates indicate surface soils in Wright Valley which are early Pliocene in age [*Ugolini and Jackson*, 1977]. This evidence suggests that a barren oasis existed in these areas during ice age maxima, when the ablation of clear lake ice, together with little annual input from meltwater streams, may have removed all the lake systems. Evaporite bands of gypsum in the sediment of Lake Vanda (DVDP 4A) indicate that this lake has dried up in the past [*Torii*, 1975].

Recent climatic records from Wright Valley show that about 60 summer days have a temperature maximum between 0° and 8°C and that about 3 days have a minimum above 0°C [*Chin*, 1973, 1974, 1975]. The summer meltwater raises lake levels, but yearly ablation rates are high, such as 2.02 m in 1972 and 1.00 m in 1974 for the frozen surface of Lake Vanda in Wright Valley. In some years there is a net gain in such a lake, but in other years with a cold summer there is a net loss. Considering these ablation rates it would seem that all lakes in the dry valleys would evaporate during an ice age. During an ice age, temperatures would drop owing to albedo effects from a larger area of pack ice in the Antarctic seas and from a global reduction in temperature. Even a small reduction in the length and intensity of the present austral summer would mean net loss in lake levels, so that a conservative estimate of −5°C average summer temperature decline in the dry valleys during an ice age (W. Budd, personal communication, 1979) would suggest the eventual removal of the existing lakes. If this is so, where then do the nonmarine lake algae survive?

Survival in Ice Shelf Pools

Today the largest nonmarine water biota in southern Victoria Land exists on the dirty ice of the western Ross Ice Shelf, which is separately identified as the McMurdo Ice Shelf. The surface is covered with terrestrial glacial moraine and with marine sediment. Anchor ice incorporates marine sediment in the base of the shelf, and ablation gradually exposes it on the surface [*Debenham*, 1920; *Dayton et al.*, 1969]. The sediment-covered shelf surface is in sharp contrast to the clear ice surface of terrestrial lakes. Initial ablation rates are higher owing to the absorption of radiation by sediment (a black body effect), but eventually a thick cover of moraine insulates the underlying ice and produces a stable desertlike pavement. One clear example of this phenomenon is The Strand Moraines, thought to be a grounded remnant of the Ross Ice Shelf stranded against the western side of McMurdo Sound as the Ross Ice Shelf retreated to its present position to the south after the last major glaciation [*Denton and Borns*, 1974]. The thick cover of moraine hides this large body of ice. Melt pools and even small streams develop on its surface during summer months. The diatom flora in these pools is similar to that in pools on the Ross Ice Shelf to the south and terrestrial mainland lakes in the dry valleys to the west.

The present-day ice shelf pool system extends throughout an area of at least 2,500 km^2 and embraces the environs of Black Island, White Island, and Brown Peninsula; the area between Brown Peninsula and the mainland to the west; the area from the north of Black Island to the edge of the shelf; and a large area to the south of Black and White islands. It is difficult to estimate the total pool area, which ranges between 5% and 50% from inspection of aerial photos. However, it is clear that the surface area of these pools is far greater (at least 3 orders of magnitude) than the surface area of pools in the dry valley oasis of Taylor, Wright, and Victoria valleys, which is less than 100 km^2.

These McMurdo Ice Shelf pools may well be the key to nonmarine algal survival in southern Victoria Land during an ice age. Morainal cover eventually would lower ablation rates, while elsewhere the clear ice of the lakes is steadily removed. The absorption of heat by the surface sediment could form annual meltwater even when the surface air temperatures are close to zero or slightly below zero. Once some water is produced within the ice, any frozen ice on the surface acts as a heat trap for incoming radiation. This author has measured temperatures in shallow 30-cm-deep pools with a frozen surface layer of ice on days when the air temperatures were between $+1°C$ and $-5°C$. In five pools the water temperature adjacent to the basal sediment and algal layers ranged from $+1°C$ to $+7°C$.

There is solid evidence to suggest that during ice age conditions the potential area for surface pools on the ice shelf will be far greater than it is at the present time. Evidence from the dry valleys shows that coastal snowfall in southern Victoria Land was less than it is today and that alpine glaciers fed from local névé fields in the ranges of the dry valleys retreated. *Denton et al.* [1968, 1970, 1974] have shown that these alpine glaciers, out of phase with polar glaciations, retreated and advanced during glacial periods and interglacial periods, respectively. During glacial periods a greater expanse of shelf and pack ice led to drier atmospheric conditions along the Antarctic coastal belt. Moraines indicate that the Ross Ice Shelf filled McMurdo Sound [*Denton and Borns*, 1974] and that its grounding line was well to the north of the present Ross ice front position [*Thomas and Bentley*, 1977]. Most areas of the Ross Ice Shelf today are zones of snow accumulation in contrast to the stagnant ablating area of its western section near McMurdo Sound. In ice age conditions, ablation areas covered by morainal material suitable for formation of surface pools would be far more extensive. Consequently, this author proposes the hypothesis that a large nonmarine algal Antarctic biota was located on flat ice shelf surfaces, particularly in local areas of slow-moving or stagnant shelf ice during Pleistocene glaciations. During interglacial periods, algal mats from ice shelf pools transported by wind [*Wilson*, 1965] could colonize main-

land lakes. Conversely, if ice shelves completely collapse during interglacials [*Wilson*, 1964], then algae from mainland lakes will recolonize them as they re-form.

MARINE DIATOMS

Since DVDP successions in DVDP 8 and 9 are contained in the longer succession of DVDP 10 drilled at the same location and since the only pure marine diatom flora of DVDP 15 is at 0 meters (i.e., the present-day sea floor), there are only three critical DVDP successions containing marine fossil diatoms (DVDP 4A, 10, and 11). This author's opinion about the diatom biostratigraphy, the age, and the interpretation of these successions has remained unchanged following the DVDP conference in Tokyo [*Brady*, 1979a, b].

DVDP 4A (Wright Valley)

This author maintains that underneath the uppermost sediments of Lake Vanda (0-3.84 m) the succession contains only marine microfossils. These also occur in a sedimentary section underlying nonmarine deltaic sediments on the northwestern shore of Lake Vanda, which this author inspected with its discoverer, G. Denton of the University of Maine [*Brady*, 1979b]. The taxa correlate with the lowermost late Miocene succession of DVDP 11 rather than with its characteristic early Pliocene section. Broken frustules of pennate diatoms, indicative of a shallow benthic water environment, occur together with planktonic forms; this author's report on these 'benthic' taxa is not completed owing to the fragmentary nature of the fossil frustules. The abundance of these fragmented marine diatom fossils throughout these sections would suggest that they were deposited by glaciofluvial processes reworking the shallow marine sediments deposited during the final times of the Wright Valley Miocene fjord.

DVDP 10 and 11 (Taylor Valley)

Diatom age determinations can only be applied to two marine intervals in DVDP 10: a Pleistocene section from 0 to 24.86 m and an early Pliocene section from 137.10 to 183.27 m.

The core obtained near the Commonwealth glacier (DVDP 11) contains only one section with age determinative marine diatoms, from 194.80 to 292 m. Within this interval the section from 194.80 to

BRADY: FOSSIL MARINE AND NONMARINE DIATOMS IN DVDP CORES

208.62 m can be confidently dated as early Pliocene because of the presence of *Nitzschia praeinterfrigidaria* McCollum, but the underlying section may be late Miocene if the results of *Brady's* [1979a] revision of the age range of *Thalassiosira torokina* Brady are considered. Since the fossil flora of the underlying section is only slightly different from the early Pliocene flora, this author would suggest that it reflects only the upper late Miocene. This does not conflict with a magnetic age which includes part of epoch 7 [*Elston and Purucker*, this volume] and so encompasses a period of approximately 1½ m.y. (following the time scale of *La Breque* [1977]).

The in situ late Miocene and early Pliocene diatoms of DVDP 10 and 11 indicate full marine fjord conditions. The diatoms are planktonic apart from rare occurrences of the genus *Diploneis*. There is no evidence of uplift or shallower conditions in the fossil diatom flora. One can only surmise that the shallow fjord sediments which could contain diatoms were not deposited owing to ice cover during uplift, or, if they were deposited, that they have subsequently been removed by Ross type glaciations or even earlier erosion. In the last Miocene diatom flora or Wright Valley there are many broken and badly preserved shallow water benthic diatoms which may reflect the last stage of the Wright Valley fjord; no such diatoms occur in DVDP 10 and 11.

While the nonmarine diatom fossils in DVDP successions shed new light on their survival and evolution in the Antarctic, the fossil marine diatoms in DVDP cores reflect the stratigraphy already developed in Deep Sea Drilling Project holes in the Antarctic seas [*McCollum*, 1975; *Schrader*, 1976]. *Thalassiosira torokina* Brady—the new taxon described by *Brady* [1977d]—which was very common in the early Pliocene and late Miocene intervals of DVDP 10 and 11, was overlooked in the DSDP and *Eltanin* reports. The uppermost time range for this taxon lies between the 'b' and 'c' events of the Gilbert paleomagnetic epoch, but its lowermost range in the late Miocene is unknown owing to large unconformities in *Eltanin* and DSDP successions throughout the cores in the Antarctic seas. The early Pliocene intervals in DVDP 10 and 11 can be confidently placed within the *Nitzschia praeinterfrigidaria* partial range zone of McCollum [1975]. Older intervals in DVDP 11 do not contain many age determinative diatoms other than *Denticula hustedtii* Simonsen and Kanaya, *Thalassiosira torokina* Brady, and flat forms

of *Actinocyclus ingens* Rattray, which were common in the Miocene of DSDP holes 266 and 274 [*McCollum*, 1975]. This flora poorly reflects that of the *Denticula hustedtii* partial range zone of the late Miocene [*McCollum*, 1975].

SUMMARY

Marine and nonmarine fossil diatoms were the most common microfossil recovered during the Dry Valley Drilling Project. The nonmarine diatoms provide the first clear evidence for the long-range survival of such floras in Antarctica through glacial and interglacial periods. The marine fossil diatoms reflect the fjord history of Taylor and Wright valleys in the late Miocene and early Pliocene. It is now clear that the diatom stratigraphy of DSDP successions (which may be further refined) can be applied to inshore sediments in future Antarctic drilling programs.

Acknowledgments. This research was supported by grant 76-29657 from the National Science Foundation, Office of Polar Programs, to P. N. Webb, Northern Illinois University. Further research assistance was provided by the School of Biological Sciences, Macquarie University, Sydney.

REFERENCES

Ahmadjian, V., Adaptations of Antarctic terrestrial plants, in *Proceedings of the SCAR Symposium on Antarctic Ecology*, vol. 2, edited by M. W. Holdgate, pp. 801-811, Academic, New York, 1970.
Baker, A., Algae from Lake Miers, a solar heated Antarctic lake, *N. Z. J. Bot.*, *5*, 453-468, 1967.
Brady, H. T., Late Neogene history of Taylor and Wright Valleys and McMurdo Sound, derived from diatom biostratigraphy and paleoecology of DVDP cores, in *Antarctic Geoscience*, edited by C. Craddock, University of Wisconsin Press, Madison, 1977a.
Brady, H. T., Freshwater lakes in Pleistocene McMurdo Sound, *Antarct. J. U.S.*, *12*(4), 117-118, 1977b.
Brady, H. T., Extraction of diatoms from glacial sediments, *Antarct. J. U.S.*, *12*(4), 123-124, 1977c.
Brady, H. T., *Thalassiosira torokina* n. sp. (diatom) and its significance in Late Cenozoic biostratigraphy, *Antarct. J. U.S.*, *12*(4), 122-123, 1977d.
Brady, H. T., Late Neogene diatom biostratigraphy and paleoecology of the dry valleys and McMurdo Sound, Antarctica, M.S. thesis, Northern Ill. Univ., DeKalb, 1977e.
Brady, H, The dating and interpretation of diatom zones in Dry Valley Drilling Project holes 10 and 11, Taylor Valley, South Victoria Land, in *Proceedings of the Seminar III on Dry Valley Drilling Project, 1978, Mem. Spec. Issue 13*, edited by T. Nagata, National Institute of Polar Research, Tokyo, 1979a.
Brady, H, A diatom report on DVDP cores 3, 4A, 12, in *Proceedings of the Seminar III on Dry Valley Drilling Project, 1978, Mem. Spec. Issue 13*, edited by T. Nagata, National Institute of Polar Research, Tokyo, 1979b.
Chapman, F., Report on the foraminifera and ostracods from el-

evated deposits on the shores of the Ross Sea, in *Reports of Scientific Investigations, British Antarctic Expedition, 1907-1909, Geology*, vol. 2, no. 2, pp. 25-52, Heinemann, London, 1916.

Chin, T. J., *Hydrological Research Report, Dry Valleys, Antarctica, 1972-35*, Water and Soil Division, Ministry of Works, Christchurch, N. Z., 1973.

Chin, T. J., *Hydrological Research Report, Dry Valleys, Antarctica, 1973-74*, Water and Soil Division, Ministry of Works, Christchurch, N. Z., 1974.

Chin, T. J., *Hydrological Research Report, Dry Valleys, Antarctica, 1974-75*, Water and Soil Division, Ministry of Works, N. A., 1975.

Dayton, P. K., G. A. Robilliard, and A. L. DeVries, Anchor ice formation in McMurdo Sound, and its biological effects, *Science, 163*, 273-274, 1969.

Debenham, F., A new mode of transportation by ice: The raised marine muds of South Victoria Land (Antarctica), *Quart. J. Geol. Soc. London, 75*(2), 51-76, 1920.

Denton, G., and R. Armstrong, Glacial geology and chronology of the McMurdo Sound region, *Antarct. J. U.S., 3*(4), 99-101, 1968.

Denton, G., and Borns, H. Jr., Former grounded ice sheets in the Ross Sea, *Antarct. J. U.S., 9*(4), 167, 1974.

Denton, G., R. Armstrong, and M. Stuiver, Late Cenozoic glaciation in Antarctica: The record in the McMurdo Sound region, *Antarct. J. U.S., 5*(1), 15-21, 1970.

Elston, D. E., and S. L. Bressler, Magnetic stratigraphy and late Cenozoic history of Taylor Valley, this volume.

Elston, D. P., M. E. Purucker, S. L. Bressler, and H. Spall, Polarity zonations, magnetic intensities, and the correlation of Miocene and Pliocene DVDP cores, Taylor Valley, Antarctica, in *Antarctic Geoscience*, edited by C. Craddock, University of Wisconsin Press, Madison, 1977.

Elston, D. P., M. E. Purucker, S. L. Bressler, and H. Spall, Polarity zonations, magnetic intensities, and correlation of Miocene-Pliocene DVDP cores, Taylor Valley, Antarctica, in *Proceedings of the Seminar III on Dry Valley Drilling Project 1978, Mem. Spec. Issue 13*, edited by T. Nagata, National Institute of Polar Research, Tokyo, 1979.

Fritsch, F., Freshwater algae in National Antarctic Expedition, in *National Antarctic Expedition Reports, Natural History*, vol. 7, 1912.

Fritsch, F., Freshwater algae, *Natur. Hist. Rep. Brit. Antarct. Terra Nova Exped. 1910, Bot., 1*, 1917.

Fukushima, H., Notes on the diatom flora of Antarctic inland waters, in *Proceedings of the SCAR Symposium on Antarctic Ecology*, vol. 2, edited by M. W. Holdgate, Academic, New York, 1970.

Greene, S. W., and R. E. Longton, The effects of climate on Antarctic plants, in *Proceedings of the SCAR Symposium on Antarctic Ecology*, vol. 2, edited by M. W. Holdgate, pp. 786-800, Academic, New York, 1970.

LaBreque, J. L., D. V. Kent, and S. C. Cande, Revised magnetic polarity time scale for the Late Cretaceous and Cenozoic time, *Geology, 5*, 330-335, 1977.

McCollum, D., Diatom stratigraphy of the Southern Ocean, in *Initial Report of the Deep Sea Drilling Project*, vol. 28, edited by D. Hayes and L. A. Frakes, pp. 515-572, U.S. Government Printing Office, Washington, D. C., 1975.

Purucker, M. E., D. Elston, and S. L. Bressler, Magnetic stratigraphy of Late Cenozoic glaciogenic sediments from drill cores, Taylor Valley, Antarctica, this volume.

Rudolph, E. D., Local dissemination of plant propagules in Antarctica, in *Proceedings of the SCAR Symposium on Antarctic Ecology*, vol. 2, edited by M. W. Holdgate, pp. 812-817, Academic, New York, 1970.

Schrader, H., Cenozoic planktonic diatom stratigraphy of the southern Pacific Ocean, in *Initial Reports of the Deep Sea Drilling Project*, vol. 35, edited by C. Craddock et al., pp. 605-672, U.S. Government Printing Office, Washington, D. C., 1976.

Thomas, R., and C. Bentley, A model for Holocene retreat of the West Antarctic Ice Sheet, in *Antarctic Geoscience*, edited by C. Craddock, University of Wisconsin Press, Madison, 1980.

Torii, T. (Ed.), *Geophysical and geochemical studies of the Dry Valleys, Victoria Land in Antarctica, Mem. Spec. Issue 4*, National Institute of Polar Research, Tokyo, 1975.

Treves, S., The Hyaloclastite of DVDP 3, Hut Point Peninsula, Ross Island, Antarctica, in *Proceedings of the Seminar III on Dry Valley Drilling Project, 1978, Mem. Spec. Issue 13*, edited by T. Nagata, National Institute of Polar Research, Tokyo, 1979.

Ugolini, F. C., and M. L. Jackson, Weathering and Mineral Synthesis in Antarctic soils, in *Antarctic Geoscience*, edited by C. Craddock, University of Wisconsin Press, Madison, 1980.

Webb, P., Wright Fjord, Pliocene marine invasion of an Antarctic dry valley, *Antarct. J. U.S., 7*(6), 227-234, 1972.

Webb, P., and J. Wrenn, Foraminifera from DVDP holes 8, 9, 10, 11, and 12, Taylor Valley, *Antarct. J. U.S., 11*(2), 85-86, 1976.

Webb, P. N., and J. H. Wrenn, Late Cenozoic micropalaeontology and biostratigraphy of Eastern Taylor Valley, Antarctica, in *Antarctic Geoscience*, edited by C. Craddock, University of Wisconsin Press, Madison, 1977.

West, W., and G. S. West, Freshwater algae, in *British Antarctic Expedition Reports, (1907-1909), Biology*, vol. 1, no. 7, pp. 263-298, Heinemann, London, 1911.

Wilson, A., Origin of ice ages: An ice shelf theory for Pleistocene glaciations, *Nature, 201*(4915), 147-149, 1964.

Wilson, A., The escape of algae from frozen lakes and ponds, *Ecology, 46*(3), 1964.

Wrenn, J., Cenozoic and subsurface micropaleontology and geology of Eastern Taylor Valley, Antarctica, M.S. thesis, Geol. Dep., Northern Ill. Univ., DeKalb, 1977.

RADIOLARIANS AND SILICOFLAGELLATES FROM DRY VALLEY DRILLING PROJECT CORE

HSIN YI LING

Department of Geology, Northern Illinois University, DeKalb, Illinois 60115

Siliceous microfossils, radiolarians and silicoflagellates, were recovered from DVDP cores and grab samples from McMurdo Sound near hole 15. Except for siliceous microfossils reported in samples collected during the Ross Ice Shelf Project, silicoflagellates and radiolarians from the sediments of DVDP holes 8, 10, and 11 constituted at that time the southernmost occurrence of these microfossils. The inferred paleo-ecological interpretation of the sediments generally agrees with studies of foraminifera and oxygen isotopes, whereas the age assignment, regarded as Pleistocene, favors the results of carbon-14 determination. Only modern and abundant silicoflagellates were observed from the surface sediments of McMurdo Sound in the vicinity of hole 15. Abundant acid resistant organic remains of unknown affinity were observed at a depth of 163-166 m in hole 12, Lake Hoare.

INTRODUCTION

Abundant sediment exposures near McMurdo Sound provide a unique opportunity for correlating geological features on land with adjacent ocean bottom sediments. Dry Valley Drilling Project (DVDP) samples represent some of the southernmost submarine sediments available to the scientific community (Figure 1).

Within this framework, the author's investigation has been directed toward the examination of siliceous microfossil assemblages of radiolarians and silicoflagellates recovered from the cored sediments. Together with other studies (such as oxygen isotope, physical stratigraphy, and micro-organic remains of both calcareous (foraminifera) and siliceous (diatoms) nature), this study provides microfossil evidence that aids in the interpretation of regional geological history of McMurdo Sound.

METHODS OF STUDY

Selection of samples was guided by (1) lithologic logs prepared by field scientists at the drill sites and (2) the megascopic examination of all but hole 15 samples by David W. McCollum and myself at the Antarctic Research Facility, Florida State University. Each sample was later divided into three parts by facility curatorial personnel, under the direction of Dennis S. Cassidy, and shipped directly to respective institutions at the University of Washington, University of South Carolina (Beaufort), and San Francisco State University.

The results from different planktonic groups (diatoms, radiolarians, and silicoflagellates) can now be compared.

A total of 284 samples from eight holes was examined for the present investigation. Boreholes and the number of samples collected from each are listed in Table 1. Samples were first treated chemically with a combination of 10% hydrochloric acid and 30% hydrogen peroxide. Chemical treatment was followed by several decantations with distilled water and washing through a 74-μm sieve. Three slides were prepared from each sample, two for the fraction larger than 74 μm and one for the fraction smaller than 74 μm. A 22 × 40 mm cover slip was used, and each sample sealed with Canada balsam as the mounting medium. All the studies examined have been permanently deposited in the micropaleontology collection of the Department of Oceanography at the University of Washington, Seattle.

TABLE 1. Boreholes and Sample Numbers Examined for Radiolarians and Silicoflagellates

Location	Number of Samples
Hole 3 (Ross Island)	19
Hole 4 and 4a (Lake Vanda)	13
Hole 8 (New Harbor)	24
Hole 9 (New Harbor)	3
Hole 10 (New Harbor)	37
Hole 11 (Taylor Valley)	139
Hole 12 (Lake Hoare)	29
Hole 15 (McMurdo Sound)	20

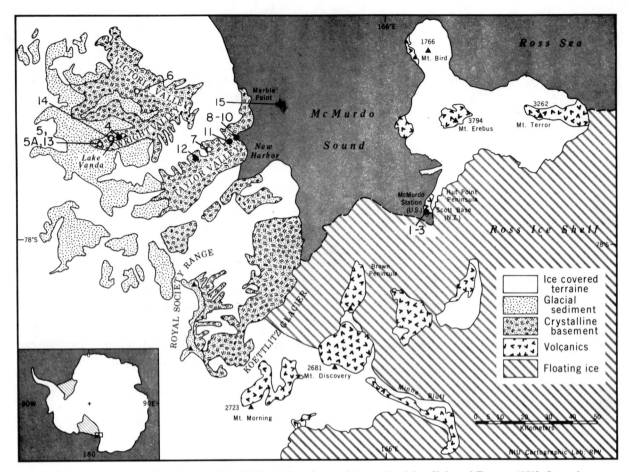

Fig. 1.   General geological map of the McMurdo Sound area of Antarctica [after Kyle and Treves, 1973]. Open circles show holes drilled during Dry Valley Drilling Project, and filled circles indicate samples examined for the present investigation.

Seven slides of bottom samples from the hole 15 area were also examined. Samples were collected with an Eckman Grab Sampler and were made available to the author for microscopic observation by Howard T. Brady, formerly of Northern Illinois University. Lithologies and sample depths are described in Table 2 [Barrett et al., 1976].

RESULTS OF ANALYSIS

Occurrence

Preservation of siliceous microfossil groups is less than ideal in sediments consisting of diamicton, sands, and gravels, which constitute the major

TABLE 2.   Lithologies and Depths of Grab Samples Examined From the Hole 15 Area [After Barrett et al., 1976]

Sample Site	Water Depth, m	Percent Gravel	Percent Sand	Percent Mud	Comments	No. of Slides Examined
2	124	trace	70	30	Yellowish green muddy fine to medium sand, with abundant organic material	4
4	184	2	69	29	Dark poorly sorted muddy coarse sand	2
8	173	trace	86	14	Dark moderately sorted medium sand	1

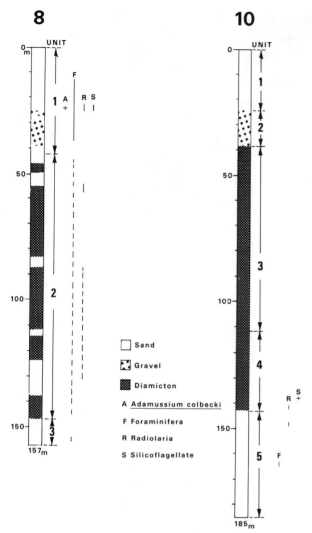

Fig. 2. Stratigraphic occurrence of radiolarians and silicoflagellates from the New Harbor holes 8 and 10. Lithology and unit (lithological) after Chapman-Smith and Luckman [1974] and McKelvey [1975].

portion of sediments cored. Nevertheless, radiolarians and silicoflagellates were recovered from samples of holes 8 and 10 at New Harbor and from grab samples collected at hole 15 in McMurdo Sound (Figure 2 and Plate 1). In addition, a few incomplete specimens were also observed from hole 11.

In hole 8, silicoflagellates and radiolarians were recovered in samples from 23- to 24-m depths (unit 1, sand and gravel). In the underlying diamicton (unit 2), scattered radiolarians occurred between the depths of 53 and 132 m. This group of siliceous microfossils was completely absent in the lowest sand (unit 3) of hole 8 and 132- to 157-m depth (Table 3).

Rare occurrence of a silicoflagellate species, *Distephanus speculum*, was observed at 137 m, and fragments of radiolarians were found at depths of 142 and 149 m in hole 10. Only rare and incomplete specimens of silicoflagellates and radiolarians were observed in one sample at a depth of 292 m in hole 11. Grab samples around hole 15 contained abundant and well-preserved silicoflagellate specimens; however, radiolarians were completely absent.

DISCUSSION

Radiolarian and silicoflagellate occurrences at 23- to 24-m depth in hole 8 indicate that deposition of unit 1 was under full marine conditions. This interpretation is in agreement with investigations on foraminifera by Webb and Wrenn [1975] and Webb [1980], on diatoms by Brady [1978], and on oxygen isotopes by Stuiver et al. [1976]. The scattered, rare occurrence of radiolarians in the underlying unit 2 may be reworked, as was the case with

TABLE 3. Occurrence of Radiolarians and Silicoflagellates From Hole 8 of the Dry Valley Drilling Project

DVDP Hole 8 Depth, m	Silicoflagellates and Radiolarians						
	Distephanum speculum	Antarctissa denticulata	Antarctissa sp. cf. A. strelkovi	Echinomma delicatum	Spongodiscus sp.	Siphocampe arachnea Group	Spongopyle osculosa
23.80-23.85	C	+	+	+	+	+	+
24.20-24.24	R				+		+
53.86-53.90		+		+	R		
56.97-57.01		R	+		R		
86.30-86.33							+
87.70-87.73		+	+		+		
103.97-104.00		+	+		+	R	+
112.13-112.16		+			+		
125.51-125.54		+					
126.17-126.21		+	+	+			

Plus sign, single specimen; R, 2-5 specimens; C, 5-10 specimens.

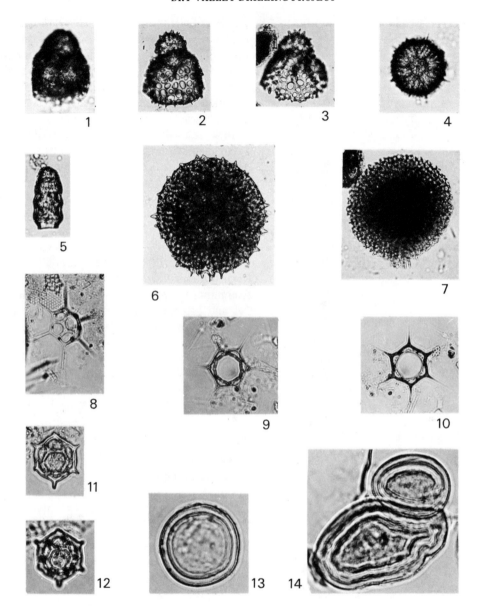

Plate 1.   Radiolarians and silicoflagellates from the McMurdo Sound region, Antarctica. Magnification ×200 unless otherwise indicated. The location of figured specimen in the slide is indicated as follows: hole number, depth interval of the sample in meters shown in parentheses, slide number, position of specimen in the slide by an England Finder.

Fig. 1.   *Antarctissa denticulata*, hole 8 (56.97-57.01 m), R-2 (P29/1).

Fig. 2.   *Antarctissa strelkovi*, hole 8 (53.86-53.90 m), R-1 (H36/0), ×300.

Fig. 3.   *Antarctissa* sp. cf. *A. strelkovi*, hole 8 (103.97-104 m), R-1 (V13/2).

Fig. 4.   *Echinomma delicatum*, hole 8 (126.17-126.21 m), R-1 (V16/1).

Fig. 5.   *Siphocampe arachnea* group, hole 8 (103.97-104 m), R-1 (C19/3), ×250.

Fig. 6.   *Spongodiscus* sp., hole 8 (112.13-112.16 m), R-1 (E3/0).

Fig. 7.   *Spongopyle osculosa*, hole 8 (103.97-104 m), R-1 (O18/2).

Figs. 8-12.   *Distephanus speculum* ×500; as fig. 8, hole 15 (site 4), 10 C (E27/4); figs. 9 and 10, hole 15 (site 4), 10 C (E20/3); figs. 11 and 12, hole 10 (137.56-137.61 m), L-2 (T23/2).

Figs. 13-14.   *Incertae sedis* ×500; as fig. 13, hole 12 (163.27-163.33 m), R-1 (H34/0); fig. 14, hole 12 (163.27-163.33 m), R-1 (N19/2).

foraminifera. On the other hand, abundant but poorly preserved foraminifera and marine diatom assemblages were found in unit 3 near the bottom of hole 8. Oxygen-isotope data indicate freshwater conditions with some admixing of marine water in unit 3. Since radiolarians and silicoflagellates are from a marine planktonic habitat, their complete absence from unit 3 may be considered as negative evidence favoring the oxygen isotope interpretation. Possible marine conditions can be inferred for sediments at approximately 137-m depth in hole 10 because of the rare occurrence of silicoflagellates.

Although neither radiolarians nor silicoflagellates were encountered in hole 15 cores, grab samples around the hole yielded abundant and well-preserved silicoflagellate microflora consisting solely of *Distephanus speculum* and its varieties.

Unfortunately, the rare occurrence of long-ranging silicoflagellate and radiolarian species did not allow any positive age determination for the sediments of the holes other than dating them provisionally as Pleistocene. A general consensus, on the basis of diatom and foraminiferal analyses, suggests the upper 153-155 m in holes 8 and 10 are Pleistocene. Webb and Wrenn [1975] and Webb [1980] suggest early Pliocene to late Miocene ages for foraminifera below 153 m in hole 8. This is in essential agreement with diatom studies by Brady [1978]. From paleomagnetic stratigraphy, Elston et al. [1978] suggest a Miocene age, whereas Stuiver et al. [1976] indicate only a Pleistocene age from carbon-14 studies.

TAXONOMIC REFERENCE LIST

Microfaunal and floral taxa observed in the samples are listed below in alphabetical order followed by the original references. Remarks are provided where necessary.

Radiolarians

Antarctissa denticulata (Ehrenberg), Petrushevskaya, 1967, p. 87, fig. 49, I–IV.— proposed as *Lithobotrys denticulata* Ehrenberg, 1844, p. 203.
Antarctissa streklovi Petrushevskaya, 1967, pp. 89-91, fig. 51, III-VI.
Echinomma delicatam (Dogiel), Petrushevskaya, 1967, pp. 22, 23, fig. 11, I-III,— proposed as *Heliosoma delicatum* Dogiel, 1952, pp. 7, 8, fig. 2.
Spongopyle osculosa Dreyer, 1889, pp. 118, 119, Plate 11, figs. 99, 100.

Spongodiscus sp.
The specimen possesses short, thick, conical spines all over the cortical shell. Several spines are located at one margin, indicating a possible pylome structure, although not enough specimens were available to confirm this.
Siphocampe arachnea (Ehrenberg) group, Nigrini, 1977, pp. 255, 256, Plate 3, figs. 7, 8.—proposed as *Lithocampe lineata* Ehrenberg, 1839, p. 130 (partim).

Silicoflagellates

Distephanus speculum (Ehrenberg), Haeckel, 1887, p. 1565.—proposed as *Dictyocha speculum* Ehrenberg, 1839, p. 129.
Included in this taxon are, in addition to the normal form, varieties with shorter radial spines with regular size of apical window, and normal length radial spines but relatively large apical window form.
Incertae sedis
These specimens are acid resistant, microorganic remains with circular to oval shape, distinct zonal pattern follows the outer margin. Maximum diameter range is 50-120 μm.

These microfossils occur abundantly only in sediments from hole 12, Lake Hoare (sand, mud sediments below the depth of 163 m to the bottom of the hole). Although the author favors a botanical origin for the microfossils, probably unicellular algal cells, no positive confirmation was reached during the present study.

Acknowledgments. The research discussed herein has been supported by National Science Foundation grant 73-05834 awarded to A. Lincoln Washburn, principal investigator, Quaternary Research Center, University of Washington. The author's grateful appreciation is extended to Dr. Washburn for his efforts in forming a DVDP research group at the University of Washington and for his continued encouragement and support. Special thanks are also due to Howard T. Brady of Macquarie University, Australia, for providing microslides of surface sediments around hole 15; Dennis S. Cassidy of the Antarctic Research Facility at Florida State University for his efficient curatorial assistance with the sediment samples; and Kozo Takahashi, currently at Woods Hole Oceanographic Institute, for his technical assistance throughout the investigation.

REFERENCES

Barrett, P. J., S. Treves, C. Barnes, H. Brady, S. McCormick, N. Nakai, J. Oliver, and K. Sillars, Initial report on DVDP 15, Western McMurdo Sound, *Dry Val. Drilling Proj. Bull.*, 7, 1-100, 1976.

Brady, H. T., The dating and interpretation of diatom zones in Dry Valley Drilling Project holes 10 and 11, Taylor Valley, South Victoria Land, Antarctica, report, 37 pp., Jap. Inst. of Polar Stud., Tokyo, 1978.

Chapman-Smith, M., and P. G. Luckman, Late Cenozoic glacial sequence at New Harbor, Victoria Land, Antarctica (DVDP 8 and 9), *Dry Val. Drilling Proj. Bull.*, *3*, 120-148, 1974.

Dogiel, V. A., and V. V. Reschetnjak, Materials on radiolaria of the northwestern part of the Pacific Ocean (investigation of the far east seas of the USSR), *Akad. Nauk SSSR Issled. Dal'nevost. Morei*, *3*, 5-36, 1952.

Dreyer, F., Die Pylombildungen in vergleichend-anatomischer und entwicklungsgeschichtlicher Beziehung bei Radiolarien und bei Protisten überhaupt, *Jena. Z. Naturwiss.*, *23*(16), 1-138, 1889.

Ehrenberg, C. G., Über die Bildung der Kreidefessen und des Kreidemergels durch unsichtbare Organismen, *Königl. Akad. Wiss. Berlin Abh.*, *1838*, 59-148, 1839.

Ehrenberg, C. G., Einige vorläufige Resultate seiner Untersuchungen der ihm von der Südpolreise des Capitain Ross, so wie von den Herren Schayer und Darwin zugekommenen Materialien über das Verhalten des kleinsten Lebens in den Ozeanen und den grössten bisher zugänglichen Tiefen des Weltmeers, *Königl. Preuss. Akad. Wiss. Berlin Ber.*, *1844*, 182-207, 1844.

Elston, D. P., M. E. Purucker, S. L. Bressler, and H. Spall, Polarity zonations, magnetic intensities and the correlation of Miocene and Pliocene DVDP cores, Taylor Valley, Antarctica, *Dry Val. Drilling Proj. Bull.*, *8*, 12-15, 1978.

Haeckel, E., Report on the radiolaria collected by H.M.S. *Challenger* during the years 1873-75, *Rep. Voyage Challenger Zool.*, *18*, 1803, 1887.

Kyle, P. R., and S. B. Treves, Review of the geology of Hut Point Peninsula, Ross Island, Antarctica, *Dry Val. Drilling Proj. Bull.*, *2*, 1-10, 1973.

McKelvey, B. C., Preliminary site reports: DVDP sites 10 and 11, Taylor Valley, *Dry Val. Drilling Project Bull.*, *5*, 16-60, 1975.

Nigrini, C., Tropical Cenozoic Artostrobiidae (radiolaria), *Micropaleontology*, *23*(3), 241-269, 1977.

Petrushevskaya, M. G., Antarctic spumelline and nasseline radiolarians, *Akad. Nauk SSSR Zool. Inst. Issled. Fauny Morei*, *4*(12), 1–186, 1967.

Stuiver, M., I. C. Yang, and G. H. Denton, Permafrost oxygen isotope ratios and chronology of three cores from Antarctica, *Nature*, *261*(5561), 547-550, 1976.

Webb, P. N., Paleogeographic evolution of the Ross sector during the Cenozoic, *Mem. Ser. C*, pp. 206-212, Nat. Inst. Polar Res., Tokyo, 1980.

Webb, P. N., and J. H. Wrenn, Foraminifera from DVDP holes 8, 9, and 10, Taylor Valley, *Antarct. J. U.S.*, *10*(4), 168-169, 1975.

MACERAL AND TOTAL ORGANIC CARBON ANALYSES OF DVDP DRILL CORE 11

J. H. Wrenn[1]

Department of Geology, Louisiana State University, Baton Rouge, Louisiana 70803

S. W. Beckman[2]

Department of Marine Sciences, Louisiana State University, Baton Rouge, Louisiana 70803

Maceral and total organic carbon (TOC) analyses were conducted on sediment samples from Dry Valley Drilling Project (DVDP) site 11, from Taylor Valley, Antarctica. These analyses were used to determine the concentration, provenance, and thermal maturation of acid resistant organic materials (macerals) contained in the sediments and to test the validity of hypothesized unconformities in DVDP 11. The applicability of TOC analysis to solving stratigraphic problems was also studied. The maceral content of all DVDP 11 samples was extremely low. Phytoclasts, indicative of a terrestrial (preglacial?) origin, were the dominant macerals. Fluctuations in protistoclast concentrations suggest varying paleoenvironmental conditions are represented in the core. The thermal maturity of macerals appears to be random throughout the core and indicates postdepositional thermal alteration has not affected the sediments in the eastern Taylor Valley basin. The results of TOC analysis confirm the low organic carbon content indicated by maceral analysis. The highest TOC concentration observed was approximately 0.3%. The distribution and concentration of TOC and macerals in DVDP 11 sediments indicate major changes occur in the core at depths of approximately 200 and 10 m. These changes probably were caused by variation in the maceral (and sediment?) provenance or depositional environmental conditions. In any event, the occurrence of these changes in proximity to the hypothesized unconformities lends strong support to those hypotheses. The favorable results of TOC analysis indicate further study of its use in subdividing sedimentary sequences is warranted.

INTRODUCTORY REMARKS

The Dry Valley Drilling Project (DVDP) core 11 was recovered from the Commonwealth Glacier site (77°35′S, 163°24′E; Figure 1) in eastern Taylor Valley, Antarctica [*McGinnis*, 1975]. The 320-m-long drill core is composed of glaciomarine and fluvioglacial sediments [*McKelvey*, 1975, 1977, 1980] that were deposited in 'Taylor Fjord' and subsequently uplifted during Late Miocene through Late Pleistocene time [*Webb and Wrenn*, 1976, 1977, 1980].

This paper reports the results of maceral and total organic carbon (TOC) analyses conducted on DVDP 11 sediments. The term 'maceral' is used here in a much broader sense than usual. The *Glos-*sary of Geology [*Gary et al.*, 1972], for example, defines maceral as 'organic units that comprise the coal mass; all petrologic units seen in microscope thin sections of coal. Macerals are to coal as minerals are to rock.'

G. F. Hart (personal communication, 1977), on the other hand, reasons that since almost all sedimentary rocks contain some organic constituents, coal is merely an organic-rich, inorganic-poor sedimentary rock. The less abundant organic clasts observed in shales and sandstones are basically the same as the macerals reported from coals.

Consequently, *Hart* [1979] interprets the definition of maceral as including all dispersed, acid resistant organic debris found in sedimentary rocks: 'Maceral analysis is the qualitative characterization of the acid resistant organic matter present in sediments and sedimentary deposits' [*Hart*, 1979]. Such analyses determine the (1) types of macerals present; (2) presence, type (biologic or chemical),

---

[1] Now at Amoco Production Company, P.O. Box 591, Tulsa, Oklahoma 74102.

[2] Now at Phillips Petroleum Company, Research and Development, Bartlesville, Oklahoma 74004.

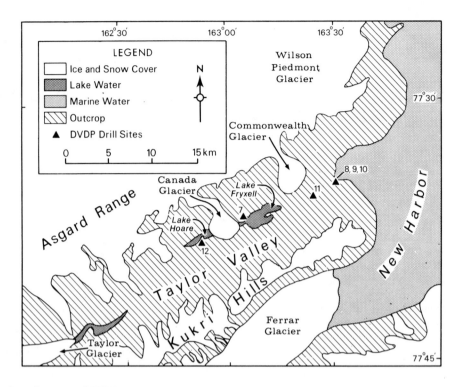

Fig. 1.   Location map of DVDP drill sites in Taylor Valley, Antarctica. Drawn from United States Geological Survey map ST57-60/6.

and degree of degradation; and (3) presence and degree of thermal alteration.

These data can be used to determine the maceral provenance, depositional paleo-environment, and postdepositional alteration processes that have affected the organic debris.

TOC analysis is a quantitative measure of the organic carbon concentration in sediments and sedimentary rocks. TOC values represent noncatabolized organic matter exclusive of carbonate material.

In this study, the absolute abundance of each type of maceral clast present in each sample was determined by combining the results of TOC and maceral analyses.

OBJECTIVES OF THIS STUDY

The main purpose of this study was to determine the maceral and TOC profiles of DVDP 11 in order to test the hypothesized unconformities in the core [*Brady*, 1977; *McKelvey*, 1977, 1980; *Webb and Wrenn*, 1977, 1980; *Elston et al.*, 1978].

In addition, the high-latitude location of the Taylor Valley drill cores, from an area characterized by extreme cold and periodic ice cover since

the Late Miocene Epoch [*McKelvey*, 1980; *Webb and Wrenn*, 1977], provided an excellent test case for determining the applicability of maceral analysis to sedimentary sequences deposited under environmental conditions adverse to most maceral producing life forms.

Finally, the relationship among the TOC distribution, biostratigraphic units based upon micropaleontological data [*Brady*, 1977, 1980; *Webb and Wrenn*, 1977, 1980] and lithostratigraphic units based upon sedimentological data [*McKelvey*, 1977, 1980], was investigated to determine the applicability of TOC analysis to stratigraphic subdivision.

METHODS AND MATERIALS

The 49 samples used in this study were selected from the DVDP 11 samples used by *Elston et al.* [1978] in paleomagnetic studies of the core. The sample depths are listed in Table 1.

TOC Analysis

Approximately 5 g of sample were disaggregated, placed in a beaker, and covered with a 1:1 solution of hydrochloric acid (37% reagent grade) and double-distilled de-ionized water until all carbonate

TABLE 1. Sample Depth and % TOC Values
of DVDP 11 Samples

Depth, m	% TOC
1.96	0.273
4.70	0.114
8.75	0.045
20.98	0.037
31.47	0.031
39.84	0.036
46.31	0.000
54.11	0.053
59.27	0.074
60.38	0.048
66.43	0.089
75.30	0.081
84.93	0.068
89.05	0.084
96.15	0.076
109.60	0.122
115.29	0.050
121.50	0.028
127.93	0.073
134.19	0.078
137.08	0.098
140.11	0.019
156.90	0.051
172.00	0.086
188.46	0.089
201.31	0.194
202.50	0.164
203.30	0.065
205.15	0.062
205.73	0.043
207.21	0.146
215.00	0.158
219.04	0.055
222.94	0.172
227.05	0.048
237.07	0.127
247.94	0.131
252.77	0.085
258.13	0.219
266.02	0.112
270.03	0.311
276.22	0.133
277.91	0.147
303.53	0.099
305.26	0.141
307.00	0.210
309.34	0.085
311.40	0.079
313.49	0.116

The % TOC value is plotted opposite the appropriate core depth of each DVDP 11 sample. Each sample depth (in meters) is the center of a 2-cm plug removed from the drill core and originally used for paleomagnetic investigations by *Elston et al.* [1978].

material had been dissolved (usually 24 hours was sufficient). The sample was then washed through a millipore filter with double-distilled de-ionized water and dried in an oven.

One-half gram of the dried residue was subsequently analyzed in a Leco Gasometric Carbon Analyzer (series 572-100).

Maceral Analysis

Approximately 5 g of sediment were placed in a 400-ml Nalgene beaker and covered with distilled water. Twenty milliliters of concentrated hydrochloric acid (37%) were added to dissolve any calcium carbonate present in the sample. When digestion of all calcium carbonate was complete, the sample was washed to neutrality with distilled water.

After decanting as much water as possible, 20 ml of concentrated hydrofluoric acid (technical grade) were added to dissolve any siliceous clasts in the sample. The samples were again washed to neutrality after all siliceous material had been removed.

The sample was then washed thru a 20-μ screen with distilled water, and both size fractions were retained. Two slides were prepared from each size fraction by using 22 × 40 mm, number 1 coverslips, and standard microscope slides. Glycerine jelly was used as the mounting medium.

Counting Procedure

Both slides of the greater than 20-μ size fraction prepared for each sample were utilized in this study. Twenty-five maceral clasts on each slide were counted and classified (Figure 2) into one of four major categories: (1) phytoclast (plant cuticles, tracheids, miospores, etc.), (2) protistoclast (clear, unstructured membranes, microforaminifera, dinoflagellates, acritarchs, etc.), (3) amorphous-infested indeterminate (chemically and/or biologically degraded material unassignable to either the phytoclast or protistoclast categories), and (4) scleratoclasts (fungal hyphae and spores, bacteria, etc.).

Miospore, dinocyst, and acritarch frequencies were recorded separately (Figure 2), but during data analysis, the frequencies were incorporated in the appropriate major category (see above). Microforams were not encountered in DVDP 11, and inertinite frequencies were not recorded during this study.

Maceral clasts were further categorized with reference to their degree of preservation (well, poor, infested, amorphous, or amorphous infested) and thermal alteration (as indicated by their color). Very pale yellow clasts (assigned a thermal altera-

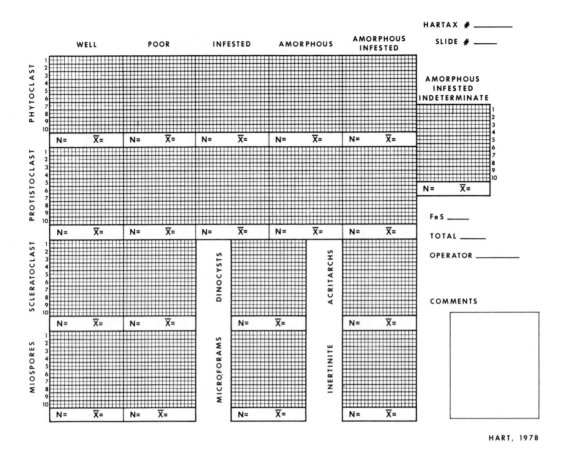

Fig. 2.   Tabulation sheet used in the collection of data for maceral analysis. Hartax number refers to the sample number; FeS is used for iron sulfide presence/absence notation; N is the total number of macerals observed in a particular maceral-type preservational cell, and $\bar{x}$ is the mean value of a cell.

tion index (TAI) number of 1) are thermally unaltered, whereas very dark brown clasts (assigned a TAI number of 9) are highly altered.

The data collected from the two slides of each sample were combined for statistical analysis, yielding a total number $N = 50$.

A detailed discussion of the maceral analysis techniques used in this study is presented in a paper by *Hart* [1979].

Method of Determining Absolute Abundance

The distribution of maceral clasts in DVDP 11 is discussed in terms of relative and absolute abundances of the various maceral types. The relative abundance (in percent) of a particular clast type is defined as its frequency (F) divided by the total number (N) of macerals counted multiplied by 100; that is,

$$(\text{percent relative abundance}) = \frac{F}{N}\ (100)$$

In this study, N is constant and equal to 50.

TOC values are recorded as weight percentage of organic carbon (% TOC).

The absolute abundance for each maceral type present in a sample was derived by multiplying F/N by the % TOC of the sample, or

$$(\text{absolute abundance}) = \%\ (\text{TOC})\ \frac{F}{N}$$

RESULTS

Maceral Analysis

The concentration and diversity of macerals were consistently low in all DVDP 11 samples ex-

COMPOSITE DIAGRAM OF
MACERALS IN DVDP II

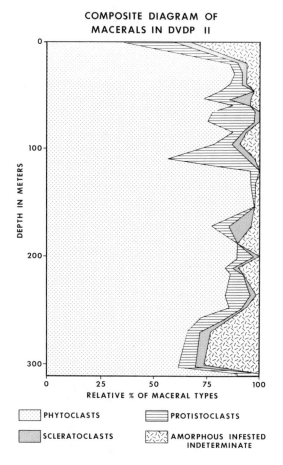

RELATIVE % OF MACERAL TYPES

PHYTOCLASTS   PROTISTOCLASTS

SCLERATOCLASTS   AMORPHOUS INFESTED
INDETERMINATE

ABSOLUTE ABUNDANCE OF
PHYTOCLASTS VS. DEPTH

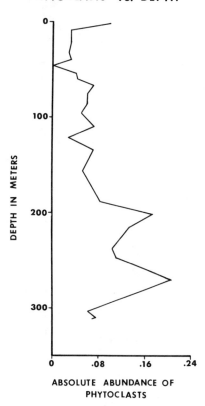

ABSOLUTE ABUNDANCE OF
PHYTOCLASTS

Fig. 3. Composite diagram depicting the relative percentage of the major maceral types present in DVDP 11, plotted against depth in meters. The elevation of DVDP 11 (80.20 m above sea level) is taken as the zero point on the depth-in-meters scale of the diagram.

Fig. 4. The absolute abundance of phytoclasts plotted against depth, in meters. The elevation of DVDP 11 (80.20 m above sea level) is taken as the zero point on the depth-in-meters scale of the diagram.

amined. The maceral content was always a mere fraction of that observed by the authors (unpublished studies) in sedimentary sequences deposited under less severe climatic conditions (e.g., the Mississippi Delta; Lower Tertiary sediments of Seymour Island, Antarctica; and off-shore California).

Phytoclasts. Phytoclasts are the dominant maceral type in DVDP 11, and their high relative abundance is characterized by only minor fluctuations (Figure 3), except between 309.34 and 247.94 m, 121.50 and 84.93 m, and above 8.75 m, where significant decreases were observed. The absolute abundance of phytoclasts (Figure 4) in the DVDP 11 core exhibits two distinct groups of values. Values are much lower and less variable above than they are below the approximately 200-m level in the core.

Most phytoclasts were unstructured lignin and epidermal terrestrial plant fragments, although structured clasts were also observed (Plate 1, figs. 5, 6, 8, 9, 11, and 12). It should be noted that spores and pollen were extremely rare in DVDP 11. The color range of phytoclasts varied from pale yellow to black; most were yellow to light orange.

Protistoclasts. The relative concentration of protistoclasts (Figure 3) was low in most samples, although noteworthy increases occurred between 109.60 and 96.15 m, and above 8.7 m. The absolute abundance of protistoclasts (Figure 5) decreased uphole, except for the same two intervals noted above, where the relative abundance increased.

The protistoclasts observed in DVDP 11 were acritarchs (Plate 1, figs. 1-3), (?) dinoflagellates, and fragments of clear, thin membranes. Protisto-

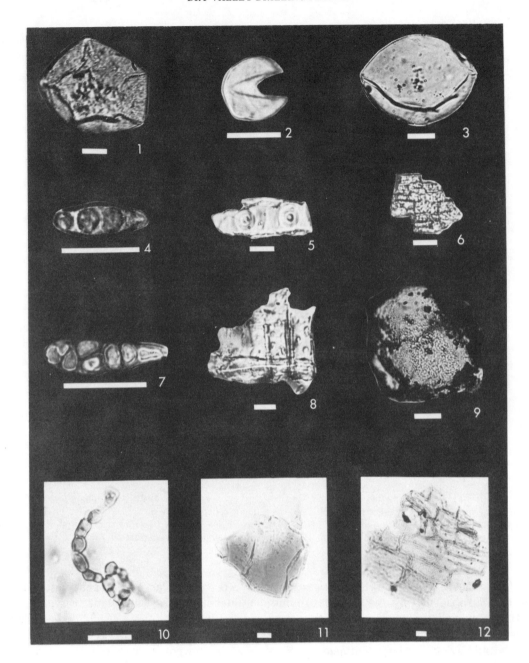

Plate 1.   Maceral types recovered from DVDP 11 sediments. Scale bar on each figure equals 20 μ.

Figs. 1-3.   Acritarchs exhibiting hyaline walls with folds. The acritarch in Figure 2 has ruptured equatorially.

Fig. 4.   A fungal spore, categorized as a scleratoclast in this paper.

Fig. 5.   Phytoclast (tracheid) exhibiting internal structures.

Fig. 6.   A well-structured phytoclast in which cell walls are still visible.

Fig. 7.   *Alternaria* sp., typical of the fungal spores (scleratoclasts) encountered in DVDP 11.

Fig. 8.   Well-structured phytoclast composed of tracheids.

Fig. 9.   Amorphous phytoclast.

Fig. 10.   Fungal chain (scleratoclast).

Fig. 11.   Amorphous phytoclast.

Fig. 12.   Well-structured phytoclast in which the original cell walls can be clearly seen.

ABSOLUTE ABUNDANCE OF PROTISTOCLASTS VS. DEPTH

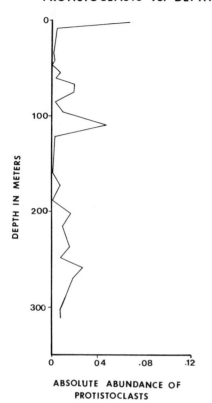

Fig. 5.   The absolute abundance of protistoclasts plotted against depth, in meters. The elevation of DVDP 11 (80.20 m above sea level) is taken as the zero point on the depth-in-meters scale of the diagram.

clasts were generally well preserved, and their color ranged from clear to orange.

Scleratoclasts.   Scleratoclasts (Figure 3) constitute a very small portion of the total maceral population. The most common scleratoclasts were well-preserved *Alternaria* sp. fungal spores and fungal hyphae (Plate 1, figs. 4, 7, and 10). The color range of these macerals was from yellow to orange-brown.

Amorphous-infested indeterminate (AII) macerals.   The relative and absolute abundances of AII macerals were high between 247.94 and 309.34 m, and above 8.7 m; all other intervals exhibited low values (Figures 3 and 6).

AII macerals consisted of amorphous organic matter of unknown origin which appeared to have undergone bacterial degradation. These macerals ranged in color from pale yellow to light orange; most clasts were yellow.

Total Organic Carbon Analysis

TOC values in DVDP 11 are characterized by high values below 200 and above 8.75 m (Figure 7). The low TOC values between these two depths display much less variation between samples than do the higher TOC values below 200 m.

Results of a student t-test comparing TOC values above and below 200 m (exclusive of those values above 8.75 m) demonstrated that a highly significant difference exists between the two sample sets ($t = -3.49$ with 46 degrees of freedom). The mean of TOC values above 200 m was 0.072%, whereas the TOC mean below 200 m was 0.129%.

Compared to the results of TOC studies of the Deep Sea Drilling Project (DSDP) core material from the Ross Sea area and the Falkland Plateau, TOC values in DVDP 11 were extremely low. The TOC average was 0.100% ($S_x = 0.009$%) in DVDP 11, whereas sediments from DSDP 272 and 273 in the Ross Sea had average TOC values of 0.44% ($S_x = 0.003$%) and 0.40% ($S_x = 0.600$%), respectively [*McIver*, 1975]. The average TOC value from DSDP 327A on the Falkland Plateau was 0.65% ($S_x = 0.163$%) [*Cameron*, 1977].

DISCUSSION

Maceral analysis indicates the organic debris in DVDP 11 is a mixture of reworked clasts and macerals produced contemporaneously with sediment deposition. Four possible sources of the macerals are indicated in Figure 6.

Reworked macerals.   Reworking of macerals from preglacial sedimentary deposits is indicated by the predominance in DVDP 11 of phytoclasts, including tracheids of woody, terrestrial plants (Plate 1, figs. 5, 6, 8, 9, 11, and 12). The existence of terrestrial floras contemporaneous with the deposition of DVDP 11 sediments is precluded because the environment in the Taylor Valley area has been characterized by polar conditions at least since the Late Miocene [*Webb and Wrenn*, 1977, 1980]. Such harsh conditions prevent the growth of vascular plants [*Plumstead*, 1964].

The Paleozoic-Mesozoic Beacon Supergroup deposits, which crop out in western Taylor Valley, may be the source of reworked macerals since they have yielded both plant macrofossils [*Plumstead*, 1962, 1964] and microfossils [*Helby and McElroy*, 1968]. The failure to recover Paleozoic or Mesozoic spores in DVDP 11 is surprising due to the position of the drill site along a major avenue for ice flow

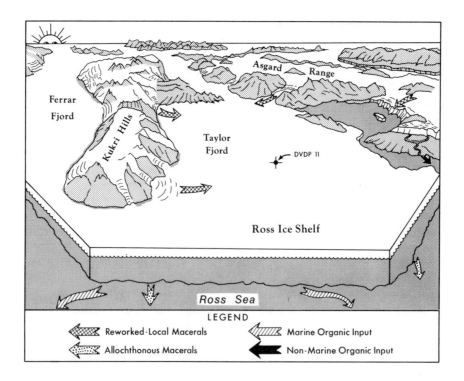

Fig. 6.   Schematic diagram of the Taylor Valley area in pre-Pliocene times, depicting the possible sources of the
macerals observed in DVDP 11.

from the polar plateau in the west. One would expect to find spores from the Beacon Supergroup in
the core if ice had transported sediments and macerals downvalley from the west. The absence of
spores suggests the DVDP 11 sediments and macerals were not derived from the Beacon Supergroup. Macerals could also have been derived from
totally unknown western deposits, currently buried by ice.

Alternately, sediments could have been moved
up the valley from eastern submarine deposits
cropping out on the floor of McMurdo Sound or under the Ross Ice Shelf. Repeated incursions of an
expanded Ross Ice Shelf into eastern Taylor Valley [*Denton et al.*, 1970; *McKelvey*, 1978, 1980]
would provide a transport mechanism for the huge
sediment volumes involved. Grounding of the ice
shelf during expansion would scrape up and transport sediment from the sea floor into Taylor Valley
for subsequent deposition.

However, the reported occurrence of Permo-
Triassic spores and Lower Tertiary dinoflagellates
[*Wilson*, 1968] in Ross Sea bottom sediments
would suggest these should also be found in DVDP
11, if the core sediments had been transported

westward, into Taylor Valley, from the Ross Sea
area.

Thus the nature of reworked macerals in DVDP
11 indicates a terrestrial origin. However, the direction of transport into Taylor Valley cannot be
established on present maceral data.

Contemporaneous maceral production.   Local
production (Figure 6) of macerals contemporaneous with sedimentation is suggested by the occurrence in DVDP 11 of hyaline, thin-walled spheroids (Plate 1, figs. 1-3). These spheroids, referred
to as acritarchs in this paper, would probably have
been oxidized or mechanically destroyed if they
had been subjected to the processes of reworking.

Similar acritarchs have also been observed
throughout DVDP 4 and 4a drill cores from Lake
Vanda, in Wright Valley (referred to as probable
algal spores by *Hall* [1975]), and DVDP 9, 10, and
12 (referred to as probable algal spores by *Wrenn
and Webb* [1976]). Whether these acritarchs are indicative of marine or freshwater paleo-ecological
conditions is unclear.

Study of the diatoms recovered from DVDP 4
and 4a indicated the core was composed of a marine sequence overlain by lacustrine sediments

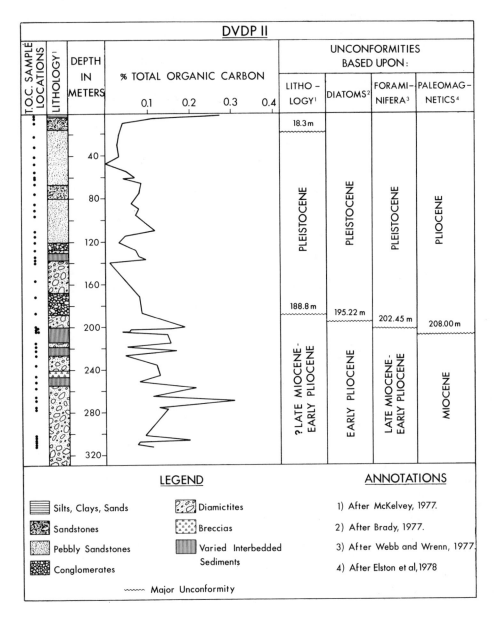

Fig. 7.   The distribution of % TOC is plotted to the left of the hypothesized locations of DVDP 11 unconformities. The basis of each unconformity, its author, and the geologic epochs separated by each unconformity are shown in individual columns in the right half of the diagram. The locations of most TOC samples are shown to the left of a generalized lithologic log and depth scale (in meters). Some TOC samples were too close together to be plotted as discrete points on the diagram (see Table 1).

[*Brady*, 1974]. Subsequently, *Brady* [1977, 1980] suggested the marine section of the core might actually have been deposited in the Vanda basin as lacustrine sediments, after having been reworked from older Wright Valley marine deposits. This implies the acritarchs which *Hall* [1975] observed throughout DVDP 4 and 4a could have been of lacustrine and/or marine origin, or transported from one environment into the other, depending upon one's interpretation of the diatom depositional history and paleo-ecologic significance.

The acritarchs observed in DVDP 12 [*Wrenn and Webb*, 1976] are of interest since *Chapman-Smith* [1975] suggested the core sediments were alternately deposited under terrestrial, fluvial, and lacustrine environmental conditions. This would indicate the acritarchs are nonmarine in origin, if it is assumed they were produced contemporaneously with sedimentation.

Finally, high relative and absolute abundances of protistoclasts, including the hyaline acritarchs, occur in two intervals of the DVDP 11 core where indicators of nonmarine conditions have been reported. H. T. Brady (personal communication, 1978) reports a high concentration of nonmarine diatoms between 89.76 and 101.90 m, which overlaps the protistoclast high between 96.15 and 109.60 m (Figures 4 and 7). This suggests the acritarchs are of nonmarine origin.

The protistoclast concentration above 8.75 m (Figure 4) occurs within *McKelvey*'s [1977, 1980] Quaternary unit 2. The lithologic composition, sedimentary structures, and textural features of this unit led McKelvey to suggest that they are glacial outwash sediments deposited under terrestrial or strandline conditions. Nonmarine diatoms recovered from the upper 20 m or so of DVDP 11 indicate fluvial or lacustrine conditions prevailed in the area of the drill site at the time of deposition [*Brady*, 1977, 1980]. Lithologic and diatom studies indicate the acritarchs are nonmarine in origin.

On the other hand, swarms of laevigate acritarchs were observed in sediments of the Early Miocene sediments penetrated at DSDP site 270 in the Ross Sea [*Hayes et al.*, 1975]. In subunits 2G and 2C, acritarchs outnumber pollen 25 to 1. The only other macerals reported from these subunits were fine fragments of terrestrial debris (phytoclasts). Sediments of these units were interpreted to be of marine origin. This implies the acritarchs are marine in origin.

The occurrence of acritarchs in the supposed marine and nonmarine deposits noted above precludes their use as a basis for definitive paleo-environmental interpretations. However, we believe the combined diatom [*Brady*, 1977, 1980] and lithologic [*McKelvey*, 1977, 1980] evidence strongly suggests the DVDP 11 acritarchs, at least, are nonmarine in origin. Their occurrence in DVDP 11 marine sediments is the result of local transport into the depositional basin, perhaps by glacial meltwater streams. Conversely, the marine origin of other types of protistoclasts observed in DVDP 11 cannot be denied.

CONCLUSIONS

The distribution of TOC (Figure 7) and phytoclast absolute abundance values (Figure 4) noted in the results above indicate that a major change occurred at approximately the 200-m depth in the core. The TOC and phytoclast values below 200 m are higher and more variable than the values above 200 m.

This suggests a change in the provenance of reworked sediments (source deposits relatively rich in macerals versus maceral-poor deposits), or the depositional regimen (such as the velocity fluctuations or the total absence of currents). A similar change is indicated above 8.75 m, where TOC and maceral values increase significantly to the top of the hole. These changes occur within portions of the core where unconformities have been hypothesized on the basis of lithologic, paleomagnetic, and paleontologic data (Figure 7) [*Brady*, 1977, 1980; *McKelvey*, 1977, 1980; *Spall and Elston*, 1977; *Webb and Wrenn*, 1977].

Although it is not possible to postulate that portions of the geologic record are missing at approximately 200 and 8.75 m based on maceral data, it is reasonable to conclude that a major change in DVDP 11 sediment provenance or depositional conditions did occur. Thus maceral data support the existence of the hypothesized unconformities to that extent. Maceral analysis indicates that the lower unconformity lies between 199.53 and 188.46 m, whereas the uppermost unconformity lies between 8.75 and 4.70 m. These intervals could not be narrowed because of the location of samples used in this study.

The absence of a downhole increase in the thermal maturity of macerals indicates thermal alteration has not occurred since deposition of the DVDP 11 sediments. Furthermore, the occurrence of totally unaltered pale yellow macerals and completely altered black macerals, often in the same sample, indicates that a multiplicity of maceral sources is involved and that reworking has occurred.

TOC analysis resulted in a subdivision of the core almost identical to that provided by more conventional stratigraphic tools. The usefulness of TOC analysis for stratigraphic subdivision deserves further study. The authors suspect it may be particularly useful for paleo-environmental subdivision of stratigraphic sequences, particularly lithologically monotonous sections.

The techniques of maceral analysis described by *Hart* [1979] are useful in studying sedimentary sequences deposited under harsh environmental conditions.

SUGGESTIONS FOR FUTURE WORK

1. The application of maceral and TOC analyses may prove to be of value in the study of DVDP 12, where the lack of biostratigraphically useful microfossils has hampered more conventional paleontological approaches [Wrenn and Webb, 1976]. Maceral and TOC analyses, shown in this study to yield supportive data for DVDP 11 unconformities hypothesized on the basis of micropaleontology, sedimentology, and paleomagnetic data, may be able to suggest the location of unconformities in DVDP 12.

2. Future maceral and TOC studies should incorporate sediment size analysis in order to determine the grain size effect on maceral distribution.

3. Acritarchs have been observed in DVDP cores 8-12 by one of the authors (J.H.W.), in DVDP 4 and 4a [Hall, 1975] and in DSDP 270 [Hayes and Frakes et al., 1975]. Their widespread vertical and horizontal distribution, usually, though not always, in association with micropaleontological and/or lithological evidence of nonmarine paleo-environmental conditions, suggests the acritarchs may be of paleo-ecologic significance. A comprehensive palynological investigation of all DVDP sedimentary cores, DSDP 270, and present-day lacustrine and fluvial sediments of the dry valley area may determine, among other things, the origin of these acritarchs.

Acknowledgments. The authors thank Donald P. Elston, George F. Hart, Thomas Whelan III, Thomas T. Lefebvre, Donald Y. Nugent, John H. Bair, Cheryl Whitehead, and Mary Crawford for assistance during this investigation. Assistance with drafting was provided by the Cartography Laboratory, School of Geoscience, Louisiana State University, and is gratefully acknowledged. Special thanks is extended to the Office of Polar Programs, National Science Foundation, and the Louisiana State University geology alumni fund for providing travel grants to one author (J.H.W.) so that this paper could be presented at the Dry Valley Drilling Project Seminar III, June 5-10, 1978, in Tokyo, Japan.

REFERENCES

Brady, H. T., Diatoms from the Lake Vanda core, *Dry Val. Drilling Proj. Bull., 3*, 181-184, 1974.

Brady, H. T., Late Neogene history of Taylor and Wright valleys and McMurdo Sound, derived from diatom biostratigraphy and paleoecology of DVDP cores, Volume of Abstracts, Third Symposium on Antarctic Geology and Geophysics, p. 20, Univ. of Wisc., Madison, 1977.

Brady, H. T., Late Neogene history of Taylor and Wright valleys and McMurdo Sound, derived from diatom biostratigraphy and paleoecology of DVDP cores, in *Antarctic Geoscience*, edited by C. Craddock, University of Wisconsin Press, Madison, in press, 1980.

Cameron, D. H., Grain size and carbon/carbonate analyses, Leg 36, in *Initial Reports of the Deep Sea Drilling Project, Leg 36*, edited by S. W. Wise, pp. 1047-1050, U.S. Government Printing Office, Washington, D.C., 1977.

Chapman-Smith, M., Geology of DVDP holes 12 and 14, *Antarct. J. U.S., 10*(4), 170-172, 1975.

Denton, G. H., R. L. Armstrong, and M. Stuiver, Late Cenozoic glaciation in Antarctica: The record in the McMurdo Sound region, *Antarct. J. U.S., 5*(1), 15-21, 1970.

Elston, D. P., M. E. Purucker, S. L. Bressler, and H. Spall, Polarity zonations, magnetic intensities and the correlation of Miocene and Pliocene DVDP cores, Taylor Valley, Antarctica, *Dry Val. Drilling Proj. Bull., 8*, 12-15, 1978.

Gary, M., R. McAfee, Jr., and C. L. Wolf (Eds.), *Glossary of Geology*, American Geological Institute, Washington, D.C., 1972.

Hall, S. A., Palynologic investigation of Quaternary sediment from Lake Vanda, *Antarct. J. U.S., 10*(4), 173-174, 1975.

Hart, G. F., Maceral analysis: Its use in petroleum exploration, *Methods Pap. 2*, Hartax Int. Inc., Baton Rouge, La., 1979.

Hayes, D. E., L. A. Frakes. P. J. Barrett, D. A. Burns, P. H. Chen, A. B. Ford, A. G. Kaneps, E. M. Kemp, D. W. McCollum, D. J. W. Piper, R. E. Wall, and P. N. Webb, Sites 270, 271, 272, in *Initial Reports of the Deep Sea Drilling Project, Leg 28*, edited by A. G. Kaneps, pp. 211-334, U.S. Government Printing Office, Washington, D.C., 1975.

Helby, R. J., and C. T. McElroy, Microfloras from the Devonian and Triassic of the Beacon Group, Antarctica, *N.Z. J. Geol. Geophys., 12*(2, 3), 376-382, 1968.

McGinnis, L. D., Dry Valley Drilling Project summaries: An overview of DVDP 1974-1975, *Dry Val. Drilling Proj. Bull., 5*, 1-4, 1975.

McIver, R. D., Hydrocarbon gases in canned core samples from leg 28 sites 271, 272, and 273, Ross Sea, in *Initial Reports of the Deep Sea Drilling Project*, vol. 28, edited by A. G. Kaneps, pp. 815-817, U.S. Government Printing Office, Washington, D.C., 1975.

McKelvey, B. C., Preliminary site reports, DVDP sites 10 and 11, Taylor Valley, *Dry Val. Drilling Proj. Bull., 5*, 16-69, 1975.

McKelvey, B. C., Upper Cenozoic marine and terrestrial glacial sedimentation in eastern Taylor Valley, southern Victoria Land, Volume of Abstracts, Third Symposium on Antarctic Geology and Geophysics, p. 112, Univ. of Wisc., Madison, 1977.

McKelvey, B. C., The Miocene-Pleistocene stratigraphy of eastern Taylor Valley—An interpretation, *Dry Val. Drilling Proj. Bull., 8*, 60-61, 1978.

McKelvey, B. C., Upper Cenozoic marine and terrestrial glacial sedimentation in eastern Taylor Valley, South Victoria Land, in *Antarctic Geoscience*, edited by C. Craddock, University of Wisconsin Press, Madison, in press, 1980.

Plumstead, E. P., Geology, 2, Fossil floras of Antarctica, *Scient. Rep. Transantarct. Exped., 9*, 154 pp., 1962.

Plumstead, E. P., Palaeobotany of Antarctica, in *Antarctic Geology*, edited by R. J. Adie, pp. 637-654, North-Holland, Amsterdam, 1964.

Spall, H. T., and D. P. Elston, Magnetic stratigraphy in DVDP hole 11, Taylor Valley, Antarctica, Volume of Abstracts, Third Symposium on Antarctic Geology and Geophysics, p. 140, Univ. of Wisc., Madison, 1977.

Webb, P. N., and J. H. Wrenn, Foraminifera from DVDP holes 8-12, Taylor Valley, *Antarct. J. U.S.*, *11*(12), 85-86, 1976.

Webb, P. N., and J. H. Wrenn, Late Cenozoic micropaleontology and biostratigraphy of eastern Taylor Valley, Antarctica, Volume of Abstracts, Third Symposium on Antarctic Geology and Geophysics, p. 159, Univ. of Wisc., Madison, 1977.

Webb, P. N., and J. H. Wrenn, Late Cenozoic micropaleontology and biostratigraphy of eastern Taylor Valley, Antarctica, in *Antarctic Geoscience*, edited by C. Craddock, University of Wisconsin Press, Madison, in press, 1980.

Wilson, G. J., On the occurrence of fossil microspores, pollen grains, and microplankton in bottom sediments of the Ross Sea, Antarctica, *N.Z. J. Mar. Freshwater Res.*, *2*(3), 381-389, 1968.

Wrenn, J. H., and P. N. Webb, Microfossils from Taylor Valley, *Antarct. J. U.S.*, *11*(2), 82-85, 1976.

GLACIAL HISTORY OF THE McMURDO SOUND AREA AS INDICATED BY THE DISTRIBUTION AND NATURE OF McMURDO VOLCANIC GROUP ROCKS

PHILIP R. KYLE

Institute of Polar Studies, Ohio State University, Columbus, Ohio 43210

Hyaloclastite at Castle Rock, Hut Point Peninsula, dated at 1.18 m.y., is believed to have been erupted subglacially into an expanded and grounded Ross Ice Sheet. The ice was at least 400 m thicker than the present-day Ross Ice Shelf and probably correlates with the Ross Sea Glaciation IV. Anorthoclase phonolite erratics which originate from Mount Erebus indicate both southward and westward movement of ice into McMurdo Sound. This can be understood when the paleogeography of the McMurdo Sound area is considered. Prior to about 5 m.y. ago, Minna Bluff, Ross Island, and White Island had not formed, so that the Ross Ice Shelf was free to move northward through McMurdo Sound. With the development of Minna Bluff, the northward flow was probably deflected and restricted. The growth of Ross Island, over the last 3 m.y., has resulted in the ice flow, during expansions of the Ross Ice Shelf, to be both southward and westward into McMurdo Sound.

INTRODUCTION

Studies of the sedimentologic characteristics [*McKelvey*, 1978; *Powell*, 1978], provenance, and depositional environments [*Porter and Beget*, 1978] of sediments in Dry Valley Drilling Project (DVDP) drill cores from the eastern Taylor Valley indicate that since the Miocene there have been frequent incursions into Taylor Valley of ice from McMurdo Sound and the Ross Sea. A full understanding of ice movement is dependent on a knowledge of the paleogeography of the region. Eruptions of alkali volcanics of the McMurdo Volcanic Group throughout the Late Cenozoic (since about 18 m.y.) have caused changes in topography, which in turn, have affected ice movement since the inception of glaciation. In this paper I wish to present several preliminary paleogeographic maps for McMurdo Sound based on the available radiometric age determinations on McMurdo Volcanic Group samples. Prior to describing these maps, several observations which constrain models for glaciation in the McMurdo Sound are described. These observations, when integrated, allow a better understanding of previous fluctuations of the Ross Ice Shelf.

Lava flows, pyroclastic deposits, scoria cones, and lava domes of the Late Cenozoic McMurdo Volcanic Group (commonly referred to as the McMurdo Volcanics) are widespread in the McMurdo Sound area (Figure 1) of Antarctica [*Warren*, 1969; *Kyle and Cole*, 1974]. Ross Island and the eruptive centers at and surrounding Mount Discovery form the bulk of these lavas, while smaller basaltic scoria cones occur in the ice-free valley area [*Armstrong et al.*, 1968] and along the front of the Royal Society Range [*McIver and Gever*, 1970] (Figure 2).

Drill cores obtained by the JOIDES project from the Ross Sea indicate that glacial calving was initiated in east Antarctica during the Late Oligocene, about 28-25 m.y. ago [*Hayes and Frakes*, 1975]. Eruption of the McMurdo Volcanic Group has therefore occurred throughout much of the period of glaciation. Volcanic rocks older than 0.5 m.y. can in most cases be dated using the K/Ar method. Therefore subglacially erupted volcanic rocks and interbedded glacial deposits and volcanic rocks can be used to establish a chronology for glacial events.

Minimum ages of 3.9 and 4.2 m.y. for valley cutting in the Taylor and Wright valleys, respectively, were established by dating basalt scoria cones erupted on the valley walls [*Armstrong et al.*, 1968; *Fleck et al.*, 1972]. *Denton et al.* [1970] used K/Ar age determinations of McMurdo volcanics from several locations in the McMurdo Sound area to establish a model for the Late Cenozoic glacial history of the area. *Armstrong* [1978] has reviewed the available K/Ar dates of McMurdo Volcanic Group rocks and commented on their sig-

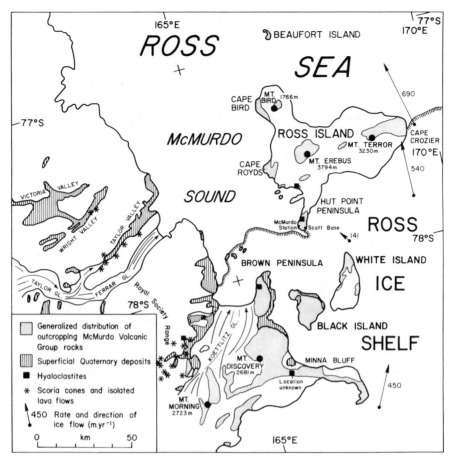

Fig. 1.   Location map of the McMurdo Sound area showing the distribution of Quaternary surficial deposits,
generalized areas of outcropping McMurdo Volcanic Group rocks, and occurrences of hyaloclastites.

nificance with regard to the glacial history of the
McMurdo Sound area.

HYALOCLASTITES

Hyaloclastites are the products of subaqueous
volcanic eruptions, which may occur in either submarine or subglacial environments. They are composed predominantly of palagonitic breccia, tuff,
and pillow lava complexes, usually of basaltic compositions. *LeMasurier* [1972] summarized three
distinctive field characteristics found in hyaloclastite deposits; however, none of these are diagnostic
of the depositional environment. At present there
are also no petrologic or geochemical criteria to
distinguish between hyaloclastites deposited in
subglacial (i.e., fresh water) and submarine environments. Indirect evidence must therefore be
used to identify subglacial hyaloclastites.

Distribution in Antarctica

Extensive deposits of hyaloclastite occur in the
Hallett volcanic province of the McMurdo Volcanic
Group, northern Victoria Land [*Hamilton*, 1972]
and in Marie Byrd Land [*LeMasurier*, 1972]. *Hamilton* [1972] considered hyaloclastites in the Hallett
area formed subglacially during the Late Cenozoic
when the Antarctic Ice Sheet was more extensive
than at present and was grounded on the Ross Sea
shelf. Hyaloclastites in Marie Byrd Land provide
evidence of a thick ice cap; unfortunately, the composition, structure, and possible submarine origin
of some of the deposits complicate interpretation of
the geologic history [*LeMasurier*, 1972].

In the McMurdo Sound area, hyaloclastites are
rare. They occur at Minna Bluff [*Hamilton*, 1972],
Turks Head, Tryggve Point, Hut Point Peninsula,
and Brown Peninsula [*Cole and Ewart*, 1968;
Eggers, 1976].

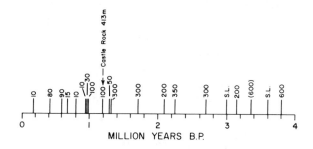

Fig. 2. Frequency of potassium-argon age determinations of subaerial cones and flows below previous higher ice levels in the McMurdo Sound area. Data compiled from *Armstrong* [1978], *Forbes et al.* [1974], *Kyle* [1976], *Muncy* [1979], and P. N. Webb (personal communication, 1978). Approximate height of sample sites above sea level are shown in meters; S. L. is sea level. See text for discussion of Castle Rock age.

Hyaloclastite was penetrated in all holes drilled by the DVDP at Hut Point Peninsula. Over 240 m of hyaloclastite was penetrated in DVDP 3, and on the basis of hydrogen and oxygen isotope measurements of permafrost samples, the deepest hyaloclastite (250 m below sea level) may be submarine whereas the upper part may be subglacial [*Lyon*, 1974; *Nakai et al.*, 1978]. *McGinnis* [1976] has suggested, however, that the ice may have resulted from freezing of groundwater and may not indicate the origin of the original regolith.

Castle Rock, Hut Point Peninsula

Castle Rock (77°48'S, 166°46'E; elevation 413 m) stands as a distinctive flat-topped turret-shaped monolith about 3.5 km north of Scott Bass and McMurdo Station on Hut Point Peninsula. It has been referred to as the eroded plug of a basaltic cone [*Debenham*, 1923] and as an autoclastic breccia [*Cole et al.*, 1971] but is now known to be composed of poorly bedded olivine-clinopyroxene basanite hyaloclastite. A dyke at the base of Castle Rock, considered to be a feeder for the hyaloclastite, has

a K/Ar age of 1.18 ± 0.05 m.y. (In this paper all K/Ar ages were calculated using the old (pre-1977) decay constants; see Table 1.) Boulder Cones that are topographically lower than Castle Rock may be part of the same deposit. Castle Rock is believed to be subglacial in origin and therefore gives an indication of previous ice levels. A submarine origin for Castle Rock is untenable for the following reasons.

1. Uplift of Hut Point Peninsula including Castle Rock is unlikely, as the required rate of over 400 m in about 1 m.y. is unrealistically high for the McMurdo Sound area. *Webb* [1972] and *Drewry* [1975] have suggested uplift rates along the Transantarctic Mountains of 7.7 × 10⁻⁵ m/yr and 7.5 × 10⁻⁵ m/yr, respectively, giving about 75 m of uplift in a million years. Raised beaches are absent at Hut Point Peninsula. There is also no evidence of faulting around Ross Island or in the seismic profiling records from McMurdo Sound [*Northey et al.*, 1975] that would be consistent with rapid uplift.

2. Subsidence rather than uplift appears to be occurring. The bathymetry of McMurdo Sound shows a large moat around Ross Island, which *McGinnis* [1973] interpreted as resulting from crustal subsidence due to isostatic loading. This is supported by seismic profiling in McMurdo Sound [*Northey et al.*, 1975] which clearly indicated depression of sediments beneath Ross Island. Unfortunately, the data are insufficient to determine quantitatively the amount of depression. The present height above sea level of Castle Rock is therefore a minimum; it may have been higher in the past.

A subglacial origin for Castle Rock is considered probable for the following reasons.

1. Evidence for higher ice level at Hut Point Peninsula is provided by the occurrence of glacial erratics of basement rocks from the Transantarctic Mountains on Observation Hill at ~200 m and on

TABLE 1. Whole-Rock K/Ar Age Determination of a Basanite Dike, Castle Rock, Hut Point Peninsula

Victoria Univ. Sample	Inst. of Nuclear Sci. Sample	K, %	⁴⁰Ar, nl/g	⁴⁰Ar, % Total	Age,† m.y.	
					Old	New
22879	R3020TR	1.54	0.072	17.5	1.18 ± 0.05	1.21 ± 0.05

Victoria University, Wellington, sample number refers to the petrological collection of the Department of Geology.

Institute of Nuclear Sciences sample age was determined by J. E. Gabites, Department of Scientific and Industrial Research, Lower Hutt.

*Radiogenic ⁴⁰Ar.

†Calculated using old and new decay constants. Old: $^{40}K\lambda_e = 0.472 \times 10^{-9}$ yr⁻¹, $\lambda_\beta = 0.584 \times 10^{-10}$ yr⁻¹. Abundance $^{40}K = 0.0119$ atom %. New: $^{40}K\lambda_e = 0.4962 \times 10^{-9}$ yr⁻¹, $\lambda_\beta = 0.581 \times 10^{-10}$ yr⁻¹. Abundance $^{40}K = 0.01167$ atom %.

Crater Hill at ~300 m. Glacial striations also occur on rocks near the top of Observation Hill [*Forbes and Ester*, 1964].

2.   Castle Rock has a shape similar to, but many times smaller than, table mountains in Iceland which are formed by subglacial eruptions [*Tazieff*, 1972].

Castle Rock, therefore, is believed to have formed subglacially 1.18 ± 0.05 m.y. ago. As uplift cannot be demonstrated and crustal depression appears to be occurring instead, a minimum thickness of ice at Castle Rock would have been over 400 m above present sea level.

Discussion

The argument is developed here that the nature of the ice mass covering Hut Point Peninsula during the formation of Castle Rock was an expanded and grounded 'Ross Ice Shelf' (here termed the Ross Ice Sheet). It is presumed that expansion of the Ross Ice Shelf occurred in response to thickening and expansion of the West Antarctic Ice Sheet. At the time of formation of Castle Rock, Mount Erebus was in a juvenile form and would have been too small to support a glacier of the dimensions necessary to cover the Castle Rock area. There was also insufficient topographic relief in the area to support a local ice dome. It is thus suggested that the eruption occurred into grounded ice originating in the Ross Sea embayment.

Denton et al. [1970, 1971] suggested that expansion of the Ross Ice Shelf, resulting in a grounded ice sheet on the Ross Sea floor, occurred on at least four separate occasions. The events were informally termed the Ross Sea Glaciations I (youngest) through IV (oldest). A maximum age of 1.2 m.y. was suggested for the Ross Sea Glaciation IV on the basis of dated lava flows from near the Walcott Glacier. The thick ice present at the time of formation of Castle Rock about 1.18 ± 0.05 m.y. ago is possibly a correlative of the Ross Sea Glaciation IV. The height of Castle Rock suggests ice thickness must have exceeded 400 m at Hut Point Peninsula at this time. Thickening of ice to this extent is consistent with the invasion of ice into the Taylor Valley and probably the Wright Valley.

AGE DISTRIBUTION OF DATED ROCKS

Most exposures of McMurdo Volcanic Group rocks in the McMurdo Sound area are scoria cones, subaerial flows, and cones. As discussed above, hy-

aloclastites possibly suggestive of subglacial eruptions are rare. The abundance of glacial erratics and glacial benches at various heights around Ross Island, and the other volcanic centers in McMurdo Sound, clearly indicates higher ice levels of from 800 m to over 1000 m above present sea level [*Vella*, 1969; *Denton et al.*, 1970] assuming no isostatic or tectonic uplift. Many samples of subaerial flows and cones have been dated by the K/Ar method [*Armstrong*, 1978; *Forbes et al.*, 1974; *Muncy*, 1979; *Kyle*, 1976, and unpublished data]. These dates therefore indicate times when these locations were ice free and an ice mass, such as the Ross Ice Sheet, was not present, at least at the eruptive sites.

Figure 2 shows the age distribution of dated rocks erupted in a subaerial environment below the level of known glacial erratics (that is, below the level of previous major ice cover). Based on Figure 2, any major period of expansion of the Ross Ice Shelf was unlikely to have been longer than 0.2-0.3 m.y. and, in the last million years, must have been even shorter. This is not a surprising conclusion, for during the last 200,000 years or so there have been fluctuations of sea level indicative of rapid changes of ice cover in the polar regions [*Veeh and Chappell*, 1970].

ICE MOVEMENT AND ERRATIC DISTRIBUTION

Porphyritic anorthoclase phonolite lavas (commonly referred to as kenyte) from Mount Erebus are distinctive in hand specimen and in thin section. They are easily distinguished from anorthoclase phonolite lavas from the upper part of Mount Discovery by their larger anorthoclase phenocrysts and different whole-rock chemistry [*Kyle*, 1976]. Erratics of Erebus-type anorthoclase phonolite can therefore be used to indicate ice movements in the McMurdo Sound area.

Some confusion has arisen because past workers have referred to porphyritic plagioclase-rich lavas of intermediate compositions as kenyte. (Intermediate is used to describe lavas with chemical compositions intermediate between basanite (44% SiO_2) and phonolite (54% SiO_2).) In hand-specimen Erebus-type anorthoclase phonolite lavas can be distinguished by the morphology of the feldspar phenocrysts. Anorthoclase is typically rhombic in outline, often lozenge shaped, whereas the intermediate lavas contain tabular or rectangular plagioclase phenocrysts. The true distribution of Ere-

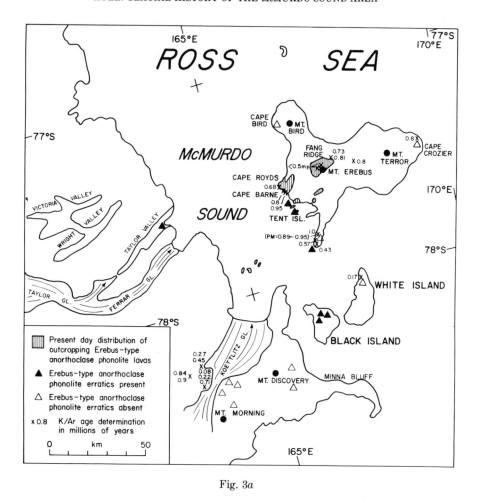

Fig. 3a

Fig. 3. Paleogeography of the McMurdo Sound area. The major volcanic piles are assumed to have accumulated in less than 1.5 m.y. The reconstructions have not been corrected for isostatic and/or tectonic uplift. (a) Present-day topography showing the locations of lavas with K/Ar age determinations of less than 1 m.y. and the distribution of anorthoclase phonolite erratics. At Hut Point Peninsula the rock sample with a K/Ar age determination of 1.0 ± 0.2 m.y. has a paleomagnetic (PM) age of 0.89-0.95 m.y. (b) Paleogeography of the McMurdo Sound area about 1.5 m.y. B.P. Location of volcanic rock samples with K/Ar age determinations of between 1 and 4 m.y. are indicated. (c) Paleogeography of the McMurdo Sound area about 5 m.y. B.P., showing the inferred directions of movement of the Ross Ice Shelf and location of volcanic rock samples with K/Ar age determinations greater than 4 m.y.

bus-type anorthoclase phonolite erratics is difficult to determine from the literature. The writer has, however, seen such erratics at Black Island and the eastern Taylor Valley. They are not known to occur on White Island, Hut Point Peninsula, Cape Bird, Cape Crozier, or on the lower slopes of Mount Morning and Mount Discovery. *Treves et al.* [1975] have reported anorthoclase phonolite erratics in sediment samples from the floor of McMurdo Sound in the vicinity of McMurdo Station. G. Denton (personal communication, 1979) has observed anorthoclase phonolite erratics in dry valleys west of the Koettlitz Glacier (Figure 3a).

The distribution of Erebus-type anorthoclase phonolite erratics indicates both southward and westward movement of ice away from Ross Island (Figure 1). *Denton and Borns* [1974] and *Treves et al.* [1975] have also suggested that an ice sheet entered McMurdo Sound around the northern tip of Ross Island. It is difficult to envisage the erratics being moved from Mount Erebus and vicinity to Black Island and Taylor Valley simultaneously. Therefore a multistage process may be necessary with initial erosion and transport of erratics into McMurdo Sound followed by transport to areas south and west of Erebus, at a later stage.

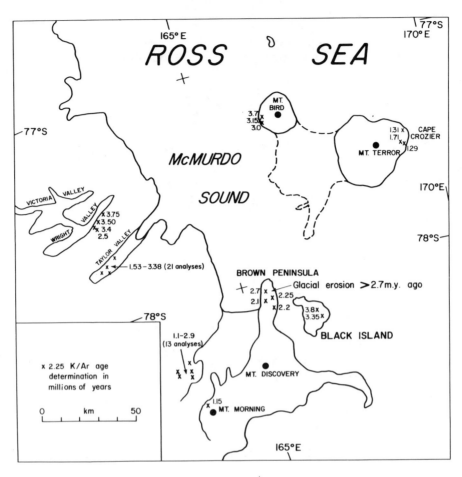

Fig. 3*b*

Ice flow models that would account for the distribution of the Erebus-type anorthoclase phonolite erratics are complicated. Present-day northward movement of ice into McMurdo Sound from the south is considerably slower than movement of the Ross Ice Shelf near Càpe Crozier and the east tip of Minna Bluff (Figure 1). White Island and, to a lesser extent, Black Island restrict ice movement into McMurdo Sound. Therefore it is likely that with an extension and thickening of the Ross Ice Shelf a larger proportion of ice would move around Ross Island and enter McMurdo Sound from the north, rather than westward between Hut Point Peninsula and White Island.

K/Ar ages of anorthoclase phonolite lavas from Mount Erebus and surrounding areas are all <1.0 m.y. (Figure 3*a*), the range of seven determinations being from 0.94 ± 0.05 to 0.15 ± 0.05 m.y. It is considered that Mount Erebus is unlikely to be older than 1.5 m.y., which places an upper time limit on the dispersal of anorthoclase phonolite erratics.

McMURDO SOUND

Paleogeography and Glacial History

Three major glacier systems are involved in the glacial history of the McMurdo Sound area. They are the Ross Ice Shelf, the east Antarctic Ice Sheet, and the alpine glaciers of the Transantarctic Mountains [*Denton et al.*, 1970, 1971]. Fluctuations of the three systems are not synchronous. Ross Ice Shelf fluctuations are examined here; the other systems, particularly in the ice-free valley area, are reviewed by *Bull and Webb* [1973], *Calkin and Bull* [1972], and *Calkin* [1973].

Denton et al. [1970] considered that the Ross Ice Shelf expanded and thickened at least four times during the last 1.2 m.y. and, on grounding, formed the Ross Ice Sheet. The expansion probably oc-

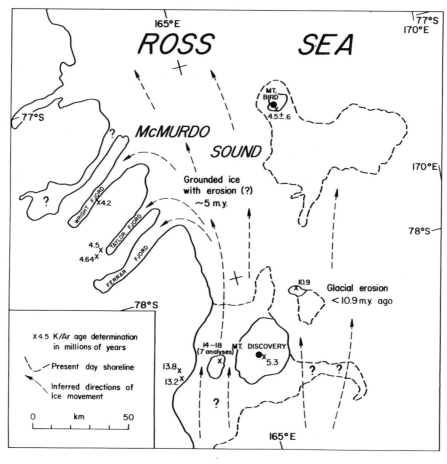

Fig. 3c

curred in response to thickening and buildup of the west Antarctic Ice Sheet. Evidence from the DVDP drill holes in the Taylor Valley suggests that other expansions occurred during the Pleistocene and Pliocene [*Porter and Beget*, 1978]. Two diagrams showing the paleogeography of McMurdo Sound are described. These show how movement of the Ross Ice Shelf and the Ross Ice Sheet through McMurdo Sound has varied due to eruptions and development of the McMurdo Volcanic Group.

Three diagrams are presented here; the first shows the present topography and location of dated McMurdo Volcanic Group samples less than 1 m.y. old (Figure 3a). Using these dates, a paleogeographic reconstruction of 1.5 m.y. B.P. is derived, along with the location of dated samples which range in age from 1 to 4 m.y. (Figure 3b). The final reconstruction shows the paleogeography about 5 m.y. ago, and the location of dated samples >4 m.y. (Figure 3c). It was assumed in compiling the maps that major volcanic centers were each active for about 1.5 m.y., unless the radiometric dates indicate otherwise. This assumption seems reasonable, for where sufficient data are available, it is apparent that many volcanic centers in the McMurdo Sound area were only active for about 1 m.y. or less. As examples, Mount Terror has dates ranging from 0.8 to 1.7 m.y., and four dates at Brown Peninsula range from 2.1 to 2.7 m.y. One million years of activity is also suggested by Mount Erebus, which commenced erupting about 0.95 m.y. ago. The volcano is now in a mature stage and unlikely to undergo any more major development. The reconstructions are based on present sea level, and no correction has been applied for subsidence of Ross Island due to isostatic loading.

Five Million Years Ago

The most significant feature of the paleogeography 5 m.y. ago (Figure 3c) was the complete absence of Ross Island, White Island, and Brown Peninsula. Most of Mount Morning was probably absent, apart from a few small outcrops at its northern base. It is tentatively suggested that Minna Bluff was also absent. No age determinations have been made on samples from Minna Bluff; however, geomorphic evidence, particularly the presence of unmodified phonolitic cones suggests an age of <5 m.y. A date of 4.5 ± 0.6 m.y. on a sample from the summit of Mount Bird has a large analytical uncertainty. Dates and paleomagnetic data [Kyle, 1976] on old cone-forming lavas at Cape Bird indicate Mount Bird began to form about 4 m.y. ago.

The dominate volcanic feature 5 m.y. ago was Mount Discovery. A date of 5.3 ± 0.14 m.y. for a sample from near the summit of the volcano indicates that the volcano was fully developed at that time. There was also a small volcanic center at the north end of Black Island. Small scoriaceous basanite cones and lava flows were erupted in the Taylor and Wright valleys between 4 and 5 m.y. B.P.

Prior to 4-5 m.y. ago, the Ross Ice Shelf, if it existed, may have moved northward directly through the present McMurdo Sound area. Lavas at Brown Peninsula and the northern end of Black Island are glacially eroded [Cole and Ewart, 1968]. Rainbow Ridge, Brown Peninsula, may have had in excess of 40 m of glacial erosion [Kyle et al., 1979]. At Black Island, erosion postdates 10.9 ± 0.4 m.y. old lavas [Armstrong, 1978], while at Rainbow Ridge an uneroded lava flow 2.7 ± 0.09 m.y. old [Armstrong, 1978] overlies glacially eroded lavas. Erosion at both of these localities may have resulted from a Ross Ice Sheet passing into McMurdo Sound without the impedance of Minna Bluff.

Seismic profiling in McMurdo Sound indicates the presence of an unconformity which, by correlation with core from the Deep Sea Drilling Project sites in the western and southern Ross Sea [Hayes and Frakes, 1975], is considered to represent an erosion surface 4-5 m.y. old [Northey et al., 1975]. Fillon [1975] believes this unconformity, as evident in piston cores from the Ross Sea, represents a hiatus lasting from 4.7 and 4.3 m.y. B.P. The erosion surface was considered to result from erosion by a grounded Ross Ice Sheet.

A major unconformity is found in DVDP boreholes 8, 10, and 11 from the lower Taylor Valley [Webb and Wrenn, 1980] and is probably correlative with the unconformity in McMurdo Sound and the Ross Sea. Paleomagnetic data [Elston et al.,

1978] on DVDP 11 drill core suggest the unconformity is about 5 m.y. old.

It is tentatively suggested here that glacial erosion at Brown Peninsula and possibly Black Island was associated with a grounded Ross Ice Sheet that moved northward through McMurdo Sound about 5 m.y. ago, prior to the formation of Minna Bluff. Subsequently, Minna Bluff formed, and effectively barred direct movement of the Ross Ice Shelf into McMurdo Sound.

Although the unconformities may be due to glacial scouring, an alternative explanation may be that when an ice shelf is present, sedimentation ceases. At the Ross Ice Shelf Project J-9 drill site, the base of the Ross Ice Shelf is melting [Jacobs et al., 1978], and there is no evidence of recent sedimentation on the sea floor [Ronan and Lipps, 1978; Webb and Brady, 1978]. According to this alternative hypothesis, about 5 m.y. ago, the Ross Ice Shelf expanded, invading the McMurdo Sound area, including the 'Taylor Fjord' and possibly 'Wright Fjord.' Major sedimentation ceased as the Ross Ice Shelf pushed northward and grounded, becoming an ice sheet. Later, as the ice sheet retreated, icebergs derived from local glaciers entered McMurdo Sound and deposited diamictons.

One and a Half Million Years Ago

About 1.5 m.y. ago (Figure 3b), Mount Erebus, Hut Point Peninsula, and White Island had not yet formed. The Ross Ice Shelf or the Ross Ice Sheet were probably able to flow into McMurdo Sound but at a much reduced rate due to the impedance of Minna Bluff. Later as White Island, Hut Point Peninsula, and Mount Erebus developed, ice flow was further restricted.

According to Fillon [1975], piston cores from the Ross Sea show a major hiatus from 2.4 to 0.7 m.y. B.P., although Kellogg et al. [1979] have recently suggested an alternative interpretation of the piston cores. The unconformity, if it exists, may have resulted from expansion and grounding of the Ross Ice Shelf [Drewry, 1976] to form an ice sheet. Drewry [1976] noted that this erosion surface is not apparent in the seismic profile records from McMurdo Sound [Northey et al., 1975]. If movement of ice during the last 1 m.y. occurred both westward between White Island and Hut Point Peninsula and southward around the northern end of Ross Island, as indicated by the anorthoclase phonolite erratics, then ice would be ponded in McMurdo Sound, thus reducing the possibility of extensive glacial scouring in this area. Therefore it is suggested that because McMurdo Sound has apparently been sheltered from the direct effects of an expanded Ross Ice Shelf, it may be the best site in the Ross Sea area to obtain a continuous sedimentary record from the present day back to about 4 m.y. B.P.

CONCLUSIONS AND RECOMMENDATIONS

1. Castle Rock on Hut Point Peninsula is a 400-m high turret-shaped deposit of hyaloclastite which is believed to have formed subglacially about 1.18 m.y. ago. The eruption probably occurred when the Ross Ice Shelf was thicker, perhaps grounded, and more extensive than at present. The event is probably correlative with Ross Sea Glaciation IV of *Denton et al.* [1970].

2. Glacial erratics of anorthoclase phonolite (kenyte) lavas are believed to have originated at Mount Erebus or associated vents on its flank. The distribution of such erratics at the mouth of Taylor Valley and on Black Island indicates southward and westward movement of ice into McMurdo Sound within the last 1.5 m.y.

3. K/Ar age determinations on subaerial flows and cones indicate periods when thick ice was not present in McMurdo Sound. Major expansions of the Ross Ice Shelf, resulting in the formation of an ice sheet, during the last 4 m.y. are unlikely to have encompassed periods longer than about 0.2-0.3 m.y. Over the last million years, expansions have probably been of even shorter duration.

4. Paleogeographic reconstructions of the McMurdo Sound area are fundamental to an understanding of movement of the Ross Ice Shelf in and into McMurdo Sound. Further geologic mapping and radiometric age determinations of McMurdo Volcanic Group rocks from the Dellbridge Islands, Erebus Bay, White Island, Mount Discovery, and areas east and southeast of Mount Morning are necessary to refine the reconnaissance maps presented here.

5. Minna Bluff probably acts as a barrier to direct movement of the Ross Ice Shelf through McMurdo Sound. It is inferred that prior to the development of the volcanic cones and flows which make up Minna Bluff, the Ross Ice Shelf moved through McMurdo Sound and glacially eroded Black Island (less than 10.9 m.y. ago) and Brown Peninsula (prior to 2.7 m.y. ago). Such movement may also be responsible for an unconformity observed in seismic profiling records from McMurdo Sound,

dated by correlation with sedimentary cores from the Ross Sea, as approximately 5 m.y. old. With the development of Minna Bluff, the flow of the Ross Ice Shelf probably was deflected, causing the eventual westward and southward movement of ice into McMurdo Sound.

Minna Bluff is undoubtedly one of the main obstacles to movement of the Ross Ice Shelf into McMurdo Sound. Unfortunately, the geological history and nature of the eruptive products of Minna Bluff are unknown. By the nature of its position, Minna Bluff is ideally situated to record fluctuations of the Ross Ice Shelf. A careful search of the area should be made for hyaloclastites; also routine geologic mapping and radiometric-age determinations are essential to an understanding of the paleogeography.

Acknowledgments. This paper was prompted by discussions with numerous people involved in the Dry Valley Drilling Project; to all of them I express my thanks. The K/Ar ages, which are important for the paleogeographic reconstructions, were made available by Dick Armstrong, to whom I extend special thanks for making his data so freely available prior to publication. Thanks also go to Janet Gabites, Institute of Nuclear Sciences, New Zealand, for providing the new K/Ar age determination on a sample from Castle Rock. Discussions with Peter Barrett, Ross Powell, and George Denton on Antarctic glaciations, were extremely illuminating. The manuscript was reviewed by George Denton, David Elliot, John Mercer, Steve Porter, and Ian Whillans. The National Science Foundation, the New Zealand University grants committee, and Victoria University of Wellington provided partial financial support; thus the author was able to attend DVDP seminars in Seattle (1974) and Tokyo (1978). This work was supported by NSF grants DPP76-23440 and DPP77-21959.

REFERENCES

Armstrong, R. L., K-Ar dating: Late Cenozoic McMurdo Volcanic Group and dry valley glacial history, Victoria Land, Antarctica, *N.Z. J. Geol. Geophys., 21,* 685-698, 1978.

Armstrong, R. L., W. Hamilton, and G. H. Denton, Glaciation in the Taylor Valley, Antarctica, older than 2.7 million years, *Science, 159,* 187-189, 1968.

Bull, C., and P. N. Webb, Some recent developments in the investigation of the glacial history of Antarctica, in *Paleoecology of Africa,* edited by E.M. van Zinderen Bakker, pp. 55-84, Scientific Committee on Antarctic Research Conference on Quaternary Studies, Balkema, Cape Town, 1973.

Calkin, P.E., Processes in the ice-free valleys of southern Victoria Land, Antarctica, in *Research in Polar and Alpine Geomorphology,* edited by B. D. Fahey and R. D. Thompson, pp. 167-186, Geo Abstracts Ltd., Norwich, 1973.

Calkin, P.E., and C. Bull, Interaction of the east Antarctic Ice Sheet, alpine glaciations and sea level in the Wright Valley area, southern Victoria Land, in *Antarctic Geology and Geophysics,* edited by R. J. Adie, pp. 435-440, Universitetsforlaget, Oslo, 1972.

Cole, J. W., and A. Ewart, Contributions to the volcanic geology of the Black Island, Brown Peninsula and Cape Bird areas, McMurdo Sound, Antarctica, *N.Z. J. Geol. Geophys.*, *11*, 793-828, 1968.

Cole, J.W., P. R. Kyle, and V. E. Neall, Contributions to the Quaternary geology of Cape Crozier, White Island and Hut Point Peninsula, McMurdo Sound region, Antarctica, *N.Z. J. Geol. Geophys.*, *14*, 528-546, 1971.

Debenham, F., The physiography of the Ross Archipelago, *Br. Antarct. Terra Nova Exped. 1910-1912*, 40 pp., 1923.

Denton, G. H., and H. W. Borns, Former grounded ice sheets in the Ross Sea, *Antarct. J. U.S.*, *10*, 167, 1974.

Denton, G. H., R. L. Armstrong, and M. Stuiver, Late Cenozoic glaciation in Antarctica: The record in the McMurdo Sound region, *Antarct. J. U.S.*, *5*, 15-21, 1970.

Denton, G. H., R. L. Armstrong, and M. Stuiver, The Late Cenozoic glacial history of Antarctica, in *The Late Cenozoic Glacial Ages*, edited by K. K. Turekian, pp. 267-306, Yale University Press, New Haven, Conn., 1971.

Drewry, D. J., Initiation and growth of the east Antarctic ice sheet, *J. Geol. Soc. London*, *131*, 255-273, 1975.

Drewry, D. J., Deep-sea drilling from *Glomar Challenger* in the Southern Ocean, *Polar Rec.*, *18*(112), 47-77, 1976.

Eggers, A. J., The Scallop Hill Formation, Brown Peninsula, McMurdo Sound, Antarctica, B.Sc. (Hons) project, 60 pp., Victoria Univ., Wellington, 1976.

Elston, D. P., M. E. Purucker, S. L. Bressler, and H. Spall, Polarity zonations, magnetic intensities, and the correlation of Miocene and Pliocene DVDP cores, Taylor Valley, Antarctica (abstract), *Dry Val. Drilling Proj. Bull.*, *8*, 12-15, 1978.

Fillon, R. H., Late Cenozoic Paleo-oceanography of the Ross Sea, Antarctica, *Geol. Soc. Amer. Bull.*, *86*, 839-845, 1975.

Fleck, R. J., L. M. Jones, and R. E. Behling, K-Ar dates of the McMurdo volcanics and their relation to the glacial history of Wright Valley, *Antarct. J. U.S.*, *7*, 244-246, 1972.

Forbes, R.B., and D. W. Ester, Glaciation of Observation Hill, Hut Point Peninsula, Ross Island, Antarctica, *J. Glaciol.*, *5*, 87-92, 1964.

Forbes, R. B., D. L. Turner, and J. R. Carden, Age of trachyte from Ross Island, Antarctica, *Geology*, *2*, 297-298, 1974.

Hamilton, W., The Hallett Volcanic Province, Antarctica, *U.S. Geol. Surv. Prof. Pap.*, *456-B*, 1972.

Hayes, D. E., and L. A. Frakes, General synthesis, deep sea drilling project leg 28, in *Initial Reports of the Deep Sea Drilling Project*, vol. 28, pp. 919-942, U.S. Government Printing Office, Washington, D. C., 1975.

Jacobs, S. S., J. Ardai, and P. Bruchhausen, Thermohaline stratification and the sea floor below the Ross Ice Shelf (abstract), *Eos Trans. AGU*, *59*(4), 308, 1978.

Kellogg, T. B., R. S. Truesdale, and L. E. Osterman, Late Quaternary extent of the west Antarctic ice sheet: New evidence from Ross Sea cores, *Geology*, *7*, 249-253, 1979.

Kyle, P. R., Geology, mineralogy and geochemistry of the Late Cenozoic McMurdo Volcanic Group, Victoria Land, Antarctica, Ph.D. thesis, 444 pp., Victoria Univ., Wellington, 1976.

Kyle, P. R., and J. W. Cole, Structural control of volcanism in the McMurdo Volcanic Group, Antarctica, *Bull. Volcanol.*, *38*, 16-25, 1974.

Kyle, P. R., J. Adams, and P. C. Rankin, Geology and petrology of the McMurdo Volcanic Group at Rainbow Ridge, Brown Peninsula, Antarctica, *Geol. Soc. Amer. Bull.*, *9*, 676-6883 1979.

LeMasurier, W. E., Volcanic record of Cenozoic glacial history of Marie Byrd Land, in *Antarctic Geology and Geophysics*, edited by R. J. Adie, pp. 251-259, Universitetsforlaget, Oslo, 1972.

Lyon, G. L., Stable isotope analyses of ice from DVDP 3, *Dry Val. Drilling Proj. Bull.*, *3*, 160-170, 1974.

McGinnis, L. D., McMurdo Sound—A key to the Cenozoic of Antarctica, *Antarct. J. U.S.*, *8*, 157, 1973.

McGinnis, L. D., Application of the Ghyben-Herzberg principle to an understanding of the physical chemistry of Ross Island and New Harbor permafrost core, *Dry Val. Drilling Proj. Bull.*, *7*, 113-114, 1976.

McIver, J. R., and T. W. Gevers, Volcanic vents below the Royal Society Range, central Victoria Land, *Trans. Geol. Soc. South Africa*, *73*, 65-88, 1970.

McKelvey, B. C., The Miocene-Pleistocene stratigraphy of eastern Taylor Valley—An interpretation (abstract), *Dry Val. Drilling Proj., Bull.*, *8*, 60-61, 1978.

Muncy, H. L., Geologic history and petrogenesis of alkaline volcanic rocks, Mt. Morning, Antarctica, M.S. thesis, Ohio State Univ., Columbus, 1979.

Nakai, N., Y. Mizutani, and H. Wada, Stable isotope studies: Past volcanic events deduced from H, O, S and C isotopic compositions of ice and salts from DVDP 3 (abstract), *Dry Val. Drilling Proj. Bull.*, *8*, 66-67, 1978.

Northey, D. J., C. Brown, D. A. Christoffel, H. K. Wong, and P. J. Barrett, A continuous seismic profiling survey in McMurdo Sound, Antarctica, *Dry Val. Drilling Proj. Bull.*, *5*, 167-179, 1975.

Porter, S. C., and J. E. Beget, Provenance and depositional environments of Late Cenozoic sediments in permafrost cores from lower Taylor Valley, Antarctica (abstract), *Dry Val. Drilling Proj. Bull.*, *8*, 74-76, 1978.

Powell, R. D., Sedimentation conditions in Taylor Valley, Antarctica, from DVDP cores (abstract), *Dry Val. Drilling Proj. Bull.*, *8*, 77-80, 1978.

Ronan, T. E., and J. H. Lipps, Sediments and biological environment under the Ross Ice Shelf, Antarctica (abstract), *Eos Trans. AGU*, *59*(4), 308, 1978.

Tazieff, H., About deep-sea volcanism, *Geol. Rundsch.*, *61*(2), 470-480, 1972.

Treves, S. B., C. G. Barnes, and D. P. Stillwell, Geology and petrography of rocks from the Ross Sea near Ross Island and the mouth of Taylor Valley, *Antarct. J. U.S.*, *10*, 297-302, 1975.

Veeh, H. H., and J. Chappell, Astronomical theory of climate changes: Support from New Guinea, *Science*, *167*, 862-865, 1970.

Vella, P.O., Surficial geological sequence, Black Island and Brown Peninsula, McMurdo Sound, Antarctica, *N.Z. J. Geol. Geophys.*, *12*, 761-770, 1969.

Warren, G., Geology of the Terra Nova Bay-McMurdo Sound area, Victoria Land, Geological Map of Antarctica, scale 1-1,000,000, Antarctic Map Folio Ser., edited by V. C. Bushnell, Amer. Geogr. Soc., New York, 1969.

Webb, P. N., Wright Fjord, Pliocene marine invasion of an Antarctic dry valley, *Antarct. J. U.S.*, *7*, 227-234, 1972.

Webb, P. N., and H. T. Brady, Cenozoic glaciomarine sediments at site J-9, southern Ross Ice Shelf (abstract), *Eos Trans. AGU*, *59*(4), 309, 1978.

Webb, P. N., and J. H. Wrenn, Late Cenozoic micropaleontology and biostratigraphy of eastern Taylor Valley, Antarctica, in *Antarctic Geoscience*, edited by C. Craddock, University of Wisconsin Press, Madison, in press, 1980.

MAGNETIC STRATIGRAPHY OF DVDP DRILL CORES AND LATE CENOZOIC HISTORY OF TAYLOR VALLEY, TRANSANTARCTIC MOUNTAINS, ANTARCTICA

DONALD P. ELSTON AND STEPHEN L. BRESSLER

U.S. Geological Survey, Flagstaff, Arizona 86001

Glaciogenic sections cored in Taylor Valley, and their magnetic polarity and susceptibility zonations, are in part temporally controlled and constrained by paleontologic and isotopic ages. When the magnetic zonations are evaluated in light of the ages, stratigraphy, provenance of the sediments, field geologic relationships, and the polarity time scale, an improved understanding emerges of the late Cenozoic glacial and structural history of Taylor Valley and adjacent McMurdo Sound. Polarity zones underlying a paleontologic datum that approximately marks the Miocene-Pliocene boundary (208 m deep in the Commonwealth Glacier drill core) suggest that the stratigraphically lowest glacial sediments cored in eastern Taylor Valley (326 m below the surface) are about 7 m.y. old. These late Miocene sediments are comparatively fine grained, in part are of Taylor Valley provenance, and from faunal data were deposited when Taylor Valley was a fjord 600-900 m deep. Uplift (perhaps preceded by a lowering of sea level) took place between about 5.1 and 4.6 m.y. ago, followed by the deposition of coarse, ice-cemented glacial sediments in a shallower Taylor Valley fjord. Uplift was accompanied by the appearance of basaltic detritus of the McMurdo Volcanic Group of Ross Sea provenance, recording the onset of volcanism in the Ross Sea and in Taylor Valley. During Pliocene time, continental glacier ice merged with shelf ice of the Ross Sea in the fjord. Renewed uplift, or perhaps a lowering of sea level, resulted in exposure of the upper ~100 m of fjord deposits in eastern Taylor Valley. Exposure occurred no earlier than late Pliocene or early Pleistocene time (<2.4 to ~1.8 m.y. ago). Since then, the fjord deposits were variously eroded and mantled by glacial drift related to late Pleistocene incursions of the Ross Ice Sheet, and subsequently further modified by Holocene stream erosion and deposition.

INTRODUCTION

Magnetic zonations and correlations of DVDP cores drilled in Taylor Valley [*Purucker et al.*, this volume, Figures 1, 8, and 9] are here related, with acknowledged uncertainties, to the polarity time scale. Because the age of the sediments is young, ranging from a few thousand years at the surface of Taylor Valley to a few million years at depths reached by the drill, estimates of minimum age can be derived solely from polarity zonations of the individual cores. Such minimum estimated ages, however, commonly are shown to be low when other geologic information is employed for temporal control. We here review paleontologic and isotopic ages in light of the polarity and susceptibility zonations. Use of the magnetic zonations has resolved some uncertainties and even apparent conflicts in age reported from other studies. However, because of hiatuses in deposition and the uncertainty of some polarity reversals (owing to the coarse-grained nature of parts of the glacial sections), no unique correlation can be proposed with the polarity time scale. Nonetheless, when the magnetic zonations are evaluated in light of the stratigraphy, paleontology, provenance, isotopic ages, and geologic framework, a provisional correlation with the polarity time scale emerges, as does an apparently improved understanding of the late Cenozoic glacial and structural history of Taylor Valley. We emphasize that additional study is needed to establish and refine the outline summarized here.

GEOLOGIC SETTING

The Dry Valley Drilling Project (DVDP) cores in Taylor Valley were drilled with the object of obtaining stratigraphic and age data bearing on the glacial history of the region from deposits that underlie the valley floor. Glacial deposits in the floor and in the lower walls of Taylor Valley characteris-

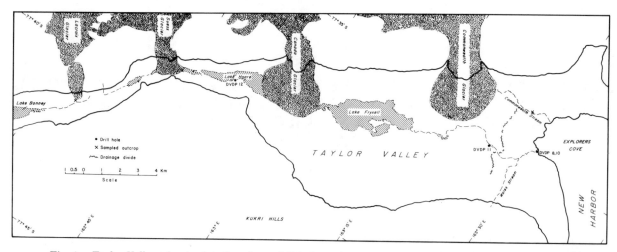

Fig. 1. Taylor Valley, showing location of core drill holes and outcrops sampled for paleomagnetic analysis. Outline of valley floor drawn on 250-m contour, U.S. Geological Survey 1:50,000 scale, Lake Fryxell (1977) and Lake Bonney (1977) quadrangles.

tically weather to a disaggregated mantle that obscures details of the stratigraphy of underlying ice-cemented sediments. Locally, sections a few meters to more than 20 m thick are exposed along and near stream courses. At most places, however, exposures are obtained only by the removal of the disaggregated overburden.

The youngest deposit in the area of study is a coarse-grained, moderately well sorted and stratified sand that forms a modern delta at the terminus of Wales Stream at Explorers Cove, New Harbor (Figure 1). DVDP holes 8 and 10 were collared in this Holocene deposit and reached its base at a depth of no less than 25 m. West (upstream) of New Harbor, similar deposits form low terraces adjacent to Wales Stream, and they rest unconformably on, and appear to wedge out against, comparatively fine-grained glacial deposits of unknown age. The older sediments form a hilly and partly dissected terrain in the southern floor of eastern Taylor Valley, in which DVDP hole 11 was collared (Figure 1). A coarse drift of Ross Sea I glaciation of *Denton et al.* [1970, 1971] caps part of the hilly terrain (chronozone IIIa of *Vucetich and Robinson* [1978, Figure 10]). This drift was deposited in late Pleistocene time from beneath the Ross Ice Sheet during its most recent occupation of Taylor Valley, between 34,800 and 9490 years ago [*Denton et al.*, 1971, Table 1]. In exposures along Commonwealth Stream about 2.7 km north of hole 11, poorly consolidated glacial sediments underlie the coarse Ross Sea I drift. These deposits, sampled for the determination of polarity (Figure 1), were found to be normally polarized, indicating that the sediments underlying the coarse drift near Commonwealth Stream are of Bruhnes age and are less than 700,000 years old.

Hole 11 was collared about 20 m below the highest deposits that underlie the coarse drift, as seen along the axis of a partially dissected transverse divide in the valley floor, about 400 m northeast of hole 11. The stratigraphy of these deposits has yet to be related to the section penetrated in the uppermost part of hole 11. *McKelvey* [1975] and *Vucetich and Robinson* [1978] have recognized that a young deposit, a few meters thick, overlies distinctly older glacial sediments at and near hole 11.

In central Taylor Valley, stratigraphic relations also have yet to be resolved between glacial deposits locally exposed low in the valley walls and deposits underlying the valley floor. The near-surface materials appear to be only a few thousand to a few hundred thousand years old on the basis of stratigraphic studies [*Denton et al.*, 1971] and isotopic ages [*Hendy et al.*, 1979]. We sampled deposits in an exposure on the valley wall adjacent to the terminus of Lacroix Glacier for the determination of polarity (Figure 1). All samples were found to be of normal polarity, supporting a young age assignment for the near-surface deposits. DVDP hole 12, located in the axis of the valley adjacent to Lake Hoare (Figure 1), was cored through the glacial fill to the basement. Although the polarity record in the cored sediments is ambiguous in some

DVDP 10, New Harbor

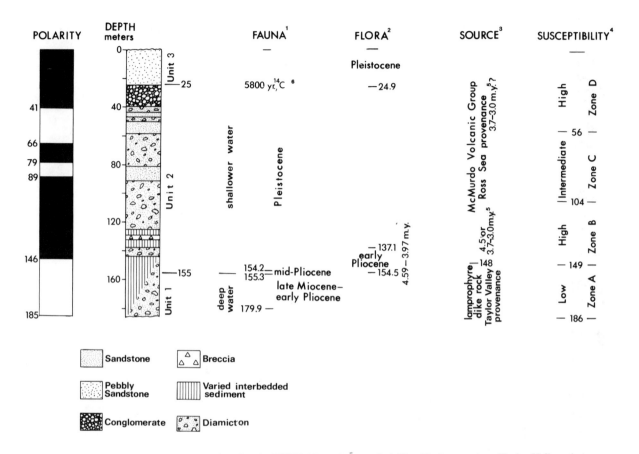

Fig. 2. Compilation of geologic data for the DVDP 10 section cored at New Harbor, eastern Taylor Valley, Antarctica. Lithologic column from *McKelvey* [1975]. Superscript numbers represent the following references: 1, *Webb and Wrenn* [1977]; 2, *Brady* [1978a, b]; 3, *Porter and Beget* [this volume]; 4, *Purucker et al.* [this volume]; 5 *Armstrong* [1978]; 6, *Stuiver et al.* [1976]. Depths are in meters.

detail, there is evidence from polarity alone that some of the glacial fill is older than 700,000 years and that the lower part of the section could be at least 2 m.y. old. However, geologic and paleomagnetic data are insufficient to allow unambiguous correlations to be made between the subsurface sections of eastern and central Taylor Valley. Such correlations are needed to clarify interrelations of glacial and glacial-marine deposits of Taylor Valley provenance and deposits derived from shelf ice of the Ross Sea.

EASTERN TAYLOR VALLEY

The polarity and susceptibility zonations for the New Harbor (DVDP 10) and Commonwealth Gla-

cier (DVDP 11) cores are summarized in Figures 2 and 3. Generalized stratigraphic sections and pertinent paleontologic, isotopic age, and provenance data are also summarized in these figures. Source areas for the sediments are interpreted from the mineralogy and the magnetic mineralogy and susceptibility. K-Ar ages [*Armstrong*, 1978] reported for the McMurdo Volcanic Group [*Nathan and Schulte*, 1968] provide constraints as to the time of introduction of volcanic material into the sedimentary record. Temporal control is obtained by comparing these stratigraphic ages with paleontologic ages and with ages inferred from referring the polarity zonation to the polarity time scale.

Lithologic variability and an apparent lack of lateral persistence of individual beds, seen in descrip-

DVDP 11, Commonwealth Glacier

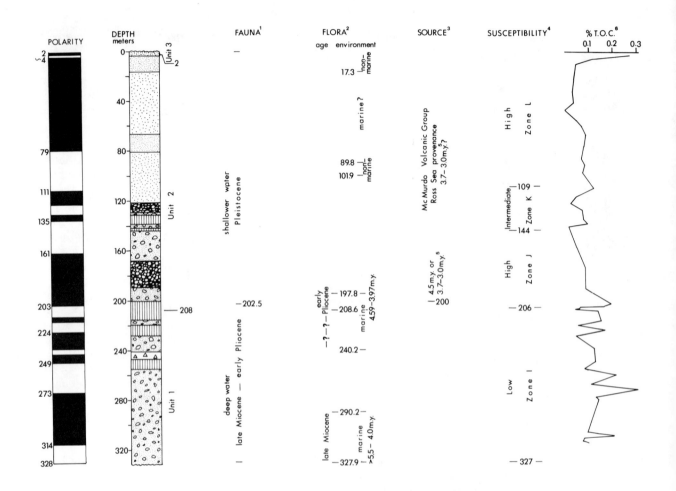

Fig. 3. Compilation of geologic data for the DVDP 11 section cored near Commonwealth Glacier, eastern Taylor Valley, Antarctica. Lithologic column from *McKelvey* [1975]. See Figure 1 caption for references to superscripts 1-6; superscript 7 is reference to *Wrenn and Beckman* [this volume]. Depths are in meters.

tive logs of sections from the closely adjacent holes 8 and 10 [*McKelvey*, 1975], preclude the use of physical stratigraphy for detailed correlations. Nonetheless, intervals of particularly coarse diamictons are found above a depth of 150 m in hole 10 and above 200 m in hole 11. Their appearance approximately marks the time of introduction of basaltic and trachytic detritus of the McMurdo Volcanic Group of Ross Sea provenance [*Porter and Beget*, this volume]. An interval marked by the deposition of abundant volcanic material is recorded magnetically by a correlative zone of high susceptibility (zones B, F, and J, holes 8, 10 and 11, respectively; Figures 4a, 4b, and 4c of

Purucker et al. [this volume]). The high susceptibility and intensity reflect the presence of titanomagnetite, which appears to characterize the eruptive rocks of Ross Sea provenance. Flows of the McMurdo Volcanic Group in central and western Taylor Valley presumably were not the source for the zones of high susceptibility in the eastern Taylor Valley sections because they appear to contain titanohematite, a weakly magnetic mineral, rather than titanomagnetite and thus would not contribute significantly to the magnetization of the sediments [*Purucker et al.*, this volume].

The McMurdo Volcanic Group includes all the volcanic rocks of Tertiary and Quaternary age in northern Victoria Land. Although a maximum K-Ar age of about 18 m.y. has been reported, most ages determined are less than 5 m.y. [*Armstrong*, 1978, Figure 4, Table 1]. The oldest volcanic rocks of Ross Island are found in the northern part of the island on Mount Bird [4.5 ± 0.6 m.y.) and on Cape Bird (3.0 to 3.1 ± 0.1 to 0.2 m.y.) [*Armstrong*, 1978, Table 1, Figure 3, and p. 694]. If the ages are taken at face value, volcanism in the Ross Sea began in early Pliocene time, followed by a second eruptive episode in middle Pliocene time. A similar sequence also is reported from basaltic rocks in central and western Taylor Valley [*Armstrong*, 1978]. On the basis of paleontologically dated sediments that contain basaltic detritus (Figures 2 and 3), the older (4.5 m.y.) K-Ar ages appear valid; however, they are few in number, and a problem of contamination by excess argon has been recognized in some of the younger rocks of the McMurdo Volcanic Group (a situation that may or may not apply for the older rocks). The lower parts of the zones of high susceptibility in holes 10 and 11, which reflect titanomagnetite-bearing detritus of Ross Sea provenance, include stratigraphic intervals that contain early Pliocene diatoms. These beds, in turn, directly overlie strata bearing a fauna assigned to the late Miocene or early Pliocene.

Paleontologic ages reported from the fauna and flora are in part mutually supporting but in part do not agree (see Figures 2 and 3). *Webb and Wrenn* [1976, 1977] identified in situ deep-water (600-900 m) benthic foraminifera to which they assigned a late Miocene and early Pliocene age; these foraminifera are found at minimum depths of 156 m in hole 10 and 202 m in hole 11. This age generally agrees with a diatom age of early Pliocene (3.97-4.59 m.y.) based on the uppermost age range of *Thallassiosira torokina* and the earliest occurrence of *Nitzschia praeinterfrigidaria* [*Brady*, 1978*a*, *b*, 1979]. However, *Webb and Wrenn* [1976, 1977, 1980] assigned a middle Pliocene age to foraminifera from a depth of 155.3-154.2 m in hole 10, and a Pleistocene age to foraminifera in strata above 154 m in hole 10 and above 170 m in hole 11. These ages conflict with the early Pliocene diatom age assigned to DVDP 10 from 154.5 to 137.1 m [*Brady*, 1978*a*, *b*] and DVDP 11 from 208.6 to 194.8 m, which Brady assigned as early Pliocene, and *Webb and Wrenn* [1976] zoned first as late Miocene and then as late Miocene and early Pliocene [*Webb and Wrenn*, 1977, 1980]. Ages for the two diatom species come from deep-sea cores that are tied to

the polarity time scale [*Brady*, 1978*a*, *b*, 1979]; thus fauna from the same beds that are assigned middle Pliocene and Pleistocene age appear to be longer ranging than previously thought.

The paleontologic age agrees with the maximum K-Ar age for the onset of volcanism in the Ross Sea and Taylor Valley. The essentially coincident appearance of detritus of the McMurdo Volcanic Group in the sediments suggests that subaerial eruptions efficiently distributed materials over shelf ice, and presumably over sea ice and open sea as well. Strata of late Miocene or early Pliocene age (~5 m.y. old) that immediately predate the onset of volcanism occur at depths of 155 m at New Harbor and 208 m at Commonwealth Glacier. From referring the polarity zonation in hole 11 to the polarity time scale, strata below 208 m would appear to be late Miocene in age, at least 6 m.y. and perhaps as much as 7 m.y. old at the base. The polarity zonation from 208-326 m indicates a greater time span for the deposition of the lower part of the DVDP 11 section than is assigned from diatoms.

Brady [1978*a*, *b*, 1979] tentatively assigned an early Pliocene age to strata from a depth 209 m to about 240 m in hole 11, and below 154 m to the bottom of hole 10, from diatom fragments that might be *Nitzschia praeinterfrigidaria*. The maximum age Brady reported for this species appears too young to accord with the polarity zonation in beds underlying 208 m in hole 11. *Brady* [1978*a*, *b*, 1979] also identified a diatom of late Miocene age *(Thallassiosira torokina)* from below 290.15 m in hole 11, which he reported has a presently known range that extends from the 'b' event of the Gilbert epoch to epoch 5 (late Miocene), or 4.0-5.5 m.y. Again, the maximum age is younger than would be inferred from the polarity zonation.

Diagnostic paleontologic evidence for age is lacking for much of the upper parts of sections cored at New Harbor and Commonwealth Glacier. However, in hole 10 at a depth of 24.86 m and above, *Brady* [1978*a*, *b*, 1979] reported diatoms that he assigned to the Bruhnes epoch (<700,000 years) of the Pleistocene. A ^{14}C age of about 5800 years (Holocene) has been reported from an in situ bivalve shell collected from a depth of about 25 m in hole 10 [*Stuiver et al.*, 1976]. Holes 8 and 10 are collared in sediment of a modern delta [*Chapman-Smith*, 1978] that is forming at New Harbor. The base of the modern deposit thus may be at a depth of 25 m, corresponding to the abrupt appearance of a modern flora, a level that also is marked by a small but abrupt change in susceptibility (Figures 4a and 4b of *Purucker et al.* [this volume]). Alternatively, the base of the modern delta could lie as deep as 39 m (a short distance above the highest reverse-to-normal polarity switch), a depth proposed by *Vucetich and Robinson* [1978]. The shallower depth of 25 m is favored here because it apparently marks an abrupt change in the environment of deposition recorded by the fauna and is supported directly by isotopic dating and indirectly by magnetic data (lower susceptibility and highly scattered and unreliable magnetic inclinations are found above 25 m).

A near-surface interval of reversed polarity was found at a depth of 1.67 to ~5.3 m in hole 11. It was determined from seven of eight samples collected in the interval. Seven of the eight samples displayed fairly steep positive (reversed polarity) inclinations prior to cleaning. Following cleaning in peak fields of 150 Oe, six of the eight samples displayed yet steeper positive inclinations. However, only two of the six samples subsequently passed the Q-test for coherence of magnetization, and these samples were from a depth of about 2 m [*Purucker et al.*, this volume, Figure 2c]. The failure of four samples to pass the Q-test was unexpected, and the reason for the failure is as yet unknown. The steep inclinations accord with the idea of a detrital remanence in the fine-grained, well-sorted, and partly thin-bedded sandstone. The clustering of steep positive inclinations following cleaning, and the fact that two of the samples passed the Q-test, indicate an interval of reversed polarity. Moreover, the core segments were examined for any evidence that the core in this interval had been inverted in handling; none was found. The samples came from four core-drilling runs, and core ends and lithology appear to match. We thus conclude that strata containing an interval of reversed polarity, possibly as much as 3.5 m thick, directly underlie the surficial deposits at the hole 11 site.

Deposits above about 2 m in hole 11 are of normal polarity. From field studies and studies of the core [*McKelvey*, 1975; *Vucetich and Robinson*, 1978], the deposits at and closely underlying the surface are considered to be very young. *Vucetich and Robinson* [1978, Figure 6] assigned deposits in the upper ~2 m of hole 11 to 'chronozone IIIa,' which they correlate with Ross Sea I glaciation of *Denton et al.* [1971] having an age range of 34,800-

9490 years. Vucetich and Robinson tentatively correlate strata below about 4 m in hole 11 with beds below a depth of 39 m in hole 10, assigning these strata to an older 'chronozone II.' Additionally, they correlate chronozone II partly with Alpine II glaciation of Taylor Valley and partly with Ross Sea II glaciation, reporting an age range of about 2.1-0.4 m.y.

From field relations, the reversed polarity beds from about 2 to 5 m in hole 11 thus appear to be bounded at the top by an unconformity marking a significant hiatus, suggesting the reversed polarity beds are at least 700,000 years old (boundary between the Matuyama reversed and Bruhnes normal polarity epochs). Because a long interval of normal polarity underlies the 2- to 3-m thick interval of reversed polarity in hole 11 (Figure 3), strata at a depth of 5 m could be as old as 2.4 m.y. (boundary between the Gauss normal and Matuyama reversed polarity epochs). In such a situation, the age for chronozone II and for the Ross Sea II glaciation would be somewhat greater than reported by *Vucetich and Robinson* [1978]. The likelihood is not great that the reversed polarity interval from 2 to 5 m in hole 11 records the very young Blake event, 110,000 years old and of a few thousand years' duration [*Smith and Foster*, 1969]. Although such an assignment does not appear to be supported by the field evidence, it cannot be completely ruled out at this time.

From the foregoing considerations the polarity zonation in hole 11, above 208 m and below 2 m, is assigned with least conflict to the polarity time scale for the Pliocene. Such an assignment seems to accord best with a scheme for the timing of uplifts and the history of emergence of the Taylor and Wright fjords deduced by various workers from geologic mapping, stratigraphic studies, and isotopic dating [*Denton et al.*, 1970, 1971; *Webb*, 1972a, b, 1979; *Vucetich and Topping*, 1972; *Wrenn and Webb*, 1980; *Armstrong*, 1978].

In summary, three discrete stratigraphic units are recognized in the drill core sections at New Harbor and Commonwealth Glacier (Figures 2 and 3). The lowest unit, unit 1, represents the accumulation of comparatively fine-grained glacial sediment deposited from ice in a fjord that was 600-900 m deep. From the paleontologic age at the top of this depositional interval (208 m in hole 11) and the polarity zonation in the underlying beds, about 2 m.y. of time, mostly late Miocene, is represented in unit 1. The intermediate unit, unit 2, records an

DVDP 12, Central Taylor Valley

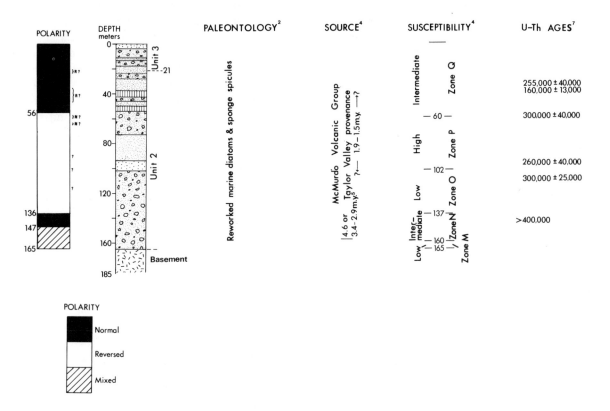

Fig. 4. Compilation of geologic data for the DVDP 12 section cored in central Taylor Valley near Lake Hoare, Antarctica. Lithologic column compiled from descriptive log of *Chapman-Smith* [1975]. Superscript 2, *Brady* [1978c]; 4, *Purucker et al.* [this volume]; 5, *Armstrong* [1978]; 8, *Hendy et al.* [1979, Table 1]. Depths are in meters.

abrupt decrease in total organic content, the introduction of basaltic detritus of the McMurdo Volcanic Group of Ross Sea provenance, a distinct shallowing of the fjord, the onset of severe glacial conditions marked by the deposition of coarse diamictons, and cementation of the sediments by ice. This unit extends to within 25 m (or alternatively 39 m) of the surface at New Harbor, and probably within 2 m of the surface at the Commonwealth Glacier drill site, 78 m above sea level. The uppermost stratigraphic unit, unit 3, records the accumulation of late Pleistocene glacial deposits and Holocene fluvioglacial deposits that unconformably overlie and irregularly mantle the exposed fjord deposits.

CENTRAL TAYLOR VALLEY

The lithology, magnetic zonations, provenance, paleontology, and U-Th ages for hole 12 are sum-

marized in Figure 4. Hole 12 was drilled adjacent to Lake Hoare on the axis of Taylor Valley about 16 km west of New Harbor. It is collared in glacial drift assigned to a late stage of glaciation in Taylor Valley having an age of about 10,000 years [*Denton et al.*, 1970, Figure 3]. The base of the young deposits in hole 12 can be no lower than the base of the interval of normal polarity (56 m), and probably lies above 38 m, above carbonate that has been dated by the U-Th method as 160,000 years [*Hendy et al.*, 1979, Table 1, sample 76-53]. We have provisionally placed the base of unit 3 at 21 m, above the highest interval of possible reversed polarity. This highest horizon of possible reversed polarity is also a place marked by a change in susceptibility within susceptibility zone Q [see *Purucker et al.*, this volume, Figure 4d].

The glacial section beneath 21 m in hole 12 is classed as unit 2 and is considered broadly equivalent in age to some part or parts of unit 2 of eastern Taylor Valley. From the polarity, the age of much of the hole 12 section is greater than 700,000 years old, the age of the (reversed to normal) Matuyama-Bruhnes polarity boundary. If some of the queried polarity reversals shown in Figure 4 are real, then strata near the bottom of the section in the interval of mixed polarity are at least 2 m.y. old. Such an age would be consistent with K-Ar ages reported for flows in Taylor Valley [*Armstrong*, 1978, Table 1], which were the source of basaltic detritus seen across the hole 12 section. Two samples of dike rock associated with basalt flows in west-central Taylor Valley have ages of about 4.6 m.y. If excess Ar has not contaminated the rock to give anomalously great ages, the 4.6 m.y. age indicates that volcanism in Taylor Valley began at the same time as that in the Ross Sea. The major period of volcanism in Taylor Valley, however, appears to have occurred about 3.4-2.9 m.y. ago, following volcanism about 3.75-3.5 m.y. ago in nearby Wright Valley [*Armstrong*, 1978, Table 1]. Detritus derived from flows of this middle Pliocene eruptive episode could have been incorporated in sediments of late Pliocene and Pleistocene age. *Armstrong* [1978] reports that polarities of the 2.9-3.4 m.y. old flows in Taylor Valley are dominantly reversed and concludes that much of the early volcanism occurred about 3 m.y. ago during two intervals of reversed polarity in the Gauss normal polarity epoch. *Armstrong* [1978] also has recognized a third, younger episode of basaltic volcanism in Taylor Valley that occurred in late Pliocene and early Pleistocene time, from about 1.9 to 1.5 m.y. ago. Flows of this episode crop out in the western part of the valley, near the terminus of Taylor Glacier.

Basaltic detritus low in the hole 12 section may have been the product of the main stage volcanism of middle Pliocene age. Products of the younger late Pliocene and early Pleistocene volcanism perhaps were incorporated within the main reversed polarity interval in core 12, but their incorporation would not necessarily correspond to high susceptibility zone P. Zone P reflects an increase in magnetite content, the most likely source of which is the Ferrar Dolerite (Jurassic). In contrast, basaltic detritus in Taylor Valley appears to be characterized by titanohematite, which is much more weakly magnetic than magnetite. Basaltic flows in Taylor Valley may have been erupted into ice to produce the titanohematite [*Purucker et al.*, this volume], although no hyaloclastite or other evi-

dence for subglacial eruption has been reported (S. Porter, written communication, 1979). An increase in detritus attributable to erosion of the Ferrar Dolerite during a time of reversed polarity accords with the idea of increased glacier activity in the early Pleistocene. Subaerial and marginal marine(?) glaciation presumably would have followed regional uplift and the emergence of Taylor and Wright valleys. Marine deposits may have once extended into Taylor Valley, suggested but not proven by the presence of reworked marine diatoms and sponge spicules reported from core 12 [*Brady*, 1978c]. Additionally, a lack of terminal moraines for the Taylor Glacier suggests that the terminus of the glacier floated in a body of water that perhaps was trapped between the Ross Sea Ice Sheet and the Taylor Glacier (G. H. Denton, personal communication, 1979). Paleontologic dating of core 12 [*Webb and Wrenn*, 1976; *Brady*, 1978c] has not proved successful.

Hendy et al. [1979, Table 1] report U-Th ages from carbonate rocks in core 12, as well as from carbonates collected from exposures of till in the wall of Taylor Valley. Although the U-Th ages when plotted on a histogram appear to agree with climate stages derived from the marine isotope record [*Hendy et al.*, 1979, Figure 3], the U-Th ages from carbonate rocks and chips and carbonate cement in the core of hole 12 do not accord with the polarity zonation. The isotopic ages appear too young, and thus may not date the time of deposition of the sediments.

The section in hole 12 cannot be correlated in any straightforward manner with the eastern Taylor Valley section. Deposition in central Taylor Valley, as concluded by *Chapman-Smith* [1978], was not related to deposition of the sections to the east. From the studies of *Denton et al.* [1970, 1971], central Taylor Valley appears to have been much more strongly affected by continental-based (Taylor) glaciation than eastern Taylor Valley [*Vucetich and Robinson*, 1978]. Fjord deposits equivalent in age to unit 1 and the early part of unit 2, if once present in central Taylor Valley, presumably were considerably eroded by Taylor Glacier after uplift and perhaps partial emergence in early Pliocene time.

Evidence in Wright Valley [*Webb*, 1972a, b, 1979; *Vucetich and Topping*, 1972; *Wrenn and Webb*, 1977, 1980] suggests that a second increment of uplift and presumably emergence, occurred at about the end of Pliocene time. Pa-

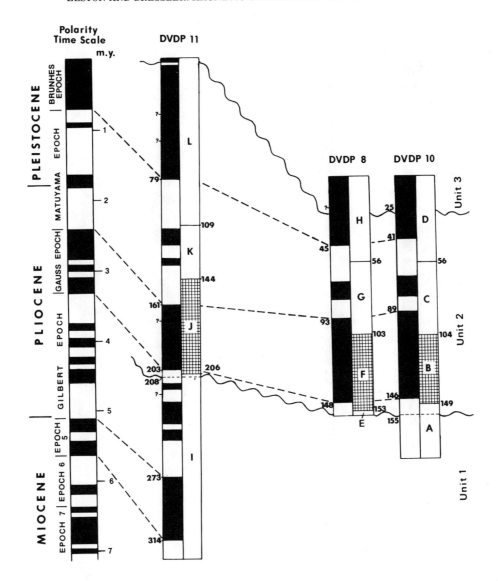

Fig. 5.   Correlation of magnetic zonations in eastern Taylor Valley sections with polarity time scale of *LaBrecque et al.* [1977]. Correlation employs the simplest visual fit and assumes that an uninterrupted depositional record extends from surface downward. Depths are in meters.

leomagnetic and radiolarian studies by *Fillon* [1972, 1974] have provided evidence for a widespread erosion surface in the Ross Sea separating late Bruhnes sediments from underlying sediments of early Matuyama and Gauss age. A minimum age for the time of onset of the erosion therefore is about 1.8 m.y., and the maximum age may be about 2.4 m.y. Onset of erosion presumably corresponded to a time of regional uplift of the Ross Sea basin and the dry valleys.

Considering the various lines of evidence, we suggest that the lower part of the hole 12 section below 147 m (but perhaps below 120 m) is no younger than late Pliocene. From the polarity zonation, Pleistocene deposits of Matuyama age may occur as high as 56 m depth. Younger deposits, related to the Taylor I and II glaciations of *Denton et al.* [1971] and partly correlative events I, II, and III of *Hendy et al.* [1979] that are less than ~300,000 years old, would lie in the normal polarity interval above 56 m. Deposits related to the older Taylor III glaciation of *Denton et al.* [1971] thus would lie below 56 m.

In this model, much of unit 2 of core 12 would be younger than strata in the upper part of unit 2 of eastern Taylor Valley.

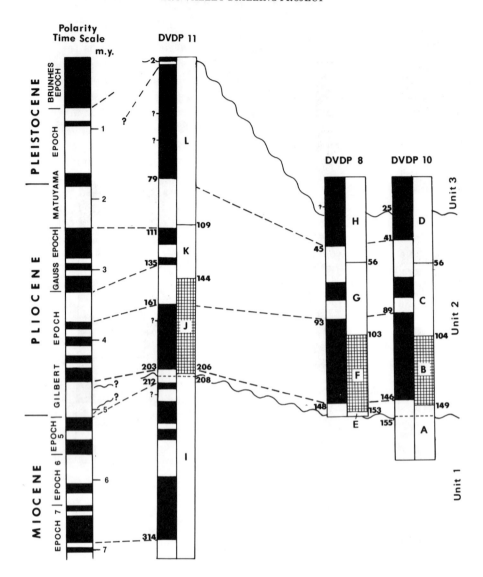

Fig. 6. Correlation of magnetic zonations with polarity time scale of *LaBrecque et al.* [1977] employing geologic data. Patterned intervals indicate zone of high susceptibility [*Purucker et al.*, this volume]. Wavy lines indicate positions of unconformities. Depths are in meters.

CORRELATION WITH POLARITY TIME SCALE

If the drill hole sections are assumed to have accumulated without significant interruption to the present time, and if only the thick polarity zones are considered to be valid, the following are minimum ages for the strata at the base of the drill hole sections. The estimated ages are derived from assigning the zonations to the polarity time scale of *LaBrecque et al.* [1977]: New Harbor, ~2.5 m.y.; Commonwealth Glacier, ~5.5 m.y.; and Lake Hoare, ~2 m.y. This rather mechanical approach,

in which few geologic and no paleontologic constraints are employed, is shown in Figure 5 for the New Harbor and Commonwealth Glacier sections. Ages assigned to purported magnetostratigraphic horizons in Figure 5 are distinctly younger than indicated when paleontologic and other geologic data are added to the analysis. A correlation using the geologic data (Figure 6) is presented here as a model for future testing and refining. Key to this model is a paleontologic and lithologic datum at approximately the Miocene-Pliocene boundary that serves as a horizon for correlation with the polarity time scale.

The oldest stratigraphic unit (unit 1) represents fairly regular deep-water deposition in which the sedimentary record may be more complete than in overlying units. In Figure 6 an age of 4.6 m.y. is assigned to a datum at a depth of 208 m and 155 m in holes 11 and 10, respectively, approximately marking the top of unit 1. In hole 11 the number of polarity reversals in unit 1 indicates a minimum age that is greater than 6 m.y., and a probable age of at least 7 m.y. at the base. The boundary between units 1 and 2 is shown as an unconformity, following the report of an unconformity at 154 m in core of hole 10 [Porter and Beget, this volume]. The unconformity perhaps represents scouring that accompanied elevation of the fjord bottom from deep water to shallower water, an event that appears to have occurred during the initial interval of reversed polarity of the Gilbert epoch.

The Miocene-Pliocene boundary lies within the reversed polarity interval that includes the lithologic, paleontologic, and magnetic datum. The age for the Miocene-Pliocene boundary is about 5.1 m.y. on the polarity time scale of LaBrecque et al. [1977]. However, a Southern Hemisphere age of about 4.6 m.y. has been proposed for the time of climatic cooling at the accepted base of the Pliocene in New Zealand [Lienert et al., 1972]. This age appears to accord better with the time of onset of basaltic volcanism, uplift of the Taylor Valley fjord, and the beginning of intense glaciation in the Ross Sea that is documented in the DVDP cores. This younger age for the Miocene-Pliocene boundary has been supported by additional but as yet unpublished studies of polarity in marine strata exposed in New Zealand (D. A. Christoffel, personal communication, 1979). However, other studies of marine sediments in New Zealand [Loutit and Kennett, 1979] have led to the proposal that the Miocene-Pliocene boundary is about 5.3 m.y. old. Such an age, from the magnetostratigraphic and paleontologic information in hole 11 referred to the polarity time scale, appears too great. The Miocene-Pliocene boundary lies within the initial reversed polarity interval of the Gilbert, the age range for which is reported to be 4.59-5.12 m.y. [LaBrecque et al., 1977]. In holes 10 and 11 the faunal boundary lies in the middle to upper parts of an interval of reversed polarity, suggesting an age closer to 4.6 than 5.1 m.y.

The lower part of unit 2 in eastern Taylor Valley records the onset of basaltic volcanism in the Ross Sea, about 4.6 m.y. ago, apparently accompanying regional uplift. Uplift was followed by increasingly severe glaciation. Perhaps tephra from subaerial eruptions deposited on sea or shelf ice allowed for a near-synchronous introduction of basaltic detritus into the geologic record at New Harbor and Commonwealth Glacier shortly after the onset of volcanism. As the glacial deposits became coarser grained, polarity zones almost certainly were less perfectly recorded. In the correlation shown in Figure 6, much of the fine polarity structure from the Gilbert epoch is assumed to have been lost, and the proposed correlation from the Gilbert and Gauss epochs is drawn on the basis of the main polarity intervals. An alternate interpretation would be to consider both the stratigraphic and the magnetic records above the datum (208 m in hole 11) to be complete. This assumption requires rapid deposition for the coarse-grained beds found in the interval from a depth of about 200-120 m, and would imply a generally older age for strata of unit 2. Although such an interpretation is permitted by the age data, we believe the correlation shown in Figure 6 is more reasonable. In it, polarity zonations of unit 2 in eastern Taylor Valley are assigned to the Gilbert and Gauss epochs, and provisionally to the early part of the Matuyama epoch. In central Taylor Valley, unit 2 of hole 12 may contain a thin section of strata of Gilbert and Gauss age and a comparatively thick section of strata of Matuyama age.

The boundary between units 2 and 3 appears to be a major unconformity developed after a second period of uplift of the fjord. In eastern Taylor Valley, deposits of unit 3 are subordinate in thickness to deposits of units 1 and 2.

Figure 7 summarizes a correlation of the DVDP drill hole sections in Taylor Valley. A lack of detailed correlation between central and eastern Taylor Valley probably results from complex stratigraphic relations between deposits of Taylor Valley origin and deposits owing to incursions of Ross Sea ice. Evidence from surface geologic mapping may not be adequate to resolve the interrelations.

GEOLOGIC HISTORY

A deep Taylor Valley fjord existed at least 6 m.y. and very likely more than 7 m.y. ago. Comparatively fine-grained diamictons and interstratified fine-grained sorted clastic sediments were deposited from ice (sea or shelf, and perhaps lobe ice)

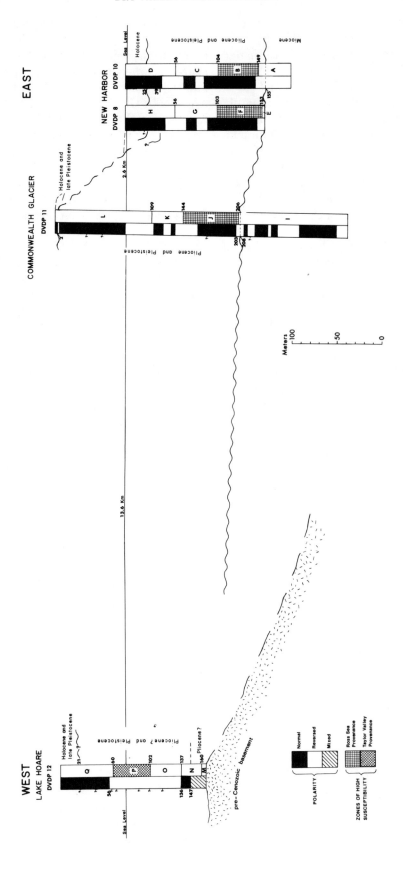

Fig. 7.   Correlation of sections between central and eastern Taylor Valley. Depths are in meters.

that occupied the fjord of eastern and central Taylor Valley. At times, ice locally may have been absent. In situ deep-water benthic foraminifera in cores from eastern Taylor Valley indicate that water depth was of the order of 600-900 m. At that time some lamprophyre dike-rock detritus was derived from western Taylor Valley, implying that erosion and transport from a throughgoing Miocene glacier were contributing sediment into the fjord. In eastern Taylor Valley the fine-grained, horizontally stratified, and well-bedded sediments and the fine-grained nature of interbedded diamictons that are assigned to the Miocene imply deposition in quiet and deep water. The fairly fine-grained nature of the diamictons perhaps also suggests a subdued glacial regime, characterized by the transport of sparse fine-grained detritus. Deposits that predate the appearance of basaltic detritus should exist between eastern and central Taylor Valley, and they should exhibit the increased effects of deposition of sediment glacially transported from western Taylor Valley. These older prebasalt deposits presumably thin from east to west and disappear by onlap on the basement erosion surface, which between Lake Hoare and Commonwealth Glacier has an eastward slope of about 1° or less.

The most pronounced stratigraphic, faunal, and magnetic breaks in the eastern Taylor Valley cores occur at a level that appears to correspond to a time of climatic cooling, and perhaps to the Miocene-Pliocene boundary. An age of about 4.6 or 5.1 m.y. can be assigned to strata near the break in the cores. The younger age comes from paleontologic ages, and from K-Ar ages marking the onset of basaltic volcanism. A 5.1 m.y. age derives if the break corresponds to the Miocene-Pliocene boundary and if the boundary is referred to the polarity time scale of *LaBrecque et al.* [1977].

An age of 4.6 m.y. for the onset of widespread glacial conditions is close to the time of increased deposition of ice-rafted debris on the Macquarie Ridge, south of New Zealand, about 4.4 m.y. ago [*Kennett and Brunner*, 1973]. To account for an unconformity and zone of oxidation reported at the Miocene-Pliocene boundary in the New Harbor section, uplift is needed to bring the Miocene section to near sea level from a depth of 600 m or more, after about 70-100 m of eustatic lowering of sea level about 5.5 m.y. ago [*Berggren and Haq*, 1976]. Major structural uplift in late Miocene and

early Pliocene time presumably was responsible for the emergence of Wright Valley and for a partial emergence of central and western Taylor Valley.

In early Pliocene time, uplift was accompanied and followed by (1) basaltic volcanism in the Ross Sea and Taylor Valley, (2) the introduction of basaltic materials of Ross Sea provenance into the sedimentary record in eastern Taylor Valley, (3) intense glaciation marked by the deposition of coarse diamictons and coarse stratified deposits, and (4) an increase in the cementation of the sediments by ice. In western and central Taylor Valley, a throughflowing glacier extended from the west into the fjord, presumably eroding as well as depositing sediment from its base along the axis of the valley. Taylor Glacier ice perhaps merged with Ross Sea ice, hindering ablation of the sea or shelf ice. Such a situation could have delayed the end of deposition of abundant basaltic detritus of Ross Sea provenance near Commonwealth Glacier with respect to deposition at New Harbor (seen in the time-transgressive upper boundary of the zone of high susceptibility revealed by the polarity zonation). Marine deposition continued in eastern Taylor Valley until late Pliocene or early Pleistocene time (about 2.4-1.8 m.y.) when renewed uplift and marine emergence, accompanied and followed by basaltic volcanism in western Taylor Valley, exposed Taylor Valley to the increased effects of continental (Taylor) and alpine glaciation, subordinating the effects of Ross Sea glaciation. In eastern Taylor Valley a major unconformity appears to separate poorly indurated glaciogenic deposits of late Pleistocene and Holocene age from better indurated deposits of Pliocene and early Pleistocene(?) age.

Acknowledgments. Research leading to this report was supported in part by National Science Foundation grants DPP-76-00941, DPP-77-13048, and DPP-79-07253 and in part by the U.S. Geological Survey. Early discussions concerning the geology and ages in the dry valleys with R. L. Armstrong, D. A. Christoffel, and J. H. Wrenn were greatly appreciated, and we thank S. C. Porter and J. H. Wrenn for providing preprints summarizing results of their work. Reviews by S. C. Porter, D. Champion, and R. Reynolds greatly improved the quality of the original manuscript. Field work in central and eastern Taylor Valley during November 1979, and illuminating reviews of geologic relations in the field with G. H. Denton, have allowed us to couch the paleomagnetic results from the DVDP cores in a framework more closely attuned to field geologic relations.

REFERENCES

Armstrong, R. L., K-Ar dating: Late Cenozoic McMurdo volcanic group and dry valley glacial history, Victoria Land, Antarctica, *N.Z. J. Geol. Geophys.*, *21*(6), 685-698, 1978.

Berggren, W. A., and B. U. Haq, The Andalusian stage (late Miocene): Biostratigraphy, biochronology, and paleoecology, *Palaeogeogr. Palaeoclimatol. Paleoecol.*, *20*, 67-129, 1976.

Brady, H. T., The dating and interpretation of diatom zones in DVDP 10 and 11, *Dry Val. Drilling Proj. Bull.*, *8*, 5-6, 1978*a*.

Brady, H. T., The dating and interpretation of diatom zones in Dry Valley Drilling Project holes 10 and 11, Taylor Valley, South Victoria Land, Antarctica, report, 37 pp., Jap. Inst. of Polar Stud., Tokyo, 1978*b*.

Brady, H. T., A diatom report on DVDP cores (3, 4A, 12, 14, 15) and other related surface sections, *Dry Val. Drilling Proj. Bull.*, *8*, 7, 1978*c*.

Brady, H. T., The dating and interpretation of diatom zones in Dry Valley Drilling Project holes 10 and 11, Taylor Valley, South Victoria Land, Antarctica, in *Proceedings of the Seminar III on Dry Valley Drilling Project, 1978, Spec. Issue 13*, edited by T. Nagata, pp. 150-163 and 10 plates, National Institute of Polar Research, Tokyo, 1979.

Chapman-Smith, M., Geologic log of DVDP-12, Lake Hoare, Taylor Valley, *Dry Val. Drilling Proj. Bull.*, *5*, 61-70, 1975.

Chapman-Smith, M., The Taylor formation (Holocene) and its macrofaunas, Taylor Valley, Antarctica (abstract), *Dry Val. Drilling Proj. Bull.*, *8*, 9-10, 1978.

Denton, G. H., R. L. Armstrong, and M. Stuiver, Late Cenozoic glaciation in Antarctica: The record in the McMurdo Sound region, *Antarct. J. U.S.*, *5*(1), 15-21, 1970.

Denton, G. H., R. L. Armstrong, and M. Stuiver, The late Cenozoic glacial history of Antarctica, in *The Late Cenozoic Glacial Ages*, edited by K. K. Turekian, pp. 267-306, Yale University Press, New Haven, Conn., 1971.

Fillon, R. H., Evidence from the Ross Sea for widespread erosion, *Nature*, *238*(81), 40-42, 1972.

Fillon, R. H., Late Cenozoic foraminiferal paleoecology of the Ross Sea, Antarctica, *Micropaleontology*, *20*(2), 129-151, 1974.

Hendy, C. H., T. R. Healy, E. M. Rayner, J. Shaw, and A. T. Wilson, Late Pleistocene glacial chronology of the Taylor Valley, Antarctica, and the global climate, *Quaternary Res.*, *11*, 172-184, 1979.

Kennett, J. P., and C. A. Brunner, Antarctic late Cenozoic glaciation: Evidence for initiation of ice rafting and inferred increased bottom-water activity, *Geol. Soc. Amer. Bull.*, *84*, 2043-2052, 1973.

LaBrecque, J. L., D. V. Kent, and S. C. Cande, Revised magnetic polarity time scale for late Cretaceous and Cenozoic time, *Geology*, *5*, 330-335, 1977.

Lienert, B. R., D. A. Christoffel, and P. Vella, Geomagnetic dates on a New Zealand upper Miocene-Pliocene section, *Earth Planet. Sci. Lett.*, *16*(2), 195-199, 1972.

Loutit, T. S., and J. P. Kennett, Application of carbon isotope stratigraphy to late Miocene shallow marine sediments, New Zealand, *Science*, *204*, 1196-1199, 1979.

McKelvey, B. C., DVDP sites 10 and 11, Taylor Valley, *Dry Val. Drilling Proj. Bull.*, *5*, 16-60, 1975.

Nathan, S., and F. J. Schulte, Geology and petrology of the Campbell-Aviator Divide, northern Victoria Land, Antarctica, *N.Z. J. Geol. Geophys.*, *11*(4), 940-975, 1968.

Porter, S. C., and J. E. Beget, Provenance and depositional environments of late Cenozoic sediments in permafrost cores from lower Taylor Valley, Antarctica, this volume.

Purucker, M. E., D. P. Elston, and S. L. Bressler, Magnetic stratigraphy of late Cenozoic glaciogenic sediments from drill cores, Taylor Valley, Transantarctic Mountains, Antarctica, this volume.

Smith, J. D., and J. H. Foster, Geomagnetic reversal in Bruhnes normal polarity epoch, *Science*, *163*, 565-567, 1969.

Stuiver, M., G. H. Denton, and H. W. Borns, Carbon-14 dates of *Adamussium colbecki* (Mollusca) in marine deposits at New Harbor, Taylor Valley, *Antarct. J. U.S.*, *11*(2), 86-88, 1976.

Vucetich, C. G., and P. H. Robinson, Quaternary stratigraphy and glacial history of lower Taylor Valley, Antarctica, *N.Z. J. Geol. Geophys.*, *21*(4), 467-482, 1978.

Vucetich, C. G., and W. W. Topping, A fjord origin for the pecten deposits, Wright Valley, Antarctica, *N.Z. J. Geol. Geophys.*, *15*(4), 600-673, 1972.

Webb, P. N., Paleontology of late Tertiary-Quaternary sediments, Wright Valley, Antarctica, *Antarct. J. U.S.*, *7*, 96-97, 1972*a*.

Webb, P. N., Wright fjord, Pliocene marine invasion of an Antarctic dry valley, *Antarct. J. U. S.*, *7*, 227-234, 1972*b*.

Webb, P. N., Paleogeographic evolution of the Ross sector during the Cenozoic, in *Proceedings of the Seminar III on Dry Valley Drilling Project, 1978, Spec. Issue 13*, edited by T. Nagata, National Institute of Polar Research, Tokyo, 1979.

Webb, P. N., and J. H. Wrenn, Interpretation of foraminiferal assemblages from lower Taylor Valley (DVDP 8-12), *Dry Val. Drilling Proj. Bull.*, *7*, 105-107, 1976.

Webb, P. N., and J. H. Wrenn, Late Cenozoic micropaleontology and biostratigraphy of eastern Taylor Valley, Antarctica (abstract), Third Symposium on Antarctic Geology and Geophysics, p. 159, Univ. of Wisc., Madison, 1977.

Webb, P. N., and J. H. Wrenn, Late Cenozoic micropaleontology and biostratigraphy of eastern Taylor Valley, Antarctica, in *Antarctic Geoscience*, edited by C. Craddock, University of Wisconsin Press, Madison, in press, 1980.

Wrenn, J. H., and S. W. Beckman, Maceral and total organic carbon analyses of DVDP drill core 11, this volume.

Wrenn, J. H., and P. N. Webb, Physiographic analysis and interpretation of the Ferrar Glacier-Victoria Valley area (abstract), Third Symposium on Antarctic Geology and Geophysics, p. 167, Univ. of Wisc., Madison, 1977.

Wrenn, J. H., and P. N. Webb, Physiographic analysis and interpretation of the Ferrar Glacier-Victoria Valley area, Antarctica, in *Antarctic Geoscience*, edited by C. Craddock, University of Wisconsin Press, Madison, in press, 1980.

GEOLOGIC HISTORY OF HUT POINT PENINSULA AS INFERRED FROM DVDP 1, 2, AND 3 DRILLCORES AND SURFACE MAPPING

P H I L I P   R .   K Y L E

Institute of Polar Studies, Ohio State University, Columbus, Ohio 43210

Three holes (DVDP 1, 2, and 3) were cored by the Dry Valley Drilling Project (DVDP) in Late Cenozoic McMurdo Volcanic Group rocks near McMurdo Station. DVDP 1 reached 201.5 m and penetrated 40 lithologic units, which consist of six petrographic sequences as shown in Table 1. Drillcores from DVDP 2 and 3 are nearly identical and consist of four petrographic sequences as shown in Table 2. Hut Point Peninsula is built of a thick pedestal of basaltic hyaloclastite which in DVDP 3 consists mainly of weakly palagonitized block lapilli tuff and lapilli tuff. The hyaloclastite at the bottom of DVDP 1 is overlain by five units with intermediate compositions; this episode of differentiation is not recognized in DVDP 2/3. A major basaltic phase of volcanism commenced with formation of hyaloclastite as DVDP 2/3 (sequence IV), which was followed by eruption of subaerial flows and a pyroclastic breccia (sequence III). The basanite in DVDP 2/3 contains the rare titanosilicate rhönite; similar rhönite-bearing flows occur in basanite sequence IV from DVDP 1. In DVDP 2/3, two porphyritic Ne-hawaiite units (sequence II) are overlain by a series of aphyric flows (sequence I) with compositions which range from Ne-hawaiite to Ne-benmoreite. In DVDP 1, two units (sequence III), which appear to be petrographically similar to the porphyritic Ne-hawaiite at DVDP 2/3, are intruded by Ne-benmoreite and basanite dikes (sequence II). These in turn are overlain by 20 units consisting mainly of Ne-hawaiite flows, although Ne-mugearite, Ne-benmoreite, and phonolite flows also occur. Ne-benmoreite flow unit 14 has been tentatively correlated with flow unit 4 at DVDP 2/3 on the basis of the major element chemistry. The Ne-hawaiite flows, in eruptive sequence I of DVDP 1, are all compositionally very similar and probably represent a series of closely spaced eruptions from a single vent. The eruptive sequences I at DVDP 1 and DVDP 2/3 were probably concomitant. The youngest events at Hut Point Peninsula are known from surface mapping and included eruption of phonolite at Observation Hill (~1.25 m.y. ago) and subglacial eruptions of hyaloclastite at Castle Rock (~1.21 m.y. ago). The final activity occurred between about 0.4 and 0.9 m.y. with basanite eruptions from numerous small vents along the northern side of Hut Point Peninsula. The drillcores reveal a more complete geological history for Hut Point Peninsula than can be inferred from surface mapping. At least three episodes of differentiation have resulted in eruptive sequences of lavas having intermediate compositions. Such rocks are rarely observed at the surface.

INTRODUCTION

The primary objective of the Dry Valley Drilling Project (DVDP) was to examine the subsurface geology of the dry valley areas (McMurdo Oasis), McMurdo Sound, and Ross Island, to obtain a detailed record of the Late Cenozoic history of south Victoria Land. DVDP 1 and 2 were drilled during January–February 1973, and as sufficient interest was generated by these two holes, a third (DVDP 3) was drilled adjacent to DVDP 2 in September–October 1973.

The purpose of this paper is to describe the volcanic history of Hut Point Peninsula as inferred from surface mapping and an examination of subsurface DVDP drillcores. The geology of DVDP 1 (Table 1), 2, and 3 (Table 2) will be reviewed, and updated summary logs will be presented for DVDP 1 and 2. Discussion of the geochemistry, mineralogy, and petrogenesis of the lava sequence is presented elsewhere [*Kyle*, 1976, 1981; *Kyle and Rankin*, 1976; *Sun and Hanson*, 1976].

Hut Point Peninsula is a 20-km long, 2- to 4-km wide en echelon line of coalescing volcanic cones, lava flows, and associated pyroclastic rocks that extend in a south-southwest direction from Mount Erebus, Ross Island (Figure 1). The linear alignment of the eruptive vents suggests their distribution may be related to a major crustal fracture. The rocks constitute part of the Erebus volcanic province [*Kyle and Cole*, 1974] of the McMurdo Volcanic Group [*Harrington*, 1958; *Nathan and*

TABLE 1. Lithologic and Eruptive Sequences, DVDP 1

Sequence	Units	Thickness, m	Lithology
I	1-20	62.88	Mainly Ne-hawaiite flows
II	23,24	7.67	Two dikes (basanite, Ne-benmoreite) intruding sequence III
III	21,22	21.80	Porphyritic Ne-hawaiite lava flow and pyroclastic unit
IV	25-34	31.03	Basanite flows and pyroclastic rocks
V	25-39	20.31	Flows and pyroclastic units of intermediate compositions
VI	40	54.43	Basanite hyaloclastite

Schulte, 1968; *Kyle et al.*, 1979a]. The latter encompasses all Late Cenozoic volcanic rocks from Victoria Land, islands in the western Ross Sea, and the Balleny Islands.

Ross Island (Figure 1) consists of three mainly basaltic eruptive centers at Mount Bird, Mount Terror, and Hut Point Peninsula distributed radially around the dominantly phonolitic Mount Erebus (Figure 1). K/Ar age determinations [*Armstrong*, 1978] show the Mount Bird eruptive center to range from 4.5 to 3 m.y. old; Mount Terror from 1.7 to 0.8 m.y., and Hut Point Peninsula from 1.3 to 0.4 m.y. Mount Erebus is the youngest eruptive center, ranging back to about 1 m.y. Active volcanism continues at Mount Erebus, where a persistent anorthoclase phonolite lava lake occurs in the summit crater and weak strombolian eruptions are common [*Kyle et al.*, 1980]. Surface mapping [*Cole and Ewart*, 1968; *Cole et al.*, 1971; *Kyle*, 1976] indicates that the eruptive history of Hut Point Peninsula is similar to that of Cape Crozier and Cape Bird and can therefore be used as a model for understanding the evolution and lava petrogenesis at these areas.

SURFACE GEOLOGY OF HUT POINT PENINSULA

Surface geological mapping of Hut Point Peninsula indicates a predominance of basanitic lavas and pyroclastic rocks with a small phonolite cone at Observation Hill and rare occurrences of lavas with intermediate compositions. (Intermediate is used as a general term to describe rocks with petrographic and geochemical compositions between basanite (42–44% SiO_2) and phonolite (54% SiO_2).) Five informal stratigraphic units termed 'sequences' (Figure 2) have been inferred from geological mapping, K/Ar age determinations (Table 3), paleomagnetic measurements, and geomorphic evidence.

Crater Hill Sequence

The Crater Hill sequence is the oldest, consisting of porphyritic olivine-clinopyroxene basanite flows and associated pyroclastic rocks. The lavas show a moderate degree of erosion, with well-developed marine(?)-cut cliffs at The Gap and near Scott Base. A large snow-filled cirque overlain by a younger flow, on which Scott Base is situated, occurs to the south of Crater Hill. Reversely magnetized, 1.21 (1.18) m.y. old phonolite lavas of the Observation Hill sequence overlie the basanite at The Gap and Cape Armitage: (All K/Ar age determinations on Hut Point Peninsula rocks (Table 3) have been recalculated using the latest decay constants [*Steiger and Jager*, 1977]. Ages calculated using the old decay constants are given in parentheses.) A single normal magnetic polarity measurement on a lava flow north of Scott Base (Figure 2) suggests that the Crater Hill sequence may extend back into the Gilsa polarity event.

TABLE 2. Lithologic and Eruptive Sequences, DVDP 2 and 3

Sequence	Units DVDP 2	DVDP 3	Thickness, m	Lithology
I	1-7	1-6	60.42	Flows of intermediate compositions
II	8,9	7,8	30.86	Porphyritic Ne-hawaiite lava flow and pyroclastic unit
III	10-14	9-14	70.28	Basanite flows and pyroclastic rocks
IV	15,16	15	214.17	Basanite hyaloclastite

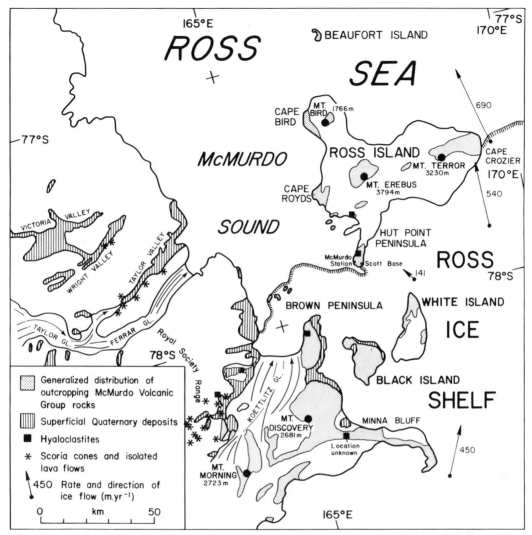

Fig. 1. Distribution of volcanic rocks within the Erebus volcanic province and location map of Ross Island.

Observation Hill Sequence

The Observation Hill sequence consists of kaersutite phonolite and lavas of intermediate compositions which crop out around the base of Observation Hill. The intermediate lavas have little surface expression but are abundant subsurface as shown by DVDP 2 and 3 cores. A mean K/Ar age of 1.21 (1.18) m.y. for the phonolite is consistent with its reversed magnetic polarity.

Castle Rock Sequence

Olivine-clinopyroxene basanite hyaloclastite at Castle Rock and Boulder Cones is the product of a subglacial eruption that occurred when the Ross

Ice Shelf was considerably thicker than at present [*Kyle*, this volume]. A K/Ar age of 1.21 ± 0.05 (1.18) m.y. was determined on a basanite feeder (Table 3).

Half Moon Crater Sequence

Clinopyroxene-kaersutite basanite flows at Half Moon Crater, and to the north, overlie the Castle Rock sequence. They have a K/Ar age of 1.0 (1.0) m.y. and normal magnetic polarity which suggests that the sequence was erupted, at least in part, during the Jaramillo magnetic event (0.89-0.95 m.y.) (Figure 2). Some of the flows are Nehawaiites (P. R. Kyle, unpublished data, 1981).

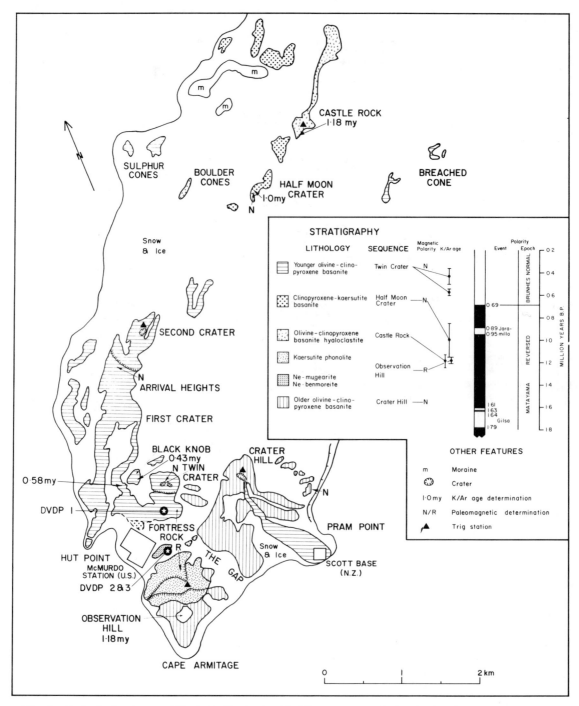

Fig. 2. Geologic sketch map of Hut Point Peninsula. All K/Ar ages were calculated using the old decay constants; see Table 3 for revised ages based on the 1977 decay constants.

TABLE 3.   Whole Rock K/Ar Age Determinations of Samples From Hut Point Peninsula

Sample Number	Location	Age,* m.y. Old	Age,* m.y. New	Reference
		Basanite Lava Flow		
22892	Black Knob	0.43 ± 0.1	0.44 ± 0.1	*Armstrong* [1978]
22900	SW of Black Knob	0.58 ± 0.06	0.60 ± 0.06	*Armstrong* [1978]
22878	Half Moon Crater	1.0 ± 0.2	1.0 ± 0.2	*Armstrong* [1978]
		Basanite Dike		
22879	Castle Rock	1.18 ± 0.05	1.21 ± 0.05	This paper‡
		Phonolite		
	Observation Hill	1.18 ± 0.03†	1.21 ± 0.04	*Forbes et al.* [1974]

Sample numbers refer to the petrology collection of the Department of Geology, Victoria University of Wellington.

*Calculated using old and new decay constants. Old: $^{40}K\lambda_\beta = 4.72 \times 10^{-10}yr^{-1}$, $\lambda_\beta = 0.584 \times 10^{-10}yr^{-1}$, $^{40}K/K_{tot} = 0.0119$ atom/atom. New: $^{40}K\lambda_e = 4.962 \times 10^{-10}yr^{-1}$, $\lambda_\beta = 0.581 \times 10^{-10}yr^{-1}$, $^{40}K/K_{tot} = 0.01167$ atom/atom.

†Mean of three samples.

‡Determined by J. E. Gabites, Institute of Nuclear Sciences, Department of Scientific and Industrial Research, Lower Hutt, New Zealand.

Twin Crater Sequence

Olivine-clinopyroxene basanites and rare clinopyroxene-kaersutite basanites which occur mainly as a series of small cones between Hut Point and Second Crater constitute the Twin Crater sequence (Figure 2). Black Knob lavas have a K/Ar age of 0.44 (0.43) m.y. and are probably the youngest volcanic rocks at the area. The paleomagnetic data and the age determinations indicate the rocks of this sequence were erupted between 0.4 and 0.69 m.y. ago.

GENERAL CORE DESCRIPTION

Detailed preliminary core descriptions of DVDP 1, 2, and 3 have been presented by *Treves and Kyle* [1973] and *Kyle and Treves* [1974]. The DVDP 1 and 2 cores have since been reexamined and new summary logs are presented in Tables 4 and 5, respectively. Preliminary descriptions of DVDP 1 and 2 were based on direct measurements of the core rather than the depth indicated by the drill stem. As a result, a 3.7-m offset was noted between units in DVDP 2 and 3 [*Kyle and Treves*, 1974]. The revised logs, based on the drill stem depths, show no discrepancies between DVDP 2 and 3 (Figure 3). The summary log for DVDP 3 (Table 6) follows *Kyle and Treves* [1974].

Petrographic classification of core samples is unsatisfactory, as most are fine grained and glassy, so instead a classification based on the major ele-ment chemistry and the CIPW normative components has been adopted (Table 7). For units in which no chemical analyses exist, core samples have been broadly classified as basaltic or of an unspecified intermediate composition. The basaltic lavas are porphyritic, contain olivine, clinopyroxene, and usually have a basic groundmass plagioclase (labradorite, bytownite). The intermediate lavas are usually fine grained (<1 mm) and contain microphenocrysts of kaersutite, clinopyroxene, and andesine plagioclase.

DVDP 1

Introduction

DVDP 1 was drilled northeast of McMurdo Station on the south-southwest flank of Twin Crater (Figure 2). Megascopic and microscopic examination of the core identified 40 lithologic units consisting of 29 flow units, 2 dike units, and 9 pyroclastic units (Table 4, Figure 3). Six lithologic sequences (Table 1) are recognized and are described with increasing depth. The following log shows a large number of changes from the original preliminary description by *Treves and Kyle* [1973].

Ne-hawaiite Sequence (I)

Aphyric intermediate rocks which form units 1-20 are 63.18 m thick and consist of 19 lava flows and one possible pyroclastic unit.

DRY VALLEY DRILLING PROJECT

TABLE 4. Lithologic Units, DVDP 1

Unit	Description	Thickness, m	Depth Below Platform, m	Elevation Above/Below Sea Level, m
1	Ne-benmoreite lava flow	3.05	6.40	63.88
2	Ne-hawaiite lava flow	3.96	10.36	59.92
3	Ne-hawaiite lave flow	2.13	12.49	57.79
4	Ne-hawaiite lava flow	4.46	16.95	53.33
5	Lapilli layer	0.33	17.28	53.00
6	Ne-hawaiite lava flow	1.92	19.20	51.08
7	Ne-hawaiite lava flow	1.39	20.59	49.69
8	Ne-hawaiite lava flow	1.92	22.51	47.77
9	Unspecified intermediate lava flow	1.48	23.99	46.29
10	Ne-hawaiite lava flow	2.05	26.04	44.24
11	Ne-hawaiite lava flow	6.25	32.29	37.99
12	Ne-hawaiite lava flow	3.21	35.50	34.78
13	Unspecified intermediate lava flow	4.04	39.54	30.74
14	Ne-mugearite lava flow	6.86	46.40	23.88
15	Ne-hawaiite lava flow	6.17	52.57	17.71
16	Ne-hawaiite lava flow	1.95	54.52	15.76
17	Unspecified intermediate lava flow	4.55	59.07	11.21
18	Ne-benmoreite lava flow	2.46	61.53	8.75
19	Phonolite lava flow	3.81	65.34	4.49
20	Unspecified intermediate lava flow	0.89	66.23	4.05
21	Ne-hawaiite pyroclast	14.80	81.03	−10.75
22a	Basanite lava flow	3.34	84.37	−14.09
23	Ne-benmoreite dike	5.63	90.00	−19.72
22b	Basanite lava flow	0.31	90.31	−20.03
24	Basanite dike	2.04	92.35	−22.07
22c	Basanite lava flow	3.35	95.70	−25.42
25	Basaltic lava flow	0.87	96.57	−26.29
26	Basanite lava flow	2.36	98.93	−28.65
27	Basaltic pyroclast	4.62	103.55	−33.27
28	Basanite lava flow	1.34	104.89	−34.61
29	Basaltic pyroclast	1.62	106.51	−36.23
30	Basanite lava flow	1.08	107.59	−37.31
31	Basanite pyroclast	6.96	114.55	−44.27
32	Basanite lava flow	0.69	115.24	−44.96
33	Basaltic pyroclast	4.44	119.68	−49.40
34	Basanite lava flow	7.05	126.73	−56.45
35	Unspecified intermediate pyroclast	4.86	131.59	−61.31
36	Ne-benmoreite lava flow	2.44	134.03	−63.75
37	Benmoreite lava flow	0.56	134.59	−64.31
38	Ne-hawaiite pyroclast	10.04	144.63	−74.35
39	Ne-hawaiite lava flow	2.41	147.04	−76.76
40	Basanite hyaloclastite	54.43	201.47	−131.19

Depths of DVDP 1 core given in the preliminary log [*Treves and Kyle*, 1973] were based on measurement of the core only. Depths and unit thicknesses have been revised to conform with depths as indicated by the drill stem. Depths listed above are from the drill platform which was 70.28 m above sea level; coring commenced 3.35 m below this level.

The flows are thin, ranging in thickness from 1.39 to 6.86 m; many are oxidized. The tops and bases of the flows are characterized by rubbly, scoriaceous zones that are ice cemented and exhibit ice-filled cavities up to 0.9 m thick. The walls of many of the cavities are lined with delicate, radiating sheafs and clusters of white secondary minerals [*Browne*, 1973, 1974].

Unit 5 is 0.39 m thick and was considered by *Treves and Kyle* [1973] to be a possible paleosol. The unit consists of rounded to angular black and dusky red clasts of vesicular lava that range up to 60 mm in diameter. Some of the clasts show evidence of surficial oxidation which apparently occurred during the emplacement of the overlying lava flow unit. The drill core resembles material

TABLE 5. Lithologic Units, DVDP 2

Unit	Description	Thickness, m	Depth Below Platform, m	Elevation Above/Below Sea Level, m
1	Unspecified intermediate lava flow	3.51	6.81	40.82
2	Ne-benmoreite lava flow	14.94	21.75	25.88
3	Unspecified intermediate lava flow	7.11	28.86	18.77
4	Ne-mugearite lava flow	8.13	36.99	10.64
5	Ne-mugearite lava flow	15.05	52.04	−4.41
6	Ne-hawaiite lava flow	4.35	56.39	−8.76
7	Ne-benmoreite lava flow	7.33	63.72	−16.09
8	Ne-hawaiite pyroclast	16.23	79.95	−32.32
9	Ne-hawaiite lava flow	14.60	94.55	−46.92
10	Basanite pyroclast	5.63	100.18	−52.55
11	Basanite lava flow	7.53	107.71	−60.08
12	Basanite lava flow	44.51	152.22	−104.59
13	Basanite pyroclast	3.35	155.57	−107.94
14	Basanite lava flow	9.26	164.83	−117.20
15	Basanite pyroclast	4.42	169.25	−121.62
16	Basanite hyaloclastite	10.16	179.41	−131.78

Depths of DVDP 2 core given in the preliminary log [*Treves and Kyle*, 1973] were based on measurement of the core only. Depths and unit thicknesses have been revised to conform with depths as indicated by the drill stem. The depths listed above are from the drill platform which was 47.63 m above sea level; coring commenced 3.30 m below this depth.

TABLE 6. Lithologic Units, DVDP 3

Unit	Description	Thickness, m	Depth Below Platform, m	Elevation Above/Below Sea Level, m
1	Ne-benmoreite lava flow	11.03	20.48	27.15
2	Unspecified intermediate lava flow	4.89	25.37	22.26
	No core	6.94	32.31	15.32
3	Ne-mugearite lava flow	3.05	35.36	12.27
	No core	10.06	45.42	2.21
4	Ne-mugearite lava flow	6.99	52.41	−4.78
5	Ne-hawaiite lava flow top	3.37	55.78	−8.15
	No core	7.62	63.40	−15.77
6	Ne-benmoreite lava flow base	1.00	64.40	−16.77
7	Ne-hawaiite pyroclast	16.10	80.50	−32.87
8	Ne-hawaiite lava flow	14.66	95.16	−47.53
9	Basanite pyroclast	4.44	99.60	−51.97
10	Basanite lava flow	8.18	107.78	−60.15
11	Basanite pyroclast	1.36	109.14	−61.51
12	Basanite lava flow	44.03	153.17	−105.54
13	Basanite pyroclast	2.25	155.42	−107.79
14	Basanite lava flow	11.41	166.83	−119.20
15	Basanite hyaloclastite; mixed volcanic breccia with lapilli tuff and blocky lapilli tuff	214.17	381.00	−333.37

The depths listed above are from the drill platform which was 47.63 m above sea level; coring commenced 9.45 m below this depth.

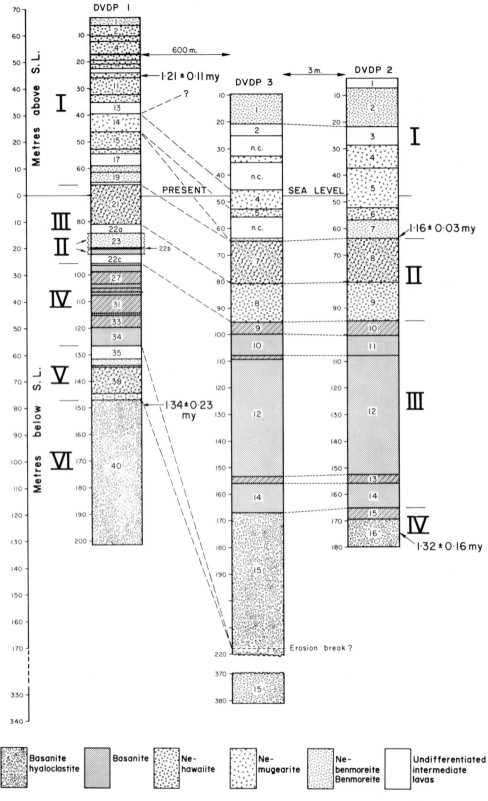

Fig. 3.   Geologic columns for DVDP 1, 2, and 3. Tentative correlations are indicated by large dashes. Units which are cross hatched are composed of pyroclastic rocks. All K/Ar ages were calculated using the new (1977) decay constants (Table 8); n.d. indicates the interval was not cored. Roman numerals indicate the eruptive sequence (see text).

TABLE 7. Nomenclature Used to Describe Hut Point Peninsula and DVDP Samples

Name	Classification Criteria
Basanite	An > 50%, Ne > 5%
Ne-hawaiite	An = 30–50%, Ne > 10%
Ne-mugearite	An = 10–30%, Ne > 10%
Ne-benmoreite	DI = 65–75, Ne > 10%
Benmoreite	DI = 65–75, Ne < 10%
Phonolite	DI > 75, Ne > 10%

An, normative $100An/An + Ab$; Ne, normative nepheline; and DI = differentiation index (normative $Q + Ab + Or + Ne + Lc$) of *Thornton and Tuttle* [1960].

Weight percent CIPW norms used throughout.

presently found at the surface in frost-heaved polygonal cracks. Chemical analyses discussed below show the three flow units above and below unit 5 to be identical in composition and thus most likely represent a single eruptive phase. The clasts of unit 5 are considered to represent lapilli-sized material that were rolled on top of a flow or in front of an advancing flow; it is not a paleosol.

Ne-hawaiite flow unit 10 has been dated by the K/Ar method as 1.21 ± 0.11 m.y. old (Table 8).

Petrographically, the lavas are aphyric; phenocrysts (arbitrarily taken to have any dimension exceeding 1 mm) are generally lacking, although rare kaersutite phenocrysts were observed. Typically, the samples are microporphyritic with microphenocrysts of kaersutite, clinopyroxene, plagioclase, and opaque oxides; the groundmass textures vary from pilotaxitic to intersertal (Table 9). Major element analyses (Table 10) of 15 of the 20 units show the majority have Ne-hawaiite compositions. Units 14, 18, and 19 are Ne-mugearite, Ne-benmoreite, and phonolite, respectively. Units 2-8, excluding 5, have identical chemical compositions within analytical uncertainties. All the remaining Ne-hawaiite flows are similar in composition and probably represent a single eruptive episode or series of closely spaced eruptions from a single vent. The eruptive

sequence commenced with evolved Ne-benmoreite, and phonolitic lavas then passed into the monotonous flows of Ne-hawaiite. Ne-mugearite unit 14 is chemically identical to a flow in DVDP 2 and is probably related to eruptive sequence I in DVDP 2 and 3.

Dike Sequence (II)

Two dikes, units 23 and 24, a Ne-benmoreite and basanite, respectively, intrude lava flow unit 22. The two dikes are considered as a separate eruptive sequence (Figure 3).

The Ne-benmoreite dike is 5.63 m thick, with a poorly defined upper intrusive contact which appears to dip very steeply at an angle of about 75-80°. The uncertainty in the upper contact is mainly a result of its vesicular nature. Megascopically, the contact looks gradational; however, the obvious steep dip is consistent with a dike. Petrographically, the Ne-benmoreite is weakly microporphyritic with rare microphenocrysts of prismatic clinopyroxene (0.3 mm) and completely or nearly completely oxidized pseudomorphs after kaersutite. The pilotaxitic groundmass consists mainly of interlocking feldspar microlites, with cubic opaque oxides, ragged anhedral clinopyroxene, and a brown unidentified mineral, possibly aenigmatite.

Basanite dike unit 24 is 2.04 m thick. The lower intrusive contact dips at 30° and shows strong flow banding parallel to the contact. The basanite is weakly porphyritic with rounded oxidized pseudomorphs after kaersutite and rare resorbed grains of opaque oxides. Microphenocrysts that consist of prismatic clinopyroxene (rarely up to 0.8 mm), olivine (generally less than 0.5 mm), plagioclase, opaque oxides, and apatite are set in an intergranular groundmass. Chemically (Table 10), the dike is a basanite; however, it is somewhat fractionated compared to the basanites of sequence IV, as it contains only 5.54 wt % MgO.

Porphyritic Intermediate Lava Sequence (III)

This sequence consists of 14.80-m thick pyroclastic unit 21 and a flow unit 22. The latter is intruded by dike units 23 and 24 and is divided into three parts, 22a, 22b, and 22c.

Unit 21, a bomb-lapilli tuff is strongly oxidized, consisting mainly of vesicular and scoriaceous blocks and lapilli, up to 0.1 m in size. One block

TABLE 8. Whole Rock K/Ar Age Determination of DVDP 1 and 2 Samples [*Kyle et al.*, 1979b]

Hole and Depth, m	Unit	Rock Type	Apparent Age, m.y. $\pm$ 1 σ
1–25.32	10	Ne-hawaiite	1.21 ± 0.11
1–148.81	40	Basanite	1.34 ± 0.23
2–62.38	7	Ne-benmoreite	1.16 ± 0.03
2–173.93	16	Basanite	1.32 ± 0.16

Calculated using $^{40}K/K_{tot} = 1.167 \times 10^{-4}$ atom/atom, $\lambda_{\beta} = 4.962 \times 10^{-10}$/yr, and $\lambda_{\epsilon} = 0.581 \times 10^{-10}$/yr.

TABLE 9.   Summary of the Petrographic Character of Representative Thin Sections From DVDP 1

Unit	Depth	Rock Type	Texture	Ol	Cpx	Kaer	Rh	Op	Fsp	Ap	Groundmass
1	4.60	Ne-benmoreite	mp(5),v	x	p	r,co		p	p	r	h
2	8.22	Ne-hawaiite	mp(5),v		p	p,po		p	p	r	i,2o
3	11.65	Ne-hawaiite	mp(10),v		p	p,po		r	p		i,2c,2o
4	16.72	Ne-hawaiite	mp(5)	x	p	p,po		g	p	g	i,2o
6	18.75	Ne-hawaiite	mp(10)	x	p	p,po		g	p	g	i,2o
7	19.60	Ne-hawaiite	mp(15),v		p,po	p,po		g	p	g	h,2o
8	21.06	Ne-hawaiite	mp(10),v		p	p,po		g	p		c
9	23.41	unspec. intermed.	mp(10),v		p,x	p		p	p	r	g,2c,2o
10	25.52	Ne-hawaiite	mp(15),v		p	p,po		r	p		i,2o
11	29.62	Ne-hawaiite	mp(10)		p	p,po		g	p		p,2o
12	34.37	Ne-hawaiite	mp(20)		p	p		g	p		1,2c,2o
13	35.70	unspec. intermed.	mp(10)		p	p		g	p		gl,2c
14	43.64	Ne-mugearite	mp(5)	x	p	x		g	p		c
15	51.57	Ne-hawaiite	mp(10),v		p	p,po		g	p		c,2c,2o
16	53.45	Ne-hawaiite	mp(25)		p	p		g	p		h,2c,2o
17	58.20	undiff. intermed.	mp(15),v		p	p		g	p		p,gl,2c,2o
18	60.49	Ne-benmoreite	mp(25),v		r	p,po		g	p		gl,2c
19	61.62	Phonolite	mp(15)		p	p,po		g	p		c,2o
20	65.59	unspec. intermed.	mp(10),v		p	p		g	p		p,gl,2c
21	74.07	Ne-hawaiite	p-mp(10)		p,gc	p,po		g	p		g,2o
22	83.60	Basanite	p(15)	p	p,gc	p,po		p	g	g	g
23	87.76	Ne-benmoreite	mp(<5)		g	r,co		g	g	g	p
24	90.28	Basanite	mp(<5)	p	p			p	g	g	g
25	95.53	Basaltic	p(20),v	p	p			g	g		g,2o
26	97.93	Basanite	p(20)	p,po	p			g	g		h,2o
27	101.21	Basaltic	p(10)	p,po	p			g	g		g,2c,2o
28	103.55	Basanite	p(15),v	p	p,gc		p	r	g		g
29	106.20	Basaltic	p(15)	p,po	p		r?	r	g	g	g,2o
30	107.59	Basanite	p(10)	p,po	p		r	r	g		g,2o
31	114.08	Basanite	p(15)	p,po	p,gc			p	g	g	g,2o
32	115.05	Basanite	p(10)	p,po	p		p	g	g		g,2c,2o
33	116.92	Basaltic	p(10)	p,co	p			r	g	g	c,2c,2o
34	124.88	Basanite	p(20)	p	p,gc		p	r	g		c
36	133.56	Ne-benmoreite	mp(10)		g	p,po		g	p		p
37	133.93	Benmoreite	mp(10)		p	p,po		g	p		p
38	140.04	Ne-hawaiite	mp(<5),v		r	r,po		r	p		i-p,c
39	145.47	Ne-hawaiite	mp(<5)		r	r,po		r	p		i-p,c
40	149.51	Basanite	p(15),v	p	p			r	g		g

Abbreviations:

Minerals: Ol-olivine, Cpx-clinopyroxene, Kaer-kaersutite, Rh-rhönite, Op-opaque oxides, Fsp-feldspar, Ap-apatite.

Texture: mp(5)-microporphyritic, % microphenocrysts to nearest 5% in parentheses; p-porphyritic, % phenocrysts and microphenocryst to nearest 5% in parentheses; v-vesicular.

Mineralogy: co-completely oxidized, g-ground mass only, gc-green sodic cores, p-present (>1%), po-partly oxidized, r-rare (<1%), x-xenocrystic.

Groundmass: c-cryptocrystalline, g-intergranular, gl-glassy, h-hyalopilitic, i-intersertal, p-pilotaxitic, 2c-secondary minerals (mainly carbonates), 2o-secondary oxides.

0.48 m in diameter was noted. In hand specimen the clasts are porphyritic, with kaersutite megacrysts (arbitrarily set as any grain exceeding 10 mm), up to 25 mm, scattered throughout the unit. Microscopically, the phenocrysts are mainly seriate, euhedral, and strongly pleochroic kaersutite. Microphenocrysts of clinopyroxene (up to 0.8 mm), some with green sodic cores, and apatite are occasionally present. The kaersutite is usually unaltered except for thin opaque oxide rims; very sparse rounded resorbed(?) grains and one strongly oxidized grain were noted. The groundmass is intergranular with plagioclase, apatite, clinopyroxene, opaque oxides, and secondary oxides.

Flow unit 22 has a combined thickness of 7.00 m. It is apparent from megascopic, microscopic, and geochemical data that unit 22 is either a multiple-flow unit or if a single-flow unit, it is strongly zoned. Some confusion may also exist because dike unit 24 was not recognized in the initial log of *Treves and Kyle* [1973] and the exact location of

TABLE 10. Reconnaissance Major Element Analyses of Selected Samples From DVDP 1

Sample Depth	4.60	8.22	11.65	16.72	18.75	19.60	21.06	25.52	29.62	34.37	42.49	51.57	53.45	60.49	62.35
Unit	1	2	3	4	6	7	8	10	11	12	14	15	16	18	19
SiO_2	50.61	48.43	48.40	48.52	48.77	48.71	48.56	48.38	48.45	49.98	50.72	48.36	48.46	53.08	53.42
TiO_2	2.03	2.58	2.59	2.57	2.57	2.57	2.59	2.70	2.72	2.41	2.20	2.60	2.65	1.58	1.40
Al_2O_3	18.52	17.70	17.59	17.69	17.72	17.73	17.74	17.74	17.69	18.47	18.90	18.13	18.36	19.77	19.89
Fe_2O_3*	7.04	8.64	8.66	8.59	8.50	8.58	8.68	8.86	8.96	8.11	7.79	8.57	8.71	6.53	6.03
MnO	0.17	0.16	0.19	0.19	0.17	0.16	0.16	0.17	0.18	0.18	0.19	0.18	0.18	0.20	0.17
MgO	2.91	3.69	3.79	3.74	3.82	3.68	3.64	3.70	3.85	3.06	2.62	3.10	3.06	1.67	1.57
CaO	5.67	7.08	7.09	7.07	6.99	7.00	7.03	7.27	7.21	6.33	5.93	6.58	6.65	4.69	4.36
Na_2O	6.79	6.26	6.45	6.49	6.40	6.40	6.11	6.32	6.34	6.69	7.08	6.51	6.51	7.58	8.21
K_2O	3.99	3.36	3.30	3.32	3.31	3.34	3.35	3.22	3.30	3.58	3.67	3.45	3.42	4.15	4.26
P_2O_5	0.46	0.59	0.56	0.61	0.58	0.59	0.59	0.59	0.56	0.43	0.48	0.58	0.51	0.42	0.36
Loss	0.82	0.62	0.69	0.63	0.51	0.65	0.85	0.83	0.51	0.65	0.28	1.68	0.67	0.34	0.27
Sum	99.01	99.11	99.31	99.42	99.34	99.41	99.30	99.78	99.77	99.89	99.86	99.74	99.18	100.01	99.94
$\frac{100\,An}{An+Ab}$†	26.7	35.6	34.0	33.7	33.7	33.8	36.1	36.2	36.1	31.2	27.6	34.2	36.1	20.8	14.7
DI	65.1	57.1	57.3	57.8	57.8	57.9	56.8	56.5	56.7	61.7	64.9	59.1	58.8	72.6	75.7

Sample Depth	74.07	83.60	87.76	90.72	93.83	97.93	103.55	107.59	114.08	115.05	124.88	133.56	134.03	137.24	143.75	192.10
Unit	21	22a	23	24	22c	26	28	30	31	32	34	36	37	38	39	40
SiO_2	48.92	43.02	52.32	42.15	42.73	42.52	41.57	41.40	41.63	41.50	41.68	55.41	55.78	49.13	50.25	43.26
TiO_2	2.50	4.01	1.62	4.66	4.51	3.75	4.03	4.02	4.12	4.15	4.34	1.25	1.29	2.49	2.40	3.60
Al_2O_3	18.36	15.86	19.55	15.84	15.85	12.71	12.75	12.64	13.18	12.99	13.32	18.86	19.00	17.29	17.77	12.93
Fe_2O_3*	8.82	12.78	6.84	13.92	13.80	11.88	12.38	12.39	12.40	12.51	12.93	6.15	6.32	10.59	10.39	12.55
MnO	0.20	0.22	0.20	0.22	0.21	0.16	0.17	0.16	0.17	0.16	0.18	0.17	0.20	0.22	0.22	0.19
MgO	3.20	6.23	1.81	5.54	5.85	12.36	11.83	12.07	11.27	11.72	11.85	1.77	1.76	3.54	3.54	11.72
CaO	6.83	9.75	4.99	10.60	10.54	10.49	11.20	11.21	11.14	11.20	11.56	4.22	4.33	7.06	7.06	11.81
Na_2O	6.62	4.63	7.48	4.51	4.19	3.18	2.93	2.48	2.92	2.69	2.90	7.65	7.00	5.84	6.26	3.00
K_2O	3.24	2.11	3.95	2.09	2.05	1.46	1.46	1.25	1.51	1.45	1.50	3.67	3.64	2.67	2.74	1.38
P_2O_5	0.67	0.87	0.47	1.30	1.13	0.82	0.78	0.79	0.81	0.81	0.89	0.37	0.40	0.94	0.98	0.78
Loss	0.53		0.65	0.10	0.11	0.23		0.29	0.15	0.16	0.06	0.34	0.09			0.10
Sum	99.71	99.48	99.88	100.93	100.97	99.56	99.10	98.70	99.30	99.34	101.21	99.86	99.81	99.77	101.61	101.32
$\frac{100\,An}{An+Ab}$†	33.9	68.2	21.5	64.3	63.1	73.2	87.1	82.0	84.4	84.8	84.7	13.8	18.3	32.5	30.7	79.1
DI	59.1	37.2	71.2	37.3	36.2	25.9	23.2	20.7	23.9	22.5	23.7	74.8	73.4	55.0	57.6	24.0

DI, differentiation index [*Thornton and Tuttle*, 1960].

*Total Fe as Fe_2O_3.

†Norms calculated using standardized Fe_2O_3 values, following convention of *Thompson et al.* [1972].

samples collected at that time relative to the new log is uncertain. In hand specimen, unit 22a is strongly porphyritic with numerous megacrysts and inclusions as well as phenocrysts of clinopyroxene. Unit 22c appears less porphyritic in its lower 1.48 m. There are some poorly defined breaks in the texture of unit 22c that could be flow banding, which marks possible flow boundaries. Note, however, that a correlative flow in DVDP 3 shows chemical zoning and no flow boundaries; it therefore seems likely that unit 22 was erupted as a mineralogically stratified flow. Four thin sections were examined; all were similar, although two sections from unit 22c contained fewer phenocrysts and in differing proportions than samples from units 22a and 22b. The samples are porphyritic with phenocrysts and microphenocrysts of purplish euhedral clinopyroxene (ranging from 0.1 to 4.5 mm), opaque oxides, and kaersutite (up to 5 mm) which varies from 50% to completely oxidized. A granular dunite nodule composed of anhedral grains of olivine (0.1 mm in diameter) and a glomeroporphyritic clump with olivine phenocrysts (4 mm in diameter) were observed; both had reaction rims. The groundmasses are intergranular with plagioclase, clinopyroxene, opaque oxides, and rare olivine.

Chemical analyses (Table 10) indicate unit 21 is a Ne-hawaiite, whereas units 22a and 22c are basanites. Units 22a and 22c are chemically distinct, however; unit 22a contains lower TiO_2, Fe_2O_3*, CaO, and P_2O_5 and higher MgO and Na_2O than unit 22c. Unit 22c is chemically very similar to dike unit 24 (Table 10); thus there is a possibility that the observed differences may be due to a sampling error.

Basanite Sequence (IV)

Units 25-34 are 31.03 m thick and consist of 6 basanite lava flows (units 25, 26, 28, 30, 32, 34) and 4 pyroclastic units (27, 29, 31, 33). The flow units are generally thin, averaging 1.27 m (excluding unit 34, which is 7.05 m thick); the combined thickness of the six flows is 13.39 m. The pyroclastic units range from 1.62 to 6.96 m and total 17.64 m in thickness.

The lava flows are mostly oxidized, red and black vesicular basanites with scoriaceous tops and bases that have ice-filled cavities. The pyroclastic units are also red and black and are mainly strongly oxidized. They are composed predomi-

nantly of lapilli tuff with minor lapilli breccia, lapilli-ash tuff, and pyroclastic breccia.

The alternating pattern of pyroclastic and lava flow units suggests that the former may be scoriaceous flow tops rather than distinct pyroclastic units. Although the contacts in most cases are gradational, the pyroclastic units are obviously airfall and can be distinguished from the coarser scoriaceous flow tops. It is considered that the basanite sequence of thin flows and thicker pyroclastic units were deposited close to their source of eruption, from a small lava fountaining scoria cone. Lava may have periodically overflowed from a crater, forming the thin lava flows.

Petrographically, the basanites are porphyritic, with seriate phenocrysts of olivine and clinopyroxene and microphenocrysts of rhönite, rare plagioclase, and opaque oxides. Samples have varying degrees of oxidation (Table 9); in many, olivine is strongly oxidized. The groundmasses are variable, ranging from intergranular to hyaloophitic and contain plagioclase, clinopyroxene, opaque oxides, olivine, rhönite, and glass. Rhönite is difficult to distinguish in the oxidized samples but has been identified in most samples except from units 25, 26, and 27.

Reconnaissance major element analyses (Table 10) of six samples show all are basanites with normative plagioclase contents (100 An/An + Ab) of An_{73} to An_{87}. The units have relatively uniform chemical compositions except unit 26, which has higher SiO_2, MgO, and Na_2O and lower TiO_2, Fe_2O_3*, and CaO relative to the other units. This chemical difference may be reflected petrographically in the absence of rhönite.

Intermediate Lava Sequence (V)

Units 35-39 are 20.31 m thick and consist of three lava flow units (36, 37, 39) and two pyroclastic units (35 and 38). They all have intermediate compositions. The two pyroclastic units are thick (4.86 and 10.04 m) and account for over 70% of the total thicknesses of the sequence. Unit 35 is a lapilli tuff and lapilli breccia, whereas unit 38 is a pyroclastic breccia with large blocks up to 0.55 m in diameter. The lava flows are thin, ranging from 0.56 to 2.44 m in thickness; they are aphyric and in some cases have scoriaceous tops and bases.

Petrographically, units 36 and 37 are similar; they contain rare oxidized kaersutite phenocrysts (up to 1.5 mm in length), microphenocrysts of plagioclase, oxidized kaersutite, and rare clinopyroxene, all set in a pilotaxitic groundmass. Units 28 and 39 are also similar; they are extremely fine grained with microphenocrysts of plagioclase, clinopyroxene, cubic opaque oxides, and rare oxidized kaersutite. The groundmass is black, dense, and cryptocrystalline; it is partly intersertal and partly pilotaxitic. No samples from unit 35 were examined petrographically; however, a large black clinopyroxene(?) megacryst was noted in hand specimen.

Chemical analyses (Table 10) of units 36 and 37 are similar except the latter is noticeably lower in Na_2O, which results in a normative nepheline (Ne) content of 9.3%. Unit 37 is therefore a benmoreite, whereas unit 36 is a Ne-benmoreite. Compared to other Ne-benmoreite lavas from DVDP 1 and 2, units 36 and 37 have distinctly different chemistries. Units 38 and 39 are Ne-hawaiites (Table 10); they have noticeably higher Fe_2O_3* and lower Al_2O_3 than Ne-hawaiite from eruptive sequence I.

Basanite Hyaloclastite Sequence (VI)

The lower 54.43 m of DVDP 1 (unit 40) consist of a basanite pyroclastic unit, the majority of which is believed to have formed subaqueously and to be a hyaloclastite. The upper 0.60 m of the unit is a yellow palagonitic tuff with well-rounded basaltic fragments. It appears as if the tuff is weathered and probably marks a disconformity. The remainder of the core consists mainly of black to grey, weakly palagonitized lapilli tuff and block lapilli tuff. A basanite clast near the top of the unit at 148.81 has a conventional whole rock K/Ar age of 1.34 ± 0.23 m.y. (Table 8).

Petrographically, the basanite is porphyritic with phenocrysts of seriate olivine and clinopyroxene. Microphenocrysts of plagioclase and opaque oxides are also present. More detailed descriptions of the hyaloclastites appear below in the section on DVDP 2 and 3.

A chemical analysis of a basanite clast from 192.10 m (Table 10) is typical of basanite from Ross Island [Goldich et al., 1975; Kyle, 1976]. In comparison to other DVDP 1 basanites, sample 192.10 has noticeably higher SiO_2 and lower TiO_2.

DVDP 2 AND 3

Introduction

DVDP 2 and 3 were drilled 3 m apart, at a site recommended by Cole and Kyle [1972], low on the northern flank of Observation Hill (Figure 2).

TABLE 11.　Major Element Analyses of DVDP 2 Core Samples

	Sample Depth										
	18.23	33.76	45.86	53.64	61.23	76.38	87.78	105.53	109.65	117.95	159.43
Unit	2	4	5	6	7	3648	9	11	12	12	14
SiO_2	51.32	49.36	50.76	49.28	51.90	47.27	47.27	41.68	41.80	41.69	41.34
TiO_2	2.03	2.49	2.23	2.68	1.85	2.95	3.00	4.06	4.15	4.17	4.12
Al_2O_3	18.82	18.18	18.88	18.41	19.31	17.61	17.65	12.92	12.84	12.96	13.18
Fe_2O_3*	3.00	2.47	2.58	2.98	3.01	3.35	3.64	3.06	4.29	3.92	5.81
FeO	3.65	5.18	4.72	5.48	3.96	6.06	6.04	8.48	7.74	8.05	6.06
MnO	0.18	0.18	0.19	0.20	0.20	0.21	0.21	0.18	0.18	0.18	0.18
MgO	2.78	3.39	2.67	3.19	2.07	3.72	3.84	12.13	12.12	11.99	11.65
CaO	5.73	6.71	5.99	6.84	5.43	7.76	7.79	11.32	11.53	11.40	10.94
Na_2O	6.84	6.58	6.78	6.37	7.26	6.07	5.82	3.16	3.02	3.09	3.00
K_2O	3.88	3.49	3.65	3.43	3.70	3.11	2.91	1.48	1.50	1.53	1.66
P_2O_5	0.52	0.65	0.49	0.65	0.60	0.81	0.80	0.84	0.83	0.85	0.83
Loss	0.47	0.31	0.27	0.31		0.40	0.09		0.06	0.08	0.24
Sum	99.22	98.99	99.21	99.82	99.29	99.32	99.06	99.31	100.06	99.91	99.01
$\dfrac{100An}{An + Ab}$†	26.4	32.2	29.6	35.1	24.2	40.2	40.9	83.0	87.3	84.6	74.5
DI	66.0	60.2	63.9	59.2	68.3	54.1	52.8	24.3	24.3	24.6	26.3

DI, differentiation index [Thornton and Tuttle, 1960].
*Total Fe as Fe_2O_3.
†Norms calculated using standardized Fe_2O_3 values, following convention of Thompson et al. [1972].

TABLE 12. Summary of the Petrographic Character of Representative Thin Sections From DVDP 3

Unit	Depth	Rock Type	Texture	Mineralogy								Groundmass
				Ol	Cpx	Kaer	Rh	Op	Fsp	Ap		
1	12.73	Ne-benmoreite	mp(<5),v		p	p,po		p	p		h
2	22.74	unspec. intermed.	mp(25),v	x	p	p,po		p	p		h
4	48.74	Ne-mugearite	mp(20)		p	p,po		p	p		h
5	55.49	Ne-hawaiite	mp(20),v		p	p,po		p	p		h,2o
6	63.70	Ne-benmoreite	mp(30)		r	p,po			p	r	h
7	67.00	Ne-hawaiite	p(5),v		p	p,p		p	p	r	h,2c
8	85.96	Ne-hawaiite	p(10)	x	p	p,co		p	p		h,2c
9	99.47	Basanite	p(20),v	p,po	p		r	r	r		gl,2c
10	102.93	Basanite	p(25)	p	p		r	p	p		g,2c
12	111.20	Basanite	p(30)	p	p,gc			p	p		g,2c
14	156.68	Basanite	p(25)	p	p			p	p	r	h
15	167.70	Basanite	p(25)	p	p			p	r	r	h
15	219.92	Basanite	p(20)	p	p				r		gl

For explanation of abbreviations, see Table 9 footnote.

DVDP 2 and 3 reached a depth of 179.41 and 381.00 m, respectively. Eleven flow units and five pyroclastic units were identified in DVDP 2, whereas in DVDP 3, ten flow units and five pyroclastic units were logged. The cores of DVDP 2 [*Treves and Kyle*, 1973] and DVDP 3 [*Kyle and Treves*, 1974] are essentially the same. Minor differences in correlation do occur (Figure 3); however, the following discussion will consider the two as a composite. The core of DVDP 2 is more complete than for the equivalent depth in DVDP 3, as three intervals in the upper part of DVDP 3 were not cored. Four lithologic sequences (Table 2) were distinguished on the basis of petrographic examination of the core. They are described from youngest to oldest.

Chemical analyses of DVDP 2 samples are given in Table 11, and a summary of the petrography of DVDP 3 samples is given in Table 12.

Aphyric Lava Sequence (I)

The first 62.73 m (units 1-7) and 64.60 m (units 1-6) of DVDP 2 and 3 (Tables 5 and 6), respectively, are aphyric Ne-hawaiite, Ne-mugearite, and Ne-benmoreite lava flows. The lava flow units of DVDP 2 range from greater than 3.51 to 15.05 m in thickness (Table 5). The tops and bases of the flows are red and black, scoriaceous, and rubbly. Irregular cooling joints in the massive interiors of the flows are covered with secondary minerals [*Browne*, 1973, 1974] and red oxide coatings. Ice is common along joints and filling cavities.

Ne-benmoreite flow unit 7 has been dated by the conventional K/Ar method as 1.16 ± 0.03 m.y. old (Table 8).

Petrographically, the lavas typically are vesicular and microporphyritic with microphenocrysts (<1 mm in diameter) of kaersutite, clinopyroxene, plagioclase (generally andesine), and opaque oxides (Table 11). The microphenocrysts show a pilotaxitic texture, whereas the groundmass usually has a hyalopilitic texture.

Chemical analyses of DVDP 2 units 2 and 4-7 (Table 11) indicate the variable nature of these intermediate rock types (Figure 4). A discussion of the chemistry is given by *Kyle* [1981], who suggests the flows may represent at least two fractionation trends.

Porphyritic Intermediate Lava Sequence (II)

A pyroclastic unit (2-8 and 3-7) and a lava flow (2-9 and 3-8) represent this sequence. (Units are prefixed by the hole number, i.e., 2-8 is unit 8 of DVDP 2.) The pyroclastic unit is 16.23 and 16.10 m thick in DVDP 2 and 3, respectively, and consists of red, brown, and yellow volcanic breccia with a lapilli and ash matrix. The flow is 14.60 m thick and black with reddish orange spots due to oxidation; oxide coatings are common on joints. Thin scoriaceous zones occur at the top and bottom of the flow, and apart from a meter of flow banding near the top of the flow, it is massive. Inclusions of gabbro, dunite, lithic fragments, and megacrysts of kaersutite and clinopyroxene were noted.

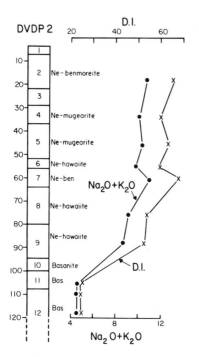

Fig. 4. Variation of total alkalis Na_2O + K_2O (in weight percent) and differentiation index (DI) [*Thornton and Tuttle*, 1960] in lava flows from the upper 120 m of DVDP 2.

Petrographically, the two units are porphyritic with phenocrysts and microphenocrysts of kaersutite, clinopyroxene, and minor opaque oxides. Rounded xenocrysts of olivine with black reaction rims are rarely observed. Kaersutite phenocrysts show varying degrees of oxidation from thin oxidized rims to completely oxidized pseudomorphs. The clinopyroxene is usually colorless; occasionally it forms glomeroporphyritic clumps and often shows some evidence of resorption. Opaque oxides, plagioclase (mainly andesine), and apatite also occur as microphenocrysts. The groundmass is hyalopilitic, usually with microlites of andesine, clinopyroxene, opaque oxides, and rarely, kaersutite.

Major element analyses of units 8 and 9 from DVDP 2 are very similar; only $Fe_2O_3^*$, Na_2O, and K_2O show differences outside analytical error. The samples are Ne-hawaiites. Samples from the lower part of the flow have lower abundances of phenocrysts, suggesting zoning of the flow. This is supported by analyses for alkalis of samples from unit 8 of DVDP 3; at depths of 89.55 and 94.72 m, total alkalis are 8.85 wt % (Na_2O = 5.75, K_2O = 3.10) and 6.65 wt % (Na_2O = 4.25, K_2O = 2.40), respectively (P. R. L. Browne, written communication, 1974). The correlative unit in DVDP 1 also shows

marked zoning; however, it has a basanite composition.

Basanite Sequence (III)

In DVDP 2 the basanite sequence is 70.28 m thick and consists of three flow units (11, 12, and 14) and two pyroclastic units (10 and 13). The flows range in thickness from 7.53 to 44.51 m, whereas the pyroclastic units are 3.35 and 5.63 m thick (Table 4). In DVDP 3 the basanite sequence (III) is 71.67 m thick and consists of three flow units (10, 12, and 14) and three pyroclastic units (9, 11, and 13). The flows range in thickness from 8.18 to 44.03 m, wheras the pyroclastics are thinner and vary from 1.36 to 4.44 m.

The flows are primarily black and dusky red with scoriaceous and fragmental tops and bases. Secondary minerals are not as abundant as in the overlying units of intermediate lava, but they do occur on the walls of ice-filled cavities and on joints. Joint surfaces are also commonly coated with thin oxide coatings. Flow units 2-12 and 3-12 are exceptional in being 44.51 and 44.03 m thick, respectively. In places, the lower third of the flow is hydrothermally altered with a streaky texture composed of veins of secondary minerals. *Browne* [1973] has described the alteration and considers some of it formed by gas discharged from the cooling lava. Thick lava flows frequently show strong oxidation toward their bases [*Haggerty*, 1976]. Direct measurement of oxygen fugacity in a thick, cooling ponded-lava lake at Makaopuhi, Hawaii [*Sato and Wright*, 1966; *Wright and Okamura*, 1977], showed a zone of intense oxidation developed at temperatures of 550-570°C. *Sato and Wright* [1966] suggested the oxidation was due to dissociation of H_2O; the small ionic size of H_2 allowed it to diffuse and escape, whereas the oxygen was retained, thus causing the oxidation. Such a model seems applicable to unit 12 in DVDP 2 and 3.

Pyroclastic unit 3-11 is 1.36 m thick and may be the scoriaceous top of flow unit 3-12. In DVDP 2 the equivalent rocks were considered to be a scoriaceous pyroclastic flow top subunit (2-12A) of the flow unit 2-12 [*Treves and Kyle*, 1973].

Petrographically, all the samples examined are very similar except for varying degrees of vesiculation and oxidation. Typically, the lavas are strongly porphyritic with seriate phenocrysts of olivine and clinopyroxene. The abundance of each is variable but usually ranges between 8 and 15%. Microphenocrysts of rhönite, opaque oxides, and plagioclase may range up to about 5% each. Groundmass textures are variable and include intergranular, subophitic, or hyalopilitic. The groundmass contains laths of labradorite, prisms of clinopyroxene, small cubic opaque oxides, minor olivine, rhönite, and glass.

Chemical analyses (Table 11) show the lavas are strongly undersaturated, usually with 7-10% normative nepheline (Ne); they are therefore basanites.

Basanite Hyaloclastite Sequence (IV)

Pryoclastic rocks, mainly hyaloclastite, occur in the lower 14.58 m of DVDP 2. The same sequence in DVDP 3 was cored for 214.17 m, and so the following discussion will be restricted to DVDP 3.

The first 20.62 m consist of a mixed volcanic breccia in which blocky palgonitic lapilli tuff predominates. Interbedded porphyritic olivine-clinopyroxene basanite flows(?) (perhaps pillows) show strong shattered bases and occasional chilled margins. Some basanites do not have chilled margins but are jointed and may be in the form of large blocks. This zone may represent a transition from subaerial to subaqueous deposition.

The remaining 139.55 m of the unit is composed predominantly of lapilli tuff and blocky lapilli tuff with minor tuff and black vitric tuff. A palagonitic matrix predominates in the lapilli tuffs. The tuff bands often show laminated bedding, which is concordant over basanite fragments and is compacted under and truncated around basanite fragments. The whole unit is only weakly to moderately lithified but is ice cemented except for the lower 10 m, where the core is uncemented and rubbly.

In DVDP 3 there is a break in the rock type of 218 m; above this depth the basanite is characterized by microphenocrysts of rhönite. Lapilli in a lapilli tuff sample from 218.47 m are composed of a large number of different lithologies. Many of the lapilli are rounded, probably indicating transportation from some distance. The diversity of the lithologies, which are not represented in the hyaloclastite below this depth, suggests that this material has been derived from a volcanic center(s) other than Hut Point Peninsula. The rock types are similar to those reported from Black Island and Brown Peninsula [Cole and Ewart, 1968]. Some of the lapilli may have been derived by erosion of units 35-39 in DVDP 1. There was probably a break in the

eruptive activity at this time, thus allowing an opportunity for the lapilli described above to be transported to the present site. A similar break has been noted in the hydrogen and oxygen isotope measurements of the included permafrost ice close to this lithologic break [Lyon, 1973; Nakai et al., 1978].

CORRELATION

Correlation of units between DVDP 2 and 3 is straightforward (Figure 3) and has been discussed above. Lithostratigraphic correlation of units in DVDP 1 with units in DVDP 3 is less certain; tentative suggestions are shown in Figure 3.

Basanite clasts in DVDP 1 hyaloclastite unit 40 do not contain rhönite and probably correlate with hyaloclastite below 218 m in DVDP 3, in which basanite clasts are also rhönite free. The DVDP 1 lava flow and pyroclastic units 28-34 are correlated with DVDP 3 units 9-14 and unit 15 to a depth of 218 m, as all are basanites and contain rhönite. The correlation is supported by chemical analyses of selected samples.

DVDP 3 units 7 and 8 are correlated with DVDP 1 units 21 and 22, respectively. Samples from the correlated units in all three drillholes are petrographically indistinguishable. Chemical analyses of DVDP 1 and 3 samples also indicate the lava flow unit (1-22 and 3-7) is zoned, a feature that is not usually observed in McMurdo Volcanic Group lava flows. The zoning supports the inferred correlations.

Ne-mugearite flow unit 1-14 is correlated with unit 2-5 on the basis of chemical analyses given in Tables 10 and 11. Analyses of samples 1-42.49 and 2-45.86 m are identical, within analytical error. Only Na_2O shows minor differences, which may be a result of different analytical techniques. Na_2O in samples 1-42.49 and 2-45.86 was analyzed by atomic absorption and X-ray fluorescence in 1973 and 1978, respectively.

Units in DVDP 1 which are correlated with those in DVDP 3 are all at higher altitudes. Assuming no fault between the two drill holes, and a reasonably uniform topography, it is likely that correlated flows in DVDP 1 originated closer to the eruptive vents. The thick basanite unit 12 in DVDP 2 and 3 is also likely to result from ponding of lava away from the vent, rather than represent a lava lake at an eruptive center. The lack of any correlation of flows in sequences I of DVDP 1 and 3, Ne-mugearite flow units 1-14 and 3-4 being an exception, shows these intermediate lavas were probably of only local extent.

GEOLOGIC HISTORY OF HUT POINT PENINSULA

The geologic history has been reconstructed using data from the drillcores combined with data obtained from mapping at the surface. Hyaloclastite was probably produced during the initial buildup of Hut Point Peninsula from the floor of McMurdo Sound. The elongate nature of the Peninsula suggests the eruptions were either from fissures or from a series of linearly aligned vents.

The oldest observed event is inferred to be the formation of Crater Hill. On the basis of a single paleomagnetic determination of an eroded flow north of Scott Base (Figure 2) and inferring that it underlies the Observation Hill sequence, the Crater Hill sequence is considered to be older than about 1.61 m.y. (Figure 2). All the DVDP core has reversed magnetic polarity [McMahon and Spall, 1974] consistent with the K/Ar ages (Figure 3, Table 8) and eruption during the Matayama Reversed Paleomagnetic Epoch (Figure 2). Therefore the Crater Hill sequence cannot be represented in the DVDP core, except perhaps in the deeper part of DVDP 3. This is a little surprising considering the close proximity of the drill sites to Crater Hill.

Erosional features are conspicuous around Crater Hill, particularly at The Gap, where there are steep cliffs, and northwest of Scott Base, where a cirque occurs, but is not partly buried by a younger flow on which Scott Base is situated. Crater Hill is strongly eroded, to a degree that has not since been repeated at Hut Point Peninsula; the erosion must have occurred prior to the emplacements of the lava flows at DVDP 1, 2, and 3. Tentatively, it is suggested the hyaloclastite below the erosion break at 218 m in DVDP 3 may be correlated with the Crater Hill sequence: note, however, that the hyaloclastite at this level has also been correlated with hyaloclastite unit 40 in DVDP 1, in which a clast has an age of 1.34 ± 0.23 m.y. This suggests that even though the hyaloclastite in DVDP 1 and that deeper than 218 m in DVDP 3 are lithologically similar, they could have been erupted from different vents at different times.

Following the basanite eruptions of the Crater Hill sequence and the hyaloclastite sequence VI in DVDP 1, intermediate lava types are found as sequence V in DVDP 1. This eruptive event is not represented at the surface or in DVDP 2/3 and represents the first evidence of a differentiating magma chamber at depth. Flow unit 37 is a benmoreite and is unusual in that intermediate rock types with <10% normative nepheline (Ne) are extremely rare in the Erebus volcanic province.

A major phase of volcanism then commenced with the eruption of rhönite-bearing basanite followed by Ne-hawaiite and then a variety of intermediate compositions. The final event was the formation of the phonolite cone at Observation Hill. This eruptive episode is excellently recorded as sequences I, II, and III in the DVDP 2/3 drill core and at Observation Hill.

Part of the eruptive event is also recorded in DVDP 1 as basanite and Ne-hawaiite eruptive sequences IV and III, respectively. Units 14-20 of sequence I in DVDP 1 may also represent part of the same eruptive event. Chemically, unit 1-14 is identical to unit 2-5. Units 1-15 and 1-16 plot on the chemical trend defined by analyses of units 2-8 and 2-6. In DVDP 2, unit 7 is not apparently related to the fractionation trend exhibited by flows 9, 8, 6, and 5 (Figure 4). The same is true for DVDP 1 units 18 and 19, which may be part of the same fractionation trend as unit 2-7. This suggests that there were several differentiating bodies involved in the overall general trend of basanite to phonolite. A detailed discussion of the differentiation process will be presented elsewhere [Kyle, 1981].

At DVDP 1, phonolitic lavas similar to those that overlie DVDP 2/3 at Observation Hill, do not occur. Instead there are a series of 11 or 12 Ne-hawaiite flows which probably represent a single eruptive episode or series of closely spaced eruptions. The DVDP 1 Ne-hawaiite flows, the phonolite at Observation Hill, and the hyaloclastite at Castle Rock (Figure 2) are all similar in age when the analytical uncertainties in the K/Ar age are considered (Tables 3 and 8). It is not possible to determine the relative age of these three events, on stratigraphic grounds.

The final chapter in the geological evolution of Hut Point Peninsula was the eruption of lavas belonging to the Half Moon Crater and Twin Crater sequences mapped at the surface (Figure 2). Volcanism ceased with the eruption of Black Knob about 0.4 m.y. ago (Table 3).

CONCLUSIONS

DVDP 1, 2, and 3 drillcores have established a more complete geologic history for Hut Point Peninsula than that inferred from surface mapping.

The record from DVDP 1 is more varied and complete, compared to DVDP 2 and 3; however, DVDP 1 core is also considerably more oxidized and altered than DVDP 2 and 3. The outstanding feature of the core is the abundance of intermediate lava types: Ne-hawaiite, Ne-mugearite, and Ne-benmoreite. Surface mapping had indicated a bimodal distribution of basanite and phonolite at Hut Point Peninsula. However, when the surface and subsurface data are combined, they reveal a near-complete differentiation sequence from basanite to phonolite, which has been termed the DVDP lava lineage [*Kyle et al.*, 1979*a*; *Kyle*, 1981]. In all, three distinct episodes of differentiation are revealed by the DVDP core.

The inability to correlate individual flow units between DVDP 1 and DVDP 2/3, which are only 600 m apart, demonstrates the weak nature of the eruptions responsible for the lava flows and pyroclastic rocks. Further drilling, particularly north of McMurdo Station, is also likely to enlarge the already complex geologic history discussed above.

In final conclusion the drilling of McMurdo Volcanic Group rocks at McMurdo Station has proved to be a resounding success. The drilling project provided the necessary stimulation to undertake detailed geochemical studies on the volcanic rocks. Our understanding of the petrogenesis of the volcanics is now well advanced [*Sun and Hanson*, 1975, 1976; *Kyle and Rankin*, 1976; *Kyle*, 1981]; however, the details of their tectonic setting still requires more detailed geophysical data.

Acknowledgments. I would like to acknowledge the New Zealand drillers, who under the supervision of Martin McGale and Jack Hoffman, went to tremendous efforts to provide the core. All the drillcores were logged in association with Sam Treves. To Sam goes my greatest respect and thanks; I will always remember his jovial cries of 'more core, more core.' To all the other scores of people who assisted in the Thiel Earth Science Laboratory, at McMurdo Station, at the core storage facility at Northern Illinois University, and at Victoria University of Wellington go many thanks. To the respective government agencies of Japan, New Zealand, and the United States go congratulations for their commitment to DVDP and all its scientific aims. My involvement in DVDP was sponsored by Victoria University of Wellington Antarctic Expeditions (VUWAE) 17 (1972-1973) and 18 (1973-1974), which were supported by grants from the New Zealand University grants committee. Financial support for this study was provided by Victoria University of Wellington, Northern Illinois University, Ohio State University, and the National Science Foundation under grants DPP77-23440 and DPP77-21590. The manuscript was reviewed by J. W. Cole and D. H. Elliot.

REFERENCES

Armstrong, R. L., K-Ar dating: Late Cenozoic McMurdo Volcanic Group and dry valley glacial history, Victoria Land, Antarctica, *N.Z. J. Geol. Geophys.*, *21*, 685-698, 1978.

Browne, P. R. L., Secondary minerals in cores from DVDP 1 and 2, *Dry Val. Drilling Proj. Bull.*, *2*, 83-93, 1973.

Browne, P. R. L., Secondary minerals from Ross Island drill holes (abstract), *Dry Val. Drilling Proj. Bull.*, *4*, 15, 1974.

Cole, J. W., and A. Ewart, Contributions to the volcanic geology of the Black Island, Brown Peninsula and Cape Bird areas, McMurdo Sound, Antarctica, *N.Z. J. Geol. Geophys.*, *11*, 793-828, 1968.

Cole, J. W., and P. R. Kyle, Stratigraphy, petrology and chemistry of rocks from Ross Island drill holes, *Dry Val. Drilling Proj. Bull.*, *2*, 8-10, 1972.

Cole, J. W., P. R. Kyle, and V. E. Neall, Contributions to the Quaternary geology of Cape Crozier, White Island and Hut Point Peninsula, McMurdo Sound region, Antarctica, *N.Z. J. Geol. Geophys.*, *14*, 528-546, 1971.

Forbes, R. B., D. L. Turner, and J. R. Carden, Age of trachyte from Ross Island, Antarctica, *Geology*, *2*, 297-298, 1974.

Goldich, S. S., S. B. Treves, N. H. Suhr, and J. S. Stuckless, Geochemistry of the Cenozoic volcanic rocks of Ross Island and vicinity, Antarctica, *J. Geol.*, *83*, 415-435, 1975.

Haggerty, S. E., Oxidation of opaque mineral oxides in basalts, chap. 4, in *Short Course Notes*, edited by D. Rumble III, Mineralogical Society of America, Washington, D. C., 1976.

Harrington, H. J., Nomenclature of rock units in the Ross Sea region, Antarctica, *Nature*, *182*, 290, 1958.

Kyle, P. R., Geology, mineralogy and geochemistry of the Late Cenozoic McMurdo Volcanic Group, Victoria Land, Antarctica, Ph.D. thesis, 444 pp., Victoria Univ., Wellington, N. Z., 1976.

Kyle, P. R., Differentiation of a basanite to phonolite sequence at Hut Point Peninsula, Antarctica, based on core from Dry Valley Drilling Project drillholes 1, 2 and 3, *J. Pet.*, in press, 1981.

Kyle, P. R., Glacial history of the McMurdo Sound area as indicated by the distribution and nature of McMurdo Volcanic Group rocks, this volume.

Kyle, P. R., and J. W. Cole, Structural control of volcanism in the McMurdo Volcanic Group, Antarctica, *Bull. Volcanol.*, *38*, 16-25, 1974.

Kyle, P. R., and P. C. Rankin, Rare earth element geochemistry of Late Cenozoic alkaline lavas of the McMurdo Volcanic Group, Antarctica, *Geochim. Cosmochim. Acta*, *40*, 1497-1507, 1976.

Kyle, P. R., and S. B. Treves, Geology of DVDP 3, Hut Point Peninsula, Ross Island, Antarctica, *Dry Val. Drilling Proj. Bull.*, *3*, 13-18, 1974.

Kyle, P. R., J. Adams, and P. C. Rankin, Geology and petrology of the McMurdo Volcanic Group at Rainbow Ridge, Brown Peninsula, Antarctica, *Geol. Soc. Amer. Bull.*, *90*, 676-688, 1979*a*.

Kyle, P. R., J. F. Sutter, and S. B. Treves, K/Ar age determinations on drill core from DVDP holes 1 and 2, in *Proceedings of the Seminar III on Dry Valley Drilling Project, Mem. Spec. Issue 13*, edited by T. Nagata, pp. 214-219, National Institute of Polar Research, Tokyo, 1979*b*.

Kyle, P. R., R. R. Dibble, W. F. Giggenbach, and J. R. Keys,

Volcanic activity associated with the anorthoclase phonolite lava lake, Mt. Erebus, Antarctica, in *Antarctic Geoscience*, edited by C. Craddock, University of Wisconsin Press, Madison, in press, 1980.

Lyon, G. R., Preliminary stable isotope analysis of drill hole ice at McMurdo Station, *Antarct. J. U.S.*, *8*, 160-162, 1973.

McMahon, B. E., and M. Spall, Paleomagnetic data from unit 13, DVDP hole 2, Ross Island, *Antarct. J. U.S.*, *9*, 229-232, 1976.

Nakai, N., Y. Mizutani, and M. Wada, Stable isotope studies: Past volcanic events deduced from H, O, S, and C isotopic compositions of ice and salts from DVDP 3 (abstract), *Dry Val. Drilling Proj. Bull.*, *8*, 66-67, 1978.

Nathan, S., and F. J. Schulte, The geology and petrology of the area between the Campbell and Aviator glaciers, *N.Z. J. Geol. Geophys.*, *11*, 940-975, 1968.

Sato, M., and T. L. Wright, Oxygen fugacity directly measured in magnetic gases, *Science*, *153*(3740), 1103-1105, 1966.

Steiger, R. H., and E. Jager, Subcommission on geochronology—Convention on the use of decay constants in geo- and cosmochronology, *Earth Planet. Sci. Lett.*, *36*, 359-362, 1977.

Sun, S. S., and G. N. Hanson, Origin of Ross Island basanitoids and limitations upon the heterogeneity of mantle sources for alkali basalts and nephelinites, *Contrib. Mineral. Petrol.*, *52*, 77-106, 1975.

Sun, S. S., and G. N. Hanson, Rare-earth element evidence for differentiation of McMurdo Volcanics, Ross Island, Antarctica, *Contrib. Mineral. Petrol.*, *54*, 139-155, 1976.

Thompson, R. N., J. Esson, and A. C. Dunham, Major element chemical variation in the Eocene lavas of the Isle of Skye, Scotland, *J. Pet.*, *13*, 219-253,1972.

Thornton, C. P., and O. F. Tuttle, Chemistry of igneous rocks, I, Differentiation index, *Amer. J. Sci.*, *258*, 664-684, 1960.

Treves, S. B., and P. R. Kyle, Geology of DVDP 1 and 2, Hut Point Peninsula, Ross Island, Antarctica, *Dry Val. Drilling Proj. Bull.*, *2*, 11-82, 1973.

Wright, T. L., and R. T. Okamura, Cooling and crystallization of tholeiitic basalt, 1965 Makaopuhi lava lake, Hawaii, *U.S. Geol. Surv. Prof. Pap.*, *1004*, 1977.

DVDP CORE STORAGE AND SAMPLE DISTRIBUTION

DENNIS S. CASSIDY

Antarctic Research Facility, Department of Geology, Florida State University
Tallahassee, Florida 32306

Cores recovered by the Japan-New Zealand-United States Dry Valley Drilling Project (DVDP) in Antarctica are stored at the Florida State University's Antarctic Research Facility. More than 1100 m of DVDP drill core are stored at −23°C, from which 3504 samples have been distributed to authorized investigators worldwide. All cores remain in excellent condition, and further research interest in them is invited.

INTRODUCTION

The purpose of this article is to introduce the role of the Antarctic Research Facility at Florida State University (FSU) within that of the Dry Valley Drilling Project (DVDP), and to provide a summary account of the status of the DVDP core collection curated at the facility, including conditions of storage, sampling methods, and sample distribution totals.

The apparent anomaly of DVDP core storage being located in the state of Florida can be attributed to the wealth of unused, refrigerated storage space available within the Antarctic Marine Geology Research Facility and Core Library at FSU. A curatorial and research activity, the facility was established by the National Science Foundation as a U.S. depository and research center for geological materials collected in the Southern Ocean. Since 1962, the permanent staff of the facility has maintained the marine geology shipboard coring program aboard the research vessels, USNS *Eltanin* and ARA *Islas Orcadas*. In addition to the DVDP materials, the facility presently houses more than 11,000 m of *Eltanin/Islas Orcadas* marine sediment cores, as well as a variety of core and dredge sediments collected in both polar regions under the auspices of the U.S. National Science Foundation.

In December 1972, James H. Zumberge, Chairman of the U.S. Academy of Sciences Committee on Polar Research, appointed Sayed Z. El-Sayed to the task of performing an on-site inspection of the facility in order to evaluate its capabilities necessary to the handling, storage, and sample distribution of DVDP core specimens. The result of this visit was that the facility was designated the domestic repository for DVDP sedimentary materials.

In retrospect, this arrangement has proven highly satisfactory and represents a unique, cooperative effort in that the facility has functioned as a satellite sample distribution and storage center for DVDP materials under the direction and guidance of the U.S. project coordinator, Lyle D. McGinnis at Northern Illinois University.

CORE STORAGE

DVDP core storage at FSU totals more than 1100 m of P, H, N, and B drill core packaged in 408 core boxes. This includes all core from the 15 drill sites, except for cores from DVDP holes 1, 2, and 3 and basement core below 10.52 m from hole 6; these are stored at Northern Illinois University. From U.S. ports of entry, the cores were shipped to the facility by refrigerated truck transportation at temperatures below −15°C. Upon receipt, they were immediately placed in a low-temperature storage vault maintained at a constant temperature of −23°C. This vault is located within a larger refrigerated storage room (2°C) and comprises 40 of the 510 m² of refrigerated storage space available at the facility. The 408 core boxes are arranged (Figure 1) on modular, bulk storage rack units with a total shelf capacity of about 700 core boxes. A duplicate, backup refrigeration unit has been installed to provide continuous service in the event of failure of the main unit.

DVDP core was received in three shipments following the termination of the last three drilling seasons. Cores from DVDP sites 4-9 were received

Fig. 1. Dry Valley Drilling Project cores stored at −23°C. (Holes 14 and 15 materials not yet received at the time this picture was taken.

Fig. 2.   Frozen Dry Valley Drilling Project core segment being dry drilled for paleomagnetic sample plug by Don Elston U.S. Geological Survey.

during May 1974, cores from sites 10-13 were received during March 1975, and cores from sites 14 and 15 during May 1976.

Essentially, the cores remain in excellent condition for further sampling, since frozen storage has preserved the structural integrity of the ice-cemented sediments. In some lithologies, a loosely consolidated, friable outer rind has developed due to the migration of ice out of the core by sublimation. In these cases, it is difficult to prevent some sediment loss in handling and sampling due to crumbling of the rind. Also, certain geochemical studies of interstitial water may no longer be feasible due to moisture loss.

SAMPLING

All sampling of DVDP core at the facility has been carried out with the direct approval of the U.S. coordinator of the project according to the terms of the official DVDP sampling policy. At the time of sampling, information concerning the hole and box number, sample interval, sample weight, proportion of core diameter sampled, and other data, including comments as may be necessitated, for example, by problems of interpretation, are recorded on a sample inventory form. Prepared in triplicate, one copy of the form is forwarded to the investigator receiving the samples, another is forwarded to the DVDP office at Northern Illinois University, which maintains a computerized inventory of all DVDP samples, and the third copy remains in the DVDP file at FSU. Additional inventory control is provided by the placement of sample identification cards at the point of sampling within each core box and by the recording of sample intervals and their locations, keyed to the investigator, on individually printed sets of core box photos which appear in Dry Valley Drilling Project bulletins 3, 5, and 7, prepared at Northern Illinois University.

All sampling is done within the 2°C storage room and is effected by chiseling and handsawing (rarely), dry sawing (no cutting fluid) by circular, diamond blade utilizing a Felker Di-Met Model 41A cut-off saw, or by diamond core drilling using a 38-cm, floor model Clausing drill press (Figure 2). The latter method makes use of a precooled drill bit using compressed air as a drilling 'fluid' and was developed by Don Elston of the U.S. Geological Survey to obtain oriented sample 'plugs' for paleomagnetic studies of DVDP cores.

Voids in the core resulting from the removal of samples are filled with cut-to-fit pieces of Dow Ethafoam rod in order to prevent shifting of core segments during box handling; thus the relative positions of the segments are preserved for further measurement of sample intervals.

SAMPLE DISTRIBUTION

Initial sampling and inspection of the DVDP materials at FSU began in August 1974 with a contingent of seven visiting investigators coordinated by Peter N. Webb. A second DVDP core sampling and inspection session was hosted by the facility on July 7, 8, and 9, 1975, involving 12 specialists in polar studies, and coordinated jointly by Lyle D. McGinnis and Mort D. Turner. For purposes of both sampling and core inspection, a total of 24 scientists have been received by the facility. An additional seven investigators have received samples on the basis of letter requests as have many of those who journeyed to the facility.

A total of 3504 samples has been distributed from the DVDP collection at FSU over a 5-year period. This total does not include samples removed from the cores prior to their arrival at the facility, nor does it include the redistribution of sample portions by investigators to whom samples were originally distributed. Figure 3 summarizes the sample distribution total according to the number of samples received per hole by investigators to whom the samples were assigned, and the number of samples distributed from each hole.

The shipment of samples to principal investigators has been by both frozen and unfrozen means of transportation. Frozen samples are packed in dry ice within commercially available styrofoam containers and have been shipped by air to points as far as Seattle without damage to or thawing of the materials.

FUTURE SAMPLING

Following publication of this volume, and the correlation of the presented data with those of related projects underway, it is anticipated that there will be a resurgence of interest in the availability of DVDP materials for further sampling. Potential investigators requiring samples, in order to place a meaningful request for them, are asked to review carefully the available literature, particularly the DVDP bulletins prepared at Northern Illinois University, in which appear the lithologic

SAMPLE RECIPIENT[1]	TOTAL NUMBER OF SAMPLES RECEIVED BY RECIPIENT PER HOLE NUMBER[2]:														SAMPLE TOTALS PER RECIPIENT
	4	4A	5	5A	6	7	8	9	10	11	12	13	14	15	
BARRETT							7	2	1	18				53	81
BRADY		89												65	154
CAMERON							42	35							77
CLARIDGE									60						60
DECKER									36	74	35		11		156
ELSTON							104	18	147	363	147				779
HENDY							10			9	25				44
JONES	12	17	2			2	2					8	22	18	83
KELLOGG		14										9	10		33
LING	2	11					24	3	37	139	29			21	266
MANDRA	2	11					20	3	37	139	29				241
McCOLLUM	2	11					25	3	37	139	29				246
McGINNIS	1	3	2		3	1	29	2	36	65	33	8	7	10	200
PORTER							33		5	35					73
STUIVER							50	15	20	150	84				319
TORII													36		36
TREVES														78	78
WEBB	5	13				7	99	17	59	268	67	5	22		562
WRENN										16					16
SAMPLE TOTALS PER HOLE NUMBER	24	169	4	0	3	10	445	98	475	1399	494	30	108	245	3,504

[1] NOTE: ALTHOUGH OTHER INVESTIGATORS ARE INVOLVED IN RESEARCH USING THESE SAMPLES, ONLY THE INDIVIDUALS TO WHOM SAMPLES HAVE BEEN OFFICIALLY ASSIGNED ARE LISTED.

[2] HOLE 4, 4A : LAKE VANDA   HOLE 7 : LAKE FRYXELL   HOLE 10 : NEW HARBOR   HOLE 13 : DON JUAN POND
HOLE 5, 5A : DON JUAN POND   HOLE 8 : NEW HARBOR   HOLE 11 : COMMONWEALTH GLACIER   HOLE 14 : NORTH FORK BASIN
HOLE 6 :   LAKE VIDA   HOLE 9 : NEW HARBOR   HOLE 12 : LAKE HOARE   HOLE 15 : McMURDO SOUND

Fig. 3.   Sample distribution from DVDP core stored at FSU.

logs, photographs, and sediment descriptions of DVDP core. Specifications for samples should indicate, in addition to routine information such as the core interval, sample size, and method of shipment, the criteria used for the determination of the desired sample interval. The latter information is particularly important in that it is extremely helpful to know whether the requested interval is based upon published data or upon perusal of the lithologic log (or both), since an element of subjectivity is often involved in the actual determination of the sample interval due to core condition, percent of core recovery, etc.

Further information concerning the DVDP cores at FSU and other aspects of the operation of the Antarctic Research Facility and its programs can be obtained by writing to the curator of the facility, as well as by reference to those articles appearing in the selected bibliography accompanying this report.

Acknowledgments. The author considers it appropriate to acknowledge the exceptional degree of cooperation and confidence afforded him by many persons throughout the course of the curatorial phase of DVDP, particularly by Peter J. Barrett, Lyle D. McGinnis, and Michael G. Mudrey, Jr. Funding for the curation of DVDP cores at the Antarctic Research Facility has been provided by National Science Foundation contracts C-564 and C-1059.

SELECTED BIBLIOGRAPHY

Cassidy, D. S., Antarctic Marine Geology Research Facility, 1972-1973, *Antarct. J. U.S.*, *8*(6), 356-357, 1973.

Cassidy, D. S., Polar programs of the Antarctic Marine Geology Research Facility and Core Library, *Die Erde*, *109*(2), 254-256, 1978.

Cassidy, D. S., Antarctic Marine Geology Research Facility and Core Library, 1978-79, *Antarct. J. U.S.*, *14*(5), 230-231, 1979*a*.

Cassidy, D. S., Dry Valley Drilling Project: Summary of core storage at Florida State University and sample distribution, *Antarct. J. U.S.*, *14*(5), 231-232, 1979*b*.

Cassidy, D. S., DVDP at the Florida State University: Core storage and sample distribution, in *Proceedings of the Seminar III on Dry Valley Drilling Project, 1978, Mem. Spec. Issue 13*, edited by T. Nagata, pp. 240-245, National Institute of Polar Research, Tokyo, 1979*c*.

Cassidy, D. S., Antarctic Marine Geology Research Facility, 1979-1980, *Antarct. J. U.S.*, *15*(5), 227-228, 1980*a*.

Cassidy, D. S., ARA *Islas Orcadas:* Core recovery, core storage, and sample distribution, *Antarct. J. U.S.*, *15*(5), 228-230, 1980*b*.

Cassidy, D. S., and G. W. DeVore, Antarctic Marine Geology Research Facility and Core Library, *Antarct. J. U.S.*, *8*(3), 120-128, 1973.

Cassidy, D. S., and S. Shepley, Core recovery: USNS *Eltanin* and ARA *Islas Orcadas*, *Antarct. J. U.S.*, *8*(4), 75-76, 1977.

Cassidy, D. S., and S. W. Wise, Jr., Antarctic Marine Geology Research Facility, 1973-1974, *Antarct. J. U.S.*, *9*(6), 319-321, 1974.

Cassidy, D. S., and S. W. Wise, Jr., Antarctic Marine Geology Research Facility, 1974-1975, *Antarct. J. U.S.*, *10*(5), 315-318, 1975.

Cassidy, D. S., and S. W. Wise, Jr., Antarctic Marine Geology Research Facility, 1975-1976, *Antarct. J. U.S.*, *11*(4), 287-290, 1976.

Cassidy, D. S., and S. W. Wise, Jr., Antarctic Marine Geology Research Facility, 1976-1977, *Antarct. J. U.S.*, *12*(4), 83-84, 1977.

Cassidy, D. S., and S. W. Wise, Jr., Antarctic Marine Geology Research Facility, 1977-1978, *Antarct. J. U.S.*, *13*(4), 221-222, 1978.

Cassidy, D. S., F. A. Kaharoeddin, I. Zemmels, and M. B. Knapp, USNS *Eltanin:* An inventory of core location data, with core location maps and cruise 55 core descriptions, *Contrib. 44*, 90 pp., Sedimentol. Res. Lab., Dep. of Geol., Fla. State Univ., Tallahassee, 1977a.

Cassidy, D. S., P. F. Ciesielski, F. A. Kaharoeddin, S. W. Wise, Jr., and I. Zemmels, ARA *Islas Orcadas* cruise 0775 sediment descriptions, *Contrib. 45*, 77 pp., Sedimentol. Res. Lab., Dep. of Geol., Fla. State Univ., Tallahassee, 1977b.

Kaharoeddin, F. A., ARA *Islas Orcadas* cruise 1176 sediment descriptions, *Contrib. 46*, 121 pp., Sedimentol. Res. Lab., Dep. of Geol., Fla. State Univ., Tallahassee, 1978.

Kaharoeddin, F. A., M. R. Eggers, R. S. Graves, E. H. Goldstein, J. G. Hattner, S. C. Jones, and P. F. Ciesielski, ARA *Islas Orcadas* cruise 1277 sediment descriptions, *Contrib. 47*, 108 pp., Sedimentol. Res. Lab., Dep. of Geol., Fla. State Univ., Tallahassee, 1979.

BIBLIOGRAPHY OF THE DRY VALLEY DRILLING PROJECT

PAULA REBERT

Northern Illinois University, DeKalb, Illinois 60115

The following publications include research resulting directly from the Dry Valley Drilling Project (DVDP) or from peripheral studies supported by the various national programs of Japan, New Zealand, and the United States under the auspices of DVDP. The major sources containing these materials are *Dry Valley Drilling Project Bulletin* (1 through 8), *Memoirs of the National Institute of Polar Research, Tokyo* (4 and 13), *Antarctic Journal of the United States*, National Science Foundation, Division of Polar Programs, Washington, D.C., and *Third Symposium on Antarctic Geology and Geophysics, Volume of Abstracts*, University of Wisconsin Press, Madison. *Antarctic Geoscience*, scheduled for publication by the University of Wisconsin Press in 1980, will comprise articles from the Third Symposium on Antarctic Geology and Geophysics. Articles in this AGU Antarctic Research Series volume are not included.

DVDP PUBLICATIONS

Publications in 1971

McGinnis, L. D., and T. E. Jensen, Permafrost-hydrogeologic regimen in two ice-free valleys, Antarctica, from electrical depth sounding, *Quaternary Res.*, *1*(3), 389-409.

Publications in 1972

Behar, J. V., L. Zafonte, R. E. Cameron, and F. A. Morelli, Hydrocarbons in air samples from antarctic dry valley drilling sites, *Antarct. J. U.S.*, *VII*(4), 94-96.

Cameron, R. E., and F. Morelli, Antarctic microbiol ecology, Dry Valley drilling sites, *Dry Val. Drilling Proj. Bull.*, *1*, 28-29.

Cameron, R. E., F. A. Morelli, and L. P. Randall, Aerial, aquatic, and soil microbiology of Don Juan Pond, Antarctica, *Antarct. J. U.S.*, *VII*(6), 254-258.

Clark, C. C., Seismic refraction and electrical resistivity investigations in the dry valleys, *Antarct. J. U.S.*, *VII*(4), 91-92.

McGinnis, L. D., and G. E. Montgomery, Aeromagnetic reconnaissance and geologic summary of the Dry Valley region—Phase I, *Dry Val. Drilling Proj. Bull.*, *1*, 61-90.

McGinnis, L. D., K. Nakao, and C. C. Clark, Geophysical identification of frozen and unfrozen ground, Antarctica, *Dry Val. Drilling Proj. Bull.*, *1*, 30-60.

McGinnis, L. D., T. Torii, and P. N. Webb, Dry Valley Drilling Project—Three nations are studying the subsurface in the McMurdo Sound region, *Antarct. J. U.S.*, *VII*(3), 53-56.

Montgomery, G. E., Aeromagnetic surveys of the Ross Island and Taylor Glacier quadrangles, *Antarct. J. U.S.*, *VII*(4), 90-91.

Morelli, F. A., R. E. Cameron, D. R. Gensel, and L. P. Randall, Monitoring of antarctic dry valley drilling sites, *Antarct. J. U.S.*, *VII*(4), 92-94.

Nakao, K., Y. Nishizaki, and K. Nakayama, Report of the Japanese summer parties in Dry Valleys, Victoria Land, 1971-1972, XI, Sedimentary structure near the saline lakes in three ice-free valleys, Victoria Land, Antarctica, inferred from electrical depth sounding, *Antarct. Rec.*, *45*, 89-104.

Nelson, C. S., and A. T. Wilson, Bathymetry and bottom sediments of Lake Vanda, Antarctica, *Antarct. J. U.S.*, *VII*(4), 97-99.

Torii, T., Japanese field surveys in the dry valleys, *Antarct. J. U.S.*, *VII*(4), 96.

Webb, P. N., Paleontology of late Tertiary-Quaternary sediments, Wright Valley, Antarctica, *Antarct. J. U.S.*, *V*(5), 96-97.

Webb, P. N., Wright Fjord, Pliocene marine invasion of an antarctic dry valley, *Antarct. J. U.S.*, *VII*(6), 226-234.

Publications in 1973

American Geophysical Union, Dry Valley Drilling Project, *Eos Trans. AGU*, *54*(8), 784-785.

Browne, P. R. L., Secondary minerals in cores from DVDP 1 and 2, *Dry Val. Drilling Proj. Bull.*, *2*, 83-93.

Browne, P. R. L., Secondary minerals from Dry Valley Drilling Project holes, *Antarct. J. U.S.*, *VIII*(4), 159-160.

Cameron, R. E., F. A. Morelli, and R. C. Honour, Aerobiological monitoring of dry valley drilling sites, *Antarct. J. U.S.*, *VIII*(4), 211-214.

Goldich, S. S., J. S. Stuckless, N. H. Suhr, and S. B. Treves, Geochemistry of volcanic rocks of the Ross Island area, Antarctica (abstract), *Geol. Soc. Amer. Abstr. Programs*, *5*(7), 638-639.

Kyle, P. R., and S. B. Treves, Review of the geology of Hut Point Peninsula, Ross Island, Antarctica, *Dry Val. Drilling Proj. Bull.*, *2*, 1-10.

Lyon, G. L., Preliminary stable isotope analysis of drillhole ice at McMurdo Station, *Antarct. J. U.S.*, *VIII*(4), 160-162.

McGinnis, L. D., Dry Valley Drilling Project, 1972-1973, *Antarct. J. U.S.*, *VIII*(4), 157.

McGinnis, L. D., McMurdo Sound—A key to the Cenozoic of Antarctica, *Antarct. J. U.S.*, *VIII*(4), 166-169.

McGinnis, L. D., K. Nakao, and C. C. Clark, Geophysical identification of frozen ground, Antarctica, in *Permafrost, 2nd International Conference, North American Contribution*, pp. 136-146, National Academy of Science, Washington, D. C.

Mudrey, M. G., Jr., Quaternary-Tertiary problems in Antarctica, Dry Valley Drilling Project—1973-1974, *Ice*, *42*(2), 10.

Mudrey, M. G., Jr., N. F. Shimp, C. W. Keighin, G. L. Oberts, and L. D. McGinnis, Chemical evolution of water in Don Juan Pond, Antarctica, *Antarct. J. U.S.*, *VIII*(4), 164-166.

Mudrey, M. G., Jr., S. B. Treves, P. R. Kyle, and L. D. McGinnis, Frozen jigsaw puzzle: First bedrock coring in Antarctica, *Geotimes*, *18*(11), 14-17.

Parker, B. C., M. G. Mudrey, Jr., R. E. Cameron, and K. Cartwright, Environmental appraisal for the Dry Valley Drilling Project Phase III (1973-74), report, 122 pp., N. Ill. Univ., DeKalb.

Stuckless, J. S., Geochemistry of McMurdo Volcanics, *Antarct. J. U.S.*, *VIII*(4), 162.

Torii, T., Japanese activities in the dry valleys, 1972-1973, *Antarct. J. U.S.*, *VIII*(4), 163-164.

Treves, S. B., and P. R. Kyle, Geology of DVDP 1 and 2, Hut Point Peninsula, Ross Island, Antarctica, *Dry Val. Drilling Proj. Bull.*, *2*, 11-82.

Treves, S. B., and P. R. Kyle, Renewed volcanic activity of Mt. Erebus, Antarctica, *Antarct. J. U.S.*, *VIII*(4), 156.

Treves, S. B., and P. R. Kyle, Geology of boreholes 1 and 2, Hut Point Peninsula, Antarctica, *Antarct. J. U.S.*, *VIII*(4), 157-159.

Treves, S. B., and J. S. Stuckless, Geology of volcanic rocks of the Ross Island area, Antarctica (abstract), *Geol. Soc. Amer. Abstr. Programs*, *5*(7), 843-844.

Wong, H. K., Aeromagnetic data from the McMurdo Sound region, *Antarct. J. U.S.*, *VIII*(4), 162-163.

Wrenn, J., and M. G. Mudrey, Jr., Disposition of core from DVDP 1 and 2, *Dry Val. Drilling Proj. Bull.*, *2*, 108-113.

Publications in 1974

Ankenbauer, G., J. L. Kilbourne, and M. G. Mudrey, Jr., Photographic record of New Harbor core, DVDP 8 and 9, *Dry Val. Drilling Proj. Bull.*, *3*, 191-225.

Antarctic Journal of the United States, DVDP 1973-1974 personnel, *IX*(4), 146.

Barrett, P. J., Prospects for the McMurdo Sound drillhole, Antarctica (abstract), *Dry Val. Drilling Proj. Bull.*, *4*, 11.

Barrett, P. J., S. A. Christoffel, D. J. Northey, and B. A. Sissons, Seismic profiles across the extension of Wright Valley into McMurdo Sound, *Antarct. J. U.S.*, *IX*(4), 138-140.

Black, R. F., Dating the Late Quaternary geomorphic events in the McMurdo Sound region (abstract), *Dry Val. Drilling Proj. Bull.*, *4*, 12-14.

Brady, H., Diatoms from the Lake Vanda core, *Dry Val. Drilling Proj. Bull.*, *3*, 181-184.

Browne, P. R. L., Secondary minerals from Ross Island drillholes (abstract), *Dry Val. Drilling Proj. Bull.*, *4*, 15.

Cameron, R. E., and F. A. Morelli, Viable microorganisms from ancient frozen cores drilled from Ross Island and Taylor Valley, Antarctica, *Dry Val. Drilling Proj. Bull.*, *3*, 171-180.

Cameron, R. E., and F. A. Morelli, Viable microorganisms from ancient Ross Island and Taylor Valley drill core, *Antarct. J. U.S.*, *IX*(4), 113-116.

Cameron, R. E., F. A. Morelli, R. Donlan, J. Guilfoyle, G. Markley, and R. Smith, DVDP environmental monitoring, *Antarct. J. U.S.*, *IX*(4), 141-144.

Cameron, R. E., F. A. Morelli, and R. C. Honour, Environmental impact monitoring of the Dry Valley Drilling Project (DVDP) (abstract), *Dry Val. Drilling Proj. Bull.*, *4*, 16-18.

Cartwright, K., Hydrogeology studies in the dry valleys from DVDP boreholes (abstract), *Dry Val. Drilling Proj. Bull.*, *4*, 19.

Cartwright, K., H. Harris, and M. Heidari, Hydrogeological studies in the dry valleys, *Antarct. J. U.S.*, *IX*(4), 131-133.

Cartwright, K., S. B. Treves, and T. Torii, Geology of DVDP 4, Lake Vanda, Wright Valley, Antarctica, *Dry Val. Drilling Proj. Bull.*, *3*, 49-74.

Cartwright, K., S. B. Treves, and T. Torii, Geology of DVDP 5, Don Juan Pond, Wright Valley, Antarctica, *Dry Val. Drilling Proj. Bull.*, *3*, 75-91.

Chapman-Smith, M., and P. G. Luckman, Late Cenozoic glacial sequence cored at New Harbor, Victoria Land, Antarctica (DVDP 8 and 9), *Dry Val. Drilling Proj. Bull.*, *3*, 120-148.

Clark, R. H., New Zealand science programs in the dry valleys—McMurdo Sound area (abstract), *Dry Val. Drilling Proj. Bull.*, *4*, 20-21.

Decker, E. R., Preliminary geothermal studies of the Dry Valley Drilling Project holes at McMurdo Station, Lake Vanda, Lake Vida, and New Harbor, Antarctica (abstract), *Dry Val. Drilling Proj. Bull.*, *4*, 22-23.

Denton, G. H., Late Cenozoic glacial history of the McMurdo Sound region as derived from surficial sediments and its relation to DVDP cores (abstract), *Dry Val. Drilling Proj. Bull.*, *4*, 24.

Fletcher, J. O., U.S. research efforts in Antarctica since the International Geophysical Year (abstract), *Dry Val. Drilling Proj. Bull.*, *4*, 25.

Friedman, I., and G. I. Smith, Studies carried out on the Lake Vanda core, *Dry Val. Drilling Proj. Bull.*, *3*, 185-186.

Friedman, I., A. Rafter, and G. I. Smith, An isotopic study of Lake Vanda, Antarctica (abstract), *Dry Val. Drilling Proj. Bull.*, *4*, 26.

Gumbley, J. W., A. T. Wilson, C. H. Hendy, and C. S. Nelson, Sedimentology of shallow cores from Lake Vanda, *Antarct. J. U.S.*, *IX*(4), 135-136.

Harris, H., and M. G. Mudrey, Jr., Core from Lake Fryxell, DVDP 7, and general geology of Lake Fryxell area, Taylor Valley, *Dry Val. Drilling Proj. Bull.*, *3*, 109-119.

Hendy, C. H., and R. Holdsworth, The physical impact of DVDP drilling at site 4, Lake Vanda, *Antarct. J. U.S.*, *IV*(4), 144-146.

Hendy, C. H., A. T. Wilson, and A. B. Field, The geochemistry of the closed drainage basins of the McMurdo Oasis, Antarctica (abstract), *Dry Val. Drilling Proj. Bull.*, *4*, 27.

Jones, L. M., J. C. Stormer, Jr., and R. E. Behling, Characterization of a volcanic ash fall, Wright Valley (abstract), *Dry Val. Drilling Proj. Bull.*, *4*, 28-29.

Kurasawa, H., Strontium isotope studies of the Ross Island volcanics (abstract), *Dry Val. Drilling Proj. Bull.*, *4*, 30-31.

Kurasawa, H., Y. Yoshida, and M. G. Mudrey, Jr., Geologic log of the Lake Vida core—DVDP 6, *Dry Val. Drilling Proj. Bull.*, *4*, 92-108.

Kusunoki, K., Japanese programs planned in the McMurdo Sound region (abstract), *Dry Val. Drilling Proj. Bull.*, *4*, 32.

Kyle, P. R., Petrology and mineralogy of DVDP 1 and 2 core samples (abstract), *Dry Val. Drilling Proj. Bull.*, *4*, 33.

Kyle, P. R., and S. B. Treves, Geology of DVDP 3, Hut Point Peninsula, Ross Island, Antarctica, *Dry Val. Drilling Proj. Bull.*, *3*, 13-48.

Kyle, P. R., and S. B. Treves, Geology of DVDP 3, Hut Point Peninsula, Ross Island, *Antarct. J. U.S.*, *IX*,(4), 127-129.

Kyle, P. R., and S. B. Treves, Geology of Hut Point Peninsula, Ross Island, *Antarct. J. U.S.*, *IX*(5), 232-234.

Lyon, G. L., Stable isotope analyses of ice from DVDP 3, *Dry Val. Drilling Proj. Bull.*, *3*, 160-170.

McGinnis, L. D., An overview of DVDP 1973-74, *Dry Val. Drilling Proj. Bull.*, *3*, 1-6.

McGinnis, L. D., Dry Valley Drilling Project, 1973-1974, *Antarct. J. U.S.*, *IX*(4), 125-127.

McGinnis, L. D., DVDP seminar held in Seattle, *Antarct. J. U.S.*, *IX*(4), 191-192.

McGinnis, L. D., C. C. Clark, D. R. Pederson, H. K. Wong, C. P. Ervin, and G. E. Montgomery, Aeromagnetic and refraction seismic studies for the Dry Valley Drilling Project (DVDP) (abstract), *Dry Val. Drilling Proj. Bull.*, *4*, 34-36.

McMahon, B., and H. Spall, Results of paleomagnetic investigation of selected cores recovered by Dry Valley Drilling Project (abstract), *Dry Val. Drilling Proj. Bull.*, *4*, 37-38.

McMahon, B. E., and H. Spall, Paleomagnetic data from unit 13, DVDP hole 2, Ross Island, *Antarct. J. U.S.*, *IX*(5), 229-232.

Morikawa, H., and J. Ossaka, The distribution

of secondary minerals at Lake Vanda (abstract), *Dry Val. Drilling Proj. Bull.*, *4*, 39-40.

Mudrey, M. G., Jr., Summary of drilling activities December 1973-February 1974, *Dry Val. Drilling Proj. Bull.*, *3*, 9-12.

Mudrey, M. G., Jr., Geology of DVDP holes 6, 7, 8, and 9, *Antarct. J. U.S.*, *IX*(4), 130.

Mudrey, M. G., Jr., A model for chemical evolution of surface waters in Wright Valley, Antarctica (abstract), *Dry Val. Drilling Proj. Bull.*, *4*, 41-42.

Mudrey, M. G., Jr., A model for the chemical evolution of surface water brines in Don Juan Pond, Wright Valley, Antarctica (abstract), *Geol. Soc. Amer. Abstr. Programs*, *6*(7), 880.

Mudrey, M. G., Jr., and J. L. Kilbourne, Disposition and meterage of DVDP core, *Dry Val. Drilling Proj. Bull.*, *3*, 226-233.

Mudrey, M. G., Jr., and L. D. McGinnis, Antarctic geologic history investigated by diamond drilling, *Geology*, *2*(6), 291-294.

Mudrey, M. G., Jr., and L. D. McGinnis, Antarctic drilling scheduled, *Geotimes*, *19*(11), 14-15.

Nagata, T., Japanese program on earth sciences in Antarctica (abstract), *Dry Val. Drilling Proj. Bull.*, *4*, 43-45.

Nakai, N., Stable isotope studies of the salts, water and ice from Ross Island core and Lake Vanda (abstract), *Dry Val. Drilling Proj. Bull.*, *4*, 46-47.

Nakao, K., Electrical resistivity studies at DVDP drilling sites (abstract), *Dry Val. Drilling Proj. Bull.*, *4*, 48-49.

Northey, D. J., and B. A. Sissons, Preliminary seismic profiling survey in McMurdo Sound, *Dry Val. Drilling Proj. Bull.*, *3*, 234-239.

Parker, B. C., Photosynthetic production and in situ concentrations of organic matter in Lake Bonney (abstract), *Dry Val. Drilling Proj. Bull.*, *4*, 50.

Parker, B. C., M. G. Mudrey, Jr., and K. Cartwright, Environmental appraisal for the Dry Valley Drilling Project Phase IV (1974-1975), report, 141 pp., N. Ill. Univ., DeKalb.

Pruss, E. F., E. R. Decker, and S. B. Smithson, Preliminary temperature measurements at DVDP holes 3, 4, 6, and 8, *Antarct. J. U.S.*, *IX*(4), 133-134.

Smith, G. I., and I. Friedman, Lacustrine deposits around Lake Vanda and Don Juan Pond, Antarctica, *Dry Val. Drilling Proj. Bull.*, *3*, 187-190.

Smith, G. I., and I. Friedman, Lacustrine deposits around Lake Vanda and Don Juan Pond,
Antarctica (abstract), *Dry Val. Drilling Proj. Bull.*, *4*, 51.

Smith, P. J., Drilling in the Antarctic, *Nature*, *251*(5476), 570-571.

Stuckless, J. S., P. W. Weiblen, and K. J. Schulz, Magma evolution for the alkalic rocks of the Ross Island area: Antarctica (abstract), *Eos Trans. AGU*, *55*, 474.

Stuckless, J. S., P. W. Weiblen, and S. S. Goldich, A petrogenetic model for the alkalic rocks from the Ross Island area: Antarctica (abstract), *Dry Val. Drilling Proj. Bull.*, *4*, 52-53.

Sun, S. S., and G. N. Hanson, Genesis of Ross Island, Antarctica volcanic rocks based on rare earth elements and Pb isotope studies (abstract), *Eos Trans. AGU*, *55*, 474.

Sun, S. S., and G. N. Hanson, Genesis of Ross Island, Antarctica volcanic rocks based on rare earth elements and Pb isotope studies (abstract), *Dry Val. Drilling Proj. Bull.*, *4*, 54.

Sun, S. S., and G. N. Hanson, Genesis of McMurdo volcanics on Ross Island, *Antarct. J. U.S.*, *IX*(5), 234-236.

Thomson, R. B., DSIR Antarctic programs (abstract), *Dry Val. Drilling Proj. Bull.*, *4*, 55-57.

Torii, T., Japanese activities in DVDP, 1973-1974, *Antarct, J. U.S.*, *IX*(4), 130-131.

Torii, T., N. Nakai, H. Morikawa, S. Ohno, and K. Nakayama, Preliminary report on the Japanese research work in Phase III of DVDP, *Dry Val. Drilling Proj. Bull.*, *3*, 149-155.

Torii, T., N. Yamagata, and S. Nakaya, Distribution of the nutrient matters in the saline lakes of the Dry Valleys (abstract), *Dry Val. Drilling Proj. Bull.*, *4*, 58-59.

Treves, S. B., Summary of drilling activities September 1973-December 1973, *Dry Val. Drilling Proj. Bull.*, *3*, 7-8.

Treves, S. B., and M. Z. Ali, Geology and petrography of DVDP 1, Hut Point Peninsula, Ross Island, Antarctica (abstract), *Dry Val. Drilling Proj. Bull.*, *4*, 60-61.

Treves, S. B., and P. R. Kyle, Volcanic activity of Mount Erebus, 1973-1974, *Antarct. J. U.S.*, *IX*(4), 147.

Treves, S. B., J. G. Rinehart, and R. Pochon, Geology and petrography of rocks from the floor of the Ross Sea near Ross Island, *Antarct. J. U.S.*, *IX*(4), 152-154.

Turner, M. D., U.S. earth science programs in Antarctica (abstract), *Dry Val. Drilling Proj. Bull.*, *4*, 156-159.

Watanuki, K., and H. Morikawa, A note on the

minerals found in DVDP core, *Dry Val. Drilling Proj. Bull.*, *3*, 156-159.

Watanuki, K., and H. Morikawa, Geochemical aspects of various minerals found at Lake Vida (abstract), *Dry Val. Drilling Proj. Bull.*, *4*, 63-65.

Webb, P. N., Synthesis of DVDP results and some views on future objectives (abstract), *Dry Val. Drilling Proj.*, *Bull.*, *4*, 66.

Weiblen, P. W., M. G. Mudrey, Jr., J. S. Stuckless, and K. J. Schulz, Reaction relations of clinopyroxenes in alkali basalts of the Ross area, Antarctica (abstract), *Eos Trans. AGU*, *55*, 475.

Weiblen, P. W., K. J. Schulz, J. S. Stuckless, W. C. Hunter, and M. G. Mudrey, Jr., Clinopyroxenes in alkali basalts from the Ross Island area, Antarctica: Clues to stages of magma crystallization (abstract), *Dry Val. Drilling Proj. Bull.*, *4*, 67-71.

Wilson, A. T., and C. H. Hendy, McMurdo dry valleys lake sediments—A record of Cenozoic climatic events (abstract), *Dry Val. Drilling Proj. Bull.*, *4*, 72.

Wilson, A. T., C. H. Hendy, T. R. Healy, A. B. Gumbley Field, and C. P. Reynolds, Dry valley lake sediments—A record of Cenozoic climatic events, *Antarct. J. U.S.*, *IX*(4), 134-135.

Wilson, A. T., R. Holdsworth, and C. H. Hendy, Lake Vanda: Source of heating, *Antarct. J. U.S.*, *IX*(4), 137-138.

Yoshida, Y., DVDP and some aspects of antarctic geomorphology (abstract), *Dry Val. Drilling Proj. Bull.*, *4*, 73-74.

Yusa, Y., The heat balance and thermal structure of Lake Vanda (abstract), *Dry Val. Drilling Proj. Bull.*, *4*, 75-76.

Publications in 1975

Antarctic Journal of the United States, DVDP drilling stopped, *10*(6), 324-325.

Cartwright, K., H. Harris, and L. R. Follmer, DVDP hydrogeological studies, *Dry Val. Drilling Proj. Bull.*, *5*, 134-138.

Chapman-Smith, M., Geologic log of DVDP 12, Lake Leon, Taylor Valley, *Dry Val. Drilling Proj. Bull.*, *5*, 61-70.

Chapman-Smith, M., Geologic Log of DVDP 14, North Fork Basin, Wright Valley, *Dry Val. Drilling Proj. Bull.*, *5*, 94-99.

Chapman-Smith, M., Geology of DVDP holes 12 and 14, *Antarct. J. U.S.*, *10*(4), 170-172.

Craig, J. R., J. F. Light, B. C. Parker, and M.
G. Mudrey, Jr., Identification of hydrohalite, *Antarct. J. U.S.*, *10*(4), 178-179.

Decker, E. R., K. H. Baker, and H. Harris, Geothermal studies in the Dry Valleys and Ross Island, *Antarct. J. U.S.*, *10*(4), 176.

Goldich, S. S., and J. L. Bodkin, Fluorine in Cenozoic volcanic rocks of Ross Island and vicinity, Antarctica—A progress report, *Dry Val. Drilling Proj. Bull.*, *6*, 6-7.

Goldich, S. S., S. B. Treves, N. H. Suhr, and J. S. Stuckless, Geochemistry of the Cenozoic volcanic rocks of Ross Island and vicinity, Antarctica, *J. Geol.*, *83*(4), 415-436.

Hall, S. A., Palynologic investigation of quaternary sediment from Lake Vanda, *Antarct. J. U.S.*, *10*(4), 173-174.

Jackson, J. K., Geophysical study of permafrost drill core from Ross Island and Victoria Valley, Antarctica, M.Sc. thesis, 60 pp., N. Ill. Univ., DeKalb.

Jackson, J. K., L. D. McGinnis, and M. G. Mudrey, Jr., Resistivity and velocity measurements on DVDP core from Ross Island and Lake Vida, Antarctica, *Dry Val. Drilling Proj. Bull.*, *6*, 10-12.

Kaminuma, K., Activities of the Japanese DVDP party in 1974-75, *Dry Val. Drilling Proj. Bull.*, *5*, 100-101.

Kato, K., Geochemical study of water and secondary minerals from cores and on the ground in the dry valley region, *Dry Val. Drilling Proj. Bull.*, *5*, 118-119.

Kurasawa, H., Strontium isotopic studies of the Ross Island volcanics, Antarctica, in *Geochemical and Geophysical Studies of Dry Valleys, Victoria Land in Antarctica, Mem. Spec. Issue 4*, edited by T. Torii, pp. 67-74, National Institute of Polar Research, Tokyo.

Kyle, P. R., Electron microprobe analyses of minerals in core samples from Dry Valley Drilling Project (DVDP), Holes 1 and 2, Ross Island, Antarctica, *Antarct. Data Ser. 4*, Dep. of Geol., Victoria Univ., Wellington, N.Z.

Kyle, P. R., and R. C. Price, Occurrences of rhonite in alkalic lavas of the McMurdo Volcanic Group, Antarctic, and Dunedin Volcano, New Zealand, *Amer. Mineral.*, *60*(7), 722-725.

McGinnis, L. D., Dry Valley Drilling Project, 1974-1975, *Antarct. J. U.S.*, *10*(4), 166-167.

McGinnis, L. D., Overview of DVDP 1974-1975, *Dry Val. Drilling Proj. Bull.*, *5*, 1-4.

McGinnis, L. D., and M. G. Mudrey, Jr., Summary of drilling activities, December, 1974–February 1975, *Dry Val. Drilling Proj. Bull.*, *5*, 11-15.

McGinnis, L. D., T. Torii, and R. Clark, Antarctic Dry Valley Project: Report of Seminar 1, *Eos Trans. AGU*, *56*(4), 217-220.

McKelvey, B. C., DVDP sites 10 and 11, Taylor Valley, *Dry Val. Drilling Proj. Bull.*, *5*, 16-60.

McKelvey, B. C., Stratigraphy of DVDP sites 10 and 11, Taylor Valley, *Antarct. J. U.S.*, *10*(4), 169-170.

Morikawa, H., I. Minato, J. Ossaka, and T. Hayashi, The distribution of secondary minerals and evaporites at Lake Vanda, Victoria Land, Antarctica, in *Geochemical and Geophysical Studies of Dry Valleys, Victoria Land in Antarctica, Mem. Spec. Issue 4*, edited by T. Torii, pp. 45-59, National Institute of Polar Research, Tokyo.

Mudrey, M. G., Jr., Basement geology of DVDP boreholes in Taylor and Wright Valleys, *Antarct. J. U.S.*, *10*(4), 172-173.

Mudrey, M. G., Jr., Basement rocks in DVDP 12, Lake Leon, Taylor Valley, *Dry Val. Drilling Proj. Bull.*, *5*, 71-77.

Mudrey, M. G., Jr., T. Torii, and H. Harris, Geology of DVDP 13, Don Juan Pond, Wright Valley, Antarctica, *Dry Val. Drilling Proj. Bull.*, *5*, 78-93.

Nakai, N., Stable isotope studies on DVDP 3, 6, and 8, and possible sources of secondary minerals and evaporites in the McMurdo Region, *Dry Val. Drilling Proj. Bull.*, *6*, 20-21.

Nakai, N., Y. Kiyosu, H. Wada, and M. Takimoto, Stable isotope studies of salts and water from dry valleys, Antarctica, I, Origin of salts and water, and the geologic history of Lake Vanda, in *Geochemical and Geophysical Studies of Dry Valleys, Victoria Land in Antarctica, Mem. Spec. Issue 4*, edited by T. Torii, pp. 30-44, National Institute of Polar Research, Tokyo.

Nishiyama, T., and H. Kurasawa, Distribution of secondary minerals from Taylor Valley, *Dry Val. Drilling Proj. Bull.*, *5*, 120-133.

Powell, R. D., and P. J. Barrett, Grain size distribution of sediments cored by the Dry Valley Drilling Project in the eastern part of Taylor Valley, Antarctica, *Dry Val. Drilling Proj. Bull.*, *5*, 150-166.

Spirakis, C. S., Pleistocene plagues, *Geology*, *3*(7), 372.

Stuckless, J. S., and R. L. Ericksen, Rubidium-strontium ages of basement rocks recovered from DVDP 6, Southern Victoria Land, *Antarct. J. U.S.*, *10*(6), 302-307.

Stuckless, J. S., and R. L. Ericksen, Strontium isotope geochemistry of the volcanic and associated mafic and ultramafic nodules from the Ross Island area, Antarctica (abstract), *Eos Trans. AGU*, *56*(4), 471.

Sun, S. S., and G. N. Hanson, Evolution of the mantle: Geochemical evidence from alkali basalt, *Geology*, *3*(6), 297-302.

Sun, S. S., and G. N. Hanson, Evolution of the mantle sources for alkali basalts and nophelenites, *Dry Val. Drilling Proj. Bull.*, *6*, 30.

Thomson, R. B., N. Z. activities in DVDP, 1974-75, *Antarct. J. U.S.*, *10*(4), 167-168.

Torii, T., Introduction, in *Geochemical and Geophysical Studies of Dry Valleys, Victoria Land in Antarctica, Mem. Spec. Issue 4*, edited by T. Torii, pp. 1-4, National Institute of Polar Research, Tokyo.

Torii, T., and O. Waguri, Preliminary report, 1974-75, *Dry Val. Drilling Proj. Bull.*, *5*, 106-117.

Torii, T., N. Yamagata, and J. Ossaka, Salt balance in the Don Juan Basin, *Dry Val. Drilling Proj. Bull.*, *6*, 33-40.

Torii, T., N. Yamagata, S. Nakaya, S. Murata, T. Hashimoto, O. Matsubaya, and H. Sakai, Geochemical aspects of the McMurdo saline lakes with special emphasis on the geophysical studies of dry valleys, Victoria Land in Antarctica, in *Geochemical and Geophysical Studies of Dry Valleys, Victoria Land in Antarctica, Mem. Spec. Issue 4*, edited by T. Torii, pp. 5-29, National Institute of Polar Research, Tokyo.

Treves, S. B., and M. A. Ali, Geology and petrography of DVDP Hole 1, Hut Point Peninsula, Ross Island, *Antarct. J. U.S.*, *10*(5), 5-10.

Treves, S. B., and B. C. McKelvey, Drilling in Antarctica, September-December, 1974, *Dry Val. Drilling Proj. Bull.*, *5*, 5-10.

Washburn, A. L., Analysis of permafrost cores from Antarctic dry valleys, *Antarct. J. U.S.*, *10*(5), 238-239.

Watanuki, K., and H. Morikawa, Geochemical studies on the minerals obtained by the Dry Valley Drilling Project, in *Geochemical and Geophysical Studies of the Dry Valleys, Victoria Land in Antarctica, Mem. Spec. Issue 4*, edited by T. Torii, pp. 60-66, National Institute of Polar Research, Tokyo.

Webb, P. N., Preliminary Micropaleontological report on samples from Dry Valley Drilling Project, hole 4-4A (Lake Vanda), Wright Valley and holes 8 and 9, New Harbor, Taylor Valley, *Dry Val. Drilling Proj. Bull.*, *5*, 139-149.

Webb, P. N., and J. H. Wrenn, Foraminifera

from DVDP holes 8, 9, and 10, Taylor Valley, *Antarct. J. U.S.*, *10*(4), 168-169.

Yusa, Y., On the water temperature in Lake Vanda, Victoria Land, Antarctica, in *Geochemical and Geophysical Studies of the Dry Valleys, Victoria Land in Antarctica, Mem. Spec. Issue 4*, edited by T. Torii, pp. 75-89, National Institute of Polar Research, Tokyo.

Publications in 1976

Barrett, P. J., S. B. Treves, C. G. Barnes, H. T. Brady, S. A. McCormick, N. Nakai, and J. Oliver, Initial report on DVDP 15, Western McMurdo Sound, *Dry Val. Drilling Proj. Bull.*, *7*, 1-99.

Barrett, P. J., S. B. Treves, C. G. Barnes, H. T. Brady, S. A. McCormick, N. Nakai, J. S. Oliver, and K. J. Sillars, Dry Valley Drilling Project, 1975-76: First core drilling in McMurdo Sound, *Antarct. J. U.S.*, *11*(2), 78-80.

Brady, H. T., and P. N. Webb, Microfossils from DVDP hole 15, Western McMurdo Sound region, *Antarct. J. U.S.*, *11*(2), 81-82.

Bucher, G. J., and E. R. Decker, Down-hole temperature measurements in DVDP 15, McMurdo Sound, *Dry Val. Drilling Proj. Bull.*, *7*, 111-112.

Bucher, G. J., and E. R. Decker, Geothermal studies in the McMurdo Sound region, *Antarct. J. U.S.*, *11*(2), 88-89.

Cartwright, K., and H. Harris, Ground water at Don Juan Pond, Wright Valley, Southern Victoria Land, Antarctica: A probable origin at the base of the East Antarctic Icecap (abstract), *Geol. Soc. Amer. Abstr. Programs*, *8*(6), 804.

Cartwright, K., and H. Harris, Hydrogeology in the dry valleys, *Antarct. J. U.S.*, *11*(2), 89-90.

Kaminuma, K., T. Torii, H. Kurasawa, K. Kato, and O. Waguri, Report of the Dry Valley Drilling Project, 1974-1975 (in Japanese with English summary), *Antarct. Rec.*, *56*, 54-69.

McGinnis, L. D., Application of the Ghyben-Herzberg principle to an understanding of the physical chemistry of Ross Island and New Harbor permafrost core, *Dry Val. Drilling Proj. Bull.*, *7*, 113-114.

McKelvey, B. C., Cenozoic marine and terrestrial glacial sedimentation in Taylor Valley, *Dry Val. Drilling Proj. Bull.*, *7*, 101-104.

Miller, G. H., and P. N. Webb, Racemization of amino acids in Late Cenozoic foraminifera from Taylor and Wright Valleys and Ross Island, *Dry Val. Drilling Proj. Bull.*, *7*, 110.

Nakai, N., Participation in the Dry Valley Drilling Project in 1975-76 (in Japanese), *Kyokuchi*, *23*, 17-21.

Stuiver, M., I. C. Yang, and G. H. Denton, Permafrost oxygen isotope ratios and chronology of three cores from Antarctica, *Nature*, *261*(5561), 547-550.

Stuiver, M., G. H. Denton, and H. W. Borns, Jr., Carbon-14 dates of *Adamussium colbecki* (Mollusca) in marine deposits at New Harbor, Taylor Valley, *Antarct. J. U.S.*, *11*(2), 86-88.

Treves, S. B., P. J. Barrett, and R. B. Thomson, Antarctic Dry Valley Drilling Project: Report on Seminar II, *Eos Trans. AGU*, *57*(8), 584-588.

Webb, P. N., and J. H. Wrenn, Foraminifera from DVDP holes 8-12, Taylor Valley, *Antarct. J. U.S.*, *11*(2), 85-86.

Wrenn, J. H., and P. N. Webb, Microfaunal reconnaissance of core material from lower Taylor Valley (DVDP 8-12), *Dry Val. Drilling Proj. Bull.*, *7*, 108-109.

Wrenn, J. H., and P. N. Webb, Microfossils from Taylor Valley, *Antarct. J. U.S.*, *11*(2), 82-85.

Publications in 1977

Brady, H. T., Extraction of diatoms from glacial sediments, *Antarct. J. U.S.*, *12*(4), 123-124.

Brady, H. T., Freshwater lakes in Pleistocene McMurdo Sound, *Antarct. J. U.S.*, *12*(4), 117-118.

Brady, H. T., Late Neogene diatom biostratigraphy and paleoecology of the dry valleys and McMurdo Sound, Antarctica, M.S. thesis, 172 pp., N. Ill. Univ., DeKalb.

Brady, H. T., Late Neogene history of Taylor and Wright Valleys and McMurdo Sound, derived from diatom biostratigraphy and paleoecology of DVDP cores, in *Third Symposium of Antarctic Geology and Geophysics, Volume of Abstracts*, p. 20, University of Wisconsin Press, Madison.

Brady, H. T., *Thalassiosira torokina* n. sp. (diatom) and its significance in Late Cenozoic biostratigraphy, *Antarct. J. U.S.*, *12*(4), 122-123.

Cartwright, K., and H. J. H. Harris, Hydrogeology of Western Wright Valley, Victoria Land, Antarctica, in *Third Symposium of Antarctic Geology and Geophysics, Volume of Abstracts*, p. 25, University of Wisconsin Press, Madison.

Cassidy, D. S., and S. W. Wise, Jr., Antarctic

marine geology research facility, 1976-1977, *Antarct. J. U.S.*, *12*(4), 83-84.

Decker, E. R., and G. J. Bucher, Geothermal studies in Antarctica, *Antarct. J. U.S.*, *12*(4), 102-104.

Decker, E. R., and G. J. Bucher, Preliminary geothermal studies in the Ross Island-Dry Valley region (review), in *Third Symposium of Antarctic Geology and Geophysics, Volume of Abstracts*, p. 41, University of Wisconsin Press, Madison.

Ervin, C. P., and M. G. Wolf, Ground magnetic studies of volcanic rocks in the Taylor and Wright Valleys region, *Antarct. J. U.S.*, *12*(4), 105.

Horiuchi, K., T. Torii, and Y. Murakami, Geochemical study on the distribution of some minor elements in deposits and water samples of the Antarctic oases, 1, The RA content of DVDP 13 core and the deposits of the Vestfold Hills, *Antarct. Rec.*, *58*, 69-80.

Kaminuma, K., T. Torii, K. Yanai, G. Matsumoto, and Y. Tanaka, Activities of Japanese party in McMurdo Sound area 1976-77, *Antarct. Rec.*, *60*, 132-146.

Kato, K., $^{18}O/^{16}O$ activity ratio at 0°C of salt water of Don Juan Pond and Lake Bonney, Dry Valleys, Antarctica, *Antarct. Rec.*, *58*, 271-275.

Koga, A., Preliminary geochemical prospecting of thermal sources around Lake Vanda, Dry Valley, Antarctica, *Antarct. Rec.*, *58*, 138-144.

Kurasawa, H., Volcanisms and volcanic rocks in Antarctica, *Antarct. Rec.*, *58*, 204-234.

Kurasawa, H., Volcanoes and volcanic rock in Antarctica, *J. Geogr.*, *86*(1), 1-19.

Matsubaya, O., H. Sakai, and T. Torii, Isotopic study of saline lakes in Antarctica, *Antarct. Rec.*, *58*, 276.

Matsumoto, G., and T. Hanya, Organic carbons and fatty acids in Antarctic saline lakes, *Antarct. Rec.*, *58*, 81-88.

McGinnis, L. D., and D. R. Osby, Logging summary of the Dry Valley Drilling Project, *Antarct. J. U.S.*, *12*(4), 115-117.

McGinnis, L. D., D. R. Osby, and F. A. Kohout, Paleohydrology inferred from salinity measurements on Dry Valley Drilling Project (DVDP) core in Taylor Valley, Antarctica, in *Third Symposium of Antarctic Geology and Geophysics, Volume of Abstracts*, p. 111, University of Wisconsin Press, Madison.

Morikawa, H., I. Minato, J. Osaka, and K. Watanuki, Secondary minerals from the drilling cores of DVDP 3 (Ross Island) and 6 (Lake Vida), *Antarct. Rec.*, *58*, 186-194.

Nagata, T., Japanese scientific activities in the McMurdo Region, 1976-1977, *Antarct. J. U.S.*, *12*(4), 94-96.

Nakai, N., and Y. Mizutani, Geological history of the dry valleys, Antarctica, based on stable isotope studies, *Antarct. Rec.*, *58*, 244-253.

Nakai, N., Y. Mizutani, and H. Wada, Volcanic events of Ross Island, Antarctica, based on stable isotope studies of drilled volcanic rocks, *Antarct. Rec.*, *58*, 277-288.

Nakaya, S., and M. Nishimura, Geochemical study of the formation process for the saline lakes in the Dry Valleys, South Victoria Land, Antarctica, *Antarct. Rec.*, *58*, 89-92.

Nakaya, S., T. Torii, and N. Yamagata, Distribution of nutrient matter in saline lakes in the dry valleys, South Victoria Land, Antarctica, *Antarct. Rec.*, *58*, 20-31.

Nishiyama, T., Studies of evaporite minerals from dry valleys, Victoria Land, Antarctica, *Antarct. Rec.*, *58*, 171-185.

Osby, D. R., Paleohydrology of Taylor Valley, Antarctica inferred from pore ice in permafrost core, M.S. thesis, 89 pp., N. Ill. Univ., De Kalb.

Parker, B. C., and R. V. Howard, First environmental impact monitoring and assessment in Antarctica: The Dry Valley Drilling Project, *Biol. Conserv.*, *12*(3), 163-177.

Spall, H., and D. P. Elston, Magnetic stratigraphy in DVDP hole 11, Taylor Valley, Antarctica, in *Third Symposium of Antarctic Geology and Geophysics, Volume of Abstracts*, p. 140, University of Wisconsin Press, Madison.

Stuiver, M., and G. H. Denton, Glacial history of the McMurdo Sound region, *Antarct. J. U.S.*, *12*(4), 128-130.

Torii, T., N. Yamagata, S. Nakaya, and S. Murata, Chemical characteristics of antarctic saline lakes, *Antarct. Rec.*, *58*, 9-19.

Treves, S. B., Geology of DVDP 1, Hut Point Peninsula, Ross Island, Antarctica, in *Third Symposium of Antarctic Geology and Geophysics, Volume of Abstracts*, p. 150, University of Wisconsin Press, Madison.

Treves, S. B., Geology of some volcanic rocks from the Ross Island, Mount Morning, and southern Victoria Land area, *Antarct. J. U.S.*, *12*(4), 104-105.

Washburn, A. L., Analysis of permafrost cores from antarctic dry valleys, *Antarct. J. U.S.*, *12*(4), 113-115.

Watanuki, K., Analysis of antarctic water systems by concentration correlation matrix, *Antarct. Rec.*, *58*, 131-137.

Watanuki, K., Geochemical studies on the minerals obtained by the Dry Valley Drilling Project, *Antarct. Rec.*, *58*, 195-196.

Watanuki, K., T. Torii, H. Murayama, J. Hirabayashi, M. Sano, and T. Abiko, Geochemical features of antarctic lakes, *Antarct. Rec.*, *59*, 18-25.

Webb, P. N., and J. H. Wrenn, Late Cenozoic micropaleontology and biostratigraphy of Eastern Taylor Valley, Antarctica, in *Third Symposium of Antarctic Geology and Geophysics, Volume of Abstracts*, p. 159, University of Wisconsin Press, Madison.

Wrenn, J. H., Cenozoic subsurface micropaleontology and geology of eastern Taylor Valley, Antarctica, M.S. thesis, 255 pp., N. Ill. Univ., DeKalb.

Yusa, Y., A study on thermosolutal convection in saline lakes, *Mem. Fac. Sci. Kyoto Univ., Ser. A, XXXV*(1), 149-183.

Yusa, Y., Thermosolutal convection in saline lakes in the dry valleys, *Antarct. Rec.*, *58*, 154-168.

Publications in 1978

Barrett, P. J., and P. C. Froggatt, Densities, porosities and seismic velocities of some rocks from Victoria Land, Antarctica, *N. Z. J. Geol. Geophys.*, *21*(2), 175-187.

Bodkin, J.B., and S. S. Goldich, Fluorine, chlorine, and sulfur in Cenozoic rocks of Ross Island and vicinity, Antarctica, *Dry Val. Drilling Proj. Bull.*, *8*, 1-4.

Brady, H. T., The dating and interpretation of diatom zones in DVDP 10 and 11, *Dry Val. Drilling Proj. Bull.*, *8*, 5-6.

Brady, H. T., A diatom report of DVDP cores (3, 4A, 12, 14, 15) and other related surface sections, *Dry Val. Drilling Proj. Bull.*, *8*, 7.

Brady, H. T., Geological studies in southern Victoria Land, on Black Island, and on the McMurdo Ice Shelf, *Antarct. J. U.S.*, *13*(4), 13-14.

Bucher, G. J., and E. R. Decker, Geothermal studies in the Ross Island-dry valley region, *Dry Val. Drilling Project Bull.*, *8*(insert).

Cartwright, K., and H. J. H. Harris, Origin of water in lakes and ponds of the dry valley region, Antarctica, *Dry Val. Drilling Proj. Bull.*, *8*, 8.

Chapman-Smith, M., The Taylor Formation (Holocene) and its macrofaunas, Taylor Dry Valley, Antarctica, *Dry Val. Drilling Proj. Bull.*, *8*, 9-10.

Christoffel, D. A., Interpretation of seismic profiling surveys in McMurdo Sound and Terra Nova Bay, Antarctica, 1975, *Dry Val. Drilling Proj. Bull.*, *8*(insert).

Decker, E. R., Geothermal models of the Ross Island-dry valley region, *Dry Val. Drilling Proj. Bull.*, *8*, 11.

Elston, D. P., M. E. Purucker, S. L. Bressler, and H. Spall, Polarity zonations, magnetic intensities, and the correlation of Miocene and Pliocene DVDP cores, Taylor Valley, Antarctica, *Dry Val. Drilling Proj. Bull.*, *8*, 12-15.

Goldich, S. S., J. S. Stuckless, J. B. Bodkin, and R. C. Wamser, Trace elements in volcanic rocks of Ross Island and vicinity, Antarctica, and implications for the origin of alkali-basalt magma, *Dry Val. Drilling Proj. Bull.*, *8*, 16-20.

Harris, H. J. H., and K. Cartwright, Hydrogeology and geochemistry of Don Juan Pond, *Dry Val. Drilling Proj. Bull.*, *8*, 21.

Hendy, C. H., The chronology of glacial sediments in the Taylor Valley, *Dry Val. Drilling Proj. Bull.*, *8*(insert).

Horiuchi, K., T. Torii, and Y. Murakami, Studies of the radium, uranium and thorium concentrations in DVDP cores of Wright Valley, *Dry Val. Drilling Proj. Bull.*, *8*, 22-28.

Kaminuma, K., The upper crustal structure under McMurdo Station, Antarctica, obtained by blasts, *Dry Val. Drilling Proj. Bull.*, *8*, 29.

Kato, K., T. Torii, and N. Nakai, Rapid concentration of saline water in Don Juan Pond, *Dry Val. Drilling Proj. Bull.*, *8*, 30-31.

Koga, A., and T. Torii, Is there no geothermal activity around Lake Vanda, the dry valleys, Antarctica?, *Dry Val. Drilling Proj. Bull.*, *8*, 32-33.

Kurasawa, H., Geochemistry of Ferrar dolerite, dikes, and Cenozoic volcanics of the dry valley region, *Dry Val. Drilling Proj. Bull.*, *8*, 34-38.

Kyle, P. R., The geology and volcanic history of Hut Point Peninsula as inferred from DVDP 1, 2 and 3 drillcores, *Dry Val. Drilling Proj. Bull.*, *8*, 39-43.

Kyle, P. R., Petrogenesis of the basanite to phonolite sequence at DVDP 2 and 3 and Observation Hill, *Dry Val. Drilling Proj. Bull.*, *8*, 44-45.

Kyle, P. R., J. F. Sutter, and S. B. Treves, K/Ar age determinations on DVDP 1 and 2 core samples, *Dry Val. Drilling Proj. Bull.*, *8*, 46-48.

Ling, H. Y., Radiolarians and silicoflagellates

from Dry Valley Drilling Project cores, *Dry Val. Drilling Proj. Bull.*, *8*, 49.

Matsumoto, G., T. Torii, and T. Hanya, Organic constituents of antarctic lake waters and sediments of the McMurdo Sound region, *Dry Val. Drilling Proj. Bull.*, *8*, 50-53.

McGinnis, L. D., Dry Valley Drilling Project Seminar III, Tokyo, *Antarct. J. U.S.*, *13*(4), 30.

McGinnis, L. D., Polar groundwater—A thermo-saline system, paper presented at the 2nd Colloquium on Planetary Water and Polar Processes, Cold Regions Res. and Eng. Lab., Hanover, N. H.

McGinnis, L. D., J. S. Stuckless, and P. N. Kyle, Significance of gamma-ray, salinity and electrical logs in DVDP boreholes, *Dry Val. Drilling Proj. Bull.*, *8*, 54-59.

McKelvey, B. C., The Miocene-Pleistocene stratigraphy of eastern Taylor Valley—An interpretation, *Dry Val. Drilling Proj. Bull.*, *8*, 60-61.

Mudrey, M. G., Jr., L. D. McGinnis, and S. E. Treves, Summary of field activities of the Dry Valley Drilling Project 1972-73 and 1973-74, in *Environmental Impact in Antarctica*, edited by B. C. Parker, pp. 179-210, Virginia Polytechnic Institute and State University, Blacksburg.

Murayama, H., S. Nakaya, S. Murata, T. Torii, and K. Watanuki, Interpretation of salt deposition in Wright Valley, Antarctica: Chemical analysis of DVDP core, *Dry Val. Drilling Proj. Bull.*, *8*, 62-63.

Nakai, N., Y. Kiyosu, H. Wada, R. Nagae, and T. Nishiyama, Stable isotope studies: The evidence of relative sea-level fluctuation and the environmental changes in Wright and Taylor valleys, *Dry Val. Drilling Proj. Bull.*, *8*, 64-65.

Nakai, N., Y. Mizutani, and H. Wada, Stable isotope studies: Past volcanic events deduced from H, O, and C isotopic compositions of ice and salts from DVDP 3, *Dry Val. Drilling Proj. Bull.*, *8*, 66-67.

Nakao, K., T. Torii, and K. Tanizawa, Interpretation of salt deposition in Wright Valley, Antarctica: Granulometric analysis of DVDP #14 core, *Dry Val. Drilling Proj. Bull.*, *8*, 68.

Nakaya, S., and H. Murayama, A note on analytical procedures for saline lake waters, *Dry Val. Drilling Proj. Bull.*, *8*, 107.

Nakaya, S., Y. Motooti, and M. Nishimura, One aspect of the evolution of saline lakes in the dry valleys area of South Victoria Land, Antarctica, *Dry Val. Drilling Proj. Bull.*, *8*, 69-70.

Nishiyama, T., Distribution and origin of evaporite minerals from dry valleys, Victoria Land, Antarctica, *Dry Val. Drilling Proj. Bull.*, *8*, 71.

Parker, B. C. (Ed.), *Environmental Impact in Antarctica*, 390 pp., Virginia Polytechnic Institute and State University, Blacksburg.

Parker, B. C., R. V. Howard, F. C. T. Allnutt, and L. D. McGinnis, Summary of environmental monitoring and impact assessment of the DVDP, in *Environmental Impact in Antarctica*, edited by B. C. Parker, pp. 211-254, Virginia Polytechnic Institute and State University, Blacksburg.

Parker, B. C., M. G. Mudrey, Jr., K. Cartwright, R. E. Cameron, and L. D. McGinnis, Environmental appraisal for the Dry Valley Drilling Project Phases III, IV, and V (1973-74, 1974-75, 1975-76), in *Environmental Impact in Antarctica*, edited by B. C. Parker, pp. 37-143, Virginia Polytechnic Institute and State University, Blacksburg.

Pederson, D. R., G. E. Montgomery, L. D. McGinnis, C. P. Ervin, and H. K. Wong, Magnetic study of Ross Island and Taylor Glacier quadrangles, Antarctica, *Dry Val. Drilling Proj. Bull.*, *8*, 72-73.

Porter, S. C., and J. E. Beget, Provenance and depositional environments of Late Cenozoic sediments in permafrost cores from Lower Taylor Valley, Antarctica, *Dry Val. Drilling Proj. Bull.*, *8*, 74-76.

Powell, R. D., Sedimentation conditions in Taylor Valley, Antarctica, from DVDP cores, *Dry Val. Drilling Proj. Bull.*, *8*, 77-80.

Stuckless, J. S., A. T. Miesch, S. S. Goldich, and P. W. Weiblen, A petrochemical model for the genesis of the volcanic rocks from Ross Island and vicinity, Antarctica, *Dry Val. Drilling Proj. Bull.*, *8*, 81-89.

Stuiver, M., G. H. Denton, T. B. Kellogg, and D. E. Kellogg, Glacial geologic studies in the McMurdo Sound region, *Antarct. J. U.S.*, *13*(4), 44-45.

Stuiver, M., I. C. Yang, and C. H. Hendy, Permafrost oxygen isotope ratios of antarctic DVDP cores, *Dry Val. Drilling Proj. Bull.*, *8*, 80-91.

Thomson, R. B., DVDP—Project organization and accomplishments, *Dry Val. Drilling Proj. Bull.*, *8*, 92-95.

Torii, T., N. Yamagata, J. Ossaka, and S. Murata, A view of the formation of saline waters in the dry valleys, *Dry Val. Drilling Proj. Bull.*, *8*, 96-101.

Treves, S. B., DVDP 1, Hut Point Peninsula, Ross Island, Antarctica, *Dry Val. Drilling Proj. Bull.*, 8, 102.

Treves, S. B., Geology of DVDP 1 and the hyaloclastite of DVDP 3, Hut Point Peninsula, Antarctica, *Antarct. J. U.S.*, 12 (4), 30.

Treves, S. B., Hyaloclastite of DVDP 3, Hut Point Peninsula, Ross Island, Antarctica, *Dry Val. Drilling Proj. Bull.*, 8, 103.

Ugolini, F. C., and W. Deutsch, Chemistry and clay mineralogy of cores 8, 9 and 10, New Harbor, Antarctica, *Dry Val. Drilling Proj. Bull.*, 8, 104.

Vocke, R. D., Jr., G. N. Hanson, and J. A. Stuckless, Ages for the vida granite and olympus gneiss, Victoria Valley, Southern Victoria Land, *Antarct. J. U.S.*, 13(4), 15-17.

Vucetich, C. G., and P. H. Robinson, Quaternary stratigraphy and glacial history of the Lower Taylor Valley, Antarctica, *N. Z. J. Geol. Geophys.*, 21(4), 467-482.

Watanuki, K., and T. Torii, Some characteristics of mineral obtained in the dry valleys, *Dry Val. Drilling Proj. Bull.*, 8, 105-106.

Webb, P. N., Miocene sediments and micropaleontology in gravity cores from RISP/J9 and comparisons with DVDP and DSDP drillcore successions, *Dry Val. Drilling Proj. Bull.*, 8, 125.

Webb, P. N., Paleogeographic evolution of the Ross Sea and adjacent montane areas during the Cenozoic, *Dry Val. Drilling Proj. Bull.*, 8, 124.

Weiblen, P. W., J. S. Stuckless, W. C. Hunter, K. J. Schulz, and M. G. Mudrey, Correlation of clinopyroxene composition with environment of formation based on data from Ross Island volcanic rocks, *Dry Val. Drilling Proj. Bull.*, 8, 108-115.

Wrenn, J. H., and S. W. Beckman, Maceral and total organic carbon analysis of DVDP core #11, *Dry Val. Drilling Proj. Bull.*, 8, 116-118.

Yoshida, Y., and K. Moriwaki, Some characteristics of antarctic coastal features, *Dry Val. Drilling Proj. Bull.*, 8, 119.

Yusa, Y., Analysis of thermosolutal phenomena observed in McMurdo saline lakes, *Dry Val. Drilling Proj. Bull.*, 8, 120-123.

Publications in 1979

Barrett, P. J., Proposed drilling in McMurdo Sound—1979, in *Proceedings of the Seminar III on Dry Valley Drilling Project, Mem. Spec. Issue 13*, edited by T. Nagata, p. 231, National Institute of Polar Research, Tokyo.

Barrett, P. J., and S. B. Treves, Sedimentology and petrology of DVDP 15, Western McMurdo Sound, Antarctica, in *Proceedings of the Seminar III on Dry Valley Drilling Project, Mem. Spec. Issue 13*, edited by T. Nagata, p. 213, National Institute of Polar Research, Tokyo.

Brady, H. T., The dating and interpretation of diatom zones in Dry Valley Drilling Project holes 10 and 11 Taylor Valley, South Victoria Land, Antarctica, in *Proceedings of the Seminar III on Dry Valley Drilling Project, Mem. Spec. Issue 13*, edited by T. Nagata, p. 150, National Institute of Polar Research, Tokyo.

Brady, H. T., A diatom report on DVDP cores 3, 4A, 12, 14, 15 and other related surface sections, in *Proceedings of the Seminar III on Dry Valley Drilling Project, Mem. Spec. Issue 13*, edited by T. Nagata, p. 165, National Institute of Polar Research, Tokyo.

Brady, H., and H. Martin, Ross Sea region in the Middle Miocene: A glimpse into the past, *Science*, 203(4379), 437-438.

Cassidy, D. S., DVDP at the Florida State University: Core storage and sample distribution, in *Proceedings of the Seminar III on Dry Valley Drilling Project, Mem. Spec. Issue 13*, edited by T. Nagata, p. 240, National Institute of Polar Research, Tokyo.

Chapman-Smith, M., The Taylor Formation (Holocene) and its macrofaunas, Taylor Dry Valley, Antarctica (extended abstract), in *Proceedings of the Seminar III on Dry Valley Drilling Project, Mem. Spec. Issue 13*, edited by T. Nagata, p. 196, National Institute of Polar Research, Tokyo.

Harris, H. J. H., and K. Cartwright, Dynamic chemical equilibrium in a polar desert pond: A sensitive index of meteorological cycles, *Science*, 204(4390), 301-303.

Hendy, C. H., The Late Quaternary chronology of the Taylor Glacier, in *Proceedings of the Seminar III on Dry Valley Drilling Project, Mem. Spec. Issue 13*, edited by T. Nagata, p. 204, National Institute of Polar Research, Tokyo.

Horiuchi, K., T. Torii, and Y. Murakami, Studies on Ra, U, Th and minor element contents in DVDP cores and deposits from Vestfold Hills, in *Proceedings of the Seminar III on Dry Valley Drilling Project, Mem. Spec. Issue 13*, edited by T. Nagata, p. 121, National Institute of Polar Research, Tokyo.

Kaminuma, K., The upper crustal structure under McMurdo Station, Antarctica, deduced from blasts during nuclear power plant removal, in *Proceedings of the Seminar III on Dry Valley Drilling Project, Mem. Spec. Issue 13*, edited by T. Nagata, p. 34, National Institute of Polar Research, Tokyo.

Kato, K., T. Torii, and N. Nakai, Dilution and concentration of saline water in Don Juan Pond in 1974, in *Proceedings of the Seminar III on Dry Valley Drilling Project, Mem. Spec. Issue 13*, edited by T. Nagata, p. 53, National Institute of Polar Research, Tokyo.

Kyle, P. R., J. F. Sutter, and S. B. Treves, K/Ar age determinations on drill core from DVDP holes 1 and 2, in *Proceedings of the Seminar III on Dry Valley Drilling Project, Mem. Spec. Issue 13*, edited by T. Nagata, p. 214, National Institute of Polar Research, Tokyo.

Matsubaya, O., H. Sakai, T. Torii, H. Burton, and K. Kerry, Antarctic saline lakes—Stable isotopic ratios, chemical compositions and evolution, *Geochim. Cosmochim. Acta, 43*(1), 7-25.

Matsumoto, G., T. Torii, and T. Hanya, Distribution of organic constituents in lake waters and sediments of the McMurdo Sound region in the Antarctic, in *Proceedings of the Seminar III on Dry Valley Drilling Project, Mem. Spec. Issue 13*, edited by T. Nagata, p. 103, National Institute of Polar Research, Tokyo.

McGinnis, L. D., The Dry Valley Drilling Project—An exercise in international cooperation—A viewpoint from the United States, in *Proceedings of the Seminar III on Dry Valley Drilling Project, Mem. Spec. Issue 13*, edited by T. Nagata, p. 1, National Institute of Polar Research, Tokyo.

McKelvey, B. C., The Miocene-Pleistocene stratigraphy of Eastern Taylor Valley—An interpretation of DVDP cores 10 and 11, in *Proceedings of the Seminar III on Dry Valley Drilling Project, Mem. Spec. Issue 13*, edited by T. Nagata, p. 176, National Institute of Polar Research, Tokyo.

Murayama, H., S. Nakaya, S. Murata, T. Torii, and K. Watanuki, Interpretation of salt deposition in Wright Valley, Antarctica: Chemical analysis of DVDP 14 core, in *Proceedings of the Seminar III on Dry Valley Drilling Project, Mem. Spec. Issue 13*, edited by T. Nagata, p. 60, National Institute of Polar Research, Tokyo.

Nagata, T., Foreword, in *Proceedings of the Seminar III on Dry Valley Drilling Project, Mem. Spec. Issue 13*, edited by T. Nagata, National Institute of Polar Research, Tokyo.

Nakao, K., T. Torii, and K. Tanizawa, Paleohydrology of Lake Vanda in Wright Valley, Antarctica inferred from granulometric analysis of DVDP 14 core, in *Proceedings of the Seminar III on Dry Valley Drilling Project, Mem. Spec. Issue 13*, edited by T. Nagata, p. 73, National Institute of Polar Research, Tokyo.

Nakaya, S., Y. Motoori, and M. Nishimura, One aspect of the evolution of saline lakes in the dry valleys of South Victoria Land, Antarctica (extended abstract), in *Proceedings of the Seminar III on Dry Valley Drilling Project, Mem. Spec. Issue 13*, edited by T. Nagata, p. 49, National Institute of Polar Research, Tokyo.

Nishiyama, T., Distribution and origin of evaporite minerals from dry valleys, Victoria Land, in *Proceedings of the Seminar III on Dry Valley Drilling Project, Mem. Spec. Issue 13*, edited by T. Nagata, p. 136, National Institute of Polar Research, Tokyo.

Powell, R. D., Conditions of sediment deposition in Taylor Valley, Antarctica, from Dry Valley Drilling Project cores 8-12 (extended abstract), in *Proceedings of the Seminar III on Dry Valley Drilling Project, Mem. Spec. Issue 13*, edited by T. Nagata, p. 187, National Institute of Polar Research, Tokyo.

Torii, T., N. Yamagata, J. Ossaka, and S. Murata, A view on the formation of saline waters in the dry valleys, in *Proceedings of the Seminar III on Dry Valley Drilling Project, Mem. Spec. Issue 13*, edited by T. Nagata, p. 187, National Institute of Polar Research, Tokyo.

Ugolini, F. C., W. Deutsch, and H. J. H. Harris, Chemistry and clay mineralogy of cores 8, 9, 10, New Harbor, Antarctica, in *Proceedings of the Seminar III on Dry Valley Drilling Project, Mem. Spec. Issue 13*, edited by T. Nagata, p. 84, National Institute of Polar Research, Tokyo.

Watanuki, K., T. Torii, S. Nakaya, H. Murayama, and S. Murata, A note on analytical method for saline lake water, in *Proceedings of the Seminar III on Dry Valley Drilling Project, Mem. Spec. Issue 13*, edited by T. Nagata, p. 227, National Institute of Polar Research, Tokyo.

Webb, P. N., Miocene sediments and micropaleontology in gravity cores from RISP/J9 and comparisons with DVDP and DSDP drillcore successions, in *Proceedings of the Seminar III on Dry Valley Drilling Project, Mem. Spec. Issue 13*, edited by T. Nagata, p. 148, National Institute of Polar Research, Tokyo.

Webb, P. N., Paleogeographic evolution of the

Ross sector during the Cenozoic, in *Proceedings of the Seminar III on Dry Valley Drilling Project, Mem. Spec. Issue 13*, edited by T. Nagata, p. 206, National Institute of Polar Research, Tokyo.

Yoshida, Y., Editorial preface, in *Proceedings of the Seminar III on Dry Valley Drilling Project, Mem. Spec. Issue 13*, edited by T. Nagata, National Institute of Polar Research, Tokyo.

Yoshida, Y., and K. Moriwaki, Some consideration on elevated coastal features and their dates around Syowa Station, Antarctica, in *Proceedings of the Seminar III on Dry Valley Drilling Project, Mem. Spec. Issue 13*, edited by T. Nagata, p. 220, National Institute of Polar Research, Tokyo.

Yusa, Y., Analysis of thermosolutal phenomena observed in McMurdo saline lakes (extended abstract), in *Proceedings of the Seminar III on Dry Valley Drilling Project, Mem. Spec. Issue 13*, edited by T. Nagata, p. 42, National Institute of Polar Research, Tokyo.

Publications in press

Brady, H. T., Late Neogene history of Taylor and Wright Valleys and McMurdo Sound, derived from diatom biostratigraphy and paleoecology of DVDP cores, in *Antarctic Geoscience*, edited by C. Craddock, University of Wisconsin Press, Madison.

Cartwright, K., and H. J. H. Harris, Hydrogeology of western Wright Valley, Victoria Land, Antarctica, in *Antarctic Geoscience*, edited by C. Craddock, University of Wisconsin Press, Madison.

Decker, E. R., and G. J. Bucher, Preliminary geothermal studies in the Ross Island–Dry Valley region, in *Antarctic Geoscience*, edited by C. Craddock, University of Wisconsin Press, Madison.

McGinnis, L. D., D. R. Osby, and F. A. Kohout, Paleohydrology inferred from salinity measurements on Dry Valley Drilling Project (DVDP) core in Taylor Valley, Antarctica, in *Antarctic Geoscience*, edited by C. Craddock, University of Wisconsin Press, Madison.

McKelvey, B. C., Upper Cenozoic marine and terrestrial glacial sedimentation in eastern Taylor Valley, Southern Victoria Land, in *Antarctic Geoscience*, edited by C. Craddock, University of Wisconsin Press, Madison.

Spall, H., Magnetic stratigraphy in DVDP hole 11, Taylor Valley, Antarctica, in *Antarctic Geoscience*, edited by C. Craddock, University of Wisconsin Press, Madison.

Treves, S. B., Geology of DVDP 1, Hut Point Peninsula, Ross Island, Antarctica, in *Antarctic Geoscience*, edited by C. Craddock, University of Wisconsin Press, Madison.

Ugolini, F. C., and M. L. Jackson, Weathering and mineral synthesis in antarctic soils, in *Antarctic Geoscience*, edited by C. Craddock, University of Wisconsin Press, Madison.

Webb, P. N., and J. H. Wrenn, Late Cenozoic micropaleontology and biostratigraphy of eastern Taylor Valley, Antarctica, in *Antarctic Geoscience*, edited by C. Craddock, University of Wisconsin Press, Madison.

Wrenn, J. H., and P. N. Webb, Physiographic analysis and interpretation of the Ferrar Glacier–Victoria Valley area, in *Antarctic Geoscience*, edited by C. Craddock, University of Wisconsin Press, Madison.